DESIGN OF PRESTRESSED CONCRETE STRUCTURES

DESIGN OF PRESTRESSED CONCRETE STRUCTURES

THIRD EDITION

T. Y. LIN,
Board Chairman, T. Y. Lin International, Consulting Engineers,
San Francisco;
Professor of Civil Engineering, Emeritus, University of California, Berkeley

NED H. BURNS,
Professor of Civil Engineering,
The University of Texas at Austin

JOHN WILEY & SONS
New York Chichester Brisbane Toronto Singapore

Library of Congress Cataloging in Publication Data:

Lin, T'ung-yen, 1911-
Design of prestressed concrete structures.

Includes bibliographies and indexes.
1. Prestressed concrete construction. I. Burns,
Ned Hamilton, 1932- joint author. II. Title.
TA683.9.L5 1980 693'.542 80-20619
ISBN 0-471-01898-8

Printed in the United States of America

10 9 8 7 6 5

To engineers who, rather than blindly following
the codes of practice, seek to apply the laws of nature

PREFACE

The development of prestressed concrete can perhaps be best described by this parody, which Lin presented before the World Conference on Prestressed Concrete in San Francisco, 1957.

All the world's a stage,
And all engineering techniques merely players:
They have their exits and their entrances.
Prestressed concrete, like others, plays a part,
Its acts being seven ages. At first the infant,
Stressing and compressing in the inventor's arms.
Then the curious schoolboy, fondly created
By imaginative engineers for wealthy customers;
Successfully built but costs a lot of dough. Then the lover,
Whose course never runs smooth, embraced by some,
Shunned by others, especially building officials. Now the soldier,
Produced en masse the world over, quick to fight
Against any material, not only in strength, but in economy as well.
Soon the justice—codes and specifications set up to abide,
Formulas and tables to help you decide. No more fun to the pioneers,
But so prestressed concrete plays its part. The sixth age
Shifts to refined research and yet bolder designs,
Undreamed of by predecessors and men in ivory towers.
Last scene of all, in common use and hence in oblivion,
Ends this eventful history of prestressed concrete
As one of engineering methods and materials
Like timber, like steel, like reinforced concrete,
Like everything else.

When Lin wrote the first edition of this book, prestressed concrete in the United States had barely entered its fourth stage—the beginning of mass production. Now it has emerged from the sixth stage and into the last. This rapid advancement has made possible a rather thorough revision of the previous two editions.

This edition presents the basic theory in prestressed concrete similar to the second edition. The method of load balancing is discussed side by side with the working-load and the ultimate-load methods, forming a tripod for prestressed concrete design.

The chapters on materials, prestressing systems, and loss of prestress have been brought up to date. An important section on moment-curvature relationships has been added for flexural analysis. The problems of shear and bond, having been codified since the second edition was written, are now based on substantial experimental findings as well as theoretical interpretations. Camber and deflection are more thoroughly discussed with respect to time-dependent properties of concrete. The coverage of analysis and design of continuous beams with draped posttensioned tendons is more complete. Load-balancing with idealized tendons is covered along with the analysis using actual tendon layout in numerical example problems.

Since building codes and bridge specifications on prestressed concrete are now pretty well standardized, the ACI Code with its latest revisions is used in design examples. Bridge designers should refer to AASHTO Bridge Specifications for some minor adjustments concerning allowable stresses and load factors.

Some selected problems, which the authors have used for their undergraduate and graduate classes, are presented in Appendix E. Solutions for these are available to faculty members upon request.

We believe that this revised edition, with its up-to-date contents, will continue to be used both as a text and a reference for engineers interested in prestressed concrete.

The assistance of Martha Burns and Stephanie Burns with manuscript typing and Sow-Wen Chang and Sung L. Lam with S.I. units is gratefully acknowledged.

Berkeley, California and Austin, Texas
1981.

T. Y. Lin
Ned H. Burns

CONTENTS

1 INTRODUCTION 1
2 MATERIALS 40
3 PRESTRESSING SYSTEMS; END ANCHORAGES 67
4 LOSS OF PRESTRESS; FRICTION 87
5 ANALYSIS OF SECTIONS FOR FLEXURE 126
6 DESIGN OF SECTIONS FOR FLEXURE 188
7 SHEAR; BOND; BEARING 241
8 CAMBER, DEFLECTIONS; CABLE LAYOUTS 290
9 PARTIAL PRESTRESS AND NONPRESTRESSED REINFORCEMENTS 325
10 CONTINUOUS BEAMS 345
11 LOAD-BALANCING METHOD 392
12 SLABS 428
13 TENSION MEMBERS; CIRCULAR PRESTRESSING 461
14 COMPRESSION MEMBERS; PILES 488
15 ECONOMICS; STRUCTURAL TYPES AND LAYOUTS 517
16 DESIGN EXAMPLES 545
APPENDIX A DEFINITIONS, NOTATIONS, ABBREVIATIONS 592
APPENDIX B DATA FOR SOME PRESTRESSING SYSTEMS 597
APPENDIX C CONSTANTS FOR BEAM SECTIONS 610
APPENDIX D PRESTRESSED CONCRETE LOSS OF PRESTRESS CALCULATIONS 614
APPENDIX E PROBLEM STATEMENTS 628
INDEX 643

1

INTRODUCTION

1-1 Development of Prestressed Concrete

The development of structural materials can be described along three different columns, as in Fig. 1-1. Column 1 shows materials resisting compression, starting with stones and bricks, then developing into concrete and more recently high-strength concrete. For materials resisting tension, people used bamboo, and ropes, then iron bars and steel, then high-strength steel. Column 3 indicates materials which resist both tension and compression, namely, bending. Timber was utilized, then structural steel, reinforced concrete and finally prestressed concrete was developed.

The main difference between reinforced and prestressed concrete is the fact that reinforced concrete combines concrete and steel bars by simply putting them together and letting them act together as they may wish. Prestressed concrete, on the other hand, combines high-strength concrete with high-strength steel in an "active" manner. This is achieved by tensioning the steel and holding it against the concrete, thus putting the concrete into compression. This active combination results in a much better behavior of the two materials. Steel is ductile and now is made to act in high tension by prestressing. Concrete is a brittle material with its tensile capacity now improved by being compressed, while its compressive capacity is not really harmed. Thus prestressed concrete is an ideal combination of two modern high strength materials.

The historical development of prestressed concrete actually started in a different manner when prestressing was only intended to create permanent compression in concrete to improve its tensile strength. Later it became clear that prestressing the steel was also essential to the efficient utilization of high-tensile steel. Prestressing means the intentional creation of permanent stresses in a structure or assembly, for the purpose of improving its behavior and strength under various service conditions.

Throughout this chapter, photographs of significant structures designed in prestressed concrete using this fundamental concept will be shown. Note that the basic idea and the high-strength materials have now become a vital part of modern structural engineering practice.

The basic principle of prestressing was applied to construction perhaps centuries ago, when ropes or metal bands were wound around wooden staves to form barrels (Fig. 1-2). When the bands were tightened, they were under tensile

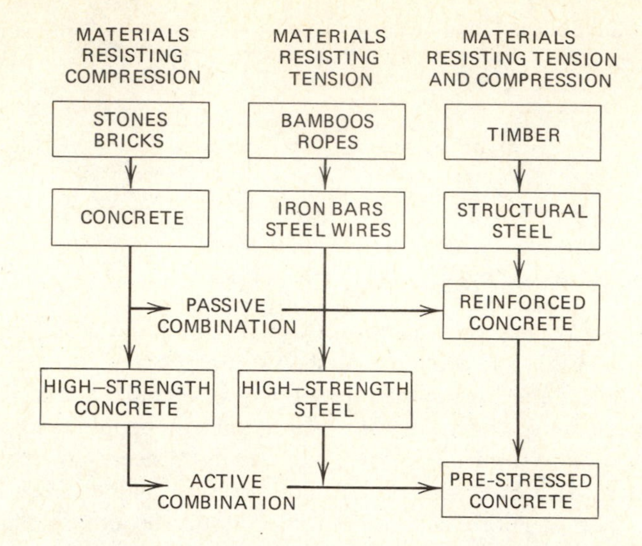

Fig. 1-1. Development of building materials.

prestress which in turn created compressive prestress between the staves and thus enabled them to resist hoop tension produced by internal liquid pressure. In other words, the bands and the staves were both prestressed before they were subjected to any service loads.

The same principle, however, was not applied to concrete until about 1886, when P. H. Jackson, an engineer of San Francisco, California, obtained patents for tightening steel tie rods in artificial stones and concrete arches to serve as floor slabs. Around 1888, C. E. W. Doehring of Germany independently secured a patent for concrete reinforced with metal that had tensile stress applied to it

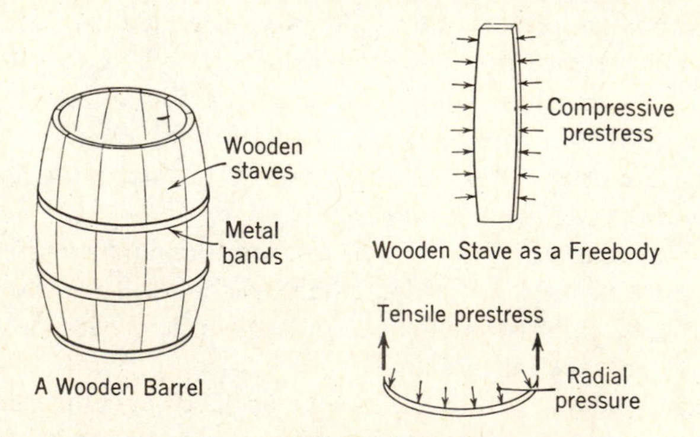

Fig. 1-2. Principle of prestressing applied to barrel construction.

before the slab was loaded. These applications were based on the conception that concrete, though strong in compression, was quite weak in tension, and prestressing the steel against the concrete would put the concrete under compressive stress which could be utilized to counterbalance any tensile stress produced by dead or live loads.

These first patented methods were not successful because the low tensile prestress then produced in the steel was soon lost as a result of the shrinkage and creep of concrete. Consider an ordinary structural steel bar prestressed to a working stress of 18,000 psi (124 N/mm^2) (Fig. 1-3). If the modulus of elasticity of steel is approximately 29,000,000 psi (200 kN/mm^2), the unit lengthening of the bar is given by

$$\delta = \frac{f}{E} = \frac{18{,}000}{29{,}000{,}000} = 0.00062$$

Since eventual shrinkage and creep often induce comparable amounts of shortening in concrete, this initial unit lengthening of steel could be entirely lost in the course of time. At best, only a small portion of the prestress could be retained, and the method cannot compete economically with conventional reinforcement of concrete.

In 1908, C. R. Steiner of the United States suggested the possibility of retightening the reinforcing rods after some shrinkage and creep of concrete had taken place, in order to recover some of the losses. In 1925, R. E. Dill of Nebraska tried high-strength steel bars coated to prevent bond with concrete. After the concrete had set, the steel rods were tensioned and anchored to the

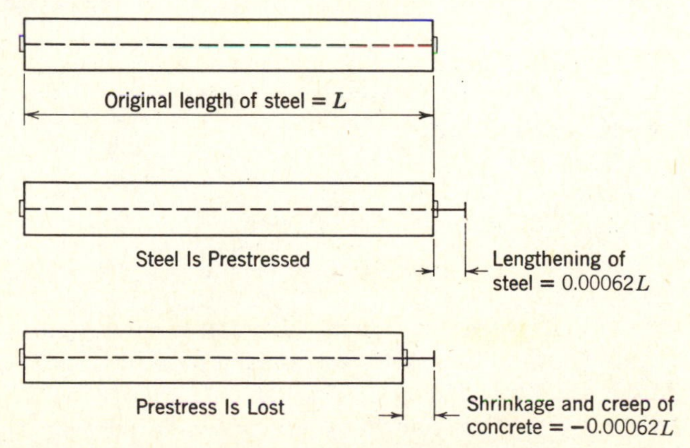

Fig. 1-3. Prestressing concrete with ordinary structural steel.

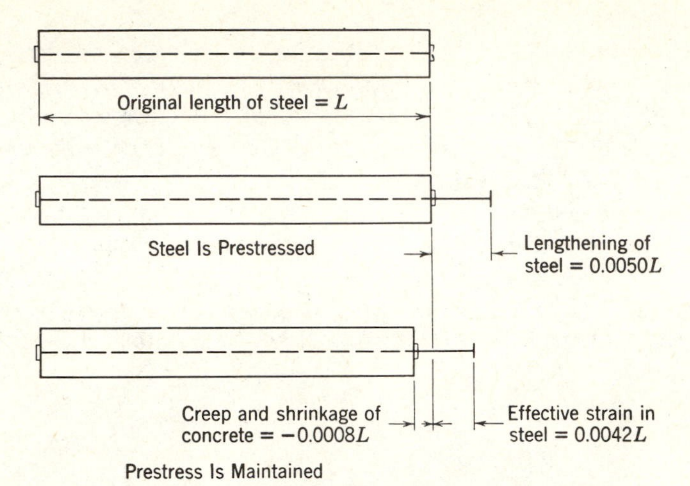

Fig. 1-4. Prestressing concrete with high-tensile steel.

concrete by means of nuts. But these methods were not applied to any apprecia-ble extent, chiefly for economic reasons.

Modern development of prestressed concrete is credited to E. Freyssinet of France, who in 1928 started using high-strength steel wires for prestressing. Such wires, with an ultimate strength as high as 250,000 psi (1725 N/mm^2) and a yield point over 180,000 psi (1,240 N/mm^2), are prestressed to about 145,000 psi (1000 N/mm^2), creating a unit strain of (Fig. 1-4)

$$\delta = \frac{f}{E} = \frac{145{,}000}{29{,}000{,}000} = 0.0050$$

Assuming a total loss of 0.0008 due to shrinkage and creep of concrete and other causes, a net strain of $0.0050 - 0.0008 = 0.0042$ would still be left in the wires, which is equilvalent to a stress of

$$f = E\delta = 29{,}000{,}000 \times 0.0042 = 121{,}800 \text{ psi } (840 \text{ N/mm}^2)$$

Although Freyssinet also tried the scheme of pretensioning where the steel was bonded to the concrete without end anchorage, practical application of this method was first made by E. Hoyer of Germany. The Hoyer system consists of stretching wires between two buttresses several hundred feet apart, putting shutters between the units, placing the concrete, and cutting the wires after the

Fig. 1-5. Modern prestressing plants produce prestressed concrete products of very high quality. Shown are very long octagonal piles being shipped out of Concrete Technology Corporation's plant in Tacoma, Washington. 36,416 ft (11,100 m) of piling went to Adak, Alaska, along with 440 deck panels to construct the Naval Supply Pier (Prestressed Concrete Institute).

concrete has hardened. This method enables several units to be cast between two buttresses.

Wide application of prestressed concrete was not possible until reliable and economical methods of tensioning and of end anchorage were devised. In 1939, Freyssinet developed conical wedges for end anchorages and designed double-acting jacks which tensioned the wires and then thrust the male cones into the female cones for anchoring them. In 1940, Professor G. Magnel of Belgium developed the Magnel system, wherein two wires were stretched at a time and anchored with a simple metal wedge at each end. About that time, prestressed concrete began to acquire importance, though it did not actually come to the fore until about 1945. Perhaps the shortage of steel in Europe during the war had given it some impetus, since much less steel is required for prestressed concrete than for conventional types of construction. But it must also be realized that time was needed to prove and improve the serviceability, economy, and safety of prestressed concrete as well as to acquaint engineers and builders with a new method of design and construction.

Although France and Belgium led the development of prestressed concrete, England, Germany, Switzerland, Holland, Soviet Russia, and Italy quickly followed. Since 1965, about 47% of all bridges built in Germany were of

Fig. 1-6. Erection of long-span precast tee section. Shipment from plant to site for these standard cross sections is usually by truck. Beams are picked from the truck and placed in the structure.

prestressed concrete.[1] Soviet Russia annually produced 25,000,000 m^3 of prestressed concrete in 1978, most of which was pretensioned products for buildings. Since the late 1960s and 1970s most medium-span bridges (100–300 ft, 30–90 m) and many long-span bridges up to about 1000 ft (305 m) were built of prestressed concrete in all parts of the world.

Prestressed concrete in the United States followed a different course of development. Instead of linear prestressing, a name given to prestressed concrete beams and slabs, circular prestressing especially as applied to storage tanks took the early lead. This was performed almost entirely by the Preload Company, which developed special wire winding machines and which, from 1935 to 1963, built about 1000 prestress-concrete tanks throughout this country and other parts of the world.

Linear prestressing did not start in this country until 1949, when construction of the famed Philiadelphia Walnut Lane Bridge was begun. The Bureau of Public Roads survey showed that for the years 1957-1960, 2052 prestressed concrete bridges were authorized for construction, totaling a length of 68 miles, with an aggregate cost of 290 million dollars and comprising 12% of all new highway bridges both in length and in cost.

Since 1960, the use of prestressed concrete bridges has become a standard practice in the United States. In many of the states almost all bridges in the span range 60 to 120 ft (18–36 m) have been constructed with prestressed concrete.

Fig. 1-7. Precast prestressed concrete girders and columns form rigid frames for 62-ft span garage, University of California, Berkeley (Architect Anshen and Allen, Structural Engineer T. Y. Lin International, Consulting Engineers).

Since the late 1970s, posttensioned bridges of medium spans (150–650 ft, 45–200 m) have gained momentum, in the form of continuous or cantilever construction.

In the United States, while there was only one precast pretensioning plant in 1950, there were 229 in 1961. The total volume of precast prestressed products was estimated to be over 2,000,000 yd^3 (1,530,000 m^3) in 1962, of which it was roughly estimated that 50% went to bridges and the remaining to buildings and other construction projects. A survey by the Prestressed Concrete Institute in 1975 indicated that 500 precasting and prestressing plants were operating in the United States.

The growth of the prestressed concrete industry in the United States and Canada is shown by the Prestressed Concrete Institute graph on dollar sales volume of precast and prestressed concrete during the 25-year period 1950–1975, Fig. 1-9. The data on sales includes posttensioned tendon assemblies but not the value of the posttensioned structures. The breakdown of industry sales is estimated by the PCI to divide approximately 50% prestressed concrete, 30% structural precast and 20% architectural precast for 1975.

The Post-tensioning Institute (PTI) was formed in 1976 with 16 member companies participating. Most of these companies had previously been part of the Post-tensioning Division of the Prestressed Concrete Institute, and the separate organization was established to permit them to cooperate in the area of

Fig. 1-8. Forty-story office building in Singapore has 8 in. posttensioned flat slabs with 30-ft spans. Reduced floor depth made possible a 40-ft reduction in the height of the building (T. Y. Lin International, Consulting Engineers).

posttensioning with clearer identity. Data shown in Table 1-1 furnished by PTI indicates that the total tonnage of prestressing steel used in the United States for posttensioning in 1974 increased to almost three times the sales in 1965. Use of posttensioning for nuclear containment structures for power plants increased to a peak in 1972 before dropping back. Note also in Table 1-1 the increasing use of posttensioning for soil and rock anchors in connection with earthwork since 1972. For 1974, the posttensioning sales were distributed as follows: buildings—59%, bridges—25%, nuclear—7%, earthwork—8%. It was estimated that 20 million square feet of flat slabs for building construction was posttensioned in the year 1974, indicating that it is a very competitive structural system. Since the

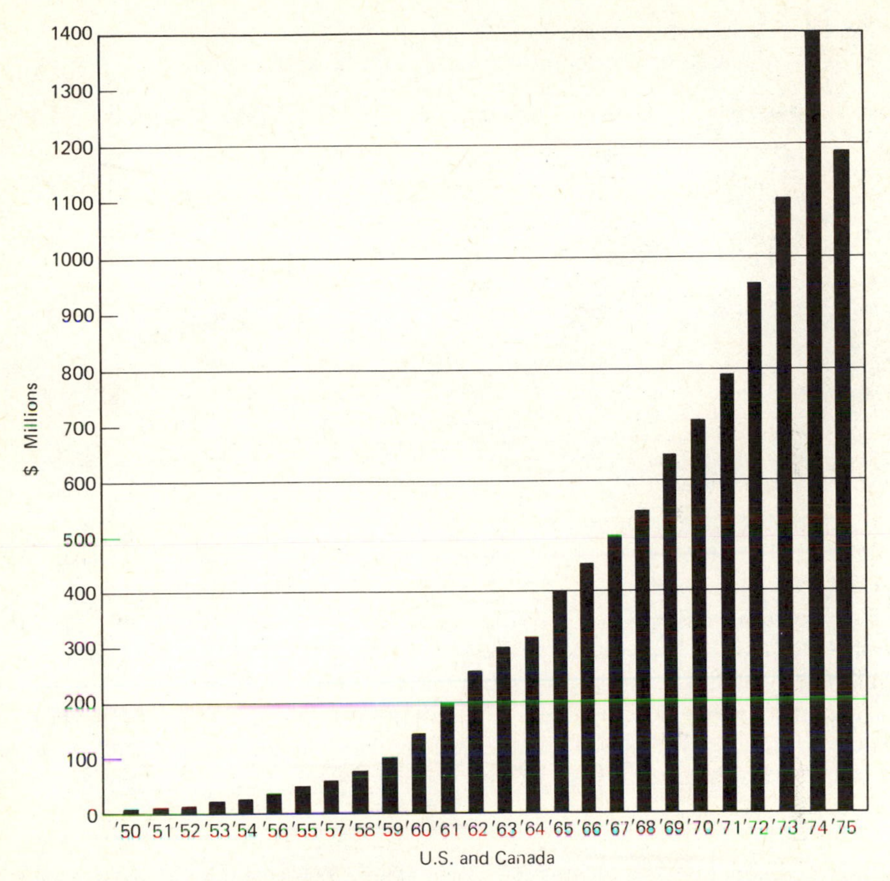

Fig. 1-9. **Estimates of dollar sales volume of precast and prestressed concrete for the United States and Canada, 1950 through 1975. No breakdown, U. S.-Canada available (Prestressed Concrete Institute).**

1970's, both precast, prestressed members and posttensioned, cast-in-place concrete are being utilized. Innovative designs using combinations of both are not uncommon.

The progress of prestressed concrete, both in research and in development, is perhaps best indicated by the growth of its technical societies and their publications. In the United States, the Prestressed Concrete Institute, formed in 1954, had a membership of 2150 by 1975. This institute publishes the *PCItems* and the *PCI Journal* and has very active technical committees. Introduction of the Prestressed Concrete Institute's *Prestressed Concrete Handbook*[8] in 1971 perhaps marked the acceptance of prestressed concrete as a structural system which engineers could use more easily and with confidence of acceptance as a viable system. The Post-Tensioning Institute published the *Post-Tensioning Manual*[9],

Table 1-1 U. S. Posttensioning Steel Tonnage—1965 Through 1975[a] (Equivalent Tonnage of $\frac{1}{2}$ in. Diameter 270 ksi Strand)

Year	Buildings	Bridges	Nuclear	Earthwork	Miscellaneous	Total
1965	10,979	2,400				13,379
1966	11,310	2,736				14,046
1967	10,335	5,148	83			15,566
1968	12,204	7,159	208			19,571
1969	17,611	7,537	718			25,866
1970	19,136	8,920	2,420			30,476
1971	22,145	8,682	4,181			35,008
1972	23,721	7,182	6,118	939	166	38,126
1973	21,809	9,228	3,244	785	422	35,488
1974	23,560	10,056	2,769	3,111	236	39,732
1975	11,994	7,954	2,134	1,942	1,893	25,917[a]

[a]1975 tonnage reports incomplete. Estimated total 1975 tonnage 29,000 tons. Information from Post-Tensioning Institute.

which was initiated by the PCI Post-Tensioning Division prior to organization of the PTI.

Proceedings of the World Conference on Prestressed Concrete in San Francisco, 1957, and Proceedings of Western Conferences on Prestressed Concrete Buildings held in California in 1960, included valuable papers on all phases of the subject. The International Federation for Prestressing (FIP) with its headquarters in London has member groups in 44 countries and FIP observers in some 25 other countries. It has published hundreds of papers of their congresses held all over the world since 1963.[11] In 1975, joint sessions were sponsored by ACI, PCI, and FIP in Philadelphia tracing the development of prestressed concrete throughout the world. These sessions indicated the extent to which prestressed concrete has been utilized in large offshore structures for storage of oil, massive containment structures for nuclear power plants, and even oceangoing barges. An excellent issue of the *PCI Journal* (September–October 1976)[10] presents papers given at the T. Y. Lin Symposium on "Prestressed Concrete, Past, Present and Future", University of California, Berkeley, June 1976.

While bridges are more easily standardized by federal and state agencies, thus helping to develop prestressed concrete construction, it took more time to develop standardized building products and designs for the individual architects and engineers. However, with the incorporation of prestressed concrete into building codes since the 1960's, and a general understanding of prestressed design and construction, a rapid rate of growth was also experienced for buildings.

Thus the development of prestressed concrete in the United States has occurred in the application of posttensioning to buildings, bridges, and pressure

Fig. 1-10. Thirteen-story apartment building with all slabs posttensioned on the ground being lifted into position; 8-in. lightweight concrete spans 28-ft; San Francisco (Owner George Belcher, Engineer August Waegeman, Consultant T. Y. Lin and Associates).

containers, including the combination of pretensioning, posttensioning, and conventional reinforcing to structures and structural components.

Outside the field of tanks, bridges, and buildings, prestressed concrete has been occasionally applied to dams, by anchoring prestressed steel bars to the foundation, or by jacking the dam against it.[2] Piles, posts, and pipes all have been constructed of prestressed concrete. In certain structures, it is possible to prestress the concrete without using prestressing tendons. For example, the Freyssinet method of arch compensation introduces compensating stresses in the arch rib by a system of hydraulic jacks inserted in the arch. Such stresses are intended to neutralize the effects of shrinkage, rib shortening, and temperature drop in the arch. The Plougastel Bridge near Brest, with three spans of 612 ft (186.5 m) each, is an example of such application.[3]

The basic principle of prestressing is not limited to structures in concrete; it has been applied to steel construction as well. When two plates are joined together by hot-driven rivets or high-tensile bolts, the connectors are highly prestressed in tension and the plates in compression, thus enabling the plates to carry tensile loads between them. The Sciotoville bridge, of 720-ft (219.5 m) spans, had its members prestressed in bending during erection in order to neutralize the secondary stresses due to live and dead loads.[4] A continuous truss prestressed with high-tensile wires was built into the airplane hangars at Brussels, Belgium, and two similar ones were tested at the University of Ghent.[5]

Whether prestressing is applied to steel or concrete, its ultimate purpose is twofold: first, to induce desirable strains and stresses in the structure; second, to counterbalance undesirable strains and stresses. In prestressed concrete, the steel is preelongated so as to avoid excessive lengthening under service load, while the concrete is precompressed so as to prevent cracks under tensile stress. Thus an ideal combination of the two materials is achieved. The basic desirability of prestressed concrete is almost selfevident, but its widespread application will be advanced by engineers' acquaintance with its principles and practice and further development of its design and construction.

1-2 General Principles of Prestressed Concrete

One of the best definitions of prestressed concrete is given by the ACI Committee on Prestressed Concrete.

> Prestressed concrete: Concrete in which there have been introduced internal stresses of such magnitude and distribution that the stresses resulting from given external loadings are counteracted to a desired degree. In reinforced-concrete members the prestress is commonly introduced by tensioning the steel reinforcement.

It might be added that prestressed concrete, in the broader sense of the term, might also include cases where the stresses resulting from internal strains are counteracted to a certain degree, such as in arch compensation. This book, however, will deal essentially with prestressed-concrete structures as defined by the ACI Committee, and will limit itself to prestressing as introduced by the tensioning of steel reinforcement, known as tendons. The tendon, as defined in Appendix A, may consist of high-strength steel strands, wires, or bars as described in Chapter 2. This is presently by far the most common form of prestressed concrete and the greater part of this chapter as well as this book will discuss this type.

Three different concepts may be applied to explain and analyze the basic behavior of this form of prestressed concrete. It is important that a designer understands all three concepts so that he or she can proportion and design prestressed concrete structures with intelligence and efficiency. These will be explained as follows.

Fig. 1-11. Ponce Coliseum, Puerto Rico. Cantilevers 138 ft with 4 in. thin hyperbolic paraboloid shell and slender edge beams (T. Y. Lin International, Consulting Engineers). See also Fig. 15-12 for more details.

First Concept—Prestressing to Transform Concrete into an Elastic Material. This concept treats concrete as an elastic material and is probably still the most common viewpoint among engineers. It is credited to Eugene Freyssinet who visualized prestressed concrete as essentially *concrete* which is transformed from a brittle material into an elastic one by the precompression given to it. Concrete which is weak in tension and strong in compression is compressed (generally by steel under high tension) so that the brittle concrete would be able to withstand tensile stresses. From this concept the criterion of no tensile stresses was born. It is generally believed that if there are no tensile stresses in the concrete, there can be no cracks, and the concrete is no longer a brittle material but becomes an elastic material.

From this standpoint concrete is visualized as being subject to two systems of forces: internal prestress and external load, with the tensile stresses due to the external load counteracted by the compressive stresses due to the prestress. Similarly, the cracking of concrete due to load is prevented or delayed by the precompression produced by the tendons. So long as there are no cracks, the stresses, strains, and deflections of the concrete due to the two systems of forces can be considered separately and superimposed if necessary.

In its simplest form, let us consider a simple rectangular beam prestressed by a tendon through its centroidal axis (Fig. 1-13) and loaded by external loads. The tensile prestress force F in the tendon produces an equal compressive force F in the concrete, which also acts at the centroid of the tendon. In this case the force is at the centroid of the cross section. Due to the prestress F, a uniform compressive stress of

$$f = \frac{F}{A} \tag{1-1}$$

Fig. 1-12. An all-precast, nine-story office building under construction, University of California, Davis; 90-ft columns, floor panels, and exterior walls are all of precast, pretensioned concrete (Architect Gardner A. Dailey, T. Y. Lin International, Consulting Engineers). See also Fig. 14-7, photo of column being erected for this structure.

will be produced across the section that has an area A. If M is the external moment at a section due to the load on and the weight of the beam, then the stress at any point across that section due to M is

$$f = \frac{My}{I} \tag{1-2}$$

where y is the distance from the centroidal axis and I is the moment of inertia of the section. Thus the resulting stress distribution is given by

$$f = \frac{F}{A} \pm \frac{My}{I} \tag{1-3}$$

as shown in Fig. 1-13.

The solution is slightly more complicated when the tendon is placed eccentrically with respect to the centroid of the concrete section, Fig. 1-14. Here the resultant compressive force F in the concrete acts at the centroid of the tendon which is at a distance e from the c.g.c. as shown in Fig. 1-14. Due to an eccentric prestress, the concrete is subject to a moment as well as a direct load. The

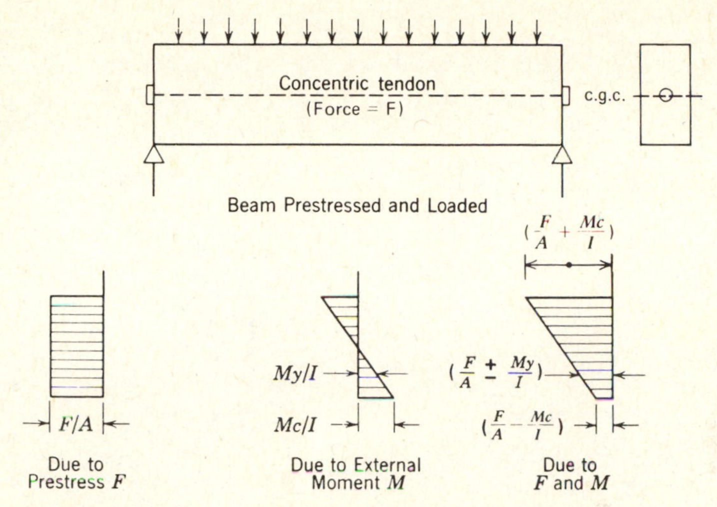

Fig. 1-13. Stress distribution across a concentrically prestressed-concrete section.

moment produced by the prestress is Fe, and the stresses due to this moment are

$$f = \frac{Fey}{I} \tag{1-4}$$

Thus, the resulting stress distribution is given by

$$f = \frac{F}{A} \pm \frac{Fey}{I} \pm \frac{My}{I} \tag{1-5}$$

as shown in the figure.

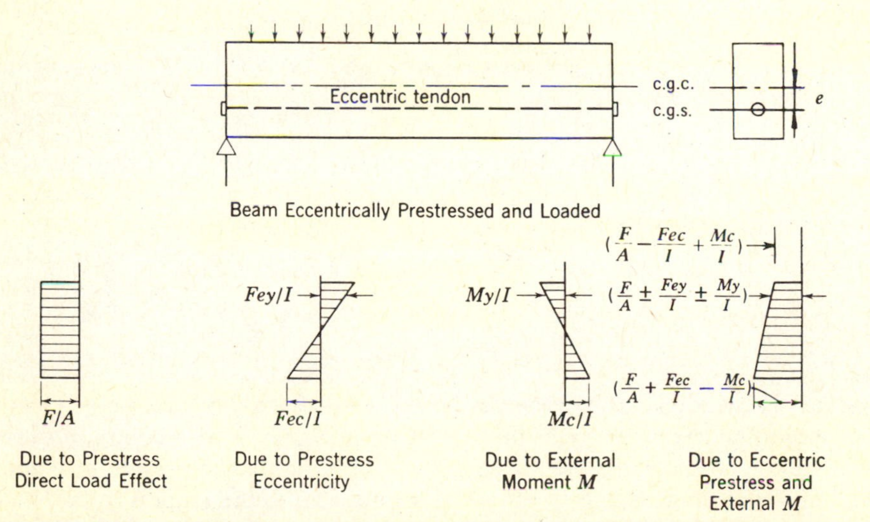

Fig. 1-14. Stress distribution across an eccentrically prestressed-concrete section.

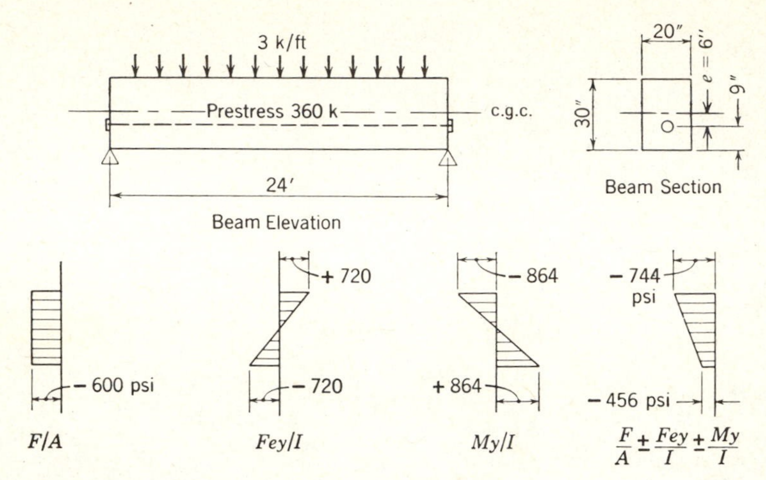

Fig. 1-15. Example 1-1.

EXAMPLE 1-1

A prestressed-concrete rectangular beam 20 in. by 30 in. has a simple span of 24 ft and is loaded by a uniform load of 3 k/ft including its own weight, Fig. 1-15. The prestressing tendon is located as shown and produces an effective prestress of 360 k. Compute fiber stresses in the concrete at the midspan section (span = 7.31 m, load = 43.8 kN/m and F = 1601 kN).

Solution Using formula 1-5, we have $F = 360$ k, $A = 20 \times 30 = 600$ in.^{2} (neglecting any hole due to the tendon), $e = 6$ in., $I = bd^3/12 = 20 \times 30^3/12 = 45{,}000$ in.^{4}; $y = 15$ in. for extreme fibers.

$$M = 3 \times 24^2/8 = 216 \text{ k-ft } (293 \text{ kN}-\text{m})$$

Therefore, assuming compressive stress negative, we have

$$f = \frac{F}{A} \pm \frac{Fey}{I} \pm \frac{My}{I}$$

$$= \frac{-360{,}000}{600} \pm \frac{360{,}000 \times 6 \times 15}{45{,}000} \pm \frac{216 \times 12{,}000 \times 15}{45{,}000}$$

$$= -600 \pm 720 \pm 864$$

$$= -600 + 720 - 864 = -744 \text{ psi } (-5.13 \text{ N/mm}^2) \text{ for top fiber}$$

$$= -600 - 720 + 864 = -456 \text{ psi } (-3.14 \text{ N/mm}^2) \text{ for bottom fiber}$$

The resulting stress distribution is shown in Fig. 1-15.

When the tendons are curved or bent, Fig. 1-16(a), it is often convenient to take either the left or the right portion of the member as a freebody in order to evaluate the effect of the prestressing force F. Note that the resultant compression on the concrete due to prestress alone is equal to the tendon force F acting at eccentricity e. Thus, in Fig. 1-16(b), equilibrium of horizontal forces indicates

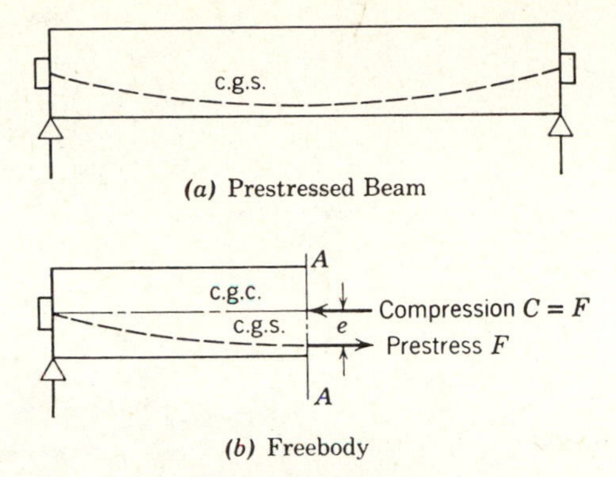

(a) Prestressed Beam

(b) Freebody

Fig. 1-16. Effect of prestress.

that the compression in the concrete equals the prestress in the steel F, and the stresses in the concrete due to eccentric force F is given by,

$$f = \frac{F}{A} \pm \frac{Fec}{I}$$

Thus, concrete stresses f at a section due to prestress are dependent only on the magnitude and location of F at that section, regardless of how the tendon profile may vary elsewhere along the beam. For example, if section A-A of the beam in Fig. 1-17 is identical with section A-A in Fig. 1-16, the concrete stresses due to prestress F with eccentricity e are identical for the two sections, regardless of variations in the shape of the beam or the cable profile away from the section. (This is true only for the statically determinate members wherein external reactions are not affected by the internal prestressing. See Chapters 10 and 11 for statically indeterminate systems.)

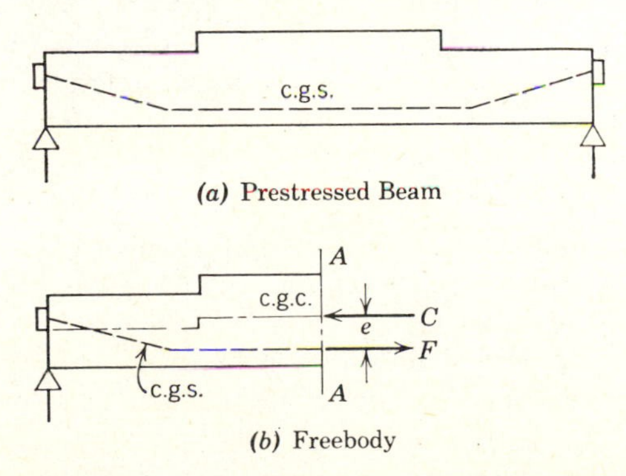

(a) Prestressed Beam

(b) Freebody

Fig. 1-17. Prestress effect is not related to variations away from section for a statically determinate member.

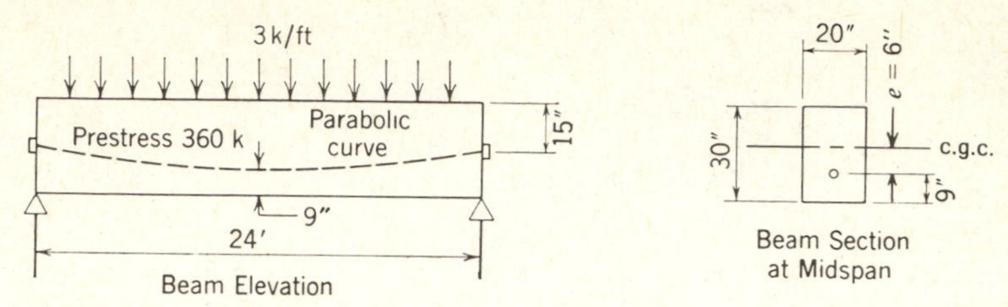

Fig. 1-18. Example 1-2.

EXAMPLE 1-2

A concrete beam with the same span, loading, section, and prestress as in example 1-1 has a parabolically curved tendon as shown, Fig. 1-18. Compute the extreme fiber stresses at midspan.

Solution The beam section at midspan is shown in the figure and is identical with the section in Fig. 1-15 for example 1-1. Hence exactly the same calculation for example 1-1 will apply, and the extreme fiber stresses are also the same at midspan,

Top fiber −744 psi (−5.13 N/mm^2) (compression)

Bottom fiber −456 psi (−3.14 N/mm^2) (compression)

Second Concept—Prestressing for Combination of High-Strength Steel with Concrete. This concept is to consider prestressed concrete as a combination of steel and concrete, similar to reinforced concrete, with steel taking tension and concrete taking compression so that the two materials form a resisting couple against the external moment, Fig. 1-19. This is often an easy concept for engineers familiar with reinforced concrete where the steel supplies a tensile force and the concrete supplies a compressive force, the two forces forming a couple with a lever arm between them. Few engineers realize, however, that similar behavior exists in prestressed concrete.

In prestressed concrete, high-tensile steel is used which will have to be elongated a great deal before its strength is fully utilized. If the high-tensile steel is simply buried in the concrete, as in ordinary concrete reinforcement, the surrounding concrete will have to crack very seriously before the full strength of

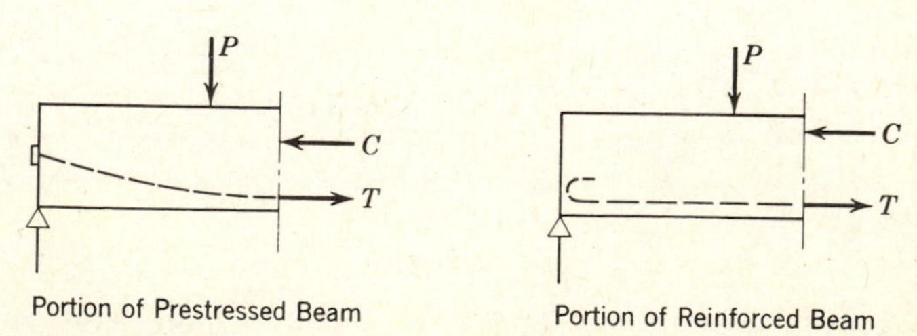

Fig. 1-19. Internal resisting moment in prestressed- and reinforced-concrete beams.

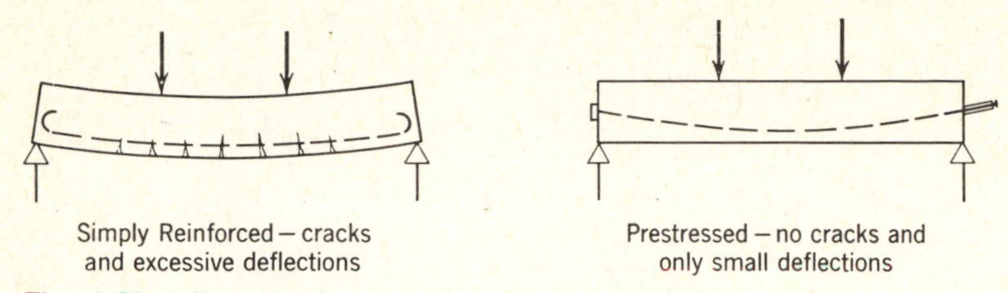

Fig. 1-20. Concrete beam using high-tensile steel.

the steel is developed, Fig. 1-20. Hence it is necessary to prestretch the steel with respect to the concrete. By prestretching and anchoring the steel against the concrete, we produce desirable stresses and strains in both materials: compressive stresses and strains in concrete, and tensile stresses and strains in steel. This combined action permits the safe and economical utilization of the two materials which cannot be achieved by simply burying steel in the concrete as is done for ordinary reinforced concrete. In isolated instances, medium-strength steel has been used as simple reinforcement without prestressing, and the steel was specially corrugated for bond, in order to distribute the cracks. This process avoids the expenses for prestretching and anchoring high-tensile steel but does not have the desirable effects of precompressing the concrete and of controlling the deflections.

From this point of view, prestressed concrete is no longer a strange type of design. It is rather an extension and modification of the applications of reinforced concrete to include steels of higher strength. From this point of view, prestressed concrete cannot perform miracles beyond the capacity of the strength of its materials. Although much ingenuity can be exercised in the proper and economic design of prestressed-concrete structures, there is absolutely no magic method to avoid the eventual necessity of carrying an external moment by an internal couple. And that internal resisting couple must be supplied by the steel in tension and the concrete in compression, whether it be prestressed or reinforced concrete. This concept has been well utilized to determine the ultimate strength of prestressed concrete beams and is also applicable to their elastic behavior.

Once the engineer sees this viewpoint, he or she understands the basic similarity between prestressed and reinforced concrete. Then much of the complexity of prestressing disappears, and the design of prestressed concrete can be intelligently accomplished and not performed by groping in the dark among a lot of complicated and confusing formulas.

The following example illustrates a simple application of the above principle in the analysis of prestressed-concrete beams; more extensive treatment will be presented in Chapter 6.

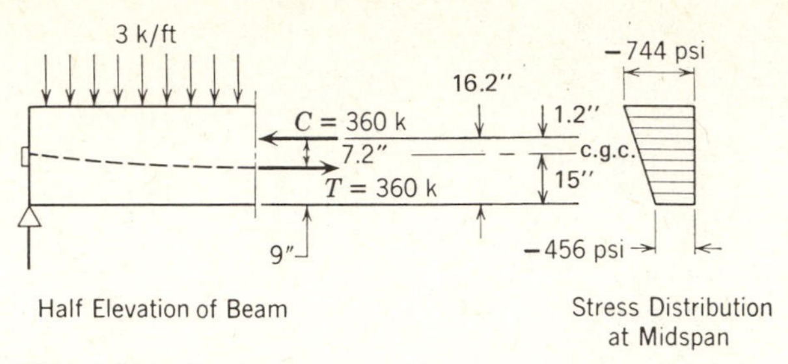

Fig. 1-21. Example 1-3.

EXAMPLE 1-3
Solve the problem stated in example 1-2 by applying the principle of the interal resisting couple.

Solution Take one half of the beam as a freebody, thus exposing the internal couple, Fig. 1-21. The external moment at the section is

$$M = \frac{wL^2}{8}$$
$$= \frac{3 \times 24^2}{8}$$
$$= 216 \text{ k-ft (293 kN-m)}$$

The internal couple is furnished by the forces $C = T = 360$ k, which must act with a lever arm of

$$\frac{216}{360} \times 12 = 7.2 \text{ in. (183 mm)}$$

Since T acts at 9 in. from the bottom, C must be acting at 16.2 in. from it. Thus the center of the compressive force C is located.

So far we have been dealing only with statics, the validity of which is not subject to any question. Now, if desired, the stress distribution in the concrete can be obtained by the usual elastic theory, since the center of the compressive force is already known. For $C = 360{,}000$ lb (1,601 kN) acting with an eccentricity of $16.2 - 15 = 1.2$ in. (30.48 mm),

$$f = \frac{F}{A} \pm \frac{Mc}{I}$$
$$= \frac{-360{,}000}{600} \pm \frac{360{,}000 \times 1.2 \times 15}{45{,}000}$$
$$= -600 \mp 144$$
$$= -744 \text{ psi } (-5.13 \text{ N/mm}^2) \text{ for top fiber}$$
$$= -456 \text{ psi } (-3.14 \text{ N/mm}^2) \text{ for bottom fiber}$$

Third Concept—Prestressing to Achieve Load Balancing. This concept is to visualize prestressing primarily as an attempt to balance the loads on a member. This concept was essentially developed by the author, although undoubtedly also utilized by other engineers to a lesser degree.

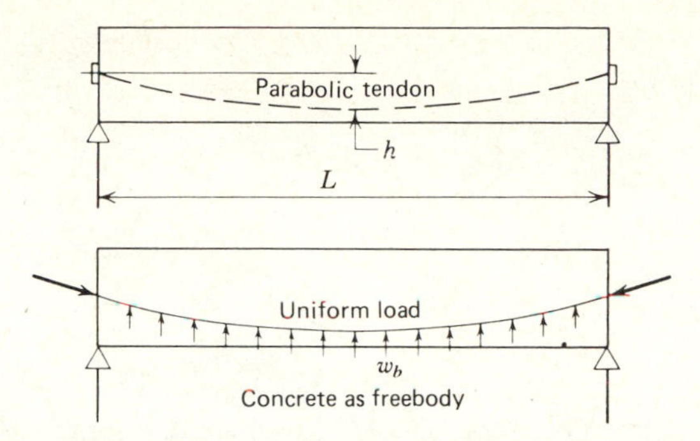

Fig. 1-22. Prestressed beam with parabolic tendon.

In the overall design of a prestressed concrete structure, the effect of prestressing is viewed as the balancing of gravity loads so that members under bending such as slabs, beams, and girders will not be subjected to flexural stresses under a given loading condition. This enables the transformation of a flexural member into a member under direct stress and thus greatly simplifies both the design and analysis of otherwise complicated structures.

The application of this concept requires taking the concrete as a freebody, and replacing the tendons with forces acting on the concrete along the span.

Take, for example, a simple beam prestressed with a parabolic tendon (Fig. 1-22) if

F = prestressing force
L = length of span
h = sag of parabola

The upward uniform load is given by

$$w_b = \frac{8Fh}{L^2}$$

Thus, for a given downward uniform load w, the transverse load on the beam is balanced, and the beam is subjected only to the axial force F, which produces uniform stresses in concrete, $f = F/A$. The change in stresses from this balanced condition can easily be computed by the ordinary formulas in mechanics, $f = Mc/I$. The moment in this case is the unbalanced moment due to $(w - w_b)$, the unbalanced load. Figure 1-23 shows load balancing applied to the Arizona State Fair Colliseum roof structure.

For a beam with bent tendon, Fig. 1-24, the load from the tendon on the concrete can easily be determined by statics. This approach, while unnecessarily cumbersome for some simple cases, often becomes very effective for complicated structures, such as continuous beams, rigid frames, flat and waffle slabs, and some thin shells, which will be explained more fully in Chapter 11. When further extended, it can be used for the design and analysis of self-anchored,

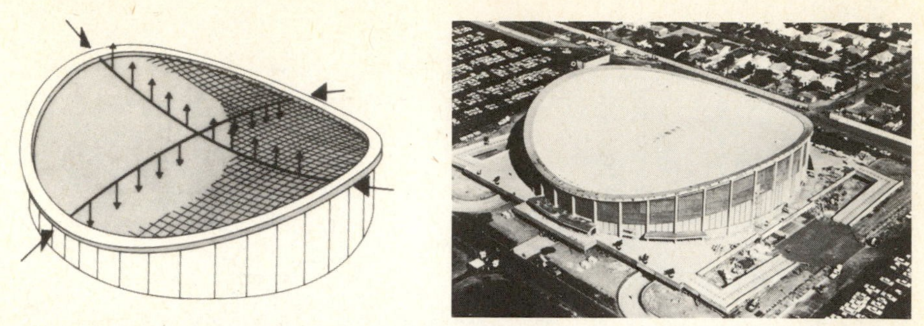

Fig. 1-23. Arizona State Fair Coliseum 380 ft diameter hyperbolic paraboloid shell of lightweight concrete waffles $2\frac{1}{2}$ in. thick (T. Y. Lin International, Consulting Engineers).

prestressed-concrete bridges, when the force from the steel cable on the concrete roadway and girders can be predetermined, and the stresses in the concrete analyzed without much difficulty.

EXAMPLE 1-4

Solve the problem in example 1-2 by the method of load balancing taking the concrete as freebody, isolated from the tendon or steel, Fig. 1-25.

Solution The upward uniform force from the tendon on the concrete is

$$w_b = \frac{8Fh}{L^2}$$

$$= \frac{8 \times 360 \times (6/12)}{24^2}$$

$$= 2.5 \text{ k/ft } (36.5 \text{ kN/m})$$

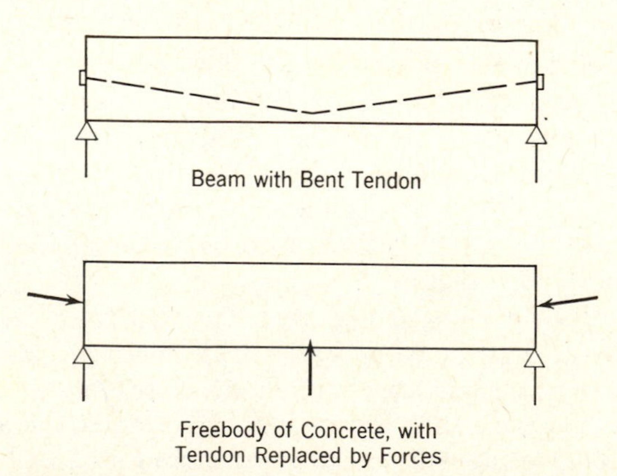

Fig. 1-24. Prestressed beam with bent tendon.

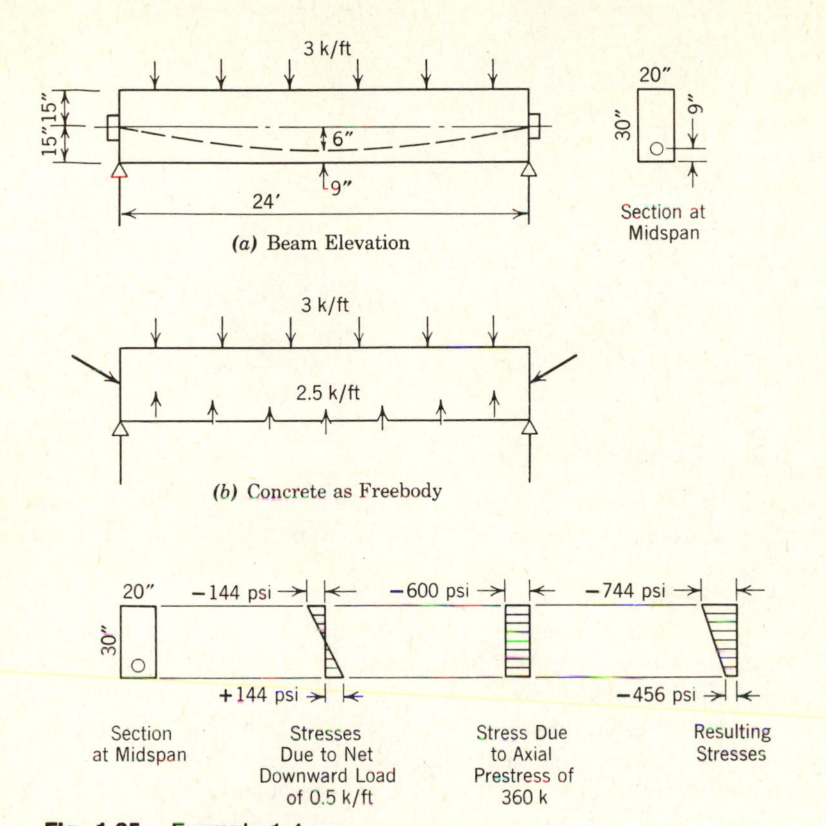

Fig. 1-25. Example 1-4.

Hence the net downward (unbalanced) load on the concrete beam is $(3-2.5)=0.5$ k/ft (7.3 kN/m), and the moment at midspan due to that load is

$$M=\frac{wL^2}{8}=\frac{0.5\times 24^2}{8}$$

$$=36 \text{ k-ft } (48.8 \text{ kN}-\text{m})$$

The fiber stresses due to that moment are

$$f=\frac{Mc}{I}=\frac{6M}{bd^2}$$

$$=\frac{6\times 36\times 12{,}000}{20\times 30^2}$$

$$=144 \text{ psi } (0.993 \text{ N/mm}^2) \text{ (compression top fiber; tension bottom fiber)}$$

The fiber stress due to the direct load effect of the prestress is very nearly

$$\frac{F}{A}=\frac{-360{,}000}{20\times 30}$$

$$=-600 \text{ psi } (-4.14 \text{ N/mm}^2) \text{ compression}$$

The resulting stresses are

$$-144-600=-744\ (-5.13 \text{ N/mm}^2) \text{ top fiber comp.}$$

$$+144-600=-456\ (-3.14 \text{ N/mm}^2) \text{ bottom fiber comp.}$$

the same as in examples 1-2 and 1-3.

1-3 Classification and Types

Prestressed-concrete structures can be classified in a number of ways, depending upon their features of design and construction. These will be discussed as follows.

Externally or Internally Prestressed. Although this book is devoted to the design of prestressed-concrete structures internally prestressed, presumably with high-tensile steel, it must be mentioned that it is sometimes possible to prestress a concrete structure by adjusting its external reactions. The method of arch compensation was mentioned previously, where a concrete arch was prestressed by jacking against its abutments. Theoretically, a simple concrete beam can also be externally prestressed by jacking at the proper places to produce compression in the bottom fibers and tension in the top fibers, Fig. 1-26, thus even dispensing with steel reinforcement in the beam. Such an ideal arrangement, however, cannot be easily accomplished in practice, because, even if abutments favorable for such a layout are obtainable, shrinkage and creep in concrete may completely offset the artificial strains unless they can be readjusted. Besides, such a site would probably be better suited for an arch bridge.

For a statically indeterminate structure, like a continuous beam, it is possible to adjust the level of the supports, by inserting jacks, for example, so as to produce the most desirable reactions, Fig. 1-27. This is sometimes practical, though it must be kept in mind that shrinkage and creep in concrete will modify

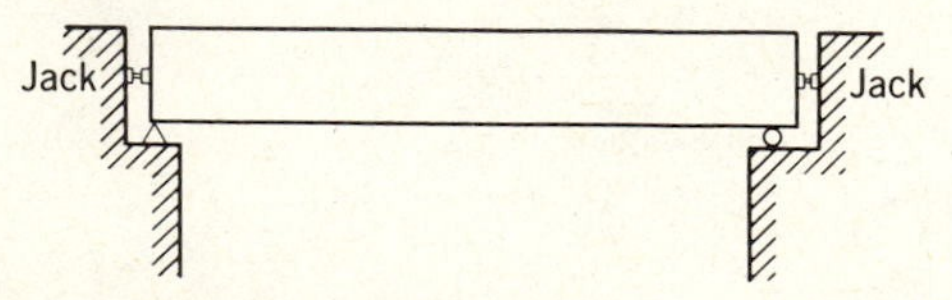

Fig. 1-26. Prestressing a simple concrete beam by jacking against abutments.

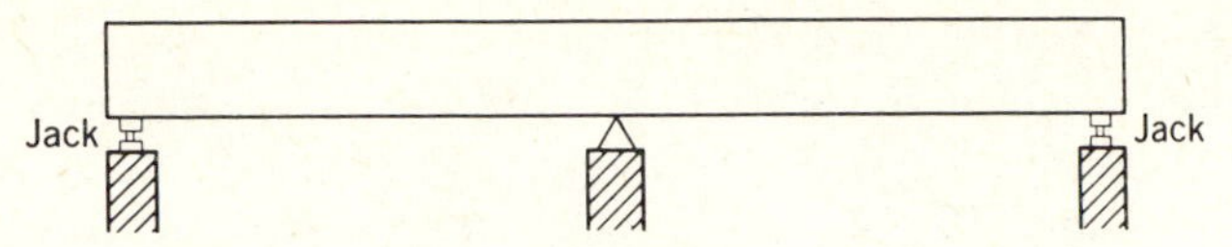

Fig. 1-27. Prestressing a continuous beam by jacking its reactions.

the effects of such prestress so that they must be taken into account or else the prestress must be adjusted from time to time.

Linear or Circular Prestressing. Circular prestressing is a term applied to prestressed circular structures, such as round tanks, silos, and pipes, where the prestressing tendons are wound around in circles. This topic is discussed in Chapter 13. As distinguished from circular prestressing, the term linear prestressing is often employed to include all other structures such as beams and slabs. The prestressing tendons in linearly prestressed structures are not necessarily straight; they can be either bent or curved, but they do not go round and round in circles as in circular prestressing.

Pretensioning and Posttensioning. The term pretensioning is used to describe any method of prestressing in which the tendons are tensioned before the concrete is placed. It is evident that the tendons must be temporarily anchored against some abutments or stressing beds when tensioned and the prestress transferred to the concrete after it has set. This procedure is employed in precasting plants or laboratories where permanent beds are provided for such tensioning; it is also applied in the field where abutments can be economically constructed. In contrast to pretensioning, posttensioning is a method of prestressing in which the tendon is tensioned after the concrete has hardened. Thus the prestressing is almost always performed against the hardened concrete, and the tendons are anchored against it immediately after prestressing. This method can be applied to members either precast or cast in place.

End-Anchored or Non-End-Anchored Tendons. When posttensioned, the tendons are anchored at their ends by means of mechanical devices to transmit the prestress to the concrete. Such a member is termed end anchored. A posttensioned member may have its tendons held by grout after mechanical end anchorage has allowed the stressing to be accomplished, but the end anchorage hardware is still required during construction. Grouting is described below in connection with bonded or unbonded classification of posttensioned members. In pretensioning, the tendons generally have their prestress transmitted to the concrete by their bond action near the ends. The effectiveness of such stress transmission is limited to wires of small size, and to larger diameter strands which possess better bond properties than smooth wires. The most common type material for pretensioning is seven-wire strand, which is also used in many posttensioning systems. Different types of end anchorages will be discussed in Chapter 3.

Bonded or Unboned Tendons. Bonded tendons denote those bonded throughout their length to the surrounding concrete. Non-end-anchored tendons are necessarily bonded ones; end-anchored tendons may be either bonded or unbonded to the concrete. In general, the bonding of posttensioned tendons is accomplished by subsequent grouting; if unbonded, protection of the tendons from corrosion must be provided by galvanizing, greasing, or some other means. Typically, the unbonded tendon is greased and wrapped with paper or plastic material to prevent bonding to the surrounding concrete. Sometimes, bonded tendons may be purposely unbonded along certain portions of their length.

Precast, Cast-in-Place, Composite Construction. Precasting involves the placing of concrete away from its final position, the members being cast either in a permanent plant or somewhere near the site of the structure, and eventually erected at the final location. Precasting permits better control in mass production and is often economical. Cast-in-place concrete requires more form and falsework per unit of product but saves the cost of transportation and erection, and it is a necessity for large and heavy members. In between these two methods of construction, there are tilt-up wall panels and lift slabs which are constructed at places near or within the structure and then erected to their final position; no transportation is involved for these. Oftentimes, it is economical to precast part of a member, erect it, and then cast the remaining portion in place. This procedure is called composite construction. The precast elements in a structure of composite construction can be more easily joined together than those in a totally precast structure. By composite construction, it is possible to save much of the form and falsework required for total cast-in-place construction. However, the suitability of each type must be studied with respect to the particular conditions of a given structure.

Partial or Full Prestressing. A further distinction between the types of prestressing is sometimes made depending on the degree of prestressing to which a concrete member is subject. When a member is designed so that under the working load there are no tensile stresses in it, then the concrete is said to be fully prestressed. If some tensile stresses will be produced in the member under working load, then it is termed partially prestressed. For partial prestressing, additional mild-steel bars are frequently provided to reinforce the portion under tension. In practice, it is often difficult to classify a structure as being partially or fully prestressed since much will depend on the magnitude of the working load used in design. For example, highway bridges in this country may be designed for full prestressing, though actually they are subject to tensile stresses during the passage of exceptionally heavy vehicles. On the other hand, roof beams designed for partial prestressing may never be subject to tensile stresses since the assumed live loads may never act on them. Partial prestressing is further discussed in section 1-6 and in Chapter 9.

1-4 Stages of Loading

One of the considerations peculiar to prestressed concrete is the plurality of stages of loading to which a member or structure is often subjected. Some of these stages of loading occur also in nonprestressed structures, but others exist only because of prestressing. For a cast-in-place structure, prestressed concrete has to be designed for at least two stages; the initial stage during prestressing and the final stage under external loadings. For precast members, a third stage, that of handling and transportation, has to be investigated. During each of these three stages, there are again different periods when the member or structure may be under different loading conditions. These will now be analyzed. Table 1-2 summarizes the permissible stresses.

Initial Stage. The member or structure is under prestress but is not subjected to any superimposed external loads. This stage can be further subdivided into the following periods, some of which may not be important and therefore may be neglected in certain designs.

TABLE 1-2 Permissible Stresses for Flexural Members (ACI Code)

Steel Stresses–not more than the following values:

1. Due to tendon jacking force,

$$0.80 f_{pu} \quad \text{or} \quad 0.94 f_{py}$$

whichever is smaller, but not greater than maximum value recommended by manufacturer of prestressing tendons or anchorages.

2. Pretensioned tendons immediately after transfer of prestress or posttensioned tendons after anchorage,

$$0.70 f_{pu}$$

Concrete Stresses–not more than the following values:

1. Immediately after transfer of prestress (before losses), extreme fiber stress

Compression–$0.60 f'_{ci}$

Tension–$3\sqrt{f'_{ci}}$ (except at ends of simply supported members where $6\sqrt{f'_{ci}}$ is permitted)

2. At service load after allowance for all prestress losses,

Compression–$0.45 f'_c$

Tension[a] –$6\sqrt{f'_c}$

[a]When analysis based on cracked sections and bilinear moment-deflection relationships show that immediate and long-time deflections satisfy Code limits, maximum tension is $12\sqrt{f'_c}$.

Before Prestressing. Before the concrete is prestressed, it is quite weak in carrying load; hence the yielding of its supports must be prevented. Provision must be made for the shrinkage of concrete if it might occur. When it is desirable to minimize or eliminate cracks in prestressed concrete, careful curing before the transfer of prestress is very important. Drying or sudden change in temperature must be avoided. Cracks may or may not be closed by the application of prestress, depending on many factors. Shrinkage cracks will destroy the capacity of the concrete to carry tensile stresses and may be objectionable.

During Prestressing. This is a critical test for the strength of the tendons. Often, the maximum stress to which the tendons will be subject throughout their life occurs at that period ($0.80 f_{pu}$ or $0.94 f_{py}$, Table 1-2). It occasionally happens that an individual wire may be broken during prestressing, owing to defects in its manufacture. But this break is seldom significant, since there are often many wires in a member. If a bar is broken in a member with only a few bars, it should be properly replaced. For concrete, the prestressing operations impose a severe test on the bearing strength at the anchorages. Since the concrete is not aged at this period while the prestress is at its maximum, crushing of the concrete at the anchorages is possible if its quality is inferior or if the concrete is honeycombed. Again, unsymmetrical and concentrated prestress from the tendons may produce overstresses in the concrete. Therefore the order of prestressing the various tendons must often be studied beforehand.

At Transfer of Prestress. For pretensioned members, the transfer of prestress is accomplished in one operation and within a short period. For posttensioned members, the transfer is often gradual, the prestress in the tendons being transferred to the concrete one by one. In both cases there is no external load on the member except its own weight. Thus the initial prestress, with little loss as yet taking place, imposes a serious condition on the concrete and often controls the design of the member. (Table 1-2 shows permissible steel and concrete stresses.) For economic reasons the design of a prestressed member often takes into account the weight of the member itself in holding down the cambering effect of prestressing. This is done on the assumption of a given condition of support for the member. If that condition is not realized in practice, failure of the member might result. For example, the weight of a simply supported prestressed girder is expected to exert a maximum positive moment at midspan which counteracts the negative moment due to prestressing. If the girder is cast and prestressed on soft ground without suitable pedestals at the ends, the expected positive moment may be absent and the prestressing may produce excessive tensile stresses on top fibers of the girder, resulting in its failure.

Decentering and Retensioning. If a member is cast and prestressed in place, it generally becomes self-supporting during or after prestressing. Thus the falsework can be removed after prestressing, and no new condition of loading is imposed on the structure. Some concrete structures are retensioned, that is, prestressed in two or more stages. Then the stresses at various stages of tensioning must be studied.

Intermediate Stage. This is the stage during transportation and erection. It occurs only for precast members when they are transported to the site and erected in position. It is highly important to ensure that the members are properly supported and handled at all times. For example, a simple beam designed to be supported at the ends will easily break if lifted at midspan, Fig. 1-28. Figure 1-6 shows a correct way to lift a prestressed simple beam.

Not only during the erection of the member itself, but also when adding the superimposed dead loads, such as roofing or flooring, attention must be paid to the conditions of support and loading. This is especially true for a cantilever layout, when partial loading may result in more serious bending than a full loading, Fig. 1-30.

Final Stage. This is the stage when the actual working loads come on the structure. (Table 1-2 shows permissible stresses.) As for other types of construction, the designer must consider various combinations of live loads on different portions of th structure with lateral loads such as wind and earthquake forces, and with strain loads such as those produced by settlement of supports and temperature effects. For prestressed-concrete sturctures, especially those of unconventional types, it is often necessary to investigate their cracking and ultimate loads, their behavior under the actual sustained load in addition to the working load. These will be discussed as follows.

Sustained Load. The camber or deflection of a prestressed member under its actual sustained load (which often consists only of the dead load) is often the controlling factor in design, since the effect of flexural creep will eventually magnify its value. Hence it is often desirable to limit the camber or deflection under sustained load.

Working Load. To design for the working load is a check on excessive stresses and strains. It is not necessarily a guarantee of sufficient strength to

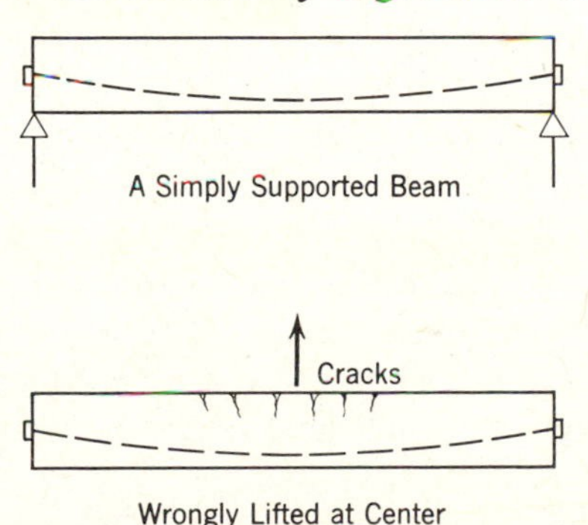

Fig. 1-28. Failure of beam due to careless handling.

Fig. 1-29. Precast Pagoda Japanese Center, San Francisco, California, wound with circular posttensioning (T. Y. Lin International, Consulting Engineers).

carry overloads. However, an engineer familiar with the strength of prestressed-concrete structures may often design conventional types and proportions on the basis of working-load computations, then check strength.

Cracking Load. Cracking in a prestressed-concrete member signifies a sudden change in the bond and shearing stresses. It is sometimes a measure of the fatigue strength. For certain structures, such as tanks and pipes, the commencement of cracks presents a critical situation. For structures subject to corrosive influences, for unbonded tendons where cracks are more objectionable, or for structures where cracking may result in excessive deflections, an investigation of the cracking load seems important.

Ultimate Load. Structures designed on the basis of working stresses may not always possess a sufficient margin for overloads. Since it is required that a

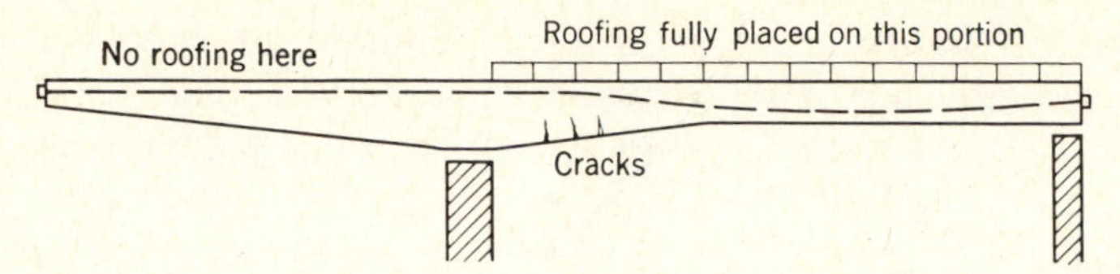

Fig. 1-30. Cracking of beam due to wrong sequence in adding superimposed load.

structure possess a certain minimum factored load capacity, it is necessary to determine its ultimate strength. In general, the ultimate strength of a structure is defined by the maximum load it can carry before collapsing. However, before this load is reached, permanent yielding of some parts of the structure may already have developed. Although any strength beyond the point of permanent yielding may serve as additional guarantee against total collapse, some engineers consider such strength as not usable and prefer to design on the basis of usable strength rather than the ultimate strength. However, ultimate strength is more easily computed and is commonly accepted as a criterion for design in prestressed concrete as with other structural systems. Table 1-3 summarizes the basic ACI Code strength requirements.

Table 1-3 Load Factors and Required Strength Under ACI Code

ACI Code Load Factors

$U = 1.4D + 1.7L$	(ACI 9-1)	Basic Required Strength[a]
where wind is included, see if one of the following is greater:		
$U = 0.75(1.4D + 1.7L + 1.7W)$	(ACI 9.2)	
or		
$U = 0.9D + 1.3W$	(ACI 9-3)	

where U = required strength
D = dead load
L = live load
W = wind load

Required Strength	$\leqslant$	**Design Strength of Member**
Req'd strength from factored loads	$\begin{Bmatrix} M_u \leqslant \phi M_n \\ P_u \leqslant \phi P_n \\ V_u \leqslant \phi V_n \end{Bmatrix}$	Member design strength is the strength reduction factor[b], ϕ, times the best estimate of member strength (nominal strength) $\phi = 0.90$ – flexure, M $\phi = 0.85$ – shear, V $\phi = 0.75$ – spiral column, P

[a]Other equations in ACI 318-77 include load factors for earthquake, lateral earth pressure, lateral liquid pressure, and temperature effects in checking strength.

[b]See ACI Code for other ϕ values where the kind of stress is different from these; such as bearing pressure, axial tension, or flexure of plain concrete.

In addition to the above normal loading conditions, some structures may be subject to repeated loads of appreciable magnitude which might result in fatigue failures. Some structures may be under heavy loads of long duration, resulting in excessive deformations due to creep, while others may be under such light external loads that the camber produced by prestressing may become too pronounced as time goes on. Still others may be subject to undesirable vibrations under dynamic loads. Under a sudden impact load or under the action of earthquakes, the energy absorption capacity of the member as indicated by its ductility may be of prime importance. These are special conditions which the engineer must consider for his individual case.

The above discussion outlines the relatively new and complex problems encountered in the design of prestressed-concrete as compared with reinforced-concrete structures. It is unfortunate that the design of prestressed concrete is more complicated, but the difficulty is by no means excessive. The new problems must be understood and solved. Ignorance of the situation might result in tragic failures such as are experienced by careless practitioners in almost any new field of endeavor.

With some experience in design, many of the loading stages mentioned above are automatically eliminated from consideration by inspection. Calculations will actually have to be made for only one or two controlling conditions. Besides, as will be shown in later chapters, calculations can be greatly simplified if the correct methods of approach and analysis are chosen. It is the observation of the authors that an engineer who belittles the complications of prestressed-concrete design will encounter problems beyond his expectations, while the majority of engineers will find it not as difficult as they may imagine.

1-5 Prestressed vs. Reinforced Concrete

As it is assumed that readers are already acquainted with reinforced concrete, it will be interesting to compare prestressed concrete with it. The most outstanding difference between the two is the employment of materials of higher strength for prestressed concrete. In order to utilize the full strength of the high-tensile steel, it is necessary to resort to prestressing to prestretch it. Prestressing the steel and anchoring it against the concrete produces desirable strains and stresses which serve to reduce or eliminate cracks in concrete. Thus the entire section of the concrete becomes effective in prestressed concrete, whereas only the portion of section above the neutral axis is supposed to act in the case of reinforced concrete.

The use of curved tendons will help to carry some of the shear in a member. In addition, precompression in the concrete tends to reduce the principal tension, increasing shear strength. Thus it is possible to use a smaller section in prestressed concrete to carry the same amount of external shear in a beam.

Hence more efficient I-shaped sections with thin webs become desirable with prestressed concrete.

High-strength concrete, which cannot be economically utilized in reinforced-concrete construction, is found to be desirable and even necessary with prestressed concrete. In reinforced concrete, using concrete of high strength will result in a smaller section calling for more reinforcement and will end with a more costly design. In prestressed concrete, high-strength concrete is required to match with high-strength steel in order to yield economical proportions. Stronger concrete is also necessary to resist high stresses at the anchorages and to give strength to the thinner sections so frequently employed for prestressed concrete.

Each material or method of construction has its own field of application. When welding was first developed in the 1930's, some engineers were overenthusiastic and believed that it would replace riveting altogether, which it has not done even yet. Prestressed concrete is likely to have a similar course of development. Not for a long time will it be used in as great quantity as reinforced concrete.

But this relatively new type of construction, basically sound in its strength and economy, has had a rapid rate of growth (Fig. 1-9) and is adaptable to new and unprecedented situations and requirements. The advantages and disadvantages of prestressed concrete as compared with reinforced concrete will now be discussed with respect to their serviceability, safety, and economy.

Serviceability. Prestressed-concrete design is more suitable for structures of long spans and those carrying heavy loads, principally because of the higher strengths of materials employed. Prestressed structures are more slender and hence more adaptable to artistic treatment. They yield more clearance where it is needed. They do not crack under working loads, and whatever cracks may be developed under overloads will be closed up as soon as the load is removed, unless the load is excessive. Under dead load, the deflection is reduced, owing to the cambering effect of prestress. This becomes an important consideration for such structures as long cantilevers. Under live load, the deflection is also smaller because of the effectiveness of the entire uncracked concrete section, which has a moment of inertia two to three times that of the cracked section. Prestressed elements are more adaptable to precasting because of the lighter weight.

So far as serviceability is concerned, the only shortcoming of prestressed concrete is its lack of weight. Although seldom encountered in practice, there are situations where weight and mass are desired instead of strength. For these situations, plain or reinforced concrete could often serve just as well and at lower cost.

Safety. It is difficult to say that one type of structure is safer than another. The safety of a structure depends more on its design and construction than on its type. However, certain inherent safety features in prestressed concrete may be mentioned. There is partial testing of both the steel and the concrete during prestressing operations. For many structures, during prestressing, both the steel and the concrete are subjected to the highest stresses that will exist in them during their life of service. Hence, if the materials can stand prestressing, they are likely to possess sufficient strength for the service loads.

When properly designed by the present conventional methods, prestressed-concrete structures have overload capacities similar to and perhaps slightly higher than those of reinforced concrete. For the usual designs, they deflect appreciably before ultimate failure, thus giving ample warning before impending collapse. The ability to resist shock and impact loads and repeated working loads has been shown to be as good in prestressed as in reinforced concrete. The resistance to corrosion is better than that of reinforced concrete for the same amount of cover, owing to the nonexistence of cracks and high quality of concrete used for prestressed members. If cracks should occur, corrosion can be more serious in prestressed concrete. Regarding fire resistance, high-tensile steel is more sensitive to high temperatures, but, for the same amount of minimum cover, prestressed tendons can have a greater average cover because of the

(a.) Completed Bridge–(with all 62 precast prestressed segments and cable stays in place).

(b.) During Construction–(lifting of precast prestressed segment in progress).

Fig. 1-31. Pasco-Kennewick Cable Stayed Bridge. The main center span is 981 ft (299 m) and the two adjacent spans are 406.5 ft (124 m). Of its total length of 2503 ft (760 m), over 1800 ft (550 m) are supported by steel stay-cables. This portion of the bridge is composed of 62 precast prestressed concrete segments 80 ft (24 m) wide and 27 ft (8.2 m) long. A uniform depth of 7 ft (2.1 m) was maintained for the continuous posttensioned girders (*Owners:* Cities of Pasco and Kennewick, Washington; *Engineer:* Arvid Grant & Associates, Inc.; *General Contractor:* Peter Kiewit Sons' Co.). (Prestressed Concrete Institute)

spread and curvature of the individual tendons. These problems are discussed in Chapter 16.

Prestressed-concrete members do require more care in design, construction, and erection than those of ordinary concrete, because of the higher strength, smaller section, and sometimes delicate design features involved. Although prestressed-concrete construction has been practiced only since the late 1940's, it is possible to conclude from experience that the life of such structures can be as long as if not longer than that of reinforced concrete.

Economics. From an economic point of view, it is at once evident that smaller quantities of materials, both steel and concrete, are required to carry the

Fig. 1-32. Ruck-a-Chucky Bridge with 1300 ft curved span to be posttensioned with cable stays from rock walls of canyon. (T. Y. Lin International, Consulting Engineers). Photo courtesy Popular Science Magazine.

same loads, since the materials are of higher strength. There is also a definite saving in stirrups, since shear in prestressed concrete is reduced by the inclination of the tendons, and the diagonal tension is further minimized by the presence of prestress. The reduced weight of the member will help in economizing the sections; the smaller dead load and depth of members will result in saving materials from other portions of the structure. In precast members, a reduction of weight saves handling and transportation costs.

In spite of the above economies possible with prestressed concrete, its use cannot be advocated for all conditions. First of all, the stronger materials will have a higher unit cost. More auxiliary materials are required for prestressing, such as end anchorages, conduits, and grouts. More complicated formwork is also needed, since nonrectangular shapes are often necessary for prestressed concrete. More labor is required to place 1 lb of steel in prestressed concrete, especially when the amount of work involved is small. More attention to design is involved, and more supervision is necessary; the amount of additional work will depend on the experience of the engineer and the construction crew, but it will not be serious if the same typical design is repeated many times.

From the above discussion, it can be concluded that prestressed-concrete design is more likely to be economical when the same unit is repeated many times or when heavy dead loads on long spans are encountered. It should also find suitable application when combined with precasting or semiprecasting such as composite or lift-slab construction. Each structure must be considered individually. The availability of good designers, of experienced crews, of pretensioning factories, and of competitive bidding often helps to tip the balance in favor of prestressed concrete.

1-6 Partial Prestressing

As previously mentioned in section 1-3, the use of partial prestressing has become common practice. Partially prestressed members are those which are designed to allow significant tensile stresses to occur at service loads, and such tensile regions are usually additionally reinforced with nonprestressed reinforcement.

Most design codes for both buildings and bridges now allow significant tensile stresses at service load (Table 1-2 shows ACI Code values), thus partial prestressing is very common. Some fully prestressed structures have developed too much upward deflection (camber), which is not desirable. Partial prestressing has accomplished the purpose of eliminating or controlling crack width at service loads by setting allowable tensile stresses which are slightly less than cracking stress for the concrete.

It is clear that most prestressed concrete design now technically falls under this classificiation rather than fully prestressing to eliminate tensile stresses as practiced earlier. But caution is still required for partial prestressing when cracking is possible at service loads, because we may get excessive live load deflection of the cracked section. Because our state of the art is such that we do have both research and actual design experience to support the widespread use of partial prestressing we can design to control stresses, and to insure deflection as well as crack control. At the same time, strength can be enhanced by the addition of supplemental flexural reinforcement. Table 1-3 gives required load factors from the ACI Code.

The special considerations described above (section 1-5) should help the reader to visualize the general comparison of reinforced concrete and prestressed concrete in various ways as we find it in practice. The designer finds the degree of prestressing which gives the desired control of stress at each of the loading stages described in section 1-4 to assure satisfactory performance for a given situation. The main concern is to produce a satisfactory structure whether the final design is "partially prestressed" or "fully prestressed" concrete. More complete discussion of partial prestressing and nonprestressed reinforcements is given in Chapter 9.

1-7 Design Codes for Prestressed Concrete

The first design guides for prestressed concrete were in the form of recommended practice, rather than building codes. In the United States, the "ACI-ASCE Joint Committee Recommendations for the Design of Prestressed Members" published in 1958, included the state-of-the-art knowledge which had developed with the limited use of prestressed concrete by the mid-1950's.

The Prestressed Concrete Institute published the first U.S. Building Code for prestressed concrete in 1961. At that time, the American Concrete Institute Building Code (318-56) contained no reference to prestressed concrete, but the inclusion of new material on this subject was being considered for the next revision.

In 1963, the ACI Code (318-63) included a chapter covering prestressed concrete, much of which was carried forward into the 1971 revision of the ACI Code (318-71). Since 1971, annual revisions have been made, and the ACI Code[6] with current revisions is the design code used throughout this book. A similar evolution occurred with the AASHTO "Standard Specification for Highway Bridges." The major provisions for prestress concrete in the current Standard Specification for Highway Bridges (AASHTO)[7] with latest revisions are very similar to those of the ACI Code. The only major differences are the allowable stress values and load factors which have been traditionally more conservative for bridges than buildings.

Both the ACI Code[6] for buildings and the AASHTO Specification[7] for bridges in the United States have evolved with information from around the world. It is

Fig. 1-33. Ekofisk Offshore Reservoir in the North Sea with combined precast and cast-in-place concrete posttensioned to form completed structure.

the feeling of the authors that design following the ACI Code (or similar provisions in AASHTO) will produce very satisfactory prestressed concrete structures, although this document may not be legally binding in some other countries. In examples throughout this book, many design provisions of the ACI Code are followed. It is noted that similar provisions are embodied in most codes throughout the world.

Since the early attempts at prestressing, outlined at the beginning of this chapter, a major industry has developed as is obvious from the photos of major structures using prestressed concrete throughout the world. We clearly have the materials and technology to be confident that out structures using this material will be safe and serviceable. We can rationally design for fire resistance and corrosion resistance following guidelines of our present codes and specifications.

Many special applications have been developed for prestress concrete in addition to buildings, bridges, and containment structures. Special products previously made with other materials have emerged; for example, railroad ties[12,13] and power line poles. Some of these situations will not necessarily fall under the Code provisions developed for buildings and bridges, but they will be helpful for guidance. There is some question whether fatigue is potentially a problem for some of these special situations. Continued research and development will solve any problems associated with fatigue and, in turn, our codes will be still further improved.

References

1. T. Y. Lin and F. Kulka, "Fifty-Year Advancement in Concrete Bridge Construction", *J. Const. Div.*, Am. Soc. of Civil Engineers, September 1975, pp. 491–510.
2. "Dams of Prestressed Concrete," *Eng. News-Rec.*, April 5, 1945, p. 456.
3. C. B. McCullough and E. S. Thayer, *Elastic Arch Bridges*, John Wiley & Sons, New York, 1931 (out of print).
4. G. A. Hool and W. S. Kinne, *Movable and Long Span Steel Bridges*, McGraw-Hill Book Co., New York, 1943.
5. G. Magnel and H. Lambotte, "Essai de deux poutres jumelées en acier précomprimé de 21.20 metres de portée," *Précontrainte Prestressing*, No. 2, 1953.
6. *Building Code Requirements for Reinforced Concrete* (ACI std. 318–77), Detroit, American Concrete Institute, 1977.
7. *Standard Specifications for Highway Bridges*, twelfth edition, American Association of State Highway and Transportation Officials (AASHTO), 1977.
8. *PCI Design Handbook*, *Precast Prestressed Concrete*, second edition, Prestressed Concrete Institute, Chicago, Illinois, 1978.
9. *Posttensioning Manual*, Post-Tensioning Institute, Phoenix, Arizona, 1976.
10. "T. Y. Lin Symposium on Prestressed Concrete, Special Commemorative Issue," *J. Prestressed Conc. Inst.*, Vol. 21, No. 5, September–October 1976.
11. Publications from Fédération Internationale de la Précontrainte (FIP):
"Recommendations for the Acceptance and Application of Posttensioning Systems, 1972.
"Recommendations for the Approval, Supply and Acceptance of Steels for Prestressing Tendons," 1974.
"Recommendations for the Design and Construction of Concrete Sea Structures," third edition, 1977.
"Guide to Good Practice: FIP/CEB Recommendations for the Design of Reinforced and Prestressed Concrete Structural Members for Fire Resistance," 1975.
"Recommendations for the Design of Aseismic Prestressed Concrete Structures," 1977.
"Recommendations for the Design of Prestressed Concrete Oil Storage Tanks," 1978.
"Proposed Recommendations for Segmental Construction in Prestressed Concrete," 1978.
12. W. J. Venuti, "Concrete Railroad Ties in North America," *Conc. Intl.*, Vol. 2, No. 1, January 1980, pp. 25–32.
13. A. N. Hanna, "Prestressed Concrete Ties for North American Railroads", State-of-the-art report, *J. Prestressed Conc. Inst.*, Vol. 24, No. 5, September/October 1979, pp. 32–61.

2

MATERIALS

2-1 Concrete, Strength Requirements

Stronger concrete is usually required for prestressed than for reinforced work. Present practice in this country calls for 28-day cylinder strength of 4000 to 8000 psi (28 to 55 N/mm^2) for prestressed concrete, while the corresponding value for reinforced concrete is around 3500 psi (24 N/mm^2). The usual cube strength specified for prestressed concrete in Europe is about 450 kg/cm^2, based on 10-, 15-, or 20-cm cubes at 28 days. If cube strength is taken as 1.25 times the cylinder strength, this would correspond to

$$450 \times 14.2/1.25 = 5100 \text{ psi cylinder strength}$$

Although the above are the usual values, strengths differing from these are occasionally specified.

Higher strength is necessary in prestressed concrete for several reasons. First, in order to minimize their cost, commercial anchorages for prestressing steel are always designed on the basis of high-strength concrete. Hence weaker concrete either will require special anchorages or may fail under the application of prestress. Such failures may take place in bearing or in bond between steel and concrete, or in tension near the anchorages. Next, concrete of high compressive strength offers high resistance in tension and shear, as well as in bond and bearing, and is desirable for prestressed-concrete structures whose various portions are under higher stresses than ordinary reinforced concrete. Another factor is that high-strength concrete is less liable to the shrinkage cracks which sometimes occur in low-strength concrete before the application of prestress. It also has a higher modulus of elasticity and smaller creep strain, resulting in smaller loss of prestress in the steel.

Experience has shown that 4000- to 5000-psi (28 to 34 N/mm^2) strength will generally work out to be the most economical mix for prestressed concrete. Although the strength of concrete to be specified for each job must be considered individually, there are some evident reasons why the economical mix usually falls within a certain range. Concrete strength of 4000 to 6000 psi (28 to 41 N/mm^2) can be obtained without excessive labor or cement. The cost of 6000-psi (41 N/mm^2) concrete averages about 15% higher than that of 3000-psi (21 N/mm^2) concrete, while it has 100% higher strength, which can be well utilized and is often seriously needed in prestressed structures. To obtain strength much greater than 6000 psi (41 N/mm^2), on the other hand, not only will cost more but also will call for careful design and control of the mixing, placing, and curing of concrete which cannot be easily achieved in the field.

Strength of 6000 to 8000 psi (41–55 N/mm^2) will sometimes be specified for precast, prestressed concrete beams. These strengths are commonly attained in plant operations where good quality control can be assured. Higher strengths, although sometimes adopted, are not in common use at this time.

To attain a strength in excess of 5000 psi (34 N/mm^2), it is necessary to use a water-cement ratio of not much more than 0.45 by weight. In order to facilitate placing, a slump of 2 to 4 in. (51 to 102 mm) would be needed, unless more than ordinary vibration is to be applied. To obtain 3-in. (76 mm) slump with water-cement ratio of 0.45 would require about 8 bags of cement per cu yd of concrete. If careful vibration is possible, concrete with $\frac{1}{2}$-in. (13 mm) or zero slump can be employed, and 7 bags of cement per cu yd may be quite sufficient. Since excessive cement tends to increase shrinkage, a lower cement factor is desirable. To this end, good vibration is advised whenever possible, and proper admixtures to increase the workability can sometimes be advantageously employed.

Not only should high-strength concrete be specified for prestressed work, but, when called for, such strength should be more closely attained in the field than for reinforced concrete. Indeed, more parts of prestressed-concrete members are subjected to high stresses than in reinforced concrete. Consider a simple prestressed beam, for example. While the top fibers are highly compressed under heavy external loads, the bottom fibers are under high compression at the transfer of prestress. While the midspan sections resist the heaviest bending moments, the end sections carry and distribute the prestressing force. Hence, in a prestressed member, it is more important to secure uniformity of strength, whereas in reinforced concrete the critical sections are relatively limited. Many engineers believe that, if the concrete is not crushed under the application of prestress, it should be able to stand subsequent loadings, since the strength of concrete increases with age and since excessive overloads are very rare for many structures. Fortunately, a 10% understrength of the concrete will result in very little change in strength of a member, but the engineer should use reasonable precautions to obtain the concrete strength specified.

It is general practice to specify a lower strength of concrete at transfer than its 28-day strength. This is desirable in order to permit early transfer of prestress to the concrete. At transfer, the concrete is not subject to external overloads, and strength is necessary only to guard against anchorage failure and excessive creep, hence a smaller factor of safety is considered sufficient. For example, in pretensioning work, a strength of 3500 psi (24 N/mm^2) at transfer is often sufficient for a specified 28-day strength of 5000 psi (34 N/mm^2).

Direct tensile strength in concrete is a highly variable item, generally ranging from $0.06f_c'$ to $0.10f_c'$, and may be zero is cracks have developed as the result of shrinkage or other reasons. Modulus of rupture in concrete is known to be higher than its direct tensile strength; ACI Code suggests $7.5\sqrt{f_c'}$ as an estimate of modulus of rupture.

Direct shearing strength, not often used in design, ranges from $0.50f_c'$ to $0.70f_c'$. Beam shear produces the principal tensile stress, whose limiting value is commonly gaged on the basis of direct tensile strength in concrete. Beam shear strength is covered in Chapter 7.

2-2 Concrete, Strain Characteristics

In prestressed concrete, it is important to know the strains produced as well as the stresses. This is necessary to estimate the loss of prestress in steel and to provide for other effects of concrete shortening. For the purpose of discussion, such strains can be classified into four types: elastic strains, lateral strains, creep strains, and shrinkage strains.

Elastic Strains. The term elastic strains is perhaps a little ambiguous, since the stress-strain curve for concrete is seldom a straight line even at normal levels of stress, Fig. 2-1. Neither are the strains entirely recoverable. But, eliminating the creep strains from consideration, the lower portion of the instantaneous stress-strain curve, being relatively straight may be conveniently called elastic. It is then possible to obtain values for the modulus of elasticity of concrete. The modulus varies with several factors,[1,2] notably the strength of concrete, the age of concrete, the properties of aggregates and cement, and the definition of modulus of elasticity itself, whether tangent, initial, or secant modulus. Furthermore, the modulus may vary with the speed of load application and with the

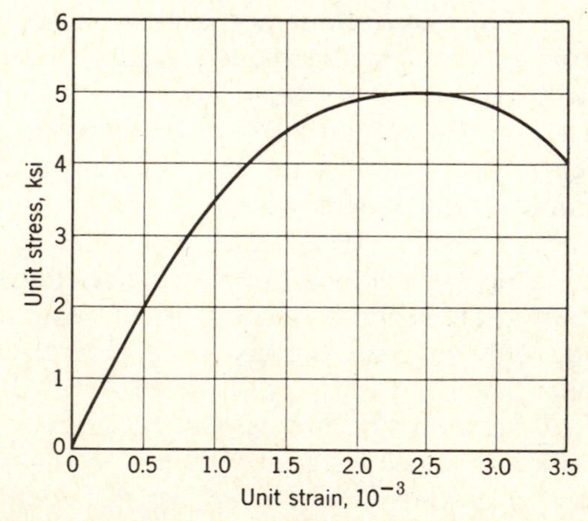

Fig. 2-1. Typical stress-strain curve for 5000 psi (34 N/mm²) concrete.

type of specimen, whether a cylinder or a beam. Hence it is almost impossible to predict with accuracy the value of the modulus for a given concrete.

As an average value for concrete at 28 days old, and for compressive stress up to about $0.40f_c'$, the secant modulus has been approximated by the following empirical formulas.

A. The ACI Code for Reinforced Concrete specifies the following empirical formula:

$$E_c = w^{1.5} 33\sqrt{f_c'} \tag{2-1}$$

where the unit weight w varies between 90 and 155 lb per ft^3 (1443 and 2485 kg/m^3). For normal weight concrete this expression may be simplified as:

$$E_c = 57{,}000\sqrt{f_c'}$$

B. Empirical formula proposed by Jensen:

$$E_c = \frac{6\times 10^6}{1+(2000/f_c')} \tag{2-2}$$

which gives more correct values for f_c' around 5000 psi (34 N/mm^2).

C. Empirical formula proposed by Hognestad:

$$E_c = 1{,}800{,}000 + 460f_c' \tag{2-3}$$

which gives results similar to the last one.

Plotting the above proposals in Fig. 2-2, we can see that those of Jensen and Hognestad come quite close to the ACI values. Some equations for E_c give values intended to represent modulus used for computing instantaneous beam deflections while the others based on measured strains from cylinder specimens.

Authorities differ on the relation between the two kinds of moduli. Some tests indicate the agreement of these two values; others tend to show that the modulus for beams is higher than that for cylinders. Not too much work has been done for the modulus of elasticity of concrete in tension, but it is generally assumed that, before cracking, the average modulus over a length of several inches is the same as in compression, although the local modulus in tension is known to vary greatly.

Lateral Strains. Lateral strains are computed by Poisson's ratio.[2] Owing to Poisson's ratio effect, the loss of prestress is slightly decreased in biaxial prestressing. Poisson's ratio varies from 0.15 to 0.22 for concrete, averaging about 0.17.

Creep Strains. Creep of concrete is defined as its time-dependent deformation resulting from the presence of stress. A great deal of work has been done in this country on the creep or plastic flow of concrete.[3,4]

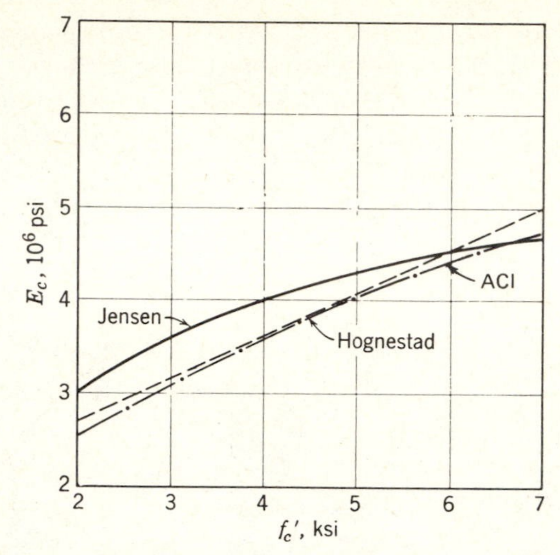

Fig. 2-2. **Empirical formulas for E_c for normal weight concrete.**

A brief summary of a comprehensive investigation carried out at the University of California extending over a period of 30 years is now presented.[5] For specimens of 4-in. (102 mm) diameter loaded in compression to 800 psi (5.52 N/mm^2) at 28 days and thereafter stored in air at 50% relative humidity and 70°F, the findings are:

1. Creep continued over the entire period, but the rate of change at the later ages was very small. Of the total creep in 20 years, 18–35% occurred in the first two weeks of loading, 40–70% within 3 months, and 60–83% within 1 year. The average values were 25, 55, and 76% respectively. Typical creep-time ratio curves with upper and lower limits are shown in Fig. 2-3 (from reference 5).
2. Creep increased with a higher water-cement ratio and with a lower aggregate-cement ratio, but was not directly proportional to the total water content of the mix.
3. Creep of concrete was appreciably greater for type IV (low-heat) than for the type I (normal) portland cement. For type IV cement the creep was greater for the coarse grind than for the fine, but the reverse was true for the type I cement.
4. Creep of concrete was greatest for crushed sandstone aggregate, followed in descending order by basalt, gravel, granite, quartz, and limestone. The creep for sandstone concrete was more than double that for limestone concrete.

For ages from 28 to 90 days at time of loading, for stresses from 300 to 1200 psi (2.07 to 8.27 N/mm^2), for storage conditions which ranged from air at 50%

relative humidity to immersion in water, and for specimen diameters from 4 to 10 in. (102–524 mm), the following statements apply.

1. The older the specimen at the time of loading, the more complete the hydration of cement, the less the creep. Those loaded at 90 days had less creep than those at 28 days, by roughly 10%.
2. The creep per unit of stress was only slightly greater at the high stresses than at the low stresses.
3. The total amount of creep strain at the end of 20 years ranged from 1 to 5 times the instantaneous deformations (averaging about 3 times) while the combined shrinkage or swelling and creep ranged from 1 to 11 times the instantaneous deformations, the low values occurring for storage in water or fog and for limestone aggregates.
4. The creep in air at 50% relative humidity was about 1.4 times that in air at 70% relative humidity and about 3 times that for storage in water.
5. Creep decreased as the size of specimen increased.

Only a limited amount of data is available concerning the creep of concrete under high stress.[6] Some of these data seem to indicate that when the sustained stress is in excess of about $\frac{1}{3}$ of the ultimate strength of concrete, the rate of increase of strain with stress tends to get higher. It is possible that this increase can become quite pronounced as the stress approaches the ultimate strength of the concrete.

Upon the removal of the sustained stress, part of the creep can be recovered in the course of time.[3] Generally, it takes a longer time to recover the creep than for the creep to take place. For the limited amount of data available, it can be stated that roughly 80 to 90% of the creep will recover during the same length of time that creep has been allowed to take place.

In Europe, the term creep coefficient C_c is employed to indicate the total strain δ_t (instantaneous plus creep strain) after a lengthy period of constant stress to the instantaneous strain δ_i immediately obtained upon the application of stress,[7] thus

$$C_c = \frac{\delta_t}{\delta_i}$$

This coefficient varies widely as reported from different tests, essentially because of the difficulty of separating shrinkage from creep. For purposes of design, it is considered safe to take C_c as around 3.0. For posttensioned members, where the prestress is applied late, the coefficient could be a little less; for pretensioned members, where the prestress is applied at an early age, the coefficient could be a little more.

This same term, creep coefficient, is sometimes used to denote the ratio of the creep strain δ_c (excluding the instantaneous strain) to the instantaneous strain δ_i, thus

$$C_c = \frac{\delta_c}{\delta_i}$$

Hence care should be exercised to find out the exact meaning of "creep coefficient" whenever the term is employed. Using the first definition, the creep coefficient is about 2.5 at the end of one year for the curve in Fig. 2-3; using the second definition, that same coefficient is only 1.5.

Of the total amount of creep strain, it can be roughly estimated that about $\frac{1}{4}$ takes place within the first 2 weeks after application of prestress, another $\frac{1}{4}$ within 2 to 3 months, another $\frac{1}{4}$ within a year, and the last $\frac{1}{4}$ in the course of many years, Fig. 2-3.

There is good reason to believe that, for smaller members, creep as well as shrinkage takes place faster than for larger members. Upon the removal of stress, part of the creep can be recovered in the course of time. Again, owing to the difficulty of separating shrinkage from creep, the amount and speed of such recovery have not been accurately measured.

Shrinkage Strains. As distinguished from creep, shrinkage in concrete is its contraction due to drying and chemical changes dependent on time and on moisture conditions, but not on stresses. At least a portion of the shrinkage resulting from drying of the concrete is recoverable upon the restoration of the lost water. The magnitude of shrinkage strain also varies with many factors, and it may range from 0.0000 to 0.0010 and beyond. At one extreme, if the concrete is stored under water or under very wet conditions, the shrinkage may be zero. There may even be expansion for some types of aggregates and cements. At the

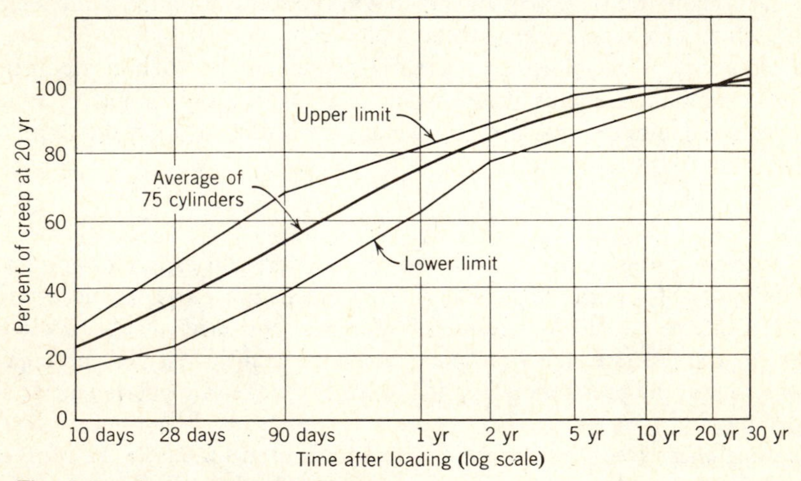

Fig. 2-3. Creep-time ratio curves.

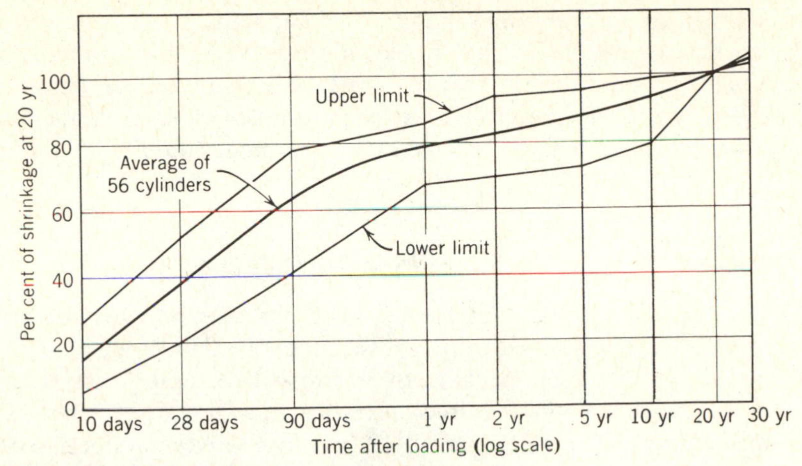

Fig. 2-4. Drying shrinkage-time ratio curves.

other extreme, for a combination of certain cements and aggregates, and with the concrete stored under very dry conditions, as much as 0.0010 can be expected. Reference 5 lists test results showing the magnitude of shrinkage and its rate of occurrence as affected by various factors. Fig. 2-4 shows some typical shrinkage-time ratio curves taken from that reference.

Shrinkage of concrete is somewhat proportional to the amount of water employed in the mix. Hence, if minimum shrinkage is desired, the water-cement ratio and the proportion of cement paste should be kept to a minimum. Thus aggregates of larger size, well graded for minimum void, will need a smaller amount of cement paste, and shrinkage will be smaller.

The quality of the aggregates is also an important consideration. Harder and denser aggregates of low absorption and high modulus of elasticity will exhibit smaller shrinkage. Concrete containing hard limestone is believed to have smaller shrinkage than that containing granite, basalt, and sandstone of equal grade, approximately in that order. The chemical composition of cement also affects the amount of shrinkage. For example, shrinkage is relatively small for cements high in tricalcium silicate and low in the alkalies and the oxides of sodium and potassium.

The amount of shrinkage varies widely, depending on the individual conditions. For the purpose of design, an average value of shrinkage strain would be about 0.0002 to 0.0006 for the usual concrete mixtures employed in prestressed construction. The rate of shrinkage depends chiefly on the weather conditions. Actual structures exposed to weather show measurable seasonal changes in the shrinkage of concrete—swelling during rainy reasons and shrinking during dry ones. If the concrete is left dry, there is reason to believe that most of the shrinkage would take place during the first 2 or 3 months. If it always wet, there may be no shrinkage at all. When stored in air at 50% relative humidity and 70°F, there were indications[5] that the rate of occurrence of shrinkage is comparable to that of creep and that the magnitude of shrinkage is often similar to that of creep produced by a sustained stress of about 600 psi (4.14 N/mm^2).

2-3 Concrete, Special Manufacturing Techniques

Most of the techniques for manufacturing good concrete, whether for plain or reinforced work, can be applied to prestressed concrete. However, they must be investigated for a few factors peculiar to prestressed concrete. First, they must not decrease the high strength required; next, they must not appreciably increase the shrinkage and creep; they must not produce adverse effects, such as inducing corrosion in the high-tensile wires.

Compacting the concrete by vibration is usually desirable and necessary. Either internal or external vibration may be used. In order to produce high-strength concrete without using an excessive amount of mortar, a low water-cement ratio and a low-slump concrete must be chosen. Such concrete cannot be well placed without compaction. There are only a few isolated applications in which concrete of high slump is employed and compaction may be dispensed with. But it will be found preferable to use at least a small amount of compaction for corners and around reinforcements and anchorages.

Good curing of concrete is most important. Too early drying of concrete may result in shrinkage cracks before the application of prestress. Besides, only by careful curing can the specified high strength be attained in concrete. In order to hasten the hardening process, steam curing is often resorted to in the precasting factory; it can also be employed in the field where the amount of work involved justifies the installation. When field work of casting must be carried out in cold weather, steam can profitably be used to raise the temperature of the ingredients and the placed concrete in order that high strength may be attained within a reasonable time.

Early hardening of the concrete is often desirable, either to speed plant production or to hasten field construction. Early high-strength concrete can be produced by any one of a number of techniques or combinations of techniques.[9] High early-strength cement or steam curing is commonly employed. Admixtures to accelerate the strength should be employed with caution. For example, calcium chloride, the most commonly used accelerator, even applied in normal amounts, will increase shrinkage. There is evidence that it will cause corrosion, which could be serious for the prestressing steel. When accelerators are used, care must also be taken not to have the initial set take place too soon.

Wetting admixtures to improve the workability of concrete may be found to be profitable, since they may permit easy placing of high-strength concrete without too high a cement content. Some of these admixtures tend to increase the shrinkage and may offset the advantage of saving cement. Each must be judged on its own merits in conjunction with the nature of the aggregates and cement. Air entrainment of 3 to 5% improves workability and reduces bleeding. When well-recognized, air-entraining agents are employed, there is no evidence of increased shrinkage or creep. Hence proper application of air entrainment is considered beneficial for prestressed concrete.

Precast segmental construction has been recently developed for prestressed bridges.[10] Breaking up a bridge superstructure into transverse segments reduces the individual weight and facilitates casting and handling. These segments can be mass produced in a plant where rigid inspection and control can be effected, or they can be poured-in-place on a traveling carriage. They are used for longer spans than could be handled with a single beam cast in one piece, thus enabling them to compete with structural steel on these larger spans. The joints for precast segments are very thin epoxy-filled space with the end surfaces being match cast for fit between segments. Posttensioning tendons are threaded through to join the segments together, thus forming the completed bridge.

2-4 Lightweight Aggregate Concrete

Since about 1955, lightweight concrete has been gaining in application to prestressed construction, especially in California. The main reason for using lightweight concrete is to reduce the weight of the structure, thus minimizing both the concrete and the steel required for carrying the load. This is especially important when the dead load is the major portion of the load on the structure, or when the weight of the member is a factor to be considered for transportation or erection.

It used to be a task to produce lightweight concrete of sufficient strength for prestressing, but this is no longer true. With experience in control and design of lightweight concrete mixes, 28-day cylinder strength of 5000 psi (34 N/mm^2) can generally be obtained with no difficulty, while 6000–7000 psi (41–48 N/mm^2) or more can be reached if desired. Strength at 1-day transfer of 3500 or 4000 psi is frequently attained by the use of high-early-strength cement and steam curing.

Data giving the physical and mechanical properties of lightweight concrete made with aggregates throughout the country are given in a paper by Shideler.[12] Those related to aggregates in the states of Texas and California are presented in two other papers.[13,14] While these test series did not yield identical values, some general observations can be made from them. When quantitative values are desired for a particular lightweight aggregate used in a given locality, it will be necessary to examine the aggregate and compare it with similar ones in the series, bearing in mind that exact values for either lightweight or regular weight concrete can be obtained only when extensive tests have been conducted for that particular aggregate, and the field conditions are under perfect control. Fortunately, a certain amount of tolerance is permissible so that when properly designed and built, prestressed, lightweight concrete will behave satisfactorily.

One objection against lightweight concrete for prestressing is its low modulus of elasticity, which indicates more elastic shortening under the same unit stress. This means that there is a slightly higher loss of prestress in the steel. It also means that for cast-in-place structures, a greater elastic movement will take place under the application of the prestress. As a rough approximation, it may be said that the E_c for lightweight concrete averages about 55% the E_c for regular weight concrete. For f_c' between 3000 and 6000 psi (21 and 41 N/mm^2), using 60% of Hognestad's formula for E_c (p. 38), we get a fairly good approximation,

$$E_c = 1{,}000{,}000 + 250 f_c'$$

where f_c' = cylinder strength of concrete at the time E_c is measured. However, E_c values may easily vary 20% either way from those given by the above formula, depending on various factors, especially the nature of the lightweight aggregate. The ACI equation (2-1) uses unit weight w directly to obtain E_c for lightweight concrete. Figure 2-2(b) simplifies use of the ACI equation for estimating E_c for lightweight concrete.

Poisson's ratio for lightweight concrete is apparently comparable to that for sand and gravel concrete; values between 0.15 and 0.25 have been reported with an average value of 0.19.[12]

The tensile strength for lightweight concrete varies with different types of manufactured material. The properties of concrete made with a particular lightweight aggregate should be available from the aggregate manufacturer, and actual results from tests should be used when available. The ACI Code uses split cylinder strength as an index for evaluating cracking and shear strength in lightweight concrete. For normal weight concrete the average splitting tensile strength, f_{ct}, is approximately $6.7\sqrt{f_c'}$, but it falls below this value for some lightweight concrete. The ACI Code provides that shear strength be modified using actual $f_{ct}/6.7$ for $\sqrt{f_c'}$ in formulas for shear strength. When the actual f_{ct} is not specified, all values of $\sqrt{f_c'}$ affecting shear strength and cracking moment are multiplied by 0.75 for "all lightweight concrete," and 0.85 for "sand-lightweight concrete." Linear interpolation between these values may be used when partial sand replacement is used. The unit weight of lightweight concrete varies considerably, between 90 and 110 pcf (1443 and 1763 kg/m^3). The addition of fine, natural sand somewhat increases the unit weight and is also known to increase the workability and strength of the mix.

The shrinkage of lightweight concrete is apparently comparable to that of similar sand and gravel concrete.[13] However, some tests showed that it was slightly higher by 6–38%;[12] while other tests indicated that it was much lower.[14] Hence it is concluded that each lightweight aggregate must be studied by itself, but the chances are that they will have no more shrinkage than sand and gravel concrete.

Total creep strain in lightweight concrete is again comparable to that in sand and gravel concrete for specimens under the same sustained stress. Some have higher creep while others have less, probably by a maximum of some 20%, one way or the other. For detailed information, readers are referred to references 12, 13, and 14. It is generally agreed that both shrinkage and creep are related to the cement paste and quite independent of the aggregates.

2-5 Self-stressing Cement

Types of cements that expand chemically after setting and during hardening are known as expansive or self-stressing cements. When these cements are used to make concrete with embedded steel, the steel is elongated by the expansion of the concrete. Thus the steel is prestressed in tension, which in turn produces compressive prestress in the concrete, resulting in what is known as chemical prestressing or self-stressed concrete.

Modern development of expansive cement started in France about 1940.[15] Its use for self-stressing has been investigated intensively in U.S.S.R. since 1953.[16] At the University of California, Berkeley, studies were directed toward the use of calcium sulfoaluminate admixtures for expansive cements in 1956, and their practical chemistry, manufacture, and potentials were analyzed and described in a paper by Klein and Troxell.[17] The physical properties of one such expansive cement were then further investigated, and results were presented in a paper by Klein, Karby, and Polivka,[18] while pilotary effort to study the structural possibilities of such expansive cement, when used for prestressing concrete, was described in another paper by Lin and Klein.[19]

When concrete made with expanding cement is unrestrained, the amount of expansion produced by the chemical reaction between the cement and water could amount to 3–5%, and the concrete would then disintegrate by itself. When restrained either internally or externally with steel or other means, the amount of expansion can be controlled. By applying restraint in one direction, the growth in the other two orthogonal directions can be limited because of the crystalline nature of the hardened paste. The Russian self-stressing cement requires hydrothermal curing resulting in a quick setting, while the component developed in California requires water or fog curing under normal temperatures.

When high-tensile steel is used to produce the prestress, say corresponding to tensile stress at 150,000 psi (1,034 N/mm^2) and an E_s of 27,000,000 psi (186 kN/mm^2), an expansion of

$$\frac{150{,}000}{27{,}000{,}000} = 0.55\%$$

is required. For other stress levels, varying amounts of expansion will be required. For proper development of the bond between steel and concrete, mechanical end-anchorages might be necessary unless the steel has sufficient corrugation to transfer the stress.

Because of the expansion in all three directions, it seems difficult to use the cement for complicated structures cast in place, such as buildings. However, for pressure pipes and pavements, where prestressing in at least two directions is desired, this type of chemical prestressing can be more economical than mechanical prestressing; this is also true for precast slabs, walls, and shells. However, no immediate economy is seen in the making of beams which require eccentric prestressing. Unless the beam soffit is precast by itself, curvature of the beam may result from steel embedded eccentrically in the beam.

Expanding cement has been successfully applied for many interesting projects, especially in France. When a concrete block of expanding cement is cast as the keystone for a concrete arch, it serves as a jack, producing the desired arch compensation to balance rib shrinkage and shortening. When used for underpinning buildings, it tends to lift the structure without jacking. It can be used for pressure grouting or for producing concrete pavements and slabs with no shrinkage joints.

While many problems remain yet unsolved concerning the use of expanding cement for self-stressing, such as the chemical and physical stability and exact control of the stresses and strains, applications for a limited amount of shrinkage compensation are found in the United States, while in the Soviet Union, it is frequently used for stressing precast elements for thin shells which require only concentric prestressing.

2-6 Steels for Prestressing

High-tensile steel is almost the universal material for producing prestress and supplying the tensile force in prestressed concrete. The obvious approach toward

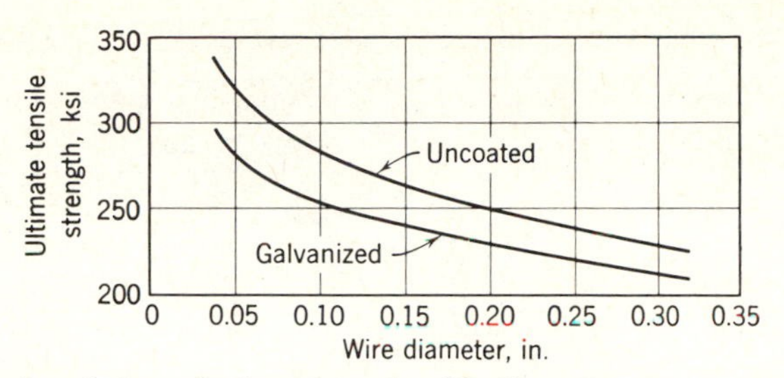

Fig. 2-5. Typical variation of wire strength with diameter.

the production of high-tensile steel is by alloying, which permits the manufacture of such steels under normal operation. Carbon is an extremely economical element for alloying, since it is cheap and easy to handle.[20] Other alloys include manganese and silicon. Other approaches are by controlled cooling of the steels after rolling and by heat treatment such as quenching and tempering. Beneficial results have been obtained by quenching from the rolling heat at a given temperature and also by interrupting the quench at a given temperature.

The most common method for increasing the tensile strength of steel for prestressing is by cold-drawing, high-tensile steel bars through a series of dyes. The process of cold-drawing tends to realign the crystals, and the strength is increased by each drawing so that the smaller the diameter of the wires, the higher their ultimate unit strength. The ductility of wires, however, is somewhat decreased as a result of cold-drawing. A curve giving the typical variation of strength with diameter is shown in Fig. 2-5. The actual strength, of course, will vary with the composition and manufacture of the steel.

High-tensile steel for prestressing usually takes one of three forms: wires, strands, or bars. For posttensioning, wires are widely employed; they are grouped, in parallel, into cables. Strands are fabricated in the factory by twisting wires together, thus decreasing the number of units to be handled in the tensioning operations. Strands, as well as high-tensile rods, are also used for posttensioning.

For pretensioning, 7-wire strands are almost exclusively used in the United States and have replaced much of wire pretensioning in other countries. Although strands cost slightly more than wires of the same tensile strength, its better bonding characteristics make it especially suitable for pretensioning.

While the ultimate strength of high-tensile steel can be easily determined by testing, its elastic limit or its yield point cannot be so simply ascertained, since it has neither a yield point nor a definite proportional limit. Various arbitrary methods have been proposed for defining the yield point of high-tensile steel, such as the 0.1% set, 0.2% set, 0.7% strain, or 1.0% strain. The more commonly accepted methods are probably the 0.2% set and the 1.0% strain.

Table 2-1 Properties of Uncoated Stress-Relieved Wire (ASTM A 421)

Nominal Diameter	Minimum Tensile Strength psi (N/mm^2)		Minimum Stress at 1% Extension psi (N/mm^2)	
in. (mm)	Type BA[b]	Type WA	Type BA[b]	Type WA
0.192 (4.88)	[a]	250,000 (1725)	[a]	200,000 (1380)
0.196 (4.98)	240,000 (1655)	250,000 (1725)	192,000 (1325)	200,000 (1380)
0.250 (6.35)	240,000 (1655)	240,000 (1655)	192,000 (1325)	192,000 (1325)
0.276 (7.01)	[a]	235,000 (1622)	[a]	188,000 (1295)

[a] These sizes are not commonly furnished in Type BA wire.

[b] Type BA wire is used for applications in which cold-end deformation is used for anchoring purposes (button anchorage), and type WA is used for applications in which the ends are anchored by wedges and no cold-end deformation of the wire is involved (wedge anchorage). Examples of tendons with button anchorages, more common in United States practice, are shown in Appendix B.

2-7 Steel Wires

Wires for prestressing generally conform to ASTM Specification A-421 for "Uncoated Stress-relieved Wire for Prestressed Concrete." They are made from rods produced by the open hearth or electric-furnace process. After cold-drawn to size, wires are stress-relieved by a continuous heat treatment to produce the prescribed mechanical properties.

The tensile strength and the minimum yield strength (measured by the 1.0% total-elongation method) are shown in Table 2-1 for the common sizes of wires.

A typical stress-strain curve for a stress-relieved $\frac{1}{4}$-in. wire conforming to the ASTM A-421 is shown in Fig. 2-6, with a typical modulus of elasticity of

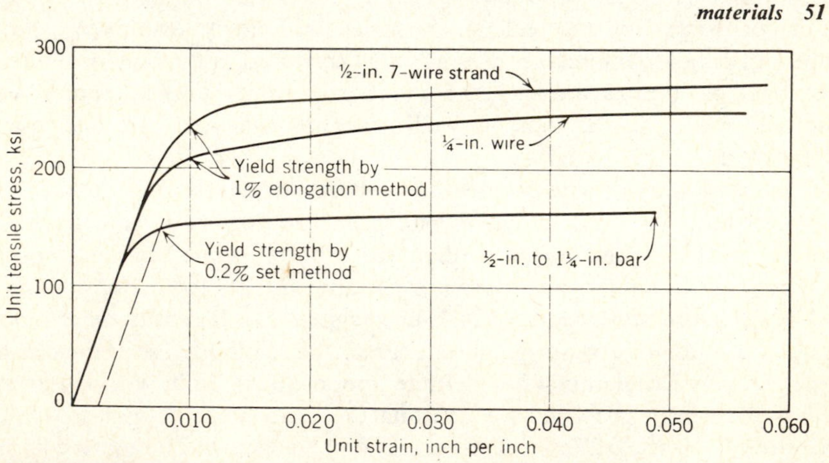

Fig. 2-6. Typical stress-strain curve for prestressing steels.

29,000,000 psi (200 kN/mm^2). The specified minimum elongation in 10 in. (254 mm) is 4.0%, while a typical elongation at rupture is more likely from 5 to 6%.

Curves for a bar and for a 7-wire strand are also shown in Fig. 2-6, and in Appendix B. Typical curves are considered to be sufficiently accurate for the purposes of structural design. For computing exact elongations, it is advised that accurate stress-strain relationships be obtained from the manufacturer or by actual testing of specimens.

Wires are supplied in reels or coils. They are cut to length and assembled either at the plant or in the field. Some steel may need a certain amount of degreasing and cleaning before placement, in order to ensure good bond with concrete. Loose rust or scale should be removed, but a firmly adherent rust film is considered advantageous in improving the bond.

In continental Europe, smooth wires 2 and 3 (sometimes 2.5) mm in diameter and corrugated wires of 4 and 5 mm have been employed in pretensioning work. Small wires possess higher unit strength and furnish better bond, which is helpful. In order to save labor and anchorage costs, larger wires are preferred for prestressing.

In England, wires are based on the British Imperial Gauge, No. 2 which has a diameter of 0.276 in., exactly 7 mm, while No. 6 has a diameter of 0.192 in., which is very close to 5 mm. Hence gage Nos. 2 and 6 are sometimes called for in posttensioning, while the exact equivalents of 7 and 5 mm (0.276 and 0.196 in., respectively) are also frequently employed. In Germany, corrugated wires with oval cross section have been used for posttensioning. These wires have areas of 20, 30, 35, and 40 mm^2, with Oval 40 having a major diameter of 11 mm and a minor diameter of 4.5 mm.

In this country, wires are manufactured according to the U.S. Steel Wire Gage, No. 2 which has a diameter of 0.2625 in. (6.668 mm) and No. 6 has a diameter of 0.1920 in. (4.877 mm). Neither of these is the exact equivalent of the millimeter counterparts. Hence, when the European types of anchorages are adopted, 0.276-in. and 0.196-in. wires are often specified. For posttensioning systems developed in the United States, $\frac{1}{4}$-in. (6.35 mm) wires have been most commonly incorporated.

2-8 Steel Strands

Strands for prestressing generally conform to ASTM Specification A-416 for "Uncoated Seven-wire Stress-relieved for Prestressed Concrete." Two grades are available, 250 ksi and 270 ksi (1,724 N/mm^2 and 1,862 N/mm^2), where the grade indicates minimum guaranteed breaking stress. While these specifications were intended for pretensioned, bonded, prestressed-concrete construction, they are also applicable to posttensioned construction, whether of the bonded or the unbonded type. These seven-wire strands all have a center wire slightly larger than the outer six wires which enclose it tightly in a helix with a uniform pitch between 12 and 16 times the nominal diameter of the strand. After stranding, all strands are subjected to a stress-relieving continuous heat treatment to produce the prescribed mechanical properties.

Seven-wire strands commonly used for prestressing conform to the ASTM A-416 Specifications, having a guaranteed minimum ultimate strength of 250,000 psi or 270,000 psi (1724 N/mm^2 or 1862 N/mm^2). Their properties are listed in Table 2-2. Since 1962, the stronger steel known as the 270K grade has been produced by various companies. For the same nomial size, the 270k grade has more steel area than the ASTM A-416 grade 250 and is about 15% stronger (see Appendix B). The 270K steel is now almost universally used for 7-wire strands in the United States, for both pretensioned and posttensioned structures. Low-relaxation strand is also available in both grades.

A typical stress-strain curve for a stress-relived $\frac{1}{2}$-in. (127 mm) 7-wire strand (ASTM A-416 is shown in Fig. 2-6 (and Appendix B), which is also typical for strands of all sizes. For calculations, a modulus of elasticity of 27,500,000 psi (189,610 N/mm^2) is often used for ASTM A-416 250K grade and 270K grade strand. The specified minimum elongation of the strand is 4% in a gage length of 24 in. (609.6 mm) at initial rupture, although typical values are usually in the range of 6%. When these strands are galvanized, they are about 15% weaker in strength and slightly lower in E_s depending on the amount of zinc coating used. The galvanized strand is not widely used in structural members and information about properties should be obtained from the manufacturer.

Table 2-2 Properties of Uncoated Seven-Wire Stress-Relieved Strand (ASTM A-416)

Nominal Diameter in. (mm)	Breaking Strength lb (kN)	Nominal Area of Strand in.^{2} (mm^2)	Minimum Load at 1% Extension lb (kN)
		Grade 250	
0.250 (6.35)	9000 (40.0)	0.036 (23.22)	7650 (34.0)
0.313 (7.94)	14,500 (64.5)	0.058 (37.42)	12,3000 (54.7)
0.375 (9.53)	20,000 (89.0)	0.080 (51.61)	17,000 (75.6)
0.438 (11.11)	27,000 (120.1)	0.108 (69.68)	23,000 (102.3)
0.500 (12.70)	36,000 (160.1)	0.144 (92.90)	30,600 (136.2)
0.600 (15.24)	54,000 (240.2)	0.216 (139.35)	45,900 (204.2)
		Grade 270	
0.375 (9.53)	23,000 (102.3)	0.085 (54.84)	19,550 (87.0)
0.438 (11.11)	31,000 (137.9)	0.115 (74.19)	26,350 (117.2)
0.500 (12.70)	41,300 (183.7)	0.153 (98.71)	35,100 (156.1)
0.600 (15.24)	58,600 (260.7)	0.217 (140.00)	49,800 (221.5)

As fabricated, 7-wire strands are several thousand feet long. When unwinding strands, care must be taken in laying them along the path to prevent kinking and permanent twisting of the strands.

2-9 Steel Bars

ASTM Specifications A-322 and A-29 are often applied to high-strength alloy steel bars. It is usually required that all such bars be proof-stressed to 90% of the guaranteed ultimate strength. Although the actual ultimate strength often reaches 160,000 psi (1,103 N/mm^2), the specified minimum is generally set at 145,000 psi (1,000 N/mm^2). A typical stress-strain curve for these bars is shown in Fig. 2-b from which it can be noticed that a constant modulus of elasticity exists only for a limited range (up to about 80,000 psi (552 N/mm^2) stress) with a value between 25,000,000 and 28,000,000 psi (172,375 and 193,060 N/mm^2).

The yield strength of high tensile bars is often defined by the 0.2% set method, as indicated in Fig. 2-b, where a line parallel to the initial tangent is drawn from the 0.002 strain, and its intersection with the curve is defined as the yield-strength point. Most specifications would call for a minimum yield strength at 130,000 psi (896 N/mm^2), though actual values are often higher. Minimum elongation at rupture in 20 diameters length is specified at 4%, with minimum reduction of area at rupture at 25%. Common sizes and properties of high-tensile rods for prestressing are listed in Appendix B.

High-tensile bars are available with length up to 80 ft. (24.4 m). Because of difficulty in shipping, the length may have to be further limited. But sleeve couplers are available to splice the bars to any desired length. These couplers have tapered threads in order to develop very nearly the full strength of the bars. They have outside diameters about twice that of the bar and a length about 4 times its diameter.

High-strength, specially deformed bars with ultimate strength of 160 ksi (1,103 N/mm^2) are available in sizes 1 to $1\frac{3}{8}$ in. (25.4 mm–34.9 mm) diameter. Ultimate strength of 230 ksi (1,586 N/mm^2) is available for these bars with $\frac{5}{8}$ in. (15.9 mm) diameter. The deformations on the bars serve as threads to fit couplers and anchorage hardward. Splicing of the bars can occur at any point where the bar may be cut since the coupler threads onto the special rolled-on deformations which are continuous along the length.

Auxiliary reinforcement using nonprestressed steel is commonly employed in prestressed construction. Steel of almost any strength will serve the purpose when properly designed. Generally, reinforcing bars conform to ASTM Specifications A-15, A-16, and A-305, and welded wire mesh conforms to ASTM Specification A-185.

2-10 Fiberglass Tendons

Fiberglass is manufactured by drawing fluid glass into fine filaments. The possible use of fiberglass for prestressing has been under investigation for some years.[21, 22, 23]

Although fiberglass has not yet been commercially applied in prestressed-concrete construction, it does possess certain superior qualities that indicate high promise for prestressing. An ultimate tensile strength of 1,000,000 psi (6,895 N/mm^2) is quite commonly obtained. Values as high as 5,000,000 psi (34,475 N/mm^2) have been reported for individual silica fibers 0.00012 in. (0.003048 mm) in diameter, it being known that the strength varies approximately inversely as the diameter of the fiber.

Fiberglass can be made in three forms: parallel chords, twisted strands, and parallel fibers embedded in plastic. The last form in the shape of fiberglass rods is considered most suitable for prestressing because of its relative simplicity for handling, gripping, and anchoring. At Princeton University, three types of resin have been tried out as bonding agents in the manufacture of fiberglass rods: polyester, epoxy, and polyamide resins. To date, the rods laminated with epoxy resins have appeared to be superior. A short-duration tensile strength has been obtained in excess of 220,000 psi (1,517 N/mm^2), based on the gross area of the rod.

Extensive research was conducted by the U. S. Army Corps of Engineers[23] on both reinforced and prestressed elements utilizing fiberglass rods and fiberglass tendons. For the prestressed test series, the fiberglass reinforced beams were found to be inferior to the prestressed beams with steel reinforcement based on an equal area of prestressing element. They recommended that fiberglass reinforcement be used in conjunction with conventional web reinforcement to avoid diagonal tension failures and that shallow depths be avoided. Other studies[24,25] have shown that the behavior of fiberglass reinforced beam and slabs is predictable, but the long-term performance of posttensioned elements with fiberglass tendons seems questionable.

An advantage of fiberglass is its low modulus of elasticity, which ranges from 6,000,000 to 10,000,000 psi (41,370 to 68,950 N/mm^2). With its high stress and low modulus, the percentage of loss of prestress would be quite small. Other advantages claimed for this material are high resistance to acids and alkalies and the ability to withstand high temperature. However, some major problems must be solved before it can be applied in practice.

1. The static fatigue limit, that is, the long-time ultimate strength of fiberglass rods as opposed to the short-time ultimate strength. This is an important problem since it is known that the duration of loading has a pronounced effect on the ultimate strength.

2. The dynamic fatigue limit of fiberglass or fiberglass rods, although there is some evidence to indicate that this may not be a serious problem.
3. The chemical stability of fiberglass such as its reaction to the surrounding concrete, especially under wet conditions.
4. The best methods of fabricating cords from fiberglass to obtain an even distribution of stress so as to increase the ratio of the strength of cords to the strength of individual fibers. The minimizing of shearing deformation of the laminating material, since such deformation could result in the breaking of the outer fibers with an inner core of fibers remaining intact.
5. The design of suitable end anchorages, since the brittle material is liable to fail in the grip under the effect of stress concentrations and combined stresses.

If these problems can be solved, there still remains a last hurdle: the economics of the application of the material in competition with high-tensile steel, which is being produced in large quantities and at relatively low cost. On the other hand, it is conceivable that the special properties of fiberglass might make it desirable in special situations, especially in very corrosive industrial environments.

2-11 Auxiliary Materials—Grouting

Among the special auxiliary materials required for prestressed concrete are those for the provision of proper conduits for the tendons. For pretensioning, no such conduits are necessary. For posttensioning, there are two types of conduits, one for bonded, another for unbonded prestressing.

When the tendons are to be bonded, generally by grouting, the conduits (ducts) are made of ferrous metal which may be galvanized. Materials commonly used for these ducts are 22 to 28 gage galvanized or bright spirally wound or longitudinally seamed steel strips with flexible or semirigid seams. Rigid tubing is sometimes provided by stiffening rods placed within the ducts, or rigid tubes. Corrugated plastic ducts have also come into limited use recently.

It is also possible to form the duct by withdrawing steel tubing or rod before the concrete hardens. More frequently, the duct is formed by withdrawing extractable rubber cores buried in the concrete. Several hours after the completion of concreting, these cores can be withdrawn without much effort, because the lateral shrinkage of the rubber under a pull helps to tear the rubber away from the surrounding concrete. In order that the rubber cores may remain straight during concreting, they are stiffened internally by inserting steel pipes or rods into axial holes provided in the rubber. To maintain the cores in position during concreting, transverse steel rods are placed under and over them at 3- to 4-ft (0.91–1.22 m) intervals. Sometimes rubber tubes inflated from their normal diameter can be substituted for the above rubber cores. These tubes can then be deflated and withdrawn.

When the tendons are to be unbonded, plastic or heavy paper sheathing is frequently used, and the tendons are properly greased to facilitate tensioning and to prevent corrosion. Rust inhibitors are usually added to the grease together with additional compounds to ensure its uniform consistency in extremes of temperature. Asbestos fibers are often added to the grease to hold it together during application. Plastic tubes of the split type should be properly overlapped and taped along the seams, so as to seal them against any leakage of mortars, which might bind the tendons to the tubes. When papers are spirally wrapped around the tendons, care should be exercised in wrapping so as to avoid jamming of the papers when the tendons are tensioned.

For bonding the tendons to the concrete after tensioning (in the case of posttensioning), cement grout is injected, which also serves to protect the steel against corrosion. Entry for the grout into the cableway is provided by means of holes in the anchorage heads and cones, or pipes buried in the concrete members. The injection can be applied at one end of the member until it is forced out of the other end. For longer members, it can be applied at both ends until forced out of a center vent.

Either ordinary portland cement or high-early-strength cement may be used for the grout with water and sometimes fine sand.[26] Commercially available additives developed to assure that sound grouting is possible. These materials increase the workability of the grout and sometimes provide for shrinkage compensation. For some situations where the grouting is vertical in walls of considerable height, special additives which prevent the segregation of water from the grout should be added as tests have shown that the water rise to the top results in a portion of the tendon being ungrouted. Such an unprotected tendon may later have severe rusting which can produce failure. To ensure good bond for small conduits, grouting under pressure is desirable; however, care should be taken to ensure that the bursting effect of the pressure on the walls of the cable enclosure can be safely resisted. Machines for mixing and injecting the grouts are commercially available.

Although sand is not used in grouting practice in the United States, it may have advantages in tendons with large void areas. Fly ash and pozzolans are occasionally added as filler material in the United States. Grouting pressure generally ranges from 80 to 100 psi (0.55 to 0.69 N/mm^2) with a maximum pressure specified as 250 psi (1.72 N/mm^2). After the grout has discharged from the far end, that end is plugged and the pressure is again applied at the injecting end to compact the grout. Historically, it has been the practice to flush the ducts with water before grouting is started, the excess water being removed with compressed air. In recent years, grouting experience has indicated that flushing may not be necessary. Grouting should not be done in cold weather because of the possibility that ice may be trapped in the duct, later leaving a void with water which can cause corrosion. The PCI grouting specification has specified concrete temperature of 35°F as the minimum temperature for grouting.

Readers are referred to a paper by Professor Milos Polivka[26] presented at the FIP-RILEM Symposium on Injection Grout for Prestressed Concrete held at Trondheim, Norway, 1961. This paper describes in detail the materials and techniques used for grouting.

The Prestressed Concrete Institute has published[27] its "Tentative Recommended Practices for Grouting Posttensioned Prestressed Concrete" in its journal of November/December 1972. The Post-Tensioning Institute has published a guide specification for grouting.[28]

2-12 Fatigue Strength

The fatigue strength of prestressed concrete can be studied from three approaches: that of concrete itself, that of high-tensile steel, and that of the combination. It may also be studied by utilizing our knowledge on the fatigue strength of reinforced concrete, since so much data has already been accumulated.[29,30,31] There are, however, some differences between prestressed and reinforced concrete. For example, in prestressed concrete, compression in the extreme fibers is frequently near zero under dead load and increases to a maximum under live load, thus varying throughout a wide range. Furthermore, high-strength steel is prestressed to a high level, while its stress range is relatively small.

For prestressed members under the action of design live loads, the stress in steel wires is seldom increased by more than 10,000 psi (69 N/mm^2) from their effective prestress of about 150,000 psi (1034 N/mm^2). It is safe to say that, so long as the concrete has not cracked, there is little possibility of fatigue failure in steel, even though the working load is exceeded. After the cracking of concrete, high stress concentrations exist in the wires at the cracks. These high stresses may result in a partial breakage of bond between steel and concrete near the cracks. Under repeated loading, either the bond may be completely broken or the steel may be ruptured.

Numerous tests have been conducted on prestressed-concrete members, giving considerable data on their fatigue strength. The results of these tests confirm the ability of the combination to stand any number of repeated loads within the working range. Failure started invariably in the wires near the section of maximum moment and often directly over the separators where the wires had a sharp change in direction or at positions where there were preformed cracks in the concrete.

Few tests are available concerning the fatigue bond strength between high-tensile steel and concrete. But, from the results of tests on prestressed-concrete beams, it seems safe to conclude that, if properly grouted, bond between the two materials can stand repeated working loads without failure. This is true because, before the cracking of concrete, bond along the length of the beam is usually low.

A rational method for predicting the fatigue strength of prestressed concrete beams in bending has been developed by Professor Ekberg.[32] It utilizes fatigue-failure envelopes for prestressing steel and concrete, and relates them to the stress-moment diagram for a beam.

A typical failure envelope for prestressing steel is shown in Fig. 2-7(*a*). This envelope indicates how the tensile stress can be increased from a given lower level to a higher level to obtain failure at one million load-cycles. Note that all values are expressed as a percentage of the static tensile strength. Thus the steel may resist a stress range amounting to $0.27f_{pu}$ if the lower stress limit is zero, but only a stress range of $0.18f_{pu}$ if the lower stress limit is increased to $0.40f_{pu}$. At a lower stress limit of $0.90f_{pu}$ or over, it takes only a negligible stress increase to fail the steel at one million cycles. While this fatigue envelope varies for different steels,[33] the curve given here may be considered a typical one.

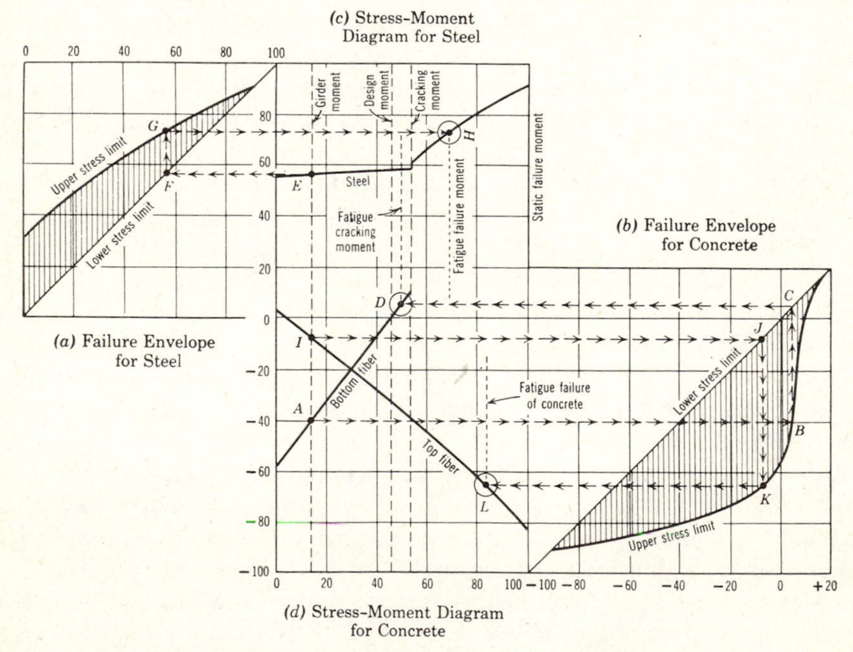

Fig. 2-7. Method for predicting fatigue strength of prestressed concrete beams.

The fatigue-failure envelope for concrete is given in Fig. 2-7(b). This is analogous to (a) for steel, except it is drawn to cover both tensile and compressive stresses. This diagram indicates that if the lower stress limit is zero, a compressive stress of $0.60f_c'$ may be repeated one million cycles. If the lower stress limit is $0.40f_c'$, the stress range can be $0.40f_c'$. If the compressive stress limit is $0.20f_c'$, the tensile stress limit, to produce cracking, is $0.05f_c'$.

A typical stress-moment diagram for steel is given in Fig. 2-7(c), which again expresses nondimensionally both the stress and the moment by relating them to the static strength and the ultimate static moment. For example, when the external moment is 70% of the ultimate static moment, the stress in the steel is shown to be $0.80f_{pu}$. Similarly, Fig. 2-7(d) gives the concrete fiber stresses relative to the moment. It is noted that under certain loading conditions, either the top or bottom fibers can be under tension rather than compression.

Combining these four portions (Fig. 2-7) it is possible to determine the fatigue-cracking moment and the fatigue-ultimate moment as limited by steel or concrete. Starting on the stress-moment diagrams at the point of dead-load stress, which represent the lowest possible stress level, we can trace three paths as follows.

Steel: *E-F-G-H*
Concrete top fiber: *I-J-K-L*
Concrete bottom fiber: *A-B-C-D*

The point H indicates that for a maximum moment of $0.68M_{ult}$, the steel will fail in tension at one million cycles. The point L indicates, for a maximum moment of $0.84M_{ult}$, the top fiber will fail in compression at one million cycles. The point D indicates that the fatigue-cracking moment is $0.50M_{ult}$.

Using this analytical approach, Ekberg studied the effect of the level of prestress, the effect of over- and underreinforcing, and the cracking characteristics. The following conclusions were reached:

1. Other conditions being equal, reducing the level of prestress considerably reduces the fatigue-failure moment. This becomes evident when it is realized that cracking would occur sooner for the lower level of prestress and a wider stress range would occur for the steel.
2. Since fatigue failure in concrete is not the controlling criterion, overreinforcing will generally increase the fatigue strength. The optimum-fatigue moment occurs for a percentage of steel higher than that indicated for a static balanced design.
3. The ratio of dead-load moment to live-load moment has very little effect on the fatigue-cracking moment. Although repetitive loading necessarily reduces the cracking moment, prestressing does delay the occurrence of cracks very substantially.

It is clear that the shape of the member and the location of the steel also have to do with the fatigue strength. When reliable fatigue resistance is to be found, it is desirable that the stress-moment curves and the fatigue-failure envelopes be obtained for the given case and the analytical method outlined above be followed for its determination.

References

1. S. Walker, "Modulus for Elasticity of Concrete," *Proc. Am. Soc. Test. Mat.*, Part II, 1919.
2. R. E. Davis and G. E. Troxell, "Modulus of Elasticity and Poisson's Ratio for Concrete and the Influence of Age and Other Factors upon These Values," *Proc. Am. Soc. Test. Mat.*, 1929. Also see A. D. Ross, "Experiments on the Creep of Concrete under Two-Dimensional Stressing," *Magazine of Concrete Research*, June 1954.
3. R. E. Davis and H. E. Davis, "Flow of Concrete under Action of Sustained Loads," *J. Am. Conc. Inst.*, March 1931 (*proc.*, Vol. 27), pp. 837–901.
4. H. R. Staley and D. Peabody, Jr., "Shrinkage and Plastic Flow of Prestressed Concrete," *J. Am. Conc. Inst.*, January 1946 (*Proc.*, Vol. 42), pp. 229–244.
5. G. E. Troxell, J. M. Raphael, and R. E. Davis, "Long-time Creep and Shrinkage Tests of Plain and Reinforced Concrete," *Proc. Am. Test. Mat.*, 1958.
6. "Creep of Concrete under High Intensity Loading," *Concrete Laboratory Report* No. C-820, Division of Engineering Laboratories, Bureau of Reclamation, U.S. Dept. of the Interior, April 10, 1956.
7. G. Magnel, "Creep of Steel and Concrete in Relation to Prestressed Concrete," *J. Am. Conc. Inst.*, February 1948 (*Proc.*, Vol. 44), pp. 485–500; also *Prestressed Concrete*, McGraw-Hill Book Co., New York, 1954.
8. R. E. Davis and G. E. Troxell, "Properties of Concrete and Their Influence on Prestressed Design," *J. Am. Conc. Inst.*, January 1954 (*Proc.*, Vol. 50), pp. 381–391.
9. P. Klieger, "Early High-strength Concrete for Prestressing," *Proceedings World Conference on Prestressed Concrete*, San Francisco, 1957.
10. N. H. Burns, G. C. Lacey, and J. E. Breen, "State of the Art for Long Span Prestressed Concrete Bridges of Segmental Construction," *J. Prestressed Conc. Inst.*, Vol. 16, No. 5, September–October 1971, pp. 52–77.
11. "Prestressed-Tile Roof and Girders Make All-Tile Building Possible," *Eng. News-Rec.*, April 15, 1954, p. 39.
12. J. J. Shideler, "Lightweight Aggregate Concrete for Structural Use," *J. Am. Conc. Inst.*, October 1957, (*Proc.*, Vol. 54), p. 299–328.
13. T. R. Jones, Jr. and H. K. Stephenson, "Properties of Lightweight Related to Prestressing," *Proceedings World Conference on Prestressed Concrete*, San Francisco, 1957.
14. C. H. Best and M. Polivka, "Creep of Lightweight Concrete," *Magazine of Concrete Research*, November 1959, pp. 129–134.

15. H. Lossier, "L'Autocontrainte des betons par les cimente expansifs," *Memoires Societe des Ingenieurs Civils de France*, March, April 1949, pp. 189–225; also, H. Lossier, "The Self-stressing of Concrete by Expanding Cement," *C.A.C.A.*, London.
16. V. V. Mikhailov, "New Developments in Self-stressed Concrete," *Proceedings World Conference on Prestressed Concrete*, San Francisco, 1957.
17. A. Klein and G. E. Troxell, "Studies of Calcium Sulfoaluminate Admixtures for Expansive Cements," *ASTM* (*Proc.*, Vol. 58), 1958, pp. 986–1008.
18. A. Klein, T. Karby, and M. Polivka, "Properties of an Expansive Cement for Chemical Prestressing," *J. Am. Conc. Inst.*, June 1961.
19. T. Y. Lin and A. Klein, "Chemical Prestressing of Concrete Elements Using Expanding Cements," *J. Am. Conc. Inst.*, September 1963.
20. Alois Legat, "Metallurgical, Metallographical and Economic Problems in the Manufacture of Prestressed Reinforcing Steels," extract of final report of *RILEM Symposium*, Liege, July 1958.
21. I. A. Rubinsky and A. Rubinsky, "A Preliminary Investigation of the Use of Fiberglass for Prestressed Concrete," *Magazine of Concrete Research*, pp. 71–78, September 1954.
22. Frank J. Maguire, III, *Report on Further Investigation Concerning the Feasibility of the Use of Fiberglass Tendons in Prestressed Concrete Construction*, M.S. Thesis, Princeton University, 1960.
23. J. C. Wines and G. C. Hoff, "Laboratory Investigation of Plastic-Glass Fiber Reinforcement for Reinforced and Prestressed Concrete," Reports 1 and 2, Miscellaneous Paper No. 6-779, U. S. Army Corps of Engineers, February 1966.
24. K. M. Gloeckner, "Investigation of Fiberglass Prestressed Concrete," Virginia Highway Research Council, June 1967.
25. E. G. Nawy and G. E. Newerth, "Fiberglass Reinforced Concrete Slabs and Beams," *J. Str. Div.*, Am. Soc. Civil Engineers, Vol. 103, No. ST2, February 1977, pp. 421–440.
26. Milos Polivka, "Grouts for Post-tensioned Prestressed Concrete Members," *J. Prestressed Conc. Inst.*, 1961.
27. "Recommended Practice for Grouting of Post-Tensioned Prestressed Concrete," PCI Committee on Post-Tensioning, *J. Prestressed Conc. Inst.*, Vol. 17 No. 6, November/December 1972.
28. *Post-Tensioning Manual*, Post-Tensioning Institute, Phoenix, Arizona, 1976, pp. 143–148.
29. G. M. Nordby, "Fatigue of Concrete—A Review of Research," *Proc. of the American Concrete Institute*, Vol. 55, 1958–1959.
30. T. Y. Lin, "Strength of Continuous Prestressed Concrete Beams Under Static and Repeated Loads," *J. Am. Conc. Inst.*, June 1955.
31. P. W. Abeles, "Fatigue Tests on Partially Prestressed Concrete Members," *Final Report, Fourth Congress, Int. Assn. Bridge and Structural Eng.*, 1953; also "Fatigue Tests of Prestressed Beams," presented at ACI Convention at Denver, 1954.
32. C. E. Ekberg, Jr., R. E. Walther, and R. G. Slutter, "Fatigue, Resistance of Prestressed Concrete Beams in Bending," *J. Str. Div. Am. Soc. Civil Engineers* (Vol. 83, No. ST4), July 1957.

33. R. E. Lane and C. E. Ekberg, Jr., "Repeated Load Tests on 7-Wire Prestressing Strands," Lehigh Univ., *Fritz Laboratory Report*, January 1959; E. J. Ruble and F. P. Drew, "Railroad Research on Prestressed Concrete," *J. Prestressed Conc. Inst.*, December 1962; R. F. Warner and C. L. Hulsbos, "Probable Fatigue Life of Prestressed Concrete Flexural Members," Fritz Engineering Laboratory Report, Lehigh Univ., July 1962.

3

PRESTRESSING SYSTEMS; END ANCHORAGES

3-1 Introduction

On account of the existence of different systems and their patents for tensioning and anchoring the tendons, the situation appears a little confusing to a beginner in the design and application of prestressed concrete. Actually, present practice in this country does not require of the designer a thorough knowledge of the details of all systems or even of the system that he intends to use for his particular job. In order to encourage competitive bidding, the engineer often specifies only the amount of effective prestressing force required so that the bid is open to all systems of prestressing. However, he should have a general knowledge of the methods available and keep it in mind while dimensioning his members so that the tendons of several systems can be well accommodated. Sometimes he must compute the actual size and number of tendons for his members in order to obtain feasible arrangements and an accurate estimate of the quantities of materials. A knowledge of the details of end anchorages and the tensioning jacks is required to design the ends of his members so as to be ready to house the anchorages and to receive the jacks.

In the United States alone, there are well over one hundred patents and patents pending on various systems of prestressing.[1] Many of these patents have never been commercially or economically applied, but many others are still being developed. It would now take the job of a specialized lawyer in addition to a specialized engineer to look into the matter, or to apply for and obtain a new patent. The situation is further complicated by the development of similar methods in other countries having reciprocal patent arrangements with ours, although, in general, any patents obtained abroad must be at least registered in this country before being effective here. These problems, are only for the inventors to worry about. The practicing engineer, who simply wants to design some structures of prestressed concrete, is free to specify and design for any system without studying the intrigues of patent rights. In fact, the owner of the structure would not have to pay any direct royalty to the patent holder. The royalty is indirectly included in the bid price for the supplying of prestressing steel and anchorages, which sometimes also includes the furnishing of equipment for prestressing and some technical supervision for jacking. Owing to the keen competition already existing in this country in the field of prestressing, the matter of patent royalty is not a serious cost item for the owner. In order to bid on a structure embodying prestressed concrete, the general contractor usually approaches the various prestressing concerns for subbids. Thus the standard method of bidding applicable to reinforcing steel and other trades is extended to prestressing work.

In some other countries, the conditions are different. In Germany, for example, there is a tendency not to accept any general contractor for a job employing prestressed concrete unless he himself has devised a system of prestressing. As a result, each contractor is forced to devise some system of his own, and he tends to charge an excessive royalty for his method if his competitors want to use it. In France and Belgium, for historic and other reasons, one or two prestressing systems are much more widely applied than the others; hence the direction of growth tends toward the development of these particular systems rather than a multiple approach.

The basic principles of prestressing cannot be patented, but the details of its application can. There are some patents on the methods of construction, such as special production of prestressed slabs or pipes, using processes different from the ordinary. Fortunately, these patents are based more on the construction procedures than on the design features and seldom affect the designing work of the engineer. Furthermore, it is not good policy for an engineer to try to hold a monopoly on his or her design. Hence engineers are seldom obstructed from using any design in prestressed concrete.

The so-called prestressing system comprises essentially a method of stressing the steel combined with a method of anchoring it to the concrete, including perhaps some other details of operation. Hence most of the patents on prestressed concrete are based on either or both of the following two operational details: (1) the methods of applying the prestress; (2) the details of end anchorages. In addition to these, sometimes the size and number of wires also form part of the patented process, although most patents contain a variety of combinations of these.

The Posttensioning Institute has a Posttensioning Manual[2] initially published by the PCI in 1972, which contains details of the various systems of posttensioning and serve as an excellent reference. Some of the more prominent and common ones will be described in this chapter.

3-2 Pretensioning Systems and End Anchorages

A simple way of stressing a pretensioned member is to pull the tendons between two bulkheads anchored against the ends of a stressing bed. After the concrete hardens, the tendons are cut loose from the bulkheads and the prestress is transferred to the concrete. Such stressing beds are often used in a laboratory and sometimes in a prestressing factory. For this set-up, both the bulkheads and the bed must be designed to resist the prestress and its eccentricity.

For mass production of pretensioned members, an extension of the above method often known as the Hoyer system is generally used. It consists of stretching the wires between two bulkheads some distance apart, say several hundred feet. The bulkheads can be independently anchored to the ground, or they can be connected by a long stressing bed. Such a bed is costly, but it can serve two additional purposes if properly designed. First, intermediate bulkheads can be inserted in the bed so that shorter wires can be tensioned. Secondly, the bed can be designed to resist vertical loads, thus permitting the prestressing of bent tendons.

With this Hoyer process, several members can be produced along one line, by providing shuttering between the members and concreting them separately. When the concrete has set sufficiently to carry the prestress, the wires are freed from the bulkheads, and the prestress is transferred to the members through bond between steel and concrete or through special pretensioning anchorages at the ends of members. This long-line production method is economical and is used in almost all pretensioning factories in the United States. Various means have been devised for deflecting the tendons up and down along the length of the bed. One method is shown in Fig. 3-1.

Devices for gripping the pretensioning wires to the bulkheads are usually made on the wedge and friction principles. There are quick-release grips, capable of many uses and effecting a saving in time due to their simplicity. Since

Fig. 3-1. Deflecting tendons (harping) in a pretensioning bed (Ben C. Gerwick).

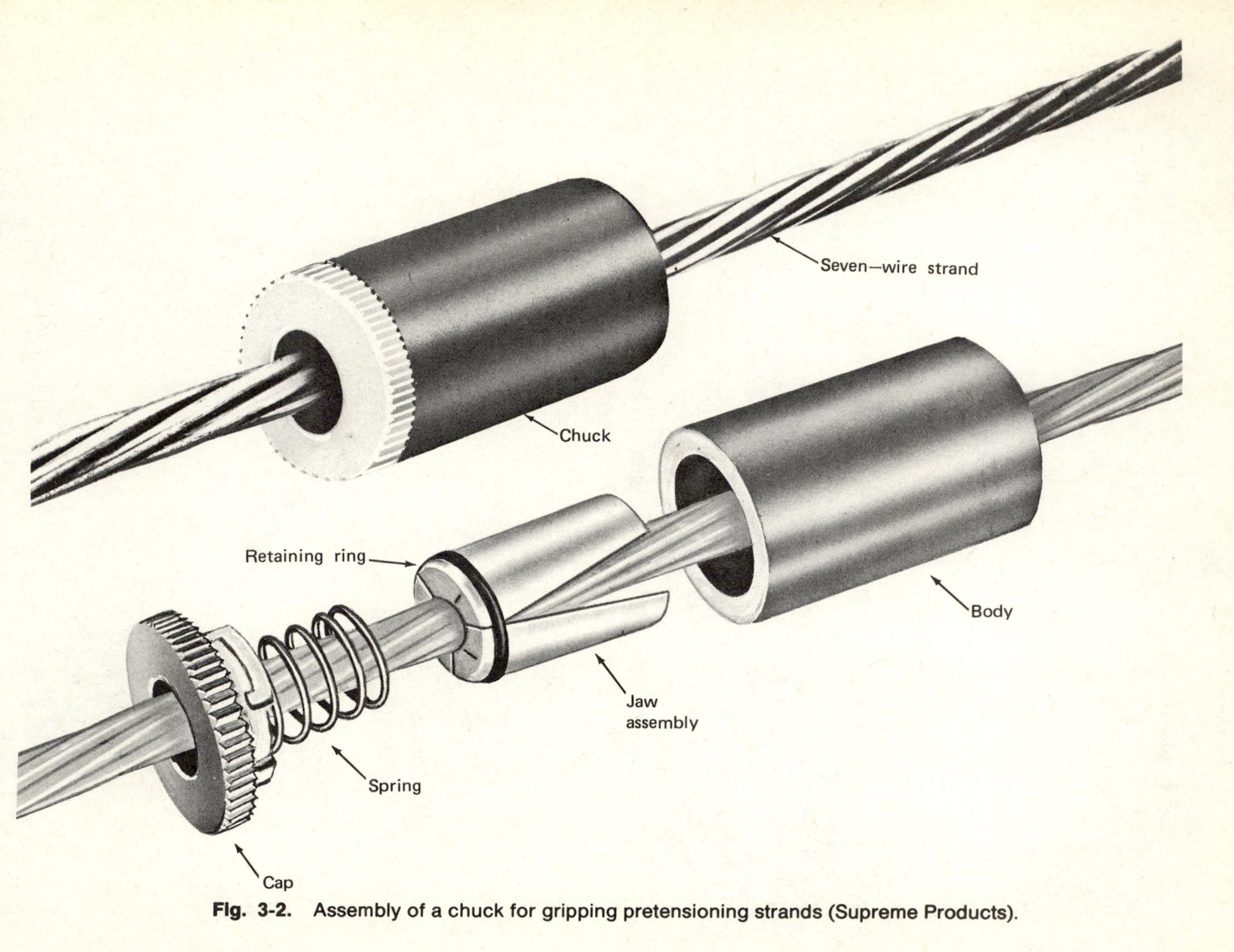

Fig. 3-2. Assembly of a chuck for gripping pretensioning strands (Supreme Products).

the wires or strands are to be held in tension only for short periods, these quick-release grips are found to be economical. Quick-release grips for holding strands are manufactured in the United States; for example, Supreme anchor grips (chucks) distributed by Supreme Products Corporation, Fig. 3.2. The grips made in England by CCL Ltd. are similar and are marketed in the United States as well as other countries.

The dependence on bond to transmit prestress between wires and concrete necessitates the use of small wires to ensure good anchorages. Wires greater than about $\frac{1}{8}$ in. (3.175 mm) are used only if they are waved along their length or if they are corrugated. In any case, a certain length of transfer is required to develop the bond. Should there be insufficient length of transfer, for example, when cracks occur near the end of a beam, the bond may be broken and the wires may slip. Addition of mechanical end anchorage to the pretensioned wires could prevent end slip, although experience with seven-wire strand up to 0.6 in. (15.2 mm) diameter has shown that end anchors are not necessary for typical pretensioned members. One method developed in San Diego, California, is the Dorland anchorage, which can be gripped on to the wires or strands at any point, thus supplying positive mechanical anchorage in addition to the bond. The clips are gripped to the tendon under high pressure, and the edges of the clips are then welded together at several points. It should be noted that such anchorage makes possible the use of bigger tendons and sometimes permits an earlier transfer of prestress.

The Shorer system[3] involves an ingenious feature, doing away with stressing beds and bulkheads. In their stead, a central tube of high-strength steel carries the prestress from the surrounding wires, and the entire assembly is placed in position and concreted. After the concrete has set and attained a certain strength, the tube is removed and the prestress is transferred to the concrete by bond. The hole left by the tube is then filled with grout. This method has not found application in this country. In France, the method is credited to Chalos and is known as the Chalos system.

As a direct contrast to the long-line method, the individual mold method for pretensioning has found application the U.S.S.R. and is occasionally used in the United States and Germany. Instead of moving the process to the product as in the case of the long-line method, the product is moved to the process when using the individual mold method, which is somewhat similar to the assembly-line method in automobile production.

The individual mold method adapts itself to relatively complicated patterns of the path of prestressing, such as prestressing in two directions for slabs or for trusses. It is also convenient for small products such as railway ties. Fig. 3-3 shows different types of continuous prestressing used in the U.S.S.R. for girders, slabs, railway ties, and trusses. The continuous prestressing is essentially carried out by two types of machines: one has a turntable with a stationary feeder head

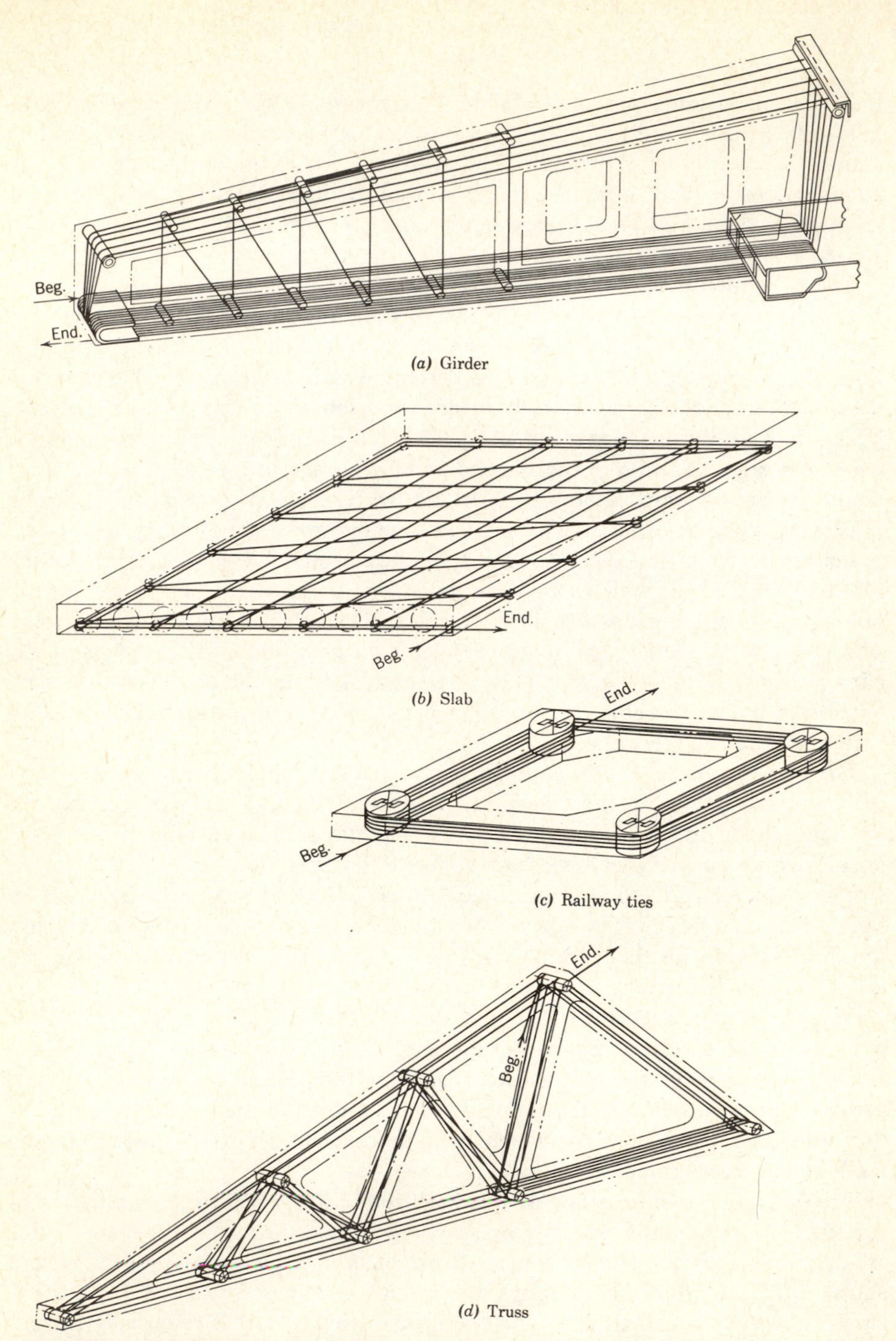

Fig. 3-3. Different types of continuous prestressing in the U. S. S. R.[4]

Fig. 3-4. Hollow-core units extruder production line and cross sections (Spancrete).

and the other consists of a stationary table with a movable feeder head. The prestressing wire is fed to the mold under controlled tension force and is laid out in a predetermined pattern as it is weaved around steel pegs fixed to the mold.[4,5]

In the United States several types of hollow-core slab units have been developed, and the equipment for their manufacture is marketed to plants which are located in many parts of the country, Fig. 3-4. Thus, the prestressed hollow-core slab has been used in many structures for floor and roof systems and is economically quite competitive. The long-line system is used with the equipment which deposits concrete and forms the hollow-core unit traveling down the prestressing bed to form long strips of the hollow-core slab. These long, continuous hollow-core slabs are later sawed into desired lengths for use in a given structure. In many structures the hollow core units may be supported by steel framing or masonry load-bearing walls.

3-3 Posttensioning, Tensioning Methods

The methods of tensioning can be classified under four groups: (1) mechanical prestressing by means of jacks; (2) electrical prestressing by application of heat; (3) chemical prestressing by means of expanding cement; (4) miscellaneous.

Mechanical Prestressing. In both pretensioning and posttensioning, the most common method for stressing the tendons is jacking. In posttensioning, jacks are used to pull the steel with the reaction acting against the hardened concrete; in pretensioning, jacks pull the steel with the reaction against end bulkheads or molds. Hydraulic jacks are used because of their high capacity and the relative ease in operating hand or electric pumps to apply the pressure. Levers may be found convenient only when very small wires are to be tensioned individually.

When hydraulic jacks are employed, one or two rams are worked by one pump unit with a control valve in the hydraulic circuit, Fig. 3-5. The capacity of hydraulic rams varies greatly, from about 3 tons up to 1000 tons. Fig. 3-6 shows the stressing equipment for a large posttensioned tendon. For some prestressing systems jacks are specially designed and rated to perform the job of tensioning particular tendons containing a given number and size of wires. For some systems, any jack of sufficient capacity can be employed, provided suitable grip for the tendon is available. Care must be taken to see that the jack can be properly mounted on the end bearing plates, and that there is enough room at the tensioning ends to accommodate the jacks.

Fig. 3-5. Electric pump operating a hydraulic jack to stress a Prescon cable (note the insertion of shims).

Fig. 3-6. Stressing of large multistressed tendon (Prescon/Freyssinet).

It is not possible to compile all the data necessary for designing each system of prestressing. New systems are being developed and existing ones are being modified from time to time. Engineers interested in a particular system should obtain the company's pamphlets or consult its representatives for particular details. In order to facilitate such consultation, the addresses of some prestressing systems in this country are listed in Appendix B.

Systems of jacking vary from pulling one or two wires up to over a hundred wires at a time. Single strand anchors such as the one shown in Fig. 3-7 are marketed by several posttensioning companies and are widely used for prestressed slabs. The anchor can be recessed to allow the end of the strand to be cut and protected by grout without any projection beyond the face of the slab.

The Freyssinet double-acting jack pulls up to 12 strands at a time, Fig. 3-8. The wires are wedged around the jack casing and are stretched by the main ram which reacts against the embedded anchorage. When the required tension is reached, an inner piston pushes the plug into the anchorage to secure the wires;

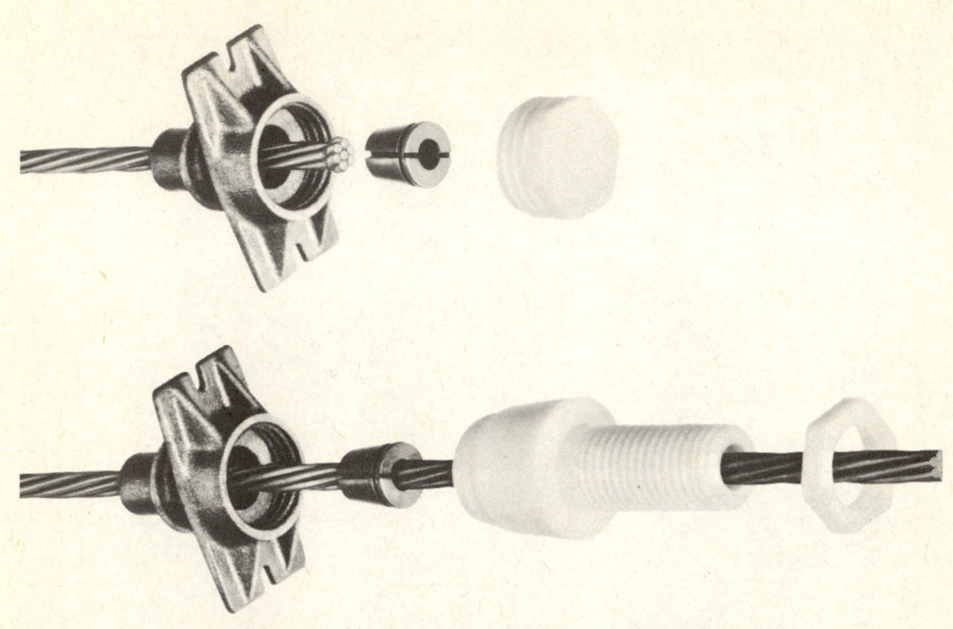

Fig. 3-7. Anchorage for single-strand tendon (Post-Tensioning Institute).

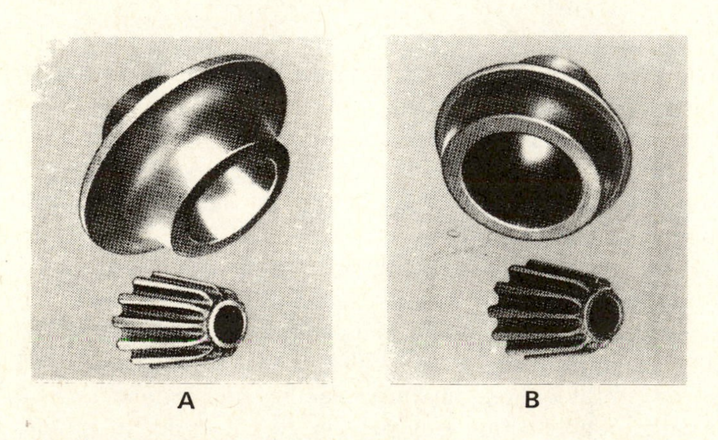

A — Freyssinet Type "BI" Internal Anchor

B — Freyssinet Type "EA" External Anchor

Fig. 3-8. Freyssinet "T" system anchorage cones (Prescon/Freyssinet).

the pressure on the main ram and that on the inner piston are then released gradually, and the jack is removed.

Dywidag threadbars (Fig. 3.9) are stressed using electrically powered hydraulic jacks. The jack nose contains a socket wrench and ratchet device which allows the nut to be tightened as the threadbar elongates. The magnitude of the prestress force applied is monitored by reading the hydraulic gage pressure and by measuring the elongation.

Some hints regarding the practice of jacking may be helpful for designers as well as for supervising engineers. In order to minimize creep in steel and also to reduce frictional loss of prestress, tendons are sometimes jacked a few per cent above their specified initial prestress. Overjacking is also necessary to compensate for slippage and take-up in the anchorage at the release of jacking pressure. When tendons are long or appreciably curved, jacking should be done from two ends. During the process of jacking, anchorage screw nuts and wedges should be run all the way home and seated moderately tight against the end plates. This may help to avoid serious damages in the event of a wire breakage or a sudden failure of the jacks.

Pressure gages for jacks are calibrated either to read the pressure on the piston or to read directly the amount of tension applied to the tendon. It is usual practice to measure the elongation of steel to be checked against the gage indications. The amount of frictional loss can be estimated from the discrepancy between the measured and the expected elongations.

When several tendons in a member are to be tensioned in succession, care should be taken to pull them in the proper order so that no serious eccentric loading will result during the process. If necessary, some tendons may have to be

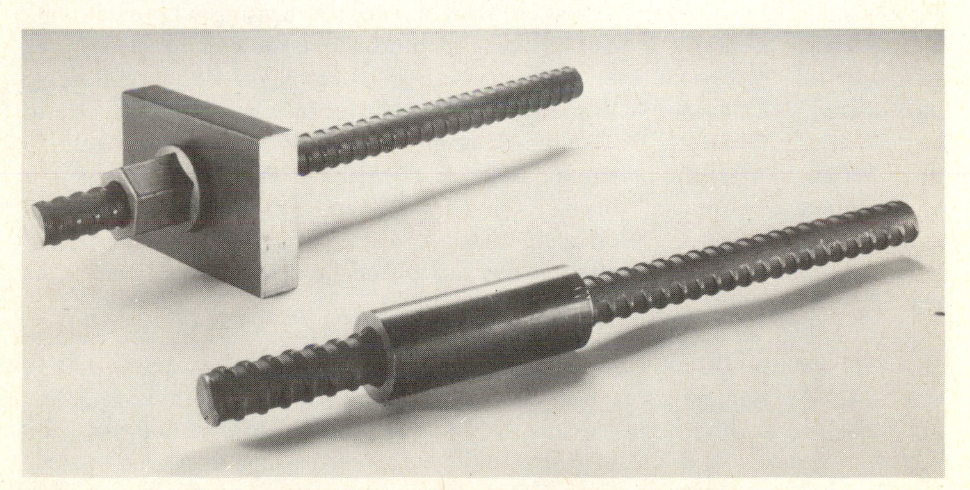

Fig. 3-9. Dywidag deformed bar anchorage (Dykerhoff and Widmann, Inc.).

tensioned in two steps so as to reduce the eccentric loading on the member during tensioning.

Instead of being attached to the steel, the jacks are sometimes inserted between two portions of concrete to force them apart, one against the other. Notably, this procedure is used in two systems: the Leonhardt system of Germany, and the Billner system of the United States. In the Leonhardt system,[6] one reinforced-concrete anchor block is poured at each end of a structural portion, and prestressing cables are wound around the blocks to be stressed all at once by hydraulic jacks inserted between the blocks and the main body of the structure. The advantage of this method lies in the reduction of the number of stressing operations. But naturally much heavier jacks are required. In order to reduce the cost of the heavy jacks, built-in ones of reinforced concrete are made on the job. They are eventually left in the structure.

In contrast to the Leonhardt system, which is specifically designed for large structures, the Billner system[7] is better suited to small ones. In the Billner system the member is cast in two portions, split at the midspan. Jacks separated by a comblike partition are inserted between the two portions. Concrete is cast on both sides of the partition, forming two separate units. The prestressing wires pass through the slots provided in the partition plate and are not bonded to the concrete except at the ends. In this system, jacking is done near the midspan and between the concrete; hence the end anchorages only have to perform the task of anchoring, which is simply achieved by looping the tendons around the concrete. In spite of the saving of the more expensive end anchorages otherwise required for posttensioning, this method has not yet been found to be commercially economical.

Electrical Prestressing. The electrical method of prestressing[8] dispenses with the use of jacks altogether. The steel is lengthened by heating with electricity. This electrical process is a posttensioning method where the concrete is allowed to harden fully before the application of prestress. It employs smooth reinforcing bars coated with thermoplastic material such as sulfur or low-melting alloys and buried in the concrete like ordinary reinforcing bars but with protruding threaded ends. After the concrete has set, an electric current of low voltage but high amperage is passed through the bars. When the steel bars heat and elongate, the nuts on the protruding ends are tightened against heavy washers. When the bars cool, the prestress is developed and the bond is restored by the resolidification of the coating.

This method, as originally developed, was intended for steel bars stretched to about 28,000 psi (193 N/mm^2), which requires a temperature of about 250°F. Owing to the high percentage of loss of prestress for steel with such a low prestress, and to other expenses involved in the process, this method has been found to be uneconomical in competition with prestressing using high-tensile steel. It has not been applied to high-tensile steel because a much higher temperature would be required for its prestressing. Such a high temperature could involve a number of complications, including possible damage to some physical properties of high-tensile steel.

In Russia, on the other hand, the electrothermal method of prestressing has found wide usage in pretensioning. Electric current is used to heat and expand prestressing steel, which is then held at the ends. As the steel cools and tends to shrink, it is stressed. A combination of electrical and mechanical stressing is also employed in the U.S.S.R.[9]

Chemical Prestressing. As described previously, the chemical reactions taken place in expansive cements can stress the embedded steel which in turn compresses the concrete. This is often termed self-stressing, but can also be called chemical prestressing. Readers are referred to section 2-5 for details.

Miscellaneous. Still another method of prestressing, not belonging to any of the above groups, was developed and applied in Belgium; it is known as the "Preflex" method.[10] The procedure consists of loading a high-tensile steel beam in the factory with a load equal to that anticipated in use. While the beam bends considerably under this load, its tensile flange is clothed with concrete of high compressive strength. After the concrete hardens, the load on the beam is removed, and the concrete is compressed as the beam regains a measure of its original shape. Then the beam is transported to the site to form a part of the structure, generally with the top flange and the web then also encased in concrete. Thus a composite section is obtained combining the strength of high-tensile steel with the rigidity of concrete. A series of tests (on Preflex beams) were carried out at Lehigh University.

3-4 Posttensioning Anchorages Utilizing Wedge Action

There are essentially three principles by which steel wires or strands are anchored to concrete.

1. *By the principle of wedge action producing a frictional grip on the wires.*
2. *By direct bearing from rivet or bolt heads formed at the end of the wires.*
3. *By looping the wires around the concrete.*

Several dependable systems have been developed based on the principles of wedge action and of direct bearing. Little can be said about the relative advantages of these two principles, the superiority of each system depending on the method of application rather than on the principle itself. The last method, looping the wires around the concrete, has not been widely applied, although it also has its advantages.

Several popular prestressing systems anchor their wires or strands by wedge action. The Freyssinet system which has been used all over the world, makes use of the wedge principle with up to 12 strands in a tendon, Fig. 3-8. Each

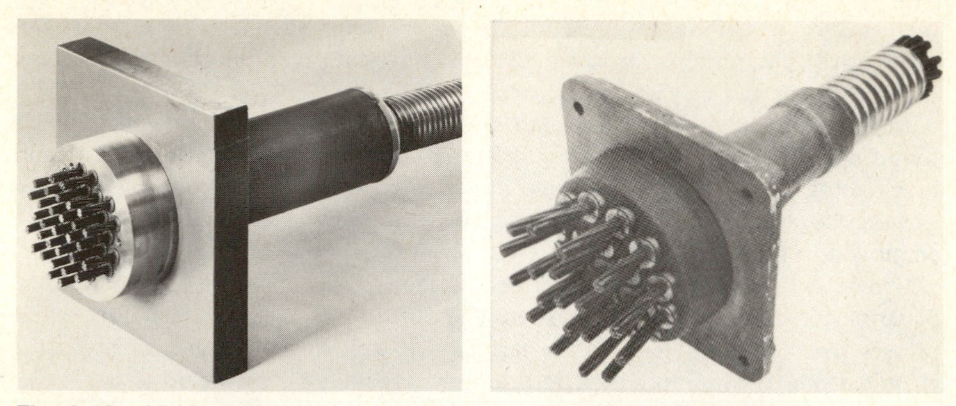

Fig. 3-10. Multistrand wedge type anchorages (Post-Tensioning Institute).

anchorage unit consists of a cone through which the wires pass, and against the walls of which the wires are wedged by a conical plug lined longitudinally with grooves to receive them. The cone is buried flush with the face of the concrete or recessed from the face if desired. It serves to transmit the reaction of the jack as well as the prestress of the wires to the concrete. After the completion of prestressing, grout is injected through a hole at the center of the conical plug. The Freyssinet cones are made for strands with 0.5 in. or 0.6 in. (12.7 mm or 15.24 mm) diameter, with the number of strands ranging from 6 to 12 per tendon.

Several companies have developed wedge-type anchors for tendons composed of several $\frac{1}{2}$-in. (12.7 mm) diameter or 0.6-in. (15.24 mm) diameter strands, Fig. 3-10. Prescon, VSL, Inryco, Freyssinet and others all utilize the friction principle with details to allow for varying numbers of strands in the tendon. These same companies market single-strand anchors which use wedge-type grips to anchor $\frac{1}{2}$-in. or 0.6-in. (12.7 mm or 15.24 mm) diameter strands, Fig. 3-7. The VSL Company's anchor uses individual wedges for various numbers of $\frac{1}{2}$-in. (12.7 mm) diameter seven-wire strands. The bearing plate has a tapered hole to receive the wedges which grip the strands individually as shown in Appendix B. Also shown in this Appendix is the Freyssinet "K" Range system for multi-strand anchorages having up to 55 strands (marketed by the Prescon Corporation in the United States).

3-5 Posttensioning Anchorages for Wires by Direct Bearing

Several systems employing cold-formed rivet heads for direct bearing at the ends of stressing wires are used throughout the United States. These systems have special head-forming machines for the purpose. One of these is the Prescon system (Appendix B). By this method, rivet heads are cold-formed at the proper place for high-tensile wires of $\frac{1}{4}$-in. diameter. Static tests on these heads have shown that the full strength of the wire can be developed. If the wires are grouted, there being practically no change in stress at the ends, no danger of fatigue failure is expected. For cable stays employed in bridges where the wires are not grouted, Prescon has developed special sockets to grip the wires. In these situations where extreme variations of stress exist at the ends, these sockets can carry loads in excess of 1000 tons/tendon. Similar sockets have also been developed in Europe by BBRV.

The Prescon system uses tendons consisting of 2 to 130 wires arranged in parallel. Figure 10-3 shows a coupling for this multiwire system. The wires are threaded through a stressing washer at each end, Fig. 3-11, before having their heads formed. A hole is provided in the stressing washer to permit grouting. The stressing jack has a special stressing collar which is screwed over the stressing washer and pumped to give the required elongation. A slight excess elongation will enable the shims to be inserted more easily. Then the jack is relieved to transmit the pressure to the shims. The height of the shims must be calculated for each particular case, depending on the length and modulus of elasticity of the wire, the amount of prestress, and the frictional force along the cable. After completion of the prestressing operations, the entire end anchorage is enclosed with concrete for protection against corrosion and fire. In order to minimize

Fig. 3-11. End anchorage (Prescon, Western Concrete, BBRV).

handling of individual wires in the field, they are made into tendons in the plant and shipped to the site ready for installation.

If the tendons are used for bonded work, metal duct is required. In order to permit the passage of grout, the inside diameter of the duct is at least $\frac{1}{4}$ in. (6.35 mm) greater than required to house the wires. For unbonded work, grease is applied to the wires, which are then wrapped with heavy paper forming a multiwire tendon. For a 6-wire unit, the tendon has a diameter of $\frac{3}{4}$ in. (19.05 mm) and the stressing washer has a diameter of about 2 in. (50.8 mm) and a thickness of $\frac{3}{4}$ in. (19.05 mm) and bears against the steel shims which rest on a 5 in. by $4\frac{1}{2}$ in. (127 mm by 114.3 mm) steel plate $\frac{1}{2}$ in. (12.7 mm) thick. For the nonstressed ends, the wire head bears directly on the steel bearing plate without the shims. Both the stressing washer and the bearing plate are made of high-strength steel, such as plow steel.

The Button Head Wire Posttensioning System, supplied by Western Concrete Structures of Los Angeles is practically identical with the Prescon System, Fig. 3-11. The BBRV system developed in Switzerland, and distributed by other companies in the United States, is also similar to the Prescon; Appendix B shows the end anchorage.

In certain systems the wires are connected to a short rod of high-tensile steel which can be anchored by nuts and washers. In Germany, three such systems were developed, differing in the method of connecting the wires to the end rod. In the Leoba system, a short tee is formed at one end of the stub, around which the wires are looped. The Monierbau system spreads the wires in a steel cone and anchors them with zinc or lead as in the case of Roebling anchorage for strands. The Huettenwerk Rheinhausen system uses a cylinder for connection. For the Leonhardt and the Billner systems, the wires loop around the concrete and bear directly on it.

3-6 Posttensioning Anchorages for Bars

A suitable end anchorage for high-tensile steel bars in prestressed concrete was developed by Donovan Lee of England where it is known as the Lee-McCall system. The ends of the bars are threaded and anchored with nuts on washers and bearing plates. The essential point is the proper threading of the ends to take a special nut capable of developing as nearly as possible the full strength of the bar. By using tapered threads, about 98% of the bar strength is developed.

Only a short length of the bar is threaded at the untensioned end, sufficient to receive the nut resting on a washer. For the jacking end, a long threaded end is required; the total length of thread is such that, after tensioning to the full value, the nut will be turned to the very bottom of the taper so as to develop the full strength of the bar. If, owing to nonuniformity of material or construction, the bar has to be lengthened more than the calculated amount in order to obtain the desired prestress, split washer shims may be inserted between the nut and the regular washer. Overtensioning will be necessary if because of friction or for other reasons the bar cannot be lengthened to the predicted amount under the desired prestress.

During jacking, an adapter from the jack screws into the threaded end of the bar to apply the pull. Since the jacking force for each bar is never more than 60 or 70% of its ultimate strength, the net section at the root of thread is not critical during jacking. However, as mentioned previously, it is a good safety measure to keep the nut near the washer at all times during the process of jacking. After completion of the prestressing operations, the protruding threaded ends can be either cut off or buried in concrete together with the anchorage plates.

The bars can be either bonded or unbonded to the concrete. If unbonded, they can be encased in flexible metal tubing or coated with grease and wrapped with heavy paper. They are then placed and supported in the forms before concreting. If bonded, the bars can be placed either before or after the pouring of concrete. For prepouring placement, flexible metal tubes with inside diameter about $\frac{1}{4}$ in. (6.35 mm) greater than the bar size are used to facilitate grouting. For postpouring placement, hole-forming cores such as inflated rubber or rubber with stiffening bars can be employed.

The hexagonal nuts for the bars have a short diameter equal to about twice the bar diameter and a thickness about 1.6 times the bar diameter. The standard and split washers are made of $\frac{3}{16}$-in. (4.76 mm) and 14-gage metal. The anchorage plates differ in size and can accommodate 1 to 3 bars per plate. The plates have a thickness of $\frac{5}{8}$ to $1\frac{1}{2}$ in. (15.875–38.1 mm) and an area per bar equal to about $(5d)^2$, where d is the bar diameter. If sleeve couplers are used to splice the bars, it is necessary to provide enough space near the couplers to permit movement during the tensioning process.

More recently, wedge anchorage has been developed for these large bars, using wedge anchors and wedge plates. The advantage of the wedge anchorage is its convenience in gripping the bar at any point along its length, whereas the threaded anchorage is limited to the threaded portion and requires the use of shims for adjustments.

In other countries, similar methods of anchorage for high-tensile bars have been devised. They differ in detail from the Lee-McCall or Stressteel system, but thread and nuts bearing against washers are employed in all methods. In Germany alone, four such methods have been developed: the Dywidag, Finsterwalder, Karig, and Polensky and Zollner systems; in Belgium, the Wets system; in Holland, the Bakker system. These will not be described here.

The Dywidag Bar (Fig. 3-9 and Appendix B) has been marketed extensively in the United States and other countries where cast-in-place segments are prestressed by posttensioning and for many other applications. The threads formed by bar deformations make it easy to couple bars cut to any length required.

Bond properties are also improved by the deformations. Dywidag bars are used in sizes from $\frac{5}{8}$-in. to $1\frac{3}{8}$ in. (15.875 mm–34.925 mm) diameter. Ultimate strength of 230 ksi (1,586 N/mm^2) is available for the $\frac{5}{8}$-in. (15.875 mm) size, but 1-in. to $1\frac{3}{8}$ in. (25.4 mm to 34.925 mm) sizes have ultimate strength of 160 ksi (1,103 N/mm^2). Couplers and anchorages with the threaded detail are designed to develop the ultimate strength of the bars.

3-7 Comparison of Systems

It is very difficult to compare the advantages of various systems of prestressing. Speaking in general, established systems that have been proved by tests and service can all be considered safe ones. That does not preclude the possibility that newer and perhaps better systems may be developed. Any new system, however, should be subjected to adequate tests before it can be adopted in practice.

Owing to the manner in which prestressed concrete has developed, any "prestressing system" generally embodies several essential features, some of which are also adopted by other systems in one form or another. For example, the methods of providing the conduits, the size, number, and arrangement of wires, and the basic principles of jacking and of anchoring are common to many methods. The essential difference between the systems, then, lies usually in the following three features: the material for producing the prestress, the details of jacking process, and the method of anchoring.

First of all, there is the choice between pretensioning and posttensioning. When a pretensioning plant is accessible, and the precast member can be conveniently transported, pretensioning is often the cheaper, because of the saving in end anchorages, in conduits, and in grouting, and because of the centralization of the production process. If a plant is too far away, the cost of transportation may be excessive. If a plant has to be established just for one job, the costs may be prohibitive unless the job is big enough to justify such an establishment. Long and heavy members can best be poured in place or cast in segments to be posttensioned at the site, and pretensioning will not likely be economical.

In the United States, pretensioning using small wires has proved to be relatively uneconomical and has given way to the widespread adoption of wire strands. Seven-wire strands up to $\frac{1}{2}$-in. (12.7 mm) diameter have been successfully employed. Since strands anchor themselves quite well through bond with concrete, anchorages for pretensioning have not been found necessary in conventional construction.

One important shortcoming in pretensioning was the fact that its application had been limited to the employment of straight wires tensioned between two bulkheads. Hence advantage could not be taken of the curving and bending of cables so beneficial to many beam layouts. In modern pretensioning plants in the United States, however, provisions are made so that pretensioning wires or strands can be bent at almost any point. Heavy girders using bent-up wires anchored to the beds at the points of bending were built for the New Northam Bridge, Southampton, England, in 1954. In the last 15 years, draping of strands has become common practice in U.S. prestressing plants (Fig. 3-1).

For posttensioning, the members can be either precast or cast in place. There is a further choice between bonded and unbonded reinforcing. Most present-day systems permit the use of either the bonded or the unbonded type. Certain systems yield a slightly better bond than others, depending on the passage provided for grouting and the bonding perimeter afforded per unit of prestressing force. For other systems, the tendons can be more easily greased and wrapped for unbonded reinforcing. When competition is keen between systems, the choice of the bonded or the unbonded type may decide the economy of one system as against the others.

Another important decision is the choice of proper materials for prestressing, whether wires, strands, or bars. Strands possess higher unit strength than the others; larger sizes of strands and bars mean fewer units for handling. The strength of strands is close to that of wires, but multiple wire tendons have been fabricated with very compact anchorage sizes. Bars possess the least strength; but they are easier to handle and cheaper to anchor for some applications.

Anchorages for strands are more costly, but the percentage cost of anchorages decreases with the length of tendons. Bars require splices for longer length, whereas strands and wires can be supplied without splice for almost any length of tendon. These are some of the inherent advantages and disadvantages of each material. Once the choice of materials is made, the choice of prestressing systems is further narrowed down. In the United States, only two or three systems are available for strands or bars, though several systems are employed for wires. Since the choice of materials almost automatically dictates the choice of prestressing systems and tends to eliminate competition, it is common practice not to commit the design to any one material. This is often achieved by specifying the amount of effective prestress instead of the material and area of the tendons.

The final decision is often an economic one, that is, which system will work out the cheapest. There are some basic advantages to each system. For example, when fewer wires are stretched per operation, smaller jacks will be needed; they are easier to handle but take more time for the total tensioning. Systems where the jacking is done all at once demand jacks of much greater capacity, which are naturally more costly and more cumbersome to move.

For any particular structure at a given time and location, one system will come out to be the most economical. This is usually the result of the surrounding conditions as much as the inherent advantages of that system. The availability of service from the system's representatives, the accessibility of materials and equipment, the acquaintance of the designing engineer with a particular system, and the desirability and ability of the system to get that job often form the deciding factors. Most surviving popular systems possess a number of merits of their own, but the economy of each system will vary with each job.

Finally, an engineer designing for any particular system should refer to the latest pamphlets issued by the respective companies for details so that his structure may be designed accordingly, (see Appendix B for addresses). It is also possible that the representatives or licensees of a system will be able to furnish special advice as to how the structure can be suitably detailed. The engineer can learn much from such advice, although he or she should always depend on his or her own judgment for the final decision.

References

1. C. Dobell, "Patents and Code Relating to Prestressed Concrete," *J. Am. Conc. Inst.*, May 1950 (*Proc.*, Vol. 46), pp. 713–724.
2. *Posttensioning Manual*, Posttensioning Institute, Phoenix, Arizona, 1976, 288 pages.
3. H. Shorer, "Prestressed Concrete, Design Principles and Reinforcing Units," *J. Am. Conc. Inst.*, June 1943 (*Proc.*, Vol 39), pp. 493–528.
4. V. V. Mikhailov, "Automation in the Production of Prestressing Units," *Proceedings World Conference on Prestressed Concrete*, San Francisco, 1957.
5. T. Y. Lin, et al, *A Report on the Visit of an American Delegation to Observe Concrete and Prestressed Concrete Engineering in the U.S.S.R.*, Portland Cement Association, 1958.
6. F. Leonhardt, "Continuous Bridge Girder Prestressed in a Single Operation," *Civil Engineering*, January 1953, pp. 42–45.
7. K. P. Billner, "Prestressing of Reinforced Concrete," *Précontrainte Prestressing*, No. 1, 1951, pp. 5–13.
8. K. P. Bilner and R. W. Carlson, "Electrical Prestressing of Reinforcing Steel," *J. Am. Conc. Inst.*, June 1943 (*Proc.*, Vol. 39), pp. 585–592.
9. V. V. Mikhailov, "Recent Developments in Automatic Manufacture of Prestressed Members in the USSR," *J. Prestressed Conc. Inst.*, September 1961.
10. L. Baes and A. Lipski, "La Poutre prëflex," *Précontrainte Prestressing*, No. 1, 1953.

4

LOSS OF PRESTRESS; FRICTION

4-1 Significance of Loss of Prestress

In Chapter 1 the concept of calculating stresses in prestressed concrete members was developed, and it was pointed out that the distinctive feature of this structural system is that these stresses may be tailored to the desired level to assure satisfactory performance. We must again note that the prestress force used in making the stress computation will not remain constant with time. Stresses during various stages of loading (section 1-4) also vary since concrete strength and modulus of elasticity increase with time.

The total analysis and design of a prestressed concrete member will involve consideration of the effective force of the prestressed tendon at each significant stage of loading, together with appropriate material properties for that time in the life history of the structure. The most common stages to be checked for stresses and behavior are the following:

1. *Immediately following transfer* of prestress force to the concrete section stresses are evaluated as a measure of behavior. This check involves the highest force in the tendon acting on the concrete which may be well below its 28-day strength, f_c'. The ACI Code designates concrete strength as f_{ci}' at this initial stage and limits the allowable and concrete stresses accordingly.
2. *At service load* after all losses of prestress have occurred and a long-term effective prestress level has been reached, stresses are checked again as a measure of behavior and sometimes of strength. The effective steel stress, f_{se}, after losses is assumed for the tendon while the member carries the service live and dead loads. Also, the concrete strength is assumed to have increased to f_c' by this later time.

Certain structures may be prestressed in stages to match loadings which may be sequentially added to the structure. While this may dictate more intermediate stages to be evaluated, the basic requirement is still to assure satisfactory behavior at each of the critical stages. The time factor remains an important consideration as the loss of prestress is related to the time-dependent material properties mentioned in Chapter 2.

The designer must consider the actual materials and individual circumstances (time elapsed, exposure conditions, dimension and size of member, etc.) which influence the amount of loss of prestress. He or she will find that this calculation cannot be "exact" but, fortunately, it does not have to be for adequate design. A reasonable estimate can assure satisfactory performance at service load. In the great majority of cases, strength will not be changed by slightly more or less loss of prestress than estimated. This is discussed in Chapter 5, but is mentioned here to assure that the estimate of loss of prestress is placed in proper perspective.

The PCI Committee statement published with the recommendations for estimating prestress losses[1] summarizes the state-of-the-art (1975) as follows:

> "*A precise determination of stress losses in prestressed concrete members is a complicated problem because the rate of loss due to one factor, such as relaxation of tendons, is continually being altered by changes in stress due to other factors, such as creep of concrete. Rate of creep in its turn is altered by change in tendon stress. It is extremely difficult to separate the net amount of loss due to each factor under different conditions of stress, environment, loading, and other uncertain factors.*
>
> "*In addition to the foregoing uncertainties due to interaction of shrinkage, creep, and relaxation, physical conditions, such as variations in actual properties of concrete made to the same specified strength, can vary the total loss. As a result, the computed values for prestress loss are not necessarily exact, but the procedures here presented will provide more accurate results than by previous methods which gave no consideration to the actual stress levels in concrete and tendons.*
>
> "*An error in computing losses can affect service conditions such as camber, deflection, and cracking. It has no effect on the ultimate strength of a flexural member unless the tendons are unbonded or the final stress after losses is less than* $0.5f_{pu}$.
>
> "*It is not suggested that the information and procedures in this report provide the only satisfactory solution to this complicated problem. They do represent an up-to-date compromise by the committee of diverse opinions, experience and research results into relatively easy to follow design formulas, parameters, and computations.*"

We can see from the PCI Committee Statement that we are faced with a complex problem, yet there may not be a need for an extremely detailed analysis of the loss of prestress in the design of many members. The results of the complex analysis would have very little bearing on design for strength. This is fortunate since we have moved toward strength design in proportioning structures in reinforced concrete, steel, and other materials. Yet the *serviceability* of prestressed concrete members at each of the two stages of behavior mentioned above is important, along with the assurance of adequate strength.

Thus the question for the designer is not *if* he will estimate loss of prestress, but rather *how* he will estimate loss of prestress because he knows some loss will occur. Some situations demand only a reasonable estimate which may be extremely easy to accomplish using recommendations mentioned below. The engineer may make an estimate based prior experience from more detailed analysis for a very similar situation. The deflections at either transfer or service load may not be very critical, and perhaps there may even be doubt that the structure will experience loading above its service design load level. We may not even be concerned about minor cracking if we should slightly underestimate the losses.

The more detailed analysis of loss of prestress is made for unusual situations where deflections could become critical. Slender beams are more sensitive to moment from prestress which balances the moment from applied loading to control deflections. Some structures have severe limits on deflection due to design tolerances which are imposed, requiring more detailed analysis. In other situations we may be allowing tension in the concrete at full service load for structures located in a very corrosive environment. The underestimate of losses might result in cracking which the designer might not want to occur in such a potentially corrosive situation.

In this chapter the lump sum estimates are reviewed briefly. In the following sections, each of the sources of loss which the designer must consider is examined individually. The collective loss of prestress is the summation of these individual losses, which will be examined using an example problem.

Reference is made throughout this chapter to two very significant recommendations for estimating prestress losses: (1) The PCI Committee recommendation[1] (1975) and (2) the ACI-ASCE Committee recommendation[2] (1979). These committee recommendations give the designer confidence that he or she is summing up the individual losses for a reasonably accurate total sum. Variation in materials with time and interactions between the various sources of loss complicate matters if we wish to make a comprehensive solution to the problem, and computer programs have been developed which can handle the solution[3,4,5].

The ACI-ASCE Committee (1979) method which is developed in this chapter accomplishes the same objectives expressed by the PCI Committee in the previously quoted statement. The two approaches differ slightly in their handling of the time-dependent losses as will be discussed in section 4-4. The use of overall total loss estimates makes the problem much simpler, and in many design problems this may be entirely adequate. Some recommendation for estimates of total loss are given in section 4-2. With experience, the designer will begin to get a feel for the order of magnitude of loss of prestress which he or she might expect for a given situation, and a very reasonable total loss estimate might be made without doing a rigorous set of individual loss calculations.

Notation used in this chapter will follow that which is generally accepted, but some terms from the Committee Recommendations[1,2] will have to be defined as they occur.

4-2 Lump Sum Estimates for Prestress Loss

The recommendation for design of prestressed concrete structures developed by ACI-ASCE Committee 423 in 1958[6] included lump sum estimates for losses. They recommended that the total loss from elastic shortening, creep, shrinkage, and relaxation (but not including friction and anchorage slip) in normal weight concrete be 35,000 psi (241 N/mm^2) for pretensioned beams and 25,000 psi (172 N/mm^2) for posttensioned. These values became widely accepted and were included in the 1963 ACI Code as well as the AASHTO Specifications for highway bridges until 1975. Most prestressed concrete highway bridge design prior to 1975 (most of the bridges now in service) was done using these lump sum estimates. These beams have generally given excellent performance. The 1977 ACI Code Commentary mentions these values, but it is now recognized that these two values cannot cover all design situations adequately.

Two current lump sum recommendations have replaced these values which probably underestimated losses in some cases. Table 4-1 shows the values adapted by AASHTO[7] in 1975 for typical prestressed concrete members. The relaxation of steel was probably underestimated[9] in the earlier values which had been used since 1958, and these revised values suggest generally higher losses. Also note that two values of F_c' were used by AASHTO, with the provision that there could be 500 psi variation more or less than the given value. The Posttensioning Institute included the recommendations for lump sum losses in Table 4-2 as part of the Posttensioning Manual (1976). These values were adapted for use in the Posttensioning Industry on projects for which the losses were not specified by the designer.

Where concrete is stressed at very low strength, where concrete is highly prestressed, or in very dry or very wet exposure conditions, they noted that there might be significant variation from these values. Friction loss was not included in the losses of either AASHTO (Table 4-1) or PTI (Table 4-2).

The use of lump sum losses such as those referred to above is recommended only for very usual cases where average conditions make it unnecessary to make

Table 4-1 AASHTO Lump Sum Losses[7]

Type of Prestressing Steel	Total Loss	
	$f_c' = 4{,}000$ psi (27.6 N/mm^2)	**$f_c' = 5{,}000$ psi (34.5 N/mm^2)**
Pretensioning strand		45,000 psi (310N/mm^2)
Posttensioning[a] wire or strand	32,000 psi (221N/mm^2)	33,000 psi (228N/mm^2)
Bars	22,000 psi (152N/mm^2)	23,000 psi (159N/mm^2)

[a]Losses due to friction are excluded. Friction losses should be computed according to Section 6.5.

Table 4-2 Approximate Prestress Loss Values[a] for Posttensioning[8]

Posttensioning	Prestress Loss-psi	
Tendon Material	Slabs	Beams and Joists
Stress relieved 270 strand and stress relieved 240 wire	30,000 (207N/mm²)	35,000 (241N/mm²)
Bar	20,000 (138N/mm²)	25,000 (172N/mm²)

[a] Losses due to friction not included. Average values of concrete strength, prestress level and exposure conditions.

an approximation of loss by sources as described in the later sections. Experience with these lump sum for such cases has been quite satisfactory, but it is recognized that some design situations require a better estimate of prestress losses.

4-3 Elastic Shortening of Concrete

This section begins the consideration of losses by each individual source. Let us first consider pretensioned concrete. As the prestress is transferred to the concrete, the member shortens and the prestressed steel shortens with it. Hence there is a loss of prestress in the steel. Considering first only the axial shortening of concrete produced by prestressing (the effect of bending of concrete will be considered later), we have

$$\text{Unit shortening } \delta = \frac{f_c}{E_c}$$

$$= \frac{F_0}{A_c E_c}$$

where F_0 is the total prestress just after transfer, that is, after the shortening has taken place. Loss of prestress in steel is

$$ES = \Delta f_s = E_s \delta = \frac{E_s F_0}{A_c E_c} = \frac{n F_0}{A_c} \tag{4-1}$$

The value of F_0, being the prestress after transfer, may not be known exactly. But exactness is not necessary in the estimation of F_0, because the loss due to this shortening is only a few per cent of the total prestress, hence an error of a few per cent in the estimation will have no practical significance. It must be further remembered that the value of E_c cannot be precisely predicted either. However, since the value of the initial prestress F_i is usually known, a theoretical solution can be obtained by the elastic theory. Using the transformed-section method, with $A_t = A_c + nA_s$, we have

$$\delta = \frac{F_i}{A_c E_c + A_s E_s}$$

$$ES = \Delta f_s = E_s \delta = \frac{E_s F_i}{A_c E_c + A_s E_s}$$

$$ES = \frac{nF_i}{A_c + nA_s}$$

$$ES = \frac{nF_i}{A_t} \tag{4-2}$$

From (4-1) and (4-2) above, we observe that the change in steel stress at transfer is simply the concrete stress at the level of steel multiplied by $n = E_s/E_c$. When bending of the member due to its own weight and moment due to eccentricity of prestress are present in a member, we can use (1-5) developed in Chapter 1 to make this calculation:

$$f_c = \frac{F}{A} \pm \frac{Fey}{I} \pm \frac{My}{I} \tag{1-5}$$

The properties of the gross concrete section may be used[1,2] here to obtain the stress in the concrete at the level of steel for the beam shown previously in Fig. 1-14. With only its own weight, w_G, acting, we will designate as M_G the moment at the section where we wish to find losses. Since we want to find stress at the level of steel, we know that $y = e$, thus

$$f_{cir} = \frac{F}{A} + \frac{Fe^2}{I} - \frac{M_G e}{I} \tag{4-3}$$

where f_{cir} = stress in the concrete at level of steel due to prestress force F

The theoretically exact analysis would use the transformed section for elastic analysis, but it can be shown (see Chapter 5) that gross section properties will give almost the same results with simpler computations. Both the ACI-ASCE and the PCI committees recommend use of the gross section properties. The prestress force, F, in (4-3) should be an estimate of the force which exists just after transfer. We probably know the initial force, F_i, for the stressing of the strands between bulkheads as shown in Fig. 4-1, but the elastic loss, ES, will reduce this immediately at transfer when we cut the strands. We will have sufficient accuracy for most situations[1,2] if we assume the loss to be 10% for a pretensioned beam where transfer of a all tendons occurs at one time. Chapter 5 shows the analysis procedure to check this more carefully, but it will not usually be necessary to make this refinement. Thus we can obtain from (4-3) the concrete stress due to $F_0 = 0.9F_i$ (pretensioned members) as follows:

$$f_{cir} = \frac{F_0}{A} + \frac{F_0 e^2}{I} - \frac{M_G e}{I} \tag{4-4}$$

where f_{cir} = stress in the concrete at c.g.s. due to prestress force F_0 which is effective immediately after prestress has been applied to concrete

The elastic shortening for the steel may be written in more general form as follows:

$$ES = \Delta f_s = nf_{cir} = \frac{E_s f_{cir}}{E_{ci}} \tag{4-5}$$

where n = modular ratio at transfer, E_s/E_{ci}
f_{cir} = concrete stress from (4-4)
E_s = 29,000,000 psi

EXAMPLE 4-1

A straight pretensioned concrete member 40 ft. long, with a cross section of 15 in. by 15 in., is concentrically prestressed with 1.2 sq. in. of steel wires which are anchored to the bulkheads with a stress of 150,000 psi (Fig. 4-1). If E_{ci} = 4,800,000 psi and E_s = 29,000,000 psi compute the loss of prestress due to the elastic shortening of concrete at the transfer of prestress. (L = 12.2 m, E_{ci} = 33.1 kN/mm², and E_s = 200 kN/mm²)

Solution

$$F_i = 150{,}000 \times 1.2 = 180{,}000 \text{ lb}$$

(a) Using elastic analysis with transformed section estimate the change in steel stress at transfer:

$$ES = \Delta f_s = \frac{nF_i}{A_c + nA_s} = \frac{6 \times 180{,}000}{223.8 + 6 \times 1.2} = 4660 \text{ psi} \tag{4-2}$$

$$\text{Steel stress} = 150{,}000 - 4660 = 145{,}340 \text{ psi}$$

(b) Using (4-4) with estimate of $F_0 \simeq 0.9\ F_i$ we find $F_0 = (0.9)(180{,}000) = 162{,}000$ lb, thus for this member with $e = 0$ and $M_G = 0$:

$$ES = \Delta f_s = \frac{E_s}{E_{ci}}\left(f_{cir} = n\frac{F_0}{A}\right) = \frac{6 \times 162{,}000}{225} = 4320 \text{ psi} \tag{4-5}$$

$$\text{Steel stress} = 150{,}000 - 4{,}320 = 145{,}680 \text{ psi}$$

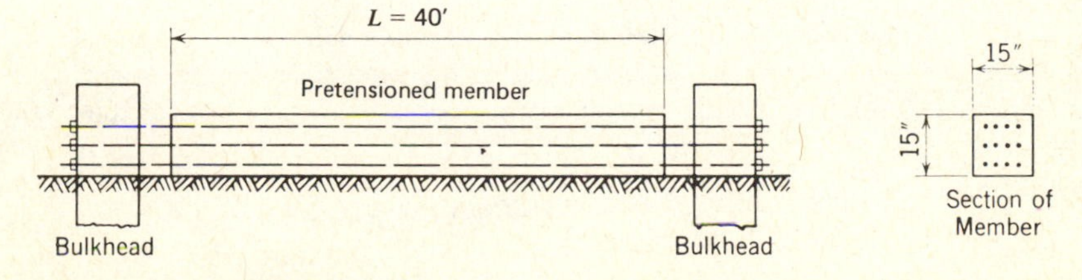

Fig. 4-1. Example 4-1.

Note that theoretically the remaining stress immediately following transfer is 145,340 psi while our approximation equation (4-5) estimated 145,680 psi, which is very adequate for design.

For posttensioning, the problem is different. If we have only a single tendon in a posttensioned member, the concrete shortens as that tendon is jacked against the concrete. Since the force in the cable is measured after the elastic shortening of the concrete has taken place, no loss in prestress due to that shortening need be accounted for.

If we have more than one tendon and the tendons are stressed in succession, then the prestress is gradually applied to the concrete, the shortening of concrete increases as each cable is tightened against it, and the loss of prestress due to elastic shortening differs in the tendons. The tendon that is first tensioned would suffer the maximum amount of loss due to the shortening of concrete by the subsequent application of prestress from all the other tendons. The tendon that is tensioned last will not suffer any loss due to the elastic concrete shortening, since all that shortening will have already taken place when the prestress in the last tendon is being measured. The computation of such losses can be made quite complicated. But, for all practical purposes, it is accurate enough to determine the loss for the first cable and use half of that value for the average loss of all the cables. This is shown in example 4-2.

EXAMPLE 4-2

Consider the same member as in example 4-1, but posttensioned instead of pretensioned. Assume that the 1.2 sq in. of steel is made up of 4 tendons with 0.3 sq in. per tendon. The tendons are tensioned one after another to the stress of 150,000 psi. Compute the loss of prestress due to the elastic shortening of concrete.

Solution The loss of prestress in the first tendon will be due to the shortening of concrete as caused by the prestress in the other 3 tendons. Although the prestress differs in the 3 tendons, it will be close enough to assume a value of 150,000 psi (1,034 N/mm^2) for them all. Hence the force causing the shortening is

$$3 \times 0.3 \times 150{,}000 = 135{,}000 \text{ lb (600 kn)}$$

The loss of prestress is given by formula 4-1

$$\Delta f_s = \frac{nF_0}{A_c} = \frac{6 \times 135{,}000}{225} = 3600 \text{ psi } (24.8 \text{ N/mm}^2)$$

Note that it is unnecessary to use the more exact formula 4.2.

Similarly, the loss due to elastic shortening in the second tendon is 2400 psi, in the third tendon 1200 psi, and the last tendon has no loss. The average loss for the 4 tendons will be

$$\frac{3600 + 2400 + 1200}{4} = 1800 \text{ psi } (12.4 \text{ N/mm}^2)$$

indicating an average loss of prestress of 1800/150,000 = 1.2%, which can also be obtained by using one-half of the loss of the first cable,

$$3600/2 = 1800 \text{ psi } (12.4 \text{ N/mm}^2)$$

The above method of computation assumes that the tendons are stretched in succession and that each is stressed to the same value as indicated by a manometer or a dynamometer. It is entirely possible to jack the tendons to different initial prestresses, taking into account the respective amount of loss, so that all the tendons would end up with the same prestress after deducting their losses. Considering the above example, if the first cable should be tensioned to a stress of 153,600 psi (1059 N/mm^2), the second to 152,400 (1051), the third to 151,200 (1,042) and the last to 150,000 (1034), then, at the completion of the prestressing process, all the tendons would be stressed to 150,000 psi (1034 N/mm^2). Such a procedure, although theoretically desirable, is seldom carried out because of the additional complications involved in the field. When there are many tendons and the elastic shortening of concrete is appreciable, it is sometimes desirable to divide the tendons into three or four groups; each group will be given a different amount of over-tensioning according to its order in the jacking sequence.

In actual practice, either of the following two methods is used.

1. Stress all tendons to the specified initial prestress (e.g., to 150,000 psi = 1,034 N/mm^2 in example 4-2), and allow for the average loss in the design (e.g., 1800 psi = 12 N/mm^2 in example 4-2).
2. Stress all tendons to a value above the specified initial prestress by the magnitude of the average loss (e.g., to 150,000 + 1800 = 151,800 psi = 1046 N/mm^2 in example 4-2). Then, when designing, the loss due to the elastic shortening of concrete is not to be considered again.

If the loss due to this source is not significant, the first method is followed. If the steel can stand some overtensioning, and if a high effective prestress is desired, the second procedure can be adopted.

The above discussion refers to the case when the prestress in the tendons is measured by manometer or dynamometer and only approximately checked by elongation measurements. At other times, prestress is measured by the amount of elongation, the gages being used merely as a check. The preference for one or the other depends on many factors: the personal practice of the engineer, the accuracy of the different instruments available, the constancy of the modulus of elasticity of steel, the amount of friction in the tendons, as well as the system of prestressing employed.

The ACI-ASCE Recommendation for elastic losses accounts for the sequence of stressing effect on elastic losses, as illustrated in example 4-2, by modifying (4-5) as follows:

$$ES = K_{es} E_s \frac{f_{cir}}{E_{ci}} \qquad (4\text{-}6)$$

Kes = 1 for pre tension,
Kes = 0.5 for post tension

where $K_{es} = 1.0$ for pretensioned members (example 4-1)
$K_{es} = 0.5$ for posttensioned members when tendons are in sequential order to the same tension

Note that in (4-4) for f_{cir} we modify the initial force to $F_0 = 0.9F_i$ for pretensioned members which have the immediate loss $ES = \Delta f_s$. Posttensioned members are stressed as described above with the elastic shortening taking place simultaneously with the application of the force F_0 from each of the tendons. The result of sequential stressing, as shown in example 4-2, is that we need only half of the force F_0 in computing f_{cir}, hence the factor $K_{es} = 0.5$ in (4-6).

4-4 Time-Dependent Losses (General)

Prestress loss due to elastic shortening at transfer is directly calculated using known or assumed material and cross section properties as discussed above. The effect of time is not directly involved in that calculation. The concrete is assumed to have the initial strength f'_{ci} in estimating E_c for use in the calculations for elastic shortening. Time from casting to release of prestress on to the concrete may be important in planning construction to assure that the initial strength will be attained. But the elastic shortening is instantaneous at the time of transfer and is independent of other sources of loss which may occur after this time.

Prestress losses due to creep and shrinkage of concrete and steel relaxation are both time dependent and interdependent. The materials have properties (described in Chapter 2) which are time dependent. The effects become interdependent in a prestressed concrete member. After transfer of prestress, a sustained stress is imposed on both steel and concrete which will change with time. To account for these changes with time, a step-by-step procedure can be used in order to account for changes which occur in successive time intervals.

A technique called rate-of-creep has been used to track the simultaneous effects of creep, shrinkage, and steel relaxation with time. This method utilizes test data for creep and shrinkage which have been obtained under laboratory conditions where constant stress was maintained. Since the stress in a prestressed concrete member is changing with time due to losses in the prestress force, the smooth curves for creep and shrinkage vs time are converted to step functions by dividing the time into short intervals over which constant strain is assumed. The effects at the beginning of the interval are known and changes over the time interval can be calculated to obtain total loss over the short time period. This leads to a step-by-step procedure in which time-dependent losses are accumulated as many short time intervals successively occur. Computer programs have been developed to handle the many calculations involved in this procedure.[3,4,5]

The PCI Committee on Prestress Losses has developed a procedure which divides the life history of a prestressed concrete member into a minimum of four time steps (see Table 4.3). The method is called the General Method.[1] The calculation of prestress losses can be made without a computer program by referring to various tables which have been developed for representative material properties reflecting time effects. The authors have used the PCI General Method for the beam shown in example 4-5 and the prestress losses which occur with time are presented. This PCI General Method adequately simplifies the actual behavior of the member to allow consideration of the important factors which influence time-dependent losses but does not require a computer program. It may easily be set up on small computers where they are available. The approach used by the PCI method is to solve the loss due to each source over a subdivided time period, the total loss being the sum of losses from the shorter periods.

This method may be necessary for special design situations where significant changes in loading occur. Such cases may require additional time intervals besides the minimum four steps of Table 4-1. For estimating losses due to time effects the long-time load should be estimated as realistically as possible. The occupancy loading may actually vary, but some estimate for the portion of the live load which will actually be present for long time effects should be made. Short applications of full-service design live load may be checked for stresses produced, but they may be computed by simple elastic procedures. Rarely would the full live load be used in computing time-dependent losses.

A Simplified Method for computing prestress losses is also included in the PCI Committee on Prestress Losses[1] recommendations. This method may save

Table 4-3 Minimum time intervals[1]

Step	Beginning time, t_1	End time, t
1	Pretensioned: anchorage of prestressing steel Postensioned: end of curing of concrete	Age at prestressing of concrete
2	End of Step 1	Age = 30 days, or time when a member is subjected to load in addition to its own weight
3	End of Step 2	Age = 1 year
4	End of Step 3	End of service life

time and yield the designer a very reasonable estimate of total loss of prestress for routine design situations. The equations are given for computing total prestress loss (TL) directly, but the appropriate equation must be selected for a given design situation.

Estimates of deflection with time accompanying the time-dependent effects which produce loss of prestress force will be discussed in Chapter 8. The deflections and prestress losses occur together and some computer programs for the rate-of-creep method handle the effects simultaneously.[3,4,5] However, in the same sense that "exact" computation for loss of prestress is not necessary or even possible, neither is an exact deflection with time calculation necessary. The estimate of final deflection is needed to assure satisfactory performance and this requires an estimate of total loss of prestress.

The loss due to elastic shortening can be compensated for, in the fabrication of pretensioned members, by tensioning the steel a bit higher than the stress desired at transfer when the strands are cut. This elastic loss may be computed and compensated for in posttensioning also. But the time dependent effects cannot be counterbalanced. It is not possible to overtension the wires or strands excessively to allow for such losses because that would mean very high initial stresses in the steel which might increase its relaxation loss or approach its yield point. Higher initial concrete stresses would also increase the creep loss in the concrete, making the total loss excessive. If the steel is unbonded it is sometimes possible to retension the steel after some losses have taken place. This has been found in practice to be expensive and undesirable, with the exception of some special situations where prestressing has been done by stages to match additional loadings. Stressing would be by steps with permanent counterbalancing load in these cases.

The equations from the ACI-ASCE Committee method[2] for estimating loss of initial tendon stress due to creep (CR) and shrinkage (SH) will be presented in the following sections and used in example 4-5 to illustrate the computation procedure. This method was developed to estimate the losses from time-dependent effects without having to break the life history of the beam into several time intervals as required by the PCI General Method described above. The paper[2] which presents the ACI-ASCE Committee Method contains comparisons of the results of the tendon stress losses computed by these relatively simple expressions with those from the test program[5]. These comparisons indicate very favorable results with a much simpler set of calculations than the PCI Method as illustrated in example 4-5 later in this chapter.

The ACI-ASCE Committee method[2] is less restricted than the PCI Simplified Method, and the authors feel that it will allow very adequate estimates of tendon stress losses for most prestressed concrete beams without the use of the time-consuming, step-by-step PCI General Method or a computer program. The interactions have been considered in setting the various coefficients which are used, thus it is significantly better than simply summing up individual estimates for loss-of-steel stress from elastic shortening, creep, and shrinkage of concrete together with relaxation of the steel.

4-5 Loss Due to Creep of Concrete (CR)

The property of concrete to experience additional strain under a sustained load was described in Chapter 2. Figure 2-3, which shows the creep ratio variation with time, gives an idea of the nature of creep. The PCI Committee assumed that the percentage of creep with time is very similar to the average curve in this figure, but it is pointed out that considerable variation has been reported by different investigators. The upper and lower curves of Fig. 2-3 reflect this variation and serve to remind us that we are simply estimating the tendon stress loss due to creep, CR, not making a precise calculation.

Many factors affect the creep ratio. The PCI General Method has modifiers to attempt to take into account the following: volume-to-surface ratio, age of concrete at time of prestress, relative humidity, and type of concrete (lightweight or normal). The ACI-ASCE Committee approximates the most important of these as will be described below. The concrete stress at the level of steel is f_{cir} immediately after transfer as described in the previous section, (4-4). The beam responds elastically to the application of the prestress force at transfer, but creep of concrete will occur over a long period of time under a sustained load.

Creep is assumed to occur with the superimposed permanent dead load added to the member after it has been prestressed. Part of the initial compressive strain induced in the concrete immediately after transfer is reduced by the tensile strain resulting from the superimposed permanent dead load. Loss of prestress due to creep is computed for bonded members from the following expression (for normal weight concrete):

$$CR = K_{cr} \frac{E_s}{E_c} (f_{cir} - f_{cds}) \tag{4-7}$$

where $K_{cr} = 2.0$ for pretensioned members
$K_{cr} = 1.6$ for posttensioned members
f_{cds} = stress in concrete at c.g.s. of tendons due to all super-imposed dead loads that are applied to the member after it is prestressed
E_s = modulus of elasticity of prestressing tendons
E_c = modulus of elasticity of concrete at 28 days, corresponding to f'_c

With sand-lightweight concrete we modify the foregoing values of K_{cr}, reducing them by 20%. There is a significantly larger amount of loss due to elastic shortening of sand-lightweight concrete because of its lower modulus of elasticity, resulting in an overall reduction of the creep coefficient. For members made of all lightweight concrete, special consideration should be given to the properties of the particular lightweight aggregate used[2].

For unbonded tendons the average compressive stress is used to evaluate losses due to elastic shortening and creep of concrete losses. The losses in the unbonded tendon are related to the average member strain rather than strain at the point of maximum moment. Thus

$$CR = K_{cr} \frac{E_s}{E_c} f_{cpa} \tag{4-8}$$

where f_{cpa} = average compressive stress in the concrete along the member length at the c.g.s. of the tendons.

4-6 Loss Due to Shrinkage of Concrete

Shrinkage of concrete is influenced by many factors, as is creep, and our calculations for loss from this source will reflect those which are most important: volume-to-surface ratio, relative humidity, and time from end of moist curing to application of prestress. Since shrinkage is time dependent (see Fig. 2-4 for curve of shrinkage ratio versus time) we would not experience 100% of the ultimate loss for several years, but 80% will occur in the first year. As with creep, there is an upper and lower variation from the average shrinkage strain value, which is taken to be 550×10^{-6} in./in. The modifying factors for volume-to-surface ratio (V/S) and relative humidity (RH) are given below:

$$\begin{aligned} \epsilon_{sh} &= 550 \times 10^{-6}\left(1 - 0.06\frac{V}{S}\right)(1.5 - 0.015RH) \\ &= 8.2 \times 10^{-6}\left(1 - 0.06\frac{V}{S}\right)(100 - RH) \end{aligned} \tag{4-9}$$

The loss of prestress due to shrinkage is the product of the effective shrinkage, ϵ_{sh}, and the modulus of elasticity of prestressing steel. For some concrete, especially lightweight concrete, the basic ultimate shrinkage may be greater than the value used above. The only other factor in the shrinkage loss equation (4-10) is the coefficient K_{sh} which reflects the fact that the posttensioned members benefit from the shrinkage which occurs prior to the posttensioning (Table 4-4). This value will be 1.0 for pretensioned beams with very early transfer of prestress and bonded tendons, but for posttensioned beams we may have a significant reduction in shrinkage. For example, if posttensioning is done 5 days after completion of moist curing we have $K_{sh} = 0.80$, or only 80% of the shrinkage for a companion pretensioned beam.

$$SH = 8.2 \times 10^{-6} K_{sh} E_s \left(1 - 0.06\frac{V}{S}\right)(100 - RH) \tag{4-10}$$

Table 4-4 Values of K_{sh} for posttensioned members

Time after end of moist curing to application of prestress, days	1	3	5	7	10	20	30	60
K_{sh}	0.92	0.85	0.80	0.77	0.73	0.64	0.58	0.45

4-7 Loss Due to Steel Relaxation

Tests of prestressing steel[9] with constant elongation maintained over a period of time have shown that the prestress force will decrease gradually as shown in Fig. 4-2. The amount of the decrease depends on both time duration and the ratio f_{pi}/f_{py}. The loss of prestress force is called relaxation. It was shown in the tests that this source of loss is more significant than had been assumed prior to 1963. We can express this loss as follows:

$$\frac{f_p}{f_{pi}} = 1 - \frac{\log t}{10}\left(\frac{f_{pi}}{f_{py}} - 0.55\right) \tag{4-11}$$

With a time interval between the moment of stressing t_1 in the pretensioning bed (such as the one shown in Fig. 4-1) and a later time t when we wish to estimate the remaining force, we can write the following equation.

$$\frac{f_p}{f_{pi}} = 1 - \left(\frac{\log t - \log t_1}{10}\right)\left(\frac{f_{pi}}{f_{py}} - 0.55\right) \tag{4-12}$$

where log t is to the base of 10 and f_{pi}/f_{py} exceeds 0.55.[9]

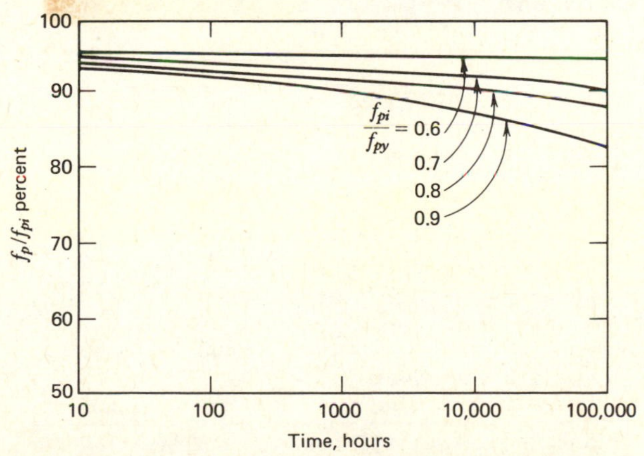

Fig. 4-2. Steel relaxation curves for stress-relieved wire and strand.

The ACI Code limit on initial prestress (immediately after anchorage) is $f_{pi}=0.7f_{pu}$. From Fig. 4-2 it is clear that higher sustained stress level will result in higher relaxation loss. This is one reason for limiting maximum initial stress, f_{pi}. The use of low-relaxation strand reduces this loss considerably and is gaining more widespread use. Some design situations may warrant the use of this material to reduce relaxation losses (about 3.5% maximum) even though its cost is slightly more than stress-relieved strand.

Prestressed beams actually have a constantly changing level of steel strain in the tendons as time-dependent creep occurs, and we must modify the calculation of relaxation loss, RE, to reflect this. The PCI Committee uses a series of steps in solving (4-12), then sums up the total. The ACI-ASCE Committee accomplishes approximately the same thing with an equation which follows:

$$RE=[K_{re}-J(SH+CR+ES)]C \tag{4-13}$$

where K_{re}, J, and C are values taken from Tables 4-5 and 4-6.

Because of elastic shortening (immediate loss at transfer) and the time dependent losses, CR and SH, there is a continual reduction in the tendon stress, and thus less relaxation loss. Equation 4-13 has the factor J from Table 4-5 to estimate this effect. The constant C from Table 4-6 shows the lesser value for low-relaxation strands than for stress-relieved strands as we would expect. Also, C accounts for the ratio f_{pi}/f_{pu}. The stress, f_{pi}, is before loss due to ES, CR, SH, and RE (but after any seating or friction loss).

Psi.

Table 4-5 Values of K_{re} and J

Type of tendon[a]	K_{re}	J
270 Grade stress-relieved strand or wire	20,000	0.15
250 Grade stress-relieved strand or wire	18,500	0.14
240 or 235 Grade stress-relieved wire	17,600	0.13
270 Grade low-relaxation strand	5,000	0.040
250 Grade low-relaxation wire	4,630	0.037
240 or 235 Grade low-relaxation wire	4,400	0.035
145 or 160 Grade stress-relieved bar	6,000	0.05

[a] In accordance with ASTM A416-74, ASTM A421-76, or ASTM A722-75.

Table 4-6 Values of C

f_{pi}/f_{pu}	Stress relieved strand or wire	Stress-relieved bar or low-relaxation strand or wire
0.80		1.28
0.79		1.22
0.78		1.16
0.77		1.11
0.76		1.05
0.75	1.45	1.00
0.74	1.36	0.95
0.73	1.27	0.90
0.72	1.18	0.85
0.71	1.09	0.80
0.70	1.00	0.75
0.69	0.94	0.70
0.68	0.89	0.66
0.67	0.83	0.61
0.66	0.78	0.57
0.65	0.73	0.53
0.64	0.68	0.49
0.63	0.63	0.45
0.62	0.58	0.41
0.61	0.53	0.37
0.60	0.49	0.33

4-8 Loss Due to Anchorage Take-Up

For most systems of posttensioning, when a tendon is tensioned to its full value, the jack is released and the prestress is transferred to the anchorage. The anchorage fixtures that are subject to stresses at this transfer will tend to deform, thus allowing the tendon to slacken slightly. Friction wedges employed to hold the wires will slip a little distance before the wires can be firmly gripped. The amount of slippage depends on the type of wedge and the stress in the wires, an average value being around 0.1 in. For direct bearing anchorages, the heads and nuts are subject to a slight deformation at the release of the jack. An average value for such deformations may be only about 0.03 in. If long shims are required to hold the elongated wires in place, there will be a deformation in the shims at transfer of prestress. As an example, a shim 1 ft long may deform 0.01 in.

With many plants the pretensioning technique is such that the anchorage take up is compensated for during the stressing operation. The chucks are seated as the tension is applied and the jacking force is calibrated to be sure that the desired tension stress is applied initially. Similarly, many of the posttensioning systems have jacking systems where a positive force pushes the wedges forward to grip the individual strands before releasing the tension onto the anchorage. Caution should be exercised in any system which depends on the wedge-type grips engaging the steel tendon simply by "grabbing" the wire or strand at release. Wide variation can occur and large anchorage take-up is possible due to the fact that the hard, smooth wires may not immediately grip the steel before it has slipped through. This source of loss in prestress can be easily minimized by exercising care in the stressing technique. A good rule is to be sure the wedges are positively engaged with the steel before release of the jacking force in the tendon. The loss from anchorage will be limited to the small movement which will occur as the wedges engage and the possibility of major slip can be eliminated.

A general formula for computing the loss of prestress due to anchorage deformation Δ_a is

$$ANC = \Delta f_s = \frac{\Delta_a E_s}{L} \tag{4-14}$$

Since this loss of prestress is caused by a fixed total amount of shortening, the percentage of loss is higher for short wires than for long ones. Hence it is quite difficult to tension short wires accurately, especially for systems of prestressing whose anchorage losses are relatively large. For example, the total elongation for a 10-ft (3.048 m) tendon at 150,000 psi (1,034 N/mm^2) is about

$$\frac{150{,}000 \times 10 \times 12}{30{,}000{,}000} = 0.6 \text{ in. } (15.24 \text{ mm})$$

and a loss of 0.1 in. (2.54 mm) would be a loss of 17% for (ANC). On the other hand, for a wire of 100 ft (30.48 m), an (ANC) loss of only 1.7% would be caused by such slippage, and it can be easily allowed for in the design, or counterbalanced by slight overtensioning.

4-9 Loss or Gain Due to Bending of Member

Loss of prestress due to a uniform shortening of the member under axial stress was discussed in section 4-3. When a member bends, further changes in the prestress may occur: there may be either a loss or a gain in prestress, depending on the direction of bending and the location of the tendon. If there are several tendons and they are placed at different levels, the change of prestress in them will differ. Then it may be convenient to consider only the centroid of all the tendons (the c.g.s. line) to get an average value of the change in prestress.

This change in prestress will depend on the type of prestressing: whether pre- or posttensioned, whether bonded or unbonded. Before the tendon is bonded to the concrete, bending of the member will affect the prestress in the tendon.

Neglecting frictional effects, and strain in the tendon will be stretched out along its entire length, and the prestress in the tendon will be uniformly modified. After the tendon is bonded to the concrete, any further bending of the beam will only affect the stress in the tendon locally but will not change its "prestress."

Consider a simple beam, where the tendon is bonded to the concrete either by pretensioning or by grouting after posttensioning (Fig. 4-3). Before any load is applied to the beam, it possesses a camber, as shown. Then an external load is applied and the beam deflects downward. The external load produces bending moment in the beam. Bending in the beam changes the unit stresses, hence the unit strains, in the tendon. The stress in the tendon near the midspan changes quite a bit, but that at the end does not change at all since there is no change in bending moment at the ends. If the "prestress" from the steel on the concrete is considered to be force applied at the ends, the change in "stress" along the length is not considered as a change in "prestress." After the tendon is bonded to the concrete, the steel and concrete form one section, and any change in stress due to bending of the section is easily computed by the transformed section method. Hence, it is convenient to say that "prestress" does not change as the result of bending of a beam after the bonding of steel to concrete, although the stress in the tendon does change.

The same is true of pretensioned members bending under prestress and their own weight. Again referring to Fig. 4-3, after the transfer of prestress, the beam bends upward and the wires shorten because of that bending. For the same reason as discussed above, that shortening of the wires due to bending is not considered a loss of prestress, just as the eventual lengthening under load is not

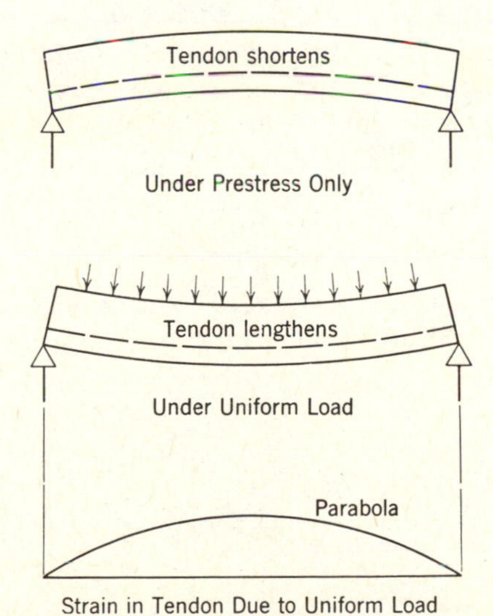

Fig. 4-3. Variation of strain in bonded tendon.

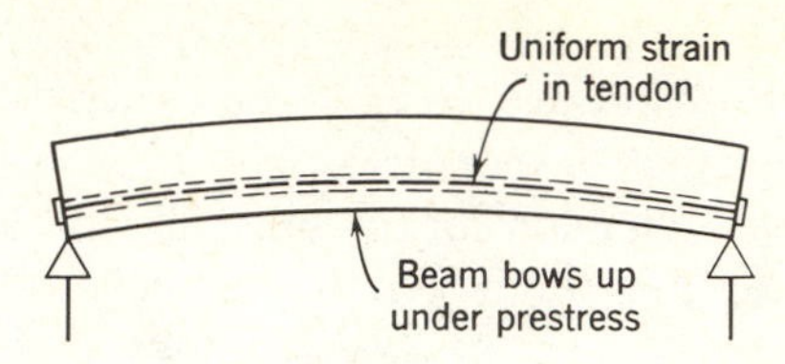

Fig. 4-4. Posttensioned member before grouting.

considered a gain in prestress. In both cases, the prestress considered is the prestress at the ends of the members, which does not change under bending.

For posttensioned bonded beams before grouting, the bending of the member will affect the prestress in the steel. Referring to Fig. 4-4, suppose that the tendons are tensioned one by one and the beam cambers upward gradually as more tendons are tensioned. Then the tendons that are tensioned first will lose some of their prestress due to this bending, in addition to the elastic shortening of concrete due to axial precompression. In general, these losses will be small and can be neglected. But when the camber is appreciable, it may be desirable to retension the tendons after completing the first round of tensioning or to allow for such losses in the design. Since it is the curvature of the beam at the time of grouting that determines the length of the tendons, the effect of creep in concrete will exaggerate the curvature and should be taken into account when allowing for such changes in prestress.

For posttensioned unbonded simple beams there may be a loss of prestress due to upward bowing caused by prestressing, and there will be a gain in prestress when the beam is fully loaded. If the tendons are permitted to slide freely within the concrete they will lengthen and shorten along their entire length as the beam bends. If a tendon does not remain at a constant distance from the c.g.c. line (center of gravity of concrete section), the computation of the change in length will be quite complicated. Fortunately, the loss or gain due to this source is ordinarily not more than 2 or 3% and for all practical purposes can be approximately estimated and allowed for.

EXAMPLE 4-3

A concrete beam 8 in. by 18 in. deep is prestressed with an unbonded tendon through the lower third point, Fig. 4-5, with a total initial prestress of 144,000 lb. Compute the loss of

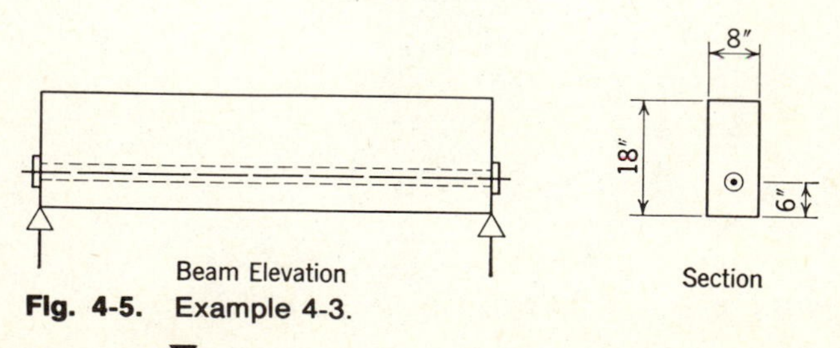

Fig. 4-5. Example 4-3.

prestress in the tendon due to the bowing up of the beam under prestress, neglecting the weight of the beam itself. $E_s = 30{,}000{,}000$ $E_c = 4{,}000{,}000$ psi. Beam is simply supported ($F_0 = 640.5$ KN, $E_s = 207$ KN/mm^2, and $E_c = 27.6$ KN/mm^2).

Solution Owing to the eccentric prestress, the beam is under a uniform bending moment of

$$144{,}000 \times 3 \text{ in.} = 432{,}000 \text{ in.-lb } (48{,}816 \text{ N-m})$$

The concrete fiber stress at the level of the cable due to this bending is

$$f = \frac{My}{I} = \frac{432{,}000 \times 3}{(8 \times 18^3)/12} = 333 \text{ psi } (2.30 \text{ N/mm}^2) \text{ compression}$$

(Note that stress due to the axial prestress of 144,000 lb is not included here; also, the gross area of the concrete is used for simplicity.)

Unit compressive strain along the level of the tendon is therefore

$$333/4{,}000{,}000 = 0.000083$$

Corresponding loss of prestress in steel is

$$0.000083 \times 30{,}000{,}000 = 2500 \text{ psi } (17.2 \text{ N/mm}^2)$$

However, if the beam is left under the action of prestress alone, the creep of concrete will tend to increase the camber and will result in further loss of prestress. On the other hand, if the prestress in the tendon is measured after the bowing of the beam has taken place, this loss due to bending of beam need not be considered.

The calculation of loss of prestress is illustrated for an I-shaped member later in this chapter, example 4-5. Note that the steel stress is estimated at a given section for this bonded member (midspan section) and the change in steel stress due to bending is included in the calculations. Since this steel stress change is easily calculated it may be combined with other effects in the individual steps. The refinement of calculating change in steel stress due to bending may thus be a part of rational calculations, but the additional moment should be only that due to the added load which was not included in the permanent, long-term loads incorporated into the calculations for creep loss. For many members where the permanent load is a large percent of the total service design load, the correction is very secondary and may be beyond the scope of accuracy in the loss calculations from other sources.

4-10 Frictional Loss, Practical Considerations

Valuable and extensive research work has been carried out to determine the frictional loss of prestress in prestressed concrete,[10,11] so that now it is possible to estimate such losses within the practical requirement of accuracy. First of all, it is known that there is some friction in the jacking and anchoring system so that the stress existing in the tendon is less than that indicated by the pressure gage. This is especially true for some systems whose wires change direction at the anchorage. This friction in the jacking and anchoring system is generally small though not insignificant. It can be determined for each case, if desired, and an overtension can be applied to the jack so that the calculated prestress will exist in the tendon. It must be remembered, though, that the amount of overtensioning is limited to stay within the yield point of the wires. The ACI Code limits the jacking force to 0.80 f_{pu}.

More serious frictional loss occurs between the tendon and its surrounding material, whether concrete or sheathing, and whether lubricated or not. This frictional loss can be conveniently considered in two parts: the length effect and the curvature effect. The length effect is the amount of friction that would be encountered if the tendon is a straight one, that is, one that is not purposely bent or curved. Since in practice the duct for the tendon cannot be perfectly straight, some friction will exist between the tendon and its surrounding material even though the tendon is meant to be straight. This is sometimes described as the wobbling effect of the duct and is dependent on the length and stress of the tendon, the coefficient of friction between the contact materials, and the workmanship and method used in aligning and obtaining the duct. Some approximate values for coefficients used to compute these losses are given in the Commentary of the ACI Code, Table 4-7.

The loss of prestress due to curvature effect results from the intended curvature of the tendons in addition to the unintended wobble of the duct. This loss is again dependent on the coefficient of friction between the contact

Table 4-7 Friction Coefficients for Posttensioning Tendons[a]

Type of Tendon	Wobble Coefficient K per Foot	Curvature Coefficient μ
Tendons in flexible metal sheathing		
Wire tendons	0.0010–0.0015	0.15–0.25
7-wire strand	0.0005–0.0020	0.15–0.25
High strength bars	0.0001–0.0006	0.08–0.30
Tendons in rigid metal duct		
7-wire strand	0.0002	0.15–0.25
Pregreased tendons		
Wire tendons and 7-wire strand	0.0003–0.0020	0.05–0.15
Mastic-coated tendons		
Wire tendons and 7-wire strand	0.0010–0.0020	0.05–0.15

[a]ACI Code Commentary.

materials and the pressure exerted by the tendon on the concrete. The coefficient of friction, in turn, depends on the smoothness and nature of the surfaces in contact, the amount and nature of lubricants, and sometimes the length of contact. The pressure between the tendon and concrete is dependent on the stress in the tendon and the total change in angle.

The Cement and Concrete Association of England has conducted extensive experiments in an effort to determine the coefficient of friction and the wobble effect for computing the frictional loss in the Freyssinet, the Magnel, and the Lee-McCall systems.[10] It is also pointed out that μ and K will depend on a number of factors: the type of steel used, whether wires, strands, or rods; the kind of surface, whether indented or corrugated, whether rusted or cleaned or galvanized. The amount of vibration used in placing the concrete will affect the straightness of the ducts; so will the overall size of the duct and its excess over the enclosed steel, and the spacing of the supports for the tendons or the duct-forming material. Individual values vary greatly, and the readers are referred to reference 10 if more accurate information is desired. Many factors cannot be predetermined, such as thin sheet metal casings which may be worn through so that the wires slide against the concrete, or the lateral binding of wires in curved sections, or the uneven movement of the separators due to the elongation of wires. Some values for coefficient of friction from the Commentary of the ACI Code are tabulated in Table 4-7, but they are intended only as a guide for the normal conditions. Values for the friction that can be expected with particular type tendons and particular type ducts can be obtained from the manufacturer of the tendons.

The coefficient of friction depends a great deal on the care exercised in construction. For unbonded reinforcement, lubricants can be used to advantage. Cables well greased and carefully wrapped in plastic tubes will have little friction, but if mortar leaks through openings in the tube, the cables may be tightly stuck. For bonded reinforcement, where lubricants may occasionally be employed, they must be applied very carefully in order not to destroy the eventual bond to be effected by grouting. Water-soluble oils have been successfully employed to reduce the friction while tensioning, and the lubricant is flushed off with water afterwards.

There are several methods of overcoming the frictional loss in tendons. One method is to overtension them. When friction is not excessive, the amount of overtension is usually made to equal the maximum frictional loss. The amount of wire lengthening corresponding to that overtension and the estimated friction can also be computed to serve as a check. This amount of overtension required for overcoming friction is not cumulative over that required for overcoming anchorage losses or of minimizing creep in steel. It is sufficient to take the greatest of the three required values and overtension for that amount. This is because in all three cases the overtensioning consists of an overstretching and a subsequent release-back. It must be noted that, if most of the friction exists near the jacking end, overtensioning to balance that friction will not produce any over-stretching of the main portion of the tendon and hence will not serve to minimize creep to any extent.

The effect of overtensioning with a subsequent release-back is to put the frictional difference in the reverse direction. Thus, after releasing, the variation of stress along the tendon takes some shape as in Fig. 4-6. When the frictional loss is a high percentage of the prestress, it cannot be totally overcome by overtensioning (curve *b*, Fig. 4-6), since the maximum amount of tensioning is limited by the strength or the yield point of the tendon. The portion of the loss that has not been overcome must then be allowed for in the design.

Jacking from both ends, of course, is another means for reducing frictional loss. It involves more work in the field but is often resorted to when the tendons are long or when the angles of bending are large. For a simple beam, where the critical point is its midspan, tensioning from both ends will not appreciably affect the controlling prestress at the midspan though it might change the beam deflection quite a bit.

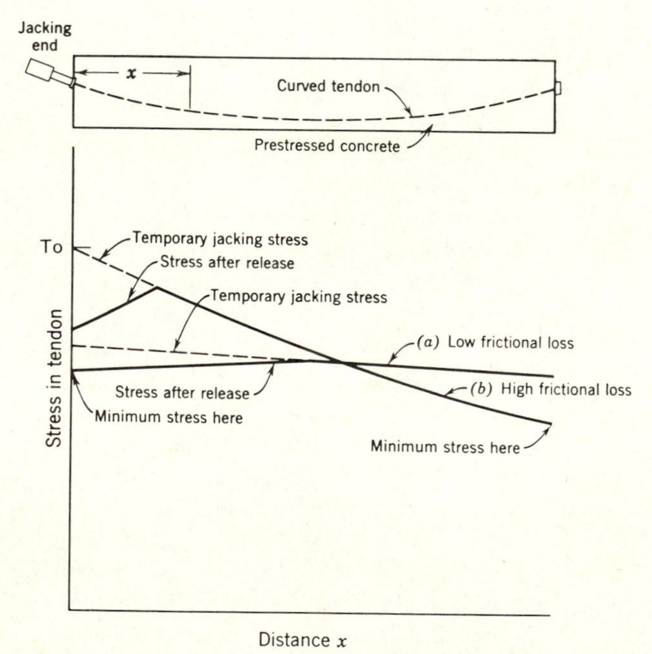

Fig. 4-6. Variation of stress in tendon due to frictional force.

4-11 Frictional Loss, Theoretical Considerations

The basic theory of frictional loss of a cable around a curve is well known in physics. In its simple form, it can be derived as follows. Consider an infinitesimal length dx of a prestressing tendon whose centroid follows the arc of a circle of radius R, Fig. 4-7, then the change in angle of the tendon as it goes around that length dx is

$$d\alpha = \frac{dx}{R}$$

For this infinitesimal length dx, the stress in the tendon may be considered constant and equal to F; then the normal component of pressure produced by the stress F bending around an angle $d\alpha$ is given by

$$N = F\,d\alpha = \frac{F\,dx}{R}$$

The amount of frictional loss dF around the length dx is given by the pressure times a coefficient of friction μ, thus,

$$\begin{aligned} dF &= -\mu N \\ &= \frac{-\mu F\,dx}{R} = -\mu F\,d\alpha \end{aligned}$$

Transposing, we have

$$\frac{dF}{F} = -\mu\,d\alpha$$

Integrating this on both sides, we have

$$\log_e F = -\mu\alpha$$

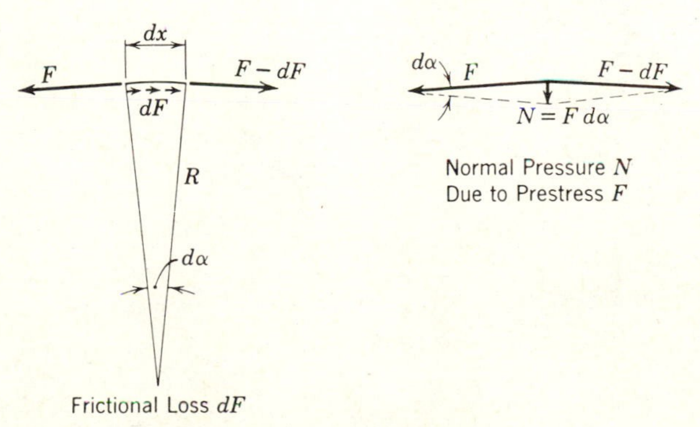

Fig. 4-7. Frictional loss along length dx.

Using the limits F_1 and F_2, we have the conventional friction formula

$$F_2 = F_1 e^{-\mu\alpha} = F_1 e^{-\mu L/R} \quad (4\text{-}15)$$

since $\alpha = L/R$ for a section of constant R.

For tendons with a succession of curves of varying radii, it is necessary to apply this formula to the different sections in order to obtain the total loss.

The above formula can also be applied to compute frictional loss due to wobble or length effect. Substituting the loss KL for $\mu\alpha$ in formula 4-4, we have

$$\log_e F = -KL \qquad F_2 = F_1 e^{-KL} \quad (4\text{-}16)$$

If it is intended to combine the length and curvature effect, we can simply write

$$\log_e F = -\mu\alpha - KL$$

For limits F_1 and F_2,

$$F_2 = F_1 e^{-\mu\alpha - KL} \quad (4\text{-}17)$$

Or, in terms of unit stresses,

$$f_2 = f_1 e^{-\mu\alpha - KL} \quad (4\text{-}17a)$$

The friction loss is obtained from this expression. Loss of steel stress is given as $FR = f_1 - f_2$, the steel stress at the jacking end is f_1, and the length to the point is L, Fig. 4-8(a). Thus, we find

$$FR = f_1 - f_2 = f_1 - f_1 e^{-\mu\alpha - KL} = f_1(1 - e^{-\mu\alpha - KL}) \quad (4\text{-}18)$$

These formulas are theoretically correct and take into account the decrease in tension and hence the decrease in the pressure as the tendon bends around the curve and gradually loses its stress due to friction. If, however, the total difference between the tension in the tendon at the start and that at the end of the curve is not excessive (say not more than 15 or 20%), an approximate formula using the initial tension for the entire curve will be close enough. On this assumption, a simpler formula can be derived in place of the above exponential form. If the normal pressure is assumed to be constant, the total frictional loss around a curve with angle α and length L is, Fig. 4-8.

$$F_2 - F_1 = -\mu F_1 \alpha = -\frac{\mu F_1 L}{R} \quad (4\text{-}19)$$

For length or wobble effect, we can again substitute KL for $\mu\alpha$ thus,

$$F_2 - F_1 = -KLF_1 \quad (4\text{-}20)$$

To compute the total loss due to both curvature and length effect, the above two formulas can be combined, giving

$$F_2 - F_1 = -KLF_1 - \mu F_1 \alpha = -F_1(KL + \mu\alpha) \quad (4\text{-}21)$$

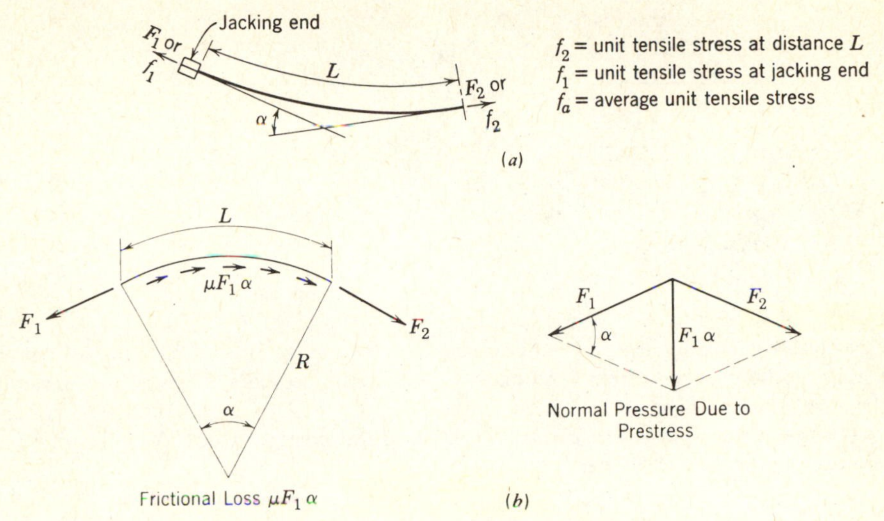

Fig. 4-8. Approximate frictional loss along circular curve.

Transposing terms, we have

$$\frac{F_2 - F_1}{F_1} = -KL - \mu\alpha = -\left(K + \frac{\mu}{R}\right)L \tag{4-22}$$

The loss of prestress for the entire length of a tendon can be considered from section to section, with each section consisting of either a straight line or a simple circular curve. The reduced stress at the end of a segment can be used to compute the frictional loss for the next segment, etc.

Since, for practically all prestressed-concrete members, the depth is small compared with the length, the projected length of tendon measured along the axis of the member can be used when computing frictional losses. Similarly, the angular change α is given by the transverse deviation of the tendon divided by its projected length, both referred to the axis of the member. Figure 4-9 shows approximation of α for this case.

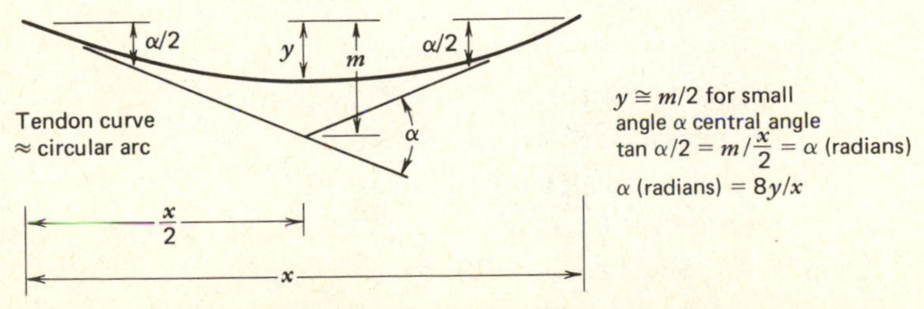

Fig. 4-9. Approximate determination of central angle for a tendon.

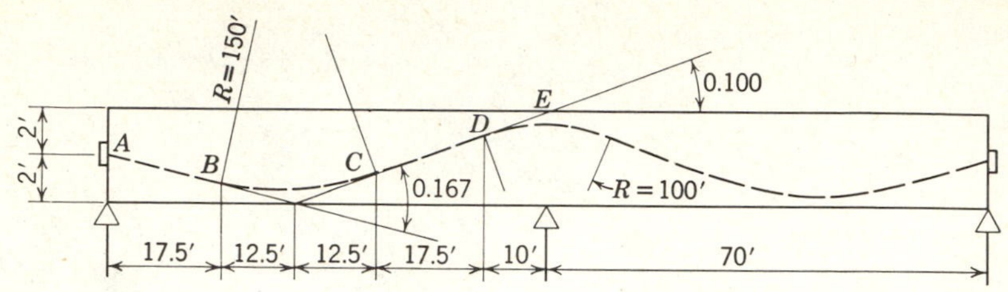

Fig. 4-10. Example 4-4.

EXAMPLE 4-4

A prestressed-concrete beam is continuous over two spans, Fig. 4-10, and its curved tendon is to be tensioned from both ends. Compute the percentage loss of prestress due to friction, from one end to the center of the beam (A to E). The coefficient of friction between the cable and the duct is taken as 0.4, and the average "wobble" or length effect is represented by $K = 0.0008$ per ft.

Solution

1. A simple approximate solution will first be presented. Using formula 4-22,

$$\frac{F_2 - F_1}{F_1} = -KL - \mu\alpha$$

$$= -0.0008 \times 70 - 0.4(0.167 + 0.100)$$

$$= -0.056 - 0.107$$

$$= -0.163$$

Solution

2. The above solution does not take into account the gradual reduction of stress from A toward E. A more exact solution would be to divide the tendon into four portions from A to E, and consider each portion after the loss has been deducted from the preceding portions. Thus, for stress at $A = F_1$,

$$AB\text{, length effect: } KL = 0.0008 \times 17.5 = 0.014$$

$$\text{Stress at } B = 1 - 0.014 = 0.986F_1$$

$$BC\text{, length effect: } KL = 0.0008 \times 25 = 0.020$$

$$\text{Curvature effect: } \mu\alpha = 0.4 \times 0.167 = 0.067$$

$$\text{Total: } 0.020 + 0.067 = 0.087$$

Using the reduced stress at B of 0.986, the loss is $0.087 \times 0.986 = 0.086$.

$$\text{Stress at } C = 0.986 - 0.086 = 0.900F_1$$

$$CD\text{, length effect: } KL = 0.0008 \times 17.5 = 0.014$$

Using the reduced stress of 0.900 at C, the loss is $0.014 \times 0.900 = 0.013$.

$$\text{Stress at } D = 0.900 - 0.013 = 0.887F_1$$

$$DE\text{, length effect: } KL = 0.0008 \times 10 = 0.008$$

$$\text{Curvature effect: } \mu\alpha = 0.4 \times 0.100 = 0.040$$

$$\text{Total: } 0.008 + 0.040 = 0.048$$

$$\text{Loss} = 0.048 \times 0.887 = 0.043$$

$$\text{Stress at } E = 0.887 - 0.043 = 0.844F_1$$

$$\text{Total loss from } A \text{ to } E = 1 - 0.844 = 0.156 = 15.6\%$$

This computation can be tabulated in order to simplify the work. It can be further noticed that this second method yields a loss only slightly less than the first approximate method.

Solution

3. A still more exact solution is to use the conventional friction formula 4-6, which takes into account not only the variation of stress from segment to segment but also that from point to point all along the cable. The solution is tabulated as shown.

Segment	L	KL	α	$\mu\alpha$	$KL+\mu\alpha$	$e^{-KL-\mu\alpha}$	Stress at End of Segment
AB	17.5	0.014	0	0	0.014	0.986	$0.986F_1$
BC	25	0.020	0.167	0.067	0.087	0.916	$0.903F_1$
CD	17.5	0.014	0	0	0.014	0.986	$0.890F_1$
DE	10	0.008	0.100	0.040	0.048	0.953	$0.848F_1$

The total frictional loss from A to E is given as

$$1 - 0.848 = 0.152 = 15.2\%$$

4-12 Total Amount of Losses

Initial prestress in steel minus the losses is known as the *effective* or the design *prestress*. The total amount of losses to be assumed in design will depend on the basis on which the initial prestress is measured. First, there is the *temporary maximum jacking stress* to which a tendon may be subject for the purpose of minimizing creep in steel or for balancing frictional losses. Then there is a slight release from that maximum stress back to the normal *jacking stress*.

As soon as the prestress is transferred to the concrete, anchorage loss will take place. The *jacking stress* minus the anchorage loss will be the stress at anchorage after release and is frequently called the *initial prestress*. For posttensioning, losses due to elastic shortening will gradually take place, if there are other tendons yet to be tensioned. This elastic shortening of concrete may be considered in two parts: that due to direct axial shortening and that due to elastic bending, as discussed in sections 4-3 and 4-9. For pretensioning, the entire amount of loss due to elastic shortening will occur at the transfer of prestress.

Depending on the definition of the term *initial prestress*, the amount of losses to be deducted will differ. If the jacking stress minus the anchorage loss is taken as the initial prestress, as described in the previous paragraph, then the losses to be deducted will include the elastic shortening and creep and shrinkage in concrete plus the creep in steel. This seems to be the most common practice. If the jacking stress itself is taken as the initial prestress, then anchorage losses must be deducted as well. If the stress after the elastic shortening of concrete is taken as the initial prestress, then the shrinkage and creep in concrete and the relaxation in steel will be the only losses. For points away from the jacking end, the effect of friction must be considered in addition. Frictional force along the tendon may either increase or decrease the stress, as discussed in section 4-10.

The magnitude of losses can be expressed in four ways:

1. In unit strains. This is most convenient for losses such as creep, shrinkage, and elastic shortenings of concrete expressed as strains.
2. In total strains. This is more convenient for the anchorage losses.
3. In unit stresses. All losses when expressed in strains can be transformed into unit stresses in steel. This is the approach used in the ACI-ASCE Committee method described in this chapter and used in example 4-5.
4. In percentage of prestress. Losses due to creep in steel and friction can be most easily expressed in this way. Other losses expressed in unit stresses can be easily transformed into percentages of the initial prestress. This often conveys a better picture of the significance of the losses.

It is difficult to generalize the amount of loss of prestress, because it is dependent on so many factors: the properties of concrete and steel, curing and moisture conditions, magnitude and time of application of prestress, and the process of prestressing. For average steel and concrete properties, cured under average air conditions, the tabulated percentages may be taken as representative of the average losses.

	Pretensioning, %	**Posttensioning, %**
Elastic shortening and bending of concrete	4	1
Creep of concrete	6	5
Shrinkage of concrete	7	6
Steel relaxation	8	8
Total loss	25	20

The table assumes that proper overtensioning has been applied to reduce creep in steel and to overcome friction and anchorage losses. Any frictional loss not overcome must be considered in addition. Allowance for loss of prestress of about 20% for posttensioning and 25% for pretensioning is seen to be not too far from the probable values for prestressed beams and girders. It must be borne in

Table 4-8 Limiting Maximum Loss (ACI-ASCE Committee[2])

	Maximum Loss psi (N/mm^2)	
Type of strand	**Normal Concrete**	**Lightweight Concrete**
Stress-relieved strand	50,000 (345)	55,000 (380)
Low-relaxation strand	40,000 (276)	45,000 (311)

mind, however, that, when conditions deviate from the average, different allowances should be made accordingly. For example, when the average prestress in a member (F_e/A_c) is high, say about 1000 psi (6.9 N/mm^2), these losses should be increased to about 30% for pretensioning and 25% for posttensioning. When the average prestress is low, say about 250 psi (1.7 N/mm^2) the above losses should be reduced to 18% for pretensioning and 15% for posttensioning. That is why an understanding and analysis of the sources of loss are of prime importance to the designing engineer.

Also noted that the above percentage losses are some 5% higher than values generally accepted in the 1950's up to the early 1970's, during which period literally hundreds of structures were designed on the basis of smaller losses and these have been only isolated cases of undesirable behavior attributable to such incorrect loss values. One explanation is the use of conservative allowable stresses in the design of these older structures which tends to compensate for the nonconservative loss of prestress assumed. Another reason is the fact that inaccurate loss assumptions (within a few %) may not result in undesirable behavior, such as larger deflections, unless the members possess a rather high span/depth ratio. However, the engineer will have to use good judgment to relate prestress loss computations to other factors and assumptions surrounding particular structures.

The ACI-ASCE Committee[2] recommended that the values shown in Table 4-8 be considered limiting maximum loss estimates. As shown in example 4-5, this may control the loss assumed in design.

EXAMPLE 4-5 (Deflection with time for this same beam is given in example 8-3.)
Estimate the change of prestress force with time for the pretensioned-prestressed concrete beam shown in Fig. 4-11. The normal weight concrete beam has only its own weight $w_G = 0.470$ k/ft acting at transfer of prestress which occurs approximately 48 h. after initially stressing the tendons to 0.75 $f_{pu} = (0.75)(270) = 202.5$ ksi in the prestressing bed. For 30 days we will assume the beam carries only $w_G = 0.470$ k/ft on a simply supported 65-ft span. Additional superimposed load $w_s = 1.0$ k/ft is added to the beam when erected at 30 days and is sustained for three years or more on the simple beam spanning 65 ft.

Assume the following material properties: $f_{ci}' = 4500$ psi, $f_c' = 6000$ psi, normal weight concrete (Type III cement, steam-cured concrete, 75% relative humidity), stress-relieved $\frac{1}{2}$ in. diameter strands with $f_{pu} = 270$ ksi. The results from analysis of this beam using a very

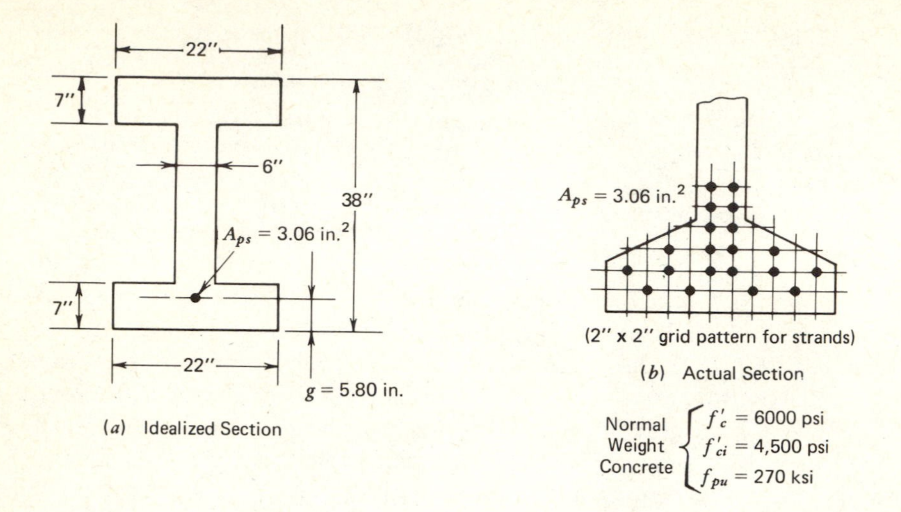

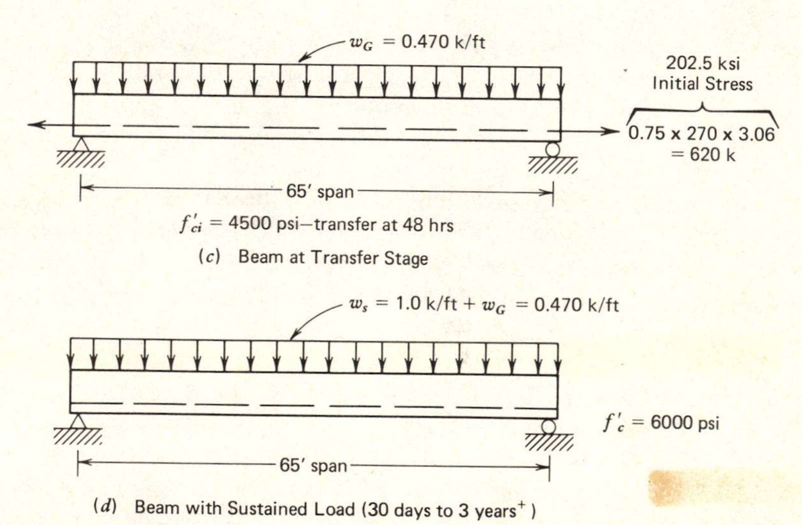

Fig. 4-11. Beam of example 4-5.

detailed computer program PBEAM (Ref. 8, Chapter 8) are shown in Table 4-9. Estimate the changes in prestress and compare results with this rather exact computation from the table at a few key stages in the life of the beam. Use the following methods for making these estimates:

(*a*) ACI-ASCE Committee 423[2] method and (*b*) PCI General Method[1] to estimate total losses. (w_G=6.68 kN/m w_s=14.59 kN/m, f_{ci}'=31 N/mm^2, f_i'=41.4 N/mm^2, and f_{pu}=1862 N/mm^2).

Table 4-9 Results of Analysis with PBEAM Program[a] (Example 4-5 and Example 8-3)

Load k/ft	Time (days)	$\Delta_{℄}$ (in.)	$f_{ps℄}$ (ksi)	$\Delta f_{ps℄}$ (ksi)	Results Example 4-5
0	2.0	1.617	175.5		
$w_G=0.470$	2.0	1.074	178.9	23.6	177.7 (PCI)
	3.0	1.245	175.3	27.2	
	7.0	1.442	170.6	31.9	
	12.0	1.560	167.3	35.2	
	20.0	1.672	163.9	38.6	
	30.0^-	1.757	161.0	41.5	161.8 (PCI)
Add'l D.L. = 1 k/ft $+w_G=0.470$ k/ft (Fig. 4.11)	30.0^+	0.752	167.6	35.9	
	33.0	0.5266	167.2	35.3	
	37.0	0.4297	166.9	35.6	
	45.0	0.3316	166.2	36.3	
	60.0	0.2449	164.0	38.5	
	80.0	0.1866	163.2	39.3	
	110.0	0.1402	161.5	41.0	
	150.0	0.1051	160.0	42.5	
	300.0	0.0603	157.0	45.5	
	1095.0	0.0169	153.4	49.1	152.4 (PCI)
	2000.0	−0.0124	152.5	50.0	152.5 (ACI)

[a] Ref. 8, Chapter 8.

Solution (*a*) ACI paper gives us total losses which are estimated as follows:

$$F_i = 0.75\ (270)(3.06) = 620\ \text{k}\ (2758\ \text{kN})$$

Elastic shortening—

$$f_{cir} = K_{cir} f_{cpi} - f_g$$

$$= (0.9)\left(\frac{620}{452} + \frac{620(13.2)^2}{82{,}170}\right) - \frac{(2979)(13.2)}{82{,}170} = 1.939\ \text{ksi}\ (13.37\ \text{N/mm}^2)$$

$$ES = K_{es} E_s \frac{f_{cir}}{E_{ci}}$$

$$= (1.0)(27.5\times10^3)\frac{1.939}{3.824\times10^3} = 13.94\ \text{ksi}\ (96.1\ \text{N/mm}^2)$$

Creep—

$$f_{cds}=\frac{M_{DL}e}{I}=\frac{(6338)(13.2)}{82{,}170}=1.018 \text{ ksi } (7.0 \text{ N/mm}^2)$$

$$CR=k_{cr}\frac{E_s}{E_c}(f_{cir}-f_{cds})$$

$$=(2.0)\frac{27.5\times10^3}{4.415\times10^3}(1.939-1.018)=11.47 \text{ ksi } (79.1 \text{ N/mm}^2)$$

Shrinkage—

$$RH=75\%,\ V/S=3$$

$$SH=8.2\times10^{-6}K_{sh}\,E_s\left(1-0.06\frac{V}{S}\right)(100-RH)$$

$$=8.2\times10^{-6}(1.0)(27.5\times10^3)(1-0.06\times3)(100-75)$$

$$=4.62 \text{ ksi } (31.9 \text{ N/mm}^2)$$

Relaxation—

$$RE=\left[K_{re}-J(SH+CR+E_s)\right]C$$

$$=\left[20.0-0.15(4.62+11.47+13.94)\right](1.45)=22.47 \text{ ksi } (154.9 \text{ N/mm}^2)$$

Total losses—

$$TL=ES+CR+SH+RE=13.94+11.47+4.62+22.47=52.5 \text{ ksi } (362 \text{ N/mm}^2)$$

$$f_{st}=0.75\times270-52.5=150.0 \text{ ksi } (1034 \text{ N/mm}^2)$$

The addition of 1 k/ft load at 30 days produces steel stress which is sometimes added back to the resulting steel stress "after losses." In this case we would get $\Delta f_{st}=6.2(1.018)=6.31$ ksi (43.5 N/mm^2).

$$f_{st}=150.0+6.31=156.31 \text{ ksi } (1078 \text{ N/mm}^2)$$ (vs. 153.4 ksi = 1058 N/mm^2 from PBEAM analysis, Table 4-9)

If we do not consider this additional steel stress in applying the method from the ACI paper, we should note that the commentary of the paper limits the design losses to 50 ksi (345 N/mm^2). In this case we would obtain the effective steel stress as follows:

Total losses = 52.5 ksi > 50 ksi maximum from Table 4-8 (section 4-12)
Use 50 ksi losses for design in this case.

$$f_{st}=202.5-50=152.5 \text{ ksi } (1051 \text{ N/mm}^2)$$ (vs. 153.4 ksi = 1058 N/mm^2 from PBEAM analysis at 3 years or 152.5 ksi = 1051 N/mm^2 at 5 years, Table 4-9)

(*b*) Shown in the following summary table are the losses computed by PCI Committee General Method (see Appendix D for detailed calculations for each of the time steps indicated in the summary table):

Time Periods	Loss by Source Each Stage				% of Total Loss All Stages	
	ES	*RE*	*CR*	*SH*	TL_i	$TL_i/TL \times 100$
Stage 1—tension strands to $0.75f_{pu}$ and maintain elongation for 48 hours (until transfer)	13.5	11.3	0	0	24.8 (171)	44
Stage 2—transfer at 48 h to 30 days with w_G load acting	0	4.68	6.29	4.97	15.94 (110)	28
Stage 3—30 days to 3 yr with $w_g + 1$ k/ft permanent load acting	0	4.78	4.04	6.86	15.68 (108)	28
Total loss at 3 yr from each source	13.5	20.76	10.33	11.83	56.42[a]	100
% of loss each source	24%	37%	18%	21%	100%	
Loss as % of initial	6.7	10.3	5.1	5.8	27.9	
$f_{si} = 0.75f_{pu} = 202.5$ ksi accumulated each source						

[a] Correction for $\Delta f_{st} = 6.31$ ksi, which changes our estimate of f_{se}. We would estimate $f_{se} = 202.5 - 56.42 + 6.31 = 152.4$ ksi. The ACI-ASCE Committee would consider the maximum loss of 50 ksi to control, which gives almost exactly the same $f_{se} = 202.5 - 50 = 152.5$ ksi.

4-13 Elongation of Tendons

It is often necessary to compute the elongation of a tendon caused by prestressing. When fabricating the anchorage parts in some systems, the expected amount of elongation must be known approximately. For the Prescon and other button-head, wire-type systems it must be known rather accurately. For all systems the measured elongation is compared to the expected value, thus serving as a check on the accuracy of the gage readings or on the magnitude of frictional loss along the length of the tendon. The computation of such elongation is discussed in two parts as follows.

Neglecting Frictional Loss along Tendon. If a tendon has uniform stress f_s along its entire length L, the amount of elongation is given by

$$\Delta_s = \delta_s L = f_s L / E_s = FL / E_s A_s \tag{4-23}$$

For prestress exceeding the proportional limit of the tendon, this formula may not be applicable, and it may be necessary to refer to the stress-strain diagram for the corresponding value of δ_s.

Before any tendon is tensioned, there almost always exists in it a certain amount of slack. For systems requiring shim plates, such as the Prescon system, that slack must be allowed for when computing the length of the shims. In addition, it may be desirable to allow for the shrinkage and elastic shortening of concrete at the time of tensioning. Hence the length of the shims must equal the elastic elongation of the tendon plus the slack in the tendon plus the shortening of concrete at transfer. Conversely, the elastic elongation of the tendon must be computed by deducting the initial slack and the elastic shortening of concrete from the apparent elongation.

It is not easy to determine the slack in a tendon accurately, hence the usual practice is to give the tendon some initial tension f_{s1} and measure the elongation Δ_s thereafter. Then, neglecting any shortening of the concrete, the total elastic elongation of the tendon can be computed by

$$\text{Elastic elongation} = \frac{f_s}{f_s - f_{s1}} \Delta_s \tag{4-24}$$

EXAMPLE 4-6

A Prescon cable, 60 ft long, Fig. 4-12, is to be tensioned from one end to an initial prestress of 150,000 psi immediately after transfer. Assume that there is no slack in the cable, that the shrinkage of concrete is 0.0002 at time of transfer, and that the average compression in concrete is 800 psi along the length of the tendon. $E_c = 3{,}800{,}000$ psi; $E_s = 29{,}000{,}000$ psi. Compute the length of shims required, neglecting any elastic shortening of the shims and any friction along the tendon (span = 18.3 m, initial prestress = 1034 N/mm^2, $E_c = 26.2$ kN/mm^2, and $E_s = 200$ kN/mm^2).

Solution From equation 4-10, the elastic elongation of steel is

$$\Delta_s = f_s L/E_s = 150{,}000 \times 60 \times 12/29{,}000{,}000 = 3.72 \text{ in. } (94.5 \text{ mm})$$

Shortening of concrete due to shrinkage is

$$0.0002 \times 60 \times 12 = 0.14 \text{ in. } (3.6 \text{ mm})$$

Elastic shortening of concrete is

$$800 \times 60 \times 12/3{,}800{,}000 = 0.15 \text{ in. } (3.8 \text{ mm})$$

Length of shims required is

$$3.72 + 0.14 + 0.15 = 4.01 \text{ in. } (101.9 \text{ mm})$$

If shims of 4.01 in. (101.9 mm) are inserted in the anchorage, there should remain an initial prestress of 150,000 psi (1034 N/mm^2) in the steel immediately after transfer.

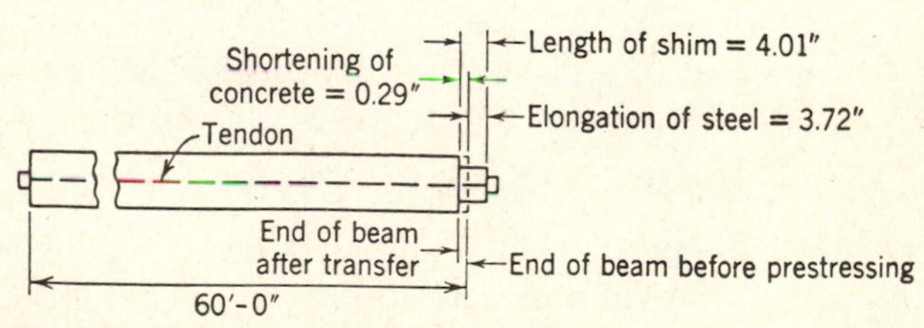

Fig. 4-12. Example 4-6.

EXAMPLE 4-7

Eighteen 0.196-in. wires in a Freyssinet cable, 80 ft (24.4 m) long, are tensioned initially to a total stress of 3000 lb. What additional elongation of the wires as measured there-from is required to obtain an initial prestress of 160,000 psi (1103 N/mm^2)? $E_s = 28{,}000{,}000$ psi (193 kN/mm^2). Assume no shortening of concrete during the tensioning process and neglect friction.

Solution

$$A_s = 18 \times 0.03 = 0.54 \text{ sq in. } (348 \text{ mm}^2)$$

$$f_{s1} = 3000/0.54 = 5500 \text{ psi } (37.9 \text{ N/mm}^2)$$

Total elastic elongation of tendon from 0 to 160,000 psi (1103 N/mm^2) is

$$f_s L/E_s = 160{,}000 \times 80 \times 12/28{,}000{,}000 = 5.48 \text{ in. } (139 \text{ mm})$$

From equation 4-11,

$$5.48 = \frac{f_s}{f_s - f_{s1}} \Delta_s$$

$$= \frac{160{,}000}{160{,}000 - 5500} \Delta_s$$

$$\Delta_s = 5.28 \text{ in. } (134 \text{ mm})$$

Thus, with zero reading taken at a total stress of 3000 lb (13.3 kN), an elongation of 5.28 in. (134 mm) must be obtained for a prestress of 160,000 psi (1103 N/mm^2).

Considering Frictional Loss along Tendon. It was shown in section 4-11 that, for a curved tendon with a constant radius R, the stress at any point away from the jacking end is

$$F_2 = F_1 e^{-(\mu\alpha + KL)}$$

The average stress F_a for the entire length of curve with stress varying from F_1 to F_2 can be shown to be

$$F_a = F_2 \frac{e^{\mu\alpha + KL} - 1}{\mu\alpha + KL} \tag{4-25}$$

This equation is solved graphically in Fig. 4-14, where the dotted lines give the values of $f_a = F_a/A_s$.

The total lengthening for length L is given by

$$\Delta_s = \frac{F_a L}{E_s A_s} = \frac{F_2 L}{E_s A_s} \frac{e^{\mu\alpha + KL} - 1}{\mu\alpha + KL} \tag{4-26}$$

If only an approximate solution is desired, the medium value of F_1 and F_2 can be used in computing the elongation; thus

$$\Delta_s = \frac{F_1 + F_2}{2} \frac{L}{E_s A_s} \tag{4-27}$$

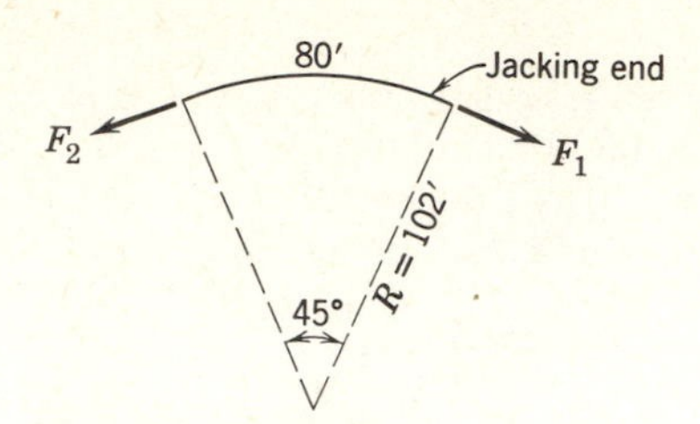

Fig. 4-13. Example 4-8.

EXAMPLE 4-8

A tendon 80 ft (24.4 m) long is tensioned along a circular curve with $R=102$ ft (31.1 m), Fig. 4-13. For a unit stress of 180,000 psi (1,241 N/mm^2) applied at the jacking end, a total elongation of 4.80 in. (122 mm) is obtained. $E_s=30{,}000{,}000$ psi (207 kN/mm^2). Compute the stress f_2 at the far end of the tendon.

Solution Approximate solution. Average stress in the tendon is given by

$$f_a=\Delta_s E_s/L=4.80\times 30{,}000{,}000/(80\times 12)$$

$$=150{,}000 \text{ psi } (1034 \text{ N/mm}^2)$$

Since the maximum stress is 180,000 psi (1241 N/mm^2), the minimum stress f_2 must be 120,000 psi (827 N/mm^2), assuming uniform decrease in the stress.

Exact solution can be shown to give $f_2=125$ ksi (862 N/mm^2).

References

1. "Recommendations for Estimating Prestress Losses," Report of PCI Committee on Prestress Losses, *J. Prestressed Conc. Inst.*, Vol. 20, No. 4, July–August 1975, pp. 43–75.
2. P. Zia, H. K. Preston, N. L. Scott, and E. B. Workman, "Estimating Prestress Losses," (ACI-ASCE Comm. on Prestressed Concrete Recommended Procedure), *Conc. Int.*, Vol. 1, No. 6, June 1979, pp. 32–38.
3. R. Sinno and H. L. Furr, "Computer Program for Predicting Prestress Loss and Camber," *J. Prestressed Conc. Inst.*, Vol. 17, No. 5, September–October 1972, pp. 27–38.
4. C. Suttikan, "A Generalized Solution for Time-Dependent Response and Strength of Non-Composite and Composite Prestressed Concrete Beams," Ph.D. Dissertation, The University of Texas at Austin, August 1978.
5. H. D. Hernandez and W. L. Gamble, "Time-Dependent Losses in Prestressed Concrete Construction," Structural Research Series No. 417, University of Illinois, Urbana, May 1975.
6. "Tentative Recommendations for Prestressed Concrete," Report by ACI-ASCE Committee 423, *J. Am. Conc. Inst.*, Vol. 54, No. 7, January 1958, pp. 548–578.
7. AASHTO Interim Specifications—Bridges—1975, Subcommittee on Bridges and Structures, American Association of Highway and Transportation Officials, Washington, 1975, pp. 41–79.

8. *Posttensioning Manual*, Posttensioning Institute, Phoenix, Arizona, 1976, p. 189.
9. D. D. Magura, M. A. Sozen, and C. P. Siess, "A Study of Stress Relaxation in Prestressing Reinforcement," *J. Prestressed Conc. Inst.*, Vol. 9, No. 2, April 1964, pp. 13–57.
10. E. H. Cooley, "Friction in Posttensioned Prestressing Systems," and "Estimation of Friction in Prestressed Concrete," Cementand Concrete Assn., London, 1953.
11. T. Y. Lin, "Cable Friction in Posttensioning," *J. Structural Div.*, Am. Soc. of Civil Engineers, November 1956.
12. F. Leonhardt, "Continuous Prestressed Concrete Beams," *J. Am. Conc. Inst.*, March 1953 (*Proc.*, Vol. 49), p. 617.

5

ANALYSIS OF SECTIONS FOR FLEXURE

5-1 Introduction and Sign Conventions

Differentiation can be made between the *analysis* and *design* of prestressed sections for flexure. By *analysis* is meant the determination of stresses in the steel and concrete when the form and size of a section are already given or assumed. This is obviously a simpler operation than the *design* of the section, which involves the choice of a suitable section out of many possible shapes and dimensions. In actual practice, it is often necessary to first perform the process of design when assuming a section, and then to analyze that assumed section. But, for the purpose of study, it is easier to learn first the methods of analysis and then those of design. This reversal of order is desirable in the study of prestressed as well as reinforced concrete.

This chapter will be devoted to the first part, the analysis; the next chapter will deal with *design*. The discussion is limited to the analysis of sections for flexure, meaning members under bending, such as beams and slabs. Only the effect of moment is considered here; that of shear and bond is treated in Chapter 7.

A rather controversial point in the analysis of prestressed-concrete beams has been the choice of a proper system of sign conventions. Many authors have used positive sign (+) for compressive stresses and negative sign (−) for tensile stresses, basing their convention on the idea that prestressed-concrete beams are normally under compression and hence the plus sign should be employed to denote that state of stress. The author prefers to maintain the common sign convention as used for the design of other structures; that is, minus for compressive and plus for tensile stresses. Throughout this treatise, plus will stand for tension and minus for compression, whether we are talking of stresses in steel or concrete, prestressed or reinforced. When the sense of the stress is self-evident, signs will be omitted.

5-2 Stresses in Concrete Due to Prestress

Some of the basic principles of stress computation for prestressed concrete have already been mentioned in section 1-2. They will be discussed in greater detail here. First of all, let us consider the effect of prestress. According to present practice, stresses in concrete due to prestress are always computed by the elastic theory. Consider the prestress F existing at the time under discussion, whether it be the initial or the final value. If F is applied at the centroid of the concrete section, and if the section under consideration is sufficiently far from the point of application of the prestress, then, by St. Venant's principle, the unit stress in concrete is uniform across that section and is given by the usual formula,

$$f = \frac{F}{A}$$

where A is the area of that concrete section.

For a pretensioned member, when the prestress in the steel is transferred from the bulkheads to the concrete, Fig. 5-1, the force that was resisted by the bulkheads is now transferred to both the steel and the concrete in the member. The release of the resistance from the bulkheads is equivalent to the application of an opposite force F_i to the member. Using the transformed section method, and with A_c = net sectional area of concrete, the compressive stress produced in the concrete is

$$f_c = \frac{F_i}{A_c + nA_s} = \frac{F_i}{A_t} \tag{5-1}$$

while that induced in the steel is

$$\Delta f_s = nf_c = \frac{nF_i}{A_c + nA_s} = \frac{nF_i}{A_t} \tag{5-2}$$

which represents the immediate reduction of the prestress in the steel as a result of the transfer.

Although this method of computation is correct according to the elastic theory, the usual practice is not to follow such a procedure, but rather to consider the prestress in the steel being reduced by a loss resulting from elastic

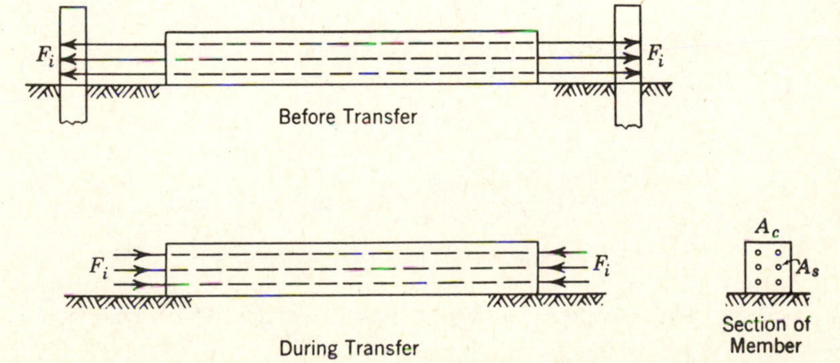

Fig. 5-1. Transfer of concentric prestress in a pretensioned member.

shortening of concrete and approximated by

$$\Delta f_s = \frac{nF_i}{A_c} \quad \text{or} \quad \frac{nF_i}{A_g} \tag{5-3}$$

which differs a little from formula 5-2 but is close enough for all practical purposes, since the total amount of reduction is only about 2 or 3% and the value of n cannot be accurately known anyway. The high-strength steel used for prestressing requires smaller area for the tension steel than would be used in reinforced concrete, thus there is not a large difference between A_c and A_g.

After the transfer of prestress, further losses will occur owing to the creep and shrinkage in concrete. Theoretically, all such losses should be calculated on the basis of a transformed section, taking into consideration the area of steel. But, again, that is seldom done, the practice being simply to allow for the losses by an approximate percentage. In other words, the simple formula $f=F/A$ is always used, with the value of F estimated for the given condition, and the gross area of concrete used for A. For a posttensioned member, the same reasoning holds true. Suppose that there are several tendons in the member prestressed in succession. Every tendon that is tensioned becomes part of the section after it is bonded by grouting. The effect of tensioning any subsequent tendon on the stresses in the previously tensioned ones should be calculated on the basis of a transformed section. Theoretically, there will be a different transformed section after the tensioning of every tendon. However, such refinements are not justified, and the usual procedure is simply to use the formula $f=F/A$, with F based on the initial prestress in the steel.

EXAMPLE 5-1

A pretensioned member, similar to that shown in Fig. 5-1, has a section of 8 in. by 12 in. (203 mm by 305 mm). It is concentrically prestressed with 0.8 sq in. (516 mm^2) of high-tensile steel wire, which is anchored to the bulkheads of a unit stress of 150,000 psi (1034 N/mm^2). Assuming that $n=6$, compute the stresses in the concrete and steel immediately after transfer.

Solution

1. An exact theoretical solution. Using the elastic theory, we have

$$f_c = \frac{F_i}{A_c + nA_s} = \frac{F_i}{A_g + (n-1)A_s}$$

$$= \frac{0.8 \times 150{,}000}{12 \times 8 + 5 \times 0.8} = 1200 \text{ psi } (8.3\ N/mm^2)$$

$$nf_c = 6 \times 1200 = 7200 \text{ psi } (49.6\ N/mm^2)$$

Stress in steel after transfer $= 150{,}000 - 7200 = 142{,}800$ psi (985 N/mm^2).

Solution

2. An approximate solution. The loss of prestress in steel due to elastic shortening of

concrete is estimated by

$$= n\frac{F_i}{A_g}$$

$$= 6\frac{120{,}000}{8\times 12} = 7500 \text{ psi } (51.7 \text{ N/mm}^2)$$

Stress in steel after loss = 150,000 − 7500 = 142,500 psi (983 N/mm²). Stress in concrete is

$$f_c = \frac{142{,}500\times 0.8}{96} = 1190 \text{ psi } (8.2 \text{ N/mm}^2)$$

Note that, in this second solution, the approximations introduced are: (1) using gross area of concrete instead of net area, (2) using the initial stress in steel instead of the reduced stress. But the answers are very nearly the same for both solutions. The second method is more convenient and is usually followed.

Next, suppose that the prestress F is applied to the concrete section with an eccentricity e, Fig. 5-2; then it is possible to resolve the prestress into two components: a concentric load F through the centroid, and a moment Fe. By the usual elastic theory, the fiber stress at any point due to moment Fe is given by the formula

$$f = \frac{My}{I} = \frac{Fey}{I} \tag{5-4}$$

Then the resultant fiber stress due to the eccentric prestress is given by

$$f = \frac{F}{A} \pm \frac{Fey}{I} \tag{5-5}$$

The question again arises as to what section should be considered when computing the values of e and I, whether the gross or the net concrete section or the transformed section, and what prestress F to be used in the formula, the initial or the reduced value. Consider a pretensioned member, Fig. 5-3. The steel has already been bonded to the concrete; the release of the force from the bulkhead is equivalent to the application of an eccentric force to the composite member; hence the force should be the total F_i, and I should be the moment of

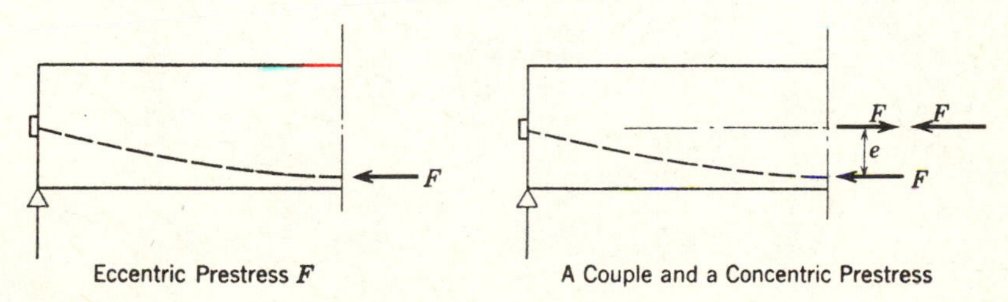

Fig. 5-2. Eccentric prestress on a section

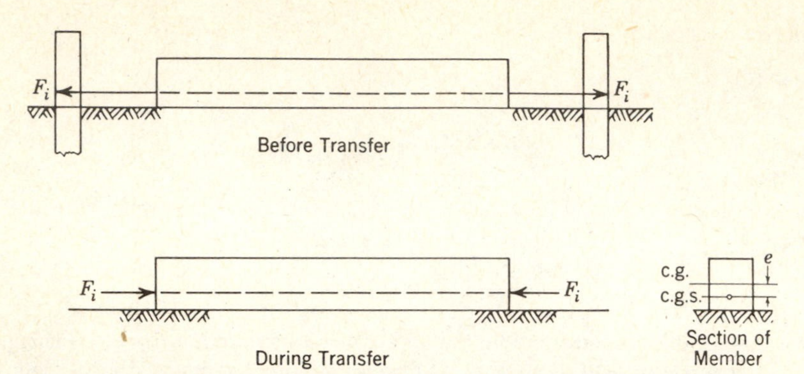

Fig. 5-3. Transfer of eccentric prestress in a pretensioned member.

inertia of the transformed section, and e should be measured from the centroidal axis of that transformed section. However, in practice, this procedure is seldom followed. Instead, the gross or net concrete section is considered, and either the initial or the reduced prestress is applied. The error is negligible in most cases.

EXAMPLE 5-2

A pretensioned member similar to that shown in Fig. 5-3 has a section of 8 in. by 12 in. (203 mm by 305 mm) deep. It is eccentrically prestressed with 0.8 sq in. (516 mm^2) of high-tensile steel wire which is anchored to the bulkheads at a unit stress of 150,000 psi (1034 N/mm^2). The c.g.s. is 4 in. (101.6 mm) above the bottom fiber. Assuming that $n=6$, compute the stresses in the concrete immediately after transfer due to the prestress only.

Solution

1. An exact theoretical solution. Using the elastic theory, the centroid of the transformed section and its moment of inertia are obtained as follows. Referring to Fig. 5-4, for $(n-1)A_s = 5\times 0.8 = 4$ sq in.,

$$y_0 = \frac{4\times 2}{96+4} = 0.08 \text{ in. (2.032 mm)}$$

$$\begin{aligned} I_t &= \frac{8\times 12^3}{12} + 96\times 0.08^2 + 4\times 1.92^2 \\ &= 1152 + 0.6 + 14.7 \\ &= 1167.3 \text{ in.}^4\ (485.9\times 10^6 \text{ mm}^4) \end{aligned}$$

$$\begin{aligned} \text{Top fiber stress} &= \frac{F_i}{A_t} + \frac{F_i e y}{I_t} \\ &= \frac{-120{,}000}{100} + \frac{120{,}000\times 1.92\times 6.08}{1167.3} \\ &= -1200 + 1200 \\ &= 0 \end{aligned}$$

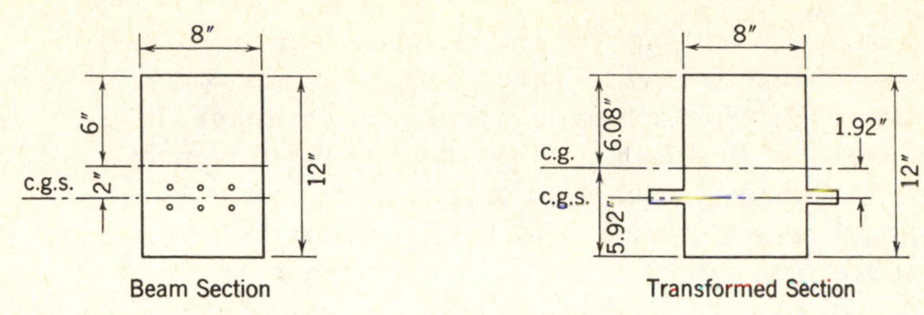

Fig. 5-4. Example 5-2.

$$\text{Bottom fiber stress} = \frac{-120{,}000}{100} - \frac{120{,}000 \times 1.92 \times 5.92}{1167.3}$$
$$= -1200 - 1170$$
$$= -2370 \text{ psi } (-16.34 \text{ N/mm}^2)$$

Solution

2. An approximate solution. The loss of prestress can be approximately computed, as in example 5-1, to be 7500 psi in the steel. Hence the reduced prestress is 142,500 psi or 114,000 lb. Extreme fiber stresses in the concrete can be computed to be

$$f_c = \frac{F}{A} \pm \frac{Fey}{I}$$
$$= \frac{-114{,}000}{96} \pm \frac{114{,}000 \times 2 \times 6}{(8 \times 12^3)/12}$$
$$= -1187 \pm 1187$$
$$= 0 \text{ in the top fiber}$$
$$= -2374 \text{ psi } (-16.37 \text{ N/mm}^2) \text{ in the bottom fiber}$$

The approximations here introduced are: using an approximate value of reduced prestress, and using the gross area of concrete. This second solution, although approximate, is more often used because of its simplicity.

Now consider a pretensioned curved member as in Fig. 5-5. If the transfer of prestress is considered as a force F_i applied at each end, the eccentricity and the

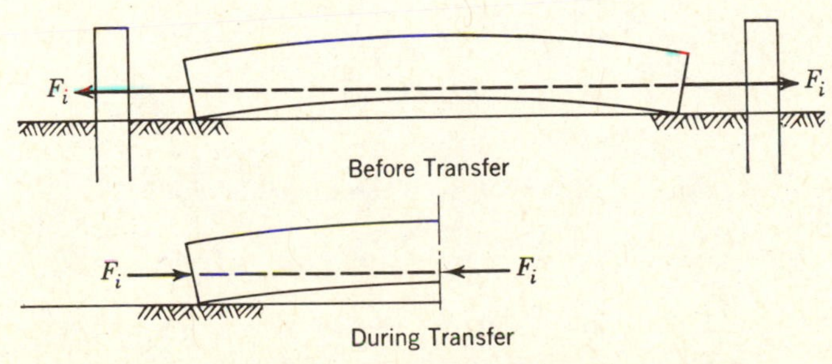

Fig. 5-5. Transfer of prestress in a curved pretensioned member.

moment of inertia will vary for each section. If the exact method of elastic analysis is preferred, different I's and e's will have to be computed for different sections. If an approximate method is permitted, a constant I based on the gross concrete area would suffice for all sections, and the eccentricity can be readily measured from the middepth of the section.

For a posttensioned member before being bonded, the prestress F to be used in the stress computations is again the initial prestress minus the estimated losses. For the value of I, either the net or the gross concrete section is used, although, theoretically, the net section is the correct one. After the steel is bonded to the concrete, any loss that takes place actually happens to the section as a whole. However, for the sake of simplicity, a rigorous analysis based on the transformed section is seldom made. Instead, the reduced prestress is estimated and the stresses in the concrete are computed for that reduced prestress acting on the net concrete section (gross concrete section may sometimes be conveniently used). Stresses produced by external loads, however, are often computed on the basis of the transformed section if accuracy is desired; otherwise, gross section is used for the computation. The permissible simplifications for each case will depend to a large degree on the degree of accuracy required and the time available for computation.

EXAMPLE 5-3

A posttensioned beam has a midspan cross section with a duct of 2 in. by 3 in. (50.8 mm by 76.2 mm) to house the wires, as shown in Fig. 5-6. It is prestressed with 0.8 sq in. (516 mm^2) of steel to an initial stress of 150,000 psi (1034 N/mm^2). Immediately after transfer the stress is reduced by 5% owing to anchorage loss and elastic shortening of concrete. Compute the stresses in the concrete at transfer.

Solution

1. Using net section of concrete. The centroid and I of the net concrete section are computed as follows

$$A_c = 96 - 6 = 90 \text{ sq in. } (58.1 \times 10^3 \text{ mm}^2)$$

$$y_0 = \frac{6 \times 3}{96 - 6} = 0.2 \text{ in. } (5.08 \text{ mm})$$

$$I = \frac{8 \times 12^3}{12} + 96 \times 0.2^2 - \frac{2 \times 3^3}{12} - 6 \times 3.2^2$$

$$= 1152 + 3.8 - 4.5 - 61.5$$

$$= 1090$$

Total prestress in steel $= 150{,}000 \times 0.8 \times 95\% = 114{,}000$ lb (507 kN)

$$f_c = \frac{-114{,}000}{90} \pm \frac{114{,}000 \times 3.2 \times 5.8}{1090}$$

$$= -1270 + 1940 = +670 \text{ psi } (+4.62 \text{ N/mm}^2) \text{ for top fiber}$$

$$f_c = -1270 - 2070 = -3340 \text{ psi } (-23.03 \text{ N/mm}^2) \text{ for bottom fiber}$$

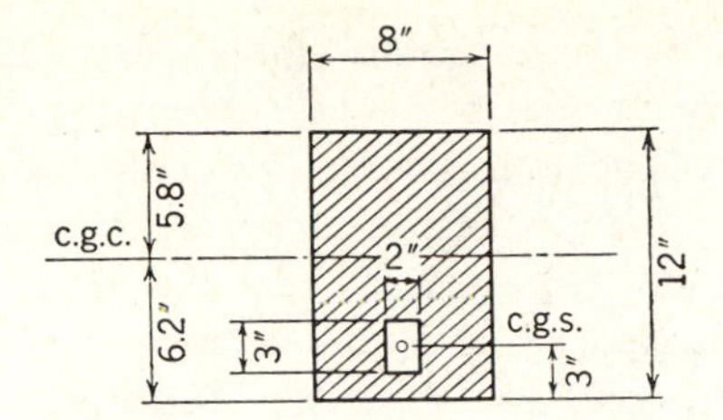

Fig. 5-6. Example 5-3.

Solution

2. Using the gross section of concrete. An approximate solutin using the gross concrete section would give results not so close in this case (11% difference):

$$f_c = \frac{-114{,}000}{96} \pm \frac{114{,}000 \times 3 \times 6}{(8 \times 12^3)/12}$$

$$= -1270 + 1940 = +1783$$

$$= +596 \text{ psi } (+4.11 \text{ N/mm}^2) \text{ for top fiber}$$

$$= -2970 \text{ psi } (-20.48 \text{ N/mm}^2) \text{ for bottom fiber}$$

If the eccentricity does not occur along one of the principal axes of the section, it is becessary to further resolve the moment into two component moments along the two principal axes, Fig. 5-7; then the stress at any point is given by

$$f = \frac{F}{A} \pm \frac{Fe_x x}{I_x} \pm \frac{Fe_y y}{I_y}$$

Since concrete is not a really elastic material, the above elastic theory is not exact. But, within working loads, it is considered an accepted form of computation. When the stresses are excessively high, the elastic theory may no longer be nearly correct.

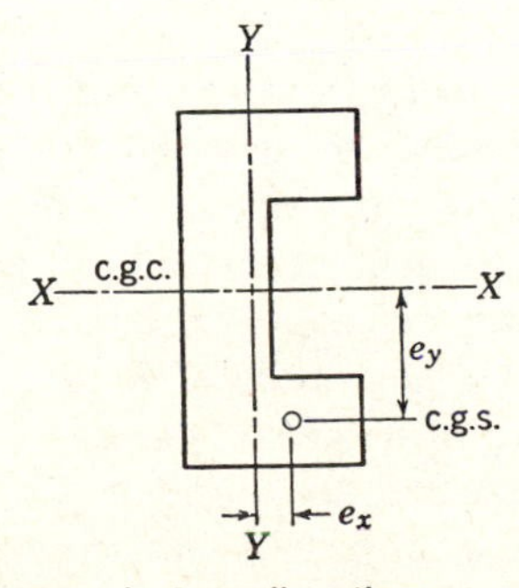

Fig. 5-7. Eccentricity of prestress in two directions.

The above method further assumes that the concrete section has not cracked. If it has, the cracked portion has to be computed or estimated, and computations made accordingly. The computation for cracked section in concrete is always complicated. Fortunately, such a condition is seldom met with in actual design of prestressed concrete. In general, any high-tensile stresses produced by prestress are counterbalanced by compressive stresses due to the weight of the member itself, so that in reality no cracks exist under the combined action of the prestress and the beam's own weight. Hence the entire concrete section can be considered as effective, even though, at certain stages of the computation, high-tensile stresses may appear on paper.

During posttensioning operations, concrete may be subjected to abnormal stresses. Suppose that there is one tendon at each corner of a square concrete section. When all four tendons are tensioned, the entire concrete section will be under uniform compression. But when only one tendon is fully tensioned, there will exist high tensile stress as well as high compressive stress in the concrete. If two jacks are available, it may be desirable to tension two diagonally opposite tendons at the same time. Sometimes it may be necessary to tension the tendons in steps, that is, to tension them only partially and to retension them after others have been tensioned. Computation for stresses during tensioning is also made on the elastic theory. It is believed that the elastic theory is sufficiently accurate up to the point of cracking, although it cannot be used to predict the ultimate strength.

Control of allowable stresses is a means of controlling serviceability, and the ACI Code continues to use limiting values of allowable stresses.

5-3 Stresses in Concrete Due to Loads

Stresses in concrete produced by external bending moment, whether due to the beam's own weight or to any externally applied loads, are computed by the usual elastic theory.

$$f = \frac{My}{I} \tag{5-6}$$

For a pretensioned beam, steel is always bonded to the concrete before any external moment is applied. Hence the section resisting external moment is the combined section. In other words, the values of y and I should be computed on the basis of a transformed section, considering both steel and concrete. For approximation, however, either the gross or the net section of concrete alone can be used in the calculations; the magnitude of error so involved can be estimated and should not be serious except in special cases.

When the beam is posttensioned and bonded, for any load applied after the bonding has taken place, the transformed section should be used as for pretensioned beams. However, if the load or the weight of the beam itself is applied before bonding takes place, it acts on the net concrete section, which should hence be the basis for stress computation. For posttensioned unbonded beams, the net concrete section is the proper one for all stress computations. It should be borne in mind, though, that when the beam is unbonded, any bending of the beam may change the overall prestress in the steel, the effect of which can be separately computed or estimated as discussed in section 4-5.

Often, only the resulting stresses in concrete due to both prestress and loads are desired, instead of their separate values. They are given by the following formula, a combination of 5-5 and 5-6.

$$f=\frac{F}{A}+\frac{Fey}{I}\pm\frac{My}{I}$$
$$=\frac{F}{A}\left(1\pm\frac{ey}{r^2}\right)\pm\frac{My}{I}$$
$$=\frac{F}{A}\pm(Fe\pm M)\frac{y}{I} \tag{5-7}$$

Any of these three forms may be used, whichever happens to be the most convenient. But, to be strictly correct, the section used in computing y and I must correspond to the actual section at the application of the force. It quite frequently happens that the prestress F acts on the net concrete section, while the external loads act on the transformed section. Judgment should be exercised in deciding whether refinement is necessary or whether approximation is permissible for each particular case.

When prestress eccentricity and external moments exist along two principal axes, the general elastic formula can be used.

$$f=\frac{F}{A}\pm\frac{Fe_x x}{I_x}\pm\frac{Fe_y y}{I_y}\pm\frac{M_x x}{I_x}\pm\frac{M_y y}{I_y}$$
$$=\frac{F}{A}\pm(Fe_x\pm M_x)\frac{x}{I_x}\pm(Fe_y\pm M_y)\frac{y}{I_y} \tag{5-8}$$

EXAMPLE 5-4

A posttensioned bonded concrete beam, Fig. 5-8, has a prestress of 350 kips (1,557 kN) in the steel immediately after prestressing, which eventually reduces to 300 kips (1334 kN)

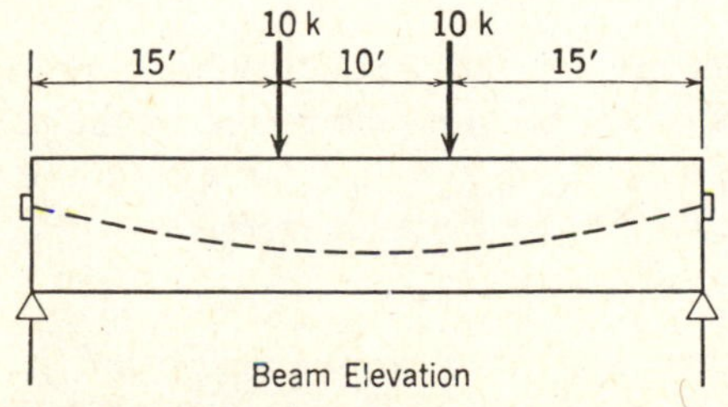

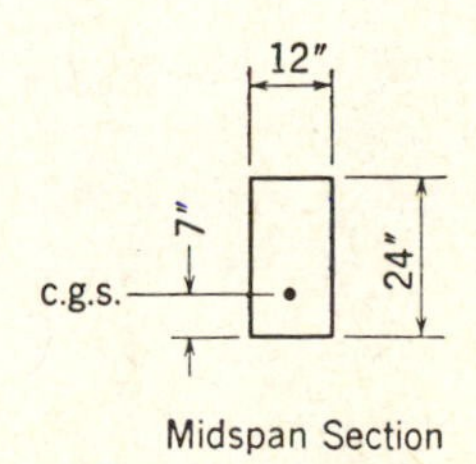

Fig. 5-8. Example 5-4.

due to losses. The beam carries two live loads of 10 kips (44.48 kN) each in addition to its own weight of 300 plf (4.377 kN/m). Compute the extreme fiber stresses at midspan, (*a*) under the initial condition with full prestress and no live load, and (*b*) under the final condition, after the losses have taken place, and with full live load.

Solution To be theoretically exact, the net concrete section should be used up to the time of grouting, after which the transformed section should be considered. This is not deemed necessary, and an approximate but sufficiently exact solution is given below, using the gross section of concrete at all times that is,

$$I = 12\times24^3/12 = 13{,}800 \text{ in.}^4\ (5744\times10^6 \text{ mm}^4)$$

1. *Initial condition*. Dead-load moment at midspan, assuming that the beam is simply supported after prestressing:

$$M=\frac{wL^2}{8}=\frac{300\times40^2}{8}=60{,}000 \text{ ft-lb } (81{,}360 \text{ N}-\text{m})$$

$$f=\frac{F}{A}\pm\frac{Fey}{I}\pm\frac{My}{I}$$

$$=\frac{-350{,}000}{288}\pm\frac{350{,}000\times5\times12}{13{,}800}\pm\frac{60{,}000\times12\times12}{13{,}800}$$

$$=-1215+1520-625=-320 \text{ psi } (-2.21 \text{ N/mm}^2), \text{ top fiber}$$

$$=-1215-1520+625=-2110 \text{ psi } (-14.55 \text{ N/mm}^2), \text{ bottom fiber}$$

2. *Final condition*. Live-load moment at midspan = 150,000 ft-lb (203,400 N-m); therefore, total external moment = 210,000 ft-lb (284,760 N-m), while the prestress is reduced to 300,000 lb (1,334 kN); hence,

$$f=\frac{-300{,}000}{288}\pm\frac{300{,}000\times5\times12}{13{,}800}\pm\frac{210{,}000\times12\times12}{13{,}800}$$

$$=-1040+1300-2190=-1930 \text{ psi } (-13.31 \text{ N/mm}^2), \text{ top fiber}$$

$$=-1040-1300+2190=-150 \text{ psi } (-1.03 \text{ N/mm}^2), \text{ bottom fiber}$$

Example 5-4 describes the conventional method of stress analysis for prestressed concrete, but it will be recalled that in section 1-2 another method of approach is described in which the center of pressure C in the concrete is set at distance a from the center of prestress T in the steel such that

$$Ta = Ca = M \tag{5-9}$$

By this method, the stresses in concrete are not treated as being produced by prestress and external moments separately, but are determined by the magnitude and location of the center of pressure C, Fig. 5-9. Most beams do not carry axial load, therefore, C equals T and is located at a distance from T.

$$a = M/T$$

Since the value of T is the value of F in a prestressed beam it is quite accurately known. Thus the computation of a for a given moment M is simply a

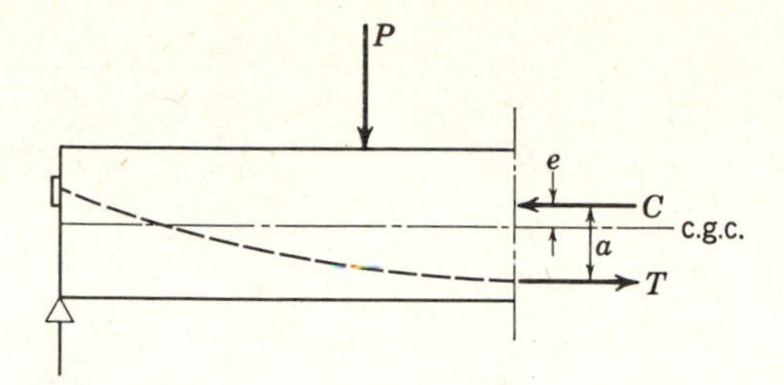

Fig. 5-9. Internal resisting couple *C-T* with arm *a*.

matter of statics. Once the center pressure C is located for a concrete section, the distribution of stresses can be determined either by the elastic theory or by the plastic theory. Generally the elastic theory is followed, in which case we have, since

$$C=T=F, \qquad f=\frac{C}{A}\pm\frac{Cey}{I}=\frac{F}{A}\pm\frac{Fey}{I} \tag{5-10}$$

where e is the eccentricity of C, not of F.

Following this approach, a prestressed beam is considered similar to a reinforced-concrete beam with the steel supplying the tensile force T, and the concrete supplying the compressive force C. C and T together form a couple resisting the external moment. Hence the value of A and I to be used in the above formula should be the net section of the concrete, and not the transformed section. If a beam has conduits grouted for bond, the stress in the grout is actually different from that in the adjacent concrete, and an exact theoretical solution would be quite involved. Under such conditions it is advisable to use the gross section of concrete for all computations for the sake of simplicity. Only when investigating the stresses before grouting should the net concrete section be used and even this refinement may not be required in most cases for design.

It can be noted that formula 5-10 is only a different form of formula 5-7, with e measured to C, thus combining the effect of M with the eccentricity of F. Although the formulas are in fact identical, the approaches are different. By following this second approach, all the inaccuracies are thrown into the estimation of the effective prestress in steel, which can generally be estimated within 5%. After that, the location of C is a simple problem in statics, and the distribution of C across the section can be easily computed or visualized. This method of approach will be further explained in the next chapter on the design of beam sections.

EXAMPLE 5-5

For the same problem as in example 5-4, compute the concrete stresses under the final loading conditions by locating the center of pressure C for the concrete section.

Solution Referring to Fig. 5-10, a is computed by

$$a=(210\times 12)/300=8.4 \text{ in. } (213 \text{ mm})$$

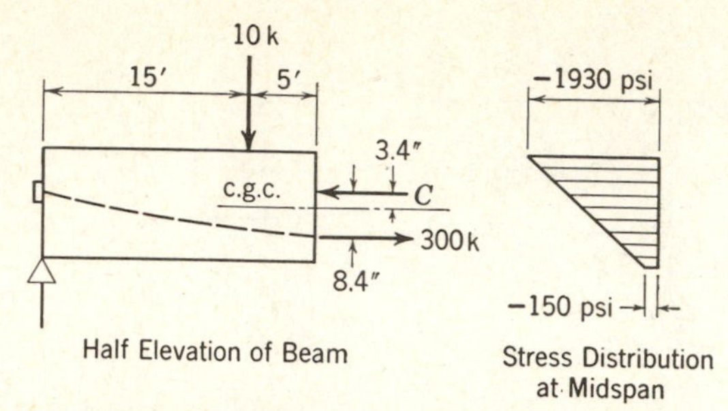

Fig. 5-10. Example 5-5.

Hence e for C is $8.4-5=3.4$ in. Since $C=F=300{,}000$ lb (1,334 kN).

$$f=\frac{C}{A}\pm\frac{Cey}{I}$$

$$=\frac{-300{,}000}{288}\pm\frac{300{,}000\times 3.4\times 12}{13{,}800}$$

$$=-1040-890=-1930 \text{ psi } (-13.31 \text{ N/mm}^2), \text{ top fiber}$$

$$=-1040+890=-150 \text{ psi } (-1.03 \text{ N/mm}^2), \text{ bottom fiber}$$

Also, by inspection, since the center of pressure is near the third point, the stress distribution should be nearly triangular as is shown. By comparing this solution with that for example 5-4, the directness and simplicity of this method seem to be evident.

5-4 Stresses in Steel Due to Loads

In prestressed concrete, prestress in the steel is measured during tensioning operations, then the losses are computed or estimated as described in Chapter 4. When dead and live loads are applied to the member, minor changes in stress will be induced in the steel. In a reinforced-concrete beam, steel stresses are assumed to be directly proportional to the external bending moment. When there is no moment, there is no stress. When the moment increases, the steel stresses increase in direct proportion. This is not true for a prestressed-concrete beam, whose resistance to external moment is furnished by a lengthening of the lever arm between the resisting forces C and T which remain relatively unchanged in magnitude.

In order to get a clear understanding of the behavior of a prestressed-concrete beam, it will be interesting to first study the variation of steel stress as the load increases. For the midspan section of a simple beam, the variation of steel stress with load on the beam is shown in Fig. 5-11. Along the X-axis is plotted the load on the beam, and along the Y-axis is plotted the stress in the steel. As prestress is applied to the steel, the stress in the steel changes from A to B, where B is at the

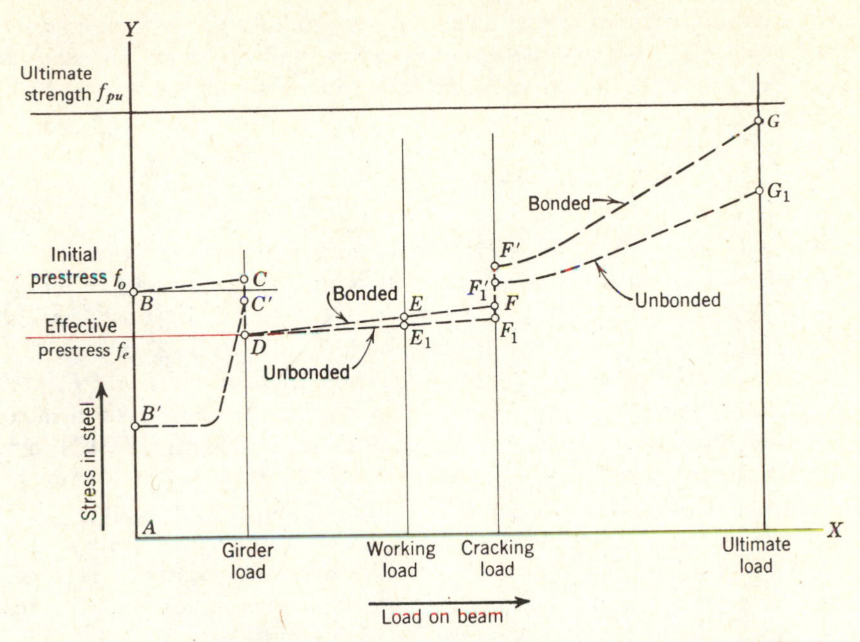

Fig. 5-11. Variation of steel stress with load.

level of f_0, which is the initial prestress in the steel after losses due to anchorage and elastic shortening have taken place.

Immediately after transfer, no load will yet be carried by the beam if it is supported on its falsework and if it is not cambered upward by the prestress. As the falsework is removed, the beam carries its own weight and deflects downward slightly, thus changing the stress in the steel, increasing it from B to C. When the dead weight of the beam is relatively light, then it can be bowed upward during the course of the transfer of prestress. The beam may actually begin to carry load when the average prestress in the steel is somewhere at B'. There may be a sudden breakaway of the beam's soffit from the falsework so that the weight of the beam is at once transferred to be carried by the beam itself, or the weight may be transferred gradually, depending on the actual conditions of support. But, in any event, the stress in steel will increase from B' up to point C'. The stress at C' is slightly lower than f_0 by virtue of the loss of prestress in the steel as caused by the upward bending of the beam. Consider now that the losses of prestress take place so that the stress in the steel drops from C or C' to some point D, representing the effective prestress f_e for the beam. Actually, the losses will not take place all at once but will continue for some length of time. However, for convenience in discussion, let us assume that all the losses take place before the application of superimposed dead and live loads.

Now let us add live load on the beam until the full-design working load is on it. The beam will bend and deflect downward, and stress in the steel will increase. For a bonded beam, such increase can be simply computed by the usual elastic theory,

$$\Delta f_s = nf_c = n\frac{My}{I}$$

where I and y correspond to the transformed section, and n is the modular ratio of steel to concrete. Since the maximum change in concrete stresses at the level of steel is not more than about 2000 psi (13.79 N/mm^2) in most cases, the corresponding change of stress in steel is limited to $2000n$, or 12,000 psi (82.74 N/mm^2) for a value of $n=6$. This stage is represented by the line DE in Fig. 5-11. It is significant to note that, in prestressed concrete, the variation in steel stress for working loads is limited to a range of about 12,000 psi (82.74 N/mm^2) even though the prestress is probably as high as 150,000 psi (1,034 N/mm^2).

If the beam is overloaded, beyond its working load and up to the point of cracking, the increase in steel stress still follows the same elastic theory. Hence the line DE is prolonged to point F. This would represent a tensile stress around 500 psi (3.45 N/mm^2) in the concrete at the level of steel indicating an increase in steel stress of about $6\times500=3000$ psi (20.68 N/mm^2) from E to F.

When the section cracks, there is a sudden increase of stress in the steel, from F to F' for the bonded beam. After cracking, the stress in the steel will increase faster with the load. As the load is further increased, the section will gradually approach its ultimate strength, the lever arm for the internal C-T couple cannot be increased any more, and increase in load is accompanied by a proportional increase in steel stress. This continues up to the point of failure. From the results of various tests, it is known that the stress in the steel approaches very nearly its ultimate strength at the rupture of the beam provided compression failure does not start in the concrete and failure of the beam is not produced by shear or bond. Hence the stress curve can be approximately drawn as from F' to G, usually slightly below f_{pu}, the steel ultimate strength.

The computation of steel stress beyond cracking and up to the ultimate load is a problem which can be rather accurately solved by analysis as shown later in this chapter. But it must be pointed out that between the two points, F' and G, there is one point when the steel ceases to be elastic, elastic in the sense that no appreciable permanent set is caused by the external load. This point is the limit to which a structure, such as a bridge or a building, could ever be subjected without permanent damage, but it will be higher than service load level. If it can be conveniently determined, it may be a more significant criterion for design than the cracking or ultimate load for some special situations.

If the beam is unbonded, the stress in the steel will be different from the bonded beam. Assuming that the same effective prestress is obtained before the addition of any external load, we can discuss the stress in an unbonded tendon as follows: Starting from point D, when load is added to the beam, the beam bends while the steel slips with respect to the concrete. Owing to this slip, the usual method of a composite steel and concrete section no longer applies. Before cracking of the concrete, stress in the concrete due to any external moment M is given by

$$f=\frac{My}{I}$$

where I and y refer to those for the net concrete section. But it must be remembered that the stress in the steel changes as load is applied, Fig. 5-12. Hence the question becomes more complicated.

At the section of maximum moment, the stress in an unbonded tendon will increase more slowly than that in a bonded tendon. This is because any strain in an unbonded tendon will be distributed throughout its entire length. Hence, as the load is increased to the working or the cracking load, the steel stress will increase from D to E_1, F_1, and F_1', below E, F, and F', respectively, Fig. 5-11. To compute the average strain for the cable, it is necessary to determine the total lengthening of the tendon due to moments in the beam. This can be done by integrating the strain along the entire length. Let M be the moment at any point of an unbonded beam; the unit strain in concrete at any point is given by

$$\delta=\frac{f}{E}=\frac{My}{E_c I}$$

The total strain along the cable is then

$$\Delta=\int\delta\,dx=\int\frac{My}{E_c I}dx$$

The average strain is

$$\frac{\Delta}{L}=\int\frac{My}{LE_c I}dx$$

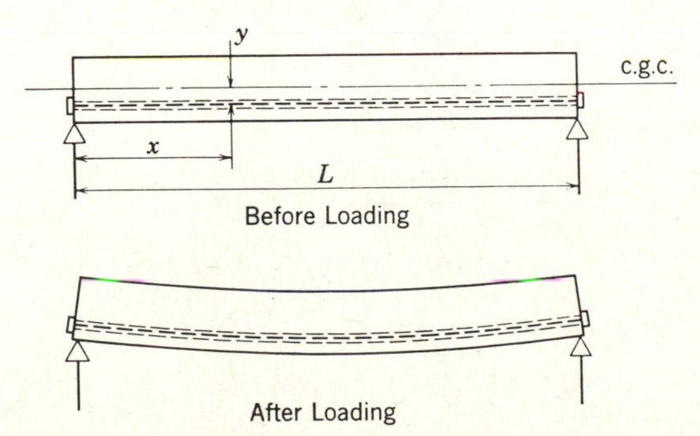

Fig. 5-12. Change of cable length in an unbonded beam.

The average stress is

$$f_s = E_s \frac{\Delta}{L} = \int \frac{M y E_s}{L E_c I} dx = \frac{n}{L} \int \frac{M y}{I} dx \tag{5-10}$$

If y and I are constant and M is an integrable form of x, the solution of this integral is simple. Otherwise, it will be easier to use a graphical or an approximate integration.

After cracks have developed in the unbonded beam, stress in the steel increases more rapidly with the load, but again it does not increase as fast as that at the maximum moment section in a similar bonded beam. In an unbonded beam, it is generally not possible to develop the ultimate strength of the steel at the rupture of the beam. Thus the stress curve is shown going up from F_1' to G_1, with G_1 below G by an appreciable amount. It is evident that the ultimate load for an unbonded beam is less than that for a corresponding bonded one, although there may be very little difference between the cracking loads for the two beams. There is a tendency for the unbonded beams to develop large cracks before rupture. These large cracks tend to concentrate strains at some localized sections in the concrete, thus lowering ultimate strength. Therefore, the strength of unbonded beams may be appreciably increased by the addition of nonprestressed bonded reinforcements, which tend to spread the cracks and to limit their size, as well as to contribute toward the tensile force in the ultimate resisting couple. The ACI Code specifies minimum amounts of such supplemental bonded reinforcement.

EXAMPLE 5-6

A posttensioned simple beam on a span of 40 ft is shown in Fig. 5-13. It carries a superimposed load of 750 plf in addition to its own weight of 300 plf. The initial prestress in the steel is 138,000 psi, reducing to 120,000 psi after deducting all losses and assuming no bending of the beam. The parabolic cable has an area of 2.5 sq in., $n = 6$. Compute the stress in the steel at midspan, assuming: (1) the steel is bonded by grouting: (2) the steel is unbonded and entirely free to slip. (Span = 12.2 m, superimposed load = 10.94 kN/m, self-weight = 4377 kN/m, initial prestress = 951.5 N/mm^2, effective prestress = 827.4 N/mm^2, and cable area = 1613 mm^2.)

Solution

1. Moment at midspan due to dead and live loads is

$$\frac{wL^2}{8} = \frac{(300 + 750)40^2}{8}$$

$$= +210{,}000 \text{ ft-lb } (+284{,}760 \text{ N-m})$$

Moment at midspan due to prestress is

$$2.5 \times 120{,}000 \times \tfrac{5}{12} = -125{,}000 \text{ ft-lb } (-169{,}500 \text{ N-m})$$

Net moment at midspan is $210{,}000 - 125{,}000 = 85{,}000$ ft-lb (115,260 N − m). Stress in concrete at the level of steel due to bending, using I of gross concrete section, is

$$=\frac{My}{I}=\frac{85{,}000\times 12\times 5}{13{,}800}=370 \text{ psi } (2.55 \text{ N/mm}^2)$$

Stress in steel is thus increased by

$$f_s=nf_c=6\times 370=2220 \text{ psi } (15.31 \text{ N/mm}^2)$$

Resultant stress in steel = 122,220 psi (842.7 N/mm²) at midspan.

Solution

2. If the cable is unbonded and free to slip, the average strain or stress must be obtained for the whole length of cable as given by formula 5-10,

$$f_s=\frac{n}{L}\int\frac{My}{I}dx$$

Using y_0 and M_0 for those at midspan and measuring x from the midspan, we can

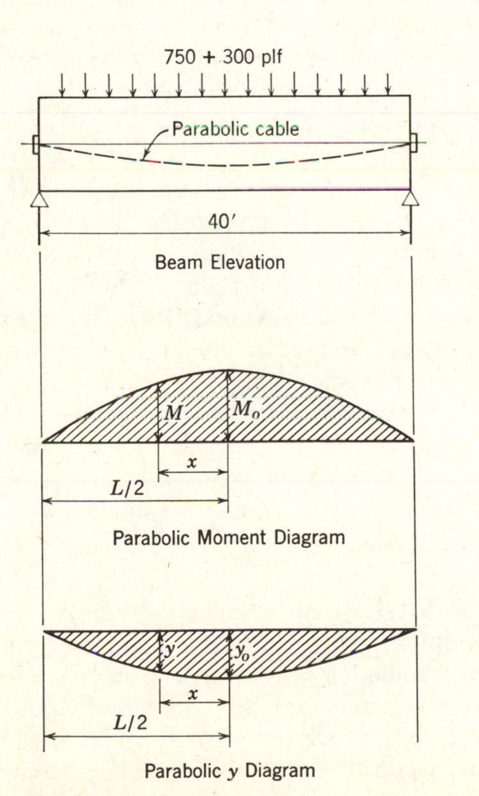

Fig. 5-13. Example 5-6.

express y and M in terms of x, thus,

$$M=M_0\left[1-\left(\frac{x}{L/2}\right)^2\right]$$

$$y=y_0\left[1-\left(\frac{x}{L/2}\right)^2\right]$$

$$f_s=\frac{n}{LI}\int_{-L/2}^{+L/2}M_0y_0\left[1-\left(\frac{x}{L/2}\right)^2\right]^2dx$$

$$=\frac{nM_0y_0}{LI}\left[x-\frac{2}{3}\frac{x^3}{(L/2)^2}+\frac{x^5}{5(L/2)^4}\right]_{-L/2}^{+L/2}$$

$$=\frac{8}{15}\left(\frac{nM_0y_0}{I}\right)$$

which is $\frac{8}{15}$ of the stress for midspan of the bonded beam, or $\frac{8}{15}(2220)=1180$ psi (8.14 N/mm^2).

Resultant stress in steel is $120{,}000+1180=121{,}180$ psi (835.5 N/mm^2) throughout the entire cable. In this calculation, the I of the gross concrete section is used and the effect of the increase in the steel stress on the concrete stresses is also neglected. But these are errors of the second order. Since the change in steel stress is relatively small, exact computations are seldom required in an actual design problem.

5-5 Cracking Moment

The moment producing first hair cracks in a prestressed concrete beam is computed by the elastic theory, assuming that cracking starts when the tensile stress in the extreme fiber of concrete reaches its modulus of rupture. Questions have been raised as to the correctness of this method. First, some engineers believed that concrete under prestress became a complex substance whose behavior could not be predicted by the elastic theory with any accuracy.[1] Then it was further questioned whether the usual bending test for modulus of rupture could give values to represent the tensile strength of concrete in a prestressed beam. However, most available test data seem to indicate that the elastic theory is sufficiently accurate up to the point of cracking, and the method is currently used. The ACI Code value for modulus of rupture, f_r, is $7.5\sqrt{f_c'}$ with units for both f_r and f_c' as psi.

Attention must be paid to the fact that the modulus of rupture is only a measure of the beginning of hair cracks which are often invisible to the naked eye. A tensile stress higher than the modulus is necessary to produce visible cracks. On the other hand, if the concrete has been previously cracked by overloading, shrinkage, or other causes, cracks may reappear at the slightest tensile stress. If the beam is made of concrete blocks, the cracking strength will depend on the tensile strength of the joining material.

Referring to formula 5-7, if f_r is the modulus of rupture, it is seen that, when

$$-\frac{F}{A}-\frac{Fec}{I}+\frac{Mc}{I}=f_r$$

cracks are supposed to start. Transposing terms, we have the value of cracking moment given by

$$M=Fe+\frac{FI}{Ac}+\frac{f_r I}{c} \tag{5-11}$$

where $f_r I/c$ gives the resisting moment due to modulus of rupture of concrete, Fe the resisting moment due to the eccentricity of prestress, and FI/Ac that due to the direct compression of the prestress.

Formula 5-11 can be derived from another approach. When the center of pressure in the concrete is at the top kern point, there will be zero stress in the bottom fiber. The resisting moment is given by the prestress F times its lever arm measured to the top kern point, (see Appendix A for definition of kern points k_t and k_b), Fig. 5-14, thus,

$$M_1=F\left(e+\frac{r^2}{c}\right)$$

Additional moment resisted by the concrete up to its modulus of rupture is $M_2=f_t I/c$. Hence the total moment at cracking is given by

$$M=M_1+M_2=F\left(e+\frac{r^2}{c}\right)+\frac{f_r I}{c} \tag{5-12}$$

which can be seen to be identical with formula 5-11.

In order to be theoretically correct when applying the above two formulas, care must be exercised in choosing the proper section for the computation of I, r, e, and c. For computing the term $f_r I/c$, the transformed section should be used for bonded beams, while the net concrete section should be used for unbonded beams (proper modification being made for the value of prestress due

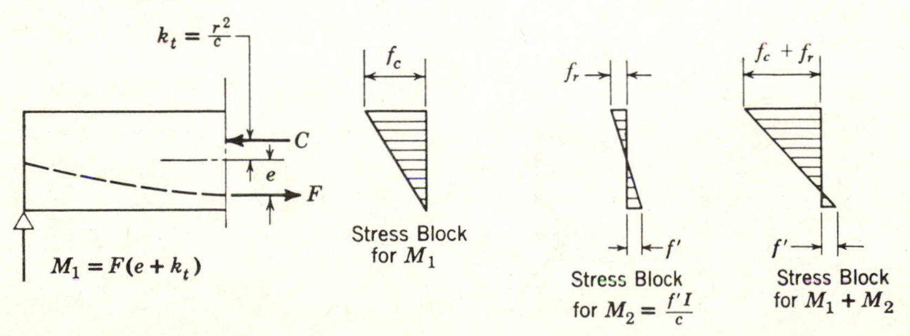

Fig. 5-14. Cracking moment.

to bending of the beam as explained in section 4-8). For the term $F\left(e+\frac{r^2}{c}\right)$, either the gross or the net section should be considered, depending on the computation of the effective prestress F. For a practical problem, these refinements are often unnecessary, and it will be easier to use one section for all the computations. In order to simplify the computations, the gross section of the concrete is most often used. If the area of holes is an important portion of the gross area, then net area may be used. If the percentage of steel is high, the transformed area may be preferred. The engineer must use discretion in choosing a method of solution consistent with the degree of accuracy required for his particular problem.

EXAMPLE 5-7

For the problem given in example 5-6, compute the total dead and live uniform load that can be carried by the beam, (1) for zero tensile stress in the bottom fibers, (2) for cracking in the bottom fibers at a modulus of rupture of 600 psi (4.14 N/mm²), and assuming concrete to take tension up to that value.

Solution

1. Considering the critical midspan section and using the gross concrete section for all computations, k_t is readily computed to be at 4 in. (101.6 mm) above the middepth, Fig. 5-15. To obtain zero stress in the bottom fibers, the center of pressure must be located at the top kern point. Hence the resisting moment is given by the prestress multiplied by the lever arm, thus

$$F(e+k_t)=300(5+4)/12=225 \text{ k-ft } (305.1 \text{ kN-m})$$

Solution

2. Additional moment carried by the section up to beginning of cracks is

$$\frac{f_r I}{c}=\frac{600\times 13{,}800}{12}$$
$$=690{,}000 \text{ in.-lb}$$
$$=57.6 \text{ k-ft } (78.1 \text{ kN}-\text{m})$$

Total moment at cracking is 225+57.6=282.6 k-ft (383.2 kN-m), which can also be obtained directly by applying formula 5-11 or 5-12.

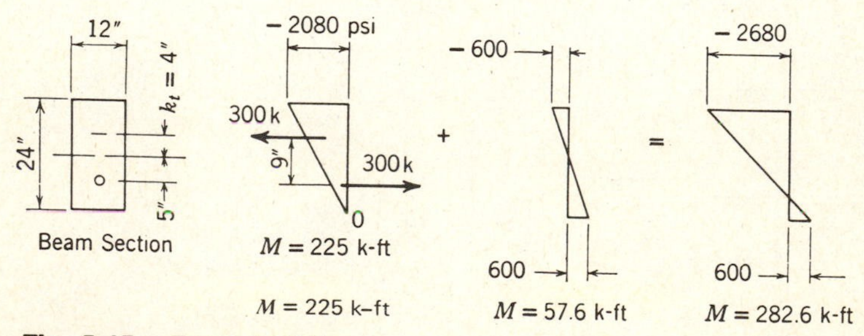

Fig. 5-15. Example 5-7.

5-6 Ultimate Moment—Bonded Tendons

Exact analysis for the ultimate strength of a prestressed-concrete section under flexure is a complicated theoretical problem, because both steel and concrete are generally stressed beyond their elastic range. The following section develops such an analysis technique for bonded beams. However, for the purpose of practical design, where an accuracy of 5-10% is considered sufficient, relatively simple procedures can be developed.

Many tests have been run, and many papers written, on the ultimate flexural strength of prestressed concrete sections. Worthy of special mention are the group of papers on this thesis[2] presented before the First International Congress on Prestressed Concrete held in London, October 1953, and another summary paper presented at the Third Congress of the International Federation for Prestressing.[3] In the United States, laboratory investigations carried out at the University of Illinois and the Portland Cement Association gave the results of extensive tests, together with definite recommendations.[4,5,6] Although formulas for ultimate strength proposed by various authors seem to differ greatly on the surface, they generally yield values within a few per cent of one another. Hence it can be concluded that the ultimate strength of prestressed concrete under flexure can be predicted with sufficient accuracy.

A simple method for determining ultimate flexural strength following the ACI Code is presented herewith, based on the results of the aforementioned tests as well as others. This method is limited to the following conditions.

1. The failure is primarily a flexural failure, with no shear bond, or anchorage failure which might decrease the strength of the section.
2. The beams are bonded. Unbonded beams possess different ultimate strength and are discussed later.
3. The beams are statically determinate. Although the discussions apply equally well to individual sections of continuous beams, the ultimate strength of continuous beams as a whole is explained by the plastic hinge theory to be discussed in Chapter 10.
4. The load considered is the ultimate load obtained as the result of a short static test. Impact, fatigue, or long-time loadings are not considered.

Of the methods proposed for determining the ultimate flexural strength of prestressed-concrete sections, some are purely empirical and others highly theoretical. The empirical methods are generally simple but are limited only to the conditions which were encountered in the tests. The theoretical ones are intended for research studies and hence unnecessarily complicated for the designer. For the purpose of design, a rational approach is presented in the following, consistent with test results, but neglecting refinements so that reasonably correct values can be obtained with the minimum amount of effort. The method is based on the simple principle of a resisting couple in a prestressed beam, as that in any other beam. At the ultimate load, the couple is made of two forces, T' and C', acting with a lever arm a'. The steel supplies the tensile force T', and the concrete, the compressive force C'.

Before going any further with the method, let us first study the modes of failure of prestressed-beam sections. The failure of a section may start either in the steel or in the concrete, and may end up in one or the other. The most general case is that of an underreinforced section, where the failure starts with the excessive elongation of steel and ends with the crushing of concrete. This type of failure occurs in both prestressed- and reinforced-concrete beams, when they are underreinforced. Only in some rare instances may fracture of steel occur in such beams; that happens, for example, when the compressive flange is restrained and possesses a higher actual strength. A relatively uncommon mode of failure is that of an overreinforced section, where the concrete is crushed before the steel is stressed into the plastic range. Hence there is only a limited amount of deflection before rupture, and a brittle mode of failure is obtained. This is similar to an overreinforced nonprestressed-concrete beam. Another unusual mode of failure is that of a too lightly reinforced section, where failure may occur by the breaking of the steel immediately following the cracking of concrete. This happens when the tensile force in the concrete is suddenly transferred to the steel whose area is too small to absorb that additional tension.

There is no sharp line of demarcation between the percentage of reinforcement for an overreinforced beam and that for an underreinforced one. The transition from one type to another takes place gradually as the percentage of steel is varied. A sharp definition of "balanced condition" cannot be made since the steel used for prestressing does not exhibit a sharp yield point. For the materials presently used in prestressed work, the reinforcement index, ω_p, which approximates the limiting value to assure that the prestressed steel (A_{ps}) will be slightly into its yield range, is given by the ACI Code as follows:

$$\omega_p = \rho_p f_{ps}/f_c' \leqslant 0.30 \quad (5\text{-}13)$$

where

$$\rho_p = A_{ps}/bd$$

There are situations where prestressing steel (A_{ps}) and ordinary reinforcing bars (A_s) are used together in a prestressed beam. In this case the total of all the tension steel is considered along with the possibility of compression steel (A_s'). The limiting reinforcement ratio is given as

$$(\omega + \omega_p - \omega') \leqslant 0.30 \quad (5\text{-}14)$$

where

$$\omega = \rho f_y/f_c' \quad \text{and} \quad \rho = A_s/bd$$

$$\omega' = \rho' f_y/f_c' \quad \text{and} \quad \rho' = A_s'/bd$$

Such ratios of reinforcement almost always end in plastic failure and can be termed as underreinforced ratios. If the ratio from equation 5-14 is over 1.0, sudden crushing of concrete without substantial elongation of steel will be likely to take place. If it is less than about 0.10, breaking of the wires following cracking of concrete may occur.

A proper definition of the percentage of steel ρ is important for prestressed sections because of their irregular shapes. For ultimate strength it is not the total concrete area or the shape of cross section but the concrete area in the compressive flange that matters; hence ρ will be more indicative of the relative strength of concrete and steel if it is expressed in terms of A_s/bd, where b is the width or average width of the compressive flange and d the effective depth as indicated above for the ACI Code expressions.

ACI Code Bonded Beams. For underreinforced bonded beams following the ACI Code, the steel is stressed to a stress level which approaches its ultimate strength at the point of failure for the beam in flexure. For the purpose of practical design, it will be sufficiently accurate to assume that the steel is stressed to the stress level, f_{ps}, given by the equation for bonded beams from the ACI Code which closely approximates test results.[6] Provided the effective prestress, f_{se}, is not less than $0.5f_{pu}$, the following approximate value for the steel stress at ultimate moment capacity for the beam is applicable for bonded beams:

$$f_{ps} = f_{pu}\left(1 - 0.5\rho_p \frac{f_{pu}}{f'_c}\right) \tag{5-15}$$

Note that as the steel ratio ρ_p is reduced, the member is increasingly underreinforced; and the steel stress f_{ps} approaches the ultimate strength of prestressing steel. In fact, there are some test data which seem to show that the steel was stressed even beyond its ultimate strength. Though this does not seem to be possible, it might perhaps be explained by the fact that the group strength of wires forced to fail together at one section of a beam might be higher than the tested strength of steel in the specimens, since, during specimen tests, only the strength of the weakest link is recorded.

The computation of the ultimate resisting moment is a relatively simple matter and can be carried out as follows. Referring to Fig. 5-16, the ultimate compressive force in the concrete C' equals the ultimate tensile force in the steel T', thus,

$$C' = T' = A_s f_{ps}$$

Let a' be the lever arm between the forces C' and T'; then the ultimate resisting moment is given by

$$M' = T'a' = A_s f_{ps} = M_n \text{ (ACI Code nominal strength)}$$

To determine the lever arm a', it is only necessary to locate the center of pressure C'. There are many plastic theories for the distribution of compressive

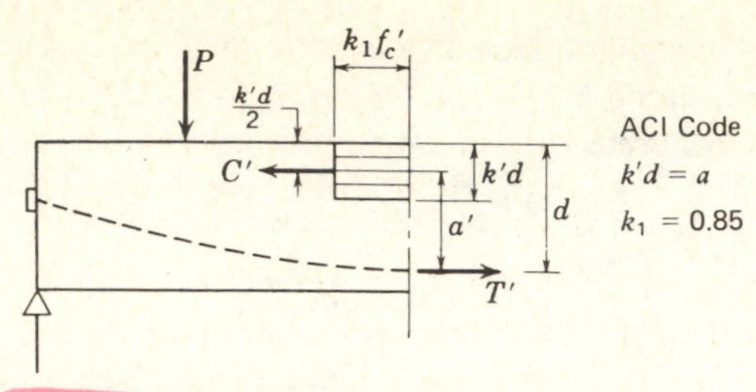

Fig. 5-16. Ultimate moment.

stress in concrete at failure,[7] assuming the stress block to take the shape of a rectangle, trapezoid, parabola, etc. Although the actual stress distribution is a very interesting problem for research, for the purpose of design, any of these methods would be sufficiently accurate, because they would yield nearly the same lever arm a', seldom differing by more than 5%.

Choosing the simplest stress block, a rectangle, for the ultimate compression in concrete, the depth to the ultimate neutral axis $k'd$ is computed by

$$C' = k_1 f_c' k' b d$$

where $k_1 f_c'$, is the average compressive stress in concrete at rupture. Hence,

$$k'd = \frac{C'}{k_1 f_c' b} = \frac{A_{ps} f_{ps}}{k_1 f_c' b}$$

$$k' = \frac{A_s f_{ps}}{k_1 f_c' b d} \tag{5-16}$$

These formulas apply if the compressive flange has a uniform width b at failure.

Locating C' at the center of the rectangular stress block, we have the lever arm

$$a' = d - k'd/2$$

$$= d\left(1 - \frac{k'}{2}\right) \tag{5-17}$$

Hence, the ultimate resisting moment is

$$M' = A_{ps} f_{ps} d\left(1 - \frac{k'}{2}\right) = M_n \text{ (ACI notation)} \tag{5-18}$$

Now the determination of the value of k_1 deserves some comments. According to Whitney's plastic theory of reinforced-concrete beams, k_1 should be 0.85, based on cylinder strength. According to some authors in Europe, k_1 should be 0.60 to 0.70 based on the cube strength; since cube strength is 25% higher than cylinder strength, this would give approximately 0.75 to 0.88 for k_1 based on the cylinder strength. The important thing for the designer to see is the fact that variation of the value of k_1 does not appreciably affect the lever arm a'. Hence it is considered accurate enough to adopt some approximate value, such as 0.85 for k_1. Using 0.85 for k_1, formula 5-16 can be written as

$$k' = \frac{A_{ps} f_{ps}}{0.85 f_c' bd} \tag{5-19}$$

By substituting this expression for k' into equation 5-18, we have

$$M' = A_{ps} f_{ps} d \left(1 - \frac{A_{ps} f_{ps}}{2 \times 0.85 f_c' bd} \right) \tag{5-20}$$

For a rectangular section for the compression area, we can let $\rho_p = A_{ps}/bd$. Then we have the following formula:

$$M' = A_{ps} f_{ps} d \left(1 - \frac{0.59 \rho_p f_{ps}}{f_c'} \right) \tag{5-21}$$

or from Fig. 5-16 with ACI notation $k'd = a$

$$M_n = A_{ps} f_{ps} \left(d - \frac{a}{2} \right) \tag{5-22}$$

which is identical to that given in the commentary of the American Concrete Institute and as first proposed by the ACI-ASCE Recommendations.[8]

The ACI Code introduces the strength reduction factor, ϕ, and writes equation 5-21 in terms of ω_p to solve the design ultimate moment as follows:

$$M_u = \phi \left[A_{ps} f_{ps} d (1 - 0.59 \omega_p) \right] \tag{5-23}$$

The alternative equation (5-22) written directly in terms of the T' and C' force couple becomes the following ACI Code design ultimate moment equation:

$$M_u = \phi \left[A_{ps} f_{ps} \left(d - \frac{a}{2} \right) \right] \tag{5-24}$$

For flexure the ACI Code uses $\phi = 0.9$ in the two equations for M_u given above, (5-23) and (5-24). These expressions apply to rectangular beams or beams which have a rectangular-shaped compression zone for the concrete cross section.

EXAMPLE 5-8

An I-shaped beam is prestressed with $A_{ps} = 2.75$ in.^{2} as prestressing steel with an effective stress, f_{se}, of 160 ksi. The c.g.s. of the strands which supply the prestress is 4.5 in. above the bottom of the beam as shown in Fig. 5-17 along with the shape of the concrete cross section. Material properties are: $f_{pu} = 270$ ksi; $f_c' = 7000$ psi. Find the ultimate resisting moment for the section for design following the ACI Code. ($A_{ps} = 1{,}774$ mm^2, $f_{se} = 1{,}103$ N/mm^2, $f_{pu} = 1{,}862$ N/mm^2, and $f_c' = 48$ N/mm^2.)

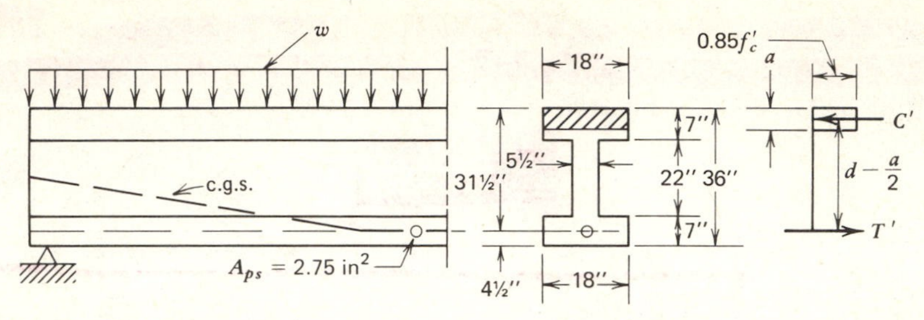

Fig. 5-17. Example 5-8.

Solution:

$$\rho_p = \frac{2.75}{(18)(31.5)} = 0.00485$$

Estimate steel stress at ultimate by the ACI equation (5-15) which is valid to use since $f_{se} = 160$ ksi (1103 N/mm²) $> 0.5f_{pu} = 135$ ksi (931 N/mm²).

$$f_{ps} = 270{,}000\left[1 - (0.5)(0.00485)\left(\frac{270{,}000}{7{,}000}\right)\right] \tag{5-15}$$

$$f_{ps} = 245{,}000 \text{ psi} = 245 \text{ ksi } (1689 \text{ N/mm}^2)$$

Check the reinforcement index

$$\omega_p = \frac{(0.00485)(245{,}000)}{7000} = 0.17 < 0.30 \tag{5-13}$$

Referring to Fig. 5-17 sketch of section

$$T' = A_{ps} f_{ps} = 2.75 \times 245 = 674 \text{ k } (2{,}998 \text{ kN})$$

$$C' = 0.85 f'_c \times 18 \times a = 674 \text{ k } (2{,}998 \text{ kN})$$

$$a = \frac{674}{(0.85)(7)(18)} = 6.29 \text{ in.} < 7 \text{ in. } O.K. \text{ rectangular section behavior}$$

$$M_n = T'\left(d - \frac{a}{2}\right) = 674\left(31.5 - \frac{6.29}{2}\right) = 19{,}100 \text{ in.-k. } (2{,}158 \text{ kN-m}) \tag{5-22}$$

$$M_u = 0.9 M_n = 17{,}200 \text{ in.-k. } (1944 \text{ kN-m}) \tag{5-24}$$

Note that even though the section in example 5-8 is I-shaped it behaves as a "rectangular section"; the compression zone of the concrete is rectangular as shown by the shaded area in Fig. 5-17. The following example illustrates the case where the compression zone is nonrectangular.

For flanged sections (nonrectangular compression zone) we may still use equation 5-15 to estimate the steel stress at ultimate, f_{ps}. The total area of prestressed steel, A_{ps}, is divided into two parts with A_{pf} developing the flanges and A_{pw} developing the web as shown in Fig. 5-18. The ultimate moment is simply computed from the two parts: the flange part has the compression resultant force acting at middepth of flange, $h_f/2$, and the arm of the moment

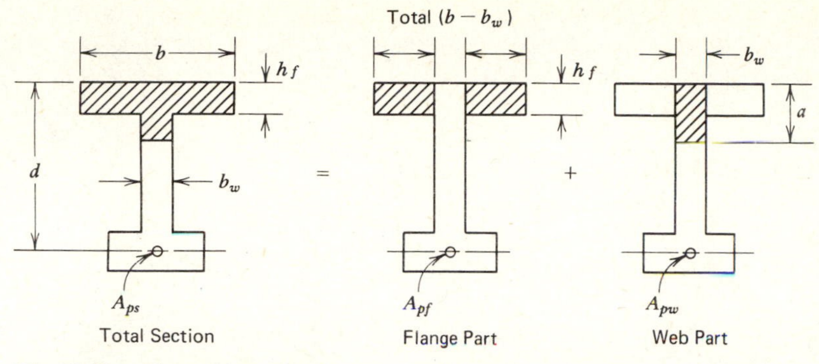

Fig. 5-18. Flanged section

couple is $\left(d-\dfrac{h_f}{2}\right)$; the web part has the compression resultant force acting at $a/2$ from the top the beam, and the arm of the moment couple is $\left(d-\dfrac{a}{2}\right)$. The equivalent rectangular stress block is assumed as before, Fig. 5-17, and the depth a is determined by the compression area required based on equal total compression and tension forces at ultimate. The commentary of the ACI Code contains equations for M_u to cover this case which it terms "flanged section."

$$M_u = \phi\left[A_{pw}f_{ps}\left(d-\frac{a}{2}\right)+0.85f_c'(b-b_w)h_f\left(d-\frac{h_f}{2}\right)\right] \tag{5-25}$$

where

$$A_{pw} = A_{ps} - A_{pf} \tag{5-26}$$

and

$$A_{pf} = 0.85f_c'(b-b_w)h_f/f_{ps} \tag{5-27}$$

EXAMPLE 5-9

The same I-shaped prestressed concrete beam as example 5-8 but the steel area is increased to $A_{ps}=3.67$ in.^{2} The effective steel stress remains 160 ksi. The c.g.s. of the strands is 4.5 in. above the bottom of the beam as shown in Fig. 5-19 along with the

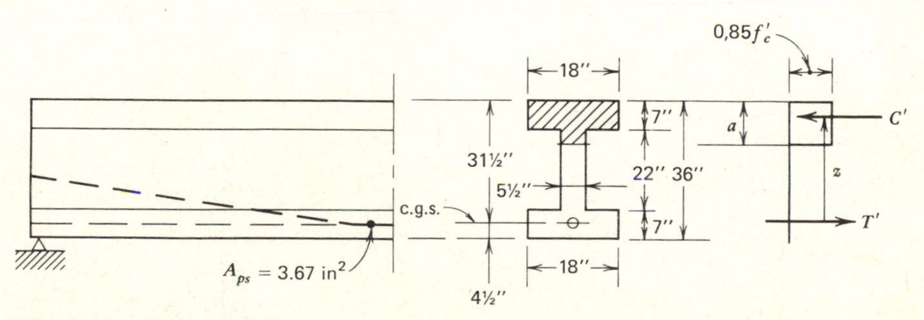

Fig. 5-19. Example 5-9.

shape of the cross section: material properties are same as example 5-8: $f_{pu}=270$ ksi, $f_c'=7000$ psi. Find the ultimate resisting moment for the section for design following the ACI Code. ($A_{ps}=2368\ \text{mm}^2$, $f_{se}=1103\ \text{N/mm}^2$, $f_{pu}=1862\ \text{N/mm}^2$, and $f_c'=48\ \text{N/mm}^2$)

Solution

$$\rho_p=\frac{3.67}{(18)(31.5)}=0.00647$$

Use equation 5-15 to estimate steel stress at ultimate.

$$f_{ps}=270{,}000\left[1-(0.5)(0.00647)\left(\frac{270{,}000}{7{,}000}\right)\right]$$

$$f_{ps}=236{,}000\ \text{psi}=236\ \text{ksi}\ (1627\ \text{N/mm}^2)$$

Check the reinforcement index after the flanged section is evaluated below.

Referring to Fig. 5-18 and 5-19 determine the extent of the compression zone

$$T'(\text{total})=(3.67)(236)=866\ \text{k}\ (3{,}852\ \text{kN})$$

$$\text{Area of compression zone}=\frac{866}{0.85f_c'}=145.5\ \text{in.}^2\ (93.87\times10^3\ \text{mm}^2)$$

$$\text{Flange area}=18\times7=126.0\ \text{in.}^2\ (81.29\times10^3\ \text{mm}^2)$$

$$\text{Web area below flange}=\overline{19.5\ in.^2}\ (12.58\times10^3\ \text{mm}^2)$$

$$a=7+\frac{19.5}{5.5}=7+3.55=10.55\ \text{in.}\ (268\ \text{mm})$$

This verifies that the section is behaving as "flanged" as shown by Fig. 5-18 and M_u can now be evaluated.

Referring to Fig. 5-18 and using ACI Commentary equations

$$A_{pf}=(0.85)(7000)(18.0-5.5)(7)/236{,}000=2.21\ \text{in.}^2\ (1426\ \text{mm}^2) \qquad (5\text{-}27)$$

$$A_{pw}=3.67-2.21=1.46\ \text{in.}^2\ (942\ \text{mm}^2) \qquad (5\text{-}26)$$

Check reinforcement index for the flanged section;

$$\rho_{pw}=A_{pw}/b_wd=1.46/(5.5)(31.5)=0.00843$$

$$\omega_{pw}=(0.00843)(236{,}000)/7000=0.284<0.30$$

$$M'\ \text{for web part}=A_{pw}f_{ps}\left(d-\frac{a}{2}\right)$$

$$M'_{\text{web}}=(1.46)(236)\left(31.5-\frac{10.55}{2}\right)=9{,}040\ \text{in.-k}\ (1{,}021.5\ \text{kN}-\text{m})$$

$$M'\ \text{for flange part}=0.85f_c'(b-b_w)h_f\left(d-\frac{h_f}{2}\right)$$

$$M'_{\text{flange}}=(0.85)(7.0)(18.0-5.5)(7)(31.5-7/2)=14{,}580\ \text{in.-k}\ (1647.5\ \text{kN}-\text{m})$$

$$M'_{\text{total}}=9040+14{,}580=23{,}620\ \text{in.-k}\ (2669\ \text{kN}-\text{m})=M_n$$

Note that the ACI commentary equation (5-25) contains these two terms, and we may write it in the form:

$$M_u = \phi[M'_{\text{web}} + M'_{\text{flange}}] = \phi[M'_{\text{total}}]$$

thus

$$M_u = (0.9)(23{,}620) = 21{,}260 \text{ in.-k } (2{,}402 \text{ kN} - \text{m})$$

The examples 5-8 and 5-9 illustrate the simplicity of analysis for M_u whether the section behaves as a "rectangular" or "flanged" section. It should be noted that the addition of more prestressed steel to the section in example 5-9 causes the section to almost reach the limit of reinforcement index, $\omega = 0.30$, which the ACI Code would allow. Further addition of tension steel would cause the beam to be overreinforced, and the beam would not have a ductile failure. Addition of compression steel might be required to be sure that ductility is assured (equation 5-14). For the flanged section we use the web of the section with b_w and the steel area required to develop the compressive strength of the web only to find $\omega_{pw} \leqslant 0.30$ as illustrated in example 5-9. The ACI Code has this requirement stated as follows: (similar to equation 5-14):

$$(\omega_w + \omega_{pw} - \omega'_w) \leqslant 0.30 \tag{5-28}$$

The terms involving ordinary reinforcement are ω_w (tension steel) and ω'_w (compression steel) for equation 5-28. In equation 5-14 the corresponding terms are ω and ω'. Thus in prestressed concrete as in reinforced concrete analysis, addition of compression steel will add ductility as the steel (unstressed bars) on the compression side of the beam carries a part of the total compressive force, relieving compression that would otherwise be carried by the concrete.

Note also in example 5-9 that addition of more prestressing strand causes the steel stress at ultimate to be reduced. The stress-strain curve for the prestressing steel, Fig. 2-7, has the characteristic of continuing slight increase in stress with strain in excess of yield. The more ductile beam of example 5-8 ($\omega_p = 0.17 < 0.30$) will fail in flexure with higher steel strain and $f_{ps} = 245$ ksi; with added A_{ps} in example 5-9 ($\omega_{pw} = 0.28 < 0.30$) we have less ductility, smaller steel strain at ultimate, and $f_{ps} = 236$ ksi. This will be discussed later in connection with the more exact moment-curvature analysis, but the trends in behavior are important to observe in the ACI Code analysis, which gives results in good agreement with tests.

If material strengths and physical dimensions are in agreement with assumed values, the ultimate moment will be quite close to M' computed by the ACI Code analysis. The $\phi = 0.9$ strength reduction factor adds safety in design, the intent being to allow for possible understrength in materials, dimensional errors in construction and errors in the assumptions made in analysis.

5-7 Moment-Curvature Analysis—Bonded Beams

A rational analysis which follows the behavior of a bonded prestressed concrete through the total load range from initial loading to failure has been developed,[6,9] and tests have shown the results of the analysis to be quite reliable. This moment-curvature analysis is derived from basic assumptions about materials and member behavior. The technique is described below, and a numerical example will show that the ultimate strength found by this more exact procedure is close to the ACI Code estimate, M'. But the added understanding of behavior which can be gained for the progressive load stages leading to failure is important to emphasize. This complete analysis is quite general, and computer programs have been written which perform the detailed calculations very rapidly. Hand calculation can be used as will be illustrated by the example problem.

The following assumptions are made in connection with the moment-curvature analysis:

1. Tendons are bonded to the concrete. *Changes* in strain in the steel and concrete after bonding are assumed to be the same.
2. The initial strain from the effective prestress in the tendon when no moment from applied loads acts on the section is illustrated by Fig. 5-20(a). At the level of steel, the concrete *compressive* strain, ϵ_{ce}, exists while the tendon has a *tensile* strain, ϵ_{se}, which corresponds to the stress f_{se}, which is initially effective.
3. Stress-strain properties for the materials are known or assumed (Fig. 5-21) for use in analysis.
4. Strains are assumed to be distributed linearly over the depth of the beam as shown in Fig. 5-20.
5. Tension and compression forces acting on the cross section must be in equilibrium for the beam which has only flexure without any applied axial load.

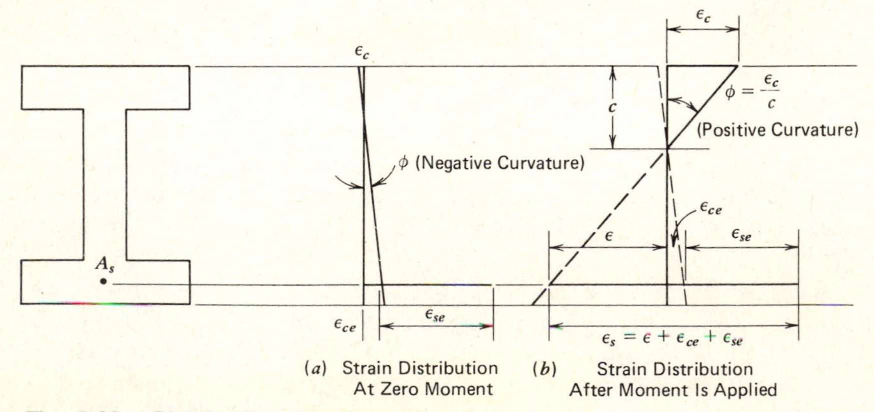

Fig. 5-20. Distribution of strain assumed.

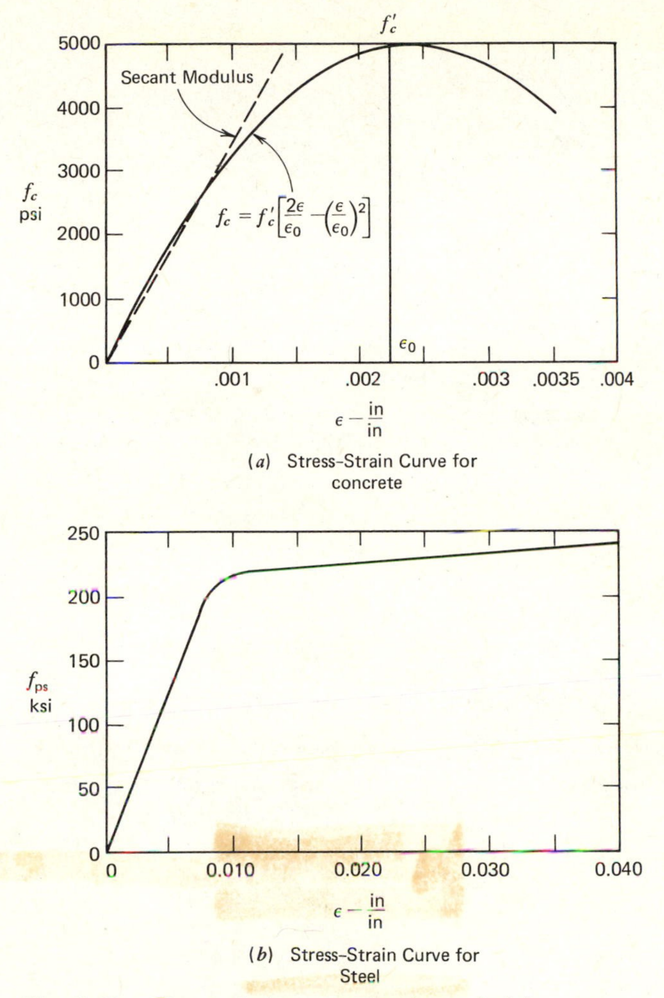

(*a*) Stress-Strain Curve for concrete

(*b*) Stress-Strain Curve for Steel

Fig. 5-21. Stress-strain properties for materials.

$$\text{Concrete stress} = f_c = f_c'\left[\frac{2\phi x}{\epsilon_0} - \left(\frac{\phi x}{\epsilon_0}\right)^2\right] \quad \text{as shown in Fig. 5-21c}$$

where $\phi x = \epsilon$ in the expression from Hognestad similar to Fig. 5-21a.

$$C_c = \int_0^c f_c b\, dx = bf_c' \int_0^c \left(\frac{2\phi x}{\epsilon_0} - \frac{\phi^2 x^2}{\epsilon_0^2}\right) dx \quad \text{(see Fig. 5-21c)}$$

solving this, the resultant compression force for a rectangular section is

$$C_c = bf_c' \frac{\phi}{\epsilon_0} c^2 \left[1 - \frac{\phi c}{2\epsilon_0}\right] \tag{5-29}$$

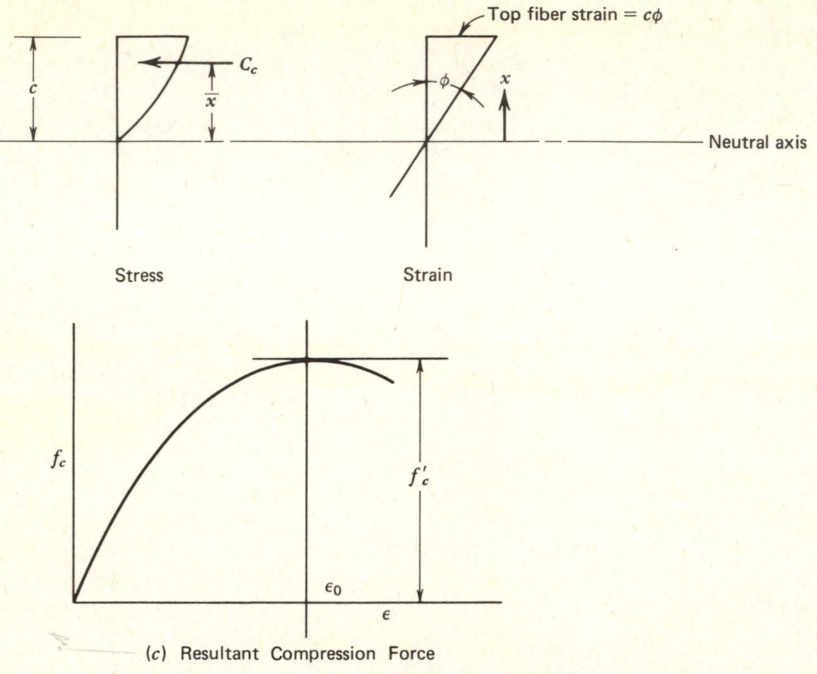

(c) Resultant Compression Force

Fig. 5-21(c). Resultant Compression Force.

$\bar{x}C_c = \int_0^c (f_c b\, dx)x$ substituting the expression above for C_c and rearranging terms, the distance from neutral axis to line of action for resultant compression force is

$$\bar{x} = c\left[\frac{8\epsilon_0 - 3\phi c}{12\epsilon_0 - 4\phi c}\right] \tag{5-30}$$

6. Ultimate moment corresponds to the occurrence of a strain in the concrete which causes crushing (usually 0.003 in./in.) or a steel strain which would fracture the tendon (for most prestressing steel about 5% strain).
7. The failure analyzed is in flexure, and it is assumed that the member will have adequate shear strength to prevent failure. Bond and anchorage of steel is assumed adequate to prevent failure prior to reaching flexural strength at the section being analyzed.

The assumptions listed above are justified by experimental data, and a few comments are in order before describing the analysis procedure. Item 1 is extremely important; bonded pretensioned beams and posttensioned beams (grouted tendons following stressing) satisfy this assumption. Unbonded tendons will slip with respect to the concrete and thus will not satisfy this assumption of strain change compatibility for steel and concrete. Item 2 relates to the initial strains which exist prior to the application of external moment. The steel will have experienced losses which would be estimated to find the effective stress, f_{se}, which exists as the starting point for analysis.

This leads to Item 3, the known stress-strain relationship for the steel, since the ϵ_{se} is simply the steel strain corresponding to f_{se} from this curve. The tests of typical materials used for tendons provide these data (Appendix B). The concrete stress-strain curve is assumed here as a parabolic form following Hognestad very closely. This is convenient because it allows integration to solve the resultant compressive force, and its location in a closed form solution as shown in Fig. 5-21(c). The secant modulus of elasticity for concrete, Fig. 5-21(a), is made to correspond to the ACI Code value for E_c, and E_s is taken from the stress-strain curve for the steel used in the beam. The initial response for the member prior to cracking (at tension in the concrete, $f_r = 7.5\sqrt{f_c'}$) is elastic, and the values of E_c and E_s relate stress to strain in the materials. As shown in Fig. 5-22 the value of f_r follows the ACI equation with scatter above or below this value.

Item 4 relates to linear strains over the depth of the member, which has been verified by tests of bonded beams where measurements were made over a gage length including cracks. Good bond of the steel results in the formation of numerous cracks as observed in tests of both pretensioned and posttensioned beams, and the average curvature is adequate to represent the beam response to moment, Fig. 5-23. This analysis deals with average curvatures with higher

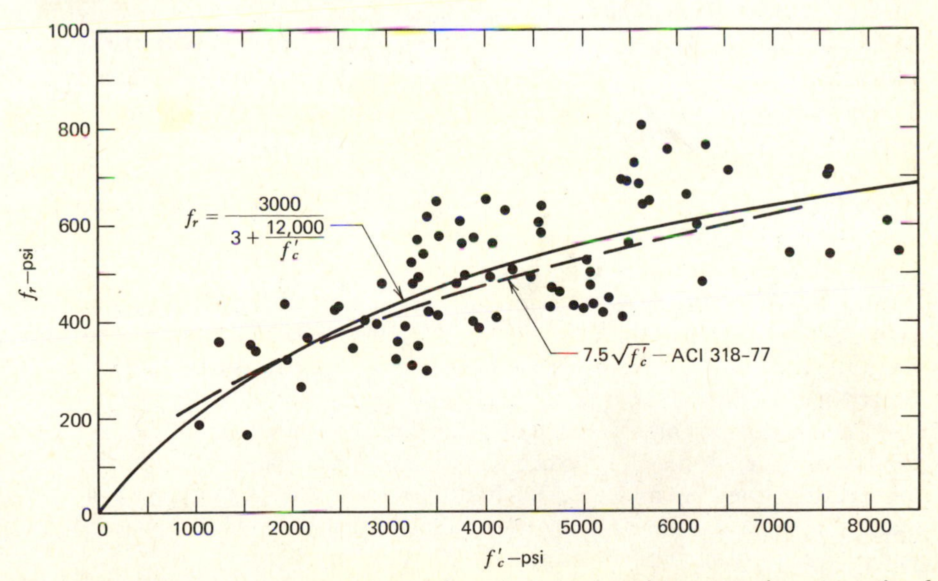

Fig. 5-22. Relationship between modulus of rupture and compressive strength of concrete.[10]

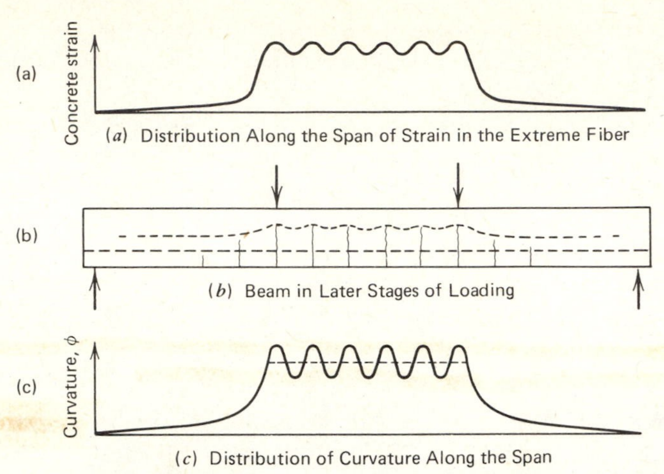

Fig. 5-23. Distribution of strain and curvature along the span.

values at a crack being averaged with lower values between cracks. The angle, ϕ, measured from the linear strains over the depth of the section is the curvature, Fig. 5-20. Note that this is initially negative as shown in Fig. 20(a), (camber results) but becomes positive curvature (downward deflection) as moment is added, Fig. 5-20(b).

The equilibrium of forces is implied from statics, but it is a key assumption here as Item 5. The total tension, T', acts together with an equal resultant compression, C'. The stress-strain curve for concrete together with the shape of the compression zone determine the point of action of C' while the tension force T' is determined from the placement of the tendons. The force is usually taken to act at the centroid of the tendon steel. Where both tendons and reinforcing bars are used in the same beam as tension steel, T' would be solved for each type of reinforcement and these two forces (acting at centroid of each type of steel) would be combined into a single total resultant, T'.

Strain at ultimate, Item 6, is based on test data. The ACI Code value for design, 0.003 in./in., is a lower-bound value from these data. Actually, the assumption of higher strain for crushing of concrete doesn't significantly change the ultimate moment calculated. Higher strain at ultimate would lead to greater deformation, thus the ACI Code value may be considered purposely conservative for safe design. Only a very lightly reinforced beam would fail by fracture of the steel prior to reaching crushing strain in the concrete at the extreme compressive fiber. As indicated by Item 7, we assume no other type of failure occurs; that is, this flexural analysis cannot assure adequate shear, bond, or anchorage strength since these must be checked separately.

The analysis procedure is carried out assuming two stages of behavior: first, the beam is elastic and uncracked; second, the beam is cracked and the actual

① Assume strain at top fiber—say $\epsilon_c = 0.0015$ for this point

② Assume depth to neutral axis c and compute internal forces.

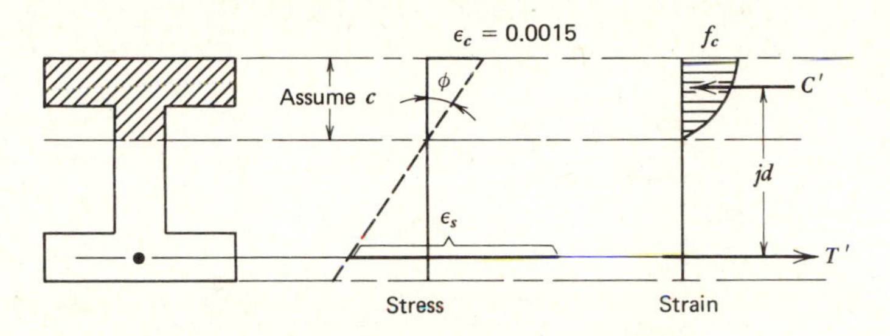

③ Check to see if assumed c yields $C' = T'$

④ Revise assumption for c until equilibrium is satisfied ($C' = T'$).

⑤ With final value of c find ϕ and moment of couple.

⑥ Assume another top fiber strain in ① and repeat ② through ⑤ to obtain ϕ and moment.

Fig. 5-24. Postcracking analysis for moment-curvature.

material properties are used for analysis of the cracked section response. The first stage is assumed elastic, but the second stage is inelastic following the response of the materials. Figure 5-24 shows the steps in the postcracking analysis. A point-to-point check is made for a series of assumed values for top fiber strain, the points collectively describing the moment-curvature relationship as shown in the numerical example, example 5-10.

EXAMPLE 5-10

The beam cross section of example 5-8 is to be analyzed to determine its moment-curvature relationship. The materials are normal weight concrete; $f_c' = 7000$ psi (48 N/mm^2), limiting strain at ultimate $= 0.003$; 7-wire strand with $f_{pu} = 270$ ksi (1.862 N/mm^2) specified. Use the *actual* curve for analysis, Fig. 5-25*a*, which has breaking strength of approximately 280 ksi (1,931 N/mm^2) from typical test. (Figure 5-25*b* shows f_c vs. ϵ_c for concrete.)

Find points for moment, M, and curvature, ϕ, for each of these stages as moment is increased:

(*a*) Initial stage—zero applied moment, $f_{se} = 160$ ksi.
(*b*) Zero strain in concrete at level of steel.
(*c*) Cracking at $f_r = 7.5\sqrt{f_c'}$.
(*d*) Top fiber strain 0.001. }
(*e*) Top fiber strain 0.002. } cracked section
(*f*) Top fiber strain 0.003. }

Make a summary of results including the steel stress at each stage, and plot the M vs. ϕ curve.

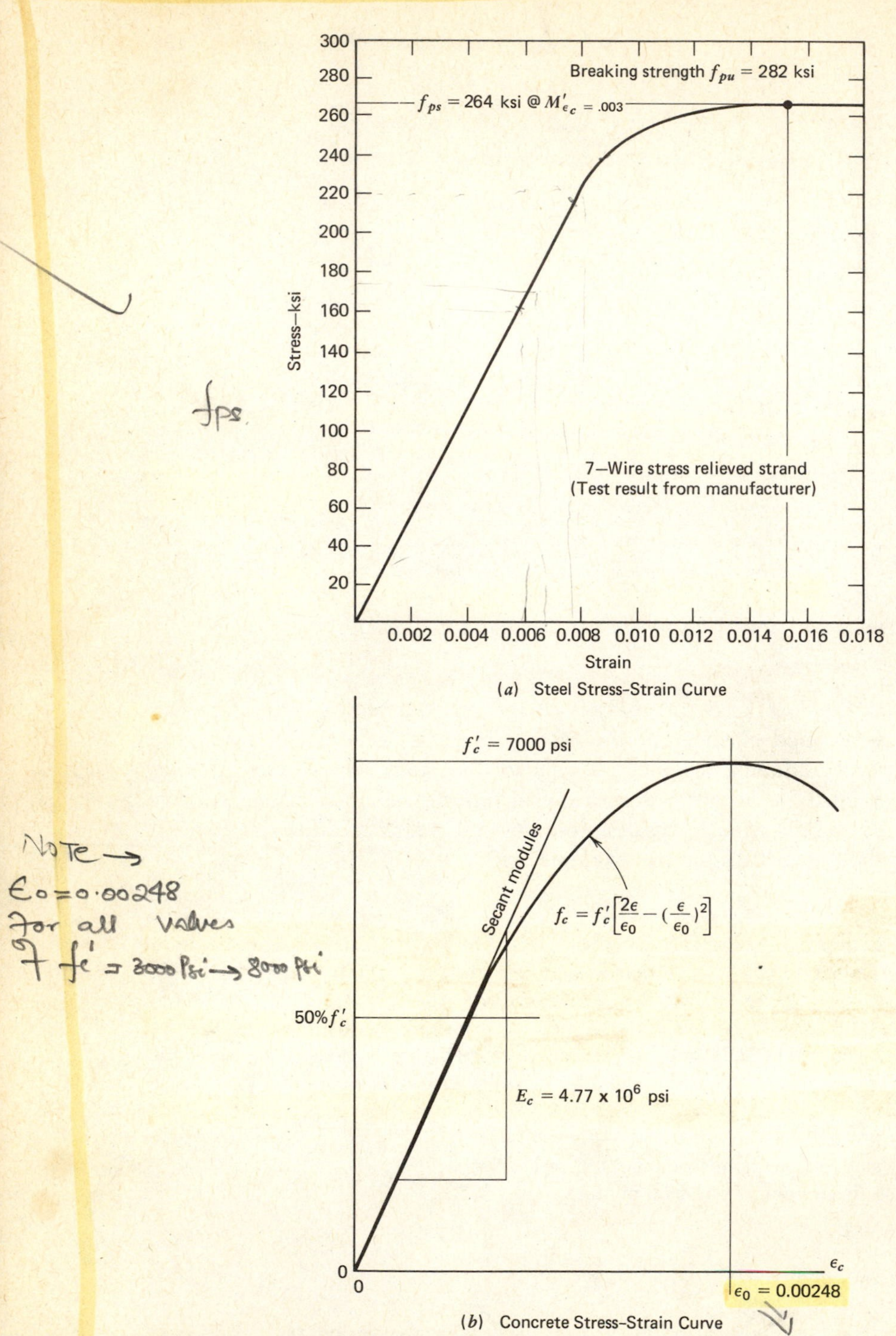

Fig. 5-25. Stress-strain curves for materials, example 5-10.

Solution (*a*) *Initial stage*: assume elastic beam may be analyzed for stresses in the concrete using gross section properties and $F = A_{ps} f_{se}$

Section properties: $A = 373$ in.^{2} (241×10^3 mm^2) $c = 18$ in. (457.2 mm)

(Fig. 5-17 shows cross-section dimensions)

$$I_x = 58{,}890 \text{ in.}^4\ (24.51 \times 10^9 \text{ mm}^4)\ e = 13.5 \text{ in. } (342.9 \text{ mm})$$

$$S_x = 3272 \text{ in.}^3\ (53.62 \times 10^6 \text{ mm}^3)$$

$$F = (2.75)(160) = 440 \text{ k } (1957 \text{ kN})$$

$$E_c = 57{,}000\sqrt{f_c'} = 57{,}000\sqrt{7000} = 4.77 \times 10^6 \text{ psi } (32.89 \text{ kN/mm}^2)$$

Compute stress and corresponding strains in section due to $F = 440$ k (1957 kN) at $e = 13.5$ in. (342.9 mm).

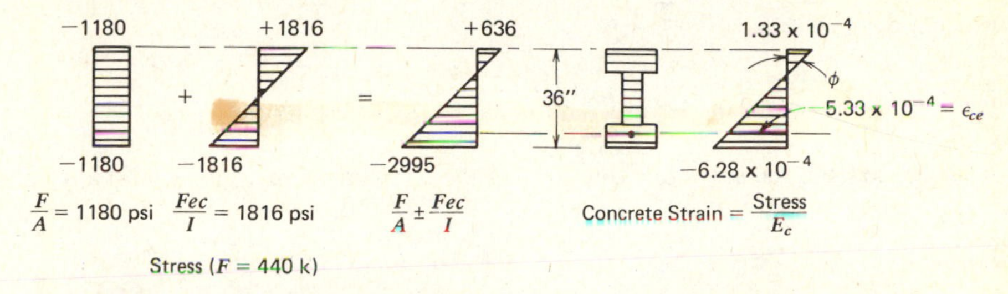

Fig. 5-26(*a*). Initial stage, example 5-10.

$$\phi = \text{curvature (slope of Fig. 5-26}(a)\text{ strain gradient)} = \frac{1.33 \times 10^{-4} + 6.28 \times 10^{-4}}{36}$$

$$\phi = -2.11 \times 10^{-5} \text{ rad/in. at } M = 0 \text{ (applied moment)}$$

Find ϵ_{se} = steel strain at $f_{se} = 160$ ksi (1103 N/mm^2)

$$\epsilon_{se} = \frac{160{,}000}{27.5 \times 10^6} = 5.82 \times 10^{-3}$$

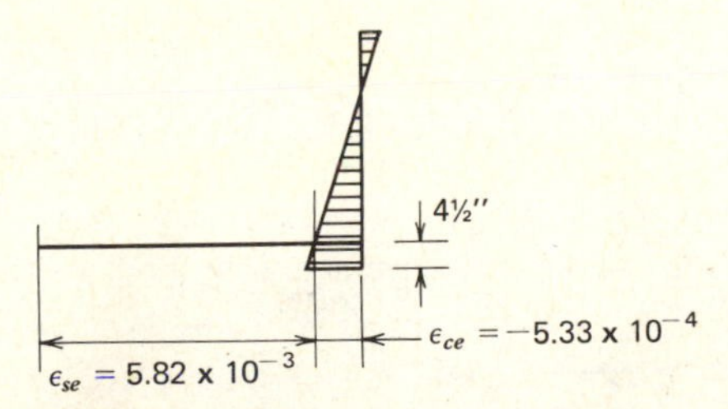

Fig. 5-26(*b*). Initial steel and concrete strains, example 5-10.

(*b*) *Zero strain in concrete at level of steel*: an applied moment which produces $\epsilon_{ce} = 5.33 \times 10^{-4}$ at the level of steel will cause the strain in the concrete to be zero as desired for this step. We note that the same change in strain will occur in the bonded steel, thus the steel strain will become

$$\epsilon_{ps} = \epsilon_{se} + \epsilon_{ce} = 5.82 \times 10^{-3} + 0.533 \times 10^{-3} = 6.35 \times 10^{-3}$$

$$f_{ps} = \epsilon_{ps} \times E_{ps} = 6.35 \times 10^{-3} \times 27.5 \times 10^{3} = 175 \text{ ksi } (1207 \text{ N/mm}^2)$$

$$F = (2.75)(175) = 481 \text{ k } (2139 \text{ kN})$$

Thus the effective stress increases from that in (a), and we can find the concrete stresses which result from this increased force as shown in Fig. 5-26(c).

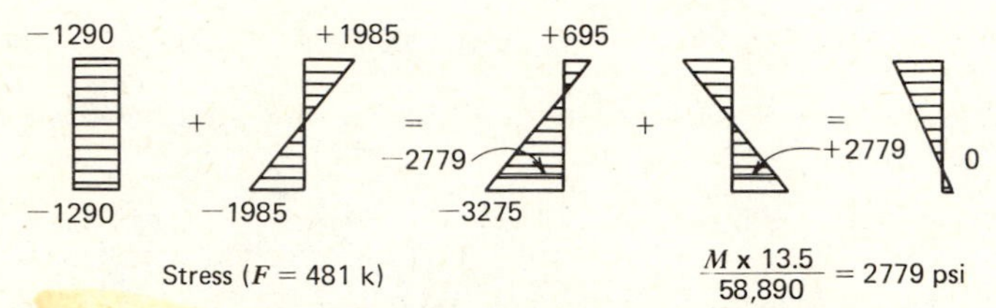

Fig. 5-26(c). Stresses for example 5-10 at stage (b).

Solving M from the stress to reduce to zero the combined stress (and strain) at level of steel as shown in Fig. 5-26(c) and using 2779 psi = 2.779 ksi:

$$M = \frac{2.779 \times 58{,}890}{13.5} = 12{,}120 \text{ in.-k } (1370 \text{ kN} - \text{m})$$

Complete the combined stress sketch with this M acting, and find corresponding strains (stress/E_c) to allow ϕ to be solved as shown in Fig. 5-26(d).

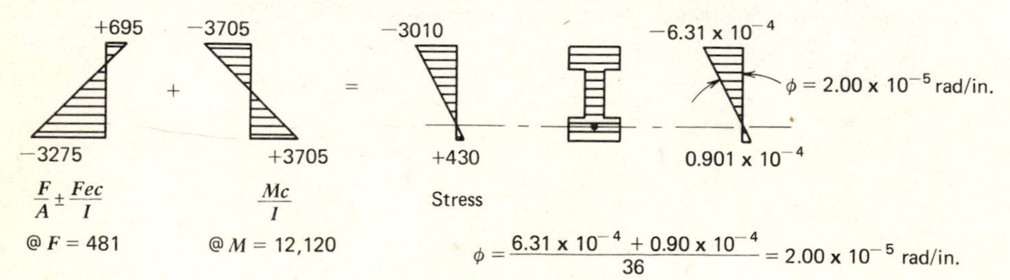

Fig. 5-26(d). Solving curvature at stage (b), example 5-10.

From these calculations we have determined a second point in the elastic range of behavior:

$$M = 12{,}120 \text{ in.-k } (1{,}370 \text{ kN} - \text{m}).$$

$$\phi = +2.00 \times 10^{-5}$$

(c) The two previous points will be used to establish the linear elastic response of the moment-curvature relationship, but we wish now to estimate the cracking moment which is the endpoint for this uncracked section analysis. M_{cr} is associated with $f_r = 7.5\sqrt{f'_c}$, the modulus of rupture for concrete.

$$f_r = 7.5\sqrt{7000} = 627 \text{ psi } (4.32 \text{ N/mm}^2)$$

Since the bottom fiber stress has 430 psi (2.96 N/mm^2) tension stress at stage (*b*) above, we can see that it will require only a rather small moment additional to pick up the additional tension stress

$$627 - 430 = 197 \text{ psi}$$
$$= 0.197 \text{ ksi } (1.36 \text{ N/mm}^2)$$

$$\Delta M = \frac{\Delta fI}{c} = \frac{(0.197)(58{,}890)}{18} = 645 \text{ in.-k } (73 \text{ kN} - \text{m})$$

$$M_{cr} = 12{,}120 + 645 = 12{,}765 \text{ in.-k } (1443 \text{ kN} - \text{m})$$

The very slight additional tension strain in the steel which accompanies this moment could be neglected. We can determine it rather easily however since

$$\Delta f_{ps} = n\frac{\Delta My}{I}$$

where y = distance to c.g.s. = 13.5 in.

$$\Delta f_{ps} = \frac{27.5}{4.77} \times \frac{645 \times 13.5}{58{,}890} = 0.85 \text{ ksi } (5.86 \text{ N/mm}^2)$$

The steel stress at M_{cr} is

$$f_{ps} = 175 + 0.85 \cong 176 \text{ ksi } (1214 \text{ N/mm}^2)$$

We can also evaluate the additional curvature, which must be associated with the additional 197 psi (1.36 N/mm^2) extreme fiber stress, Fig. 5-26(*e*), and obtain the cracking curvature, ϕ_{cr}.

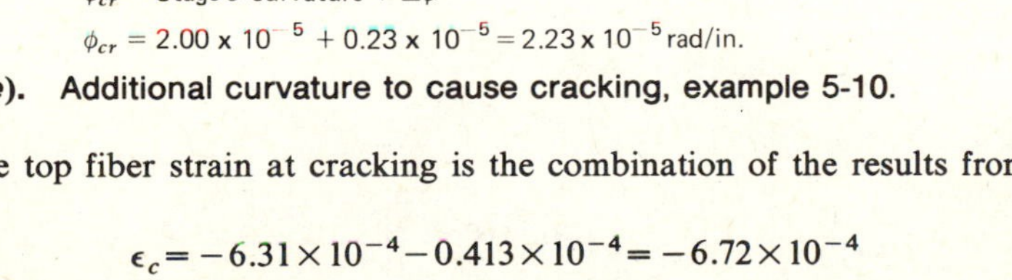

ϕ_{cr} = Stage *b* curvature + $\Delta\phi$

$\phi_{cr} = 2.00 \times 10^{-5} + 0.23 \times 10^{-5} = 2.23 \times 10^{-5}$ rad/in.

Fig. 5-26(*e*). Additional curvature to cause cracking, example 5-10.

(*d*) The top fiber strain at cracking is the combination of the results from (*b*) and (*c*):

$$\epsilon_c = -6.31 \times 10^{-4} - 0.413 \times 10^{-4} = -6.72 \times 10^{-4}$$
$$\epsilon_c = -0.000672 \text{ in/in.} < -0.001 \text{ in./in.}$$

We know that the cracked section analysis is valid for the next point asked for where $\epsilon_c = 0.001$.

The step-by-step procedure of Fig. 5-24 will now be followed for $\epsilon_c = 0.001$ (top fiber strain). The expression from Fig. 5-21(c) for resultant compression force and its point of action will be used in this solution. The stress-strain curve for the steel, Fig. 5-25, is used as well as the prior strain in the steel at stage (b) when zero strain existed in the concrete at level of steel. From (b) we know that

$$\epsilon_{se} + \epsilon_{ce} = 6.35 \times 10^{-3}$$

The first trial-neutral axis assumed 12 in. (304.8 mm) below the top fiber as shown in Fig. 5-26(f).

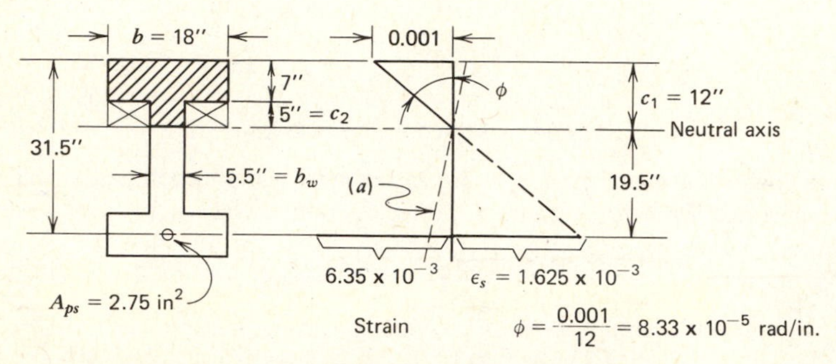

Fig. 5-26(f). First trial stage (d), example 5-10.

From the sketch above we note the neutral axis falls in the web, thus the compression zone is not rectangular as assumed in the derivation of the expression for resultant compression in Fig. 5-21(c). We will first assume the width $b = 18$ in. (457.2 mm) extends to the neutral axis as shown on the sketch and solve the resultant compressive force C_1. Next we will correct this by using the width $(b - b_w)$ for the 5 in. (127 mm) above the neutral axis as shown blocked out on the sketch. This C_2 force will be given a negative sign, and the resultant compression C_c will be the algebraic sum of C_1 and C_2.

$$C_c = bc^2 f'_c \frac{\phi}{\epsilon_0}\left[1 - \frac{\phi c}{3\epsilon_0}\right] \tag{5-29}$$

where $\epsilon_0 = 0.00248 = 2.48 \times 10^{-3}$ in./in. from Fig. 5-26 gives secant modulus of elasticity assumed in previous parts.

$$C_{c_1} = (18)(12)^2(7000)\frac{8.33 \times 10^{-5}}{2.48 \times 10^{-3}}\left[1 - \frac{(8.33 \times 10^{-5})(12)}{(3)(2.48 \times 10^{-3})}\right] \tag{5-29}$$

$$C_{c_1} = 527{,}500 \text{ lb (2346 kN)}$$

Using $b = 18 - 5.5 = 12.5$ in. (317.5 mm) for C_{c_2} correction:

$$C_{c_2} = -(12.5)(5)^2(7000)\left(\frac{8.33 \times 10^{-5}}{2.48 \times 10^{-3}}\right)\left[1 - \frac{(8.33 \times 10^{-5})(5)}{(3)(2.48 \times 10^{-3})}\right] \tag{5-29}$$

$$C_{c_2} = -69{,}360 \text{ lb } (-308.5 \text{ kN})$$

$$C_c = 527{,}500 - 69{,}360 = 458{,}140 \text{ lb} = 458 \text{ k (2,037 kN)}$$

From Fig. 5.26(f) find strain in the prestressing steel

$$\epsilon_{ps} = 6.35 \times 10^{-3} + 1.625 \times 10^{-3} = 7.98 \times 10^{-3}$$

From Fig. 5-25 we determine stress corresponding to this strain to be $f_{ps} = 218$ ksi (1503 N/mm²). For $A_{ps} = 2.75$ in.² (1774 mm²) the tension is

$$T = (2.75)(218) = 600 \text{ k } (2669 \text{ kN}) > C_c = 458 \text{ k } (2037 \text{ kN})$$

The neutral axis is too high, resulting in a steel strain which is too high, making $T > C_c$. For the second trial make the assumed distance to the neutral axis larger.

Second trial—neutral axis assumed 16.5 in. (419.1 mm) below top fiber as shown in Fig. 5-26(g).

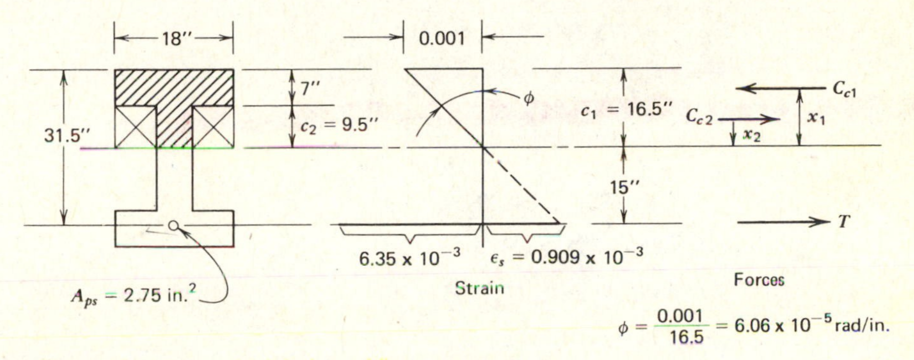

Fig. 5-26(g). Second trial stage (d), example 5-10.

$$C_{c_1} = (18)(16.5)^2(7000)\left(\frac{6.06 \times 10^{-5}}{2.48 \times 10^{-3}}\right)\left[1 - \frac{(6.06 \times 10^{-5})(16.5)}{(3)(2.48 \times 10^{-3})}\right] \qquad (5\text{-}29)$$

$C_{c_1} = 725{,}700$ lb (3228 kN)

$$C_{c_2} = -(12.5)(9.5)^2(7000)\left(\frac{6.06 \times 10^{-5}}{2.48 \times 10^{-3}}\right)\left[1 - \frac{(6.06 \times 10^{-5})(9.5)}{(3)(2.48 \times 10^{-3})}\right] \qquad (5\text{-}29)$$

$C_{c_2} = -178{,}060$ lb (-792 kN)

$C_c = 725{,}700 - 178{,}060 = 547{,}640$ lb $= 548$ k (2436 kN)

From Fig. 5-26(g) find strain in prestressing steel

$$\epsilon_{ps} = 6.35 \times 10^{-3} + 0.909 \times 10^{-3} = 7.26 \times 10^{-3}$$

From Fig. 5-25 we determine the steel stress, f_{ps}, corresponding to this strain to be 200 ksi (1379 N/mm²) and $A_{ps} = 2.75$ in.² (1774 mm²) is known, thus

$$T = (2.75)(200) = 550 \text{ k } (2{,}225 \text{ kN}) \cong 548 \text{ k } (2436 \text{ kN}) = C_c$$

Referring to Fig. 5-26(g) and using expression (5-30) from Fig. 5-21(c) to locate resultant C_c forces:

$$\bar{x}=c\left[\frac{8\epsilon_0-3\phi c}{12\epsilon_0-4\phi c}\right] \text{ measured from neutral axis} \tag{5-30}$$

$$\bar{x}_1=16.5\left[\frac{(8)(2.48\times10^{-3})-(3)(6.06\times10^{-5})(16.5)}{(12)(2.48\times10^{-3})-(4)(6.06\times10^{-5})(16.5)}\right]=10.8 \text{ in. (274 mm)}$$

$$\bar{x}_2=9.5\left[\frac{(8)(2.48\times10^{-3})-(3)(6.06\times10^{-5})(9.5)}{(12)(2.48\times10^{-3})-(4)(6.06\times10^{-5})(9.5)}\right]=6.25 \text{ in. (159 mm)}$$

Summing moments about the tension steel location

$$M=C_{c1}(15+\bar{x}_1)-C_{c_2}(15+\bar{x}_2)$$

$$=(725.7)(15+10.8)-(178.1)(15+6.25)$$

$$M=14{,}940 \text{ in.-k } (1{,}688 \text{ kN}-\text{m}) \text{ at } \phi=6.06\times10^{-5}$$

(e) With top fiber strain = 0.002 the solution is carried out as above. The results are

$c=10$ in. (254 mm)—top fiber to neutral axis

$C_c=T=688$ k (3060 kN)

$f_{ps}=250$ ksi (1724 N/mm^2)

$M=19{,}200$ in.-k. (2170 kN − m)

$\phi=20.0\times10^{-5}$ rad/in.

(f) With top fiber strain = 0.003 the solution will be shown. This is the ultimate moment corresponding to the limiting strain specified by the ACI Code.

Try $c=8$ in. (203.2 mm) with top fiber strain 0.003 as shown in Fig. 5-26(h):

$$C_{c_1}=(18)(8)^2(7000)\left(\frac{37.5\times10^{-5}}{2.48\times10^{-3}}\right)\left[1-\frac{(3.75\times10^{-5})(8)}{(3)(2.48\times10^{-3})}\right]$$

$$C_{c_1}=727{,}000 \text{ lb (3234 kN)} \tag{5-29}$$

$$C_{c_2}=-(12.5)(1)^2(7000)\left(\frac{37.5\times10^{-5}}{2.48\times10^{-3}}\right)\left[1-\frac{(3.75\times10^{-5})(1)}{(3)(2.48\times10^{-3})}\right]$$

$$C_{c_2}=-12{,}500 \text{ lb } (-55.6 \text{ kN}) \tag{5-29}$$

$$C_c=727{,}000-12{,}500=714{,}500 \text{ lb}=715 \text{ k (3178 kN)}$$

From Fig. 5-26(h) find strain in the prestressing steel

$$\epsilon_{ps}=6.35\times10^{-3}+8.81\times10^{-3}=15.16\times10^{-3}$$

From Fig. 5-25 we find the stress corresponding to this strain

$$f_{ps}=264 \text{ ksi } (1820 \text{ N/mm}^2)$$

$$T=(2.75)(264)=726 \text{ k } (3{,}229 \text{ kN}) \text{ vs. } C_c=715 \text{ k (3178 kN)}$$

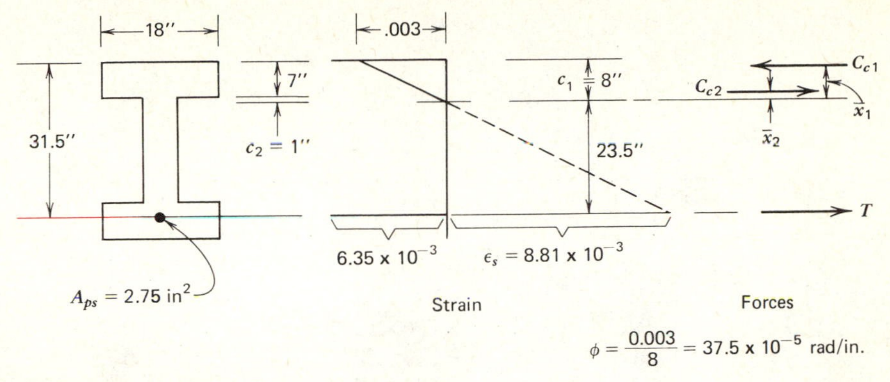

Fig. 5-26(*h*). Stage (f), example 5-10.

This is acceptable within 1.5%.

$$\bar{x}_1 = 8\left[\frac{(8)(2.48\times10^{-3})-(3)(37.5\times10^{-5})(8)}{(12)(2.48\times10^{-3})-(4)(37.5\times10^{-5})(8)}\right]$$

$$\bar{x}_1 = 4.86 \text{ in. } (123.4 \text{ mm}) \tag{5-30}$$

$$\bar{x}_2 = 1\left[\frac{(1)(2.48\times10^{-3})-(3)(37.5\times10^{-5})(1)}{(12)(2.48\times10^{-3})-(4)(37.5\times10^{-5})(1)}\right]$$

$$\bar{x}_2 = 0.66 \text{ in. } (16.8 \text{ mm}) \tag{5-30}$$

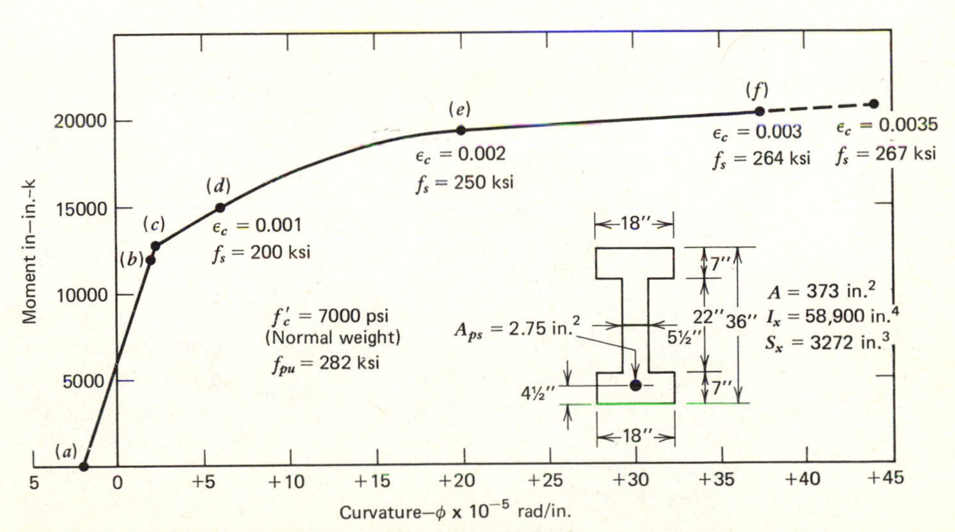

Fig. 5-27. Moment-curvature results, example 5-10.

Summing moments about the tension steel:

$$M' = (727)(23.5 + 4.86) - (12.5)(23.5 + 0.66)$$

$$M' = 20{,}320 \text{ in.-k } (2296 \text{ kN} - \text{m}) \text{ at } \phi = 37.5 \times 10^{-5}$$

Summary of results for example 5-10 (plotted Fig. 5-27):

Stage	**Moment** (in.-k)	**Curvature** (rad/in.)	**Steel Stress** (ksi)
(*a*)	0	-2.11×10^{-5}	160
(*b*)	12,120	$+2.00 \times 10^{-5}$	175
(*c*)	12,765	$+2.23 \times 10^{-5}$	176
(*d*)	14,940	$+6.06 \times 10^{-5}$	200
(*e*)	19,200	$+20.0 \times 10^{-5}$	250
(*f*)	20,320	$+37.5 \times 10^{-5}$	264

The use of the actual test curve results in $M' = 20{,}320$ in.-k (2,296 kN − m) as the ultimate moment in example 5-10. We must make two observations in order to compare this result with the ACI Code M_u for this same section in example 5-8. First, the estimate in example 5-8 was made with $f_{pu} = 270$ ksi (1862 N/mm^2), the guaranteed strength rather than $f_{pu} = 282$ ksi (1944 N/mm^2), the actual strength in example 5-10 (Fig. 5-25, stress-strain curve). Second, the M_u in example 5-8 was that for ultimate design moment under the ACI Code, $M_u = \phi M' = 0.9M'$. We can make corrections for both factors and then compare M_u values.

We will estimate that the ultimate moment is increased in proportion to f_{pu} for example 5-10. Making this correction we have

$$M' = \left(\frac{270}{282}\right)(20{,}320) = 19{,}460 \text{ in.-k } (2199 \text{ kN} - \text{m})$$

$$M_u = (0.9)(19{,}460) = \underset{\text{(example 5-10)}}{17{,}510 \text{ in.-k } (1979 \text{ kN} - \text{m})} \text{ vs. } \underset{\text{(example 5-8)}}{17{,}220 \text{ in.-k } (1944 \text{ kN} - \text{m})}$$

This excellent agreement indicates that the ACI Code estimate is certainly adequate for analysis when performing strength design.

The moment-curvature analysis allows the total range of performance to be examined, and the last point on Fig. 5-27 is the ultimate moment. We will find this total curve useful later in making estimates of ultimate deflection. It is generally true, however, that the load-deflection curve for a beam will have the same form as the $M - \phi$ curve. Thus, the curve of Fig. 5-27 indicates ductile behavior which we want in structural design. The curvature at the ultimate moment is much larger than that at cracking, and most designs of prestressed concrete beams limit stress at service load to keep the beam uncracked.

Another observation from the moment vs. curvature curve of Fig. 5-27 is that the ratio of ultimate moment to cracking moment is 1.59. The ACI Code requires that this ratio be at least 1.2 to insure that a beam will not have the possibility of collapse at the instant of cracking. Increase in strength after cracking as observed in Fig. 5-27 is the behavior we desire in the design of prestressed concrete beams.

The versatility and reliability of the moment-curvature analysis are apparent in the observed test results shown compared to calculated response shown in Fig. 5-28. Note that this curve is for a member with both prestressed strands and unstressed reinforcing bars. The procedure for finding moment and corresponding curvature for points defining the M vs. ϕ curve for this beam would be done in the step-by-step manner described in example 5-10.

The only special consideration here which was not a part of the previous example is illustrated by Fig. 5-29, which shows the internal forces. Note that the two types of steel are handled separately, and the difference in their strain histories is taken into account in the analysis. The force T'_{ps} in the prestressed strands would be calculated from the area A_{ps} for the strands and the stress for the strand material which corresponds to the total strain $(\epsilon_{ce} + \epsilon_{se} + \epsilon_{si})$, Fig.

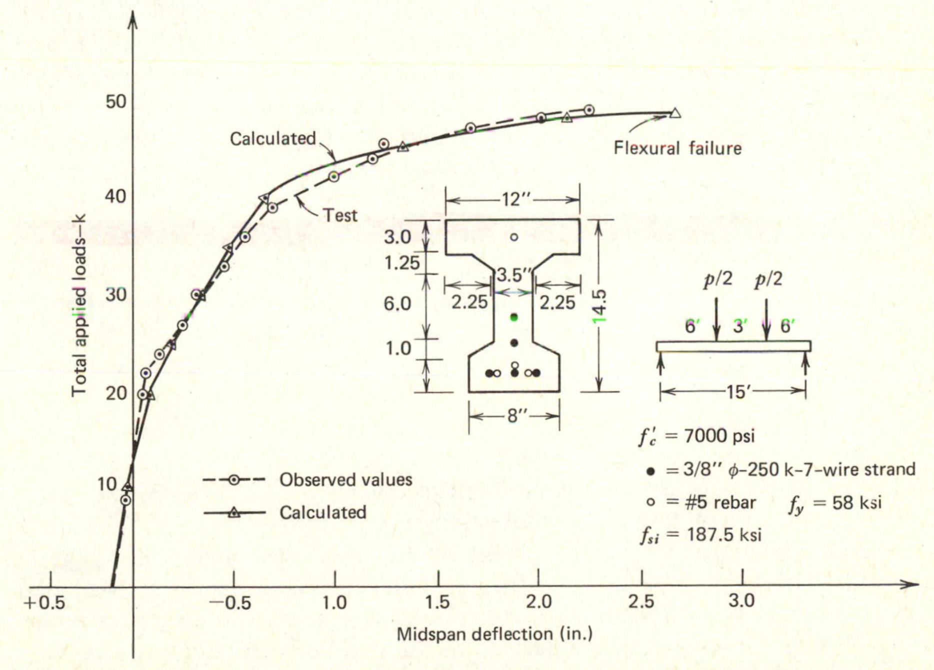

Fig. 5-28. Comparison of load-deflection observed (test) and calculated (analysis using $M-\phi$ curve).

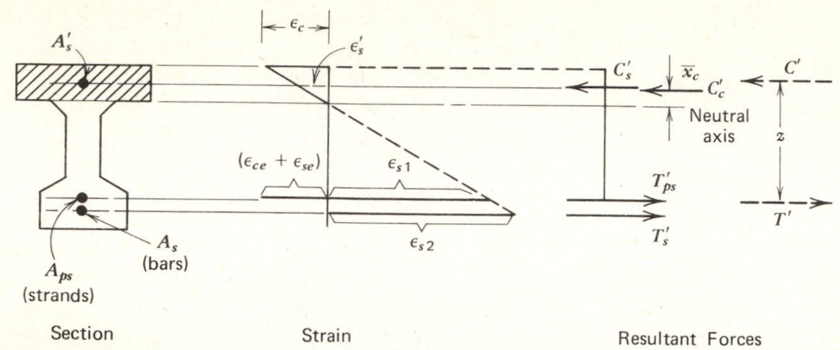

Fig. 5-29. Forces acting for beam of Fig. 5-28.

5-29. The force T_s' in the unstressed bars, A_s, would be calculated from the area A_s and the stress for the reinforcing bar material which corresponds to the strain ϵ_{s_2}, Fig. 5-29. These two tension forces combine to form the total resultant T', which must be equal to the total resultant compression C', and the moment M' would be $M' = T'z = C'z$.

The compression force C_c' in the concrete would be computed for an assumed neutral axis position and top fiber strain, ϵ_c, as was done in example 5-10. If the neutral axis were in the web, this calculation might be done in two parts as shown previously with C_c' being the resultant compression force acting at a distance $\bar{x}_c$ above the neutral axis, Fig. 5-29. The compression steel reinforcing bar, A_s', has a stress corresponding to ϵ_s' from the material stress-strain curve; thus the force C_s' may be solved. The total compression C' is the resultant of the steel and concrete compressive forces, Fig. 5-29.

Fig. 5-28 shows the load-deflection response which has the same form as the $M-\phi$ relationship for the cross section. The deflection for load levels after cracking must be computed by utilizing the known moment diagram for the simple beam together with the $M-\phi$ relationship solved by analysis of the cross section. Figure 5-30 shows the changing form of the distribution of curvature along the span at various load levels. As load approaches ultimate, note that the ultimate curvature, ϕ, at midspan is much larger than the curvature at this section at cracking. As shown by the load-deflection curve of Fig. 5-28, the deflection at ultimate is much larger than at cracking. The deflection is calculated from the distribution of curvature along the span, shown in Fig. 5-30 as the ϕ diagram. We must sum the moment about A of area under the diagram between A and B (shaded in Fig. 5-30) to obtain the deflection at B. Note that this calculation would reflect the large contribution to deflection which results from the large curvatures which develop in the middle portion of the span with flexural cracking. The end regions of the beam remain uncracked in flexure (Fig. 5-30) since the moment is less than the cracking moment in these regions, and

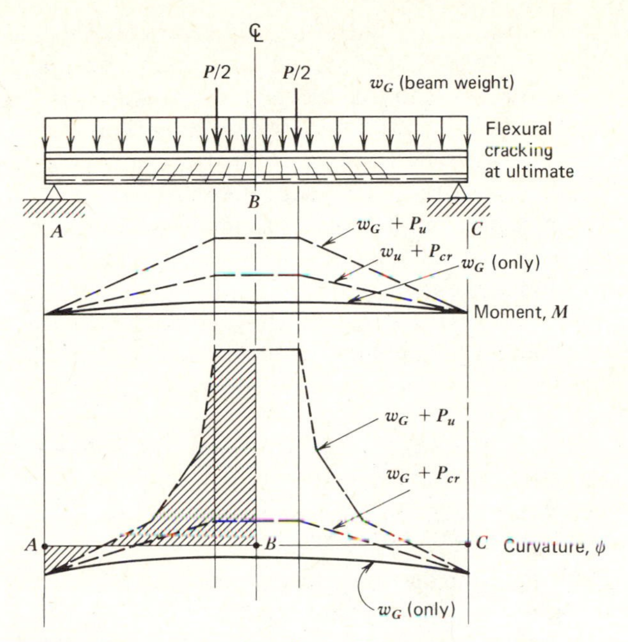

Fig. 5-30. Moment and curvature at various load levels.

they contribute insignificantly to the deflection after extensive cracking has developed.

More will be said about deflections in Chapter 8, but this tie between curvature along the span and the resulting deflection should be thought of as a part of analysis for flexure. As discussed here, the flexural behavior of a simple beam with bonded reinforcement may be analyzed for the whole range of applied load. This analysis may be coded for the computer, but only a few points from hand calculation can give reasonable accuracy. Normally, we want the character of the response and not an exact prediction of ultimate deflection.

5-8 Ultimate Moment—Unbonded Beams

An accurate calculation for the ultimate strength of unbonded beams is more difficult than for that of bonded ones, because the stress in the steel at rupture of the beam cannot be closely computed. Also there have not been sufficient data on the ultimate strength of unbonded beams to establish definitely a reliable method of computation. It is agreed, however, that unbonded beams are weaker than the corresponding bonded ones in their ultimate strength, the difference being placed at 10–30%.

Explanations can be offered for the lower strength of unbonded beams. First, since the tendon is free to slip, the strain in a tendon is more or less equalized along its length, and the strain at the critical section is lessened. Hence the stress in the tendon is increased only slowly so that, when the crushing strain has been reached in the concrete, stress in the steel is often far below its ultimate strength. When there are no cracks in the beam, stress in steel can be computed as in solution 2, example 5-6. As soon as part of the beam cracks or is stretched into the plastic range, the stress cannot be conveniently calculated. For the purpose of design, however, it may be possible to estimate the stress in the steel at the rupture of the beam and to compute the corresponding lever arm so as to approximate the ultimate resisting moment. Until further test data are available, such estimation may often err by 10–15%. Fortunately, unbonded beams are not often used where ultimate strength is a controlling factor, and they are generally designed for the working loads by the elastic theory rather than for the ultimate load.

Another reason for the lower ultimate strength of unbonded beams is the appearance of a few large cracks in the concrete instead of many small ones well distributed. Such wide cracks tend to concentrate the strains in the concrete at these sections, thus resulting in early failure.

Tests prove that the ultimate strength of unbonded beams can be materially increased by the addition of nonprestressed steel. Such increase is attributed to the resistance of the nonprestressed steel itself as well as to its effect in distributing and limiting the cracks in the concrete. This will be discussed in Chapter 11. The ACI Code requires minimum amounts of bonded reinforcement to assure that cracking will be distributed along the span rather than allowing only one or two cracks in the unbonded member at ultimate.

A general formula for f_{ps}, the stress in steel at ultimate load, in an unbonded beam is

$$f_{ps} = f_{se} + \Delta f_s$$

where f_{se} is the effective prestress in the steel, and Δf_s is the additional stress in the steel produced as a result of beam bending up to the ultimate load. Tests at the University of Illinois (Fig. 117, reference 6) showed Δf_s varying from about 10,000 to 80,000 psi (68.95–551.6 N/mm^2); those at the Portland Cement Association showed Δf_s between 40,000 and 60,000 psi (275.8 and 413.7 N/mm^2) (p. 615, reference 5). Limited tests at the University of California indicated Δf_s varying from 30,000 psi to 80,000 psi (206.85–551.6 N/mm^2), with the higher values of Δf_s occurring for the curved tendons, where frictional force probably restricted the free slipping of the wires, and for the beams which had a sizable amount of nonprestressed steel. Test results from simple and continuous beams at the University of Texas at Austin and the University of Washington show similar results.

Tests on simple and continuous beams by Mattock[11,12] at the University of Washington resulted in the correlation of f_{ps} as shown in Fig. 5-31. The

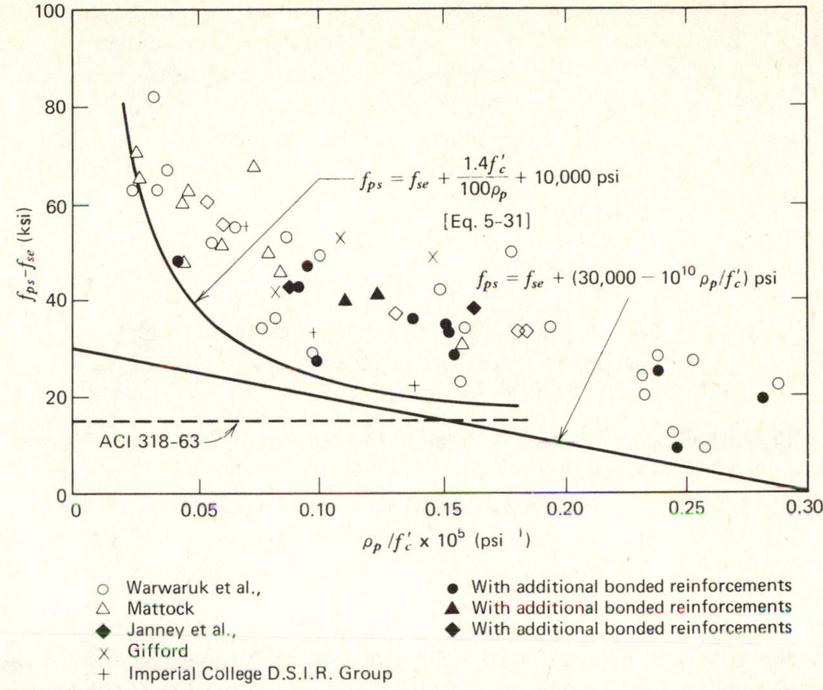

Fig. 5-31. Posttensioned beams without bond, increase in tendon stress during loading to ultimate.[12]

recommended equation for unbonded members which resulted from beam tests was as follows:

$$f_{ps}=f_{se}+10{,}000+\frac{1.4f'_c}{100\rho_p} \tag{5-31}$$

The equation was slightly modified by ACI-ASCE Committee #423 to make it slightly more conservative, giving the following ACI Code equation:

$$f_{ps}=f_{se}+10{,}000+\frac{f'_c}{100\rho_p} \tag{5-32}$$

where

$$f_{ps}\leqslant f_{py}$$

$$f_{ps}\leqslant f_{se}+60{,}000$$

In general, simple beams with unbonded tendons would be conservatively analyzed by this expression for f_{ps}. Shallow slabs (span/depth = 45) have been observed to develop slightly less than this f_{ps}, but the error is not significant. The geometry of the tendon layout enters into the elongation which will develop in the unbonded tendon. Almost no increase above f_{se} will result until after cracking. The Δf_s increase will develop as deflections become large, and equation 5-30 is adequate for analysis requiring the ultimate steel stress.

The analysis for ultimate moment M' for an unbonded beam would proceed as outlined above using the simplified ACI Code equation (5-30) for f_{ps}. The minimum bonded reinforcement which must be included for beams and one-way slabs to distribute cracking and also contribute to strength is given by the following expression:

$$A_s = 0.004A \qquad (5\text{-}31)$$

where

A = area of that part of the cross section between the flexural tension face and the center of gravity of the cross section

The steel which supplies this A_s acts at the specified yield point for the bars supplied, but not over 60,000 psi (414 N/mm^2) may be used. Tests show that these bars will yield as assumed in analysis. Ample ductility is assured by the limitation on ω as required by the ACI Code, which assures that beams will be under-reinforced.

EXAMPLE 5-11

Assume the beam of Example 5-8 is provided with unbonded tendons but is otherwise the same as shown in Fig. 5-17. How much bonded reinforcement must be provided to satisfy ACI Code requirements? What is the estimated ultimate moment capacity of the section with $A_{ps} = 2.75$ in.^{2} and minimum bonded reinforcement supplied with deformed bars having $f_y = 60$ ksi? Assume $f_{se} = 160$ ksi for the unbonded tendons, and $f_{py} = 230$ ksi; $f_c' = 7000$ psi as before. ($A_{ps} = 1.774$ mm^2, $f_y = 414$ N/mm^2, $f_{se} = 1{,}103$ N/mm^2, $f_{py} = 1{,}586$ N/mm^2, and $f_c' = 48$ N/mm^2).

Solution

$$\rho_p = 0.00485 \text{ (same as example 5-8)}$$

for analysis of the beam use equation (5-31) here rather than the ACI Code equation (5-32) which is used for design.

$$f_{ps} = f_{se} + 10{,}000 + \frac{1.4 f_c'}{100\rho_p}$$

$$= 160{,}000 + 10{,}000 + \frac{(1.4)(7000)}{(100)(0.00485)}$$

$$f_{ps} = 160{,}000 + 10{,}000 + 20{,}200 = 190{,}200 \text{ psi } (1311 \text{ N/mm}^2) < f_{py}$$

$$= 230{,}000 \text{ psi } (1586 \text{ N/mm}^2)$$

$$\Delta f_s = 10{,}000 + 20{,}200 = 30{,}200 \text{ psi } (208 \text{ N/mm}^2) < 60{,}000 \text{ psi } (414 \text{ N/mm}^2)$$

Bonded reinforcement required by ACI Code:

$$A_s = 0.004A \quad (5\text{-}31)$$

$A=$ half of cross section area since neutral axis is at middepth of symmetrical I-shaped beam

$$A=(18-5.5)(7)+(5.5)(18)=87.5+99=186.5 \text{ in.}^2 \ (120\times10^3 \text{ mm}^2)$$

$$A_s=(0.004)(186.5)=0.75 \text{ in.}^2 \ (480 \text{ mm}^2) \quad (5\text{-}31)$$

Use four #4 bars—$A_s=0.80$ in.^{2} (516 mm^2)

For this problem assume this steel is placed 2 in. (50.8 mm) above bottom of section as shown in Fig. 5-32.

$$T_p'=f_{ps}A_{ps}=(190.2)(2.75)=523 \text{ k } (2326 \text{ kN})$$

$$T_s'=f_yA_s=(60.0)(0.80)=48 \text{ k } (213.5 \text{ kN})$$

$$T'(\text{total})=523+48=571 \text{ k } (2540 \text{ kN})$$

$$g=\frac{(523)(4.5)+(48)(2)}{571}=4.29 \text{ in. } (109 \text{ mm})$$

$$T'=C'=571 \text{ k}=(0.85)(f_c')(b)(a)$$

$$a=\frac{571}{(0.85)(7.0)(18)}=5.33 \text{ in. } (135 \text{ mm})$$

$$z=36-\frac{5.33}{2}-4.29=29.04 \text{ in. } (738 \text{ mm})$$

$$M'=T'z=(571)(29.04)=16{,}580 \text{ in.-k } (1.874 \text{ kN}-\text{m})$$

The ultimate moment is less than $M'=19{,}100$ in.-k for the bonded beam of example 5-8 as anticipated. If the ACI Code were followed, the steel stress for the unbonded tendon from equation 5-30 is a little less ($f_{ps}=184$ ksi), and this results in slightly smaller $M'=16{,}130$ in.-k. In design this would be further reduced by the strength reduction

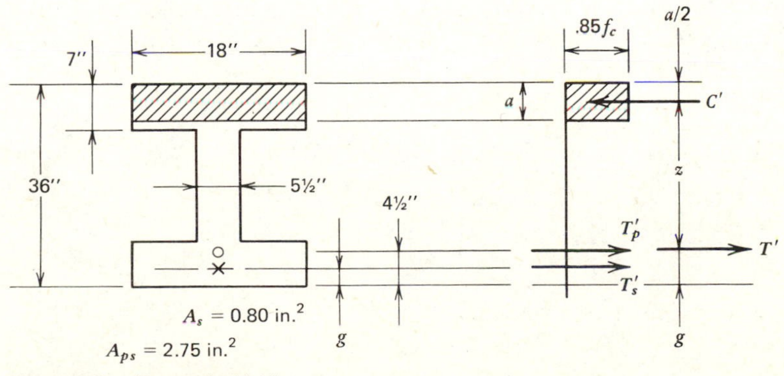

Fig. 5-32. Example 5-11.

factor $\phi=0.9$. This beam would have $M_u=(0.9)\ (16{,}130)=14{,}520$ in.-k following the ACI Code for design. Additional A_{ps} or A_s could be used to increase this to the $M_u=17{,}200$ in.-k found for the bonded beam in example 5-8 if the design loads required this strength.

5-9 Composite Sections

In prestressed-concrete construction it is often advantageous to precast part of a section (either by pretensioning for by posttensioning), lift it to position, and cast the remainder of the section in place. The precast and cast-in-place portions thus act together (with stirrups if necessary) and form a composite section. Members of composite sections laid side by side may be eventually connected together by transverse prestressing, while such members laid end to end may be further prestressed longitudinally in order to attain continuity. These points will be discussed in later chapters. We shall describe here the basic method of analysis commonly employed for such composite sections.

Figure 5-33 shows a composite section at the midspan of simply supported beam, whose lower stem is precast and lifted into position with the top slab cast in place resting directly on the stem. If no temporary intermediate support is furnished, the weight of both the slab and the stem will be carried by the stem acting alone. After the slab concrete has hardened, the composite section will carry any live or dead load that may be added on to it.

In the same figure, stress distributions are shown for various stages of loading. These are discussed as follows.

(*a*) Owning to the initial prestress and the weight of the stem, there will be heavy compression in the lower fibers and possibly some small tension in the top fibers. The tensile force T in the steel and the compressive force C in the concrete form a resisting couple with a small lever arm between them.

(*b*) After losses have taken place in the prestress, the effective prestress together with the weight of the stem will result in a slightly lower compression in the bottom fibers and some small tension or compression in the top fibers. The C-T couple will act with a slightly greater level arm.

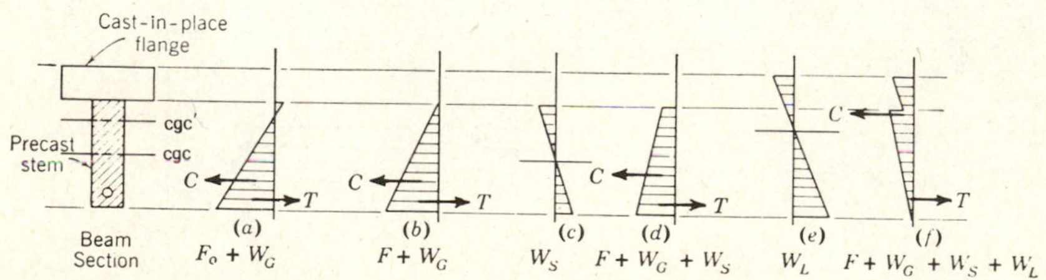

Fig. 5-33. Stress distribution for a composite section.

(*c*) Owning to the addition of the slab, its weight produces additional moment and stress as shown.
(*d*) Owning to the effective prestress plus the weight of the stem and slab, we can add (*b*) to (*c*) and a somewhat smaller compression is found to exist at the bottom fibers and some compression at the top fibers. The lever arm for the *C*-*T* couple further increases.
(*e*) Stresses resulting from live load moment are shown, the moment being resisted by the composite section.
(*f*) Adding (*d*) to (*e*), we have stress block as in (*f*), with slight tension or compression in the bottom fibers, but with high compressive stresses in the top fibers of the stem and the slab. The couple *T* and *C* now acts with an appreciable lever arm.

The above shows the stress distribution under working load conditions. For overloads, the stress distributions are shown in Fig. 5-34. For the load producing first cracks, it is assumed that the lower fibers reach a tensile stress equal to the modulus of rupture. This is obtained when the live-load stresses shown in Fig. 5-33(*e*) are big enough to result in a stress distribution as shown in Fig. 5-34(*a*), computed by the elastic theory.

Under the ultimate moment, however, the elastic theory is no longer valid. As an approximation, the ultimate resisting moment is best represented by a tensile force T' computed by the ACI equation for estimating f_{ps} acting with a compressive force C' supplied by the concrete. If failure in bond and shear is prevented, the ultimate strength of a composite section can be estimated by a method similar to that previously described for a simple prestressed section. It must be emphasized, however, that a composite section may fail in horizontal shear between the precast and the cast-in-place portions, unless proper stirrups or connectors are provided.

The above describes a simple case of composite action; there are many possible variations. First, the precast portion may be supported on falsework while the cast-in-place slab is being poured or placed, the falsework being removed only after the hardening of the slab concrete. This will permit the entire composite section to resist the moment produced by the weight of the slab. It is

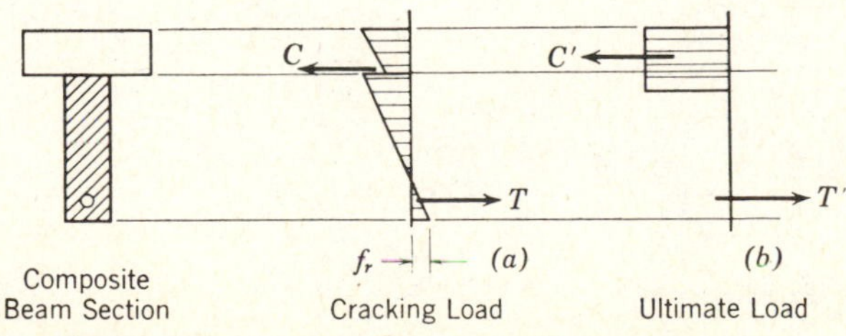

Fig. 5-34. Stress distributions for cracking and ultimate loads.

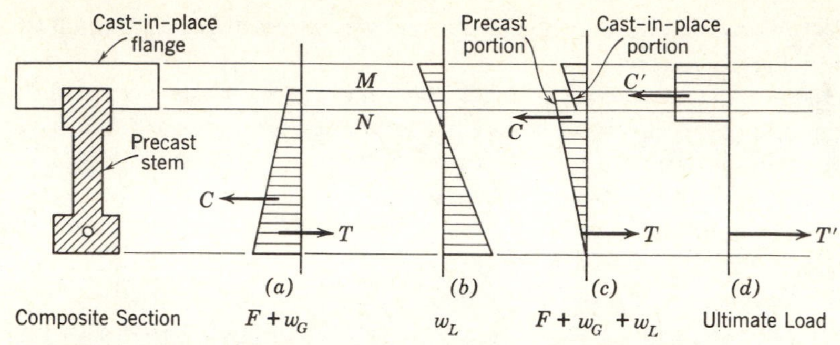

Fig. 5-35. Stress distribution for a special composite section.

also possible to prop up the falsework so that the stem will carry practically no moment by itself. Then the moments due to the weight of the stem will also be carried by the composite section. Since the composite section has a greater section modulus than the stem alone, the resulting stresses will be more favorable. The desirability of such methods depends on the cost of falsework for the particular structure.

Another variation happens when the cast-in-place slab overlaps with the precast portion as shown in Fig. 5-35. Here, the stresses in the concrete between levels M and N will follow two different variations, as shown in (c); one for the precast and another for the cast-in-place portion. At the ultimate range, however, they will all be stressed to the maximum and the difference will be hardly noticeable. Then the section can be analyzed as if it were a simple one, Fig. 5-35(d).

If the precast portion is only a small part of the whole section, it may be prestressed for direct tension only, or with a slight eccentricity of prestress. One method used in England (known as the Udall system), Fig. 5-36, employs both prestressed and nonprestressed wires in the groove of precast blocks, with the major top portion cast in place so as to be well bonded to the wires. For such a construction, high tension may exist in the bottom fibers of the cast-in-place

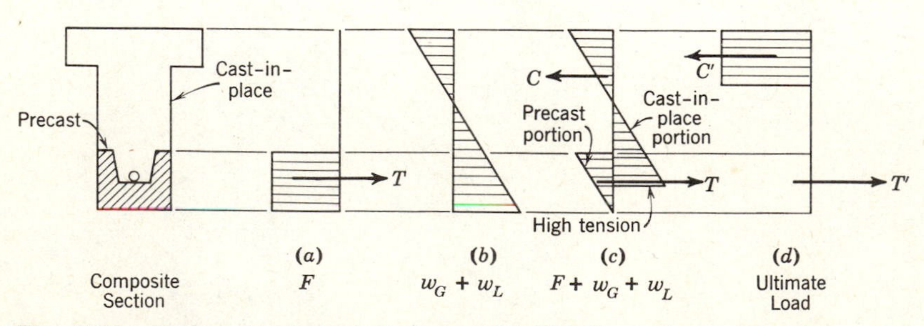

Fig. 5-36. Stress distribution for a special (Udall) composite section.

portion (c), resulting in cracks under working load. But the ultimate strength in flexure is not affected by the tensile stresses (d).

In other instances, the section is prestressed in two stages. Only part of the tendons are prestressed first in order to hold the stem together. The remaining tendons are prestressed after the slab has been cast and has hardened; otherwise the tendons may be partially prestressed first, to be fully prestressed later. If the process of retensioning is not too costly, this may result in an economical design. The stress distribution must be studied for the various stages, but the allowable stresses need not be the same as for an ordinary simple section. In certain instances, considerable tension may be permitted without adversely influencing, performance of the member.

When differential shrinkage and creep between the precast and the in-place portions are considered, high stresses are obtained. The usual practice of neglecting such stresses can be justified on the grounds that the ultimate strength of the section is seldom affected by these stresses. However, the elastic behavior, such as camber and deflection, may be seriously modified. In practice, the in-place portion will have more shrinkage, since shrinkage of the precast portion has mostly taken place; but the precast portion will have more creep because it is usually under higher compression due to prestress. If the higher shrinkage in the in-place portion is just about balanced by the higher creep in the precast portion, it would be possible to neglect both. It often happens, however, that the shrinkage of the in-place portion is more serious, especially when the concrete has a high water-cement ratio. In this case the in-place concrete may crack, or the entire composite member may be forced to deflect downward.

EXAMPLE 5-12

The midspan section of a composite beam is shown in Fig. 5-37. The precast stem 12 in. by 36 in. (305 mm by 915 mm) deep is posttensioned with an initial force of 550 kips (2446 kN), Fig. 5-37(a). The effective prestress after losses is taken as 480 kips (2135 kN). Moment due to the weight of that precast section is 200 k-ft (271.2 kN-m) at midspan. After it is erected in place, the top slab of 6 in. by 36 in. (152 mm by 915 mm) wide is to be cast in place producing a moment of 100 k-ft (135.6 kN-m). After the slab concrete

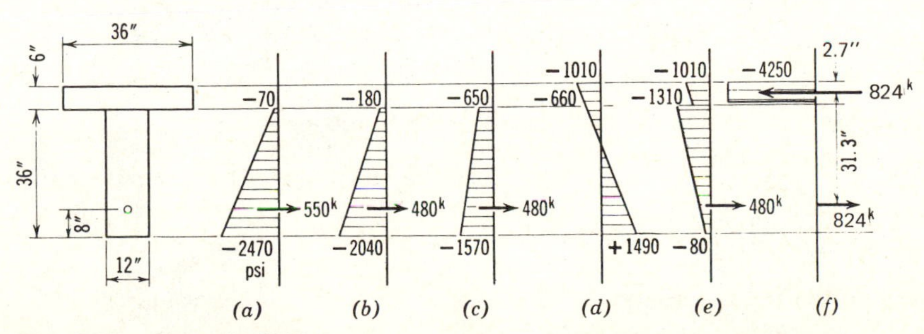

Fig. 5-37. Example 5-12.

has hardened, the composite section is to carry a maximum live load moment of 550 k-ft (745.8 kN-m). Compute stresses in the section at various stages. $A_{ps}=3.7$ sq in. (2387 mm^2) $f_{pu}=240{,}000$ psi (1655 N/mm^2). $f_c'=5000$ psi. (34 N/mm^2) Estimate the ultimate moment.

Solution C.g.c. of the composite section is located at 25 in. (635 mm) from the bottom fiber. The area and moment of inertia of the rectangular and the composite sections are computed and listed below:

	Rectangular Section	Composite Section
Area, in.^{2}	432	648
I, in.^{4}	46,600	111,000

(*a*) Immediately after prestressing, the stresses in the rectangular section will be

$$f=\frac{F}{A}\pm\frac{(M-Fe)c}{I}$$

$$=\frac{-550{,}000}{432}\pm\frac{(200{,}000\times12-550{,}000\times10)18}{46{,}600}$$

$$=-1270\pm1200$$

$$=-70\text{ psi }(-0.48\text{ N/mm}^2)\text{ top fiber}$$

$$=-2470\text{ psi }(-17.03\text{ N/mm}^2)\text{ bottom fiber}$$

(*b*) After loss of prestress, the stresses will be

$$f=\frac{-480{,}000}{432}\pm\frac{(2000{,}000\times12-480{,}000\times10)18}{46{,}600}$$

$$=-1110\pm930$$

$$=-180\text{ psi }(-1.24\text{ N/mm}^2)\text{ top fiber}$$

$$=-2040\text{ psi }(-14.07\text{ N/mm}^2)\text{ bottom fiber}$$

(*c*) After pouring of top slab, the stresses will be

$$f=\frac{-480{,}000}{432}\pm\frac{(300{,}000\times12-480{,}000\times10)18}{46{,}000}$$

$$=-1110\pm460$$

$$=-650\text{ psi }(-4.48\text{ N/mm}^2)\text{ top fiber}$$

$$=-1570\text{ psi }(-10.83\text{ N/mm}^2)\text{ bottom fiber}$$

(*d*) The live load acts on the composite section, producing stresses,

$$f=\frac{-550{,}000\times12\times17}{111{,}000}=-1010\text{ psi }(-6.96\text{ N/mm}^2)\text{ top fiber of composite section}$$

$$f=\frac{550{,}000\times12\times25}{111{,}000}=+1490\text{ psi }(+10.27\text{ N/mm}^2)\text{ bottom fiber}$$

By proportioning, the stress at top fiber of the rectangular portion is found to be -660 psi (-4.55 N/mm^2) due to this live load.

(e) The combined stresses due to prestress and dead and live loads are given in Fig. 5−37(e), which yields −80 psi (−0.55 N/mm²) for bottom fiber and -1310 psi (−9.02 N/mm²) for top fiber of the rectangular section.

(f) The ultimate moment capacity of the section can be estimated as follows.

$$\rho_p = \frac{3.7}{(36)(34)} = 0.00302$$

$$f_{ps} = f_{pu}\left(1 - 0.5\rho_p \frac{f_{pu}}{f_c'}\right) = (240{,}000)\left[1 - (0.5)(0.00302)\left(\frac{240{,}000}{5{,}000}\right)\right]$$

$$f_{ps} = 222{,}600 \text{ psi } (1{,}535 \text{ N/mm}^2)$$

$$\text{Total tensile force at ultimate} = f_{ps}A_{ps}.$$

$$T' = 222{,}600 \times 3.7 = 824{,}000 \text{ lb } (3{,}665 \text{ kN})$$

Area of compression concrete, for an average stress of $0.85f_c' = 4250$ psi (29.30 N/mm²), is

$$\frac{824{,}000}{4250} = 194 \text{ sq in. } (125 \times 10^3 \text{ mm}^2)$$

or a depth of 194/36 = 5.4 in. (137 mm) The center of compressive force is about 5.4/2 = 2.7 in. (68.5 mm) from top; hence the lever arm for the resisting moment is 42 − 2.7 − 8 = 31.3 in. (795 mm), and the ultimate moment capacity is

$$824{,}000 \times 31.3/12{,}000 = 2150 \text{ k-ft } (2915 \text{ kN-m})$$

The total applied dead and live load moment is only 850 k-ft (1153 kN-m), indicating a load factor (factor of safety) of 2150/850 = 2.5.

The ACI Code introduces the strength reduction factor $\phi = 0.9$, which would change the indicated load factor to 0.9 × 2150/850 = 2.3. This is more than adequate for factored loads of $1.4D + 1.7L$ required by the ACI.

5-10 Flexural Behavior and Ultimate Strength at Transfer

While extensive studies have been made, both analytically and experimentally, of the behavior of prestressed-concrete beams under the final conditions of loading, only a limited amount of investigation has been carried out to determine the behavior at transfer.[13] The term "at transfer" is used here in a broad sense to mean that the beam is under little or no external positive moment or under an external negative moment which increases the eccentricity of prestress. Such knowledge is of importance when a beam is subjected to reversal of moments during stressing and erecting operations, or under service conditions. When appreciable cracking may occur, reinforcing bars should be provided to control the cracks or to increase the strength. The design for these bars will be discussed later in Chapter 9.

The elastic stresses due to prestress have been discussed in section 5-2, while those due to external loads were discussed in section 5-3. The combined stresses due to prestress and loading are given by the well-known formula 5-7,

$$f=\frac{F}{A}\pm\frac{Fey}{I}\pm\frac{My}{I}$$

When the moment is negative, acting to increase the eccentricity of prestress, it is only necessary to insert the proper sign for the third term My/I in formula 5-7, as illustrated in example 5-13.

EXAMPLE 5-13

A posttensioned bonded beam with a transfer prestress of $F_t=350$ kips is being wrongly picked up at its midspan point, Fig. 5-38. Compute the critical fiber stresses. Check for cracking; $f_r=7.5\sqrt{f_c'}$. $f_c'=5000$ psi. ($F_t=1{,}557$ kN and $f_c'=34$ N/mm²).

Solution Compute the external moment due to beam's own weight of $w=300$ plf (4.377 kN/m), on a cantilever of 20 ft (6.10 m) span,

$$-M=\frac{wL^2}{2}=\frac{300\times 20^2}{2}=60{,}000 \text{ ft-lb } (81.36 \text{ kN-m})$$

The fiber stresses at midspan are computed as follows and shown in Fig. 5-38.

$$f=\frac{F}{A}\pm\frac{Fey}{I}\pm\frac{My}{I}$$

$$=\frac{-350{,}000}{288}\pm\frac{350{,}000\times 5\times 12}{13{,}800}\pm\frac{-60{,}000\times 12\times 12}{13{,}800}$$

$$=-1215+1520+625=+930 \text{ psi } (+6.41 \text{ N/mm}^2) \text{ top fiber} > f_r$$

$$f_r=7.5\sqrt{f_c'}=530 \text{ psi } (3.65 \text{ N/mm}^2)$$

(cracked section)

$$=-1215-1520-625=-3360 \text{ psi } (-23.17 \text{ N/mm}^2) \text{ bottom fiber}$$

These stresses indicate possible cracking for the top fibers and high compression for the bottom fibers. If cracking does occur, the stress distribution would be modified and the bottom compression further aggravated. Also note that as the bottom fiber is compressed, the prestress $F_t=350$ k (1557 kN) is somewhat reduced, but this will not be discussed here as the calculations shown verify cracking.

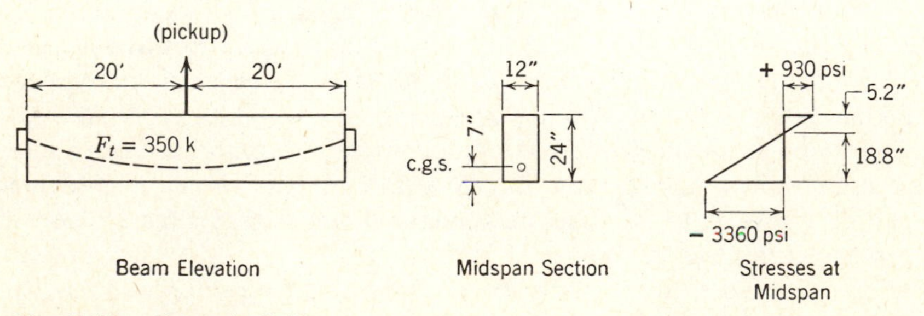

Fig. 5-38. Example 5-13.

It is sometimes more convenient to compute the fiber stresses by locating the center of pressure C, and then apply formula 5-10, using the increased eccentricity e.

$$f=\frac{F}{A}\pm\frac{Fey}{I}$$

Thus example 5-13 can be solved in a manner similar to example 5-5. When cracking occurs, locating the center of pressure C will yield a simple solution as illustrated in the next example.

EXAMPLE 5-14

For the beam picked up at midspan point in example 5-13, if the top fiber cracks and the concrete is assumed not to take any tension, compute the bottom fiber stress.

Solution Assuming prestress F_t to remain at 350 k (1557 kN), with an external negative moment of 60 k-ft (81.36 kN−m), the center of pressure C will be moved downward by

$$\frac{M}{F_t}=\frac{60\times 12}{350}=2.06 \text{ in. } (52.3 \text{ mm})$$

locating it at 4.94 in. (125.5 mm) above the bottom fiber, Fig. 5-39. Assuming a triangular stress block, the height of the triangle is

$$3\times 4.94=14.82 \text{ in. } (376 \text{ mm})$$

and the bottom fiber stress is

$$\frac{-350{,}000}{14.82\times 12}\times 2=-3930 \text{ psi } (-27.10 \text{ N/mm}^2)$$

which exceeds $f_c=0.45(5000)=2250$ psi compression allowed by the ACI Code.

When the center of pressure C moves further downward, the bottom portion of the beam will be stressed into the plastic range. Then it will be a better approximation to assume a rectangular or a trapezoidal stress block for the determination of the stress values. This is illustrated in example 5-15:

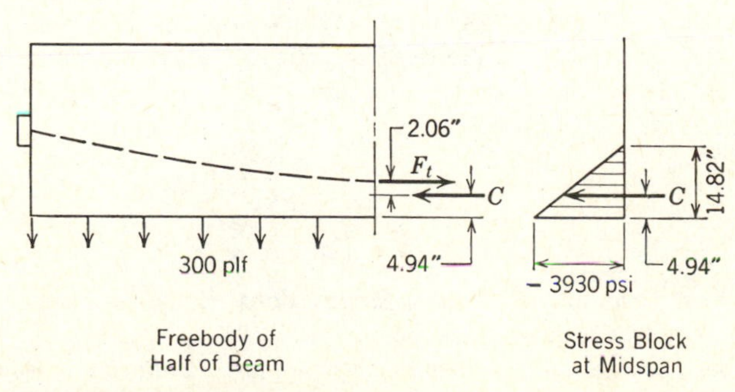

Fig. 5-39. Example 5-14.

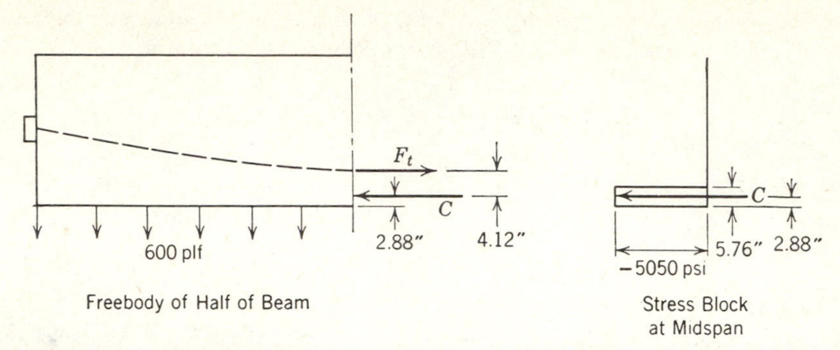

Fig. 5-40. Example 5-15.

EXAMPLE 5-15

Assuming the beam in example 5-15 is picked up suddenly at midspan so that an impact factor of 100% is considered, compute the maximum stress: $f_c' = 5000$ psi (34 N/mm^2).

Solution The external moment will be doubled as a result of 100% impact, thus,

$$-M = 2 \times 2.06 \text{ in.} = 4.12 \text{ in.} \ (104.6 \text{ mm})$$

or located 2.88 in. (73.2 mm) above the bottom fiber. A triangular stress block will yield a high maximum stress of $2 \times 350{,}000/(8.64 \times 12) = 6750$ psi (46.54 N/mm^2). Assuming a rectangular stress block, we have, Fig. 5-40,

$$\frac{350{,}000}{5.76 \times 12} = 5050 \text{ psi } (34.82 \text{ N/mm}^2) > 0.85 \times 5{,}000 = 4250 \text{ psi } (29.30 \text{ N/mm}^2)$$

Thus the beam would fail when picked up suddenly by a midspan pickup point. Note that trapezoidal stress block, using Jensen's theory, will give a more accurate answer,[13] but will not be attempted here as it is clear from the calculation above that the midspan pickup point exceeds the stress which the concrete could carry with $f_c' = 5000$ psi (34 N/mm^2).

We would revise the pickup arrangement to use two points equidistant from midspan to avoid any possibility of damage to the beam during handling.

The three foregoing examples illustrate stress distributions in beams at transfer before cracking, after cracking, and at ultimate. The permissible stress values both in tension and in compression will depend on many factors, such as the shape of the section, the magnitude and location of the prestress, the chances of misplacement of the tendons, the probability of adverse moments, and the serious ness of cracking. Values specified in the ACI Code may be used as a reference.

References

1. G. L. Rogers, "Validity of Certain Assumptions in the Mechanics of Prestressed Concrete," *J. Am. Conc. Inst.*, December 1953 (*Proc.*, Vol. 49), pp. 317–330.
2. *International Federation of Prestressing*, *Preliminary Publications*, First International Congress, London, October 1953.

3. G. S. Ramaswamy and S. K. Narayana, "The Ultimate Flexural Strength of Posttensioned Grouted Rectangular Beams," *Papers, Third Congress of the International Federation of Prestressing*, Berlin, 1958.
4. D. F. Billet and J. H. Appleton, "Flexural Strength of Prestressed Concrete Beams," *J. Am. Conc. Inst.*, June 1954 (*Proc.*, Vol. 50), pp. 837–854.
5. J. R. Janney, E. Hognestad, and D. McHenry, "Ultimate Flexural Strength of Prestressed and Conventionally Reinforced Concrete Beams," *J. Am. Conc. Inst.*, February 1956 (*Proc.* Vol. 52), pp. 601–620.
6. J. Warwaruk, M. A. Sozen, C. P. Siess, "Strength and Behavior in Flexure of Prestressed Concrete Beams," Engineering Experiment Station *Bull. No. 464*, University of Illinois, 1962.
7. E. Hognestad, H. W. Hanson, and D. McHenry, "Concrete Stress Distribution in Ultimate Strength Design," *J. Am. Conc. Inst.*, December 1955 (*Proc.* Vol. 52), pp. 455–479.
8. "Tentative Recommendations for Prestressed Concrete," *J. Am. Conc. Inst.*, January 1958 (*Proc.* Vol. 54), pp. 545–578.
9. N. Burns, "Moment-Curvature Relationships for Partially Prestressed Concrete Beams," *J. Prestressed Conc. Inst.*, Vol. 9 No. 1, February 1964, pp. 52–63.
10. N. H. Burns, and C. P. Siess, "Load-Deformation Characteristics of Beam-Column Connections in Reinforced Concrete," Civil Engineering Studies, Structural Research Series No. 234, University of Illinois, January 1962.
11. Jun, Yamazaki, Basil T. Kattula, and Alan H. Mattock, "A Comparison of the Behavior of Posttensioned Prestressed Concrete Beams With and Without Bond," *Report* SM69-3, University of Washington, College of Engineering, Structures and Mechanics, December 1969, 94 pp.
12. Alan H. Mattock, Jun Yamazaki, and Basil T. Kattula, "Comparative Study of Prestressed Concrete Beams, With and Without Bond," *J. Am. Conc. Inst.* (*Proc.* Vol. 68) February 1971, pp. 116–125.
13. A. C. Scordelis, T. Y. Lin, and H. R. May, "Flexural Strength of Prestressed Concrete Beams at Transfer," *Proceedings World Conference on Prestressed Concrete*, San Francisco, 1957.

6

DESIGN OF SECTIONS FOR FLEXURE

6-1 Preliminary Design

Preliminary design of prestressed-concrete sections for flexure can be preformed by a very simple procedure, based on a knowledge of the internal C-T couple acting in the section. In practice the depth h of the section is either given, known, or assumed, as is the total moment M_T on the section. Under the working load, the lever arm for the internal couple could vary between 30 to 80% of the overall height h and averages about $0.65h$. Hence the required effective prestress F can be computed from the equation

$$F = T = \frac{M_T}{0.65h} \tag{6-1}$$

if we assume the lever arm to be $0.65h$, Fig. 6-1. If the effective unit prestress is f_s for the steel, then the area of steel required is

$$A_{ps} = \frac{F}{f_{se}} = \frac{M_T}{0.65hf_{se}} \tag{6-2}$$

The total prestress $A_{ps}f_{se}$ is also the force C on the section. This force will produce an average unit stress on the concrete of

$$\frac{C}{A_c} = \frac{T}{A_c} = \frac{A_{ps}f_{se}}{A_c}$$

The top fiber stress, f_c, under working loads following ACI Code is $0.45f_c'$, Fig. 6-1. Table 1-2, Chapter 1, summarizes permissible stresses in steel and concrete for prestressed concrete members. For preliminary design, the average stress can be assumed to be about 50% of the maximum allowable stress f_c, under the working load. Hence,

$$\frac{A_{ps}f_{se}}{A_c} = 0.50f_c$$

$$A_c = \frac{A_{ps}f_{se}}{0.50f_c} \tag{6-3}$$

Note that in the above procedure the only approximations made are the

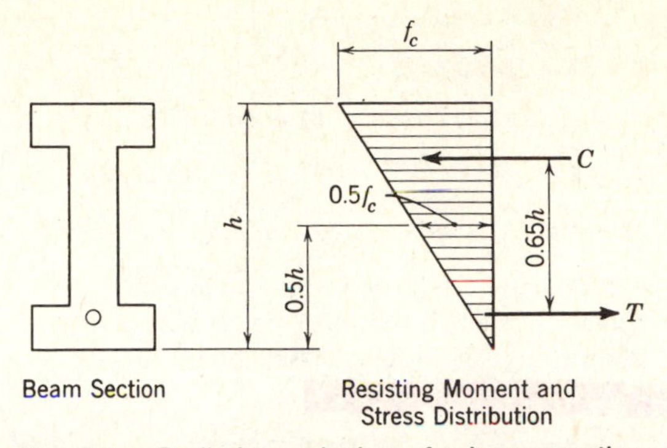

Fig. 6-1. Preliminary design of a beam section.

coefficients of 0.65 and 0.50. These coefficients vary widely, depending on the shape of the section. However, with experience and knowledge, they can be closely approximated for each particular section, and the preliminary design can be made rather accurately.

The above procedure is based on the design for working load, with little or no tension in the concrete. Preliminary designs can also be made on the basis of ultimate strength theories with proper load factors. Such an alternative procedure will be discussed in section 6-7.

EXAMPLE 6-1

Make a preliminary design for section of a prestressed-concrete beam to resist a total moment of 320 k-ft (434 kN-m). The overall depth of the section is given as 36 in (914.4 mm). The effective prestress for steel is 125,000 psi (862 N/mm^2), and allowable stress for concrete under working load is -1600 psi (-11.03 N/mm^2).

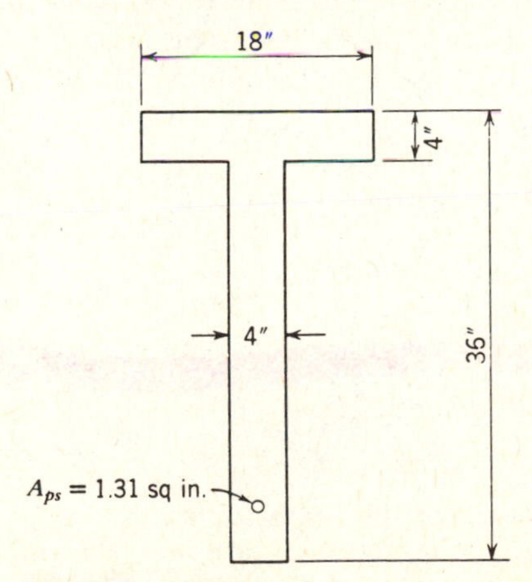

Fig. 6-2. Example 6-1.

Solution From equations 6-1, 6-2, and 6-3,

$$F = T = M_T/0.65h$$

$$= (320\times 12)/(0.65\times 36) = 164 \text{ k } (729.5 \text{ kN})$$

$$A_{ps} = F/f_{se} = 164/125 = 1.31 \text{ sq in } (845 \text{ mm}^2)$$

$$A_c = 164/(0.5\times 1.60) = 205 \text{ sq in. } (132\times 10^3 \text{ mm}^2)$$

Now a preliminary section can be sketched with a total concrete area of about 205 sq in. (132×10^3 mm^2), a height of 36 in. (914.4 mm), and a steel area of 1.31 sq in. (845 mm^2) Such a section is shown in Fig. 6-2. A T-section is chosen here because it is an economical shape when M_G/M_T ratio is large.

In estimating the depth of a prestressed section, an approximate rule is to use 70% of the corresponding depth for conventional reinforced-concrete construction. Some other empirical rules are also available. For example, the thickness of prestressed slabs may vary from $L/35$ for heavy loads to $L/55$ for light loads. The depth of beams of the usual proportions can be approximated by the following formula.

$$h = k\sqrt{M}$$

where h = depth of beam in inches
M = maximum bending moment in k-ft
k = a coefficient varying from 1.5 to 2.0

It is needless to add that such empirical rules apply only under the average conditions and should be used merely as a preliminary guide.

A more accurate preliminary design can be made if the girder moment M_G is known in addition to the total moment M_T. When M_G is much greater than 20 to 30% of M_T, the initial condition under M_G generally will not control the design, and the preliminary design needs be made only for M_T. When M_G is small relative to M_T, then the c.g.s. cannot be located too far outside the kern point, and the design is controlled by $M_L = M_T - M_G$. In this case, the resisting lever arm for M_L is given approximately by $k_t + k_b$, which averages about $0.50h$. Hence the total effective prestress required is

$$F = \frac{M_L}{0.50h} \tag{6-4}$$

When M_G/M_T is small, this equation should be used instead of equation 6-1. Equation 6-3 is still applicable.

EXAMPLE 6-2

Make a preliminary design for the beam section in example 6-1, with M_T = 320 k-ft, M_G = 40 k-ft, h = 36 in., f_{se} = 125,000 psi, and f_c = − 1600 psi (M_T = 434 kN-m, M_G = 54 kN-m, h = 914.4 mm, f_{se} = 862 N/mm^2, and f_c = − 11.03 N/mm^2).

Solution Since M_G is only 12% of M_T, it is not likely that the c.g.s. can be located much outside the kern. Hence it will be more nearly correct to apply equation 6-4. Thus,

$$M_L = M_T - M_G = 320 - 40$$
$$= 280 \text{ k-ft } (380 \text{ kN-m})$$
$$F = M_L/0.50h = 280 \times 12/(0.50 \times 36)$$
$$= 187 \text{ k } (832 \text{ kN})$$

Applying the first part of formula 6-2 and also formula 6-3, we have

$$A_{ps} = F/f_{se} = 187/125$$
$$= 1.50 \text{ sq in. } (968 \text{ mm}^2)$$
$$A_c = A_{ps}f_{se}/0.50f_c = 187/(0.50 \times 1.60)$$
$$= 234 \text{ sq in. } (151 \times 10^3 \text{ mm}^2)$$

Now a preliminary section can be sketched with a total concrete area of about 234 sq in. (151×10^3 mm^2), a height of 36 in. (914.4 mm), and a steel area of 1.50 sq in. (968 mm^2), as shown in Fig. 6-3. An I-section is chosen because it is a suitable form when the M_G/M_T ratio is small.

When it is not known whether M_T or M_L should govern the design, one convenient way is to apply both equations 6-1 and 6-4, and use the greater of the two values of F. For example, if $M_G = 80$ k-ft (108 kN-m) in example 6-1, we have, from equation 6-1,

$$F = M_T/0.65h$$
$$= (320 \times 12)/(0.65 \times 36)$$
$$= 164 \text{ k } (730 \text{ kN})$$

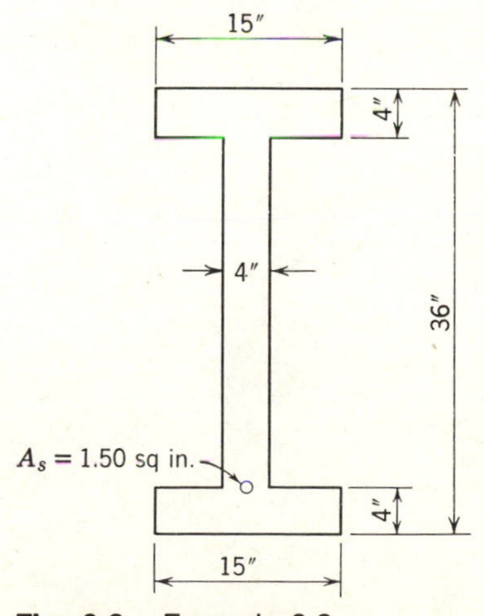

Fig. 6-3. Example 6-2.

From equation 6-4, we have

$$F = M_L/0.50h$$
$$= [(320-80)12]/(0.50\times 36)$$
$$= 160\text{ k }(712\text{ kN})$$

$F = 164$ k (730 kN) controls the design.

6-2 Elastic Design, General Concepts

There is a prevailing impression that the design of prestressed-concrete sections is much more complicated than that of reinforced ones. This is not true if the procedure recommended in this chapter is followed. However, the design of a section is based on a knowledge of its analysis. Hence readers must be familiar with the methods of analysis discussed in the previous chapter before they can master the methods of design.

The method of preliminary design presented in section 6-1 is based on the fact that the section is governed by two controlling values of external bending moment: the total moment M_T, which controls the stresses under the action of the working loads; and the girder load moment M_G, which determines the location of the c.g.s. and the stresses at transfer.

It is desirable to reiterate here the basic concept of a resisting couple in a prestressed-concrete-beam section. From the law of statics, the internal resisting moment in a prestressed beam, as in a reinforced-concrete beam, must equal the external moment. That internal moment can be represented by a couple, C–T, for either the prestressed- or the reinforced-concrete-beam section, Figs. 6-4 and 6-5. T is the centroid of the prestress or tensile force in the steel; and C is the center of pressure or the center of compression on the concrete.

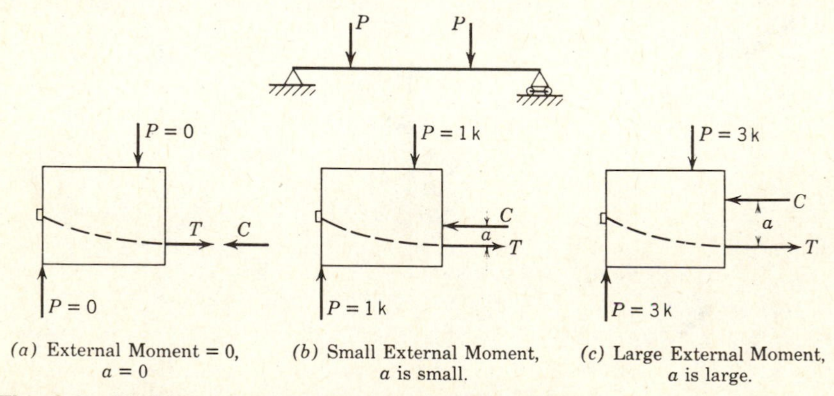

(*a*) External Moment = 0, $a = 0$

(*b*) Small External Moment, a is small.

(*c*) Large External Moment, a is large.

Fig. 6-4. Variable *a* in a prestressed-concrete beam.

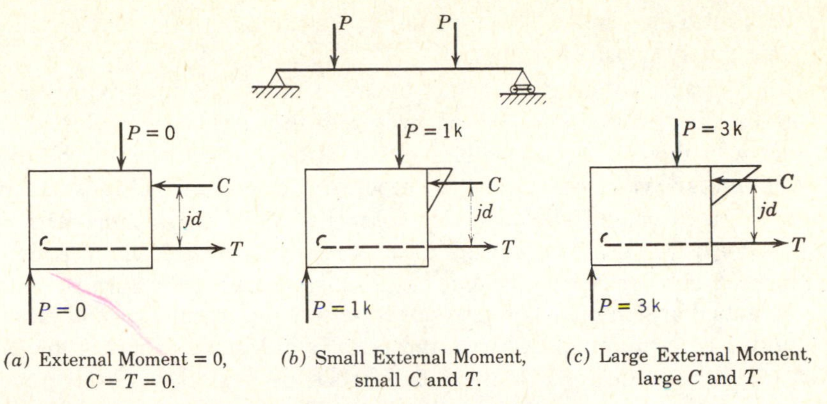

(*a*) External Moment = 0, $C = T = 0$.

(*b*) Small External Moment, small C and T.

(*c*) Large External Moment, large C and T.

Fig. 6-5. Constant *jd* in a reinforced-concrete beam.

There is, however, an essential difference between the behavior of a prestressed- and of a reinforced-concrete-beam section. The difference is explained as follows:

1. In a reinforced-concrete-beam section, as the external bending moment increases, the magnitude of the forces C and T is assumed to increase in direct proportion while the lever arm jd between the two forces remains practically unchanged, Fig. 6-5.
2. In a prestressed-concrete-beam section under working load, as the external bending moment increases, the magnitude of C and T remains practically constant while the lever arm a lengthens almost proportionately, Fig. 6-4.

Since the location of T remains fixed, we get a variable location of C in a prestressed section as the bending moment changes. For a given moment M, C can be easily located, since

$$Ca = Ta = M \tag{6-5}$$

$$a = M/C = M/T \tag{6-5a}$$

Thus, when $M=0$, $a=0$, and C must coincide with T, Fig. 6-4(a). When M is small, a is also small, Fig. 6-4(b). When M is large, a is also large, Fig. 6-4(c).

In a prestressed-concrete beam, the amount of initial prestress F_0 is measured and is rather accurately known. At the time of transfer of prestress, $T = F_0$. After all losses have taken place, $T = F$. Although the value of T does change as the beam bends under loading, the change is small within the working range and can either be taken into account or neglected in design.

Once the magnitude of T is known, the value of a can be computed from equation 6-5a for any value of M. The location of C can thus be determined. With the position and magnitude of C known, stress distribution across the concrete section can be obtained by the elastic or the plastic theory, although the elastic theory is usually followed.

It will be well to mention some of the simple relations between stress distribution and the location of C, according to the elastic theory, Fig. 6-6. If C coincides with the top or bottom kern point, stress distribution will be triangular, with zero stress at bottom or top fiber, respectively. If C falls within the kern, the entire section will be under compression; if outside the kern, some tension will exist. If C coincides with c.g.c., stress will be uniform over the entire concrete section. (See Appendix A for k_t and k_b, defining the kern.)

In the actual design of prestressed-concrete sections, similar to any other type of section, a certain amount of trial and error is inevitable. There is the general layout of the structure which must be chosen as a start but which may be modified as the process of design develops. There is the dead weight of the member which influences the design but which must be assumed before embarking on the moment calculations. There is the approximate shape of the concrete section, governed by both practical and theoretical considerations, which must be assumed for the trial. Because of these variables, it has been found that the

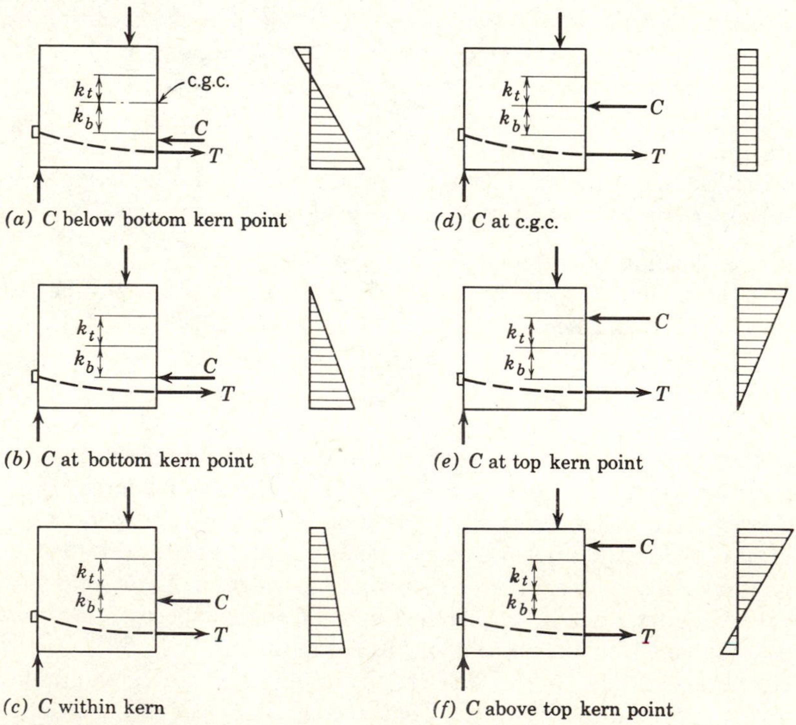

Fig. 6-6. Stress distribution in concrete by the elastic theory.

best procedure is one of trial and error, guided by known relations which enable the final results to be obtained without excessive work.

When the shape of the section and the loadings are known, the amount and location of prestress can be solved by programs set up for digital computers to meet stress requirements at different points along a beam, resulting in saving of manual computation.

6-3 Elastic Design, No Tension in Concrete

In this section will be discussed the final flexural design of sections based on the elastic theory and allowing no tension in the concrete both at transfer and under working load. While allowing no tension is generally regarded as too conservative a criterion, it does help to simplify computation and hence will be discussed first. Two cases will be considered, one for small and one for large ratios of M_G/M_T.

Small Ratios of M_G/M_T. For the section obtained from the preliminary design, the values of M_G, k_t, k_b, A_c are computed. When the ratio of M_G/M_T is small, c.g.s. is located outside the kern just as much as the M_G will allow. Since no tension is permitted in the concrete, c.g.s. will be located below the kern by the amount of Fig. 6-7(b).

$$e-k_b=M_G/F_0 \tag{6-6}$$

If c.g.s. is so located, C will be exactly at the bottom kern point for the given M_G, and the stresses at the top and bottom fibers will be

$$f_t=0$$

$$f_b=\frac{F_0}{A_c}\frac{h}{c_t}\leqslant 0.6f'_{ci} \tag{6-7}$$

Hence,

$$A_c=\frac{F_0 h}{f_b c_t} \tag{6-7a}$$

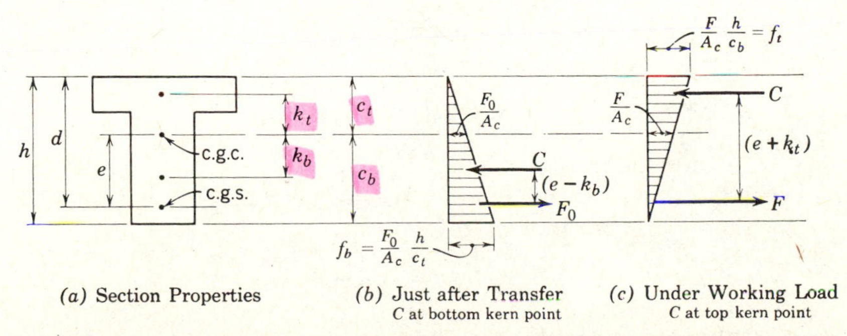

(a) Section Properties (b) Just after Transfer, C at bottom kern point (c) Under Working Load, C at top kern point

Fig. 6-7. Stress distribution, no tension in concrete (small ratios of M_G/M_T).

If c.g.s. is located father up, C will fall within the kern; then the top fibers will be under some compression, and the bottom fibers will be stressed less than given by equation 6-7. If c.g.s. is located farther below, C will fall outside the kern; then there will be some tension in the top fibers, and the bottom fibers will be stressed higher than given by equation 6-7. The bottom fiber allowable stress is $0.60 f'_{ci}$ following the ACI Code and F_0 is the prestress force acting at transfer.

With c.g.s. located as above, the available lever arm for the resisting moment is given by $e+k_t$, and the effective prestress F is given by

$$F=\frac{M_T}{e+k_t} \tag{6-8}$$

Under the action of this effective prestress F and the total moment M_T, C will be located at the top kern point, and the top and bottom fiber stresses are Fig. 6-7 (c),

$$f_b=0$$

$$f_t=\frac{F}{A_c}\frac{h}{c_b}\leqslant 0.45 f'_c \tag{6-9}$$

Hence,

$$A_c=\frac{Fh}{f_t c_b} \tag{6-9a}$$

If F is smaller than the value given by equation 6-8, there will be tension in the bottom fibers, and the compressive stress in the top fibers will be greater than that given by equation 6-9; if F is greater, there will be some residual compression in the bottom fibers, and the compressive stress in the top fibers will be less than that indicated by equation 6-9.

If f_b or f_t exceeds the allowable value, it will be necessary to increase the area of concrete A_c, or to decrease the ratio of h/c_t or h/c_b, respectively. If f_b and f_t are both lesss than the respective allowable values, A_c can be decreased accordingly. Slight changes in the dimensions of the section may not affect the k_t, k_b, and other values. But if major changes are made, it may be desirable to go over the procedure once more to obtain a new location for the c.g.s. and compute new values for F and check over the required A_c.

To summarize the procedure of design, we have:

Step **1.** From the preliminary design section, locate c.g.s. by

$$e-k_b=M_G/F_0$$

Step **2.** With the above location of c.g.s., compute the effective prestress F (and then the initial prestress F_0) by

$$F=\frac{M_T}{e+k_t}$$

Step **3.** Compute the required A_c by

$$A_c=F_0 h/f_b c_t$$

and

$$A_c = Fh/f_t c_b$$

Step **4.** Revise the preliminary section to meet the above requirements for F and A_c. Repeat steps 1 through 4 if necessary.

From the above discussion, the following observations regarding the properties of a section can be made.

1. $e+k_t$ is a measure of the total moment-resisting capacity of the beam section. Hence the greater this value, the more desirable is the section.
2. $e-k_b$ locates the c.g.s. for the section, and is determined by the value of M_G. Thus, within certain limits, the amount of M_G does not seriously affect the capacity of the section for carrying M_L.
3. h/c_b is the ratio of the maximum top fiber stress to the average stress on the section under working load. Thus, the smaller this ratio, the lower will be the maximum top fiber stress.
4. h/c_t is the ratio of the maximum bottom fiber stress to the average stress on the section at transfer. Hence, the smaller this ratio, the lower will be the maximum bottom fiber stress.

To facilitate design computations, properties of different sections are listed in Appendix C, Tables 1 through 6. Values of A_c, I, k_t, k_b, c_t, c_b, etc., are given in these tables. Properties for a rectangular section are included in Table 1 under the headings $b'/b = t/h = 1$, that is, section 1-q. By the use of these equations and the tables, it is possible to develop formulas which will give directly the required section modulus for a given shape. But for a practical design, it is generally preferable to follow a method of trial and error as we just outlined, because dimensioning and other practical considerations do not often permit keeping to an assumed shape of section.

EXAMPLE 6-3

For the preliminary section obtained in example 6-2, make a final design, allowing $f_b = -1.80$ ksi, $f_0 = 150$ ksi. Other given values were: $M_T = 320$ k-ft; $M_G = 40$ k-ft; $f_t = -1.60$ ksi; $f_{se} = 125$ ksi; $F = 187$ k. And the preliminary section is the same as in Fig. 6-3 ($f_b = -12.41$ N/mm², $f_0 = 1034$ N/mm², $M_T = 434$ kN-m, $M_G = 54$ kN-m, $f_t = -11.03$ N/mm², $f_{se} = 862$ N/mm², and $F = 832$ kN).

Solution For the trial preliminary section, compute the properties as follows

$$A_c = 2\times4\times15 + 4\times28 = 232 \text{ sq in. } (150\times10^3 \text{ mm}^2)$$

$$\begin{aligned} I &= \frac{15\times36^3}{12} - \frac{11\times28^3}{12} \\ &= 58{,}200 - 20{,}100 \\ &= 38{,}100 \text{ in.}^4 \ (15.86\times10^9 \text{ mm}^4) \end{aligned}$$

$$\begin{aligned} r^2 &= 38{,}100/232 \\ &= 164 \text{ in.}^2 \ (106\times10^3 \text{ mm}^2) \end{aligned}$$

$$k_t = k_b = 164/18 = 9.1 \text{ in. } (231 \text{ mm})$$

Step 1. For an assumed

$$F = 187 \text{ k } (832 \text{ kN})$$

$$F_0 = \frac{150}{125} 187 = 225 \text{ k } (1001 \text{ kN})$$

c.g.s. should be located at $e - k_b$ below the bottom kern, where

$$e - k_b = \frac{M_G}{F_0} = \frac{40 \times 12}{225} = 2.1 \text{ in. } (53 \text{ mm})$$

$$e = 9.1 + 2.1 = 11.2 \text{ in. } (285 \text{ mm})$$

Step 2. Effective prestress required is recomputed as

$$F = \frac{M_T}{e + k_t} = \frac{320 \times 12}{11.2 + 9.1}$$

$$= 189 \text{ k } (841 \text{ kN})$$

$$F_0 = \frac{150}{125} 189 = 227 \text{ k } (1010 \text{ kN})$$

Step 3. A_c required is

$$A_c = \frac{F_0 h}{f_b c_t}$$

$$= \frac{227 \times 36}{1.80 \times 18}$$

$$= 252 \text{ sq in. } (163 \times 10^3 \text{ mm}^2) \text{ controlling}$$

$$A_c = \frac{Fh}{f_t c_b}$$

$$= \frac{189 \times 36}{1.60 \times 18}$$

$$= 236 \text{ sq in. } (152 \times 10^3 \text{ mm}^2)$$

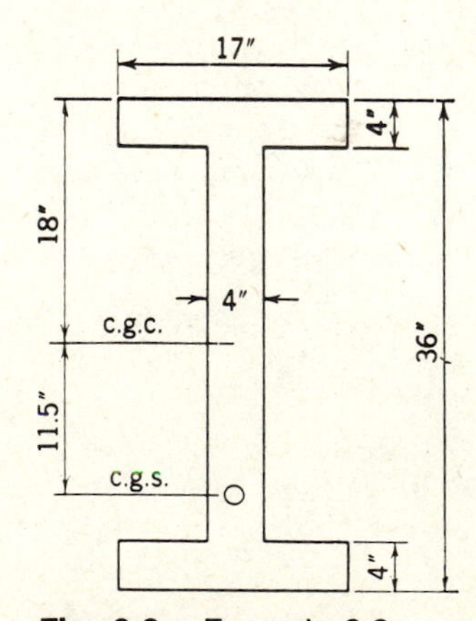

Fig. 6-8. Example 6-3.

Step 4. Try a new section as shown in Fig. 6-8, with $A_c = 248$ sq in. (160×10^3 mm^2). For this new section, $I = 42{,}200$ in.^{4} (17.57×10^9 mm^4); $k_t = k_b = 9.4$ in. (239 mm $e - k_b = 2.1$ in (53 mm); $F = 320 \times 12/(11.5 + 9.4) = 184$ k (818 kN); $F_0 = 221$ k (983 kN); A_c required for bottom fiber = 246 sq in. (159×10^3 mm^2), for top fiber = 230 sq in. (148×10^3 mm^2). Hence the section seems to be quite satisfactory. And no further revision is needed.

Large Ratios of M_G/M_T. When the ratio of M_G/M_T is large, the value of $e - k_b$ computed from equation 6-6 may place c.g.s. outside of the practical limit, for example, below the section of the beam. Then it is necessary to place the c.g.s. only as low as practicable and design accordingly.

For such a condition, the bottom fiber stress is seldom critical. Under the initial condition, just after transfer, the bottom fiber stress is shown in Fig. 6-9(b) and is given by the formula

$$f_b = \frac{F_0}{A_c} + \frac{(F_0 e - M_G)C_b}{I}$$

$$= \frac{F_0}{A_c}\left(1 + \frac{e - (M_G/F_0)}{k_t}\right)$$

from which the required area A_c can be computed as

$$A_c = \frac{F_0}{f_b}\left(1 + \frac{e - (M_G/F_0)}{k_t}\right) \tag{6-10}$$

The top fiber is always under some compression and does not control the design under this condition.

Under the working load, the stress distribution is the same as for the first case (small ratios), and is pictured in Fig. 6-9(c). The design is practically the same as before except that equation 6-10 should be used in place of equation 6-7a. For convenience, the procedure will be outlined as follows.

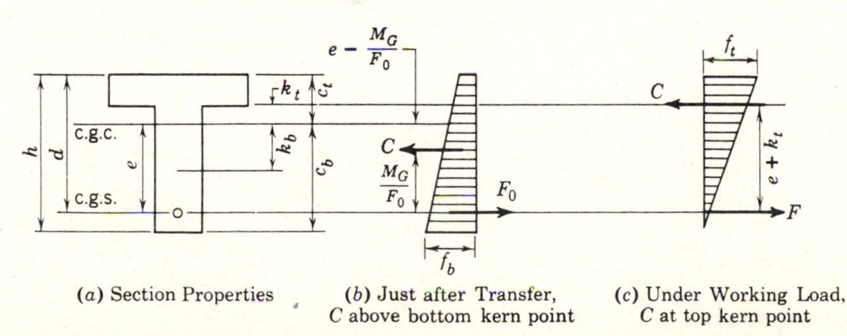

Fig. 6-9. Stress distribution, no tension in concrete (large ratios of M_G/M_t).

Step **1.** From the preliminary section, compute the theoretical location for c.g.s. by

$$e-k_b=M_G/F_0$$

If it is feasible to locate c.g.s. as indicated by this equation, follow the first procedure. If not, locate c.g.s. at the practical lower limit and proceed as follows.

Step **2.** Compute F (and then F_0) by

$$F=\frac{M_T}{e+k_t}$$

Step **3.** Compute the required area by equations 6-9*a* and 6-10.

$$A_c=Fh/f_t c_b$$

$$A_c=\frac{F_0}{f_b}\left(1+\frac{e-(M_G/F_0)}{k_t}\right)$$

Step **4.** Use the greater of the two A_c's and the new value of F, and revise the preliminary section. Repeat steps 1 through 4 if necessary.

EXAMPLE 6-4

Make final design for the preliminary section obtained in example 6-1, $M_G=210$ k-ft, allowing $f_b=-1.80$ ksi, $f_0=150$ ksi. Other values given were $M_T=320$ k-ft; $h=36$ in.; $f_{se}=125$ ksi; $f_t=-1.60$ ksi. The preliminary section is shown in Fig. 6-10, with $A_c=200$ sq in., $c_t=13.5$ in., $c_b=22.5$ in., $I=26{,}000$ in.^{4}, $k_t=5.8$ in., $k_b=9.6$ in., $F=164$ k, $F_0=164(150/125)=197$ k ($M_G=285$ kN-m, $f_b=-12.41$ N/mm^2, $f_0=1034$ N/mm^2; $M_T=434$ kN-m, $h=914.4$ mm, $f_{se}=862$ N/mm^2; $f_t=-11.03$ N/mm^2, $A_c=129\times10^3$

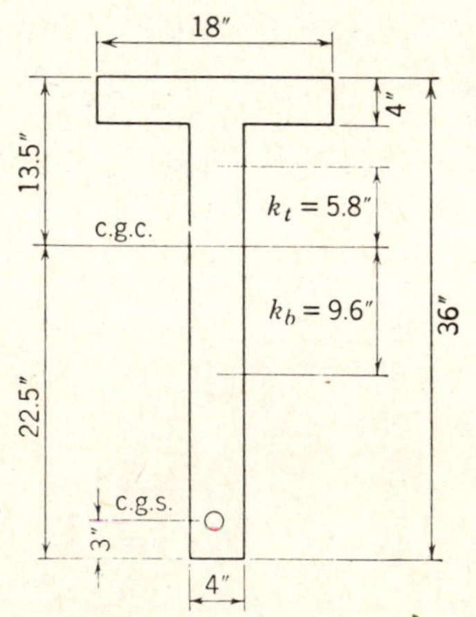

Fig. 6-10. Example 6-4. Trial section.

mm, $c_t = 343$ mm; $c_b = 572$ mm, $I = 10.82 \times 10^9$ mm^4, $k_t = 147$ mm; $k_b = 244$ mm, $F = 730$ kN, and $F_0 = 876$ kN).

Solution *Step 1.* Theoretical lowest location for c.g.s. is given by

$$\begin{aligned} e - k_b &= M_G/F_0 \\ &= (210 \times 12)/197 \\ &= 12.8 \text{ in. } (325 \text{ mm}) \end{aligned}$$

indicating 12.8 in. (325 mm) below the bottom kern, or 0.1 in. (2.54 mm) above the bottom fiber, which is obviously impossible. Suppose that for practical reasons the c.g.s. has to be kept 3 in. (76.2 mm) above the bottom fiber to provide sufficient concrete protection. This problem then belongs to the second case, and we proceed as below.

Step 2. The effective prestress required is, corresponding to a lever arm of $e + k_t = 22.5 - 3 + 5.8 = 25.3$ in. (643 mm),

$$F = (320 \times 12)/25.3 = 152 \text{ k } (676 \text{ kN})$$

$$F_0 = 152(150/125) = 182 \text{ k } (810 \text{ kN})$$

Step 3. Compute the area required by

$$\begin{aligned} A_c &= \frac{Fh}{f_t c_b} \\ &= \frac{152 \times 36}{1.60 \times 22.5} \\ &= 152 \text{ sq in. } (98 \times 10^3 \text{ mm}^2) \end{aligned}$$

$$\begin{aligned} A_c &= \frac{F_0}{f_b}\left(1 + \frac{e - (M_G/F_0)}{k_t}\right) \\ &= \frac{182}{1.80}\left(1 + \frac{19.5 - 210 \times 12/182}{5.8}\right) \\ &= 199 \text{ sq in. } (128 \times 10^3 \text{ mm}^2) \end{aligned}$$

which indicates that the trial preliminary section with $A_c = 200$ sq in. (129×10^3 mm^2) is just about right for the stress in the bottom fibers, but much more than enough as far as the top fibers are concerned. In other words, if practical conditions permit, it may be desirable to reduce the concrete area in the top flange and to add concrete area to the bottom flange, to obtain a more economical section. The reader may try this out to see whether a better section is obtainable for this example.

6-4 Elastic Design, Remarks on Allowing Tension

In the preceding section we discussed the design of prestressed-concrete sections allowing no tensile stresses. This requirement may often be an extravagance that cannot be justified. When compared to reinforced concrete, where high tensile stresses and cracks are always present under working load, it seems only logical that at least some tensile stresses should be permitted in prestressed concrete. On the other hand, there are several reasons for limiting the tensile stresses in prestressed concrete. These are:

1. The existence of high tensile stress in prestressed concrete may indicate an insufficient factor of safety against ultimate failure. When high tensile stress exists in prestressed concrete, the working lever arm a for the resisting couple is a large ratio of h, Fig. 6-11, so that no substantial increase in the lever arm can take place in case of overloads. Thus the margin of safety may not be sufficient. The actual check on strength is a separate part of the design after stress requirements are satisfied.
2. The existence of tensile stress may indicate an insufficient factor of safety against cracking and may easily result in cracking if the concrete has been previously cracked. Although cracking may not be significant under static load, it could be an important criterion when a member is subject to repeated loads. Cracking also signifies a change in the nature of bond and shearing stressing stresses. Furthermore, it is sometimes believed that the small wires in prestressed concrete are more susceptible to corrosion in the event of permanent cracks, although opening of cracks under passing loads is seldom held as contributing to corrosion to any significant extent.

Since the original idea of prestressing concrete was to produce a new material out of concrete by putting it permanently under compression, it was more or less arbitrarily concluded that tensile stresses were not permitted under working loads. As more experience and knowledge was gained with the behavior of prestressed concrete, more engineers have now shifted to the opinion that a certain amount of tension is permissible. Thus the ACI Building Code Requirements permit tensile stresses as follows.

1. *Stresses at transfer:*

   Tension in members without auxiliary reinforcement—$3\sqrt{f'_{ci}}$ ($6\sqrt{f'_{ci}}$ at ends of precast simple beams).

   Tension in members with properly designed auxiliary reinforcement—no limit.

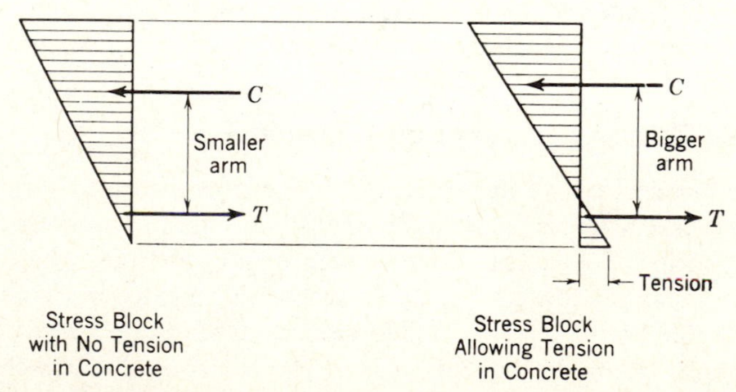

Fig. 6-11. Bigger arm for steel when allowing tension in concrete.

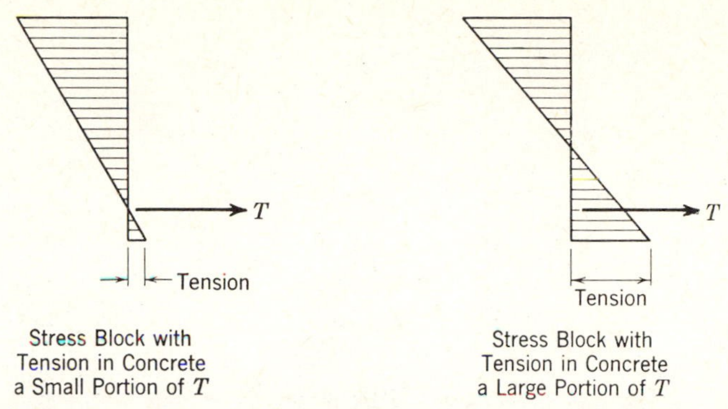

Fig. 6-12. Relative significance of tension in concrete.

2. *Stresses at service loads:*
   Tension in precompressed tensile zone.
   Tension in excess of above limiting values may be permitted when shown to be not detrimental to proper structural behavior ($12\sqrt{f_c'}$).

It is clear from the above that, while empirical limits are often specified for convenience in design and checking, the magnitude of permissible tensile stresses should vary with the conditions and cannot easily be fixed at one or two definite values.

When tensile stresses are permitted under working loads, the term "partial prestressing" is often employed, indicating that the concrete is only partially compressed by the prestress. It is the opinion of the authors that there is really no basic difference between partial and full prestress. The only difference is that, in partial prestressing, there exists a certain amount of tension in concrete under working loads. Since most structures are subject to occasional overloads, tensile stresses will actually exist in both partial and full prestressing, at one time or another. Hence there is no basic difference between them.

It is sometimes argued that the allowing of tension in concrete is a dangerous procedure, since the concrete might have cracked previously and could not take any tension. This is true if the tensile force in concrete is a significant portion of the tensile force in the steel, in which event it will be necessary to neglect tensile force furnished by the concrete, Fig. 6-12. On the other hand, if the tensile force in the concrete is only a small proportion of that in steel, the calculations will not be very different whether it is neglected or included.

6-5 Elastic Design, Allowing and Considering Tension

This method should be used with the understanding that the stresses calculated are not exactly correct if the tension stress exceeds the cracking stress for the concrete. It is a convenient method and yields results comparable to those of the

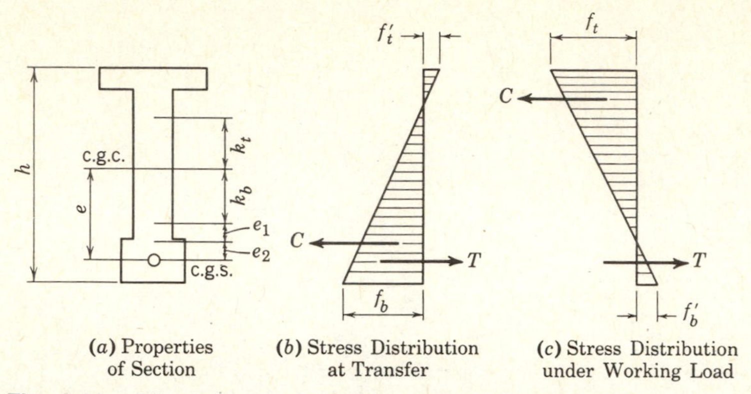

Fig. 6-13. Allowing and considering tension in concrete.

method which neglects the tension in the concrete when the tensile force in concrete considered is only a small portion of the total tension (see Fig. 6-13). This method is usually followed in design using ACI Code allowable stresses. The method will be explained as follows.

Small Ratios of M_G/M_T. If tensile stress f_t' is permitted in the top fibers, the center of compression C can be located below the bottom kern by the amount of

$$e_1 = f_t' I/F_0 c_t = f_t' A k_b / F_0 \tag{6-11}$$

For a given moment M_G, the c.g.s. can be further located below C by the amount of

$$e_2 = M_G/F_0 \tag{6-12}$$

Hence the maximum total amount that the c.g.s. can be located below the kern is given by

$$e_1 + e_2 = \frac{M_G + f_t' A k_b}{F_0} \tag{6-13}$$

The c.g.s. having been located at some value e below c.g.c., the lever arm a under working load is known. For an allowable tension in the bottom fiber, the moment carried by the concrete is

$$f_b' I/c_b = f_b' A k_t$$

The net moment $M_T - f_b' A k_t$ is to be carried by the prestress F with a lever arm acting up to the top kern point; hence the total arm is (Fig. 6-13).

$$a = k_t + e \tag{6-14}$$

and the prestress F required is

$$F = \frac{M_T - f_b' A k_t}{a} \tag{6-15}$$

The bottom fiber stress at transfer is given by

$$f_b = \frac{F_0 h}{A_c c_t} + f_t' \frac{c_b}{c_t} \tag{6-16}$$

from which we have

$$A_c = \frac{F_0 h}{f_b c_t - f_t' c_b} \tag{6-16a}$$

Similarly, the top fiber stress under working load is given by

$$f_t = \frac{Fh}{A_c c_b} + f_b' \frac{c_t}{c_b} \tag{6-17}$$

from which

$$A_c = \frac{Fh}{f_t c_b - f_b' c_t} \tag{6-17a}$$

EXAMPLE 6-5

Redesign the beam section in example 6-3, allowing and considering tension in concrete. $f_t' = 0.30$ ksi, $f_b' = 0.24$ ksi. Other given values were: $M_T = 320$ k-ft; $M_G = 40$ k-ft; $f_t = -1.60$ ksi; $f_b = -1.80$ ksi; $F = 184$ k; $F_0 = 221$ k ($f_t' = 2.07$ N/mm^2, $f_b' = 1.65$ N/mm^2, $M_T = 434$ kN-m, $M_G = 54$ kN-m, $f_t = -11.3$ N/mm^2, $f_b = -12.41$ N/mm^2, $F = 818$ kN, and $F_0 = 983$ kN).

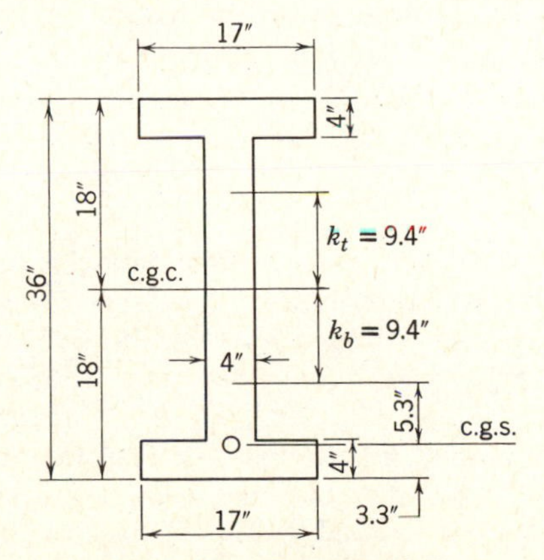

Fig. 6-14. Example 6-5.

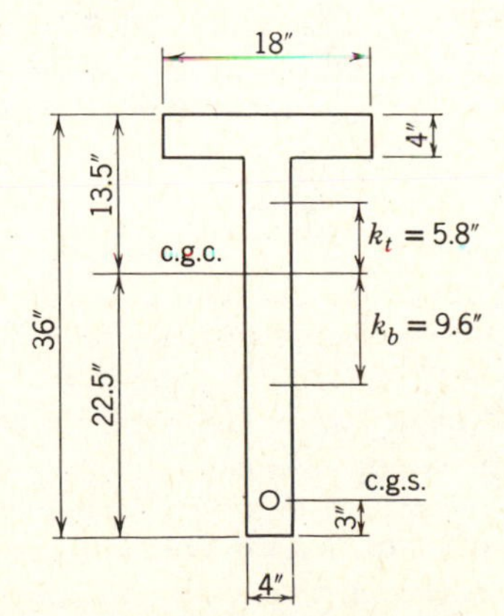

Fig. 6-15. Example 6-6.

Solution *Step* 1. From example 6-3, we have $k_t = k_b = 9.4$ in. (239 mm); $A_c = 248$ sq in. (160×10^3 mm^2). Using equation 6-13, we have

$$e_1 + e_2 = \frac{40 \times 12 + 0.3 \times 248 \times 9.4}{221} = 5.3 \text{ in. } (135 \text{ mm})$$

Hence c.g.s. can be located 5.3 in. (135 mm) below the bottom kern, or 3.3 in. (84 mm) above the bottom fiber, Fig. 6-14.

Step 2. The net moment to be carried by the prestress is

$$M_T - f_b' A k_t = 320 \times 12 - 0.240 \times 248 \times 9.4$$
$$= 3840 - 560 = 3280 \text{ k-in. } (371 \text{ kN-m})$$

For a resisting lever arm of $9.4 + 9.4 + 5.3 = 24.1$ in. (612 mm), the prestress required is

$$F = 3280/24.1 = 136 \text{ k } (605 \text{ kN})$$
$$F_0 = 136 \times 150/125 = 163 \text{ k } (725 \text{ kN})$$

Step 3. To limit the bottom fibers to -1.80 ksi (-12.41 N/mm^2), we need

$$A_c = \frac{163 \times 36}{1.80 \times 18 - 0.30 \times 18}$$
$$= 218 \text{ sq in. } (141 \times 10^3 \text{ mm}^2)$$

To keep the top fibers to -1.60 ksi (-11.03 N/mm^2), we need

$$A_c = \frac{136 \times 36}{1.60 \times 18 - 0.24 \times 18}$$
$$= 200 \text{ sq in. } (129 \times 10^3 \text{ mm}^2)$$

which indicates that the trial section can be appreciably reduced and a new section tried over again.

Large Ratios of M_G/M_T. When M_G/M_T is large, C will be within the kern at transfer, and the allowing of tension on top fiber will have no effect on the design. The c.g.s. has to be located within practical limits. Otherwise, the design is made as for the first case. This is illustrated in the next example.

EXAMPLE 6-6

Revise the design for the section in example 6-4 allowing and considering tension in concrete. Other values given were: $M_T = 320$ k-ft; $M_G = 210$ k-ft; $F = 152$ k; $F_0 = 182$ k; $A_c = 200$; $c_t = 13.5$ in.; $c_b = 22.5$ in.; $k_t = 5.8$ in.; $k_b = 9.6$ in. (Fig. 6-15) ($M_T = 434$ kN-m, $M_G = 285$ kN-m, $F = 676$ kN, $F_0 = 810$ kN, $A_c = 129 \times 10^3$ mm^2, $c_t = 343$ mm, $c_b = 572$ mm, $k_t = 147$ mm, and $k_b = 244$ mm).

Solution *Step* 1. Referring to example 6-6, since the possible theortical location for c.g.s. is 13.8 in. (351 mm) below the bottom kern (0.9 in. (23 mm) below bottom fiber) without producing tension in top fiber, whereas the practical location of c.g.s. has to be 3 in. (76.2 mm) above bottom fiber, no tension will exist in top fiber.

Step 2. Net amount to be carried by prestress is

$$M_T - f_b' A k_t = 320 \times 12 - 0.240 \times 200 \times 5.8$$
$$= 3840 - 280 = 3560 \text{ k=in. (402 kN-m)}$$

The resisting lever arm is

$$36 - 3 - 13.5 + 5.8 = 25.3 \text{ in. (643 mm)}$$

The required prestress is

$$F = 3560/25.3 = 141 \text{ k (627 kN)}$$
$$F_0 = 141(150/125) = 169 \text{ k (752 kN)}$$

To keep the bottom fiber stress within limits, we can apply equation 6-10,

$$A_c = \frac{F_0}{f_b}\left(1 + \frac{e - (M_G/F_0)}{k_t}\right)$$
$$= \frac{169}{1.80}\left(1 + \frac{19.5 - (210 \times 12/169)}{5.8}\right)$$
$$= 168 \text{ sq in. } (108 \times 10^3 \text{ mm}^2)$$

To keep the top fiber stress within limit, we have, from equation 6-17*a*,

$$A_c = \frac{141 \times 36}{1.60 \times 22.5 - 0.24 \times 13.5}$$
$$= 155 \text{ sq in. } (100 \times 10^3 \text{ mm}^2)$$

The area furnished is 200 sq in. (129×10^3 mm^2), which can be reduced if desired.

6-6 Elastic Design, Composite Sections

As described previously, a composite section consists of a precast prestressed protion to be combined with another cast-in-place portion which usually forms part or all of the top flange of the beam. The design of composite sections is more complicated than that of simple ones because there are many possible combinations in the make-up of a composite section. Only a very common case will be treated here, leaving the possible variations to the designer after he or she has mastered the principles here presented.

In the case considered here the precast portion forms the lower flange and the web while part or the whole of the top flange is cast in place. Tension is usually permitted in the top flange at transfer and often also in the bottom flange under working load. Hence, formulas will be derived to include tensile stresses. These can be easily simplified when tensile stresses are not permitted. For such composite sections compressive stress in the cast-in-place portion is seldom critical and hence will be checked only at the end of the design. When the cast-in-place portion becomes the major part of the web, or when falseworks are employed to support the precast portion during casting, the method presented here has to be modified accordingly.

The procedure of design here presented follows closely the basic approach previously adopted for noncomposite sections. It is essentially a trial-and-error process, simplified by a systematic and fast converging procedure and assisted by the use of some simple relations and formulas. One additional concept introduced for composite action is the reduction of moments on the composite section to equivalent moments on the precast portion. This is accomplished by the ratio of the section moduli of the two sections. Steps in the design and the formulas employed will now be explained.

Step **1.** Location of c.g.s. For a given trial section, the c.g.s. must be so located that the precast portion will not be overstressed and yet will possess the optimum capacity in resisting the applied external moments. Thus, the c.g.s. must be situated as low as possible but not lower than given by the following value of eccentricity, Fig. 6-16(a),

$$e = k_b + e_1 + e_2$$

where

$$e_1 = \frac{f_t' I}{c_t F_0}$$

$$e_2 = \frac{M_G}{F_0}$$

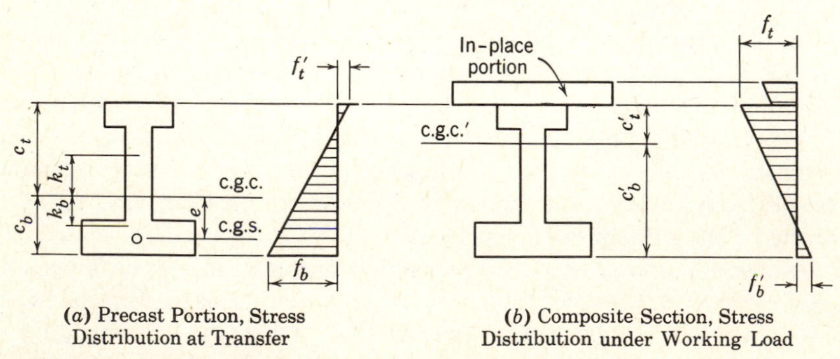

(a) Precast Portion, Stress Distribution at Transfer

(b) Composite Section, Stress Distribution under Working Load

Fig. 6-16. Elastic design of composite sections.

where $f_t' =$ allowable tension stress on top fiber of precast portion at transfer
$I =$ moment of inertia of precast portion
$c_t =$ distance to top fiber from c.g.c. of precast portion

Step **2.** Compute the equivalent moment on the precast portion. For any moment M_C acting on the composite section, it will produce stresses on the precast portion as follows, Fig. 6-16(b),

$$f_t = \frac{M_C c_t'}{I'}$$

$$f_b = \frac{M_C c_b'}{I'}$$

where $I' = I$ of composite section, $c_b' =$ distance to extreme fibers of the precast portion measured from c.g.c.′ of the composite section. Let

$$m_t = \frac{I/c_t}{I'/c_t'}$$

and

$$m_b = \frac{I/c_b}{I'/c_b'}$$

We have

$$f_t = \frac{m_t M_C c_t}{I} = \frac{m_t M_C}{A_c k_b}$$

$$f_b = \frac{m_b M_C c_b}{I} = \frac{m_b M_C}{A_c k_t}$$

where $A_c =$ area of the precast section
$k_b =$ bottom kern distance of precast section
$k_t =$ top kern distance of precast section.

The above indicates that M_C can be modified by the coefficients m_t and m_b so that it can be reduced to equivalent moments for computation based on the precast-portion properties.

Step **3.** Compute the amount of prestress required for the moments as follows. If $M_P =$ the total moment acting on the precast portion, and $f_b' =$ allowable tensile stress at the bottom fiber, we have

$$\frac{F}{A_c}\left(-1-\frac{e}{k_t}\right)+\frac{M_P}{A_c k_t}+\frac{m_b M_C}{A_c k_t}=f_b'$$

$$F=\frac{M_P+m_b M_c-f_b' k_t A_c}{e+k_t} \qquad (6\text{-}18)$$

or

$$F = \frac{M_P + m_b M_C}{e + k_t} \tag{6-18a}$$

if

$$f_b' = 0$$

from which compute the required initial prestress F_0. Revise the location of c.g.s. by this new value of F_0 if necessary.

Step **4.** In order to limit the bottom fiber stress to the allowable value at transfer, we have

$$f_b = \frac{F_0}{A_c} + \frac{(F_0 e - M_G)}{A_c k_t}$$

from which

$$A_c = \frac{1}{f_b}\left[F_0 + \frac{F_0 e - M_G}{k_t}\right] \tag{6-19}$$

In order to limit the top fibers of the precast portion to within allowable compressive stress f_t under working load, we have

$$f_t = \frac{F}{A_c} + \frac{M_P + m_t M_C - Fe}{A_c k_b}$$

$$A_c = \frac{1}{f_t}\left[F + \frac{M_P + m_t M_C - Fe}{k_b}\right] \tag{6-20}$$

The greater of the two formulas will control the A_c required for the precast portion. The top fiber of the cast-in-place top flange can be computed by the formula $f = Mc/I$, using the applicable values.

EXAMPLE 6-7

The top flange of a composite section is given as a slab 4 in. thick and 60 in. wide cast in place. Design a precast section with a total depth of 36 in. (including the slab thickness) to carry the following moments: $M_T = 320$ k-ft, $M_G = 40$ k-ft, $M_P = 100$ k-ft, $M_C = 220$ k-ft. Allowable stresses are: $f_t = -1.60$ ksi, $f_b = -1.80$ ksi, $f_t' = 0.30$ ksi, $f_b' = 0.16$ ksi. Initial prestress = 150 ksi, effective prestress = 125 ksi. (Slab: 102 mm thick and 1524 mm wide, total depth = 914.4 mm, $M_T = 434$ kN-m, $M_G = 54$ kN-m, $M_p = 136$ kN-m, $M_c = 298$ kN-m, $f_t = -11.03$ N/mm², $f_b = -12.41$ N/mm², $f_t' = 2.07$ N/mm², $f_b' = 1.10$ N/mm², initial prestress = 1034 N/mm², and effective prestress = 862 N/mm².)

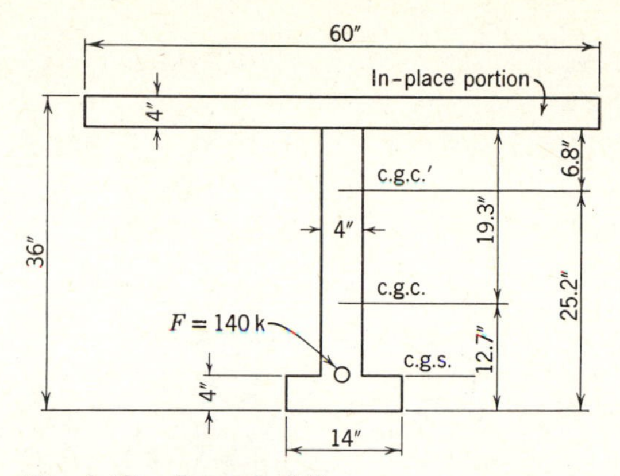

Fig. 6-17. Example 6-7.

To assume the section, make a preliminary design, assuming a lever arm of $0.65h$ for the prestressing force in resisting the total moment; we have

$$F = \frac{M_T}{0.65h}$$

$$= \frac{320 \times 12}{0.65 \times 36}$$

$$= 164 \text{ k } (730 \text{ kN})$$

$F_0 = 164 \times 150/125 = 197$ k (876 kN). For an inverted T-section, the concrete area required can be approximated by

$$A_c = 1.5 \frac{F_0}{f_b}$$

$$= 1.5 \frac{197}{1.8}$$

$$= 164 \text{ sq in. } (106 \times 10^3 \text{ mm}^2)$$

From this preliminary section, a sketch of a trial section is shown, Fig. 6-17, and the design proceeds as follows.

For the precast portion, the section properties are

$$4 \times 14 = 56 \times 2 = 112$$

$$28 \times 4 = 112 \times 18 = 2016$$

$$A_c = \overline{168} \qquad \overline{2128} \div 168 = 12.7 \text{ in.} = c_b$$

$$56(4^2/12 + 10.7^2) = 6{,}500$$

$$112(28^2/12 + 5.3^2) = 10{,}450$$

$$\overline{16{,}950} \div 168 = 101 = r^2$$

$$k_t = 101/12.7 = 8.0 \text{ in. } (203 \text{ mm})$$

$$k_b = 101/19.3 = 5.2 \text{ in. } (132 \text{ mm})$$

For the composite section, the properties are

$$4\times 60=240\times 2=480$$
$$168\times 23.3=3920$$
$$4400\div 408=10.8 \text{ in.}$$

$$240(4^2/12+8.8^2) \quad =18{,}800$$
$$168(12.5^2) \quad =26{,}200$$
$$I \text{ of precast portion}=16{,}950$$
$$\overline{62{,}000}$$

$$m_t=\frac{I/c_t}{I'/c_t'}=\frac{16{,}950/19.3}{62{,}000/6.8}=0.10$$

$$m_b=\frac{I/c_b}{I'/c_b'}=\frac{16{,}950/12.7}{62{,}000/25.2}=0.54$$

Step 1. Location of c.g.s.

$$e_1=\frac{f_t'I}{c_tF_0}$$
$$=\frac{0.30\times 16{,}950}{19.3\times 197}=1.3 \text{ in. (33 mm)}$$
$$e_2=M_G/F_0$$
$$=(40\times 12)/197=2.4 \text{ in. (61 mm)}$$
$$k_b=5.2 \text{ in. (132 mm)}$$
$$e=1.3+2.4+5.2=8.9 \text{ in. (226 mm)}$$

Thus c.g.s. can be located at $12.7-8.9=3.8$ in. (97 mm) above bottom fiber.

Step 2. As computed above,

$$m_t=0.10$$
$$m_b=0.54$$

Step 3. Compute the required F,

$$F=\frac{M_P+m_bM_C-f_b'k_tA_c}{e+k_t}$$
$$=\frac{(100+0.54\times 220)12-0.16\times 8.0\times 168}{8.9+8.0}$$
$$=\frac{2430}{16.9}$$
$$=144 \text{ k (641 kN)}$$
$$F_0=144\times 150/125=173 \text{ k (770 kN)}$$

For $F_0 = 173$ k instead of 197 k, revise e_1 and e_2 as follows;

$$e_1 = 1.3 \times 197/173 = 1.5 \text{ in. (38 mm)}$$

$$e_2 = 2.3 \times 197/173 = 2.7 \text{ in. (69 mm)}$$

$$e = 5.2 + 1.5 + 2.7 = 9.4 \text{ in. (239 mm)}$$

which indicates that c.g.s. can be located at $12.7 - 9.4 = 3.3$ in. (84 mm) above bottom fiber. With new $e + k_t = 9.4 + 8.0 = 17.4$ in. (442 mm), F can be revised to be $144 \times 16.9/17.4 = 140$ k (623 kN), $F_0 = 140 \times 150/125 = 168$ k (747 kN).

Step 2. To keep bottom fiber within allowable stress f_b,

$$A_c = \frac{1}{f_b}\left(F_0 + \frac{F_0 e - M_G}{k_t}\right)$$

$$= \frac{1}{1.80}\left(168 + \frac{168 + 9.4 - 40 \times 12}{8.0}\right)$$

$$= 170 \text{ sq in. } (110 \times 10^3 \text{ mm}^2)$$

To keep top fiber within allowable f_t,

$$A_c = \frac{1}{f_t}\left(F + \frac{M_P + m_t M_C - Fe}{k_b}\right)$$

$$= \frac{1}{1.60}\left(140 + \frac{(100 + 0.10 \times 220)12 - 140 \times 9.4}{5.2}\right)$$

$$= 106 \text{ sq in. } (68 \times 10^3 \text{ mm}^2)$$

The top fiber is not controlling in this case, and the A_c required for abottom fiber stress is 170 sq in. (110×10^3 mm^2), which is very close to the A_c of 168 sq in. (108×10^3 mm^2) furnished by the trial section. The design is considered satisfactory.

As pointed out in section 5-7, differential shrinkage and creep between the precast and the in-place portions of a composite section may produce high stresses. However, the usual practice is to ignore them in the process of design. Such practice can be justified on the basis that sections obtained by following the conventional procedure and using the allowable stresses generally result in fairly reasonable proportions. Another justification lies in the fact that the ultimate strength of composite sections is seldom much affected by shrinkage and creep. When deflections and camber are the controlling factors, the effect of differential shrinkage and creep should be carefully considered in design.

6-7 Ultimate Design

Only the ultimate design for simple sections with bonded tendons will be discussed here. Basically, the procedure is also applicable to the ultimate design of composite sections, the details of which will be left to the reader, however.

Preliminary Design. The amount of mathematics involved in the design of prestressed-concrete sections is less in ultimate design than in elastic design, since the ultimate flexural strength of sections can be expressed by simple semiempirical formulas. For preliminary design, it can be assumed that the ultimate resisting moment of bonded prestressed sections is given by the ultimate strength of steel acting with a lever arm. This arm lever varies with the shape of section and generally ranges between $0.6h$ and $0.9h$, with a common value of $0.8h$. Hence the area of steel required is approximated by

$$A_s = \frac{M_T \times m}{0.80h \times f_{ps}} \tag{6-21}$$

where m = factor of safety or the load factor.

Assuming that the concrete on the compressive side is stressed to $0.85f'_c$, then the required ultimate concrete area under compression is

$$A'_c = \frac{M_T \times m}{0.80h \times 0.85f'_c} \tag{6-22}$$

which is supplied by the compression flange (occasionally with the help of part of the web). The web area and the concrete area on the tension side are designed to provide the shear resistance and the encasement of steel, respectively. In addition, concrete on the precompressed tension side has to stand the prestress at transfer. For a preliminary design, these areas are often obtained by comparison with previous designs rather than by making any involved calculations.

The chief difficulty in ultimate design lies in the proper choice of the factor of safety or the load factor, which will depend on the Code being followed in the design. For the present it will be assumed that a load factor of 2 will be sufficient for steel and one of 2.5 for concrete. The application of the method is illustrated in the following example.

EXAMPLE 6-8

Make a preliminary design for a prestressed-concrete section 36 in. (914.4 mm) high to carry a total dead and live load moment of 320 k-ft (434 kN), using steel with an ultimate strength of 220 ksi (1517 N/mm^2) and concrete with $f'_c = 4$ ksi (28 N/mm^2). Use ultimate design, and assume a bonded beam.

Solution Using a load factor of 2 for steel, we have, from equation 6-24,

$$A_s = \frac{320 \times 12 \times 2}{0.80 \times 36 \times 220} = 1.21 \text{ sq in. } (781 \text{ mm}^2)$$

Using a load factor of 2.5 for concrete, from equation 6-22,

$$A'_c = \frac{320 \times 12 \times 2.5}{0.80 \times 36 \times 0.85 \times 4} = 98 \text{ sq in. } (63 \times 10^3 \text{ mm}^2)$$

Thus a preliminary section can be sketched as in Fig. 6-18, providing an ultimate area of 98 sq in. (63×10^3 mm^2) under compression, assuming the ultimate neutral axis to be 10 in. (254 mm) below the top fiber. Note that the exact location of the ultimate neutral axis cannot and need not be obtained for a preliminary design but can be assumed to be about 30% of the effective depth of section.

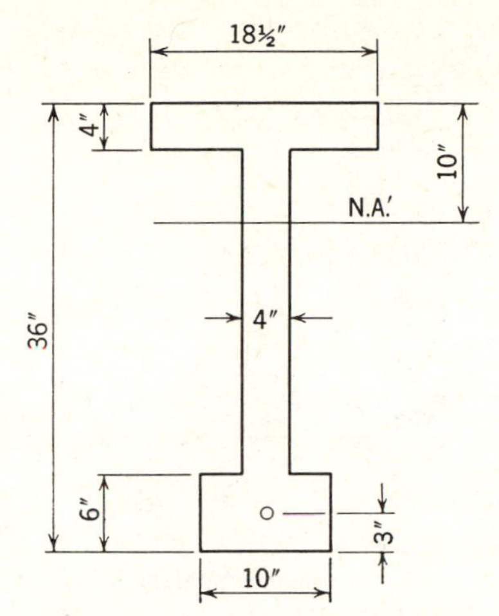

Fig. 6-18. Example 6-8 and 6-9.

Final Design. Although the above illustrates a preliminary design based on ultimate strength, a final design is more complicated in that the following factors must be considered.

1. Proper and accurate load factors must be chosen for steel and concrete, related to the design load and possible overloads for the particulat structure.
2. Compressive stresses at transfer must be investigated for the tensile flange, generally by the elastic theory. In addition, the tensile flange should be capable of housing the steel properly.
3. The approximate location of the ultimate neutral axis may not be easily determined for certain sections.
4. Design of the web will depend on shear and other factors.
5. The effective lever arm for the internal resisting couple may have to be more accurately computed.
6. Checks for excessive deflection and overstresses may have to be performed.

In spite of these factors, a reasonably good final design for flexure can be made for bonded sections based on ultimate strength considerations. This is illustrated in the following example.

EXAMPLE 6-9

Make a final design for the beam in example 6-8, based on its ultimate strength.

Solution A trial-and-error procedure is considered convenient for the purpose. Using the preliminary section obtained in example 6-8 as the first trial section, Fig. 6-18, we can proceed as follows.

With the ultimate axis 10 in. below top fiber, the centroid of the ultimate compressive force is located by

$$\frac{74\times2+24\times7}{74+24}=3.2 \text{ in. } (81 \text{ mm})$$

or 3.2 in. (81 mm) from top fiber. With the c.g.s. located 3 in. above bottom fiber, the ultimate lever arm for the resisting moment is

$$36-3.2-3=29.8 \text{ in. } (757 \text{ mm})$$

Now the area of steel required may be recomputed as

$$A_s=\frac{320\times12\times2}{29.8\times220}=1.17 \text{ sq in. } (755 \text{ mm}^2)$$

which is very near to the preliminary value of 1.21 sq in. (781 mm^2), and no further trial is necessary. Design of the top flange may be done as in example 6-8. Tensile stress should also be checked in the top flange at transfer. Since the bottom flange is seldom controlled by ultimate strength considerations, it is usually checked for elastic stresses. The web, of course, has to be designed for shear, which will be discussed in Chapter 7.

Ultimate vs. Elastic Design. At the present time, both the elastic and the ultimate designs are used for prestressed concrete, the majority of designers still following the elastic theory. It is difficult to state exact preference for one or the other. Each has its advantages and shortcomings. But, whichever method is used for design, the other one must often be applied for checking. For example, when the elastic theory is used in design, it is the practice to check for the ultimate strength of the section in order to find out whether it has sufficient reserve strength to carry overloads. When the ultimate design is used, the elastic theory must be applied to determine whether the section is overstressed under certain conditions of loading and whether the deflections are excessive. Overstressing is objectionable because it may result in undesirable cracks and creep and fatigue effects. When the design is of conventional types and proportions, such checking becomes unnecessary, because it is then generally known that designing by one method will yield safe results when checked by the other. This is, in fact, the reason why such checking is not required of reinforced-concrete structures designed by the usual codes. When we delve into new types and proportions, it is possible that elastic design alone might not yield a sufficiently safe structure under overloads, while the ultimate design by itself might give no guarantee against excessive overstress under working conditions. It is therefore deemed desirable to apply both the elastic and the ultimate methods, especially for structures of unusual proportions.

An understanding of both theories of design is also essential in forming judgment when designing structures. Sometimes, design based on one method will yield different proportions from those based on the other. In order to illustrate the point, let us compare a symmetrical I-section and its circumscribing rectangular section, example 6-10. Based on elastic design, allowing no tension in the concrete, the I-section will carry greater moment than the rectangular section; the rectangular section, however, has a higher ultimate strength. If strength is a more important consideration, the design can be based on ultimate strength. If tensile stress, cracking, creep, or deflection is a critical limit, elastic design should be followed. If both strength and stress are controlling criteria for a structure, we are forced to apply both methods to ensure safety and proper behavior, often at the sacrifice of some economy.

EXAMPLE 6-10

An I-section and another circumscribing rectangular section are both prestressed with 0.9 sq in. of steel, Fig. 6-19. $f_c' = 5$ ksi, $f_s = 125$ ksi, $f_{pu} = 250$ ksi. Compute (a) the resisting moment capacity of each section by the elastic theory, allowing no tension in concrete; (b) the ultimate moment capacity of each section ($A_{ps} = 581$ mm^2, $f_c' = 34$ N/mm^2, $f_s = 862$ N/mm^2, and $f_{pu} = 1724$ N/mm^2).

Solution (a) For no tension in concrete, using formula 6-8, we have

	I-Section	Rectangular Section
Area, sq in.	128	240
I, in.^{4}	6170	8000
k_t, in.	4.82	3.33
Lever arm between c.g.s. and k_t, in.	13.32	11.83
Effective prestress, kips	112.5	112.5
Resisting moment, k-ft	125	111

(b) For the ultimate capacity, following the method in section 5-6, we have, assuming $k_1 f_c' = 4.5$ ksi,

	I-Section	Rectangular Section
$k'd$, in.	6.5	4.2
Ultimate distance of centroid of compression force from top fiber, in.	2.4	2.1
Ultimate lever arm a' for resisting couple, in.	16.1	16.4
Ultimate tension in steel, kips	225	225
Ultimate resisting moment, k-ft	302	307

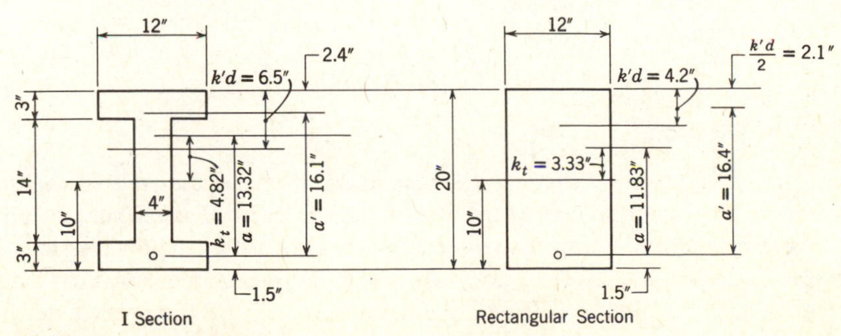

Fig. 6-19. Example 6-10.

The above example illustrates that, when designed by the elastic theory, the I-section can carry greater moment; when designed by ultimate strength, the rectangular section carries the greater moment. It is obvious from this example that the I-shaped section is more efficient in utilizing the concrete area. The rectangular section has almost twice as much area but carries less resisting moment in (a) above and only slightly more ultimate moment in (b). It will be shown later that the thin web can carry the shear stress, using stirrups if required.

There is another case where the application of the elastic and the ultimate design yields radically different results. Consider two sections of exactly the same steel and concrete dimensions, but one with bonded steel and the other unbonded (see section 5-6). By the elastic design, both sections will carry the same moment; but by the ultimate design, the unbonded section will carry much less moment. Which method should be used in design will depend on the particular conditions of the structure. When ultimate strength is an important consideration, the bonded section must be given preference. On the other hand, when excessive overloads are not likely, the bonded and unbonded sections can be considered equally satisfactory.

6-8 Shapes of Concrete Sections

Having studied both the elastic and ultimate designs, we are now ready to discuss the selection of the best shapes for prestressed concrete sections under flexure. The simplest form is the rectangular shape possessed by all solid slabs and used for some short-span beams. As far as formwork is concerned, the rectangular section is the most economical. But the kern distances are small, and the available lever arm for the steel is limited. Concrete near the centroidal axis and on the tension side is not effective in resisting moment, especially at the ultimate stage. As observed in the previous section, the rectangular section is not as efficient in the use of the concrete section as is the I-shaped section.

Hence other shapes are frequently used for prestressed concrete, Fig. 6-20:

1. The symmetrical I-section.
2. The unsymmetrical I-section.
3. The T-section.
4. The inverted T-section.
5. The box section.

The suitability of these shapes will depend on the particular requirements. The I-section has its concrete concentrated near the extreme fibers where it can most effectively furnish the compressive force, both at transfer of prestress and under working and ultimate loads. The more the concrete is concentrated near the extreme fibers, the greater will be the kern distances, and the greater will be the

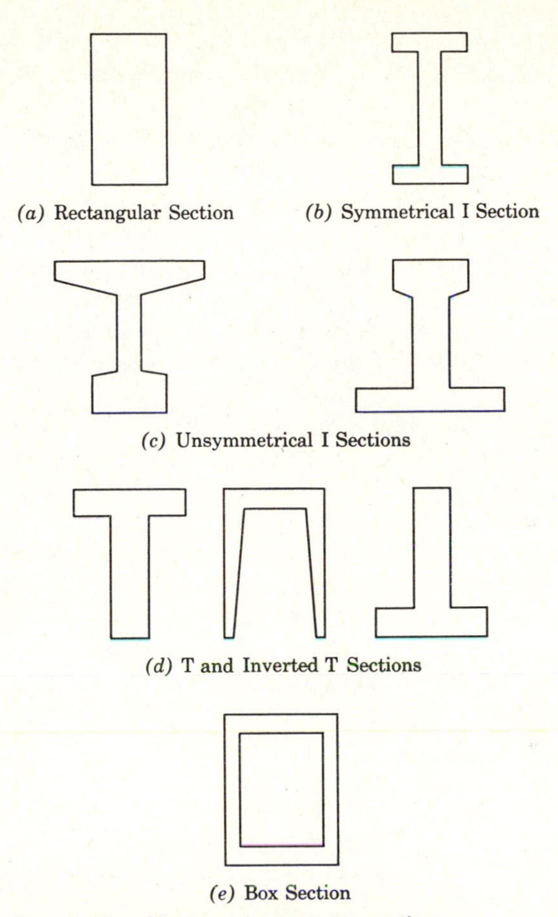

(*a*) Rectangular Section (*b*) Symmetrical I Section

(*c*) Unsymmetrical I Sections

(*d*) T and Inverted T Sections

(*e*) Box Section

Fig. 6-20. Shapes of concrete sections.

lever arm furnished for the internal resisting couple. However, this principle of concentrating the concrete in the extreme fibers cannot be carried too far, because the width and thickness of the flanges are governed by practical considerations, and the web must have a minimum thickness to carry the shear, to avoid buckling, and to permit proper placement of concrete.

If the M_G/M_T ratio is sufficiently large, there is little danger of overstressing the flanges at transfer, and concrete in the bottom flange can be accordingly diminished. This will result in an unsymmetrical I-section which when carried to the fullest extent becomes a T-section. A T-section, similar to that for reinforced beams, is often most economical, since the concrete is concentrated at the top flange where it is most effective in supplying the compressive force. It may not be economically used, however, where the M_G/M_T ratio is small, because the center of pressure at transfer may lie below the bottom kern point. Then tensile stresses may result in the top flange and high compressive stresses in the bottom flange.

The unsymmetrical I-section with a bigger bottom flange, like a rail section, is not an economical one in carrying ultimate moment, since there is relatively little concrete on the compression flange. However, there is a great deal of material to resist the initial prestress. It can be economically used for certain composite sections, where the tension flange is precast and the compression flange is poured in place. This section requires very little girder moment to bring the center of pressure within the kern and hence is suitable when the M_G/M_T ratio is small. When carried to the extreme, this section becomes an inverted T-section.

The box section has the same properties as the I-section in resisting moment. In fact, their section properties are identical and both are listed in Table 6 of Appendix C. The adoption of one or the other will depend on the practical requirements of each structure.

The above discussion can be summarized as follows. For economy in steel and concrete it is best to put the concrete near the extreme fibers of the compression flange. When the M_G/M_T ratio is small, more concrete near the tension flange may be necessary. When the M_G/M_T ratio is large, there is little danger of overstressing at transfer, and concrete in the tension flange is required only to house the tendons properly.

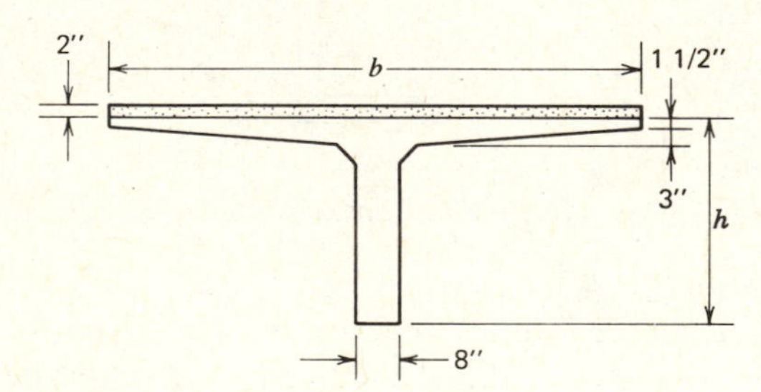

Fig. 6-21(*a*). Typical single-tee section (Lin Tee).

Table of Properties

b, ft	h, in.	A, in.^{2}	I, in.^{4}	y_t, in.	y_b, in.	V/S, in.	
8	36	570	68,917	9.99	26.01	2.16	without 2″ topping
10	48	782	168,968	11.36	36.64	2.33	without 2″ topping
8	36+2		88,260		29.09		with 2″ topping

PCI Design Handbook gives tables of safe superimposed loads for these sections.

In choosing the shapes, prime importance must be given to the simplicity of formwork. When the formwork is to be used only once, it may constitute the major cost of the beam, so that any irregular shapes for the purpose of saving concrete or steel may not be in the interest of overall economy. On the other hand, when the forms can be reused repeatedly, more complicated shapes may be justified.

For plants producing precast elements, it is often economical to construct forms that can be easily modified to suit different spans and depths. For example, by filling up the stems for the section in Fig. 6-21(a) and 6-21(b), several depths can be obtained. Again, by ommitting the center portions of a tapered beam or decreasing the distance between the side forms, one set of forms can be made to fit many shorter spans.

Sections must be further designed to enable proper placement of concrete around the tendons and the corners. This is especially true when proper vibration cannot be ensured. The use of fillets at corners is often desirable. It is also common practice to taper the sides of the flanges. Such tapering will permit easier stripping of the formwork and easier placement of concrete.

Examples of some sections commonly used in the United States are shown in Fig. 6-21. A typical single tee section (often known as Lin Tee) for which a set of forms can be adjusted to suit several variables is shown in Fig. 6-21(a). The depth of the section can be set by raising the soffit. The width of the stem can be

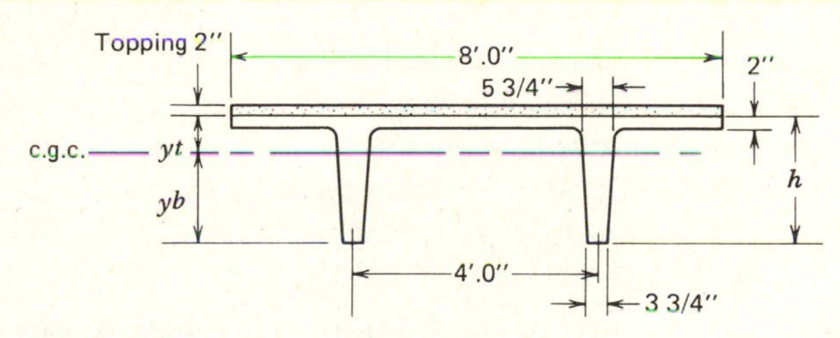

Fig. 6-21(b). Typical double-tee section.

Table of Properties

	h, in.	A, in.^{2}	I, in.^{4}	y_t, in.	y_b, in.	V/S, in.
Without	14	306	4,508	3.49	10.51	1.25
topping	18	344	9,300	4.73	13.27	1.32
	24	401	20,985	6.85	17.15	1.41
With	14+2		7,173		12.40	
topping	18+2		13,799		15.51	
2" thick	24+2		29,853		19.94	

Other sections listed in *PCI Design Handbook* 12 in., 16 in., 20 in., and 32 in. deep with tables of safe superimposed load.

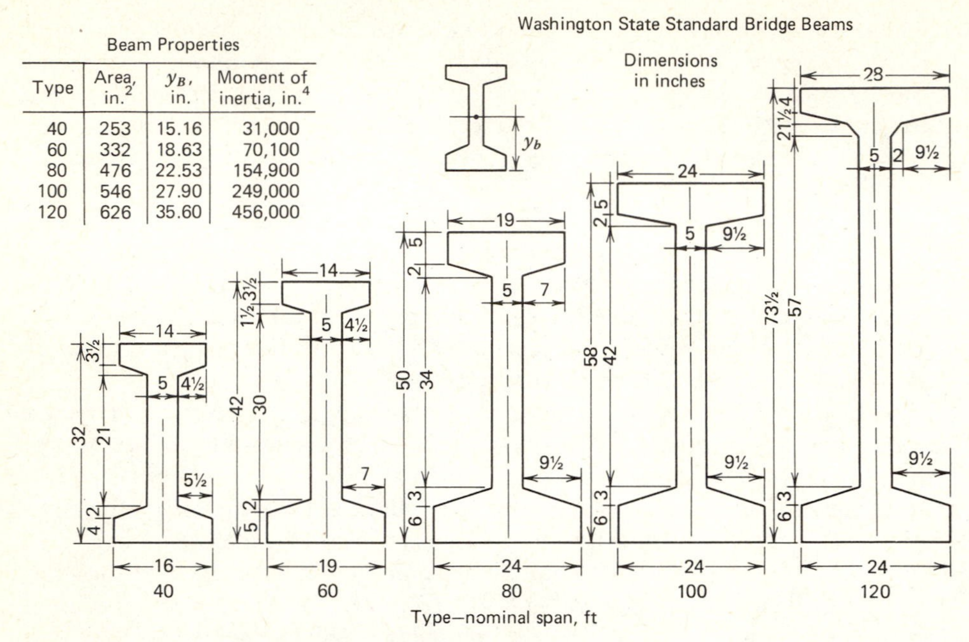

Beam Properties

Type	Area, in.^{2}	y_B, in.	Moment of inertia, in.^{4}
40	253	15.16	31,000
60	332	18.63	70,100
80	476	22.53	154,900
100	546	27.90	249,000
120	626	35.60	456,000

Fig. 6-21(*c*). Washington State standard beams.

varied by spreading the two halves of the forms. The width and the thickness of the flange can be modified by changing the outside strips. When carefully designed, the 36-inch (914.4 mm) deep section can be used for floors up to 100 ft (30 m), roofs up to 120 ft (36 m) and highway bridges up to 70 ft (21 m). Some plants have steel forms to produce these tees up to a depth of 48 in. (1,219 mm) and a width of 10 ft (3.0 m).

Slab sections of about 8-in. (203 mm) thick, with round cores of about $5\frac{1}{2}$-in. (140 mm) diameter spaced at some 8-in. (203 mm) centers, are frequently used for short span roofs and floors where a flat soffit is desired. Several companies have standardized methods of production and shapes in depths up to about 12 in. (305 mm) for hollow core slabs.

Figure 6.21(*b*) shows the double tee section commonly used for roofs and floors of buildings. Lightweight or normal weight concrete may be used and the 2 in. (50.8 mm) concrete topping is added to the section in many cases, forming a composite structure. The section properties, especially the moment of inertia, change very significantly with the addition of this topping. Many parking structures have used these double tees with spans in the order of 60 ft. (18.3 m), and deeper sections of this type have been used for spans up to 100 ft. (30.5 m). The PCI Design Handbook includes design aids for selection of the depth of

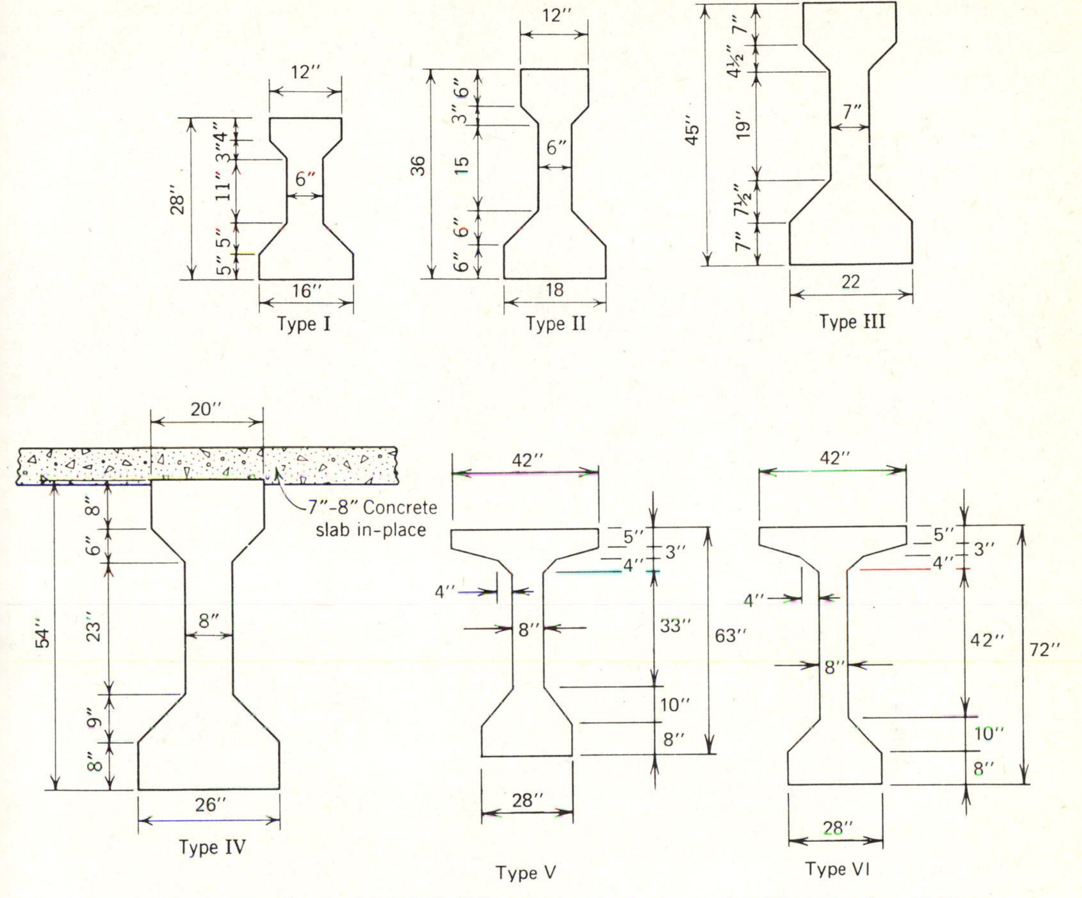

Fig. 6-21(*d*). Standard AASHTO-PCI prestressed concrete I-beams for highway bridges.[2]

Table of Properties
(Without in-place slab)

Beam Type	Area, in.^{2}	I, in.^{4}	c_b, in.	Recommended Span Limits, ft
I	276	22,750	12.59	30–45
II	369	50,980	15.83	40–60
III	560	125,390	20.27	55–80
IV	789	260,730	24.73	70–100
V	1013	521,180	31.96	90–120
VI	1085	733,320	36.38	110–140

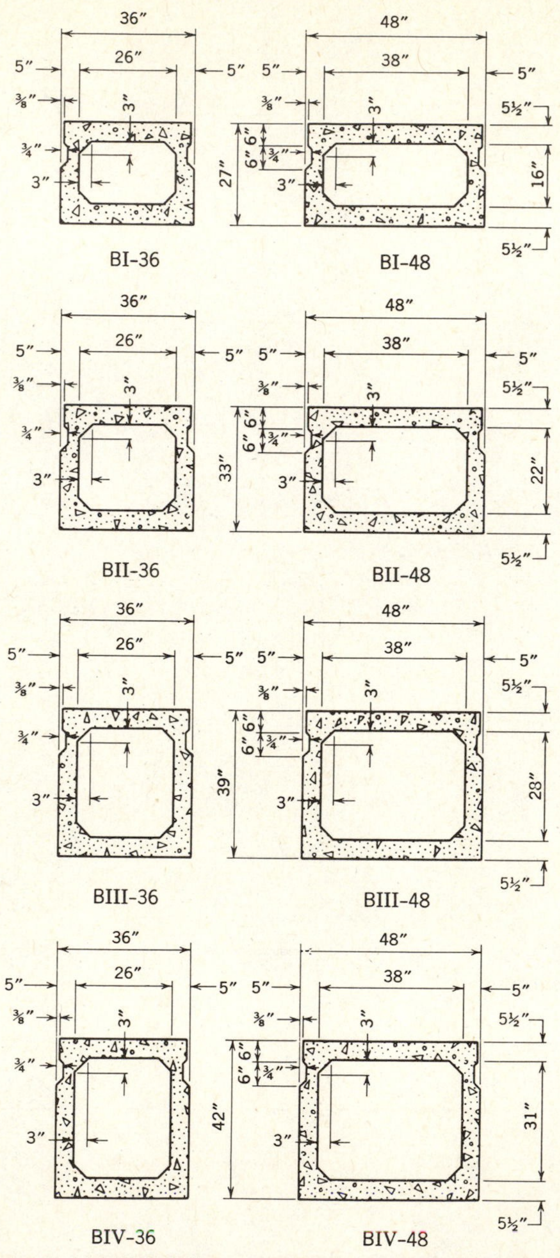

Fig. 6-21(*e*). Standard AASHTO-PCI prestressed concrete box beams for highway bridges.

Table of Properties

Beam Type	Area, in.^{2}	I, in.^{4}	c_b, in.	Recommended Span Limits, ft	
				Draped Strand	Straight Strand
BI-36	560.5	50,334	13.35	74	62
BI-48	692.5	65,941	13.37	73	63
BII-36	620.5	85,153	16.29	86	73
BII-48	752.5	110,499	16.33	86	74
BIII-36	680.5	131,145	19.25	97	83
BIII-48	812.5	168,367	19.29	96	83
BIV-36	710.5	158,644	20.73	103	87
BIV-48	842.5	203,088	20.78	103	88

standard double tee section and the strands needed for carrying known service loads on a specified span.

The Washington State bridge beams shown in Fig. 6.21(c) are used for composite structures, the cast-in-place slab being added to the top of these sections in a manner similar to the ASSHTO-PCI standard beams shown in Fig. 6.21(d). The Washington State sections have thinner webs than the AASHTO-PCI beams, which are standard in many states. The thin web makes the sections more efficient in flexure and stirrups are added as required to assure that the shear strength is adequate in these beams.

I-sections of the same sizes standardized for bridges have been used for buildings with a poured-in-place slab so as to get composite action. The AASHTO-PCI have also standardized 8 sections for prestressed box beams, Fig. 6-21(e).

6-9 Design with Standard Precast Sections

Noncomposite Sections. Many designs of standard single tees, double tees, and hollow core sections are made using design aids which allow selection of a trial section from load tables. These tables make the selection much simpler than the successive trial procedure outlined earlier in this chapter. The design aids are made up for these products assuming certain loss of prestress which should be checked to be certain that it is reasonable. The ACI Committee Recommendations presented in Chapter 4 make this rather easy (or, the Simplified Method might be used) and the designer should verify that the section is adequate at all load stages.

Both serviceability and strength must be satisfactory. Calculations will indicate whether stresses are satisfactory at critical sections both at transfer and after losses. The camber and deflection with time may be estimated using the rather simple procedures described in Chapter 8. Strength is analyzed by the ACI Code procedures discussed in Chapter 5.

These checks on the member selected are much easier than the general design procedure since the cross section is known without going through the successive trial procedure. The designer is responsible for checking the member and should not depend completely on the design aids without verifying that the section selected is totally satisfactory.

Composite Sections. The design of composite members is also very much simplified by the availability of standard sections. Usually experience will indicate the section which may be used for a given span and the major thing to be determined is the spacing of the composite beams. Again, as with noncomposite precast beams, the known section properties for the precast beam will avoid the successive trial approach to design given in Section 6-7.

It may be desirable to determine the prestress force which would fully precompress the bottom fiber of the precast beam, then solve the spacing which the beam can accommodate. This is basically an analysis procedure which can be done rather quickly. A heavier section would be selected for the precast beam if the spacing is too close, or a lighter beam if the spacing is too great. The composite system makes efficient use of the precast beams and the design would be made with the "shored" or "unshored" condition being clearly specified and considered in the design calculations. Example 6-11 illustrates this design process.

The concrete used in the slab often will be specified with smaller f_c' than that in the prestressed, precast beam. This can be accounted for in the design by using the modification of effective width as illustrated in Fig. 6-22. The ACI rule for tee beams is widely used for design of composite systems with 8t as the effective width wither side of the beam as illustrated. Note that this cannot exceed the beam spacing, s, as shown in Fig. 6-22. We can modify the effective width, b, by the modular ratio E_c (slab)/E_c (beam) to account for the difference in concrete strength as shown in Fig. 6-22(b). The elastic analysis is done for a beam which is of one material, in this case concrete with $f_c' = 6000$ psi (41 N/mm^2).

This modification applies only to the elastic analysis for stresses at various stages. Note that as stated in section 6-6, the effect of differential shrinkage and creep between the precast and the in-place concrete may seriously change the real stress distribution, although it is not often computed in analysis.

When the strength is checked, the full effective width, b, is appropriate with the f_c' being that of the slab concrete (Fig. 6-22 example, $f_c' = 4000$ psi $= 28$ N/mm^2 for slab). The M_u is evaluated for the composite section exactly as for a noncomposite section and the compression zone will be within the slab in almost all practical designs. The analysis following ACI Code procedure is simple and

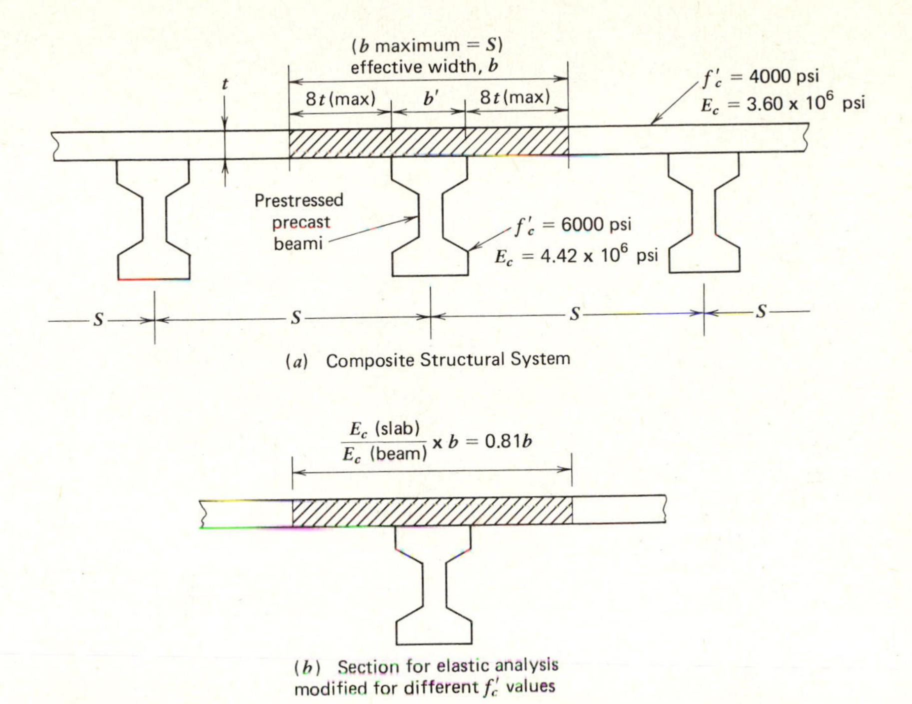

Fig. 6-22. Design of composite system with standard precast beams.

entirely adequate for this check on strength. Example 6-11 includes this check on strength.

EXAMPLE 6-11

Design a composite prestressed concrete structural system for a simply supported 90 ft. span. The beams will be Type III standard secions, Fig. 6-21*d*, with a 6-in.-thick composite slab. Design loads for the structure are: dead load of the composite system (normal weight concrete), w_G; additional dead load $w_D = 20$ lb/ft^2; live load, $w_L = 50$ lb/ft^2. Normal weight woncrete for the precast beams is specified: $f'_{ci} = 4500$ psi, $f'_c = 6000$ psi and for the slab $f'_c = 4000$ psi. Design for unshored beams in construction following the ACI Code for the critical section at midspan. Assume total loss of prestress is 20 percent for this section. ($w_D = 0.958$ kN/m^2, $w_L = 2.395$ kN/m^2, for precast beam $f'_{ci} = 31$ N/mm^2, $f'_c = 41$ N/mm^2, for slab $f'_c = 28$ N/mm^2)

(*a*) Determine the pretensioned strand arrangement to fully utilize the precast beams.

(*b*) Find the maximum spacing of the beams for this structural system based on allowable stresses at service load.

(*c*) Check strength following ACI Code.

Solution (*a*) Assume $g = 5$ in. (127 mm), Fig. 6-23, and solve limiting force, F_0, allowed at transfer stage. $f_{top} \leqslant 3\sqrt{f'_{ci}} = 201$ psi (1.39 N/mm²), $f_{bot} \leqslant 0.6 f'_c = 2700$ psi (18.62 N/mm²).

Sections properties for precast beam:

$A = 560$ in.² (361×10^3 mm) $\qquad w_g = 583$ lb/ft (8.51 kN/m)

$I = 125,390$ in.⁴ (52×10^9 mm⁴) $\qquad S_b = 6186$ in.³ (101×10^6 mm³)

$c_b = 20.27$ in. (515 mm) $\qquad S_t = 5070$ in.³ (83×10^6 mm³)

$c_t = 24.73$ in. (628 mm)

$e = 20.27 - 5.0 = 15.27$ in. (388 mm) $\qquad M_G = 7083$ in.-k (800 kN-m)

$$f_{top} = \frac{F_0}{560} + \frac{F_0(15.27)}{5070} - \frac{7083}{5070} = 0.201 \text{ ksi } (1.39 \text{ N/mm}^2)$$

$F_0 = 1303$ k (5796 kN) based on $3\sqrt{f'_{ci}}$ at top fiber

$$f_{bot} = -\frac{F_0}{560} - \frac{F_0(15.27)}{6186} + \frac{7083}{6186} = -2.70 \text{ ksi } (18.6 \text{ N/mm}^2)$$

Use $\rightarrow F_0 = 904$ k (4021 kN) based on $0.6 f'_{ci}$ at bottom fiber.

No. strands $= \dfrac{904}{28.9} = 31.3$ Try $32 - \dfrac{1''}{2}\ \phi$ strands

Check c.g.s. for pattern, Fig. 6-23.

$8 \times 2 = 16$

$8 \times 4 = 32 \qquad g = \dfrac{164}{32} = 5.13$ in. (130 mm)

$8 \times 6 = 48 \qquad e = 20.27 - 5.13 = 15.14$ in. (385 mm)

$6 \times 8 = 48 \qquad F_0 = 32 \times 28.9 = 925$ k (4114 kN)

$2 \times 10 = 20$

$\overline{32} \qquad \overline{164}$

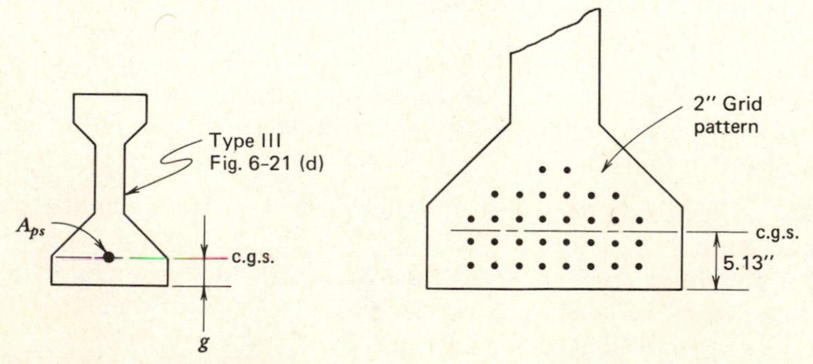

Fig. 6-23. Example 6-11.

$$\text{Actual } f_{\text{bot}} = -\frac{925}{560} - \frac{(925)(15.14)}{6186} + \frac{7083}{6186} = -2.771 \text{ ksi } (-19.1 \text{ N/mm}^2)$$

$$\text{Note:} \left\{ \begin{array}{l} f_{\text{top}} \text{ not critical since} \\ F_0 = 925 \text{ k} < 1303 \text{ k} \end{array} \right\}$$

2.6% over -2.70 ksi (-18.6 N/mm²) allowable stress say o. k.

(b) Beam spacing will depend on service loads which determine moments. We will assume that the spacing is sufficiently large that the composite section, Fig. 6-24, is valid; that is, spacing $\geqslant$ 112 in. (2845 mm)

$$c_b' = \frac{(6)(91)(48) + (560)(20.27)}{(6)(91) + 560} = 33.96 \text{ in. } (863 \text{ mm})$$

$$c_t' = 11.04 \text{ in. } (280 \text{ mm})$$

$$A' = 1106 \text{ in.}^2 \ (714 \times 10^3 \text{ mm}^2)$$

$$I' = 125{,}390 + (560)(13.69)^2 + \frac{(91)(6)^2}{12} + (6)(91)(14.04)^2$$

$$I' = 339{,}610 \text{ in.}^4 \ (141 \times 10^9 \text{ mm}^4)$$

Each composite beam carries loads for spacing, s, allowing us to solve the moments for midspan as follows (s=ft):

$$M_D = \frac{(0.020)(s)(90)^2}{8} \times 12 = 243s \text{ in.-k}$$

$$M_L = \frac{(0.050)(s)(90)^2}{8} \times 12 = 607s \text{ in.-k}$$

$$M_{\text{slab}} = \frac{(0.075)(s)(90)^2}{8} \times 12 = 911s \text{ in.-k}$$

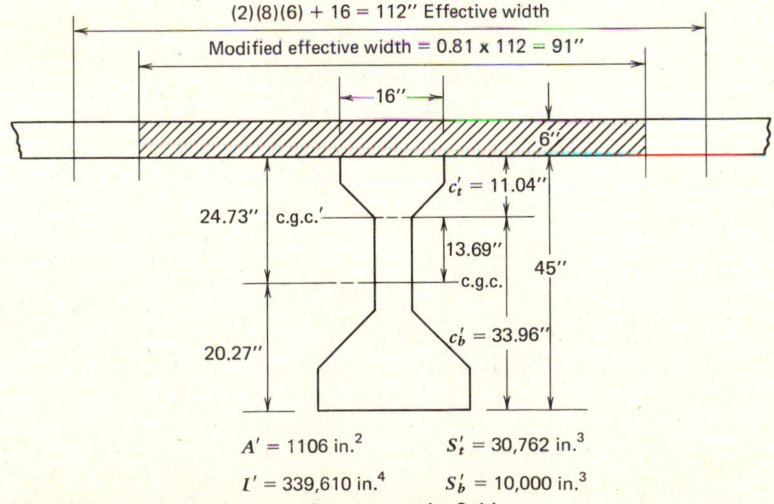

Fig. 6-24. Composite section, example 6-11.

After losses, the effective prestress is:

$$F = 0.8F_0 = 0.8 \times 925 = 740 \text{ k } (3292 \text{ kN})$$

Since the construction is unshored, the moment from slab weight is carried by the precast beam alone. The stresses at top and bottom of the precast beam combined from all effects are as follows:

$$f_{top} = -\frac{740}{560} + \frac{(740)(15.14)}{5070} - \frac{(7083 + 911s)}{5070} - \frac{(243+607)s}{30{,}762} = -2.70 \text{ ksi}$$

$$s = 10.6 \text{ ft.} = 127 \text{ in. } (3{,}226 \text{ mm})$$

$$f_{bot} = -\frac{740}{560} - \frac{(740)(15.14)}{6186} + \frac{(7083+911s)}{6186} + \frac{(243+607)s}{10{,}000} = 0.465 \text{ ksi}$$

$$s = 10.5 \text{ ft} = 126 \text{ in. } (3226 \text{ mm})$$

The maximum spacing $s = 126$ in. (3200 mm) exceeds the slab effective width of 112 in. (2845 mm) (Fig. 6-24) as assumed and the bottom fiber tension for the precast beam controls the spacing here.

We could *use* s = 10 *ft*–6 *in*. spacing (126 in.) without overstress at bottom fiber of beam and just slightly less than the allowable stress for the top fiber.

Check stress at top of composite slab which results from loads acting on the composite section only:

$$f_{top\,(slab)} = -\frac{(243+607)(10.5)(17.14)}{339{,}610} = -0.450 \text{ ksi } (-3.10 \text{ N/mm}^2)$$

$$< -1.8 \text{ ksi } (-12.41 \text{ N/mm}^2)$$

(*c*) Check strength using effective width of slab (serving as compression flange) as 112 in. (2845 mm), Fig. 6-24.

$$A_{ps} = 32 \times 0.153 = 4.90 \text{ in.}^2 \ (3{,}161 \text{ mm}^2) \qquad d = 51 - 5.13 = 45.87 \text{ in. } (1{,}165 \text{ mm})$$

$$\rho_p = \frac{A_{ps}}{p\,d} = \frac{4.90}{(112)(45.87)} = 0.0009537$$

$$f_{ps} = f_{pu}\left[1 - 0.5\rho_p \frac{f_{pu}}{f'_c}\right] = 270{,}000\left[1 - 0.5 \times 0.0009537 \frac{270{,}000}{4{,}000}\right]$$

$$f_{ps} = 261{,}000 \text{ psi } (1800 \text{ N/mm}^2)$$

$$a = \frac{4.9 \times 261{,}000}{(0.85)(4000)(112)} = 3.36 \text{ in. } (85 \text{ mm}) < 6 \text{ in. } (152 \text{ mm})$$ (compression zone entirely in slab as assumed)

$$M_u = 0.9M' = (0.9)(261)(4.9)\left(45.87 - \frac{3.36}{2}\right)$$

$$M_u = 50{,}863 \text{ in.-k} = 4{,}239 \text{ ft-k } (5{,}748 \text{ kN-m})$$

Required M_u for $1.4D + 1.7L$ following ACI Code:

$$(1.4)[7083 + (911+243)(10.5)] + (1.7)(607)(10.5) = M_u$$

$$M_u = 26{,}880 + 10{,}835 = 37{,}715 \text{ in.-k} = 3{,}143 \text{ ft-k } (4{,}262 \text{ kN-m})$$

$$M_u = 4239 \text{ ft-k } (5748 \text{ kN-m}) > 3143 \text{ ft-k } (4262 \text{ kN-m})$$

Strength is more than required.

When cast in place, simplicity of formwork is of prime importance. Thus solid slabs with or without cores, and T-shapes with vertical or tapered sides, are often convenient. Only when the forms can be re-used many times would the I-shape and other complicated shapes be considered economical.

6-10 Arrangement of Steel—Prestressing in Stages

The arrangement of steel is governed by a basic principle: in order to obtain the maximum lever arm for the internal resisting moment, it must be placed as near the tensile edge as possible. This is the same for prestressed as for reinforced sections. But, for prestressed concrete, one more condition must be considered: the initial condition at the transfer of prestress. If the c.g.s. is very near the tensile edge, and if there is no significant girder moment to bring the center of pressure near or within the kern, Fig. 6-25, the tension flange may be overcompressed at transfer while the compression flange may be under high tensile stress. Hence this brings up a special condition in prestressed concrete: a heavy moment is desirable at transfer so that the steel can be placed as near the edge as possible. However, no economy is achieved by adding unnecessary dead weight to the structure in order to enable a bigger lever arm for the steel, because whatever additional moment capacity was thus obtained would be used in carrying the additional dead load, although some additional reserve capacity is obtained for the ultimate range. Loads that will eventually have to be carried by the beam can be more economically put on the structure before transfer rather than after, because moments produced by such loads will permit the placement of steel nearer the tensile edge.

Another method sometimes used in order to permit placement of steel near the tensile edge is to prestress the structure in two or more stages; this is known as retensioning. At the first stage, when the moment on the beam is small, only a portion of the prestress will be applied; the total prestress will be applied only when additional dead load is placed on the beam producing heavier moment on

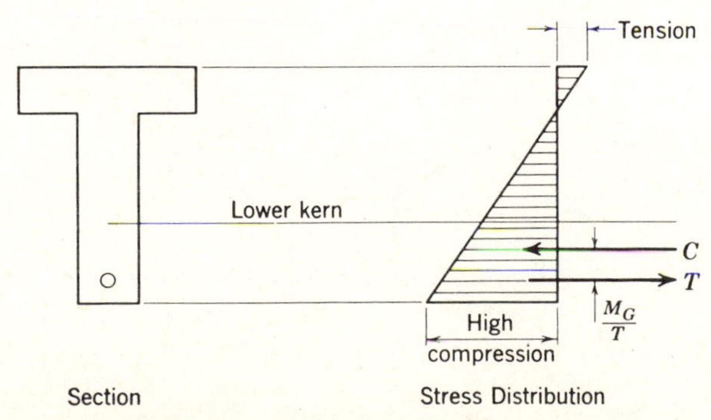

Fig. 6-25. Girder moment insufficient to bring *C* within kern.

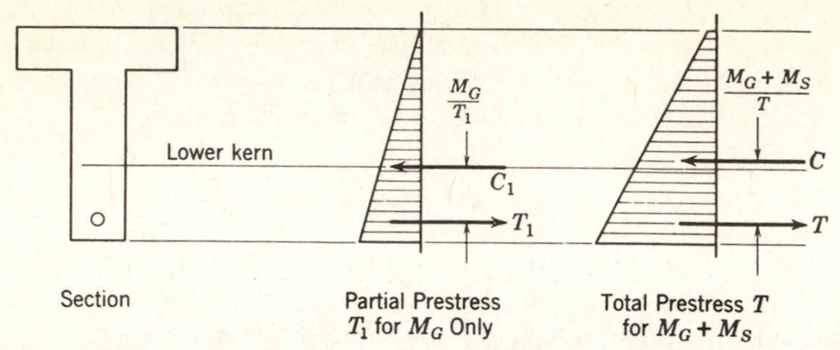

Fig. 6-26. Prestressing in two stages to keep C within kern.

the section, Fig. 6-26. Thus the center of pressure can be kept within the kern at all times, and excessive tension in the compression flange, as well as high compression in the tension flange, can be avoided.

EXAMPLE 6-12

Design the symmetrical I-shape section with $h=45$ in. (Fig. 6-27) for a simply supported 90 ft. span carrying the following service loads: weight of beam itself, w_G; additional dead load, $w_D = 700$ lb/ft; live load, $w_L = 400$ lb/ft. Normal weight concrete with $f'_c = 6000$ psi is specified. Prestressing will be done with 270 k grade $\frac{1}{2}$ in. diameter strand; $f_{pu} = 270$ ksi and $A_{ps} = 0.153$ in.^{2}/strand. Assume $0.7f_{pu}$ at transfer (28.9 k/strand) with total losses

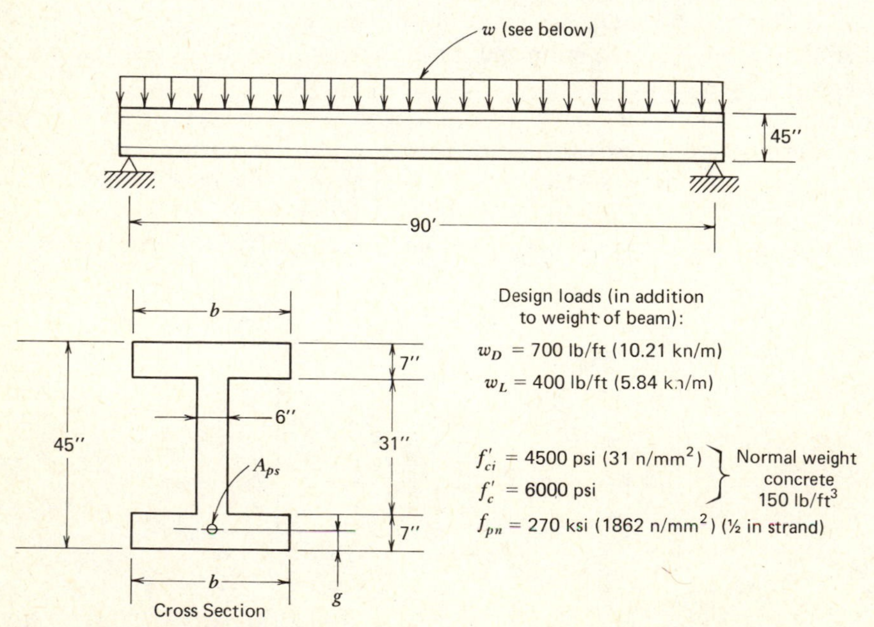

Fig. 6-27. Example 6-12.

estimated as 20% for this design situation. Use two stages of prestressing if a lighter section results.

(a) limit the design to no tension allowed in the concrete assuming the surrounding environment to be very corrosive. Assume the second stage posttensioned strands will be grouted, forming a bonded beam.

(b) allow $6\sqrt{f_c'}$ tension in the concrete. Check the strength after stresses are satisfied based on ACI Code allowable values. ($w_D=10.21$ kN/m, $w_L=5.84$ kN/m, $f_{ci}'=31$ N/mm², $f_c'=41$ N/mm², $f_{pu}=1862$ N/mm², and $A_{ps}=98.7$ mm²/strand)

Solution (a) Assume $w_G=500$ lb/ft (7.30 kN/m) for beam weight initially.

$$M_G=w_G\frac{l^2}{8}=\frac{0.50\times 90^2}{8}\times 12=6{,}075 \text{ in.-k } (687 \text{ kN-m})$$

$$M_D=w_D\frac{l^2}{8}=\frac{0.70\times 90^2}{8}\times 12=8{,}505 \text{ in.-k } (961 \text{ kN-m})$$

$$M_L=w_L\frac{l^2}{8}=\frac{0.40\times 90^2}{8}\times 12=\underline{4{,}860 \text{ in.-k}}\ \underline{(549 \text{ kN-m})}$$

$$M_T=19{,}440 \text{ in.-k } (2.197 \text{ kN-m})$$

$$\text{Estimate } F=\frac{19{,}440}{(0.65)(45)}=665 \text{ k } (2958 \text{ kN}) \qquad (6\text{-}1)$$

$$F_0=\frac{655}{0.80}=831 \text{ k } (3696 \text{ kN})$$

Estimate concrete area required if we are designing for no tension allowed: $f_t=0.45\times 6000=2700$ psi $=2.7$ ksi (18.62 N/mm²) compression allowed top fiber; $f_b=0.60\times 4500$ psi $=2700$ psi $=2.7$ ksi (18.62 N/mm²) compression allowed bottom fiber.

$$A_c=\frac{F_0 h}{f_b c_t}=\frac{(831)(45)}{(2.7)(22.5)}=616 \text{ in.}^2 \ (397\times 10^3 \text{ mm}^2) \qquad (6\text{-}7a)$$

$$A_c=\frac{Fh}{f_t c_b}=\frac{(665)(45)}{(2.7)(22.5)}=493 \text{ in.}^2 \ (318\times 10^3 \text{ mm}^2) \qquad (6\text{-}9a)$$

With two stages of prestressing we can use the lighter section $A_c=493$ in.² (318×10^3 mm²) required by service load, avoiding the $A_c=616$ in.² (397×10^3 mm²) which would be required with single stage transfer of total F_0 as shown above.

$$(2)(7)(b)+(31)(6)=493 \text{ in.}^2=A_c$$

$$b=22 \text{ in. } (559 \text{ mm})$$

$w_G=515$ lb/ft (7.51 kN/m)	$A=494$ in.² (319×10^3 mm²)
$M_G=6257$ in.-k (707 kN-m)	$I=127{,}342$ in.⁴ (53×10^9 mm⁴)
$M_T=19{,}620$ in.-k (2217 kN-m)	$k_t=k_b=11.46$ in. (291 mm)
(full service load)	$c_b=c_t=22.5$ in. (572 mm)
	$S_x=5660$ in.³ (93×10^6 mm³)

Assume $e \cong 18.5$ in. (470 mm)($g \cong 4$ in.)

$$F = \frac{M_T}{e + k_t} = 19{,}620/(18.5 + 11.5) = 654 \text{ k (2909 kN)}$$

$$F_0 = \frac{654}{0.80} = 818 \text{ k (3638 kN)}$$

$$A_c = \frac{Fh}{f_t c_t} = \frac{(654)(45)}{(2.7)(22.5)}$$

$$A_c = 484 \text{ in.}^2\ (312 \times 10^3 \text{ mm}^2) < 494 \text{ in.}^2\ (319 \times 10^3 \text{ mm}^2) \text{ provided}$$

use $b = 22$ in. (559 mm)

Check pretension strands transfer $F_{0_1} = ??$

$$A_c = \frac{f_0 h}{f_b c_t} \tag{6-7a}$$

$$494 = \frac{F_0(45)}{(2.7)(22.5)}$$

$F_0 = 667$ k (2967 *kN*), $n = 667/28.9 = 23$ *strands* (max.)

Try $22 - \frac{1}{2}$ in. ϕ strands—$F_{0_1} = 636$ k (2829 kN)

Posttension $F_{0_2} = 818 - 636 = 182$ k (810 kN)

Try $6 - \frac{1}{2}$ in. ϕ strands—$F_{0_2} = 173$ k (770 kN)

Pretensioned strands: first stage stressing

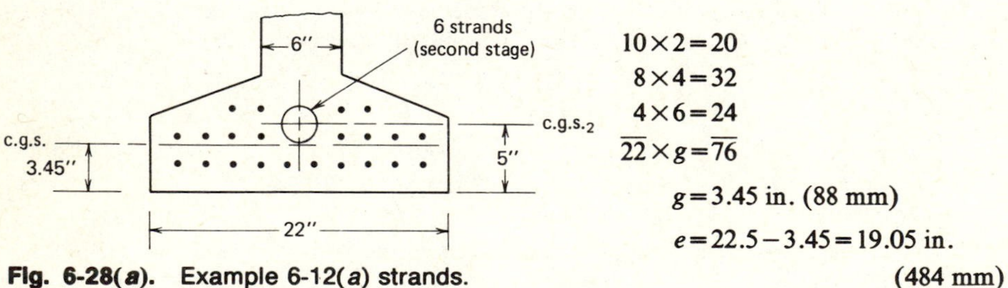

$10 \times 2 = 20$
$8 \times 4 = 32$
$4 \times 6 = 24$
$\overline{22} \times g = \overline{76}$

$g = 3.45$ in. (88 mm)

$e = 22.5 - 3.45 = 19.05$ in. (484 mm)

Fig. 6-28(*a*). Example 6-12(*a*) strands.

Check midspan at transfer ($F_{0_1} = 636$ k): ($M_G = 6257$ in.-k)

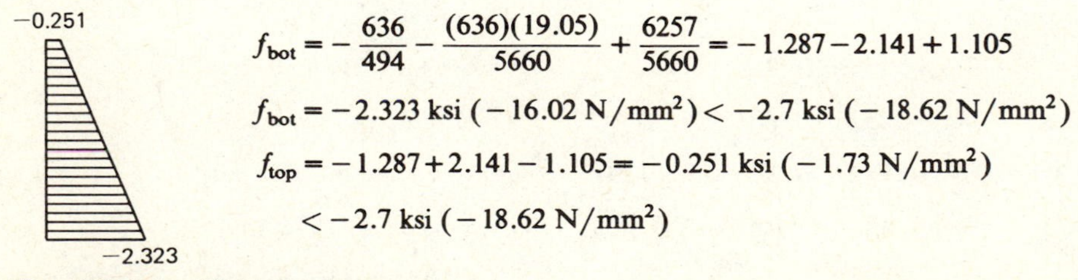

$$f_{\text{bot}} = -\frac{636}{494} - \frac{(636)(19.05)}{5660} + \frac{6257}{5660} = -1.287 - 2.141 + 1.105$$

$$f_{\text{bot}} = -2.323 \text{ ksi } (-16.02 \text{ N/mm}^2) < -2.7 \text{ ksi } (-18.62 \text{ N/mm}^2)$$

$$f_{\text{top}} = -1.287 + 2.141 - 1.105 = -0.251 \text{ ksi } (-1.73 \text{ N/mm}^2)$$

$$< -2.7 \text{ ksi } (-18.62 \text{ N/mm}^2)$$

Fig. 6-28(*b*). Midspan stresses.

Now adding posttensioned strands:

Since $g=5$ in. (127 mm) as shown above, Fig. 6-28, at $e=22.5-5.0=17.5$ in. try $6-\frac{1}{2}$ in. strands ($F_{0_2}=173$ k). Prior to stressing F_{0_2}: (w_D added is assumed 500 lb/ft of design $w_D=700$ lb/ft.

$$M_{DL}=6075 \text{ in.-k } (w_D=500 \text{ lb/ft})$$

$$f=\frac{6075}{5660}=1073 \text{ psi}=1.073 \text{ ksi } (7.40 \text{ N/mm}^2)$$

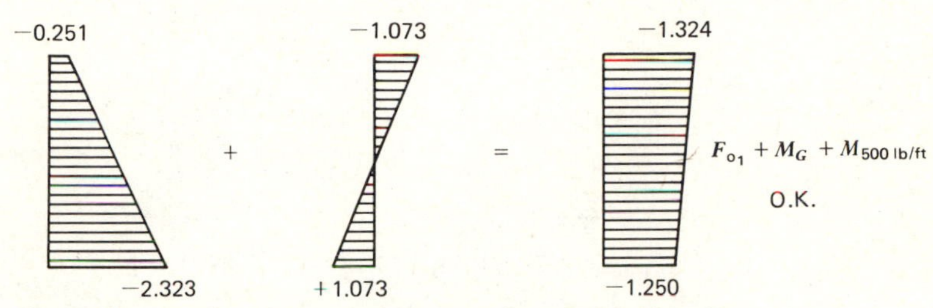

Fig. 6-28(*c*). Stresses at midspan prior to stressing F_{0_2}.

After adding $F_{0_2}=173$ k (770 kN) at $e=17.5$ in. (444 mm) (with $w_D=500$ lb/ft acting):

$$f=\frac{173\times 17.5}{5660}=0.535 \text{ ksi } (3.69 \text{ N/mm}^2)$$

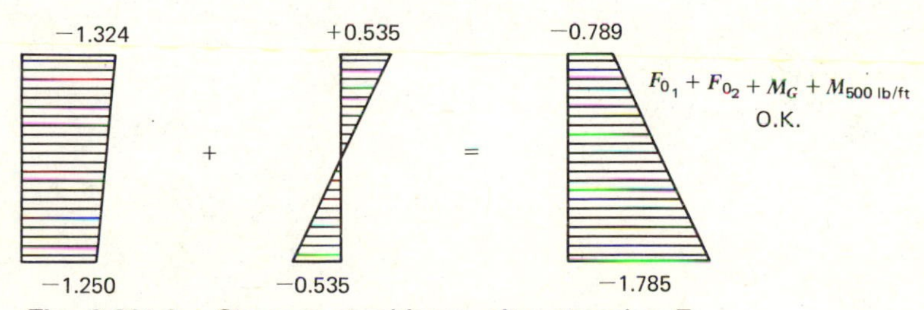

Fig. 6-28(*d*). Stresses at midspan after stressing F_{0_2}.

Section is satisfactory at transfer using the two stages.

Checking service loads after losses:

$$F_1=0.8\times 636=509 \text{ k } (2264 \text{ kN}) \text{ at } e=19.05 \text{ in. } (484 \text{ mm}) \text{ (first stage)}$$

$$F_2=0.8\times 173=\underline{138} \text{ k } (614 \text{ kN}) \text{ at } e=17.5 \text{ in. } (444 \text{ mm}) \text{ (second stage)}$$

$$F=647 \text{ k } (2878 \text{ kN})$$

$$f_{top}=-\frac{647}{494}-\frac{(509)(19.05)}{5660}-\frac{(138)(17.5)}{5660}+\frac{19{,}620}{5660}$$

$$f_{top}=-1.310+1.713+0.427-3.466=-2.363 \text{ ksi } (-18.18 \text{ N/mm}^2)$$
$$<-2.70\ (-18.62 \text{ N/mm}^2)$$

$$f_{bot}=-1.310-1.713-0.427+3.466=0.016 \text{ ksi} \cong \text{zero}$$

Use of six posttensioned strands is satisfactory at service load.
See last of (*b*) following for check on strength of cross section to carry factored loads following ACI Code. This section exceeds strength required for $U = 1.4D + 1.7L$.

(*b*) *Allowing* $6\sqrt{6000} = 465$ *psi tension at service load*: From section 6-6, $f_b' I/c_b = (0.465)(5660) = 2632$ in.-k (297 kN-m)

Net moment = 19620 − 2632 = 16,988 in.-k (1920 kN-m) From Fig. 6-28 estimate g:

$$\begin{array}{rcl} 22\times 3.45 & = & 76 \\ \underline{6\times 5} & = & \underline{30} \\ 28 & & 106 \end{array}$$

$$g = 3.79 \text{ in.} \cong 3.8 \text{ in. (96 mm)}$$

$$e = 22.5 - 3.8 = 18.7 \text{ in. (475 mm)}$$

$$a = e + k_t$$

$$a = 18.7 + 11.5 = 30.2 \text{ in. (767 mm)}$$

$$F \cong \frac{16{,}988}{30.2} = 563 \text{ k (2504 kN)}$$

$$A_c = \frac{Fh}{f_t c_b - f_b' c_t} = \frac{(563)(45)}{(2.7)(22.5) - (0.465)(22.5)} = 504 \text{ in.}^2 \; (325\times 10^3 \text{ mm}^2) \qquad (6\text{-}20a)$$

Section with $b = 22$ in. (559 mm), $A_c = 494$ in.^{2} (319×10^3 mm^2)

$\cong 504$ in.^{2} (325×10^3 mm^2)

(only 2% difference, try this same section here)
$F = 563$ k (2504 kN) $\rightarrow F_0 = 704$ k (3,131 kN) (total)

Try $24 - \frac{1}{2}$ in. strands $F_0 = 694$ k (3087 kN)

$F = 555$ k (2,469 kN)

Check layout with: 18 straight at $e = 19.6$ in. (498 mm) $F_1 = 416$ k (1851 kn)

6 draped at $e = 17.5$ in. (444 mm) $F_2 = 139$ k (618 kN)

555 k (2469 kN)

From previous calculations transfer is satisfactory.
Check service load after losses:

$$f_{top} = -\frac{554}{494} + \frac{(416)(19.6)}{5660} + \frac{(139)(17.5)}{5660} - \frac{19{,}620}{5660}$$

$$f_{top} = -1.121 + 1.441 + 0.430 - 3.466 = -2.716 \text{ ksi } (-18.73 \text{ N/mm}^2)$$

$$\cong -2.70 \text{ ksi } (-18.62 \text{ N/mm}^2)$$

(less than 1% difference from allowable)

$$f_{bot} = -1.121 - 1.441 - 0.430 + 3.466 = 0.474 \text{ ksi (3.27 N/mm}^2)$$

$$\cong 0.465 \text{ ksi (3.21 N/mm}^2)$$

(less than 2% difference from allowable)

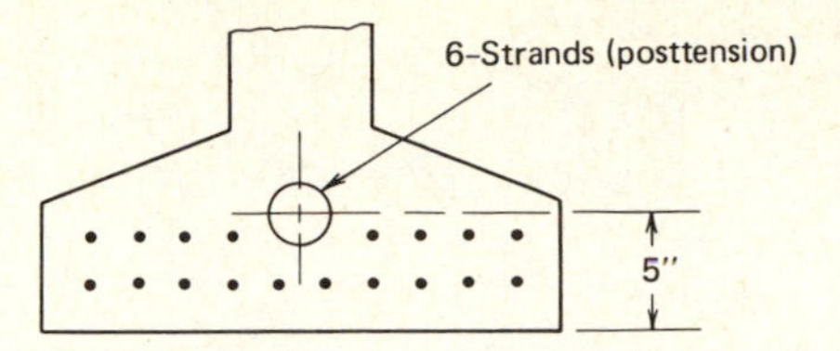

Fig. 6-28(e). Example 6-12(b) strands.

$$10\times 2=20$$
$$8\times 4=32$$
$$\underline{6\times 5=30}\qquad \underline{g=3.42\text{ in. (87 mm)}}$$
$$24\times g=82\qquad e=19.08\text{ in. (485 mm)}\quad d=41.6\text{ in.}$$

Stresses satisfy ACI Code.

Check strength: $b=22$ in. (559 mm) $\quad d=41.6$ in. (1057 mm)

$24-\frac{1}{2}$ in. ϕ strands

$$\rho_{ps}=\frac{0.153\times 24}{(22)(41.6)}=0.004012$$

$$f_{ps}=270\left[1-(0.5\times 0.004012)\left(\frac{270}{6}\right)\right]=246\text{ ksi (1696 N/mm}^2\text{)}$$

$$a=\frac{0.153\times 24\times 246}{(0.85)(6)(22)}=8.05\text{ in. (204 mm)}>7\text{ in. (178 mm)}$$

flanged section behavior

Area required $=177$ in.^{2}(114×10^3 mm^2)

$\underline{154}$ in.^{2}(99×10^3 mm^2)—flange

23 in.^{2}(15×10^3 mm^2)—web

Fig. 6-28(f) shows compression zone.

$$a=7+\frac{23}{6}=10.83\text{ in. (275 mm)}\qquad z=\frac{154\times 3.5+23\times 8.92}{177}=4.20\text{ in. (107 mm)}$$

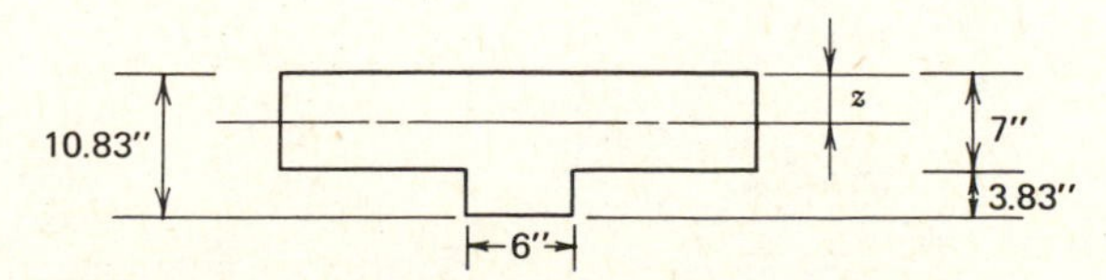

Fig. 6-28(f). Compression zone.

Required $M_u=(1.4)(14762)+(1.7)(4860)=20{,}667+8262$
Required $M_u=28{,}929$ in.-k (3269 kN-m) $<30{,}395$ in.-k (3435 kN-m)

Strength, Fig. 6-28(g) satisfies ACI Code.

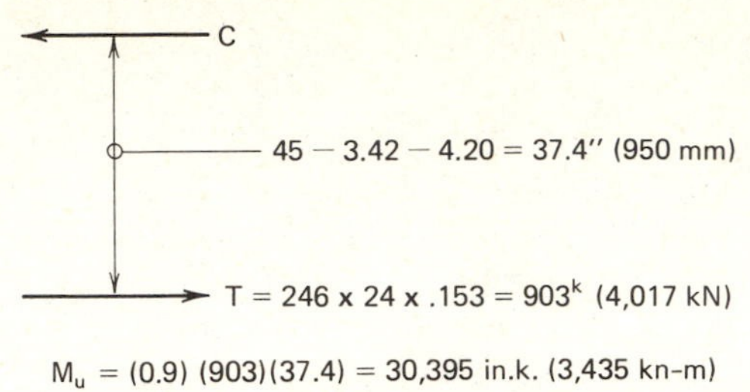

M_u = (0.9) (903)(37.4) = 30,395 in.k. (3,435 kn-m)

Fig. 6-28(*g*). Ultimate moment force couple.

For certain sections, the tendons are placed in the compression flange as well as in the tension flange, Fig. 6-29. Generally speaking, this is not an economical arrangement, because it will move the c.g.s. nearer to the c.g.c. and thereby decrease the resisting lever arm. At the ultimate range, tendons in the compressive flange will neutralize some of its compressive capacity, whereas only those in the tension flange are effective in resisting moment. However, under certain circumstances it may be necessary to put tendons in both flanges in spite of the resulting disadvantages. These conditions are:

1. When the member is to be subject to loads producing both $+M$ and $-M$ in the section.
2. When the member might be subject to unexpected moments of opposite sign, during its handling process.
3. When the M_G/M_T ratio is small and the tendons cannot be suitably grouped near the kern point. Then the tendons will be placed in both the tension and the compression flanges with the resulting c.g.s. lying near the kern.

The minimum concrete protection for tendons is governed by two requirements: first for fire resistance which is discussed in Chapter 16, then by corrosion protection which has been determined more or less by experience and practice. ACI Building Code Requirements specifies the following minimum thickness of concrete cover for prestressing steel, ducts, and nonprestressed steel.

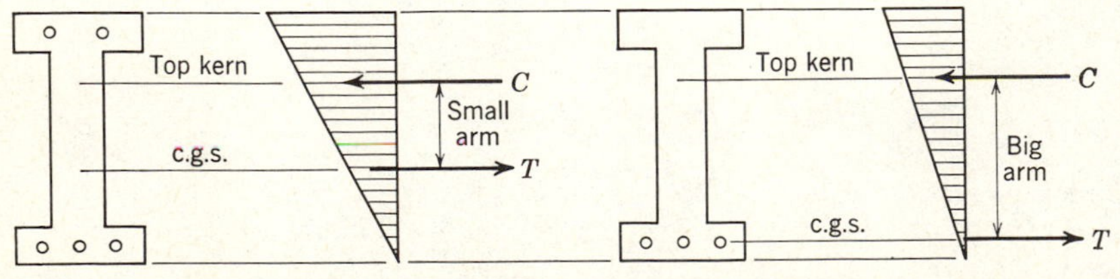

Fig. 6-29. Prestressing steel in both flanges reduces lever arm for resisting moment.

(ACI) 7.14.1.3 Prestressed concrete members—Prestressed and nonprestressed reinforcement, ducts, and end fittings

Cast against and permanently exposed to earth	3
Exposed to earth or weather:	
Wall panels, slabs, and joists	1
Other members	$1\frac{1}{2}$
Not exposed to weather or in contact with the ground:	
Slabs, walls, joists	$\frac{3}{4}$
Beams, girders, columns:	
Principal reinforcement	$1\frac{1}{2}$
Ties, stirrups, or spirals	1
Shells and folded plate members:	
Reinforcement $\frac{5}{8}$ in. and smaller	$\frac{3}{8}$
Other reinforcement	d_b but not less than $\frac{3}{4}$

(ACI) 7.14.2 The cover for nonprestressed reinforcement in prestressed concrete members under plant control may be that given for precast members.

Cover specified in section 7.14.1.3 is for prestressed members with stresses less than or equal to the limits of section 18.4.2b. When tensile stresses exceed this value for members exposed to weather, earth, or corrosive environment, cover shall be increased 50%.

It is generally believed that the above protection is quite adequate if the cover concrete is prestressed and not subjected to cracking under sustained loading.

The minimum spacing of tendons is governed by several factors. First, the clear spacing between tendons, or between tendons and side forms, must be sufficient to permit easy passage of concrete. Here we may apply the general rule for reinforced concrete, which limits the clear distance to a minimum of $1\frac{1}{3}$ times the size of the maximum aggregates. This requirement may be reduced for prestressed concrete when good vibration can be ensured. Second, to properly develop the bond between steel and concrete, we follow the ACI Code Requirements for reinforcing bars: the clear distance between bars should be at least the diameter of the bars for special anchorage and $1\frac{1}{2}$ times the diameter for ordinary anchorage, with a minimum of 1 in. These limitations may not be necessary for small wires and strands used in prestressed work, and they are often bundled together, especially at sections near the middle portion of the span.

The ACI Building Code Requirements call for a minimum clear spacing at each end of the member of four times the diameter of individual wires or three times the diameter of strands, in order to properly develop the transfer bond in pretensioning steel. Along the middle portion of the span, bundling is always permitted.

Ducts may be arranged closely together vertically when provision is made to prevent the tendon from breaking through into an adjacent duct. When the tendons are sharply curved, the radial thrusts exerted on the concrete along the bends may be considerable. Horizontal disposition of ducts shall allow proper placement of concrete. When the tendons are placed outside of the concrete to be eventually covered with mortar, then only the problem of fire and corrosion protection need be considered.

To help dimension a beam section, the sizes of some tendons and their conduits are listed in Appendix B.

References

1. T. Y. Lin and A. C. Scordelis, "Selection and Design of Prestressed Concrete Beam Sections," *J. Am. Conc. Inst.*, November 1953 (*Proc.*, Vol. 49), pp. 209–224.
2. "Tentative Standards for Prestressed Concrete Piles, Slabs, I-beams, and Box Beams, and Interim Manual for Inspection of Such Construction," American Association of State Highway Officials, 1963.

7

SHEAR; BOND; BEARING

7-1 Shear, General Considerations

The strength of prestressed-concrete beams in flexure is quite definitely known, but their strength in resisting shear or the combination of shear and flexure cannot be predicted with corresponding precision. Previous to about 1955, numerous prestressed-concrete beams had been tested for their strength in flexure, but very few in shear. Between 1955 and 1961, however, hundreds of specimens were actually tested to determine their strength in resisting shear or moment and shear, with or without web reinforcement.[1-8] Unfortunately, on account of the complexity of the problem and our inability to isolate the variables in our experiments and analyses, we cannot claim that we have completely solved the problem, even with this wealth of information now available to us.

This rather dim view of our knowledge concerning shear does not stop us from designing prestressed-concrete beams, just as our lack of knowledge concerning shear strength of reinforced-concrete beams does not stop us from designing them. In fact, it may be stated that prestressed-concrete beams possess greater reliabililty in shear resistance than reinforced-concrete beams, because prestressing will usually prevent the occurrence of shrinkage cracks which could conceivably destroy the shear resistance of the reinforced-concrete beams, especially near the point of contraflexure. Since so many prestressed-concrete beams have been designed and built on the basis of assumed theories for shear resistance, and since beams are not failing in shear, it may be concluded that our present method of design (basically unchanged since the ACI Code of 1963) is a safe one.

The degree of safety for prestressed concrete structures designed for shear by the conventional methods used prior to 1963 varied widely. Some of the designs were too safe in shear, others not so safe, even though all seemed to function safely under the usual service loads. The shear strength contributed by the concrete was computed from a design equation which was an extension of our experience with reinforced concrete members. While the design code equation was simple to use, it did not really take into account the influence of prestressing on the actual behavior of prestressed members. The present ACI Code equations for shear strength are more complex than the previous simple expression, but they are based on theory and tests which show clearly that we are reflecting actual behavior. A more uniform factor of safety in shear strength is the result of our additional effort in design.

The approach now contained in the ACI Code has a rational basis for considering the ways in which shear cracking may develop in prestressed concrete members. Cracking mechanisms assumed are those which have been verified by test results for both simple and continuous beams. Yet the design method is semiempirical in that some terms in the code equations are based on test data. In the following development of the background for the present design approach for shear strength, the tie with test data will be emphasized.

A general picture of the shear in a prestressed-concrete beam will first be presented. Consider three beams carrying transverse loads as shown in Fig. 7-1. Beam (*a*) is prestressed by a straight tendon. Taking an arbitrary section *A–A*, the shear V at that section is carried entirely by the concrete, none by the tendon which is stressed in a direction perpendicular to the shear. Beam (*b*) is prestressed with an inclined tendon. Section *B–B* shows that the transverse component of the tendon carries part of the shear, leaving only a portion to be carried by the concrete, thus,

$$V_c = V - V_p$$

This may be compared to reinforced-concrete beams with bent-up bars where the inclined portion of the steel carries some of the shear. It must be noted that a horizontal tendon, though inclined to the axis of the beam, does not carry any vertical shear, as illustrated by section *C–C* in beam (*c*). Whenever the tendon is not perpendicular to the direction of shear, then it does assist in carrying the shear, for example, section *D–D*. It is interesting to note that, in some rare

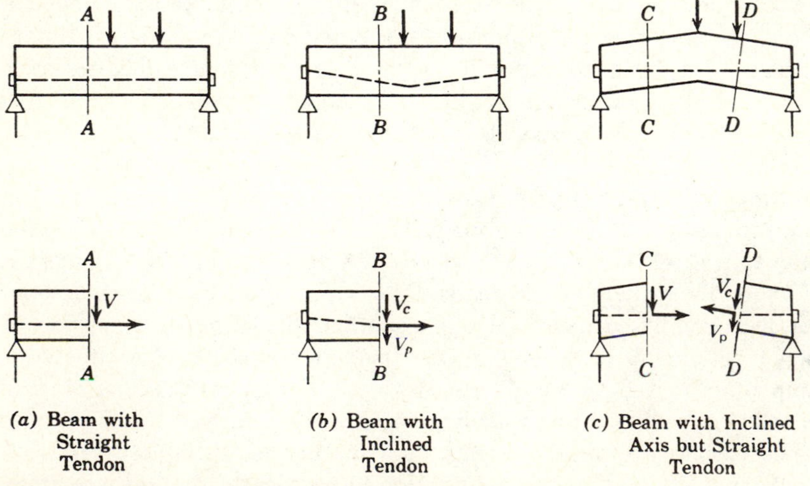

(*a*) Beam with Straight Tendon

(*b*) Beam with Inclined Tendon

(*c*) Beam with Inclined Axis but Straight Tendon

Fig. 7-1. Shear carried by concrete and tendons.

instances, the transverse component of the prestress increases the shear in concrete.

By following the balanced load approach for prestressed concrete, it is possible to design a beam with no shear in the concrete under a given condition of loading. Take Fig. 7-2, for example; if the simple beam carrying a uniform load is prestressed by a parabolic cable with a sag equal to

$$y_0 = \frac{wL^2}{8F}$$

where F is the prestress in the cable, then the transverse component of the cable equals the shear at any point, and there is no shear to be carried by the concrete. The balanced load w is in the service load range in usual cases, thus the concrete would carry shear stress at factored load (ACI design load level). For beams carrying concentrated loads, or for continuous beams over the intermediate supports, Fig. 7-2(b), the problem is more complicated since the tendon cannot be bent sharply to conform with the theoretically sudden change in shear at the points of load concentration.

The amount of shear acting on the concrete having been determined, the next step is to compute the shear resistance of the concrete. It is generally believed that prestressed beams, similar to reinforced ones, practically never fail under direct shear or punching shear. They fail as a result of tensile stresses produced by shear, known as diagonal tension in reinforced concrete and as principal tension in prestressed concrete. Before cracking, prestressed concrete can be considered as made up of a homogeneous material; the computation of principal tensile stresses can thus be made by the usual method in strength of materials for the state of stress in a homogeneous body. Although the principal tensile stresses can be computed, the strength of concrete in resisting such stresses is not definitely known, since for concrete there are many theories of failure, of which the maximum tensile stress theory is only one. After the cracking of concrete, and with the addition of web reinforcement, the problem becomes even more complicated.

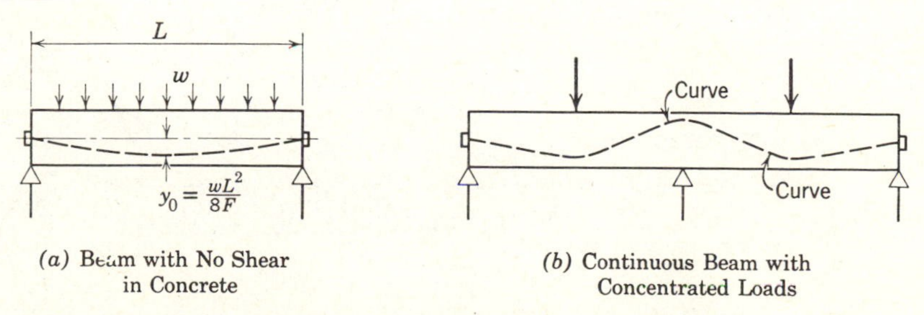

(a) Beam with No Shear in Concrete

(b) Continuous Beam with Concentrated Loads

Fig. 7-2. Varying inclination of tendons to carry shear.

The formation of web cracking in a previously uncracked section 1-1 of the beam is shown in Fig. 7-3. Note that both this web cracking at section 1-1 and the flexure-shear cracking at section 2-2, discussed below, may occur in the same beam. Since the moment is very low at the section 1-1, we can use the classical theory for computing principal tensile stress in the web to predict formation of these web-shear cracks. This is discussed in the following section. Fig. 7-4 shows the appearance of the web cracking of this type in a test beam.

Tests on both reinforced- and prestressed-concrete beams indicate that, when shear failure occurs at a section, not only the shear but also the moment at that

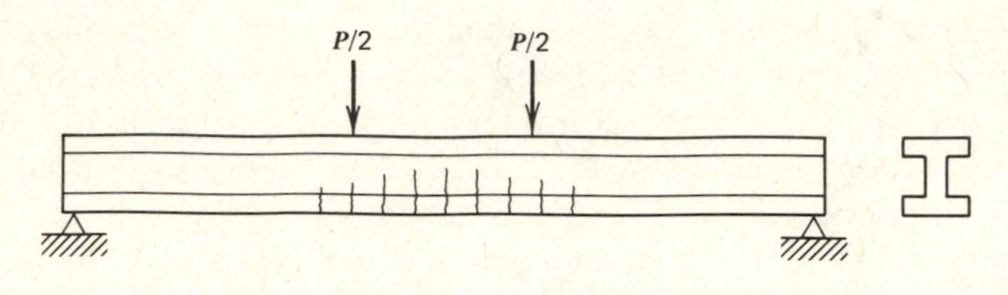

(*a*) Initial Flexural Cracking (*P* slightly more than service load for typical beam)

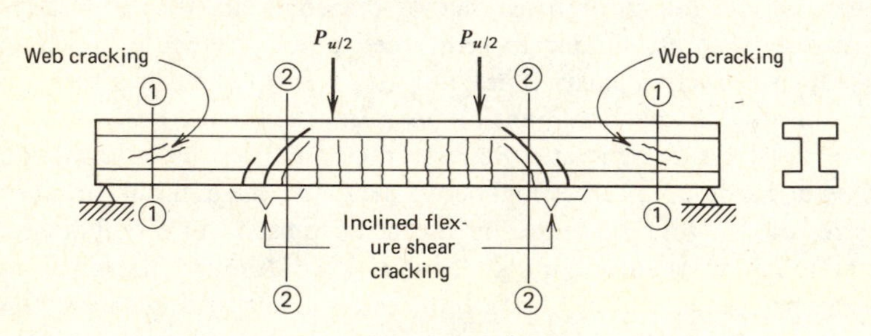

(*b*) Flexure–Shear Cracking at Factored Load

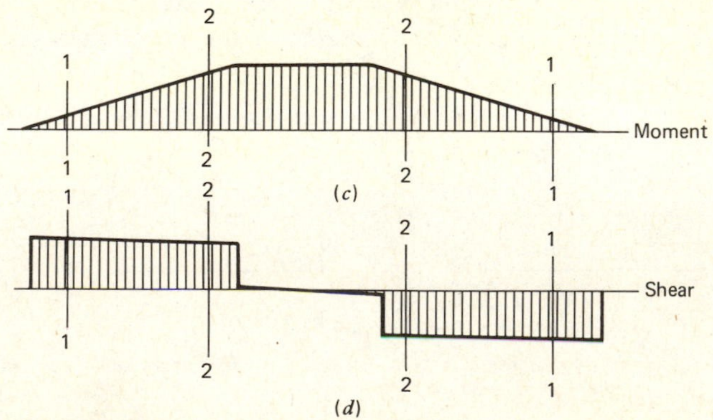

Fig. 7-3. Development of shear cracking.

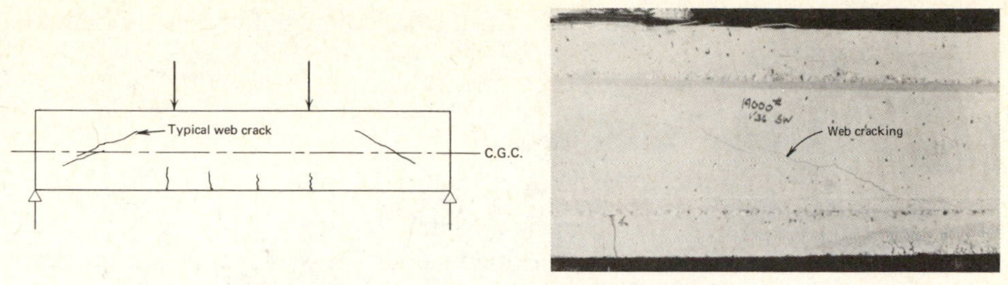

Fig. 7-4. Inclined tension crack originating in web.[3]

section has an effect on its ultimate strength. Flexural cracking may initiate due to moment, Fig. 7-3(*a*); but the combined effect of shear and moment at the critical section results in failure at section 2-2, Fig. 7-3(*b*). The flexural cracking develops in response to moment, Fig. 7-3(*c*); but the shear, Fig. 7-3(*d*), causes the initially vertical flexural cracks to become inclined cracks in the web where shear stress is high.

Failure of the beam in a very sudden and complete fashion may result from the inclined flexure-shear cracking shown in Fig. 7-3(*b*) and Fig. 7-5. Precise analysis is complicated here by the fact that the shear produces a propagation of the initial flexural cracks. The classical mechanics theory for an elastic, previously uncracked section would not be appropriate to apply here for analysis of the principal tensile stress. We must associate this flexure-shear cracking with formation of the initiating flexural cracking in regions where shear is also high. Section 2-2 in Fig. 7-3 illustrates this case where flexure-shear cracking is critical.

For shear strength design following the ACI Code (as well as other codes), semiempirical methods are employed. They are relatively simple and are presented in the following sections. For further investigation, refer to the references listed at the end of this chapter.

There are essentially two types of shear failures: one in which the cracking starts in the web as a result of high principal tension, Fig. 7-4, and another in which vertical flexural cracks occur first and gradually develop into inclined shear cracks, Fig. 7-5. The principal stresses associated with the first type of

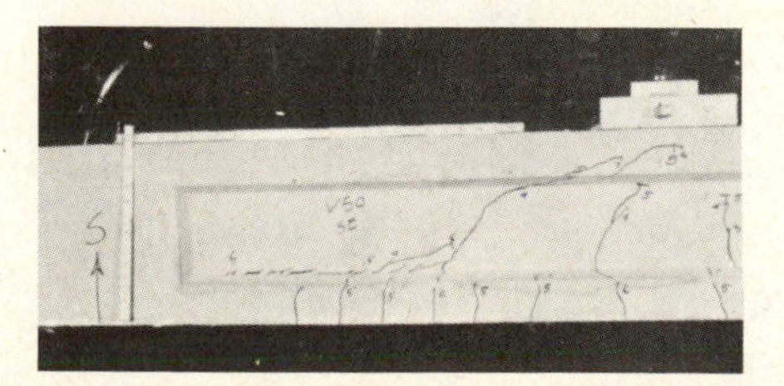

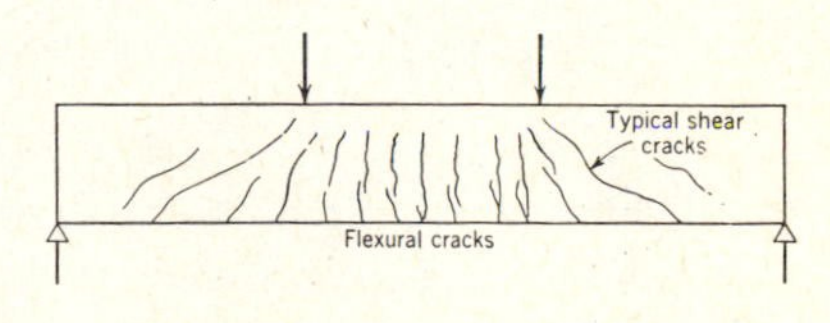

Fig. 7-5. Inclined tension crack originating from flexure crack.[3]

cracking will be treated in section 7-2, and the design for both types is discussed in section 7-3.

7-2 Shear, Principal Tensile Stress

Conventional design for web shear cracking in prestressed-concrete beams is based on the computation of the principal tensile stress in the web and the limitation of that stress to a certain specified value. The first part of this method, the computation of the principal tension based on the classical approach, is a correct procedure so long as the concrete has not cracked. The second part of this method, limiting the principal tension to a definite value, is not always an accurate approach, because there is evidence to show that the resistance of concrete to such principal tension is not a consistent value but varies with the magnitude of the axial compression.[9] It seems, however, that when the axial compression is not too high, say less than about $0.50f'_c$, the resistance of concrete to principal tensile stress is relatively consistent. Typical prestressed concrete beams have axial compression less than $0.5f'_c$. Hence, the computation of principal tensile stress can be regarded as a proper criterion for the stress conditions to determine when the concrete has cracked.

The conventional method of computing principal tensile stress in a prestressed concrete-beam section is based on the elastic theory and on the classical method for determining the state of stress at a point as explained in any treatise on mechanics of materials. The method can be outlined as follows.

1. From the total external shear V across the section, deduct the shear V_p carried by the tendon to obtain the shear V_c carried by the concrete, thus,

$$V_c = V - V_p \tag{7-1}$$

   Note again that occasionally, though rarely, $V_c = V + V_p$; this happens when the cable inclination is such that it adds to the shear on the concrete.

2. Compute the distribution of V_c across the concrete section by the usual formula, Fig 7-6,

$$v = V_c Q / Ib$$

   where v = shearing unit stress at any given level
   Q = statical moment of the cross-sectional area above (or below) that level about the centroidal axis.
   b = width of section at that level.

3. Compute the fiber stress distribution for that section due to external moment M, the prestress F, and its eccentricity e by the formula

$$f_c = \frac{F}{A} \pm \frac{Fec}{I} \pm \frac{Mc}{I}$$

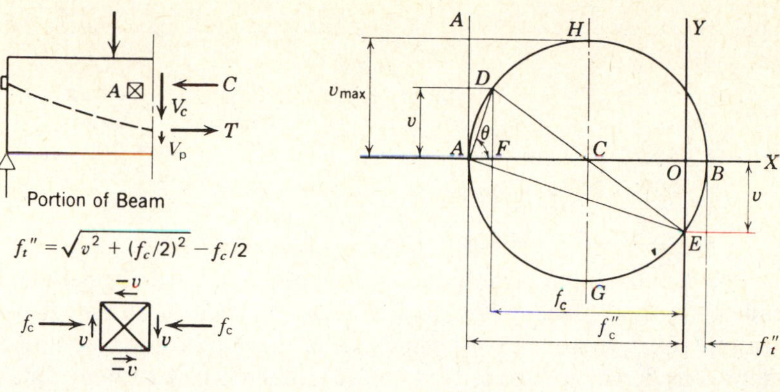

(a) Typical Beam with Usual Prestressing

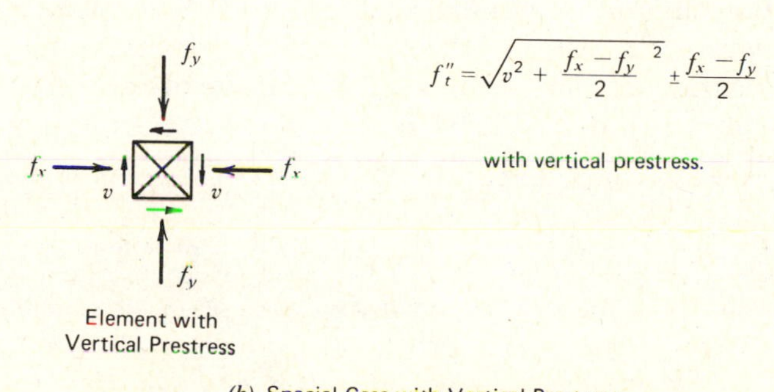

(b) Special Case with Vertical Prestress

Fig. 7-6. State of stress in concrete.

4. The maximum principal tensile stress f_t'' corresponding to the above v and f_c is then given by the formula

$$f_t'' = \sqrt{v^2 + (f_c/2)^2} - (f_c/2) \tag{7-2}$$

Graphically it can be solved by Mohr's circle of stress* as shown in Fig. 7-6. One advantage of this graphical method lies in the indication of the plane of

*Mohr's circle is constructed as follows:

Choose a pair of rectangular axes X–Y with origin at O. Measure OE equal to v. Measure OF equal to f_c and FD equal to v. Draw a circle with DE as diameter. Then OB is the principal tension, OA the principal compression, and $CG = CH =$ principal shear.

principal tension as shown in Fig. 7-6 and listed in the table. (Note: AA = plane perpendicular to AB.)

Plane	Shearing Stress	Normal Stress
AD = vertical plane	v	f_c
AE = horizontal plane	$-v$	0
AB = principal tensile plane	0	f_t''
AA = principal compressive plane	0	f_c''

It can be seen from the table that the angle between the principal tensile plane AB and the vertical plane AD is greater than 45°. Also note that the principal compressive stress, although somewhat greater than the compressive fiber stress, is seldom considered in design. It is considered sufficient to limit the compressive fiber stress to an allowable value. Similarly, no account is taken of the maximum shearing stresses which occur on planes at 45° to the principal planes, since it is the tension rather than shear that produces ultimate failure.

The greatest principal tensile stress does not necessarily occur at the centroidal axis, where the maximum vertical shearing stress exists. At some point, where f_c is diminished, equation 7-2 will often yield a higher principal tension even though v is not a maximum. For I-sections, the junction of the web with the tensile flange is frequently a critical point for computing the greatest principal tension. This is illustrated in the following example.

EXAMPLE 7-1

A prestressed-concrete beam section under the action of a given moment has a fiber stress distribution as shown in Fig. 7-7. The total vertical shear in the concrete at the section is 520 k (2313 kN). Compute and compare the principal tensile stresses at the centroidal axis N–N and the junction of the web with the lower flange M–M.

Solution I of the section about its centroidal axis is computed as 3,820,000 in.[4] Other values are listed separately for the two levels M–M and N–N as tabulated.

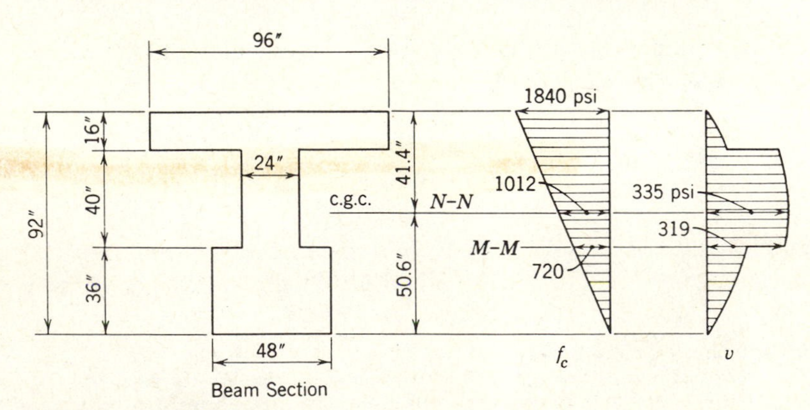

Fig. 7-7. Example 7-1.

Section	M–M	N–N
Q, in.^{3}	$36\times48\times32.6$ $=56{,}200$	$14.6\times24\times7.3$ $+56{,}200=58{,}800$
$v=\dfrac{V_cQ}{Ib}$, psi	$\dfrac{520{,}000\times56{,}200}{3{,}820{,}000\times24}$ $=319$	$\dfrac{520{,}000\times58{,}800}{3{,}820{,}000\times24}$ $=335$
f_c, psi	720	1012
$f_t''=\sqrt{v^2+\left(\dfrac{f_c}{2}\right)^2}-\dfrac{f_c}{2}$	$\sqrt{319^2+\left(\dfrac{720}{2}\right)^2}-\dfrac{720}{2}$ $=121$	$\sqrt{334^2+\left(\dfrac{1012}{2}\right)^2}-\dfrac{1012}{2}$ 100

Instead of using equation 7-2 as above, f_t'' can be obtained directly from Fig. 7-8, using the curves plotted therein, or it can be measured graphically by constructing Mohr's circles as in Fig. 7-6. The same answers will be obtained by any of these methods.

In this example, the greatest principal tension occurs at M-M rather than the centroidal axis N-N where v is a maximum. As in most prestressed sections, the principal tension is much smaller than the vertical shearing stress.

The following example illustrates how principal tension may increase much faster than the increase in shear. This indicates why shear design for working

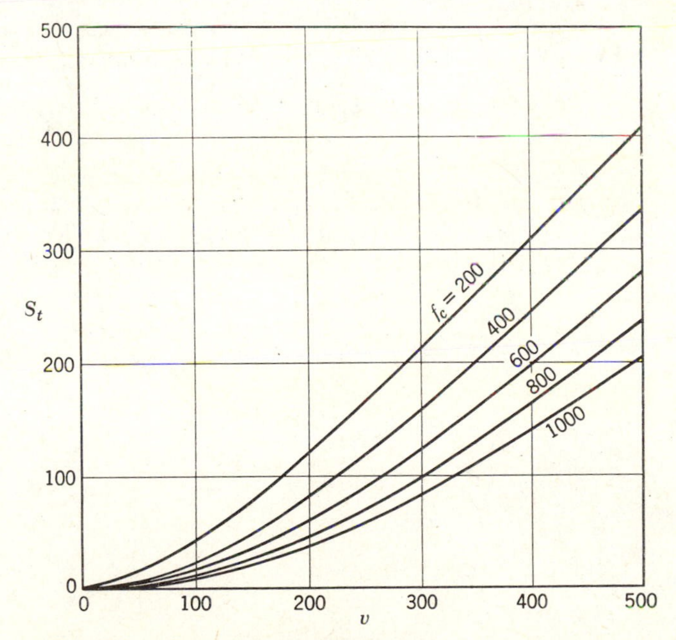

Fig. 7-8. Graph for principal tension.

load alone is not used by the ACI Code. The principal tension as well as the effect of flexural cracking under ultimate load should be investigated.

EXAMPLE 7-2

For the beam section in example 7-1, suppose that the external load is increased by 25% so that the fiber stress distribution is shown in Fig. 7-9 and the total vertical shear in the concrete is $520 \times 1.25 = 650$ k (2,891 kN). Compute the principal tensile stress at *M-M*.

Solution The compressive fiber stress at *M-M* is computed to be 560 psi (3.86 N/mm²), assuming that cracks have not occurred at the bottom fiber for the tensile stress of 561 psi. THe unit vertical shearing stress is

$$v = \frac{V_c Q}{Ib} = \frac{650{,}000 \times 56{,}200}{3{,}820{,}000 \times 24} = 399 \text{ psi } (2.75 \text{ N/mm}^2)$$

Hence the principal tensile stress at *M-M* is

$$f_t'' = \sqrt{v^2 + \left(\frac{f_c}{2}\right)^2} - \frac{f_c}{2}$$

$$= \sqrt{399^2 + \left(\frac{560}{2}\right)^2} - \frac{560}{2} = 208 \text{ psi } (1.43 \text{ N/mm}^2)$$

By comparing this value with $f_t'' = 121$ psi (0.83 N/mm²) in example 7-1, it is seen that, for this particular point, an increase of 72% in principal tension has taken place corresponding to a 25% increase in loading. Note that, in this example, V_p is assumed to be zero. If V_p is not zero, the percentage increase in the value of f_t'' will be still higher.

Some large bridges have utilized vertical or inclined tendons which cause the element to have both f_x and f_y as shown in Fig. 7-6(*b*). Note that the principal stress, f_t, will be dependent on $(f_x - f_y)$ in this case as is easily seen from comparison of the two equations for f_t'' in Fig. 7-6. For usual prestressed concrete beams, $f_y = 0$ (no vertical prestressing of the web), and we have the case which has been discussed above for the elastic analysis of web shear stress.

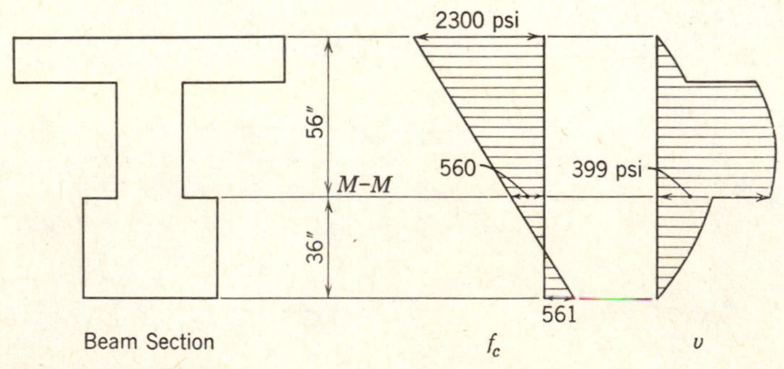

Fig. 7-9. Example 7-2.

In composite construction, shearing stress v between precast and in-place portions computed on the basis of the ordinary elastic theory,

$$v=\frac{VQ}{Ib}$$

where V = the total shear in lb applied after the in-place portion has been cast

Q = statical moment of the cross-sectional area of in-place portion taken about the centroidal axis of the composite section

I = moment of inertia of the composite section

b = width of the contact area between the precast and the in-place portions

Obviously, the shearing strength along the surface will depend on many factors. However, for purpose of design, empirical values have been set by various authorities. For example the ACI Code allows at factored load (design load) a value of 80 psi (0.55 N/MM^2) for a smooth surface with ties, and 350 psi (2.41 N/MM^2) for a roughened surface with adequate ties. These recommendations are considered adequate for design.

Generally speaking, no ties are required for composite slabs or panels where large contact areas are provided. For beams with a narrow strip of top flange to be made composite with in-place slabs, ties are almost always required. Although designed on the basis of shear resistance, a main function of the ties is to prevent the separation of the component elements in a direction normal to the contact surface. Hence ties should be provided if there is a danger of separation, regardless of the shearing stress.

Push-off tests at the Portland Cement Association[7] yielded some interesting values. It was indicated that the ultimate shearing stress for composite action (between 3000 psi = 21 N/mm^2 slab concrete and 5000 psi = 34 N/mm^2 girder concrete) was about 500 psi (3.45 N/mm^2) for a rough bonded surface and 300 psi (2.07 N/mm^2) for a smooth bonded surface. In addition to these values, approximately 175 psi shear capacity may be added for each per cent stirrup reinforcement crossing the joint. These tests also indicated, contrary to the traditional belief, that shear keys used with a rough bonded surface do not change the strength of the connection. This is explained by the fact that the slip movements required to develop the keys are greater than the movements for a bonded surface; therefore, one will fail before the other and the effects of the two are not additive.

7-3 Shear, Ultimate Strength for Beams

As discussed in the previous section, the conventional method for "analyzing" the principal tension, based on the state of stress in a homogeneous material, is a rational method of analysis as long as the concrete has not cracked. However, if service load limiting values are set for "designing," the members so proportioned will possess different factors of safety. Slight increases in loads may produce appreciable and varying increases in the principal tension while the resistance of concrete to principal tension may also change with the magnitude of the compressive fiber stress. Furthermore, after the cracking of concrete, whether produced by flexural or principal tension, the method of analysis is no longer applicable. Hence it is evident that service load shear design by stress analysis is not a satisfactory one, especially if the member is to be subjected to overloads.

In order to obtain more logical and accurate design for shear, particularly in structures of unusual proportions, it is necessary that design be based on their strength under factored design load. Their stresses under the working loads might be checked to avoid cracking at this stage of loading if this should be important in a given design.

In prestressed concrete beams, cracks may be of either inclined flexural or principal tension (web crack) type (Figs. 7-3, 7-4, 7-5). For certain beams, generally those with low percentages of longitudinal reinforcement and with high moment-to-shear ratios, the flexure cracks will develop faster than the web principal tension cracks, the steel will be highly stressed in the region of high bending moment, and final failure will occur by crushing of concrete above the flexural cracks. When the beams are overreinforced but still subject to high moment rather than high shear, failure may occur by crushing of concrete while the steel is still in the elastic range. Such beams are not allowed in design following the ACI Code limiting reinforcement indices.

When the shear is heavy, the principle tension web cracks will develop faster than the flexural cracks; the presence of principal tension cracks will tend to reduce the compressive depth of concrete, and the beam will fail at a load lower than its capacity under pure flexure. Flexure cracks are not necessarily objectionable, unless they combine with and develop into principal tension cracks. Existing by themselves, flexure cracks do not indicate any imminent failure of the beam unless it is highly overreinforced. On the other hand, when flexural cracks develop into inclined tension cracks, sudden and violent failure could result.

The design for ultimate shear strength following the ACI Code is based on extensive laboratory testing and has yielded safe structures since its adoption in 1963. Flexural design of the member is completed first, and the shear strength is then evaluated. Thus, the procedure is one of analysis to determine the shear strength of the concrete v_c compared to the ultimate shear stress at a given section v_u. Since the shear strength may be governed by either of the two types of cracking mechanisms described above, both must be considered. The one yielding lesser shear strength governs v_c at the given section.

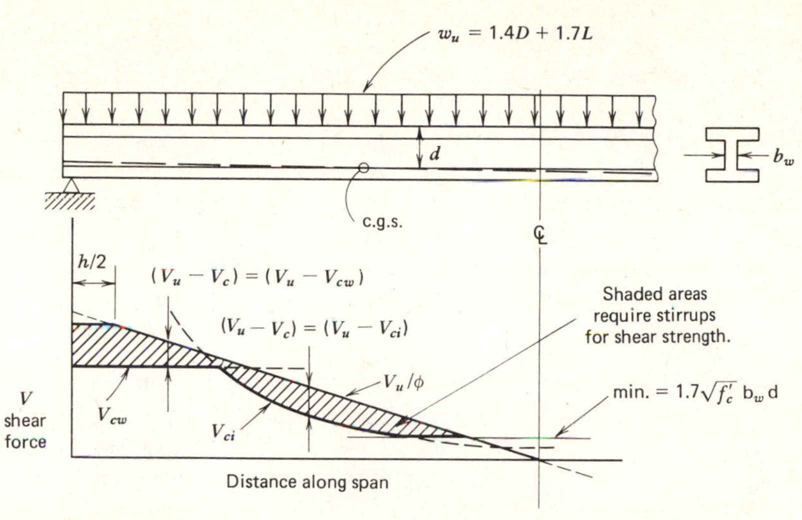

Fig. 7-10. ACI analysis for shear strength—distribution of shears along span.

From this point on, all notation in connection with shear design will follow the ACI Code. The shear stress at formation of principal tension web cracking is designated v_{cw}; shear stress at formation of inclined flexural cracking is designated v_{ci}. Stirrups will be designed for the difference, $(v_u - v_c)$, where v_c is the lesser of v_{cw} or v_{ci}, as discussed above. Figure 7-10 shows the type of distribution which we might expect for a uniformly loaded simple beam. Given in Table 7-1

Table 7-1 ACI Equations for shear strength evaluation: Shear stress $= \dfrac{\text{shear force}}{b_w d}$

Shear Stress Equations (following 1971 ACI Code)		Shear Force Equations (following 1977 ACI Code)	
$v_u = \dfrac{V_u}{\phi b_w d}$	(7-3)	V_u = Factored shear force at section	
		$V_u \leqslant \phi V_n$	(7-6)
where V_u in design is factored shear force at section. Use $\phi = 0.85$ for shear.		Design for V_u/ϕ as the nominal shear strength, V_n. Use $\phi = 0.85$ for shear.	
$v_{ci} = 0.6\sqrt{f'_c} + \dfrac{V_d + \left(\dfrac{V_i M_{cr}}{M_{max}}\right)}{b_w d} \geqslant 1.7\sqrt{f'_c}$	(7-4)	$V_{ci} = 0.6\sqrt{f'_c}\, b_w d + V_d + \dfrac{V_i M_{cr}}{M_{max}} \geqslant 1.7\sqrt{f'_c}\, b_w d$	(7-7)
where $M_{cr} = (I/y_t)(6\sqrt{f'_c} + f_{pe} - f_d)$	(7-9)	where $M_{cr} = (I/y_t)(6\sqrt{f'_c} + f_{pe} - f_d)$	(7-9)
$v_{cw} = 3.5\sqrt{f'_c} + 0.3 f_{pc} + \dfrac{V_p}{b_w d}$	(7-5)	$V_{cw} = (3.5\sqrt{f'_c} + 0.3 f_{pc}) b_w d + V_p$	(7-8)

are the expressions for computing v_u, v_{ci}, and v_{cw} along with the corresponding shear force equations for V_u, V_{ci} and V_{cw} from the 1977 ACI Code for later reference.

Evaluation of V_{cw} (Web Shear Cracking). The ACI Code equation for evaluation of v_{cw} may be shown to be an approximation of the more complex expression which would be used to compute the principal tensile stress at the centroid of a noncomposite prestressed concrete beam. The research[5] on beams of this type showed that the initiation of web shear cracking correlated with a principal tensile stress at the centroid of the section equal to $4\sqrt{f_c'}$. The ACI Code specifies that:

> "Alternatively, V_{cw} may be taken as the shear stress corresponding to a multiple of dead load plus live load which results in a computed principal stress of $4\sqrt{f_c'}$ at the centroidal axis of the member or at the intersection of the flange and the web when centroidal axis is in the flange. In a composite member, the principal tensile stress shall be computed using the cross section which resists live load."

We can show the basis for the v_{cw} equation following the ACI Code Commentary. Even though it was shown in Example 7-1 that larger principal tensile stress may be computed at sections away from the centroidal axis, the difference is not excessive. Tests show that the maximum principal tensile stress (and thus web cracking) will occur near the centroid of the cross section, Fig. 7-4. The calculated stress at centroidal axis of the member equal to $4\sqrt{f_c'}$ correlates with formation of the web shear cracking. This gives a basis for computing concrete shear strength when initial cracking is in the web ($V_{cw} = v_{cw} b_w d$).

The capacity of the member is reached if:

$$f_t'' = \sqrt{v_{cw}^2 + \left(\frac{f_{pc}}{2}\right)^2} - \frac{f_{pc}}{2}$$

where f_t'' = tensile strength of concrete, v_{cw} = shear stress, and f_{pc} = compressive stress due to prestress. This relation yields:

$$\left(f_t'' + \frac{f_{pc}}{2}\right)^2 = v_{cw}^2 + \left(\frac{f_{pc}}{2}\right)^2$$

or,

$$v_{cw} = f_t'' \sqrt{1 + \frac{f_{pc}}{f_t''}}$$

A value for f_t'' of $4\sqrt{f_c'}$ appears to be substantiated by tests, but since v_{cw} is the nominal shear stress and not the actual maximal one, $f_t'' = 3.5\sqrt{f_c'}$ is used in

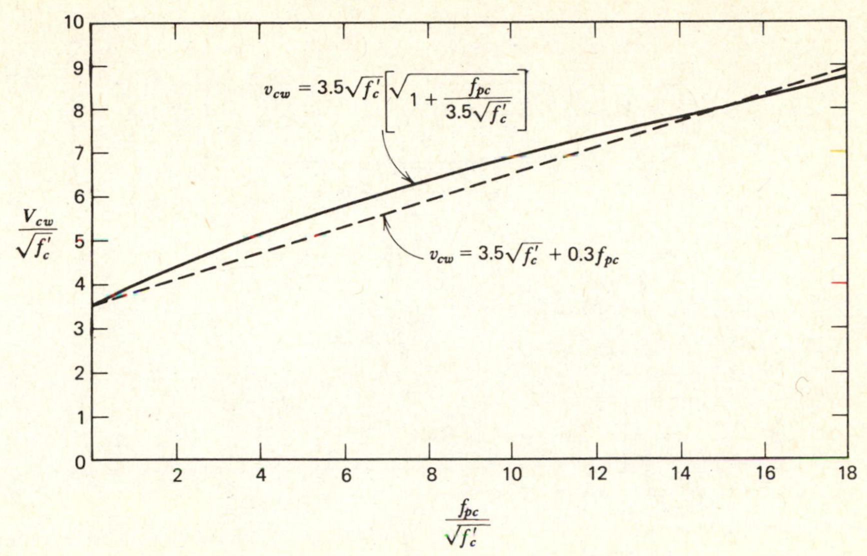

Fig. 7-11. Relationship between nominal shear stress at (web) cracking and compressive stress at centroid.[cw]

the expression for v_{cw}:

$$v_{cw} = 3.5\sqrt{f'_c}\left(\sqrt{1 + \frac{f_{pc}}{3.5\sqrt{f'_c}}}\right) \tag{7-11}$$

The curve representing this equation is plotted on Fig. 7-11, as a solid line. The equation may be simplified to the form:

$$v_{cw} = 3.5\sqrt{f'_c} + 0.3f_{pc}$$

which is shown by the dashed lines.

EXAMPLE 7-3

Check shear strength for the beam shown in Fig. 7-12 at station 1-1 which is $h/2$ from support. Given that this section is adequate for $w_u = 4.6$ k/ft (67 kN/m) on the basis of its flexural strength computed in example 5-8. $f'_c = 7000$ psi (48.3 N/mm^2)

Solution (*a*) Using equation 7-5 for v_{cw}, Table 7-1:

As shown in Fig. 7-10, we know that near the support at station 1-1 we must check v_{cw} only. By inspection, the section will not have flexural cracking at station 1-1, thus v_{ci} will not be critical for shear strength of concrete.

$$f_{pc} = \frac{440}{373} = 1.180 \text{ ksi} = 1180 \text{ psi } (8.136 \text{ N/mm}^2)$$

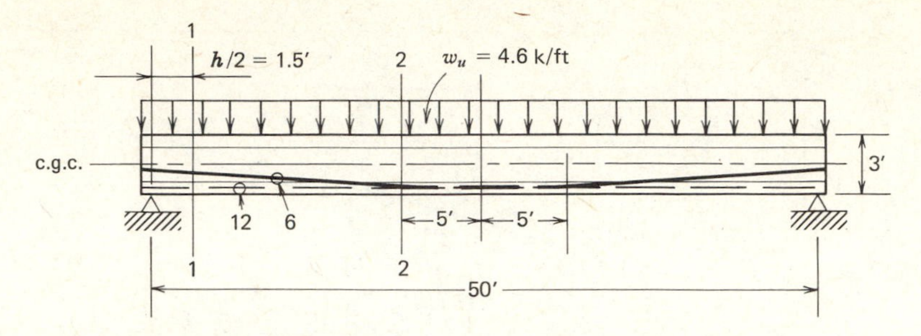

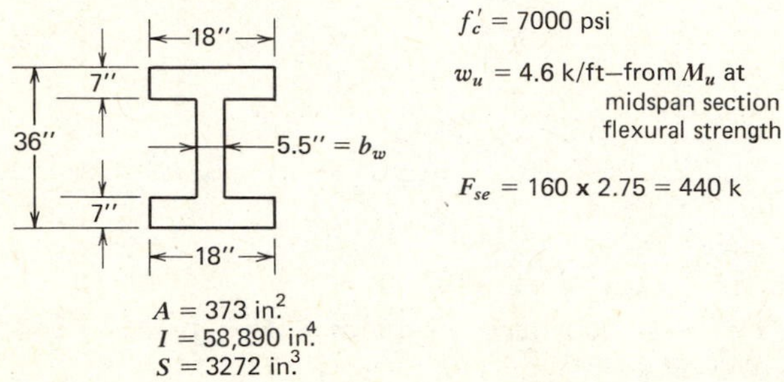

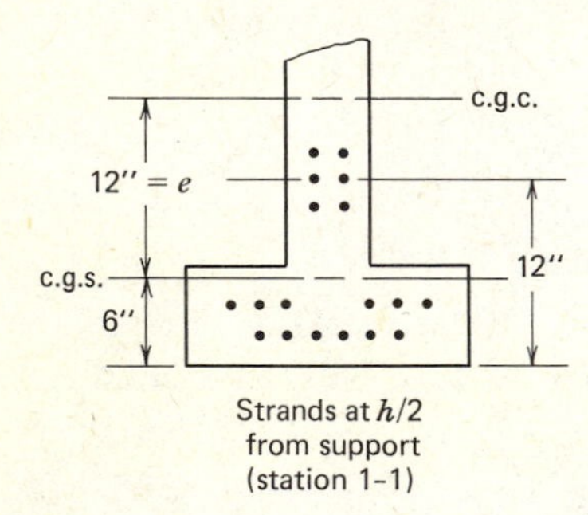

Note: at station 2-2 c.g.s. is located 4.5 in. above bottom fiber.

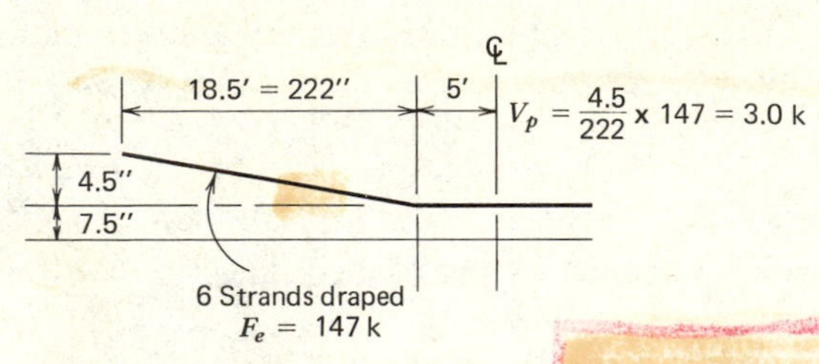

Fig. 7-12. Example 7-3.

Note that for noncomposite sections, f_{pc} at the centroid will not be changed by moment resulting from eccentricity of prestress ($y=0$ in the equation Fey/I).

$$v_{cw}=3.5\sqrt{7000}+(0.3)(1180)+\frac{3000}{(5.5)(30)} \tag{7-5}$$

$$v_{cw}=293+354+18=\underline{665\text{ psi}}\ (4.59\text{ N/mm}^2)=v_c$$

$$V_u=(25-1.5)(4.6)=108.1\text{ k }(481\text{ kN})$$

$$v_u=\frac{108{,}100}{(0.85)(5.5)(30)}=\underline{771\text{ psi}}\ (5.32\text{ N/mm}^2)>v_c \tag{7-3}$$

thus, stirrups are required to carry $(v_u - v_c)$.

$$v_u - v_c = 771 - 665 = 106 \text{ psi } (0.73 \text{ N/mm}^2)$$

(*b*) Using 1977 ACI Code equation for V_{cw}, Table 7-1:

$$V_u = 108.1 \text{ k } (481 \text{ kN}) \text{ from } (a) \text{ above.}$$

$$V_{cw} = \left(3.5\sqrt{f_c'} + 0.3f_{pc}\right)b_w d + V_p$$

$$= 109.7 \text{ k } (488 \text{ kN}) \tag{7-8}$$

thus, stirrups are required to carry $\left(\frac{V_u}{\phi} - V_{cw}\right) = 17.5 \text{ k } (77.7 \text{ kN})$

(*c*) Using ACI Code alternate procedure with principal stress equal $4\sqrt{f_c'}$ at centroid, check v_{cw}.

$$Q = (18)(7)(18-3.5) + (5.5)(11)(5.5)$$

$$Q = 1827 + 333 = 2160$$

$$v = \frac{V_c Q}{I b_w} = \frac{V_c(2160)}{(58{,}890)(5.5)} = 0.006669 V_c$$

$$f_{pc} = \frac{F_e}{A} = 1180 \text{ psi } (8.14 \text{ N/mm}^2)$$

$$f_t'' = \sqrt{v^2 + \left(\frac{f_c}{2}\right)^2} - \left(\frac{f_c}{2}\right) \tag{7-2}$$

$$4\sqrt{7000} = \sqrt{(0.006669V_c)^2 + \left(\frac{1180}{2}\right)^2} - \frac{1180}{2}$$

$$(334.7 + 590)^2 = (0.006669V)^2 + (590)^2$$

$$V_c = 106{,}765 \text{ lb } (475 \text{ kN})$$

$$v_{cw} = \frac{V_c + V_p}{b_w d}$$

$$v_{cw} = \frac{106{,}765 + 3000}{(5.5)(30)} = \underline{665 \text{ psi}} \, (4.59 \text{ N/mm}^2) < v_u = 771 \text{ psi } (5.32 \text{ N/mm}^2)$$

Note that for this case the alternate procedure for v_{cw} yields the same shear stress of 665 psi as we obtained with equation 7-5 in part (*a*). The ACI equation is simpler to apply and would thus normally be used in design. Of course, we could use either v_{cw} or V_{cw} as the designer prefers and there is no difference between (*a*) and (*b*) results except units used in the equations (Table 7-1).

Evaluation of V_{ci} (Inclined Flexural Cracking). The type cracking associated with this potential shear failure mode was shown in Figs. 7-3 (*b*) and 7-5. When this approach to shear strength analysis was introduced as part of the ACI Code in 1963, the total shear V_{ci} at inclined flexural cracking was shown to correlate

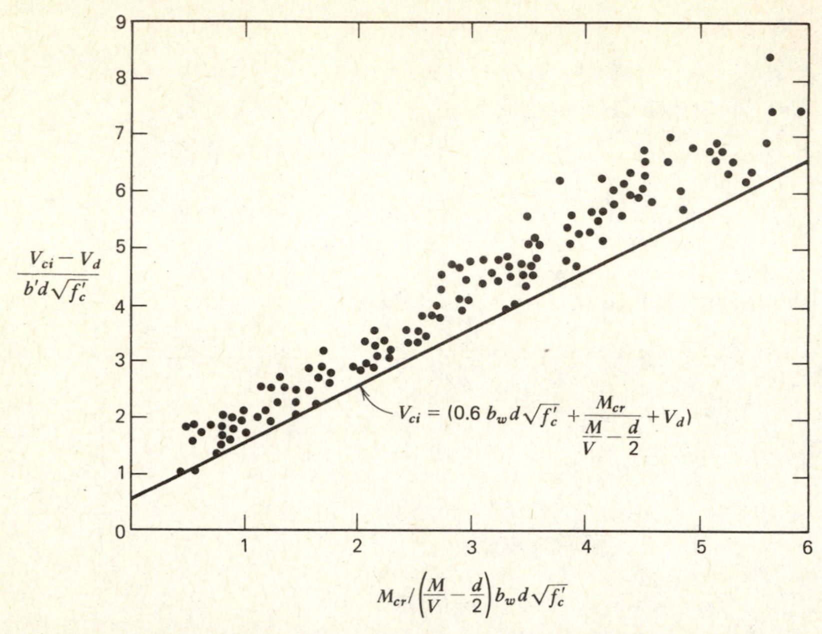

Fig. 7-13. Diagonal cracking in those regions of beams previously cracked in flexure.

(see Fig. 7-13) with formation of a flexural crack at a distance $d/2$ from the section under consideration plus a shear which is a function of the dimensions of the cross section and the strength of the concrete. This was expressed as follows:

$$V_{ci} = 0.6 b_w d\sqrt{f'_c} + \frac{M_{cr}}{\frac{M}{V} - \frac{d}{2}} + V_d \tag{7-10}$$

Figure 7-13 shows the close agreement between this equation from ACI Code (318-63) and the available experimental data. The second term involves M_{cr}, the moment due to applied loads when flexural cracking occurs. This term actually gives the shear due to applied loads when flexural cracking occurs, even though expressed in terms of moment. The cracking moment was shown by tests to correlate conservatively with $6\sqrt{f'_c}$ as modulus of rupture for concrete:

$$M_{cr} = \frac{I}{y_t}\left(6\sqrt{f'_c} + f_{pe} - f_d\right) \tag{7-9}$$

The net precompression of the expreme fiber with dead load of the beam itself (L.F. = 1.0) is $(f_{pe} - f_d)$ in this expression. Cracking occurs when this stress has been overcome by the applied load moment and the extreme fiber tension (y_t is measured to this fiber) becomes $6\sqrt{f'_c}$. The total shear V_{ci} due to applied loads and dead load when the flexural crack occurs is:

$$V_{ci} = \frac{M_{cr}}{\frac{M}{V} - \frac{d}{2}} + V_d$$

Adding the small additional shear $0.6 b_w d \sqrt{f'_c}$ completes the V_{ci} equation (7-10), and this represents the small additional shear which causes the initial tension crack to become an inclined flexural crack, as illustrated in Fig. 7-3(b).

The value of M/V at a section is a function of the distribution of the loading as shown in Fig. 7-14, and it can be rather easily evaluated for a given section where V_{ci} is being checked. For a noncomposite simple beam with distributed loading for both dead load and applied load, we do not have to treat the dead load separately. Thus the V_d term in equation 7-10 and the f_d term in equation

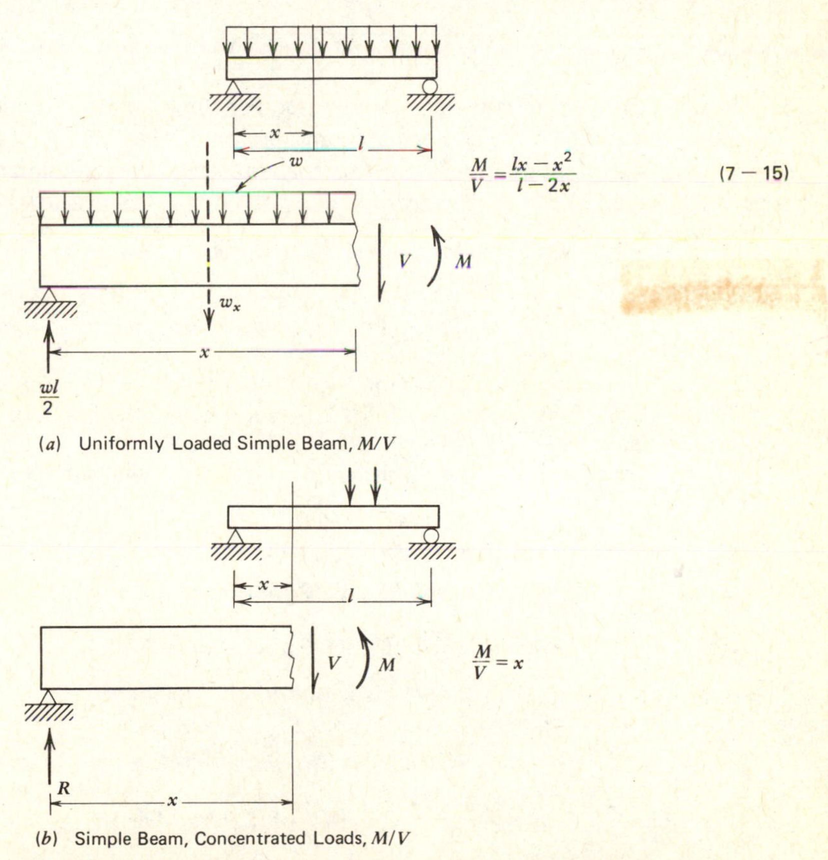

(a) Uniformly Loaded Simple Beam, M/V

(b) Simple Beam, Concentrated Loads, M/V

Fig. 7-14. Ratio of M/V for uniform and concentrated loads on simple beams.

7-9 become zero, and calculation of the V_{ci} (corresponding to inclined flexural cracking) is simplified as illustrated in example 7-4.

In the ACI Code 318-77 a simplifying change has been made in the equation for V_{ci}. The denominator of equation 7-10 was simplified to drop the distance $d/2$. The resulting equation for V_{ci} is:

$$V_{ci}=0.60b_w d\sqrt{f_c'}+V_d+V_i M_{cr}/M_{max} \quad (7\text{-}7)$$

As mentioned above, for uniformly loaded noncomposite simple beams this becomes simplified:

$$V_{ci}=0.6b_w d\sqrt{f_c'}+\frac{V_i M_{cr}}{M_{max}} \qquad \text{ACI Commentary}$$

where

$$M_{cr}=\frac{I}{y_t}\left(6\sqrt{f_c'}+f_{pe}\right)$$

For composite beams the M_{cr} is computed for the composite section, thus the I/y_t term becomes I'/y_t'. The net precompression of the precast beam $(f_{pe}-f_d)$ is for the precast section alone. For this case the complete equation (7-7) is used where V_d is the shear at the section considered due to precast beam weight (L.F. = 1.0).

EXAMPLE 7-4

Check shear strength for station 2-2 for the beam shown in Fig. 7-12. The I-shaped, noncomposite section spans 50 ft, and the c.g.s. is as 4.5 in. from the bottom fiber at station 2-2. Other information is given in Fig. 7-12 for this beam: $F_e=440$ k, $e=13.5$ in., $d=31.5$ in., $f_c'=7000$ psi, and $w_d=389$ lb/ft (beam weight). (Span = 15.12 m, $F_e=1.957$ kN, $e=343$ mm, $d=800$ mm, $f_c'=48$ N/mm², and $w_d=5.68$ kN/m.)

Solution Use equation 7-7 for V_{ci} since, by inspection, V_{ci} will control shear strength.

$$f_{pe}=\frac{F}{A}+\frac{Fe}{S}=\frac{440}{373}+\frac{(440)(13.5)}{3272}=1.180+1.815$$

$$f_{pe}=2.995 \text{ ksi}=2995 \text{ psi } (20.65 \text{ N/mm}^2)$$

Consider dead load of beam (L.F. = 1.0). Solving for station 2-2,

$$M_d=\frac{wlx}{2}-\frac{wx^2}{2}=\frac{(389)(50)(20)}{2}-\frac{(389)(20)^2}{2}$$

$$M_d=116{,}700 \text{ ft-lb } (158 \text{ kN}-\text{m})$$

$$f_d=\frac{(116{,}700)(12)}{3272}=425 \text{ psi } (2.93 \text{ N/mm}^2)$$

$$V_d=(389)(5)=1945 \text{ lb } (8651 \text{ N})$$

Using relationship from Fig. 7-14(a) for M/V, equation 7-15,

$$\frac{M_{\max}}{V_i}=\frac{lx-x^2}{l-2x}=\frac{(50)(20)-(20)^2}{50-(2)(20)}=60\text{ ft}=720\text{ in.}$$

$$M_{cr}=\frac{I}{y_t}\left(6\sqrt{f'_c}+f_{pe}-f_d\right)$$

$$M_{cr}=\frac{(3272)(6\sqrt{7000}+2995-425)}{1000}$$

$$M_{cr}=10{,}050\text{ in.-k }(1136\text{ kN-m})$$

Using the ACI Code equation for V_{ci},

$$V_{ci}=0.6\sqrt{f'_c}\,b_w d+V_d+\frac{V_{ci}M_{cr}}{M_{\max}} \qquad (7\text{-}7)$$

$$V_{ci}=\frac{0.6\sqrt{7000}}{1000}(5.5)(31.5)+1.945+\frac{10{,}050}{720}$$

$$V_{ci}=24.6\text{ k }(109.4\text{ kN})$$

Checking ultimate shear force at station 2-2,

$$V_u=(5)(4.6)=23.0\text{ k }(102\text{ kN})$$

$$V_u/\phi=23.0/0.85=27.1\text{ k }(120.4\text{ kN})$$

Stirrups are required at station 2-2 for $V_u/\phi-V_{ci}=2.5$ k (11 kN).

Note that we might use the simpler expression for V_{ci} suggested by the ACI Commentary for this noncomposite, uniformly loaded beam. Both f_d and V_d are zero, and all load is uniformly distributed applied load for $M_{\max}/V_i$ at station 2-2.

$$M_{cr}=\frac{I}{y_t}\left(6\sqrt{f'_c}+f_{pe}\right)=(3272)(6\sqrt{7000}+2995)/1000$$

$$M_{cr}=11{,}440\text{ in.-k }(1293\text{ kN-m})$$

$$M_{\max}/V_i=720\text{ in. station 2-2 from above calculation}$$

$$V_{ci}=0.6b_w d\sqrt{f'_c}+\frac{M_{cr}}{M_{\max}/V_i}=\frac{(0.6)(5.5)(31.5)\sqrt{7000}}{1000}+\frac{11{,}440}{720}$$

$$V_{ci}=24.6\text{ k (at station 2-2)}$$

This value corresponds to the result obtained above. We must check stirrup requirements following ACI Code as shown in section 7-4 and example 7-5.

Figures 7-15 and 7-16 show the results of the shear strength analysis for the beam of Examples 7-3 and 7-4. The tabular form shown in Fig. 7-15 is very convenient and allows stations 3 and 4 to be evaluated easily following Example 7-4 as a sample calculation. Note the plot of these results in Fig. 7-16 gives assurance as to the most critical regions where stirrups are required for shear strength.

Shear Strength Calculations

Summary of Results

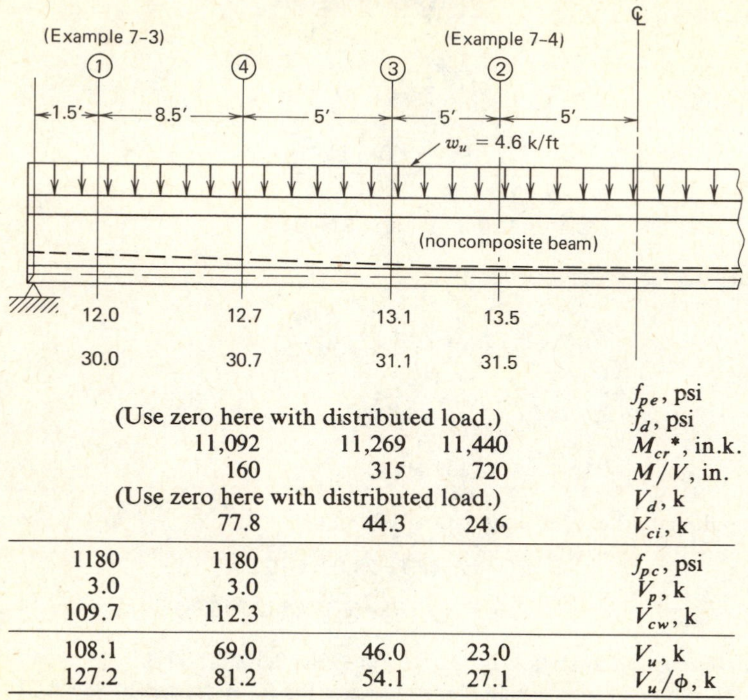

①	④	③	②	
				f_{pe}, psi
(Use zero here with distributed load.)				f_d, psi
	11,092	11,269	11,440	M_{cr}*, in.k.
	160	315	720	M/V, in.
(Use zero here with distributed load.)				V_d, k
	77.8	44.3	24.6	V_{ci}, k
1180	1180			f_{pc}, psi
3.0	3.0			V_p, k
109.7	112.3			V_{cw}, k
108.1	69.0	46.0	23.0	V_u, k
127.2	81.2	54.1	27.1	V_u/ϕ, k

*M_{cr} from simplified equation of ACI Commentary used here since distributed loading is carried.

Fig. 7-15. Shear strength calculations for beam of examples 7-3 and 7-4.

7-4 Shear, Web Reinforcement

Since we do not understand completely the behavior of prestressed or reinforced concrete beams without web reinforcement, it is obviously more difficult to analyze those with web reinforcement. Fortunately, fairly extensive series of tests have now been conducted,[5] and semiempirical methods have been formulated which allow design for shear to be made conservatively.

The expression for design of stirrups following the ACI Code may be expressed as follows:

$$V_s = A_v f_y d/s \tag{7-11}$$

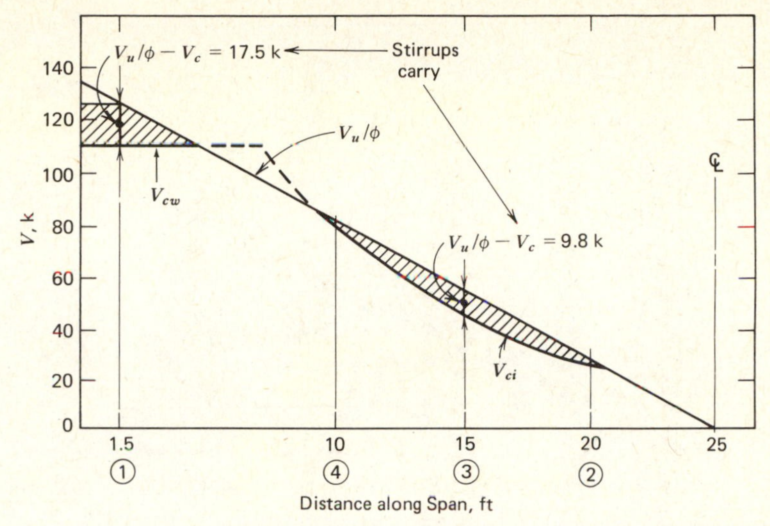

Fig. 7-16. Plot of shear force vs. distance along span beam of examples 7-3 and 7-4.

where $\quad V_s = V_n - V_c = V_u/\phi - V_c$

(Excess of nominal shear force at ultimate $V_n = V_u/\phi$ over that which the concrete can carry, V_c. Here V_c is the lesser value V_{cw} or V_{ci} as discussed in section 7-3.)

A_v = area of unstressed vertical stirrups with yield stress of f_y

s = stirrup spacing

d = depth of beam

The development of this expression will be discussed below. Note that the V_{cw} and V_{ci} calculations yield V_c, the capacity of the concrete to carry shear at the time of shear crack development. Stirrups must carry the excess of V_n above this concrete shear capacity, V_c. The shaded areas in Fig. 7-10 and 7-16 show this as a graphical display. Tests verify that the additional capacity of the beam with stirrups is actually developed, Fig. 7-17.

Combined Moment and Shear. Under combined high moment and shear, flexural cracks develop into inclined tension cracking which could reduce the moment capacity and result in sudden and violent failure of the section if proper web reinforcement is not provided. The design for combined moment and shear can be best based on the ultimate strength behavior as illustrated in Fig. 7-18. Owing to high moment and shear, a flexural crack starting at point A will progress upward and eventually turn into an inclined crack. The web stirrups, each supplying a force of $A_v f_y$ will be called into action as the crack progresses. At the ultimate load, the inclined crack will likely extend over a projected distance of at least d, and the number of stirrups intercepted by the crack will be

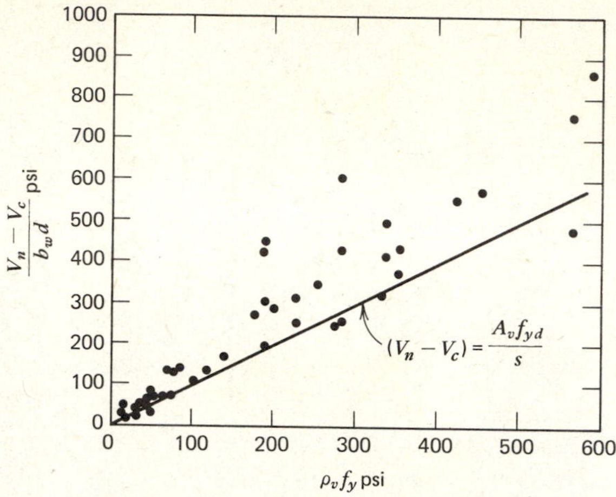

Fig. 7-17. Increase in shear strength of a prestressed member due to web reinforcement.[5]

d/s. Thus, static equilibrium requires that

$$V_n - V_c = A_v f_y d/s \tag{7-12}$$

where V_n = design ultimate external shear on the section from factored loads (V_u/ϕ)
V_c = ultimate shear carried by the concrete in the compressive flange
A_v = area of each stirrup
f_y = yield point stress of the stirrups
d = effective depth of the section
s = spacing of stirrups

Note in Fig. 7-17 that the trend of test results follows this formula rather well, with the line representing equation 7-12 falling below almost all the available

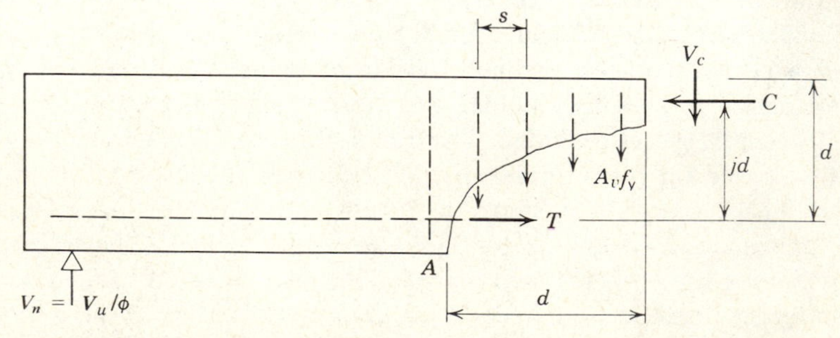

Fig. 7-18. Ultimate design for combined moment and shear.

data shown. This is on the conservative side since $(V_n - V_c)$ values of test points fall above that counted on by the stirrup design equation shown in Fig. 7-17 as the solid line.

The major difficulty in the above lies in the determination of V_c. While V_c will depend on the ultimate compressive area of the concrete and the amount of compression C it can be approximated by the shear load producing flexural crack at a point d from the load.[5] It can be further assumed that any shear load in addition to that producing the flexural crack is to be carried by the web reinforcement. This does not take into account the shear capacity lost as a result of web cracking; nor does it consider the additional shear carried by the top flange after the inclined cracking. Equation 7-12 makes two further assumptions. One assumption is that all the stirrups are stressed to the yield point. This may not be exactly correct, since some of the intercepted stirrups may be stressed below the yield point. The other is that the projected distance of the inclined crack is d. This is a safe assumption since it has been found experimentally[5] that the distance is generally greater than d, and probably at least equal to $1.1d$.

ACI Code Method for Stirrup Dcsign. An ultimate strength method originally proposed for the 1963 ACI Code follows an empirical approach based primarily on the test results of the University of Illinois.[5] The method indicates that the yield strength in the stirrups extending over a distance d could be considered effective in transmitting the shear of $V_n - V_{ci}$; thus,

$$A_v f_y d/s = V_n - V_{ci} \tag{7-13}$$

where V_n is the design ultimate load shear, and V_{ci} is the shear at the section when the vertical flexural crack starts to develop into an inclined one. A similar criterion is recommended for the case of web cracking initiating by itself without flexural cracking; thus,

$$A_v f_y d/s = V_n - V_{cw} \tag{7-14}$$

where V_{cw} is the shear at the section when web cracking starts by itself, without any flexural cracking.

The ACI Code determines the shear strength V_{cw} and V_{ci} using the equations developed in section 7-3. We assume the nominal shear force as $V_n = V_u/\phi$ where $\phi = 0.85$ for shear strength design. This adds safety to the design in addition to that which is introduced by the load factors used for obtaining V_u. The shear force carried by stirrups, V_s, is expressed as follows:

$$V_s = \frac{V_u}{\phi} - V_c = \frac{A_v f_y d}{s} \tag{7-12}$$

Example 7-5 shows the stirrup design procedure. The shaded areas of Fig. 7-16 show the regions where stirrups are required for shear strength.

The ACI Code has two expressions for the minimum A_v required, and the maximum spacing of stirrups is also limited. This maximum spacing is $0.75h$ in prestressed concrete, but not more than 24 in. Unless V_u is less than one-half of ϕV_c the minimum A_v requirement applies. If it is shown by test that the required ultimate flexural and shear capacity can be developed when shear reinforcement is omitted, the minimum A_v requirement can be waived under the ACI Code. For some standard precast concrete products this testing has been done, and the concrete alone carries the shear stress without stirrups.

The minimum A_v in square inches is given in the ACI Code by two expressions:

$$A_v = 50\frac{b_w s}{f_y} \tag{7-16}$$

or

$$A_v = \frac{A_{ps} f_{pu} s}{80 f_y d}\sqrt{\frac{d}{b_w}} \tag{7-17}$$

Equation 7-17 may be used for prestressed members having an effective prestress force at least equal to 40% of tensile strength of the flexural reinforcement. Equation 7-16 may be used for prestressed and nonprestressed members. The use of these equations is illustrated in example 7-5.

EXAMPLE 7-5

Design stirrups following ACI Code for the beam of examples 7-3 and 7-4. Use the plot of shear stress, Fig. 7-16, in this solution. Assume stirrups perpendicular to the longitudinal axis with $f_y = 50$ ksi (345 N/mm²).

Solution Try #3U stirrups—$A_v = 0.22$ in² (142 mm²) From Fig. 7-16, $V_s = \frac{V_u}{\phi} - V_c =$ 17.5 k (77.7 kN) at ends.

$$V_s = \frac{A_v f_y d}{s} \tag{7-12}$$

$$s = \frac{(0.22)(50)(30)}{17.5}$$

$s = 18.9$ in. (480 mm) (spacing required near ends) Use $s = 16$ in.

From Fig. 7-16, at 15 ft. from support $V_s = V_u/\phi - V_c = 9.8$ k,(43.5 kN).

$$s = \frac{(0.22)(50)(31.1)}{9.8}$$

$s = 34.9$ in. (886 mm) > 24 in. (588 mm) max. spacing

Maximum spacing controls except at ends.

$$\text{Check minimum } A_v = \frac{50\, b_w s}{f_y} = \frac{(50)(5.5)(24)}{50{,}000} \tag{7-16}$$

min. $A_v = 0.132$ in² (85 mm²) < 0.22 in² (142 mm²)

Use #3U stirrups at 24 in. (610 mm) spacing for entire beam except two spaces of 16 in. (406 mm) at each end. Place first stirrup no more than 8 in. (203 mm) from end.

7-5 Flexural Bond at Intermediate Points

For posttensioned concrete, bond is supplied by grouting. For pretensioned concrete, bond is secured directly when placing the concrete. When a bonded beam is subject to shear, bond stresses are produced. In order to design against bond failure, it is necessary to determine two things: first, the amount of bond stress existing between steel and concrete; second, the bond resistance between the two materials. Prestress transfer bond in pretensioning, where the end anchorage of the tendons is secured solely by bond, is discussed in the next section. Bond at intermediate points along the length of beam, whether pretensioned or posttensioned, will now be discussed.

To determine the bond stress existing between concrete and the tendons, two stages have to be considered: before and after cracking of concrete. Before the cracking of concrete, bond stress can be calculated using conventional elastic analysis. The wires or strands are found to have very low bond stress when this problem is solved. The flexural bond stress between the prestressing steel and the surrounding concrete (or grout, in posttensioned construction) has not been a problem in practice with members which are uncracked at service load. The current design codes, as a result, do not require the engineer to check flexural bond as a result of this favorable experience. Use of 7-wire strand improves bond strength compared to smooth wires; thus the low-bond stresses are of even less concern with this very common type of tendon.

After cracking, the problem is a little more complicated for calculation of bond stress. The bond stress changes suddenly at the cracks owing to the abrupt transfer of concrete to steel at such points[11, 12]. Solutions made with reasonable assumptions do show that there is a significantly higher bond stress in the regions adjacent to cracks, but there is no evidence that we have a problem with this high flexural bond stress in laboratory test beams or in actual structures. Slight local bond failure is not significant in the overall safety or serviceability of the beam. For members which experience fatigue loading we should avoid cracking at service load as it may make the bond problem more serious. We do not find any flexural bond difficulty with partially prestressed beams with static loading which causes very minor cracking at service load.

The ACI Code deals with bond in both reinforced and prestressed concrete beams in terms of development of reinforcement rather than bond stress. Most prestressed concrete beams are designed to be uncracked at service load. The above discussion shows that flexural bond stress is very low at this stage and need not be checked to assure serviceability. Design will thus be based on development length (transfer bond) as discussed in the following section.

7-6 Prestress Transfer Bond in Pretensioned Concrete

Nature of Bond and Length of Transfer. When tendons are pretensioned, their stress is often transferred to the concrete solely by bond between the two materials. Thus there is a length of transfer at each end of the tendons to perform the function of anchorage, when mechanical end anchorages are not provided. The condition of bond stress existing at these ends is radically different from that along the intermediate length of a beam. At intermediate points, the bond stress is produced by the external shear or by the existence of cracks. Where there are no cracks and no shear, the bond stress is zero. At anchorage, bond stress exists immediately after transfer. The stress in the tendons varies from zero at the exposed end to a full prestress at some distance inside the concrete. That distance is known as the length of transfer; and such bond stress is termed prestress transfer bond.[12] A rather complete survey of the literature on transfer length and development bond is contained in reference 13.

The nature of prestress transfer bond is entirely different from the flexural bond stress produced by shear or cracks. At intermediate points along a beam, the bond stress is resisted by adhesion between steel and concrete, aided by mechanical resistance provided by corrugations in the steel when deformed bars are used. At end anchorages, the pre-tensioned tendons almost always slip and sink into the concrete at the moment of transfer. This slippage destroys most of the adhesion for the length of transfer and part of the mechanical resistance of the corrugations, leaving the bond stress to be carried largely by friction between steel and concrete.

Immediately after transfer, at end A, Fig. 7-19, the wire will have zero stress and its diameter will be restored to the unstressed diameter. At B, the inner end of the length of transfer, the wire will have almost full prestress, and, owing to Poisson's ratio effect, its diameter will be smaller than the unstressed diameter. Thus along the length of transfer, there is an expansion of the wire diameter which produces radial pressure against the surrounding concrete. Frictional force resulting from this pressure serves to transmit the stress between steel and concrete. In other words, a sort of wedging action takes place within that length of transfer.

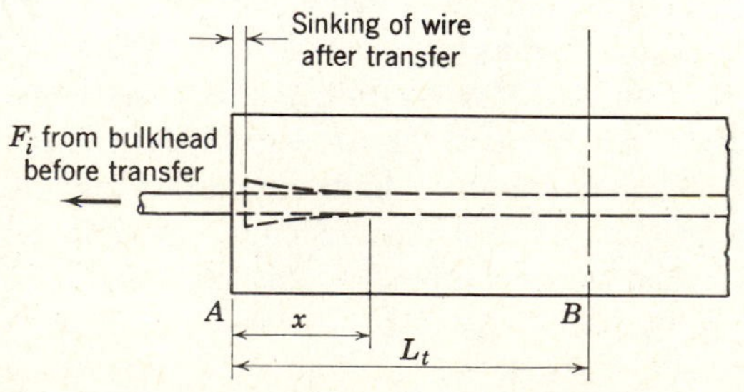

Fig. 7-19. Prestress transfer at end of pretensioned beam.

On the supposition of lateral expansion, Hoyer[14] has derived an equation giving the length of transfer, L_t, as

$$L_t = \frac{d}{2\mu}(1+m_c)\left(\frac{n}{m_s} - \frac{f_i}{E_c}\right)\frac{f_e}{2f_i - f_e} \tag{7-18}$$

where
m_c = Poisson's ratio for concrete
m_s = Poisson's ratio for steel
$n = E_s/E_c$
E_c = modulus of elasticity for concrete
f_i = initial prestress in steel
f_e = effective prestress in steel
μ = coefficient of friction between steel and concrete
d = diameter of wire

If we assume that $m_c = 0.1$, $m_s = 0.3$, $n = 6$, $E_c = 5{,}000{,}000$ psi (34.5 kN/mm^2), $f_i = 150{,}000$ psi (1034 N/mm^2), and $f_e = 125{,}000$ psi (862 N/mm^2), we will reduce equation 7-18 to

$$L_t = 8d/\mu$$

It can be observed, therefore, that the length of transfer varies directly as the diameter of wire and inversely as the coefficient of friction, assuming this wedging action to be the only bonding force. Figure 7-20 shows the form of stress variation over the transfer length.

An elastic analysis has been made by Janney,[12] also based on the elastic theory of a thick-walled cylinder, considering only the frictional bonding phenomenon and neglecting the adhesion or mechanical bond due to corrugation. The results given by this analysis are necessarily approximate because only the frictional force is considered and because concrete is not truly elastic.

Actual tests by Janney[12] to measure the variation of steel stress along the length of transfer verified the analysis in a qualitative manner, indicating that prestress transfer bond for wires is largely a result of friction between concrete and steel. However, the high stresses produced in the surrounding concrete indicated plastic behavior and the inaccuracy of predicted transfer lengths derived on the assumption of the elastic theory was noted.

Since the bond studies by Hoyer[14] in 1939, more than 30 such investigations have been reported in the literature (see Ref. 13 for listing). Most of the early tests dealt with transfer length of small wires of different sizes—either plain, twisted, crimped, indented, or deformed. Only more recent bond studies in the United States and Great Britain have dealt with multi-wire strands. One study in England by Base[15], tests at the Portland Cement Association,[16, 17, 18, 19] and a study by Anderson[20] have dealt with the question of flexural bond and development length as a combined consideration in the behavior of beams.

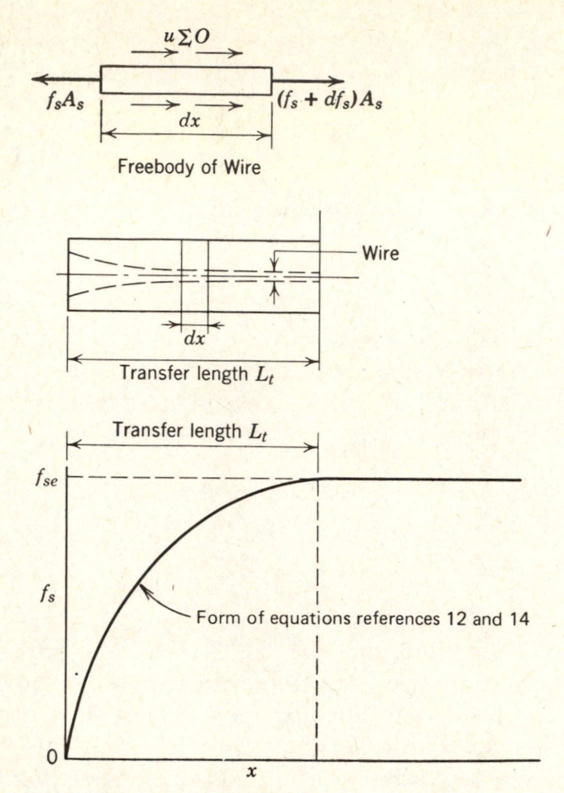

Fig. 7-20. Transfer length.

The transfer length for prestressing steel is affected by many parameters[13], the most important of them being:

1. Type of steel (e.g., wire, strand).
2. Steel size (diameter).
3. Steel stress level.
4. Surface condition of steel—clean, oiled, rusted.
5. Concrete strength.
6. Type of loading (e.g., static, repeated, impact).
7. Type of release [e.g., gradual, sudden (flame cutting, sawing)].
8. Confining reinforcement around steel (e.g., helix or stirrups).
9. Time-dependent effect.
10. Consolidation and consistency of concrete around steel.
11. Amount of concrete coverage around steel.

It is generally agreed that transfer length is longer for larger steel sizes, higher prestress levels, and lower concrete strengths. Strands provide a certain amount of mechanical resistance in addition to friction; thus their transfer length is shorter than smooth wires of comparable diameter.[21] Under repeated loading applied outside the transfer zone, no significant effect on the transfer length is observed. However, if applied within the transfer zone, repeated loading can cause early bond failure if a crack develops with near the transfer length.[13]

The ACI Code makes special provision for checking shear strength near the end of a pretensioned member. When the section at a distance h/2 from the face of the support is closer to the end of the beam than the development length of the tendon, a reduced prestress is considered when calculating V_{cw}.

The prestress force may be assumed to vary linearly from zero at the end of the tendon to a maximum at a distance from the end of the tendon equal to the transfer length, assumed 50 *diameters for strand and* 100 *diameters for single wire.*

Figure 7-20 indicates that the actual variation of f_s over the transfer length is not linear, but the ACI simplified linear design assumption is conservative.

A careful review of the data from transfer length tests[12] led to the following equation for L_t to be used in design (gradual or sudden release at transfer):

$$L_t = 1.5\frac{f_{si}}{f'_{ci}}d_b - 4.6 \tag{7-19}$$

The ACI Code suggests that $L_t = 50d_b$, which is simple but not quite so conservative as equation 7-19. Table 7-2 shows that the difference is quite small for a small-size strand but becomes appreciable for the large-size strands, especially if the concrete strength at transfer is relatively low.

Design for Prestress Transfer by Bond. The design of prestressed members as affected by the transfer bond may be described in two parts. The first involves the distribution of prestress at the ends of members, since the prestress is not concentrated at the free end but is transferred gradually along a certain length.

Table 7-2 Comparison of equation 7-19 with ACI Code requirement for transfer length L_t (in.)

Strand	250-k Grade $f_{si} = 175$ ksi, $f_{se} = 140$ ksi			270-k Grade $f_{si} = 189$ ksi, $f_{se} = 151$ ksi		
Size	Equation 7-19			Equation 7-19		
in.	$f'_{ci} = 3500$ psi	$f'_{ci} = 4000$ psi	ACI	$f'_{ci} = 3500$ psi	$f'_{ci} = 4000$ psi	ACI
$\frac{1}{4}$	14	12	12	16	13	13
$\frac{5}{16}$	19	16	15	21	18	16
$\frac{3}{8}$	24	20	18	26	22	19
$\frac{7}{16}$	28	24	21	31	26	22
$\frac{1}{2}$	33	28	24	36	31	25

Hence the stress concentration at the free end is somewhat relieved, while the stress a few inches inside the free end may be more critical. Since a certain portion of the end of the member is not fully prestressed, its resistance to flexure, cracking, and shear must be determined accordingly, and additional reinforcing steel provided as necessary.

In prestressed beams, it is significant to avoid flexural cracking near the end portions. When concrete cracks under flexure, the bond stress in the vicinity of the cracks rises, and slip occurs over a small portion of the strand adjacent to the cracks. With continued increase in load, the high bond stress progresses as a wave from the original cracks toward the beam ends. If the peak of the high bond stress wave reaches the prestress transfer zone, the increase in steel stress resulting from the bond slip decreases the strand diameter, reduces the frictional bond resistance in wires which have little mechanical resistance, and general bond failure could result. For strands, the helical shape of the individual wires will provide mechanical resistance so that the beam can support additional load even after slip of the strand at the beam ends.

Hence it seems clear that if flexural cracks cannot occur near the end portion of the beam, there is no danger of the high bond stress wave reaching the prestress transfer zone, and there is no danger of beam failure resulting from bond slip. The failure of a beam resulting from bond slippage may be illustrated in Fig. 7-21, which shows the end portion of a pre-tensioned beam loaded to cracking near the end. The solid line in Fig. 7-21 indicates the stress in the

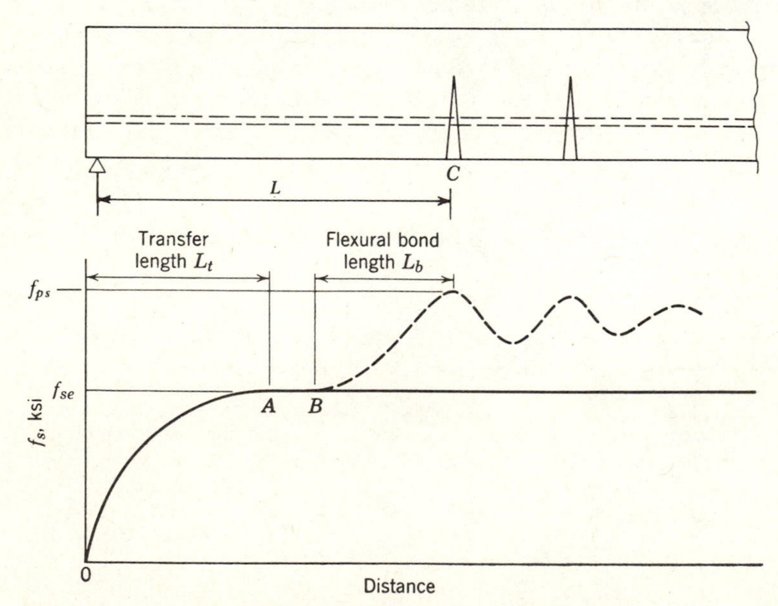

Fig. 7-21. Flexural bond overlapping with transfer length.

tendons under normal loads before cracking, having a transfer length L_t. Under increased loading, when a crack occurs at C, the stress in the tendon is raised to a maximum stress of f_{ps} as shown by the dotted line. Thus there is a bond stress developed from C to B. If B reaches past point A, that is, if the bond length overlaps the transfer length, then the tendon could be pulled through the concrete.

The bond length L_b is a function of the stress differential $f_{ps} - f_{se}$, of the bond stress u, the tendon diameter d_b, and other factors. Assuming uniform bond stress along the length L_b, and equating the total stress differential to the total bond stress, we have

$$(f_{ps} - f_{se})\frac{\pi d_b^2}{4} = u\pi d_b L_b$$

$$L_b = \frac{(f_{ps} - f_{se})d_b}{4u} \qquad (7\text{-}20)$$

To determine the value of u, it is noted that the bond stress at slippage is lower than that for the transfer length. Within the transfer length, the strand tends to expand and anchor to the concrete, while when strand is stressed above f_{se}, it tends to contract and pull away from the concrete. The ACI Code L_b is based on $u = 250$ psi from tests, thus the required flexural bond length is,

$$L_b = (f_{ps} - f_{se})d_b \qquad (7\text{-}21)$$

The design recommendation from reference 13 is that this length be increased 25% based on further review of the test results.

General bond slip is in most cases arrested by mechanical interlock between the strand and the surrounding concrete, so that additional bond can be developed before the strand is pulled through. The use of end anchors for strands could stop the pulling through, but it does not become effective until slip has occurred along the entire embedded length. Then the beam acts as a post-tensioned one without bond, and the ultimate moment resistance may not be much higher than those without anchors.

Based on the results of tests at the Portland Cement Association Laboratories[16] together with some additional data from the Association of American Railroads, the 1977 ACI Code has the following empirical formula for the minimum embedment length L in pretensioned flexural members using 7-wire strands with nominal diameter d_b:

$$L = \left(f_{ps} - \tfrac{2}{3}f_{se}\right)d_b \qquad (7\text{-}22)$$

where f_{ps} and f_{se} are in ksi, while L and d_b are in inches. The embedment length is the distance from the flexural cracking to the end of the beam, or to the end of the bonding between the strand and the concrete if a portion of the strand is unbonded.

Assuming $f_{ps}=250$ ksi (1724 N/mm^2) and $f_{se}=150$ ksi (1034 N/mm^2), we have,

$$L=\left(250-\tfrac{2}{3}\times 150\right)d_b=150d_b \tag{7-23}$$

The "pulling in" of 7-wire strand at the end of hollow core precast units has been used as one evidence of bond slip in the transfer region. These products have small depth and are cast with a very stiff concrete mix. A possible difficulty is that the failure bond stress may be less than that for strands cast in the bottom of beams with greater depth using concrete with slightly higher slump.

For hollow core and other precast units used on short spans the critical section for moment (similar to point C of Fig. 7-21) may be closer to the end of the member than the ACI development length (equation 7-22). Such a situation could be unsafe and must be avoided. In fact, it might be best to assure that more development length is provided than this minimum amount. The total length to the critical section recommended in reference 13 would be as follows:

$$L=\left(1.5\frac{f_{si}}{f'_{ci}}d_b-4.6\right)+1.25(f_{pu}-f_{se})d_b \tag{7-24}$$

This is slightly more length than the current ACI Code equation (7-22) would require, and some studies[20] have not found the need for this additional development length.

When bonding of the strand does not extend to the end of the member, the bonded development length shall be doubled following ACI Code. This provision applies to situations where some strands are unbonded in the end region of pretensioned members to avoid overstress in the concrete.

7-7 Bearing at Anchorage

For tendons with end anchorages, where the prestress is transferred to the concrete by direct bearing, various designs may be used for transmitting the prestress: steel plates, steel blocks, or reinforced-concrete ones.

The design of an anchorage consists of two parts: determining the bearing area required for concrete, and designing for the strength and detail of the anchorage itself. Stress analysis for any anchorage is a very complicated problem, because not only the elasticity but also the plasticity of concrete enters into the picture. As a result, anchorages are designed by experience, tests, and usage rather than by theory. Since anchorages are generally supplied by the prestressing companies which have their own standards for different tendons, the engineer does not have to design for them. Anchorages that have been successfully adopted are usually considered reliable, and no theoretical check on their stresses is necessary. For a new type of anchorage, the most reliable check is to run a test to determine its ultimate strength. A proper safety factor can then be applied to obtain the allowable load.

Sometimes it is necessary to design or to check the bearing areas for end anchorage, as governed by the allowable bearing in concrete. Since the cost of anchorage increases greatly if the allowable bearing stress is low, it has been the practice to use as high a bearing stress as is consistent with safety, much higher than permitted in reinforced concrete. This is true for practically all systems of prestressing. Besides reasons of economy, such high bearing stress can be justified on the following grounds.

1. The highest bearing stress that will ever exist at the anchorage occurs at transfer. As loss of prestress takes place, the bearing stress gradually diminishes.
2. The strength of concrete increases with time. Hence, if failure does not take place immediately at transfer, there is little possibility that it will happen later.
3. For bonded tendons with anchorages at the end of members, externally applied load will not increase the force on the anchorage. For unbonded tendons, the force on the anchorage will increase with load; but the increase is limited, hence a high factor of safety is not required.

The allowable bearing stress depends on several factors, such as the amount of reinforcement at the anchorage, the ratio of bearing to total area, and the method of stress computation.

Previously used equations were found to be too conservative[25] for design. Both the Post-Tensioning Institute Guide Specifications[26] and the 1977 ACI Commentary use the following equations for average bearing stress on the concrete, f_{cp}:

At service load—

$$f_{cp} = 0.6 f_c' \sqrt{A_b'/A_b} \tag{7-25}$$

but not greater than f_c'

At transfer load—

$$f_{cp} = 0.8 f_{ci}' \sqrt{(A_b'/A_b) - 0.2} \tag{7-26}$$

but not greater than $1.25 f_{ci}'$

where
f_{cp} = permissible compressive concrete stress
f_c' = compressive strength of concrete
f_{ci}' = compressive strength of concrete at time of initial prestress
A_b' = maximum area of the portion of the concrete anchorage surface that is geometrically similar to and concentric with the area of the anchorage
A_b = bearing area of the anchorage

As used in the above equation f_{cp} is the average bearing stress, P/A, in the concrete computed by dividing the force P of the prestressing steel by the net projected area, A_b, between the concrete and the bearing plate or other structural element of the anchorage which has the function of transferring the force to the concrete.

Special reinforcement, required for the performance of the anchorage, shall be indicated by the tendon supplier.

The PTI Guide Specification[26] also provides for static and dynamic test requirements. Manufacturers do not provide results of tests for every individual project, but they are usually in a position to show the results of prior tests which have been made on their products following these PTI Guidelines.

If the anchorage is rigid, the variation of pressure will be small over the contact area; if only a thin plate is used, high bearing pressure may exist near the tendons. If the anchorage simply bears on the end of concrete without being buried in it, the prestress is transferred entirely through bearing. If the anchorage is buried in the concrete, then part of the prestress may be transferred through bond along the sides of the anchorage. Take the Freyssinet cone, for example; it is believed that about a third of the prestress is transmitted through the sides.

Because of the strict economy followed in the design of end anchorages, it has not been unusual that when the concrete is poor it actually crushes under the application of the prestress. Hence it is important the concrete for prestressed work should be of high quality and should be carefully placed around the anchorages.

Besides supplying strength and rigidity, anchorage must be detailed to suit the dimensions of the jack and the ends of the beam. When both anchorages and jacks are supplied by the prestressing company, the designer will not have to worry about such details. If the thickness of the bearing plate has to be designed, a procedure similar to that used for designing column bearing plates may be followed. This consists of designing the critical section for bending at an allowable stress. Here, again, the allowable stress can be somewhat higher than ordinary, since there is no danger of overload or fatigue effect.

EXAMPLE 7-6

Determine the bearing plate area required for a tendon consisting of $12-\frac{1}{2}$ in. diameter, 7-wire strands, Fig. 7-22. At time of posttensioning assume that f'_{ci} is approximately 4000 psi (28 N/mm^2) and at service load after losses $f'_c = 5500$ psi (38 N/mm^2). The tendon forces for design are: 397 k (1,766 kN) due to maximum jacking force and 297 k (1321 kN) at service load after losses. Follow the Guide Specification of the Post-Tensioning Institute (PTI) for allowable bearing stresses on the concrete.

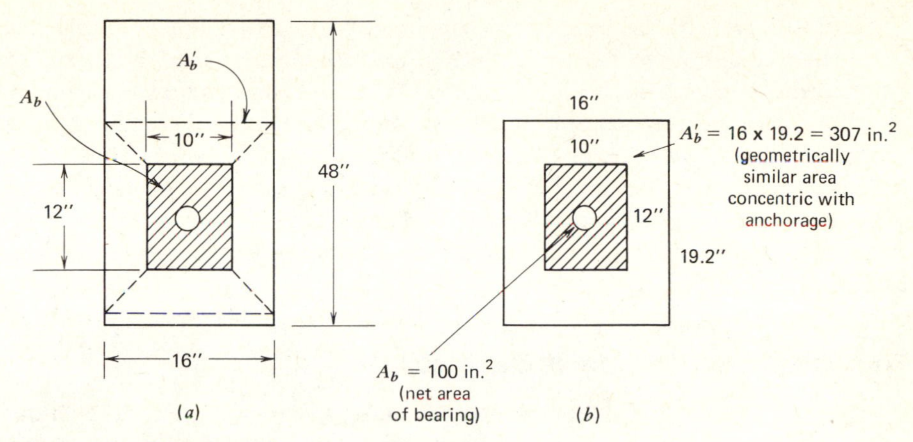

Fig. 7-22. Example 7-6.

Solution Note that we may assume $A_b'/A_b > 1.0$ as illustrated in Fig. 7-22 and estimate size of plate for service load requirements:

$$f_{cp} \cong 0.6 f_c' \sqrt{\frac{A_b'}{A_b}} = (0.6)(5500)(1) = 3300 \text{ psi} = 3.3 \text{ ksi } (22.75 \text{ N/mm}^2) \qquad (7\text{-}25)$$

Bearing area required $= \dfrac{297}{3.3} = 90$ in.^{2} (58×10^3 mm^2)

Assume an area of approximately 20 in.^{2} (13×10^3 mm^2) (5 in. diameter circular area) is lost as a bearing area for the tendon to pass through the plate.

Gross plate area $\cong 90 + 20 = 110$ in.^{2} (71×10^3 mm^2) Try plate 10 in.$\times$12 in. ($A_b = 120 - 20 = 100$ in.$^2 > 90$ in.^{2}) Check bearing pressure at transfer load. Assume maximum jacking load is 397 k (1766 kN) and $f_{ci}' = 4{,}000$ psi. (28 N/mm^2)

$$f_{cp} = 0.8 f_{ci}' \sqrt{\left(\frac{A_b'}{A_b}\right) - 0.2} \qquad (7\text{-}26)$$

$$f_{cp} = (0.8)(4{,}000)\sqrt{\frac{307}{100} - 0.2} = 5420 \text{ psi} = 5.42 \text{ ksi } (37.4 \text{ N/mm}^2)$$

$$A_b = \frac{397}{5.42} = 73.2 \text{ in.}^2 \ (47 \times 10^3 \text{ mm}^2) < 100 \text{ in}^2 \ (65 \times 10^3 \text{ mm}^2) \text{ provided}$$

Use 10 in.$\times$12 in. Plate (O.K. at both stages)

Note that the plate size could be reduced if necessary and remain within the bearing stress, f_{cp}, allowable by the PTI specification for the concrete strengths specified. Another possibility is that we might permit transfer at f_{ci}' less than 4000 psi (28 N/mm^2) as originally assumed. If space were available, the larger plate might be used, providing additional safety in the anchorage.

Elastic analysis of the bearing plate and supporting concrete resulting from the posttensioning force has been done with finite element modeling, and also photoelastic studies have been carried out. The problem requires experimental data to determine the inelastic effects which are important here. As a result, empirical expressions such as those given above from PTI will continue to be widely used for most design. Tests are carried out on large or unusual tendon anchors to verify that their performance will be satisfactory. A rather extensive study currently underway at The University of Texas at Austin should yield very useful results for design.

7-8 Transverse Tension at End Block

The portion of a prestressed member surrounding the anchorages of the tendons is often termed the end block. Throughout the length of the end block, prestress is transferred from more or less concentrated areas and distributed through the entire beam section. The theoretical length of the end block is the distance through which this change takes place and is sometimes called the lead length. It

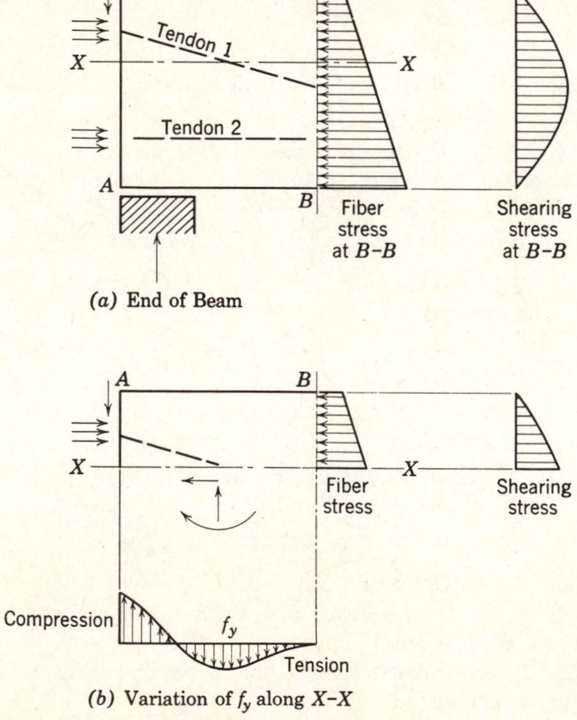

(a) End of Beam

(b) Variation of f_y along X–X

Fig. 7-23. Stresses at end block.

is known from theoretical and experimental investigations that this lead is not more than the height of the beam and often is much smaller except for pretensioned beams with long transfer length.

Referring to Fig. 7-23, the prestress at section *A–A*, whether horizontal or inclined, is applied as concentrated or somewhat distributed loads. At section *B–B*, the end of the lead length, the resistance from the beam consists of linearly distributed fiber stresses and corresponding shearing stresses as calculated by the usual beam theory. For the portion between sections *A–A* and *B–B*, the stress distribution is rather complicated. If we cut a longitudinal section *X–X*, and take a freebody as in Fig. 7-23(*b*), there will exist moment, shear, and a direct load on that section. These components of forces can be simply computed from statics, but their distribution along *X–X* cannot be easily determined. It is not possible to apply the usual beam theory assuming a plane section remaining plane, because that theory is far from being correct when applied to a short block like *A–A–B–B*. It can only be solved by the advanced theory of elasticity, which is complicated even for the simplest conditions of loading.

In order to simplify the solution, an assumption is made that the load is uniformly distributed across the width of the beam; thus, instead of a load concentrated at one point, we can assume a knife-edge load extending the entire width of the beam. Then the problem is reduced from a three to a two-dimensional problem. On this assumption, and on the theory of elasticity, stress distributions within the end block have been solved, and tables and graphs are available for certain conditions of loading.[21] For these graphs, Fig. 7-24, it is convenient to express the stresses in terms of the average direct compression f, where

$$f = F/A$$

F = total axial prestress at end of beam, and A = cross-sectional area of beam.

In general, along any longitudinal section, such as *X–X*, in Fig. 7-23, the shearing stress is small and does not cause any trouble; only the transverse tensile stress f_y can be serious. Hence we are interested only in the variation of f_y.

Graphs in Fig. 7-24 are for rectangular sections and are intended to indicate the general nature of the tensile stresses in end blocks. Lines of equal f_y, also termed "isobars," are shown in the graphs. From these isobars, it can be observed that there are two general areas of tension. One area in the center of the section is termed the "bursting zone." It has a maximum tension along the line of the load and at some distance from it. Another area is on the sides of the load close to the end surface, termed the "spalling zone." This zone is subject to high tensile stresses but only over a small area.

Additional graphs are available in Guyon's *Prestressed Concrete*,[22] to which readers are referred. Even though these graphs are theoretically correct and

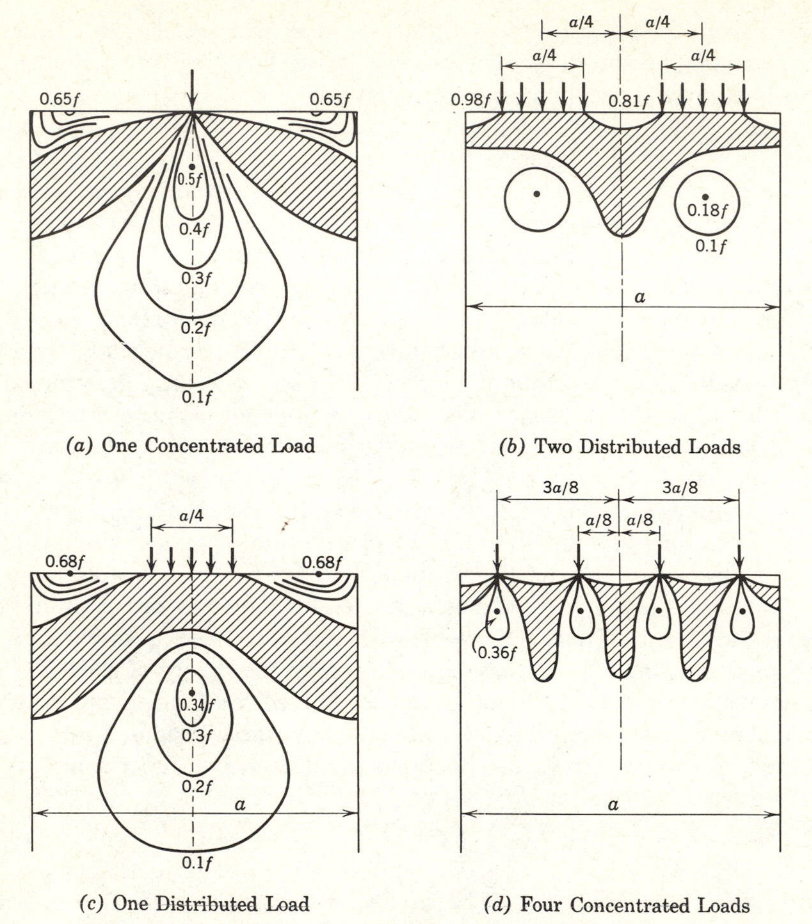

(*a*) One Concentrated Load (*b*) Two Distributed Loads

(*c*) One Distributed Load (*d*) Four Concentrated Loads

Fig. 7-24. **Isobars for transverse tension in end block (in terms of average compression *f*). Shaded areas represent compressive zones. (From Guyon's *Prestressed Concrete*.)**

some of them have been confirmed by photo-elasticity, their application to design is another problem. First, concrete is not a perfectly elastic material and will act plastically especially when part of it is over-stressed. Second, what should be the allowable tension in the concrete? Third, if the allowable tension is exceeded, how shall we design the reinforcing steel? Fourth, the pattern of forces applied at the end is often more complicated than can be handled by the theory of elasticity. Hence, though theory is needed in analysis, judgment must be exercised in design.

Guyon recommends that the allowable tensile stress be set at about a tenth of that for compression, that is, about $0.04f_c'$. Wherever the tension exceeds that value, steel reinforcement should be designed to take the entire amount of tension on the basis of the usual allowable stress in steel.

In the spalling zone, the tensile stresses are very high and will generally exceed the allowable value. However, these stresses act on only a small area, and the total tensile force is therefore small. For most cases, it has been found sufficient to provide steel for a total transverse tension of $0.03F$. For posttensioning, this steel is placed as close to the end as possible. Either wire mesh or steel bars may be used.

To carry the tension in the bursting zone, either stirrups or spiral steel may be used. For local reinforcement under the anchorage, $\frac{1}{4}$-in. (6.35 mm) spirals at 2-in. (50.8 mm) pitch or $\frac{3}{8}$-in. (9.52 mm) spirals at $1\frac{1}{2}$-in. (38.1 mm) pitch are sometimes adopted. For overall reinforcement, stirrups can be efficiently employed. A design of these stirrups is illustrated in example 7-7. It is fortunate that under ordinary conditions the number of stirrups required to resist the transverse tension is not excessive. Hence nominal amounts of reinforcement will suffice. However, longitudinal cracks have been produced by transverse tension when reinforcement was not provided or was insufficient.

EXAMPLE 7-7

The end of a prestressed beam is rectangular in section and is acted on by two prestressing tendons anchored as shown, Fig. 7-25. The initial prestress is 170 k (756 kN) per tendon. $f_c'=4000$ psi (28 N/mm^2). Design the reinforcement for the end block, allowing a maximum of 120 psi (0.83 N/mm^2) for the tension in the concrete.

Solution Tensile stresses for the bursting zone can be obtained from Fig. 7-24(*b*). Critical tensile stresses exist through sections *C–C* and *D–D* of Fig. 7-25(*b*), and their variation is plotted in Fig. 7-25(*c*). The greatest tensile stress is given as $0.18f$,

$$0.18f=0.18\left(\frac{2\times 170{,}000}{10\times 40}\right)$$
$$=0.18\times 850$$
$$=153 \text{ psi } (1.05 \text{ N/mm}^2)$$

Suppose that reinforcement is required for the portion whose tension exceeds 120 psi (0.83 N/mm^2); then the shaded portion of about 5 in. (127 mm) would require reinforcement. Assuming an average tension of 140 psi (0.97 N/mm^2) for the 5-in. (127 mm) length, the total tensile force to be resisted by steel is

$$140\times 5\times 10=7000 \text{ lb } (31.14 \text{ kN})$$

For an allowable stress of 20,000 psi (138 N/mm^2) in the steel, the area of steel required is

$$A_s=\frac{7000}{20{,}000}=0.35 \text{ sq in. } (226 \text{ mm}^2)$$

Four #3 U-stirrups will be provided as shown, giving a total area of 0.88 sq in. (568 mm^2). Note that computation such as this simply serves as a guide. Judgment must be

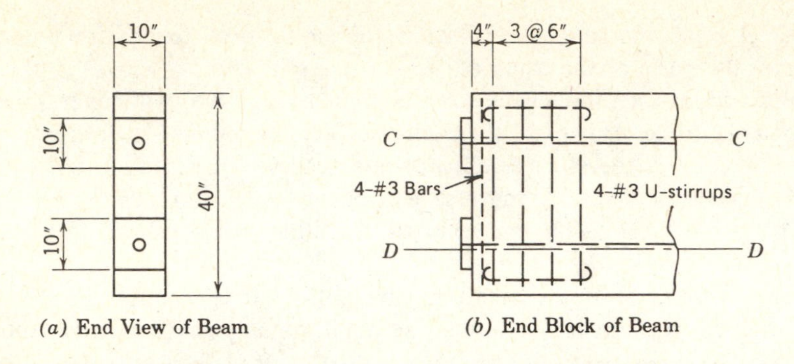

(a) End View of Beam (b) End Block of Beam

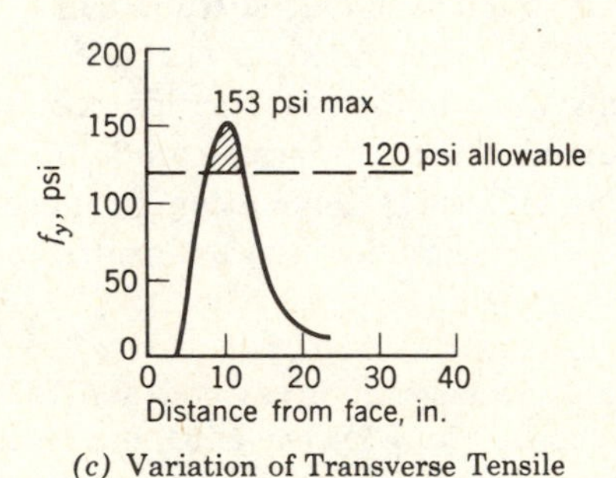

(c) Variation of Transverse Tensile Stress along C–C or D–D

Fig. 7-25. Example 7-9.

exercised in design. Since the tension is not excessive, liberal provision of steel is possible without much additional cost.

If concrete around the anchorages is thin, it is desirable to add some spiral steel, such as $\frac{1}{4}$-in. (6.35 mm) wires at 2-in. (50.8 mm) pitch.

For the spalling zone, stresses as high as $0.98f = 0.98 \times 850 = 830$ psi (5.72 N/mm^2) exist. The total force, however, is small and can be approximated by an average of 400 psi (2.76 N/mm^2) over a length of 2 in. (50.8 mm), which amounts to

$$400 \times 2 \times 10 = 8000 \text{ lb } (35.58 \text{ kN})$$

Using the value of 0.03 F as suggested, we would get

$$0.03 \times 340{,}000 = 10{,}200 \text{ lb } (45.37 \text{ kN})$$

showing not too bad an agreement in this problem. Steel required to resist 8000 lb (35.6 kN) is

$$8000/20{,}000 = 0.40 \text{ sq in. } (258 \text{ mm}^2)$$

which is adequately provided by the four No. 3 bars close to the end face of the beam.

Gergely and Sozen[27] suggested an approach to the design of end-block reinforcement which was addressed to the question "What stops a crack?" while the elastic approach more commonly used in the above discussion has been based on answering the question, "What starts a crack?" Without going into the details of this design method, it is based on the equilibrium of an assumed partial region of the end block. The authors state that the method is best suited for the design of anchorage-zone reinforcement for loads of high eccentricity.

While certain recent tests[23] confirmed the correctness of the several classical theories for anchorage zone stresses, including Guyon's method as outlined above, other tests[24] seem to indicate that the actual tensile stresses and total transverse tensile forces could be two to three times higher than the above described Guyon's method. Because of the many variables involved in a problem of this kind, it is most difficult to be exact. The AASHTO Bridge Committee, after consultation with the PCI, revised the AASHTO Bridge Specifications to read as follows (1977 edition).

ARTICLE 1.6.15—ANCHORAGE ZONES

For beams with posttensioning tendons, end blocks shall be used to distribute the concentrated prestressing forces at the anchorage. Where all tendons are pretensioned wires or 7-wire strand, the use of end blocks will not be required. End blocks shall have sufficient area to allow the spacing of the prestressing steel as specified in Article 1.6.16. Preferably, they shall be as wide as the narrower flange of the beam. They shall have a length at least equal to three-fourths of the depth of the beam and in any case 24 in. (0.610 m). In posttensioned members a closely spaced grid of both vertical and horizontal bars shall be placed near the face of the end block to resist bursting stresses. Amounts of steel in the end grid should follow recommendations of the supplier of the anchorage. Where such recommendations are not available the grid shall consist of at least No. 3 bars on 3-in. (0.076 m) centers in each direction placed not more than $1\frac{1}{2}$ in. (0.038 m) from the inside face of the anchor bearing plate.

Closely spaced reinforcement shall be placed both vertically and horizontally throughout the length of the end block in accordance with accepted methods of end block stress analysis.

In pretensioned beams, vertical stirrups acting at a unit stress of 20,000 psi (137.895 MPa) to resist at least 4% of the total prestressing force shall be placed within the distance of $d/4$ of the end of the beam, the end stirrups to be as close to the end of the beam as practicable. For at least the distance d from the end of the beam, nominal reinforcement shall be placed to enclose the prestressing steel in the bottom flange. For box girders, transverse reinforcement shall be provided and anchored by extending the leg into the web of the girder.

In pretensioned beams, vertical stirrups acting at a unit stress of 20,000 psi to resist at least 4% of the total prestressing force shall be placed within the distance of $d/4$ of the end of the beam, the end stirrup to be as close to the end of the beam as practicable.

Studies at the Portland Cement Association Laboratories[28] indicated an empirical equation for the design of stirrups to control horizontal cracking in the ends of pretensioned I-girders,

$$A_t = 0.021 \frac{T}{f_s} \cdot \frac{h}{l_t} \tag{7-27}$$

where A_t = required total cross-sectional area of sitrrups at the end of girder, to be uniformly distributed over a length equal to one-fifth of the girder depth

T = total effective prestress force, lb

f_s = allowable stress for the stirrups steel, psi

h = depth of girder, in.

l_t = length of transfer taken approximately assumed to be 50 times the strand diameter, in.

The above formula shows that the amount of end stirrups should vary directly with the depth of the section and inversely with the length of transfer. These conclusions are apparently quite logical and can be qualitatively justified.

7-9 Torsional Strength

Because of the high shear strength of concrete coupled with its low tensile strength, the failure of concrete beams in torsion seldom results from shearing stresses as such, but rather from principal tensile stress produced by the shearing stress. When the shearing stress v is combined with direct stress f_c, the principal tensile stress is given by the familiar formula (refer to section 7-2),

$$f_t'' = \sqrt{v^2 + (f_c/2)^2} - f_c/2 \tag{7-2}$$

Consider the simple case of a round plain concrete bar subject to pure torsion. Like other brittle materials, it fails along the plane of principal tension, at a spiral line 45° to the axis. This failure can be delayed by nonprestressed spiral and longitudinal reinforcement. However, such steel does not begin to act until the concrete has cracked, hence the elastic torsional strength of the concrete beam is not affected, although the ultimate failure resistance and the postcracking resilience of the member are increased.

Since the value of f_c in formula 7-2 can be considerably increased by prestressing along the length of the member, the principal tensile stress f_t'' is reduced for the same value of v. Hence the torsional resistance of a concrete member can be increased several-fold by prestressing, as has been proved experimentally by several investigators.[29, 30] It is also evident that prestressing the member along its depth and width will further increase its torsional resistance, although the economics of such two- and three-dimensional prestressing may limit its practicality for the present.

Confining our discussion to one-dimensional prestressing and rectangular sections, we can simply apply the general theory of torsion developed by St. Venant. According to St. Venant's theory, the maximum shearing stresses occur on the periphery at the middle of the sides, the absolute maximum v_{max} being at the middle of the longer sides of the rectangle,

$$v_{max} = \gamma b G \theta \tag{7-28}$$

and

$$T_M = \beta b^3 D G \theta \tag{7-29}$$

where b = the width (smaller dimension) of the rectangle
D = the depth (greater dimension) of the rectangle
G = the shear modulus of the concrete
θ = angle of twist in radians
$\gamma\, \alpha$ and β = constant, depending on the proportion of the rectangle (Table 7-3)

Letting $\alpha = \beta/\gamma$, Table 7-3 we have

$$T_M = \alpha b^2 D v_{max} \tag{7-30}$$

or

$$v_{max} = \frac{T_M}{\alpha b^2 D} \tag{7-31}$$

Then the principal tensile stress can be calculated from formula 7-2. When this principal tensile stress reaches the ultimate tensile strength of the concrete, cracking starts and the section may fail immediately without much warning. The addition of closed ties and longitudinal reinforcement can add strength and

Table 7-3[29] St. Venant's Constants for the Design of Rectangular Sections Subject to Torsion

D/b	α	β	γ
1.0	0.208	0.141	0.675
1.2	0.219	0.166	0.759
1.4	0.227	0.187	0.822
1.6	0.234	0.204	0.869
1.8	0.240	0.217	0.904
2.0	0.246	0.229	0.930
2.5	0.258	0.249	0.968
3.0	0.267	0.264	0.985
5.0	0.292	0.291	0.999
10.0	0.312	0.312	1.000
110.0	0.331	0.331	1.000
∞	0.333	0.333	1.000

ductility, but the formation of the torsional crack drastically influences the beam response to any further increase in the torsional moment, T_M.

In contrast to the mode of failure in torsion, a prestressed concrete beam under bending generally fails gradually and possesses much reserve strength and ductility after the appearance of the first cracks. This becomes evident when it is realized that the bending failure is dependent on the tensile stress and strain of steel, together with the compressive stress and strain of concrete, whereas the torsional strength of a beam without web reinforcement for torsion is exhausted when the tensile limit is reached in concrete, and there is no ductility in concrete under tension.

The behavior of concrete beams subject to combined bending and torsion is intermediate between that of beams subject to pure torsion and pure bending. For prestressed beams with relatively low ratios of bending moment M_B to torsional moment T_M, the formation of the first crack results in failure of the beam in a sudden and destructive manner, such as beams subject to pure torsions. With higher ratios of M_B/T_M, the failure becomes more gradual, and the ultimate load becomes higher than the load causing the formation of the first cracks.

The 1977 ACI Code provision for torsion design are for reinforced concrete and no provisions have been adopted for design to utilize the benefits of

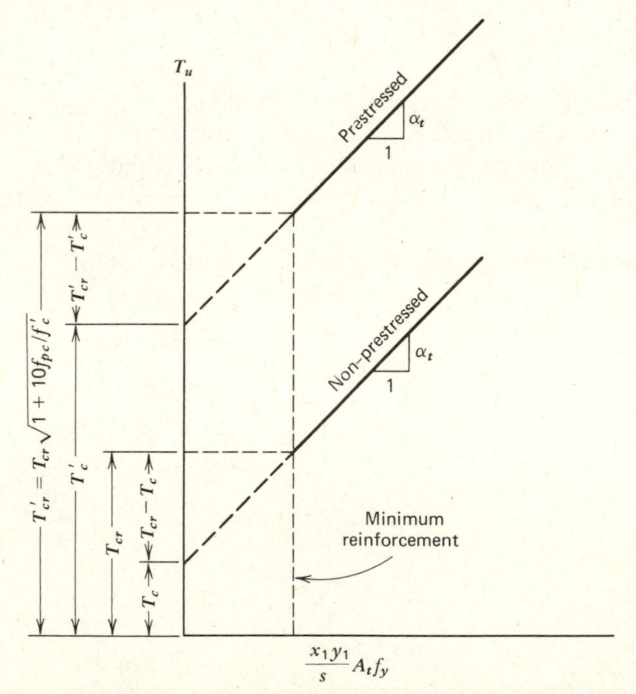

Fig. 7-26. Ultimate torque for reinforced and prestressed concrete members.[31]

prestress mentioned above. Tests show that we may design for torsional strength as the sum of the strengths contributed by the concrete and the web reinforcement as is also the case with nonprestressed members. The prestress increases the contribution of the concrete to the ultimate torsional strength compared to nonprestressed members as shown in Fig. 7-26 from reference 31. We can express the summation as

$$T_n = T_c' + T_s \tag{7-32}$$

where T_n = nominal torsional strength

T_c' = concrete contribution which is reduced from T_{cr} as shown in Fig. 7-26 for a similar reinforced concrete member

T_s = same stirrup contribution to torque as for nonprestressed beam

When a member is subject to simultaneous flexural shear and torsion we may use the circular interaction diagram of Fig. 7-27 to represent the strength of the member.

$$\left(\frac{V_n}{V_{cr}}\right)^2 + \left(\frac{T_n}{T_{cr}}\right)^2 = 1 \tag{7-33}$$

where V_n = shear force at failure under combined loading

T_n = torsional moment at failure under combined loading

V_{cr} = the lesser of V_{ci} and V_{cw}

$T_{cr} = 6\sqrt{f_c'}\sqrt{1+10f_{pc}/f_c'}\,\Sigma\eta x^2 y$

f_{pc} = average longitudinal prestress, F/A

$$n = \frac{0.35}{0.75+(b/d)}$$

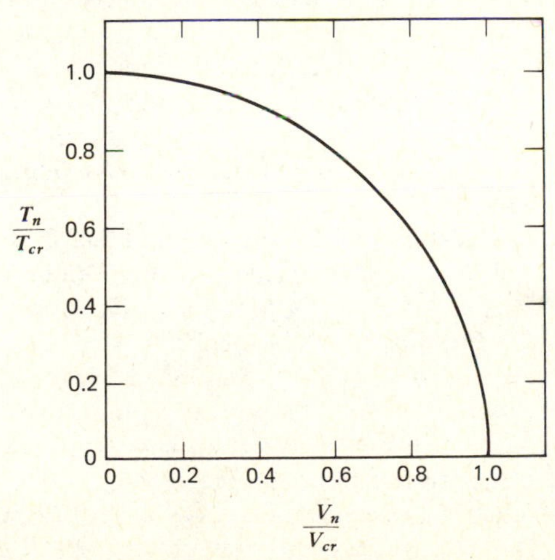

Fig. 7-27. Interaction curve for combined torsion plus flexural shear.[31]

A provision for torsion design of beams with web reinforcement will probably be added to the ACI Code in the near future. It will follow the strength approach outlined above where concrete and steel (closed ties are needed for torsion) combined carry the torque and shear from applied loads. The design approach of reference 31 is well formulated for design along these lines.

References

1. Rene Walther, "The Shear Strength of Prestressed Concrete Beams," *Proceedings Third Congress of the International Federation for Prestressing*, Berlin, 1958.
2. R. H. Evans and A. H. H. Hosny, "The Shear Strength of Post-tensioned Prestressed Concrete Beams," *Proceedings Third Congress of the International Federation for Prestressing*, Berlin, 1958.
3. M. A. Sozen, "Strength in Shear of Prestressed Concrete Beams without Web Reinforcement," *Structural Research Series No. 139*, Univ. of Illinois, August 1957; also see J. G. MacGregor, "Effect of Draped Reinforcement on Behavior of Prestressed Concrete Beams," *Structural Research Series No. 154*, Univ. of Illinois, May 1958.
4. Rene E. Walther and Robert F. Warner, *Ultimate Strength Tests of Prestressed and Conventionally Reinforced Concrete Beams in Combined Bending and Shear*, Fritz Engineering Laboratory, Lehigh University, Institute of Research, September 1958.
5. J. G. MacGregor, M. A. Sozen, and C. P. Siess, "Strength of Concrete Beams with Web Reinforcement," *J. Am. Conc. Inst.*, Vol. 62, No. 12, December 1965, pp. 1503–1519. See also *Structural Research Series No. 201*, University of Illinois, August 1960.
6. A. C. Scordelis, T. Y. Lin, and H. R. May, "Shearing Strength of Prestressed Lift Slabs," *J. Am. Conc. Inst.* October 1958, pp. 485–506.
7. Norman W. Hanson, "Precast-prestressed Concrete Bridges—2. Horizontal Shear Connections," *Journal of the PCA Research and Development Laboratories*, Vol. 2, No. 2, 1960.
8. N. M. Hawkins, M. A. Sozen, and C. P. Siess, "Strength and Behavior of Two-span Continuous Prestressed Concrete Beams," *Structural Research Series No. 225*, Univ. of Illinois, September 1961.
9. B. Bresler and K. S. Pister, "Strength of Concrete under Combined Stresses," *J. Am. Conc. Inst.*, September 1958.
10. "The Structural Use of Prestressed Concrete in Buildings," *British Standard Code of Practice*, The Council for Codes of Practice, British Standards Institution, 1959.
11. R. M. Mains, "Measurement of the Distribution of Tensile and Bond Stresses along Reinforcing Bars," *J. Am. Conc. Inst.*, November 1951 (*Proc.*, Vol. 47), pp. 225–252.
12. J. R. Janney, "Nature of Bond in Pre-Tensioned Prestressed Concrete," *J. Am. Conc. Inst.*, May 1954 (*Proc.* Vol. 50), pp. 717–736. Also E. Hognestad and J. R. Janney, "*The Ultimate Strength of Pre-Tensioned Prestressed Concrete Failing in Bond*," Magazine of Concrete Research, June 1954.
13. P. Zia and T. Mostafa, "Development Length of Prestressing Strands," *J. Prestressed Conc. Inst.* Vol. 22, No. 5, September/October 1977, pp. 54–65.

14. E. Hoyer and E. Friedrich, "Beitrag zur Frage der Hafspannung in Eisenbetonbauteilen," *Beton und Eisen*, Berlin, 1939 (Vol. 38, No. 6), pp. 107–110. Also K. Billig, *Prestressed Concrete*, Van Nostrand Co., New York, 1953.
15. G. D. Base, "An Investigation of Transmission Length in Pre-tensioned Concrete," *Papers of Third Congress* FIP, Berlin, 1958.
16. N. W. Hanson and P. H. Kaar, "Flexural Bond Tests of Pretensioned Prestressed Beams," *J. Am. Conc. Inst.*, January 1959, pp. 783–802. Also, Development Department Bulletin D28, Portland Cement Association, Skokie, Illinois.
17. W. T. Marshall and A. H. Mattock, "Control of Horizontal Cracking in the Ends of Pretensioned Prestressed Concrete Girders," *J. Prestressed Conc. Inst.* Vol. 7, No. 5, October 1962, pp. 56–74.
18. J. R. Janney, "Report of Stress Transfer Length Studies on 270K Prestressing Strand," *J. Prestressed Conc. Inst.* Vol. 8, No. 1, February 1963, pp. 41–43.
19. P. H. Kaar, R. W. Lafraugh, and M. A. Mass, "Influence of Concrete Strength on Strand Transfer Length," *J. Prestressed Conc. Inst.* Vol. 8, No. 5, October 1963, pp. 47–67. Also, Developement Department Bulletin D71, Portland Cement Association, Skokie, Illinois.
20. A. R. Anderson and R. G. Anderson, "An Assurance Criterion for Flexural Bond in Pretensioned Hollow Core Units," *J. Am. Conc. Inst.* August 1976 (*Proc.* Vol. 73), pp. 457–464.
21.. R. W. Kenning, M. A. Sozen, and C. P. Siess, "A Study of Anchorage Bond in Prestressed Concrete," Univ. of Illinois, *Structural Research Series No.* 251. June 1962.
22. Y. Guyon, *Prestressed Concrete*, John Wiley & Sons, New York, 1960, see pp. 127–174.
23. J. Zielinski and R. E. Rowe, "An Investigation of the Stress Distribution in the Anchorage Zones of Post-tensioned Concrete Members," *Report No.* 9, *C.A.C.A.*, London, September 1960.
24. S. Ban, H. Muguruma, and Z. Ogaki, "Anchorage Zone Stress Distributions in Post-tensioned Concrete Members," *Proceedings World Conference on Prestressed Concrete*, San Francisco, 1957.
25. K. H. Middendorf, "Anchorage Bearing Stresses in Post-Tensioned Concrete," *J. Am. Conc. Inst.*, November 1960, pp. 580–584.
26. Post Tensioning Institute, *Post-Tensioning Manual*, Post-Tensioning Institute, Glenview, Illinois, 1976.
27. P. Gergely and M. A. Sozen, "Design of Anchorage Zone Reinforcement in Prestressed Concrete Beams," *J. Prestressed Conc. Inst.*, April 1967, pp. 63–75.
28. W. T. Marshall and Allan H. Mattock, "Control of Horizontal Cracking in the Ends of Pre-tensioned Prestressed Concrete Girders," J. Prestressed Conc. Inst., October 1962.
29. H. J. Cowan and S. Armstrong, "The Torsional Strength of Prestressed Concrete," *Proc. World Conference on Prestressed Concrete*, 1957, San Francisco.
30. P. Zia, "Torsional Strength of Prestressed Concrete Members," *J. Am. Conc. Inst.*, Vol. 32, No. 10, April 1961, pp. 1337–1359.
31. P. Zia and W. D. McGee, "Torsion Design of Prestressed Concrete," *J. Prestressed Conc. Inst.*, Vol. 19, No. 2, March/April 1974, pp. 46–65.

8

CAMBER, DEFLECTIONS; CABLE LAYOUTS

8-1 Camber; Deflections

Before cracking, the deflections of prestressed-concrete beams can be predicted with greater precision than that of reinforced-concrete beams. Under working loads, prestressed-concrete beams do not crack; reinforced ones do. Since prestressed concrete is a more or less homogeneous elastic body which obeys quite closely the ordinary laws of flexure and shear, the deflections can be computed by methods available in elementary strength of materials.

As usually encountered for any concrete member, two difficulties still stand in the way when we wish to get an accurate prediction of the deflections. First, it is difficult to determine the value of E_c within an accuracy of 10% or even 20%. Tests on sample cylinders may not give the correct value of E_c, because E_c for beams may differ from that for cylinders. Besides, the value of E_c varies for different stress levels and changes with the age of concrete. The second difficulty lies in estimating the effect of creep on deflections.[1,2] The value of the creep coefficient as well as the duration and magnitude of the applied load cannot always be known in advance. However, for practical purposes, an accuracy of 10% or 20% is often sufficient, and that can be attained if all factors are carefully considered.

Deflections of prestressed beams differ from those of ordinary reinforced beams in the effect of prestress. While controlled deflections due to prestress can be advantageously utilized to produce desired cambers and to offset deflections due to loadings, there are also known cases where excessive cambers due to prestress have caused serious troubles.

The computation of moment-curvature relationship for a cross section and determination of load-deflection response was discussed in section 5-7 of Chapter 5. Using the principles described in that section, the response of a member to prestress forces and applied loadings can be determined. For a simple span with uniform cross section, this can be done by hand calculations for the total range of loading from the elastic range through ultimate. While hand computation is somewhat time consuming for even this simple case, it is prohibitively complex for a continuous member or even for a simple beam with variable cross sections. As a result, computer programs have been written which use these same principles of moment-curvature relationships to calculate load-deflection response for prestressed concrete beams, both bonded[3, 4] and unbonded tendons[5]. These programs do not include time-dependent effects and they are rather complex to use if one is only interested in estimating deflections for usual design calculations. For very special structures, such programs might be extremely helpful, especially if the loading produces cracking at some sections of the member or if the ultimate load capacity is of special interest.

Other computer programs have been developed which allow the consideration of time-dependent effects for prestressed concrete members in the service load range, assuming no cracking in the concrete[6]. Complex programs have been developed for use with multiple span beams[7, 8] which apply the principles of superposition of load and time-dependent effects to obtain curvatures at every section along the member at a given time, and to compute the deflection response to sustained loading. The complexity of these programs is greater than those previously referred to, but they represent the latest development of the state-of-the-art to handle this complex calculation for general problems using the computer.

To illustrate the type of deflection response we might expect for a simple span prestressed concrete beam, the computer program PBEAM developed by Suttikan in reference 8 was utilized to do a time-dependent deflection analysis. Figure 8-1 (*a* and *b*) show the details of the simple, pretensioned beam studied and Fig. 8-1 (*c*) shows the straight strands assumed for the simple beam which spans 65 ft. Table 8-1 shows data from the analysis with sustained loading.

The analysis indicates that the steel stress at midspan changes from 203 ksi (1400 N/mm^2) when initially tensioned in the prestressing bed to 179 ksi (1234 N/mm^2) after 48 hours. This change in the stress in the 20 stress-relieved strands ($\frac{1}{2}$ in. diameter) results from steel relaxation plus elastic shortening at transfer of the prestressing force to the concrete. Figure 8-1(*e*) shows the immediate centerline camber at prestress transfer is 1.1 in. (28 mm) due to prestress plus beam weight of $w_G = 0.470$ k/ft. (6.86 kN/m) on the 65-ft. (19.8 m) span. With the prestress and w_G loading sustained for 30 days, we find the camber increases to about $1\frac{3}{4}$ in. (44 mm) for the beam with only strands but is only a little over $1\frac{1}{2}$ in. (38 mm) with the four No. 5 bars included as shown in Fig. 8-1(*b*).

Addition of 1 k/ft. (14.59 kN/m) load to the beam as shown in Fig. 8-1(*d*) results in an immediate downward deflection of 1.0 in. (25.4 mm), Fig. 8-1(*e*). The steel stress in the strands is changing with time over the period of sustained loading from 30 days (167 ksi = 1151 N/mm^2) to 3 years (153 ksi = 1055 N/mm^2). Time-dependent deflection under the beam weight plus 1 k/ft. (total 1.470 k/ft) from 30 days to 3 years is shown in Fig. 8-1(*e*). It is interesting to observe that this beam with no rebar has essentially no deflection after 3 years under this loading history while the companion beam with addition of four No.

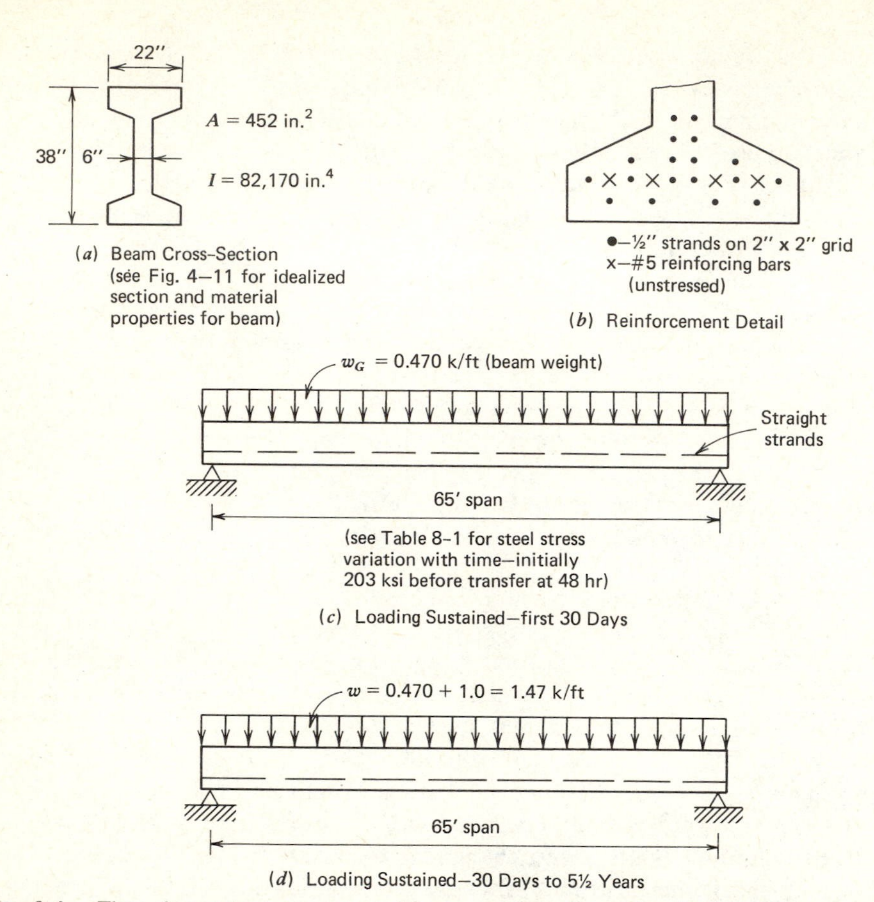

Fig. 8-1. Time-dependent response of beam subjected to sustained loading.

5 bars has a small downward deflection (slightly over $\frac{1}{4}$ in.). The final steel stress of 153 ksi (1055 N/mm^2) is approximately 24.7% less (all losses) than the initial 203 ksi (1400 N/mm^2) stress.

The results of sophisticated analysis, Table 8-1 and Fig. 8-1(e), which are illustrated by the example help one visualize the deflection with time history we might expect for prestressed concrete beams with changes in sustained loading. Approximate methods discussed later in this chapter consider the first 30 days under only beam weight to be the period up to the time of erection. The sustained dead load is added at approximately 30 days and we might consider the deflection after 3 years with this additional load acting to be the final deflection for a typical member. While the example above illustrates the influence of additional unstressed reinforcing bars on the deflection history we might find this refinement to be beyond the scope of many of the approximate

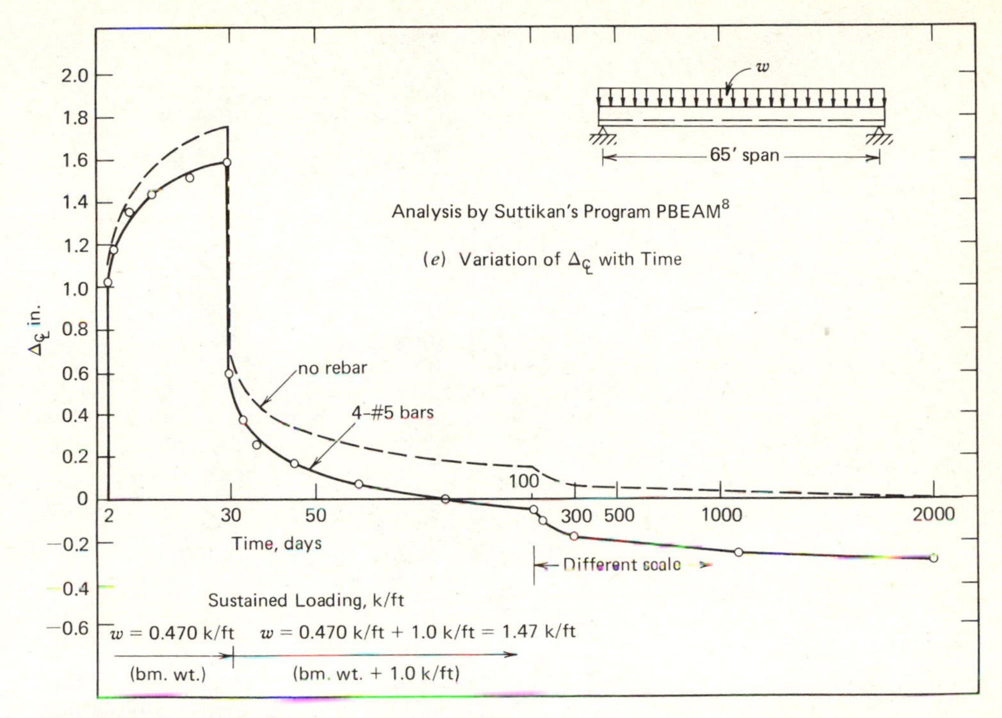

Fig. 8.1. (*continued*)

methods and even some of the computer programs which have been developed to handle the analysis.

In this chapter we present only an approximate computation procedure for estimating camber or deflection of prestressed concrete beams. A frequent design requirement is to estimate what camber may be expected at the time of prestressing the beam, as well as the deflection at some later time after it has carried its service loads for several months. Frequently no precise answers are needed since allowable deflections are not defined precisely. Thus, the rather crude and certainly approximate procedures for obtaining design estimates of deflection are usually sufficient, and the more complex procedures will not be covered (although the reader should be aware of the existence of computer programs such as reference 8).

Cambers or deflections due to prestress can be computed by two methods. The first method is to take the concrete as a freebody separated from the tendons, which are replaced by a system of forces acting on the concrete, Fig. 8-2. This would necessitate the computation of proper components of forces at the end anchorages plus transverse or radial forces at every bend of the tendons. This method is applicable to both simple and continuous beams. For the sake of

Table 8-1 Results of Analysis with PBEAM Program[a] (beam of Fig. 8-1 and example 8-1)

Load (k/ft)	time (days)	Beam with 20 Strands (initial $f_{si}L = 203$ ksi) Δ_L (in.)	f_{sL} (ksi)	Δf_{sL} (ksi)	Beam with 20 Strands Plus Four No. 5 Bars Δ_L (in.)	f_{sL} (ksi)	Δf_{sL} (ksi)
0	2^-	1.617	175.5		1.549	176.1	
$w_G = 0.47$	2^+	1.074	178.9	23.6	1.017	178.9	23.6
↓	3	1.245	175.3	27.2	1.169	176.0	26.5
	7	1.442	170.6	31.9	1.339	171.4	31.1
	12	1.560	167.3	35.2	1.434	168.4	34.1
	30	1.672	163.9	38.6	1.521	165.1	37.4
	30^-	1.757	161.0	41.5	1.583	162.4	40.1
add'l D.L. = 1 k/ft.	30^+	0.752	167.6	35.9	0.597	167.8	30.7
$+w_G = 0.47$ k/ft.	33	0.527	167.2	35.3	0.374	168.4	34.1
↓	37	0.430	166.9	35.6	0.275	168.1	34.4
	45	0.332	166.2	36.3	0.171	167.4	35.1
	60	0.245	164.0	38.5	0.0731	166.0	36.5
	80	0.187	163.2	39.3	0.0022	164.6	37.9
	110	0.140	161.5	41.0	−0.0586	163.0	39.5
	150	0.105	160.0	42.5	−0.108	161.6	40.9
	300	0.060	157.0	45.5	−0.179	158.8	43.7
	1095	0.017	153.4	49.1	−0.259	155.3	47.2
	2000	−0.012	152.5	50.0	−0.303	154.5	48.0
@ 2000 days: $\frac{\Delta f_{sqL}}{f_{siL}}$		= 24.7% loss total			= 23.7% loss total		

[a] Time-dependent deflection analysis—see reference 8.

simplicity, the following assumptions are usually made:

1. The gross section of concrete can often be used in computing the moment of inertia, although the net concrete section would be a more correct value.
2. The prestress producing deflection is somewhere between the initial and the final effective value. It is considered sufficiently accurate to assume a reasonable value for the purpose of computation.

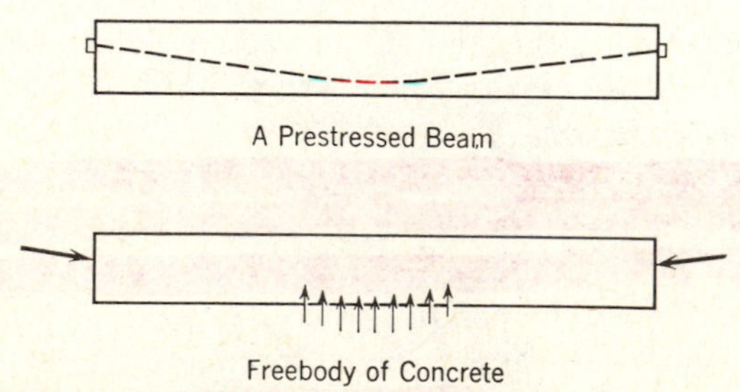

Fig. 8-2. Computation for deflection due to prestress.

3. The component of the prestress along the beam axis is assumed constant unless the inclination of the tendons becomes excessive. The component transverse to the beam is computed by the prestress times the tangent of the angle of bending unless the angle becomes unusually large.
4. Where the tendons bend suddenly, the transverse components may be assumed to be concentrated; where they form a flat curve, the transverse load may be assumed to be uniformly distributed along the bend.
5. All computations may be based on the c.g.s. line, the tendons being treated as a whole instead of individually.
6. Shearing deflections are small for ordinary proportions of prestressed beams and can be neglected.

The second method of computation is based on the same assumptions as above, but, without calculating the forces from the tendons on the concrete, a moment diagram produced by the tendons is directly drawn from c.g.s. profile. For statically determinate beams the moment diagram is similar to the eccentricity profile of the c.g.s. line; hence it is only necessary to plot the eccentricity profile to another scale to obtain the moment diagram. Then the computation of deflections from the moment diagram is performed by any method given in elementary strength of materials. This procedure is often simpler than the first, since it does away with the computation of forces from the tendons. But when applied to statically indeterminate beams, it has to be modified because of moments produced by the redundant reactions as a result of prestressing, which will be explained in Chapter 10.

Acting simultaneously with the prestress is the weight of the beam itself, which will produce deflections depending on the conditions of support. Such deflections can again be computed by the usual elastic theory. The resultant deflections of the beam at transfer are obtained by summing algebraically the deflections due to prestress and those due to beam weight.

EXAMPLE 8-1

A concrete beam of 32-ft simple span, Fig. 8-3, is posttensioned with 1.2 sq in. of high-tensile steel to an initial prestress of 140 ksi immediately after prestressing. Compute the initial deflection at midspan due to prestress and the beam's own weight, assuming $E_c = 4{,}000{,}000$ psi. Estimate the deflection after $1\frac{1}{2}$ months, assuming a creep coefficient of $C_c = 1.8$ and an effective prestress of 120 ksi at that time (span $= 9.75$ m, $A_{ps} = 774$ mm^2, $E_c = 27.58$ kN/mm^2, initial prestress $= 965$ N/mm^2, and effective prestress $= 827$ N/mm^2).

Solution Using the first method, take the concrete as a freebody and replace the tendon with forces acting on the concrete. The parabolic tendon with 6-in. (152.4 mm) midordinate is replaced by a uniform load acting along the beam with intensity

$$w = \frac{8Fh}{L^2} = \frac{8 \times 140{,}000 \times 1.2 \times 6}{32^2 \times 12} = 655 \text{ plf } (9.56 \text{ kN/m})$$

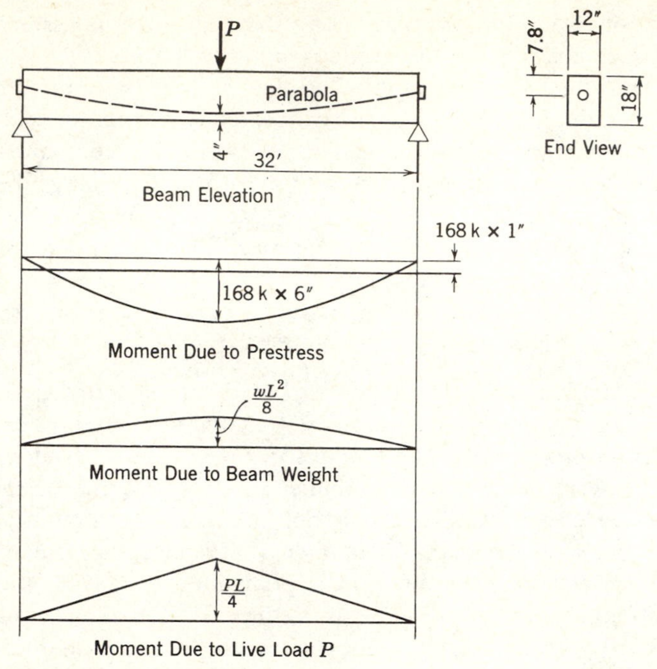

Fig. 8-3. Examples 8-1 and 8-2.

In addition, there will be two eccentric loads acting at the ends of the beam, each producing a moment of $140{,}000\times 1.2\times\frac{1}{12}=14{,}000$ ft-lb. (19.0 kN-m)

Since the weight of the beam is 225 plf (3.28 kN/m), the net uniform load on concrete is $655-225=430$ plf (6.28 kN/m), which produces an upward deflection at midspan given by the usual deflection formula

$$\Delta=\frac{5wL^4}{384EI}$$

$$=\frac{5\times 430\times 32^4\times 12^3}{384\times 4{,}000{,}000\times(12\times 18^3)/12}$$

$$=0.434\text{ in. }(11.02\text{ mm})$$

The end moments produce a downward deflection given by the formula

$$\Delta=\frac{ML^2}{8EI}$$

$$=\frac{140\times 1.2\times 1\times 32^2\times 12^2}{8\times 4{,}000{,}000\times(12\times 18^3)/12}$$

$$=0.133\text{ in. }(3.38\text{ mm})$$

Thus the net deflection due to prestress and beam weight is

$$0.434-0.133=0.301\text{ in. }(7.64\text{ mm})\text{ upward}$$

If we follow the second method, it will not be necessary to compute the forces between the tendon and the concrete. Instead, the moment diagram is drawn from the eccentricity curve of the tendon, and the deflection computed therefrom. For convenience is computation, the moment diagram can be divided into two parts, a parabola and a rectangle (Fig. 8-3). By area-moment principles or any similar method, the upward deflection due to prestress can be computed to be

$$\Delta = \frac{5FhL^2}{48EI} - \frac{ML^2}{8EI}$$

$$= \frac{5\times 140\times 1.2\times 6\times 32^2\times 12^2}{48\times 4{,}000{,}000\times (12\times 18^3)/12} - \frac{140\times 1.2\times 1\times 32^2\times 12^2}{8\times 4{,}000{,}000\times (12\times 18^3)/12}$$

$$= 0.661 - 0.133$$

$$= 0.528 \text{ in. } (13.41 \text{ mm})$$

Downward deflection due to beam weight of 225 plf is given by

$$\Delta = \frac{5wL^4}{384EI} = \frac{5\times 225\times 32^4\times 12^3}{384\times 4{,}000{,}000\times (12\times 18^3)/12} = 0.227 \text{ in. } (5.77 \text{ mm})$$

The resultant deflection is $0.528 - 0.227 = 0.301$ in. (7.64 mm) upward, the same answer as by the first method.

While the above gives the initial deflection, the eventual deflection should be modified by two factors: first, the loss of prestress, which tends to decrease the deflection; and second, the creep effect, which tends to increase the deflection. Since the prestress is reduced from 140 to 120 ksi (965–827 N/mm^2), the deflection due to prestress can be modified by the factor 120/140. Then, for the creep effect, the net deflection should be increased by the coefficient 1.8. Thus, if the beam is not subject to external loads the eventual deflection after $1\frac{1}{2}$ months can be estimated as

$$\left(0.528\times \frac{120}{140} - 0.227\right)1.8 = 0.407 \text{ in. } (10.34 \text{ mm}) \text{ upward}$$

The calculation for deflections due to external loads is similar to that for nonprestressed beams. So long as the concrete has not cracked, the beam can be treated as a homogeneous body and the usual elastic theory applied to it for deflection computations.

If the beam is bonded at the time of application of the load, the transformed section including steel should be used in computing the moment of inertia. If it is unbonded, to be theoretically correct, the net section of the concrete should be used and the effect of the change in prestress in the tendons under loading should be taken into account. For practical purposes, however, it will be close enough to consider the gross section of concrete in the computations and to neglect the change in prestress. This will simplify the procedure a great deal and will yield practically the same results. It must always be remembered that the greatest difficulties in arriving at correct deflections are the proper choice of a value for E_c and an accurate allowance for creep effect.

When the beam is loaded beyond its working load (or near its working load, for some cases), tensile stresses will exist in the beam. So long as the beam has not cracked, the elastic theory can still be applied for the computation of deflections. Although the tensile modulus of elasticity may be different from the compressive, the difference is not significant enough to alter the nature of deflection, since, at that stage, tension exists only in a small portion of the beam.

When cracks begin to occur in the beam, the nature of deflection will start to change. Even at the beginning of cracks, when they are still hair cracks hardly visible to the unaided eye, the effective section in resisting moment will be the cracked section instead of the entire concrete section. As the cracks extend deeper and deeper, the moment of inertia of the section will become smaller and smaller until eventually the cracked section may have a moment of inertia about one-half or one-third that of the uncracked section. Besides, the concrete will be under higher average stress and therefore will possess a lower average value of E_c. Hence the deflection of the section will increase much faster than before cracking. It must be noticed, however, that only the part of the beam subjected to higher moment has cracked, while the remaining portion under lower moments may still remain intact. Thus the deflection of the beam will increase faster as more cracks develop. This is shown graphically in Fig. 8-4.

Upon the removal of the applied load, the beam will return to its original position even though cracks have already developed, provided that the prestress

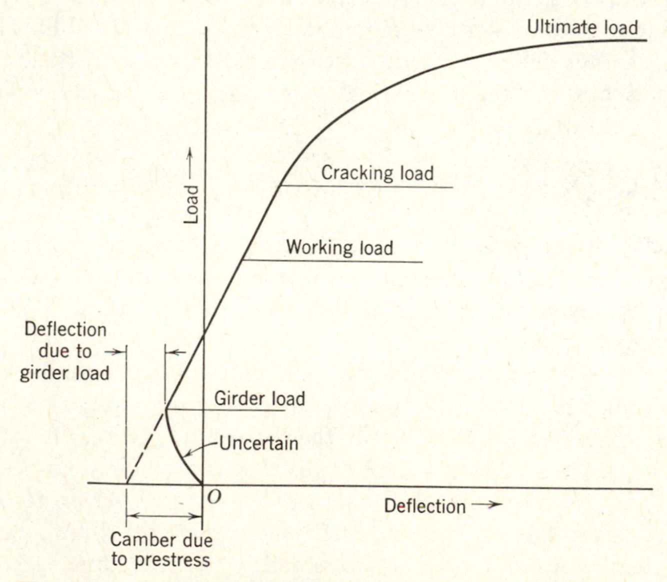

Fig. 8-4. Load deflection curve of a prestressed beam.

in the steel has not suffered any losses due to the overload. There will, in general, be some residual deflection left in the beam, depending on the degree and duration of loading. Such residual deflection is often attributed to the plasticity of concrete and can amount to a few per cent of the total deflection upon the first application of the load, but will be hardly noticeable for the second and third similar applications. If the loading is sustained for some time, residual deflections will be produced as a result of creep but most of this can be recovered in the course of time after load is removed.

When cracks have developed to an appreciable degree, portions of the steel near and across the cracks may be stressed beyond the elastic or creep limit. In such cases, there will be loss of prestress upon the removal of the load. The amount of loss naturally depends on the degree of overload; if the amount of permanent deformation equals or exceeds the prestressed strain, the prestress can be entirely lost. Then, upon reloading, that section of the beam will behave like an unprestressed one reinforced with high-tensile steel. Cracks will appear much earlier, even though the ultimate load of rupture may not be decreased.

The above description of the deflections of beams applies to both bonded and unbonded beams. It is believed that unbonded beams are almost as strong as the bonded ones in so far as the elastic limit of steel is concerned. Bonded beams, however, can carry higher load before the eventual crushing of concrete.

After concrete has cracked, the cracks will reappear as soon as tensile stresses again exist in that portion. The tensile stress does not have to approach the modulus of rupture for the cracks to reappear. Hence, between the working load and the cracking load, the beam will deflect slightly more after it has been previously cracked.

EXAMPLE 8-2

For the beam in example 8-1, compute the center deflection due to a 10-k (44.48 kN) concentrated load applied at midspan, when the beam is $1\frac{1}{2}$ months old after prestressing. Assume camber is 0.407 in. (10.34 mm) at this time prior to application of the 10-k (44.48 kN) load as computed in example 8-1.

Solution If the beam is bonded, the moment of inertia for the section should be computed on the basis of the transformed section including steel, but it can be approximated by using the gross concrete section. Also note that the modulus of elasticity E_c may be greater at the time of application of load than at transfer, but will be assumed to be 4,000,000 psi (27.58 kN/mm^2) for simplicity. Using the usual formula for deflection, we have

$$\Delta = \frac{PL^3}{48EI}$$

$$= \frac{10{,}000 \times 32^3 \times 12^3}{48 \times 4{,}000{,}000 \times (12 \times 18^3)/12}$$

$$= 0.505 \text{ in. } (12.83 \text{ mm})$$

which is the instantaneous downward deflection due to a load of 10 k (44.48 kN). Since the deflection before the application of load was 0.407 in. (10.34 mm) upward, the resultant deflection is $0.505-0.407=0.098$ in. (2.49 mm) downward. If the load is kept on for a time, the creep effect due to that load must be considered. Also, if the load is heavy enough to produce cracking, then the elastic theory for computing deflection can be used only as guidance for an approximation.

Experience has shown that it is difficult to predict cambers in prestressed concrete beams with any precision, because they vary not only with the E_c and creep of concrete, but also with age of concrete, actual support conditions, temperature and shrinkage differential between top and bottom fibers, and variation in properties between top and bottom concrete. It is generally necessary to get sufficient experience with the product of a particular plant before accurate deflection predictions can be made.

For pretensioned beams, the magnitude of prestress varies along the length of transfer at the ends. This variation can usually be neglected in camber computations, but it may need to be taken into account when greater accuracy is desired. However, the modification with time of the average effective prestress force in the member along the span is an important factor. As illustrated in example 8-1, the initial camber due to prestressing is in response to an initial prestress force. Reduction in the prestress force at a later time is accounted for by using a ratio of effective prestress force at the later time divided by the initial prestress force to modify the initial deflection calculation for prestress.

Use of a creep coefficient, C_c, for time-dependent deflections was illustrated in example 8-1. In that problem it was assumed that $C_c=1.8$ at $1\frac{1}{2}$ months was appropriate to modify the elastic deflections. The concept of reduced modulus of elasticity, E_c', for concrete has also been used to account for creep of concrete effect on deflections using elastic equations. We should be aware that by either computation method only an estimate of additional deflection with time under a sustained loading is obtained. The percent of creep which will occur with time was given in Fig. 2-3 in discussing conctete material behavior. Also, the value of C_c was taken to be 3.0 for usual design where

$$C_c=\frac{\delta_t}{\delta_i}=\frac{\delta_i+\delta_c}{\delta_i}$$

where δ_t = total strain
δ_i = instantaneous (elastic) strain
δ_c = creep strain

Figure 8-5 shows two levels of stress at which a sustained load might produce creep strain. Note from this figure that using $C_c=3$ for total creep after a long period of sustained loading (δ_c equal twice the strain δ_i) will result in a reduced modulus which is approximately $\frac{1}{3}$ that of the elastic modulus for concrete. If

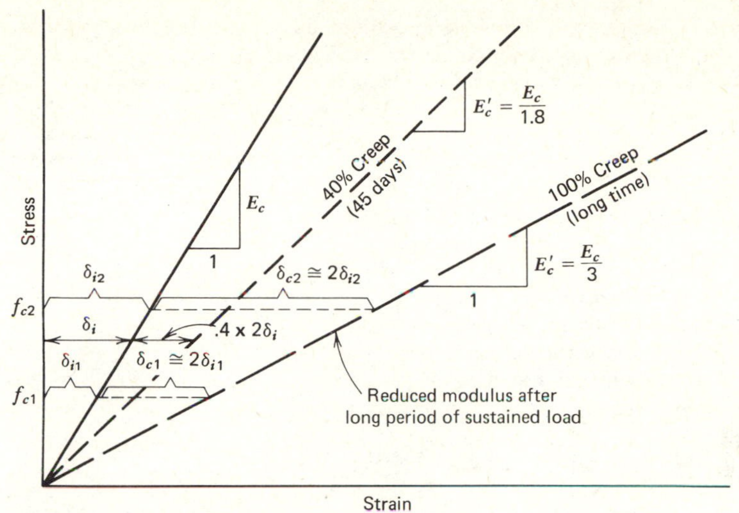

Fig. 8-5. Reduced modulus of elasticity for concrete.

information is available about creep properties of the concrete in a given beam, the value of this reduced modulus might be estimated more accurately than simply assuming $E_c' = \frac{1}{3} E_c$ as from Fig. 8-5.

The important factor is the creep strain δ_c for the material. If we wish to use a reduced modulus for some period of sustained loading less than that which would result in 100% of the creep, Fig. 2-3 may be used to estimate an average value of δ_c for the desired time period. For example, a 45-day period of sustained loading would give about 40% of the total creep, resulting in a reduced modulus $E_c' = E_c/1.8$ as illustrated in Fig. 8-5. Elastic equations for deflections would be used for the sustained load deflections, substituting the reduced E_c' for E_c. Note that this gives 1.8 times the elastic deflections and is identical to the use of a $C_c = 1.8$ modifier as used in example 8-1 for deflection at 45 days.

An approximate method of estimating the camber of a member after a period of time is to use multipliers which allow us to include both the creep effect (reduce modulus) mentioned above and other factors. Table 8-2 shows the multipliers which were derived by Martin.[9] This table is used in the *PCI Design Handbook* for estimating long-time cambers and deflections for typical members. The deflection "at erection" and "final" stages correspond generally to the time history discussed earlier in this chapter and presented in Fig. 8-1 from a more rigorous analysis. We might observe that the Table 8-2 multipliers differ with the time duration of sustained loading. The factor 1.8 for the camber (upward) component due to prestress corresponds to the creep coefficient, $C_c = 1.8$, used in example 8-1 for deflection after $1\frac{1}{2}$ months. Figure 8-4 showed how the time effect may be handled by use of a reduced modulus, E_c', for deflection calculations.

Table 8-2 Suggested multipliers to be used as a guide in estimating long-time cambers and deflections for typical members[a]

	Without Composite Topping	With Composite Topping
At erection		
1. Deflection (downward) component—apply to the elastic deflection due to the member weight at release of prestress.	1.85	1.85
2. Camber (upward) component—apply to the elastic camber due to prestress at the time of release of prestress.	1.80	1.80
Final		
3. Deflection (downward) component—apply to deflection calculated in (1) above.	2.7	2.4
4. Camber (upward) component apply to camber calculated—in (2) above.	2.45	2.2
5. Deflection (downward)—apply to elastic deflection due to superimposed dead load only.	3.0	3.0
6. Deflection (downward)—apply to elastic deflection caused by the composite topping.	—	2.30

[a] From reference 9. *PCI Design Handbook* uses same multipliers to find approximate dependent deflections.

We find that $C_c = 1.8$ (or $E_c' = E_c/1.8$) corresponds to about 40% of the creep effect which might occur with long-time loading. The creep coefficient of 3.0 for long-time deflection (or $E_c' = E_c/3.0$) is shown in Table 8-1 as the multiplier for final downward deflection due to superimposed dead load. This table also suggests multipliers for use with composite topping. The multipliers of Table 8-2 may require modification with experience as we get more actual data to compare with predicted time dependent deflections, but they should be accurate enough to allow us to avoid serious time-dependent deflection problems in the design of prestressed concrete members.

EXAMPLE 8-3

(loss of prestress with time for this same beam is estimated in example 4-5)

Use the deflection multipliers of Table 8-2 to estimate the deflection of the beam described in Fig. 8-1. Section properties are known: $A = 452$ in.^{2}, $I = 82{,}170$ in.^{4}, $S_x = 4325$

in.^{3}, $w_G = 0.47$ k/ft, and $M_G = 2979$ in-k. ($A = 292 \times 10^3$ mm^2, $I = 34.2 \times 10^9$ mm^4, $S_x = 70.9 \times 10^6$ mm^3, $w_G = 6.86$ kN/m, and $M_G = 336.6$ kN-m.)

Solution Solving deflection using multipliers of Table 8-1 for the stage after transfer (without rebar):

(*a*) At transfer:

$$\Delta(\text{prestress})\frac{PeL^2}{8EI} = \frac{(177.7 \times 3.06)(13.2)(65 \times 12)^2}{(8)(3824)(82{,}170)} = 1.74 \text{ in. } (44.2 \text{ mm})(\uparrow)$$

$$\Delta(\text{load } w_G)\frac{5w_G L^4}{384EI} = \frac{(5)(0.470/12)(65 \times 12)^4}{(384)(3824)(82{,}170)} = 0.60 \text{ in. } (15.2 \text{ mm})(\downarrow)$$

Estimate of camber at transfer $= 1.14$ in. (29.0 mm)($\uparrow$)

(*b*) At erection: (assume 30 days corresponds approximately, using multipliers at erection from Table 8-2)

$$\Delta(\text{prestress})\frac{PeL^2}{8EI} = (1.74)(1.80) = 3.13 \text{ in. } (79.5 \text{ mm})(\uparrow)$$

$$\Delta(\text{load } w_G)\frac{5w_G L^4}{384EI} = (0.60)(1.85) = 1.11 \text{ in. } (28.2 \text{ mm})(\downarrow)$$

Estimate of camber at erection $= 2.02$ in. (51.3 mm)($\uparrow$)

(*c*) At end of 3 years: (1 k/ft sustained load added at 30 days using multipliers Table 8-2, final)

$$\Delta(\text{prestress})\frac{PeL^2}{8EI} = (1.74)(2.45) = 4.26 \text{ in. } (108.2 \text{ mm})(\uparrow)$$

$$\Delta(\text{load } w_G)\frac{5w_G l^4}{384EI} = (0.60)(2.70) = 1.62 \text{ in. } (41.1 \text{ mm})(\downarrow)$$

$$\Delta(\text{load} w)\frac{5wl^4}{384EI} = \frac{(5)(1.012)(65 \times 12)^4}{(384)(4415)(82{,}170)} \times 3.0 = 3.31 \text{ in. } (84.1 \text{ mm})(\downarrow)$$

Estimated final Δ $= 0.67$ in. (17.0 mm)($\downarrow$)

Comparing results of estimated deflection for the beam without rebar with results of more exact analysis by PBEAM as shown in Fig. 8-1(*e*):

	Δ (PCI multipliers, Table 8-2)	**Δ (PBEAM)**
Transfer	1.14 in. (29.0 mm)(↑)	1.07 in. (27.2 mm)(↑)
Erection (30 days)	2.02 in. (51.3 mm)(↑)	1.76 in. (44.7 mm)(↑)
Final ($5\frac{1}{2}$ years)	0.67 in. (17.0 mm)(↓)	0.012 in. (0.30 mm)(↓)

The estimate of erection deflection using the PCI multipliers is slightly higher than the result from more exact analysis (PBEAM). If erection were at more than 30 days the camber would increase but the plot of Δ versus time in Fig. 8-1(e) indicates that even at 50 days the camber might still be less than 2 in. (50.8 mm). For this beam the estimate of camber at erection is slightly high using the multipliers suggested. Note that we could modify the multipliers to give better agreement with measured values if they showed this to be a usual trend. The same observation can be made about the final deflection. The multipliers give about $\frac{5}{8}$ in. (15.9 mm) more deflection than PBEAM predicted. This downward deflection, though conservative when compared to the PBEAM result, is still less than $\frac{1}{360} \times \text{span} = 2.2$ in. (55.9 mm). The deflection limitation following Table 9.5(b) of ACI Code should not be a problem. The reason we make this estimate of deflection is to assure satisfactory service ability and the deflection predictions with the PCI multipliers should give enough accuracy to make this check for most cases encountered in design.

Deflection response to a transient live load would be made using the elastic modulus, E_c, for concrete. A realistic estimate of the actual strength for concrete at the time of loading must be made, and the ACI equation for E_c will give a reasonable estimate of the modulus of elasticity to use in deflection calculations.

$$E_c = w^{1.5} 33 \sqrt{f'_c} \tag{2-1}$$

where w = unit weight of concrete
f'_c = concrete strength
E_c = elastic modulus

Note that this equation is valid for lightweight concrete as well as normal weight concrete. For typical lightweight concrete the value of E_c will be about 75% of the value for normal weight concrete.

It is good practice to balance the deflection resulting from dead load by the camber produced by the prestress whenever possible. When this is achieved, the flexural creep and the highly variable value of E_c will have little effect on the camber or deflection. Frequently, a designer can put a slight camber in the beam so that flexural creep tending to camber the beam upward will just about balance the downward deflection resulting from the loss of prestress. We should also note that if the initial camber is too small, sagging may eventually take place. This downward deflection in flat roof structures may result in the ponding of water during heavy rains, particularly if insufficient or poorly placed drains are provided, or if the drains are plugged. Additional loading from ponded water has been known to produce failure of slender roof members sensative to deflections, since large quantities of water may collect from very heavy rains. Positive provision for drainage is obviously more reliable than depending on camber to drain water from roofs.

Some simple formulas are listed in Fig. 8-6 to help the computation of camber due to prestress. They are derived from the well-known moment-area principles.

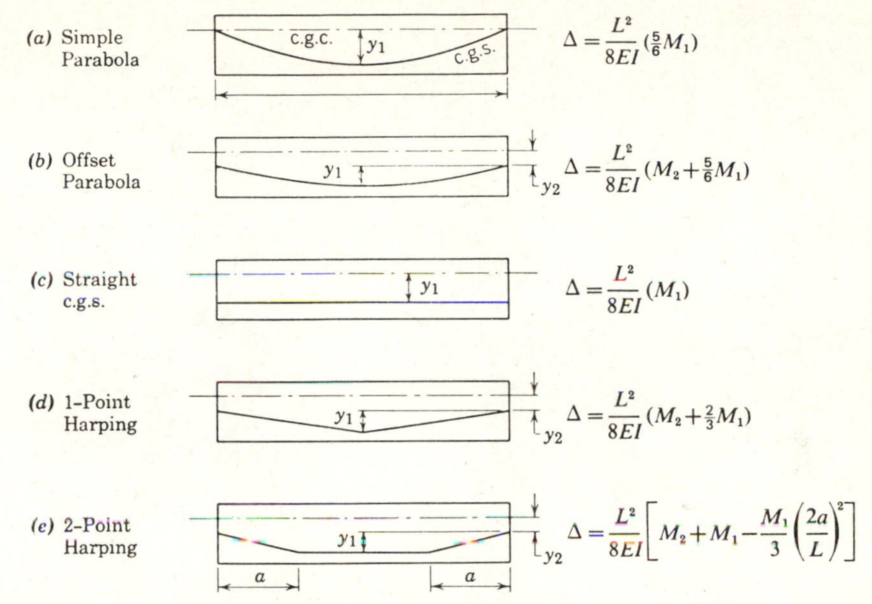

Fig. 8-6. Formulas for computing midspan camber due to prestress (simple beams).

By isolating the value of M and grouping the terms L, E, and I together, arithmetical work is reduced, and checking of the answers made easier. In these formulas, the moment M at each section is computed by the prestress F (or more accurately the horizontal component of F) multiplied by the corresponding ordinate y as marked, thus

$$M_1 = Fy_1$$

and

$$M_2 = Fy_2$$

Accurate data are lacking concerning the deflection of prestressed concrete beams after cracking. Because of the slenderness of most of these beams, they can deflect considerably before ultimate collapse. Hence they are quite resilient in the plastic as well as in the elastic range and possess a high energy-absorption capacity. The ultimate angular rotation of a beam section may be approximated by locating the ultimate neutral axis and using a maximum unit compressive strain in the concrete of 0.003. The ultimate deflection of a beam can be accurately determined by a summation procedure if the moment-curvature relationship is known for all sections of a beam (Chapter 5, section 5-7). It is noted that when a beam fails, only a limited portion develops its full rotational capacity while most other portions will be subjected to smaller moments and hence smaller rotation.

8-2 Simple Beam Layout

The layout of a simple prestressed-concrete beam is controlled by two critical sections: the maximum moment and the end sections. After these sections are designed, intermediate ones can often be determined by inspection but should be separately investigated when necessary. The maximum moment section is controlled by two loading stages, the initial stage at transfer with minimum moment M_G acting on the beam and the working-load stage with maximum design moment M_T. The end sections are controlled by the area required for shear resistance, bearing plates, anchorage spacings, and jacking clearances. All intermediate sections are designed by one or more of the above requirements, depending on their respective distances from the above controlling sections. A common arrangement for posttensioned members is to employ some shape, such as I or T, for the maximum moment section and to round it out into a simple rectangular shape near the ends. This is commonly referred to as the end block for posttensioned members. For pretensioned members, produced on a long line process, a uniform I, double-T, or cored section is employed throughout, in order to facilitate production. The design for individual sections having been explained in Chapters 5, 6, and 7, the general cable layout of simple beams will now be discussed.

The layout of a beam can be adjusted by varying both the concrete and the steel. The section of concrete can be varied as to its height, width, shape, and the curvature of its soffit or extrados. The steel can be varied occasionally in its area but mostly in its position relative to the centroidal axis of concrete. By adjusting these variables, many combinations of layout are possible to suit different loading conditions. This is quite different from the design of reinforced-concrete beams, where the usual layout is either a uniform rectangular section or a uniform T-section and the position of steel is always as near the bottom fibers as is possible.

Consider first the pretensioned beams, Fig. 8-7. Here straight cables are preferred, since they can be more easily tensioned between two abutments. Let us start with a straight cable in a straight beam of uniform section, (*a*). This is simple as far as form and workmanship are concerned. But such a section cannot often be economically designed, because of the conflicting requirements of the midspan and end sections. At the maximum moment section generally occurring at midspan, it is best to place the cable as near the bottom as possible in order to provide the maximum lever arm for the internal resisting moment. When the M_G at midspan is appreciable, it is possible to place the c.g.s. much below the kern without producing tension in the top fibers at transfer. The end section, however, presents an entirely different set of requirements. Since there is no external moment at the end, it is best to arrange the tendons so that the c.g.s. will coincide with the c.g.c. at the end section, so as to obtain a uniform stress distribution. In any case, it is necessary to place the c.g.s. within the kern if

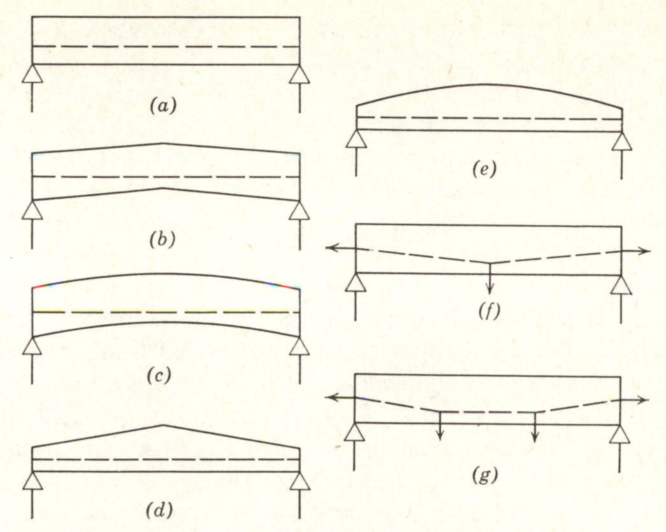

Fig. 8-7. Layouts for pretensioned beams.

tensile stresses are not permitted at the ends, and not too far outside the kern to avoid tension stress in excess of allowable values.

It is not possible to meet the conflicting requirements of both the midspan and the end sections by a layout such as (*a*). For example, if the c.g.s. is located all along the lower kern point, which is the lowest point permitted by the end section, a satisfactory lever arm is not yet attained for the internal resisting moment at midspan. If the c.g.s. is located below the kern, a bigger lever arm is obtained for resisting the moment at midspan, but stress distribution will be more unfavorable at the ends. Besides, too much camber may result from such a layout, since the entire length of the beam is subjected to negative bending due to prestress. In spite of these objections, this simple arrangement is often used, especially for short spans.

For a uniform concrete section and a straight cable, it is possible to get a more desirable layout than (*a*) by simply varying the soffit of the beam, as in Fig. 8-7(*b*) and (*c*); (*b*) has a bent soffit, while (*c*) has a curved one. For both layouts, the c.g.s. at midspan can be depressed as low as desired, while that at the ends can be kept near the c.g.c. If the soffit can be varied at will, it is possible to obtain a curvature that will best fit the given loading condition; for example, a parabolic soffit will suit a uniform loading. While these two layouts are efficient in resisting moment and favorable in stress distribution, they possess three disadvantages. First, the formwork is more complicated than in (*a*). Second, the curved or bent soffit is often impractical in a structure, for architectural or functional reasons. Third, they cannot be easily produced on a long-line pretensioning bed.

When it is possible to vary the extrados of concrete, a layout like Fig. 8-7(*d*) or (*e*) can be advantageously employed. These will give a favorable height at midspan, where it is most needed, and yet yield a concentric or nearly concentric prestress at end sections. Since the depth is reduced for the end sections, they must be checked for shear resistance. For (*d*), it should also be noted that the critical section may not be at midspan but rather at some point away from it where the depth has decreased appreciably while the external moment is still near the maximum. Beam (*d*), however, is simpler in formwork than (*e*), which has a curved extrados.

Most pretensioning plants in the United States have buried anchors along the stressing beds so that the tendons for a pretensioned beam can be bent, Fig. 8-7(*f*) and (*g*). It may be economical to do so, if the beam has to be of straight and uniform section, and if the M_G is heavy enough to warrant such additional expense of bending. Means must be provided to reduce the frictional loss of prestress produced by the bending of the tendons. For example, the tendons may be tensioned first from the ends and then bent at the harping points.

It is evident from the above discussion that many different layouts are possible. Only some basic forms are described here, the variations and combinations being left to the discretion of the designer. The correct layout for each structure will depend upon the local conditions and the practical requirements as well as upon theoretical considerations.

Most of the layouts for pretensioned beams can be used for posttensioned ones as well. But, for posttensioned beams, Fig. 8-8, it is not necessary to keep the tendons straight, since slightly bent or curved tendons can be as easily tensioned as straight ones. Thus, for a beam of straight and uniform section, the tendons are very often curved as in Fig. 8-8(*a*). Curving the tendons will permit favorable positions of c.g.s. to be obtained at both the end and midspan sections, and other points as well.

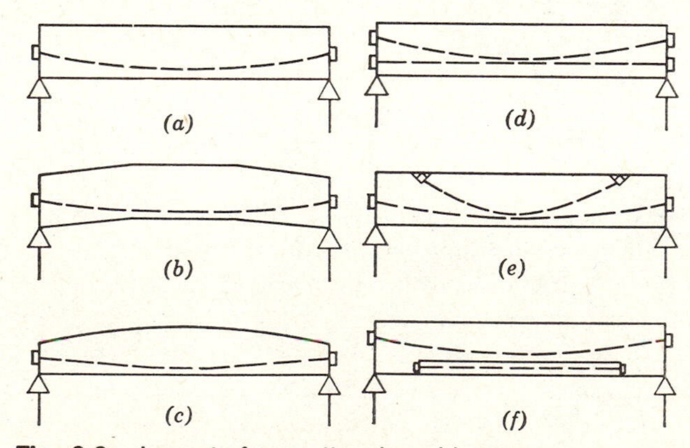

Fig. 8-8. Layouts for posttensioned beams.

A combination of curved or bent tendons with curved or bent soffits is frequently used, Fig. 8-8(*b*), when straight soffits are not required. This will permit a smaller curvature in the tendons, thus reducing the friction. Curved or bent cables are also combined with beams of variable depth, as in (*c*). Combinations of straight and curved tendons are sometimes found convenient, as in (*d*).

Variable steel area along the length of a beam is occasionally preferred. This calls for special design of the beam and involves details which may offset its economy in weight of steel. In Fig. 8-8(*e*), some cables are bent upward and anchored at top flanges. In (*f*), some cables are stopped part way in the bottom flange. These arrangements will save some steel but may not be justified unless the saving is considerable as for very long spans carrying heavy loads.

8-3 Cable Profiles

We stated in the previous section that the layout of simple beams is controlled by the maximum moment and end sections so that, after these two sections are designed, other sections can often be determined by inspection. It sometimes happens, however, that intermediate points along the beam may also be critical, and in many instances it would be desirable to determine the permissible and desirable profile for the tendons. To do this, a limiting zone for the location of c.g.s. is first obtained, then the tendons are arranged so that their centroid will lie within the zone.

The method described here is intended for simple beams, but it also serves as an introduction to the solution of more complicated layouts, such as cantilever and continuous spans, where cable location cannot be easily determined by inspection. The method is a graphical one; giving the limiting zone within which the c.g.s. must pass in order that no tensile stresses will be produced. Compressive stresses in concrete are not checked by this method. It is assumed that the layout of the concrete sections and the area of prestressing steel have already been determined. Only the profile of the c.g.s. is to be located.

Referring to Fig. 8-9, having determined the layout of concrete sections, we proceed to compute their kern points, thus yielding two kern lines, one top and one bottom, (*c*). Note that for variable sections, these kern lines would be curved, although for convenience they are shown straight in the figure representing a beam with uniform cross section.

For a beam loaded as shown in (*a*), the minimum and maximum moment diagrams for the girder load and for the total working load respectively are marked as M_G and M_T in (*b*). In order that, under the working load, the center of pressure, the *C*-line, will not fall above the top kern line, it is evident that the c.g.s. must be located below the top kern at least a distance

$$a_1 = M_T/F \tag{8-1}$$

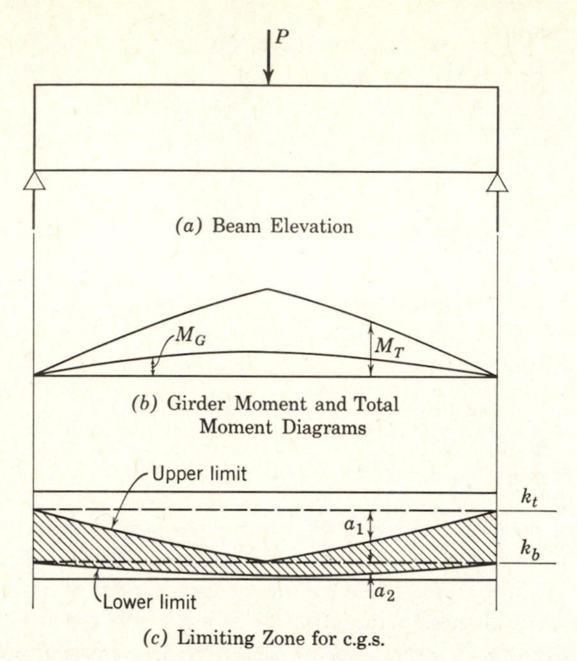

Fig. 8-9. Location of limiting zone for c.g.s.

If the c.g.s. falls above that upper limit at any point, then the *C*-line corresponding to moment M_T and prestress F will fall above the top kern, resulting in tension in the bottom fiber.

Similarly, in order that the *C*-line will not fall below the bottom kern line, the c.g.s. line must not be positioned below the bottom kern by a distance greater than

$$a_2 = M_G / F_0 \tag{8-2}$$

which gives the lower limit for the location of c.g.s. If the c.g.s. is positioned above that lower limit, it is seen that the *C*-line will be above the bottom kern and there will be no tension in the top fiber under the girder load and initial prestress F_0.

Thus, it becomes clear that the limiting zone for c.g.s. is given by the shaded area in Fig. 8-9(*c*), in order that no tension will exist both under the girder load and under the working load. The individual tendons, however, may be placed in any position so long as the c.g.s. of all the cables remains within the limiting zone.

The position and width of the limiting zone are often an indication of the adequacy and economy of design, Fig. 8-10. If some portion of the upper limit falls outside or too near the bottom fiber, in (*a*), either the prestress F or the depth of beam at that portion should be increased. On the other hand, if it falls

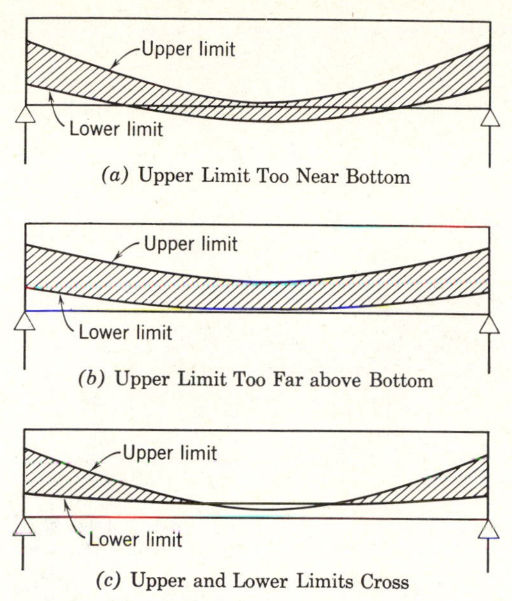

Fig. 8-10. Undesirable positions for c.g.s. zone limits.

too far above the bottom fiber, in (*b*), either the prestress or the beam depth can be reduced. If the lower limit crosses the upper limit, in (*c*), it means that no zone is available for the location of c.g.s., and either the prestress F or the beam depth must be increased or the girder moment must be increased to depress the lower limit if that can be done. On the other hand, as will be discussed later, the case shown in Fig. 8.10(*c*) may be very satisfactory when we are allowing tensile stress in concrete.

The application of the above graphical method is illustrated in example 8-4.

EXAMPLE 8-4

Preliminary design for a 50-ft pretensioned beam gives a layout with tapered top flange and symmetrical I-sections as shown in Fig. 8-11. Ten steel wires of $\frac{3}{8}$-in. diameter with anchorages are used for prestressing. $f_0 = 130$ ksi, $f_{se} = 110$ ksi, $f'_c = 5000$ psi. Determine the position for the c.g.s. line. Live and superimposed dead load on the beam totals 450 plf, in addition to the weight of the beam itself. $M_G = 58$ k-ft and $M_T = 194$ k-ft at midspan (span = 15.24 m, $f_0 = 896\ \text{N/mm}^2$, $f_{se} = 758\ \text{N/mm}^2$, load = 6.57 kN/m, $M_G = 78.6$ kN-m, and $M_T = 263.1$ kN-m).

Solution To get an accurate graphical solution, 4 or 5 points should be calculated for half of the span, but only calculations for the midspan section will be shown here. Note that the sections near the end are of rectangular shape; hence there is a sudden jump in the kern lines at the junction. Also, theoretically, there is no external moment for the portions directly over the supports.

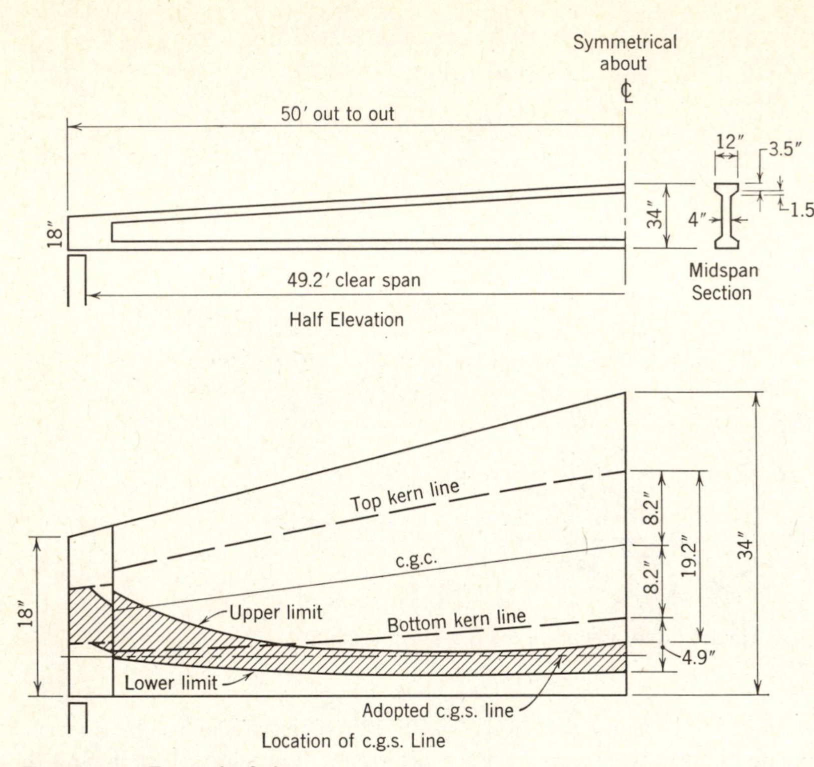

Fig. 8-11. Example 8-4.

First, locate the kern lines. Values for the midspan section are as follows:

$$I \text{ of section} = 28{,}200 \text{ in.}^4 \ (11.7 \times 10^9 \text{ mm}^4)$$

$$A \text{ of section} = 204 \text{ in.}^2 \ (132 \times 10^3 \text{ mm}^2)$$

$$r^2 \text{ of section} = 140 \text{ in.}^2 \ (90 \times 10^3 \text{ mm}^2)$$

$$k_t \text{ and } k_b = 140/17 = 8.2 \text{ in. } (208 \text{ mm})$$

With $F = 10 \times 0.11 \times 110 = 121$ kips (538 kN), minimum resisting arm required for M_T is, from equation 8-1,

$$M_T/F = (194 \times 12)/121 = 19.2 \text{ in. } (488 \text{ mm})$$

which is measured down from the top kern line and located as shown.

For initial prestress $F_0 = 10 \times 0.11 \times 130 = 143$ k (636 kN), the lever arm corresponding to $M_G = 58$ k-ft (78.6 kN-m) is, from equation 8-2,

$$M_G/F_0 = (58 \times 12)/143 = 4.9 \text{ in. } (124 \text{ mm})$$

These limiting points are calculated for several other sections, and the limiting zone is indicated by the shaded area. Straight tendons are preferred for pretensioning, while it is impossible to get a straight c.g.s. line within this shaded area. The best recourse in this design is to permit some tension near the supports and to reinforce the ends with some mild steel if the tensile stresses exceed allowable values. Then it is possible to adopt a c.g.s. line as shown which will result in no tension in the bottom fibers under working loads but will have some tension in the top fibers near the supports. If such tension is to be avoided, it will be necessary to use a greater prestressing force, thus raising the upper limit of the zone and enabling a c.g.s. line to be located at about 6 in. (125.4 mm) from the bottom. Deflection of this beam at transfer should be computed to see whether the camber is excessive, but it will not be illustrated in this example.

The location of the c.g.s. line as just described is based on the elastic theory, allowing no tensile stress both at transfer and under the working loads. If some tension is permitted, then it is possible to place the c.g.s. line slightly outside the previous limiting zone. Referring to Fig. 8-12, for an allowable tensile stress of f_t' in the top fibers at transfer, we have

$$f_t' = \frac{Mc_t}{I}$$

$$= \frac{F_0 e_b c_t}{I} \tag{8-3}$$

where e_b = the amount c.g.s. may fall below the lower limit. For an allowable tensile stress of f_b' in the bottom fibers under the working load, we have

$$f_b' = \frac{Fe_t c_b}{I} \tag{8-4}$$

where e_t = the amount c.g.s. may rise above the upper limit. From equations 8-3 and 8-4, we can write

$$e_b = \frac{f_t' I}{F_0 c_t} = \frac{f_t' A k_b}{F_0} \tag{8-5}$$

and

$$e_t = \frac{f_b' I}{F c_b} = \frac{f_b' A k_t}{F} \tag{8-6}$$

Hence, the limiting zones for no tensile stresses can be extended to lines 1-1 and 2-2 if some tensile stresses are permitted, Fig. 8-12.

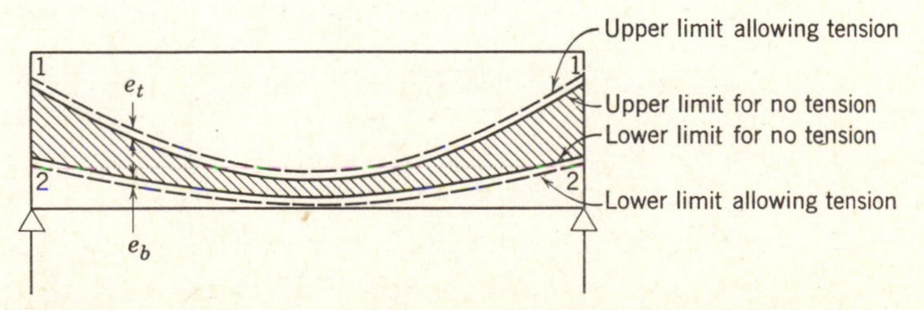

Fig. 8-12. Limiting zone for c.g.s. allowing tension in concrete.

The above graphical method can also be applied when there are changes in the cross-sectional area of steel. It is only necessary to use the corresponding value of prestress existing at the particular point when computing the position of the limiting zone. Thus, at points of change in the steel area, there will be sudden jumps in the limiting lines. If the prestress is applied in two stages, two lower limits should be computed, each based on its own prestressing force. However, if too many complications are involved, the graphical method may not be efficient.

If ultimate-strength design is to be used, the c.g.s. line can also be located by a graphical method, Fig. 8-13. But, since ultimate design applies only to the maximum loading stage, the lower limit for the c.g.s. still has to be determined by the elastic theory or some other method. The upper limit, however, can be obtained by the ultimate theory as follows. If M_T is the total moment, and m the load factor, then the ultimate moment is mM_T, which is to be resisted by the ultimate strength of the steel (in the case of bonded reinforcement) with a level arm,

$$a' = \frac{mM_T}{A_s f_s'}$$

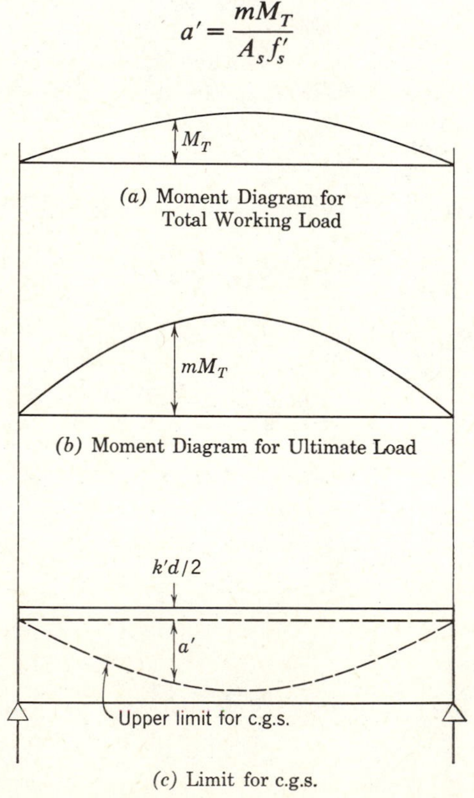

Fig. 8-13. Location of c.g.s. by ultimate design.

The line of pressure at ultimate load is located at $k'd/2$ below top fiber, where $k'd$ is obtained by

$$k'd = \frac{A_s f_s'}{k_1 f_c' b}$$

if a uniform width b is obtained for the top flange at the ultimate load.

8-4 Cantilever Beam Layout

Because of the balancing and reduction of moments, cantilever beams can be economically utilized in prestressed-concrete structures, especially for certain favorable span ratios and for long and heavy beams. The basic theories and methods for the design of cantilever beams are the same as those for simple beams. But the work of designing is more complicated, because of several factors which must be more carefully considered. These are

1. Certain portions of a cantilever are subjected to both positive and negative moments, depending on the position of live loads.
2. To obtain most severe loading conditions, partial loading of the spans must sometimes be considered.
3. In a cantilever, moments produced by loads on a certain portion are often counterbalanced by loads on other portions. Hence the moments are sensitive to changes in external load. Because of this, the sequence of the application of superimposed loads on the beam must be carefully considered and executed.
4. If the beam is precast, care must be exercised during erection and transportation of the beam. At all times the supporting conditions assumed in design must be realized for the beam. Even slight changes in the position of supports may affect the moments seriously.
5. Cantilever beams are more sensitive to temperature changes which might result in excessive deflections.
6. The ultimate capacity of cantilever beams may be relatively low if heavy partial loading is a possibility. The coexistence of high moment and shear at certain critical sections may also tend to reduce the ultimate strength in a cantilever.

In spite of the complications, cantilever beams are often used, because of their economy and their adaptability to certain structures. In fact, the above-mentioned complications should not be held against the use of cantilever beams. They only indicate that greater care must be exercised in their design and construction.

Two general layouts are possible for cantilevers: the single and the double cantilevers. Some typical layouts for the single cantilevers are shown in Fig.

8-14. Part (*a*) shows the layout for a short span with a short cantilever, where a straight and uniform section may be the most economical. In such a design, it is only necessary to vary the c.g.s. profile so that it will conform with the requirements of the moment diagrams. When the cantilever span becomes longer, it is advisable to taper the beam as in (*b*). If the anchor span is short compared to the cantilever, it may be entirely subjected to negative moments, and the c.g.s. may have to be located above the c.g.c. at all points.

For longer anchor spans, it may be desirable to haunch them as in (*c*) and (*d*). Then the c.g.s. profile can be properly curved as in (*c*) or may remain practically straight as in (*d*) where conditions permit.

For short double cantilevers, a straight and uniform section can be adopted as shown in Fig. 8-15(*a*). When the cantilevers are long, they may be tapered as in (*b*). If the anchor span is long, it may be haunched as in (*c*). If the anchor span is short compared with the cantilevers, the c.g.s. line may lie near the top of the beam at all points, as in (*d*).

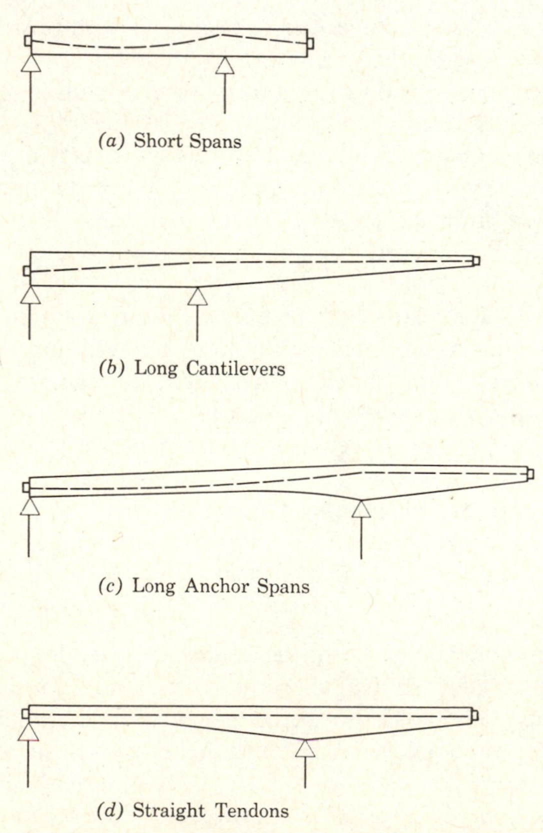

Fig. 8-14. Typical layouts for single cantilevers.

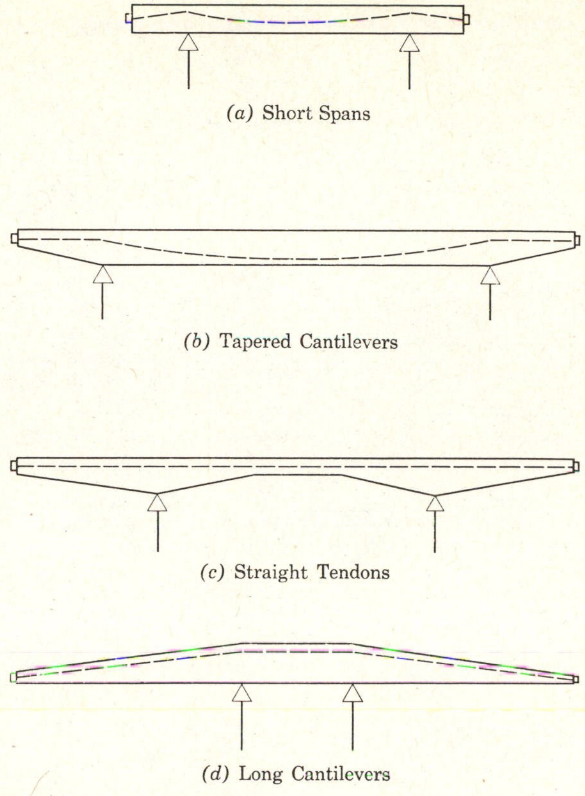
(a) Short Spans

(b) Tapered Cantilevers

(c) Straight Tendons

(d) Long Cantilevers

Fig. 8-15. Typical layouts for double cantilevers.

Cable location for cantilevers can be obtained graphically as for simple beams, except that more thought should be given to the possibilities of partial live loads and the reversal of moments. Figure 8-16(*a*) shows a cantilever beam. Assume that the beam is under the action of its own weight and the action of uniform live load on any portion. Moment due to dead weight of the beam is pictured in (*b*). Moment due to live load on the anchor span is shown in (*c*); that due to live load on the cantilever is shown in (*d*). For convenience in discussion, maximum moment will signify the greatest positive or the smallest negative moment, while minimum moment will mean the smallest positive or the greatest negative moments. So, for this beam, the maximum moment will be given by $(b)+(c)$, and the minimum moments by $(b)+(d)$. Both are plotted in (*e*).

In order to obtain the limiting zone for the c.g.s. line, first plot the top and bottom kern lines for the beam, k_t and k_b lines in (*f*). If no tension is permitted in the concrete, one limiting line is obtained by plotting from each kern line the

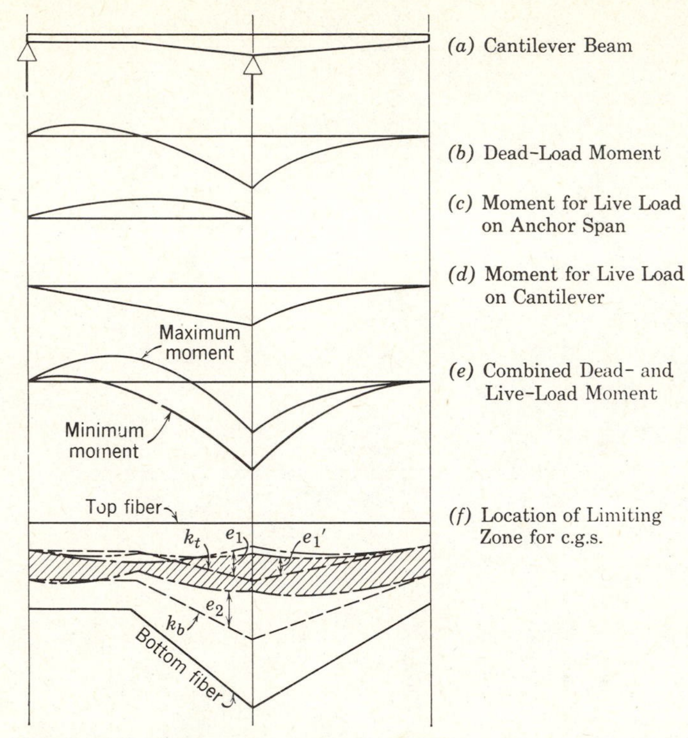

Fig. 8-16. Graphical method for location c.g.s.

permissible eccentricity e, with

$$e = M/F$$

Note that e may be plotted from either the k_t or the k_b line, whichever gives the more critical limit. But e due to $+M$ is always plotted downward, since it tends to shift the required c.g.s. line downward. By similar reasoning, e due to $-M$ is always plotted upward. In general the upper limit for the zone is plotted from the k_t line with a distance

$$e_1 = M_{max}/F$$

The lower limit for the zone is plotted from the k_b line with a distance

$$e_2 = M_{min}/F$$

Consideration should also be given to the action of dead load alone, since in this case we may have the initial prestress which is greater than the effective prestress and may impose a more critical situation. With the dead load acting alone, another limit is obtained by plotting from the k_t line a distance

$$e_1' = M_G/F_0$$

again plotting the $+M$ downward and the $-M$ upward. In this figure it is not necessary to plot e_1' from the k_b line, because evidently it will not be controlling. When plotted from the k_t line, it is seen that, for certain portions of the beam, e_1' will be controlling rather than e_1. The resulting limiting zone is shaded as in (*f*).

For long cantilevers carrying heavy loads, it is sometimes economical to cut off some of the prestressing wires at intermediate points. The number and location of cut-offs can also be established by a graphical method, which is the reverse of the above procedure and will be illustrated in example 8-5.

EXAMPLE 8-5

Compute the variation of steel area required along the 140-ft (42.67 m) length of the cantilever roof girder having a layout as shown in Fig. 8-17. Given the following data.

1. Concrete: $f_c' = 5000$ psi (34 N/mm²), allowable $f_c = 2250$ psi (15,5 N/mm²) for working load, and 2500 psi (17 N/mm²) under initial prestress, allowable tension in concrete = 0 under working load.

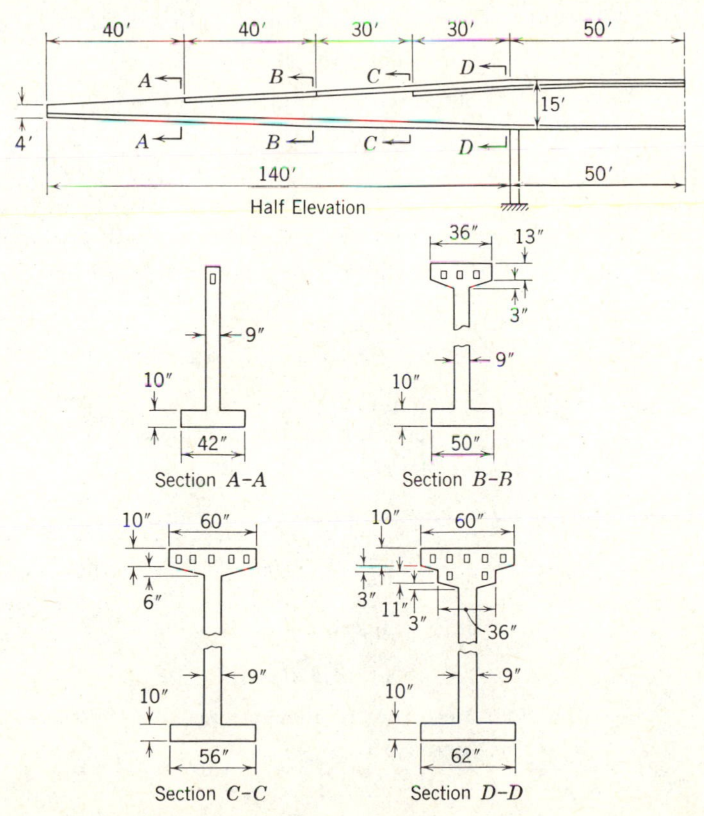

Fig. 8-17. Example 8-5. Girder sections.

2. Steel: $f_s = 240{,}000$ psi (1655 N/mm^2), initial prestress = 150,000 psi (1034 N/mm^2), final effective prestress = 125,000 psi (862 N/mm^2).

3. Live load and superimposed dead load = 1.60 k per linear foot of girder (23.34 kN/m), producing moment at support

$$wL^2/2 = 1.60 \times 140^2/2 = 15{,}700 \text{ k-ft } (21{,}290 \text{ kN-m})$$

4. Moments due to weight of girder: For the trial layout, the girder load moments are computed for various points on the cantilever, with a maximum of 15,200 k-ft (20,610 kN-m) at the support.

Solution After some preliminary investigation, it is found that, owing to the relatively heavy girder load moment, there will exist compressive stresses along most of the bottom flange. Hence it is not necessary to check for any tensile stress in the bottom flange except near the cantilevering end. For the given layout, the c.g.s. line is computed and plotted in Fig. 8-18. The resisting lever arm a_1 available for the internal resisting couple is measured from the c.g.s. line to the k_b line at each point. Corresponding to a_1, the minimum amount of prestress required is

$$F = M_T/a_1$$

and the steel area required is

$$A_s = F/125 = M_T/125a_1$$

Similarly, the maximum steel permitted without producing tension in the bottom fiber is

$$A_s = M_G/150a_2$$

where a_2 is the distance between the c.g.s. and the k_t lines.

First, the M_T and M_G moment diagrams are drawn. Next the distances a_1 and a_2 between the c.g.s. line and the k_b and k_t lines are measured. Then the minimum and maximum A_s lines are computed by the above formulas. Several points may be necessary for an accurate determination of these curves, but only some sample computations will be illustrated here. At the support, the total moment is

$$M_T = 15{,}200 + 15{,}700 = 30{,}900 \text{ k-ft } (41{,}900 \text{ kN-m})$$

The c.g.s. is located 9 in. (229 mm) from the top, and the k_b is located 42.6 in. (1082 mm) above the bottom fibers; hence the available lever arm a_1 is

$$a_1 = 180 - 42.6 - 9 = 128.4 \text{ in. } (3261 \text{ mm})$$

The minimum area of steel required at the support is, therefore,

$$\min A_s = \frac{M_T}{125a_1} = \frac{30{,}900 \times 12}{125 \times 128.4} = 23.1 \text{ sq in. } (14.9 \times 10^3 \text{ mm}^2)$$

At 35 ft (10.67 m) from the cantilevering end,

$$M_G = 600 \text{ k-ft } (814 \text{ kN-m})$$

a_2 is measured to be 10.5 in. (267 mm). The maximum steel area permitted at this point is

$$\max A_s = \frac{M_G}{150a_2} = \frac{600 \times 12}{150 \times 10.5} = 4.57 \text{ sq in. } (2.95 \times 10^3 \text{ mm}^2)$$

Similar computations are made for other points, and the curves are drawn as shown. It

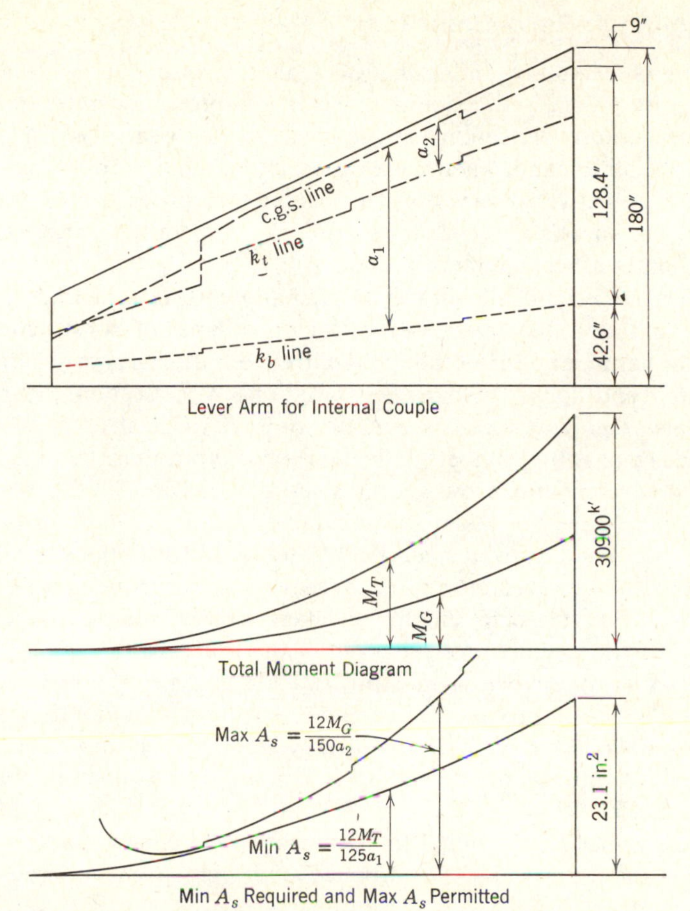

Fig. 8-18. Example 8-4. Computation for A_s.

will be seen that the maximum A_s curve is actually required only for a short portion of the beam near the cantilevering end. Keeping as close as possible to the minimum A_s curve, but without crossing the maximum A_s curve, the adopted steel area may be tailored and cut off as desired.

The checking of compressive stresses in concrete and other design features will not be discussed here. Note that all moments are obviously negative in this solution, hence no particular attention has been paid to the signs of the moments.

One advantage of such a graphical solution is that is gives a visual presentation. The variations of the lever arm a_1 and a_2 and of the A_s curves both follow certain simple laws so that necessary modifications to suit changes in design can be easily made, either by shifting the c.g.s. location or by varying the steel areas.

8-5 Span-Depth Ratio Limitations

For reasons of economy and esthetics, higher span-depth ratios are almost always used for prestressed concrete than for reinforced concrete. Higher ratios are possible because deflection can be much better controlled in prestressed design. On the other hand, when these ratios get too high, camber and deflection become quite sensitive to variations in loadings, in properties of materials, in magnitude and location of prestress, and in temperature. Furthermore, the effects of vibration become more pronounced.

It is difficult to establish a simple set of span-depth ratio limitations because the proper limitation, for prestressed as for other types of construction, should vary with the nature and magnitude of the live load, the damping characteristics, the boundary conditions, the shape and variations of the section, the modulus of elasticity, and the length of span itself. In fact, if a structure is carefully investigated for possible camber, deflections, and vibrations, there is no reason to adhere to any set ratio. However, as a result of accumulated experience, the values in Table 8-3 may be taken as a preliminary guide for building design.

For cantilever solid slabs, a span-depth ratio of 18 for floors and 20 for roofs has been found to be satisfactory. But cantilevers are sensitive to deflections and vibrations, and greater care should be taken. For example, a camber in the anchor span would usually produce a dip in the cantilever.

Generally speaking, when span-depth ratios are some 10% below the values tabulated in Table 8-3, problems of camber, deflection, and vibration should not occur unless the loadings are extremely heavy and vibratory in nature. Occasionally, these ratios can be exceeded by 10% or more, if careful study would justify and ensure proper behavior.

The above values are intended for both hard-rock concrete and light-weight concrete, but should be reduced by about 5% for lightweight concrete having E_c

Table 8-3 Approximate Limits for Span-Depth Ratios

	Continuous Spans		Simple Spans	
	Roof	**Floor**	**Roof**	**Floor**
One-way solid slabs	52	48	48	44
Two-way solid slabs (supported on columns only)	48	44	44	40
Two-way waffle slabs (3 ft waffles)	40	36	36	32
Two-way waffle slabs (12 ft waffles)	36	32	32	28
One-way slabs with small cores	50	46	46	42
One-way slabs with large cores	48	44	44	40
Double tees and single tees (side by side)	40	36	36	32
Single tees (spaced 20-ft centers)	36	32	32	28

less than 3,000,000 psi (20.69 kN/mm^2). For long spans (say, in excess of about 70 ft i.e. 21 m) and for heavy loads (say, live loads over 100 psf, that is, 4.79 kN/m^2) the above values should be reduced by 5 to 10%. For in-place concrete in composite action with the precast elements, the total depth may be considered in computing the above span-depth ratios.

It should be emphasized that the above table is intended as a guide and should not be applied blindly, without considering the local conditions. For example, the degree of continuity (whether full or partial), the existence of rigid frame action, the reliability of the production and construction control, and the local temperature differential will all affect the ratio limitations.

The problem of objectionable vibration can be studied by determining the amplitude and the natural frequency of the structure.[10] Human sensitivity to vibration increases with the frequency and the amplitude. When vibration is produced by mechanical loading, the degree of damage is also proportional to the frequency and the amplitude. The synchronization of the applied loading with the natural frequency of the structure will, of course, amplify the vibration, whereas damping would help to minimize the effects.

Little experience has been obtained for railway bridges of prestressed concrete to justify any limitation on their span-depth ratios; the usual ratios have been in the range of 10 to 14 for box sections up to 100 ft (30 m) or more. For simple-span highway bridges of the I-beam type, up to about 200 ft (61 m), a span-depth of 20 is considered conservative, 22 to 24 is normal, while 26 to 28 would be the critical limit. Box sections can have ratios about 5 to 10% higher than I-beams, while T-sections spaced far apart should have ratios about 5 to 10% lower than I-beams. Again, there is no reason to believe that a fixed span-depth ratio will apply to all cases. The effect of continuity, of varying moment of inertia, etc., should be considered. The Union Oil Pedestrian Bridge, Fig. 11-24, has a structural depth of 27 in. (686 mm) spanning 102 ft (31 m) (a span-depth ratio of 45), but was carefully designed to take care of camber, deflection, and vibrations. It has behaved in a very satisfactory manner since its completion in 1956.

References

1. W. G. Corley, M. A. Sozen, and C. P. Siess, "Time-Dependent Deflections of Prestressed Concrete Beams," *Bulletin No. 307*, Highway Research Board, Washington, D.C., 1961.
2. S. L. Bugg, "Long-time Creep of Prestressed Concrete I-beams," Technical Report R-212. U.S. Naval Civil Engineering Laboratory, Port Hueneme, California, 1962.
3. W. D. Atkins, "A Generalized Numerical Solution for Prestressed Concrete Beams," unpublished Master's thesis supervised by N. H. Burns, The University of Texas, Austin, Texas, Aug. 1965.

4. D. C. Chang, "A Numerical Method of Analyzing Composite Prestressed Concrete Members, "unpublished Master's thesis supervised by N. H. Burns, The University of Texas, Austin, Texas, May 1969.
5. D. M. Pierce, "A Numerical Method of Analyzing Prestressed Concrete Members Containing Unbonded Tendons," Ph.D. Dissertation supervised by N. H. Burns, The University of Texas, Austin, Texas, June 1968.
6. R. Sinno, and H. L. Furr, "Computer Program for Predicting Prestress Loss and Camber," *V. Prestressed Conc. Inst.*, Vol. 17, No. 5, September–October 1972.
7. A. I. Fadl, W. L. Gamble, and B. Mohraz, "Tests of a Precast Posttensioned Composite Bridge Girder Having Two Spans of 124 Feet," University of Illinois Engineering Experiment Station, Structural Research Series No. 439, April 1977.
8. C. Suttikan, "A Generalized Solution for Time-Dependent Response and Strength of Noncomposite and Composite Prestressed Concrete Beams," Ph.D. dissertation supervised by N. H. Burns, The University of Texas, Austin, Texas, Aug. 1978.
9. L. D. Martin, "A Rational Method for Estimating Camber and Deflection of Precast Prestressed Members," *V. Prestressed Conc. Inst.*, Vol. 22 No. 1, January/February 1977, pp. 100–108.
10. "Vibrations in Buildings," *Building Research Station Digest*, No. 78, London, England, June 1955.

9

PARTIAL PRESTRESS AND NONPRESTRESSED REINFORCEMENTS

9-1 Partial Prestress and Beam Behavior

When prestressed concrete was introduced in the 1930's, the philosophy of design was to create a new material by putting concrete under compression so that there would never be any tension in it, at least not under working loads. In the later 1940's, observations on those earlier structures indicated that often extra strength existed in them. Therefore some engineers believed that a certain amount of tensile stresses could be permitted in design.[1, 2, 3, 5, 6, 7, 8] In contrast to the earlier criterion of no tensile stress, which may be called "full prestressing," this later method of design allowing some tension is often termed "partial prestressing." As discussed in section 6-4, there is no basic difference between the two, because, although a structure may be designed for no tension under working loads, it will be subjected to tension under overloads. Therefore the difference is rather a matter of degree; tensile stresses will be higher and occur more frequently for the same structure if designed for partial prestressing rather than full prestressing. We can predict the moment-curvature response of partially prestressed members[3,8] using the same analysis as was discussed in Chapter 5. This allows us to analyze load-deflection performance to be certain that it is satisfactory.

In order to provide additional safety for partially prestressed concrete, nonprestressed reinforcements are often added to give higher ultimate strength to the beam, and to help carry the tensile stresses in the concrete. For these beams, some of the reinforcements are prestressed and others are not. This situation also lends itself to the use of the term "partial prestress," so that, sometimes, "partial prestress" may mean either or both of the following two conditions, although more frequently it is employed to denote the first condition only.

1. Tensile stresses are permitted in the concrete under working loads. Design on that basis is described in sections 6-4 through 6-6.
2. Nonprestressed reinforcements are employed in the member. This condition will be described in the following sections of this chapter.

An important advantage of partial prestressing is the decrease in the amount of camber. Minimizing camber is important especially when the girder load or the dead load is relatively small compared to the total design load. Minimizing the initial camber also means decreasing the effect of flexural creep and easier control of the uniformity of camber.

In order to understand the design of partially prestressed beams, it is necessary to study the behavior of such beams with varying amount of reinforcement and subjected to varying amount of prestress. The difference in behavior of overreinforced and underreinforced beams is seen by comparing curves (*a*) and (*b*) of Fig. 9-1. The difference in behavior of overprestressed and underprestressed beams is seen by comparing curves (*a*), (*b*), (*c*), and (*d*) in Fig. 9-2.

When a section is overreinforced, Fig. 9-1, it will fail by compression in concrete before the steel is stressed beyond its elastic limit. Thus, the ultimate deformation of the steel and the ultimate deflection of the beam are rather small, and the failure is brittle. When seriously overreinforced, even if the steel is not prestressed, the deflection of the beam before rupture will still be limited. When a section is underreinforced, its deflection increases very appreciably before failure, thus giving ample warning of impending collapse. Failure starts in the excessive elongation of steel and ends in the gradual crushing of concrete on the compressive side.

In order to avoid sudden and brittle failures, and also for general economy in design, beams designed following the ACI Code (see Chapter 5) are underreinforced. When an underreinforced section is designed for full prestressing, allowing no tension in concrete under the working loads, the load-deflection

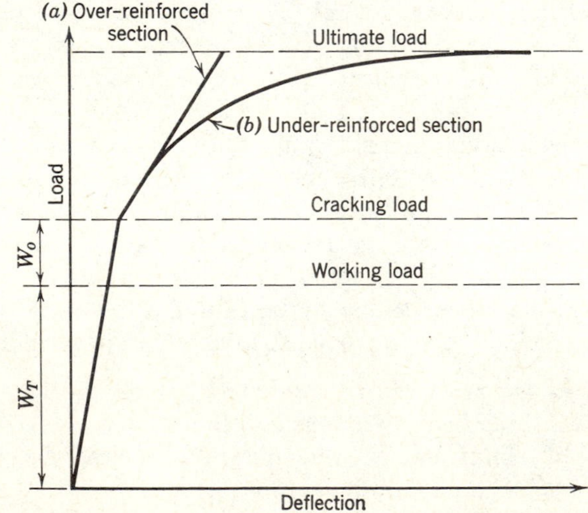

Fig. 9-1. Load-deflection curves, over- and underreinforced sections.

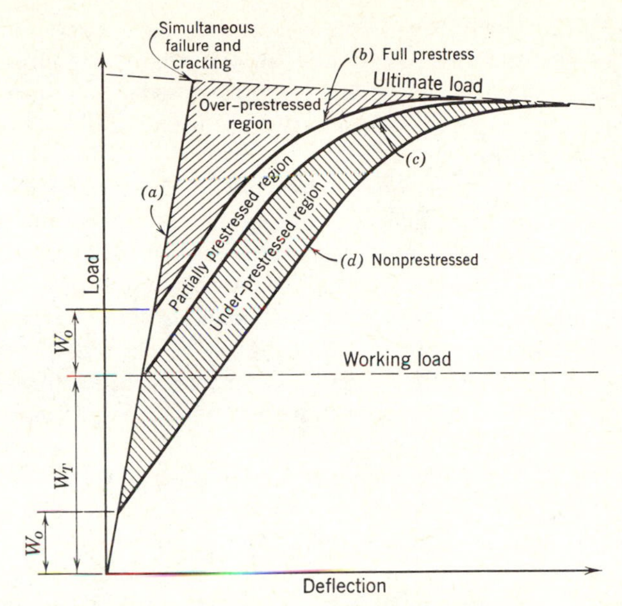

Fig. 9-2. Load-deflection curves for varying degrees of prestress (for underreinforced sections of bonded beams).

relation is given by curve (b) of Fig. 9-2. Before cracking, the section will carry an additional load W_0 above the working load W_T, the magnitude of that additional load being

$$W_0 = k\frac{f'I}{c_b}$$

where k is a constant depending on the span length and end conditions, f' is the modulus of rupture, and c_b the distance from c.g.c. to the tensile extreme fibers.

If the same underreinforced section with the same amount of steel is given somewhat smaller prestress so that cracking is reached just at the working load, the tensile stress being equal to the modulus of rupture under working load, the load-deflection relation will be given by curve (c), where the deflection corresponding to the cracked section starts at the working load. If the beam is not prestressed at all, but still reinforced with the same amount of steel, provided that the steel is bonded to the concrete, the beam will behave as in curve (d). It will start cracking as soon as load W_0 is reached, although its ultimate strength may not be greatly reduced.

If the beam is overprestressed it will crack only after the load exceeds $W_T + W_0$, and its load-deflection curve will fall between curves (a) and (b), Fig. 9-2. In the extreme case, when the beam is very much underreinforced but highly overprestressed, cracking and failure may take place simultaneously so that brittle failure occurs with sudden rupture in the steel. In principle, a partially prestressed beam may have a load-deflection curve lying anywhere between curves (b) and (d), depending on the amount of prestress. But, in practice, seldom is cracking permitted for prestressed concrete under working load; hence the actual load-deflection curve will usually fall between curves (b) and (c), and seldom below curve (c).

The desired amount of prestress will depend on the type of service to which the structure is to be subjected. For structures in which the possibility of cracking under working loads must be avoided and where overload might occur rather frequently, full prestress yielding load-deflection curve (b) is preferred. For structures which are seldom overloaded, such as certain types of buildings, partial prestress between curves (b) and (c) may be permitted. Some prestressing steel is saved by designing for partial prestress, but, if the same ultimate strength is desired, at least the same amount of total reinforcement must be used.

The area under the load-deflection curve is a measure of the ability of a beam to stand impact load and to absorb shocks. Hence it is seen that both the fully and the partially prestressed beams will supply a fairly large amount of resilience, while both the overprestressed and the underprestressed ones will supply appreciably less. The overprestressed beams will possess less plastic energy, while the underprestressed ones will absorb less elastic energy.

Partial prestress may be obtained by any of the following measures.

1. By using less steel for prestressing; this will save steel, but will also decrease the ultimate strength, which is almost directly proportional to the amount of steel.
2. By using the same amount of high-tensile steel, but leaving some nonprestressed; this will save some tensioning and anchorage, and may increase resilience at the sacrifice of earlier cracking and slightly smaller ultimate strength.
3. By using the same amount of steel, but tensioning them to a lower level; the effects of this are similar to those of method 2, but no end anchorages are saved.
4. By using less prestressed steel and adding some mild steel for reinforcing; this will give the desired ultimate strength and will result in greater resilience at the expense of earlier cracking.

The engineer must judge which method is desirable for a particular structure.

The advantages and disadvantages of partial prestress as compared to full prestress[5,6,7,8] may be now summarized.

Advantages

1. Better control of camber.
2. Saving in the amount of prestressing steel.

3. Saving in the work of tensioning and of end anchorages.
4. Possible greater resilience in the structure.
5. Economical utilization of mild steel.

Disadvantages

1. Earlier appearance of cracks.
2. Greater deflection under overloads.
3. Higher principal tensile stress under working loads.
4. Slight decrease in ultimate flexural strength for the same amount of steel.

Naaman has carefully summarized the procedures for design of partially prestressed beams in reference 8. His curve showing the load-deflection response with varying levels of prestress is essentially the same as Fig. 9-2. With the present state-of-the art as described by references 6, 7, and 8, we can now design partially prestressed members with confidence that we are assuring both adequate serviceability (service-load behavior) and sufficient strength (factored load behavior). The ACI Code now allows concrete tensile stresses which would possibly result in cracking at service load, but we must limit the steel ratio to values which will assure that members we design are underreinforced. All of the disadvantages listed above can be checked to assure that we have a serviceable and safe design. Partial prestressing has been used with very satisfactory results, and we will see increasing utilization of this approach to design in the future.[6,7,8]

9-2 Uses of Nonprestressed Reinforcements

One of the more recent developments in prestressed concrete is the use of nonprestressed flange reinforcements. These reinforcements can be made up of high-tensile wires, wire strands, bars, or merely ordinary mild-steel bars. When used in conjunction with prestressed steel, they form an effective combination, one supplementing the other. The prestressed steel balances a portion of the load, reduces the deflection, and supplies the major part of the strength, while the nonprestressed steel distributes the cracks, increases the ultimate strength, reinforces those portions not readily reached by prestressed steel, and provides additional safety for unexpected conditions of loading. With unbonded tendons, frequently it is necessary to add some bonded nonprestressed reinforcement. With proper design, both economy and safety can be attained in many cases. The experiments available regarding the behavior of such designs indicate the combination of prestressed and nonprestressed steel gives good results, whether bonded or unbonded tendons are used. The general nature of behavior is known and numerous structures have been built utilizing such combinations.

Nonprestressed reinforcements can be placed at various positions of a prestressed beam to serve different purposes and to help carry the loading at different stages. Often, one set of these reinforcements can serve to strengthen the beam in several ways. This will become evident upon examining the following functions performed by them.

1. To provide strength immediately after transfer of prestress:

A. When the compressive flange may be under some tension at transfer, nonprestressed steel will help to reinforce that flange against any possible fracture, Fig. 9-3(*a*). This design is often desirable when the beam's own weight is small compared to its live load. The use of such nonprestressed steel will permit the placing of the prestressing steel nearer to the extreme tensile fibers, thus gaining a bigger lever arm for the resisting moment.

B. When straight tendons are used for straight beams, top flange at the ends of the beam may be subjected to tensile stresses. Nonprestressed reinforcements can be placed therein for reinforcement, Fig. 9-3(*b*).

C. When high compressive stresses are produced in the tensile flange by high prestressing, steel bars may be added to reinforce that flange, Fig. 9-3(*c*). Such bars will also tend to minimize creep in the concrete. On

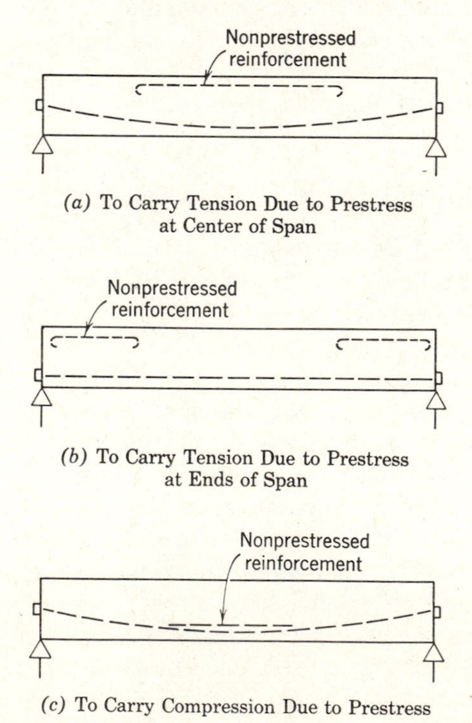

Fig. 9-3. Nonprestressed reinforcements to strengthen beam just after transfer of prestress.

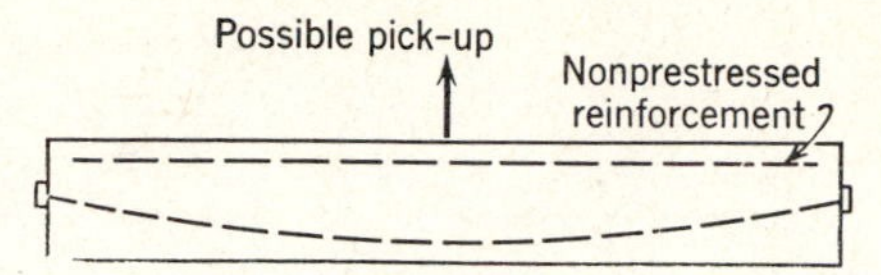

Fig. 9-4. Nonprestressed reinforcement to strengthen precast beam during handling and erection.

the other hand, when these bars are subjected to high compressive stresses, especially when considering the effect of creep and shrinkage, lateral expansion of the bars due to Poisson's ratio effect may have a tendency to split the surrounding concrete. German specifications call for a minimum cover of about 3 times the bar's diameter to prevent such splitting. The use of proper stirrups may also be helpful.

2. To reinforce certain portions of precast beams so as to be able to carry special or unexpected loads during handling and erection, Fig. 9-4. This may either permit easier handling of the beams or may prevent serious rupture in case of careless handling.
3. To reinforce the beam under working loads:
   A. Either high-tensile or ordinary steel can be placed side by side with prestressed steel, Fig. 9-5(a). This will help to distribute cracks when they occur and also to increase the ultimate strength, especially when the prestressed steel is not bonded to the concrete. By preventing the formation of concentrated big cracks, both the flexural and shear

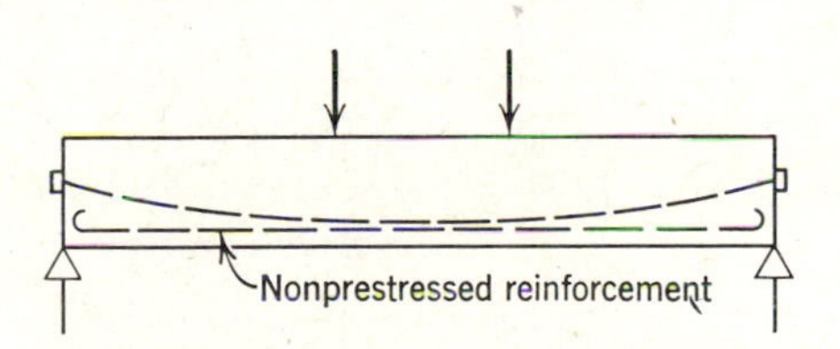

(a) To Distribute Cracks and Increase Ultimate Strength

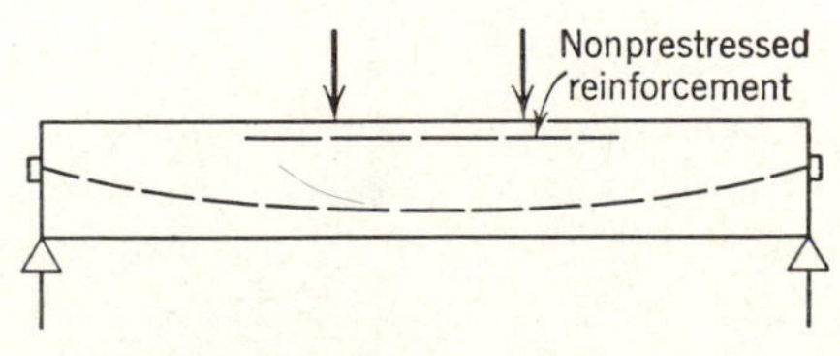

(b) To Reinforce Compression in Concrete

Fig. 9-5. Nonprestressed reinforcements to reinforce beams under working and ultimate loads.

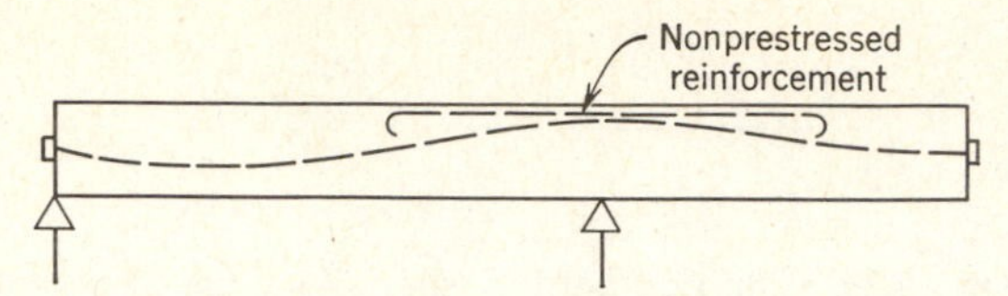

Fig. 9-6. Nonprestressed reinforcement to reinforce moment peaks in cantilevers.

resistance of the beams may be increased. Nonprestressed steel can often be economically employed because it has to be placed only over certain critical portions, while the prestressed steel generally has to extend the whole length of the beam.

B. Ordinary steel bars can be added to the compressive flange to reinforce it against high compression, Fig. 9-5(*b*). This is generally uneconomical but may be required under certain conditions.

The use of nonprestressed reinforcements is certainly not limited to simple spans. For cantilevers and continuous spans, where peak moments exist it is often economical to reinforce such portions with nonprestressed steel, Fig. 9-6. Here, again, the use of some short length of ordinary steel may save some long prestressed steel, and economy is thereby achieved.

When prestressed and nonprestressed reinforcements are combined in a structure, the cooperation of the two should be carefully investigated. Most of the time, the nonprestressed steel will not be acting effectively until the cracks have formed. Its effect on the start of hair cracks and on the elastic deflection of the beam will be small. But after cracking occurs, such steel will distribute the cracks and prevent the formation of big ones which may sometimes be detrimental in producing diagonal tension cracks and compression failures. The ultimate strength of beams under both static and repeated loads can be materially increased by proper employment of nonprestressed steel.[2, 3]

9-3 Nonprestressed Reinforcement–Elastic Stresses

It is difficult, if not impossible, to design nonprestressed reinforcement by the elastic theory, because, within the elastic range, the tensile stresses in the reinforcements are very small and the reinforcements are consequently ineffective, although in the ultimate range they are usually stressed to the yield point and function effectively. However, a study of the elastic stresses is significant in helping to understand the behavior of such beams and to design them properly. As discussed in section 9-2, nonprestressed steel can be placed on either or both sides of the beam: the tension side which is counter-compressed by prestressing, and the compression side which could be under tension before the application of

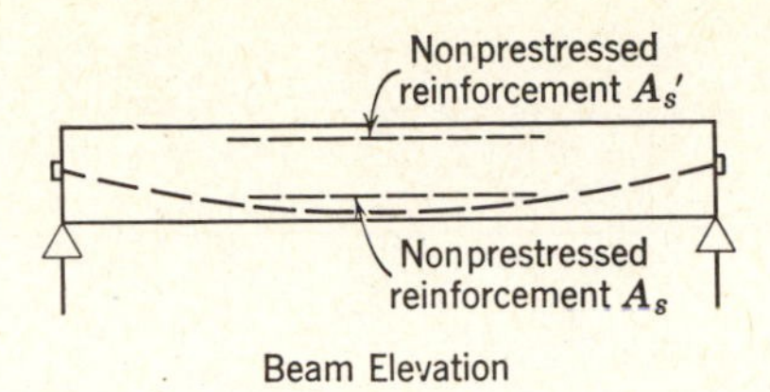

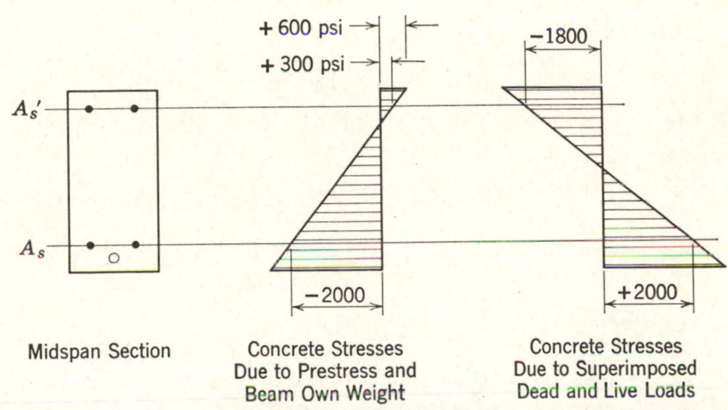

Fig. 9-7. Concrete stresses at levels of nonprestressed reinforcements.

external loads. Let us investigate the stresses in both sides together, as in Fig. 9-7.

Assume that there is no shrinkage of concrete; then there is no stress in the nonprestressed steel until prestress is transferred. At the transfer of prestress, the nonprestressed steel will have strains corresponding to the adjacent concrete, and stresses can be computed by the elastic theory with the usual formula

$$f_s = n\left[\frac{(M_G + F_0 e)y}{I_t} + \frac{F}{A}\right]$$

Owing to creep in concrete, stresses in the steel will be modified by the creep coefficient so that they will increase from f_s to $C_c f_s$.

The above method can be assumed to apply equally to the tensile and to the compressive stresses in the steel. Thus both the tensile and the compressive stresses in the steel will be increased by the effect of creep in concrete, and both can be modified by the coefficient of creep applicable for the given duration of time.

Next, let us consider the effect of shrinkage in concrete due to which compressive stresses will be produced in steel to the amount of

$$f_s = \delta E_s$$

where δ is the unit shrinkage strain in the concrete. Thus the resulting stresses in the steel before the application of external loads is given by the formula

$$f_s = C_c n\left(\frac{(M_G + F_0 e)y}{I} + \frac{F}{A}\right) + \delta E_s \tag{9-1}$$

using proper signs for each of the items.

The approximate magnitudes of these stresses can be shown as below. Assuming that the concrete fiber at the level of the nonprestressed reinforcement A_s, Fig. 9-7, is stressed to -2000 psi (13.8 N/mm^2), for a value of $n=6$ and $C_c = 1.5$, the compressive stress in steel will be

$$-2000 \times 6 \times 1.5 = -18{,}000 \text{ psi } (-124 \text{ N/mm}^2)$$

Add to this the effect of a shrinkage strain of -0.0002, which will induce a compressive stress in the steel of

$$-0.0002 \times 30{,}000{,}000 = -6000 \text{ psi } (-41.4 \text{ N/mm}^2)$$

Hence the total stress in the steel A_s on the tension flange may be around $-18{,}000 - 6000 = -24{,}000$ psi (-165 N/mm^2). Thus the steel will be stressed appreciably in compression.

For the steel A'_s on the compression flange the stresses cannot be very high. Assuming a tensile stress of 600 psi (4.14 N/mm^2) in the extreme top fibers of concrete, there may be only about 300 psi (2.07 N/mm^2) tension at the level of steel. For the same creep and shrinkage as above, the resulting stress in the steel A'_s will be

$$(1.5 \times 300 \times 6) - 6000 = -3300 \text{ psi } (-22.8 \text{ N/mm}^2)$$

Hence this nonprestressed steel in the compression flange, which is intended to carry tension in that flange under girder loads, will probably be under compression instead. In other words, before the occurrence of cracks, such reinforcements may not serve their intended purpose at all.

Now, when the external load is applied on the beam, steel A'_s on the compression flange will be further compressed, while the steel A_s on the tensile side will be decompressed. These stresses can again be computed by the elastic theory, the effect of shrinkage and creep being taken into account if necessary. To get an idea of the magnitude of the elastic stresses produced by loading, assume a compression of about 1800 psi in the concrete fiber near the A'_s, and a decompression of about 2000 psi in the concrete fiber at the A_s; we have, Fig. 9-7,

$$f_s = -3300 - 1800 \times 6 = -14{,}100 \text{ psi } (-97.2 \text{ N/mm}^2) \text{ in } A'_s$$

$$f_s = -24{,}000 + 2000 \times 6 = -12{,}000 \text{ psi } (-82.7 \text{ N/mm}^2) \text{ in } A_s$$

which indicates that the tensile steel A_s, which is intended to carry the tension under working loads, may actually still be under compression instead. Hence it is impossible to design such nonprestressed steel for working loads. Previous to cracking, they will not serve the intended function at all. They will increase the ultimate strength of the beam and minimize its deflection after cracking. For members whose serviceability is impaired by cracking, nonprestressed reinforcements cannot suitably be employed.

9-4 Nonprestressed Reinforcements, Ultimate Strength

It is shown in the previous section that nonprestressed reinforcements, when used in conjunction with prestressed ones, do not function effectively in carrying any tension within the elastic working range. Similar to bars in ordinary reinforced concrete beams, they act efficiently in tension only after the concrete has cracked. Before the cracking of concrete, their tensile stresses, if any, are limited. Since almost all prestressed beams are designed for no cracks within the working loads, the nonprestressed reinforcements are apparently useless under such conditions. The interesting phenomenon is that, though they do not serve within the working range, they are often as effective as the prestressed ones near the ultimate load. Thus, if the ultimate strength is of prime importance rather than the elastic strength, nonprestressed reinforcements can be profitably employed.

Figure 9-8 shows, for various reinforcements, the variation of stresses with strains produced by external loads. Consider first a prestressed wire with effective prestress of 125 ksi (862 N/mm^2), elastic limit of 180 ksi (1,241 N/mm^2) and ultimate strength of 250 ksi (1,724 N/mm^2). As the load on the beam increases, the strain and hence the stress increases as shown in curve (*a*). Next consider a nonprestressed wire of the same qualities embedded at the same level. Its stress-strain variation is given by curve (*b*). The wire will actually be precompressed during the transfer of prestress, so that, before any external load is applied to the beam, the wire will be under compression. If the amount of precompression is 20 ksi (138 N/mm^2), the total difference in stress between this wire and the prestressed one is of the order of 145 ksi (1,000 N/mm^2).

Under working load, producing a strain in the steel of about 0.05%, stress in the prestressed wire will be increased to 140 ksi (965 N/mm^2), while that in the nonprestressed one will be changed to 5 ksi (34.5 N/mm^2) compression. Hence the nonprestressed wire is still not functioning at all.

Now, going into the ultimate range, let us refer to Fig. 9-9, which shows the conditions of strain in a prestressed bonded beam section at the ultimate load. In (*a*), we see that, for an average overreinforced beam, the strain in steel at failure is about 0.34%. Figure 9-8 shows that, at this strain, the stress in the prestressed wire is about 207 ksi (1,427 N/mm^2) while that in the nonprestressed one is only 80 ksi (552 N/mm^2). This means that the nonprestressed wire is

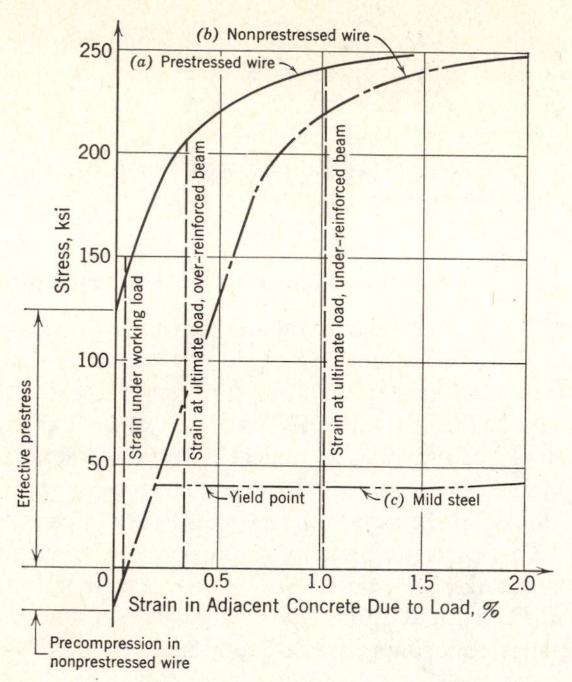

Fig. 9-8. Stress-strain diagrams of prestressed and non-prestressed steel.

picking up some stress but still has been worked only to about a third of its capacity, even at the ultimate load.

Figure 9-9(*b*) shows the strain relations of an average underreinforced beam, with a strain in the steel amounting to about 1.02% at the ultimate load. Corresponding to this strain, it is seen from Fig. 9.8 that the prestressed wire will be stressed to about 243 ksi (1,675 N/mm^2) and the nonprestressed one to 222 ksi (1,531 N/mm^2), the two values being quite close. This means that the nonprestressed wire is now almost as effective as the prestressed one. It can thus be concluded that, for an underreinforced beam, the nonprestressed wires will be quite efficient in resisting the ultimate load, although under ordinary working loads it is hardly functioning at all.

The stress in a nonprestressed ordinary mild-steel bar can be studied by referring to curve (*c*), Fig. 9-8. Under ordinary working loads, the bar may be under some compression similar to the nonprestressed wire. But at the ultimate load, it will be stressed to its yield point for either an overreinforced or an underreinforced beam. In the latter case, it is possible that the bar may sometimes be stressed even beyond its yield point.

The above discussion has been confirmed by many tests, such as mentioned in the references for this chapter. The role of bonded reinforcement for beams with

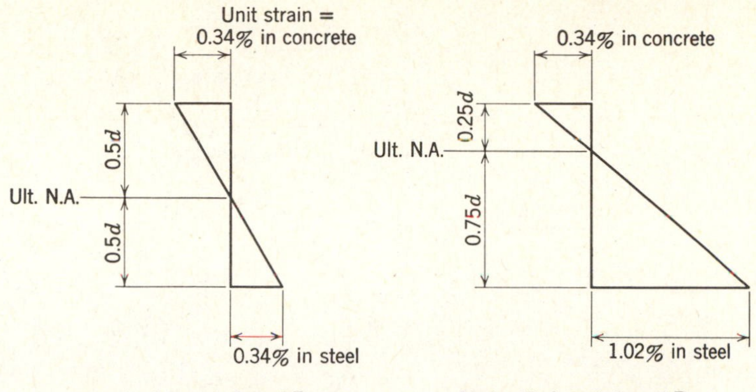

(*a*) An Overreinforced Beam (*b*) An Underreinforced Beam

Fig. 9-9. Strain relations at ultimate load for a bonded beam.

unbonded tendons is clearly shown in test results.[4] These bars assure distribution of flexural cracking and make a significant contribution to strength. Although the detailed behavior of nonprestressed reinforcements may still need experimental investigation before they can be definitely formulated, enough data are on hand to permit designs made within the usual range of accuracy desired in practice.

Having described the general behavior of nonprestressed steel in a prestressed beam, we can now proceed to the design of such beams on the basis of ultimate strength. It must again be remembered that ultimate strength is only one measure of the adequacy of a structure. High stresses and local strains, which may be detrimental if repeated often enough, and deflections and cracks, which may impair the serviceability of the structure long before the ultimate strength is reached, should be carefully studied in each case before a design can be adopted.

It is difficult to formulate a proper basis for the design of nonprestressed reinforcements in the compression flange, where they are needed to strengthen the beam during handling, since there is no way to predict exactly how the beam might be handled. A method for determining the ultimate strength of such prestressed sections, however, has been analyzed and tested. Briefly speaking, it can be assumed that the nonprestressed steel will be worked to the yield point at the failure of the beam, with a lever arm which can be approximately estimated, if not carefully analyzed. For example, Fig. 9-10(*a*) shows one half of a beam which is being lifted at the midspan point. With half of the beam as a freebody and taking moments about point A, the c.g.s. of the prestressed steel, we can write, for conditions at failure.

$$\text{Moment of tension in nonprestressed steel about } A = \text{Moment of weight of member about } A$$

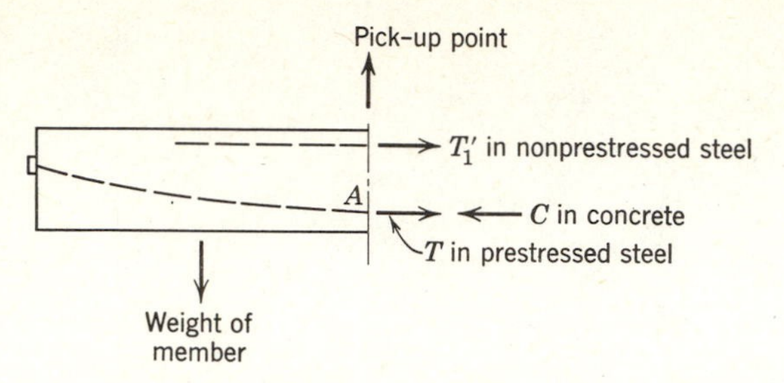

(a) Nonprestressed Steel in Compression Flange, an Arbitrary Assumption for Design

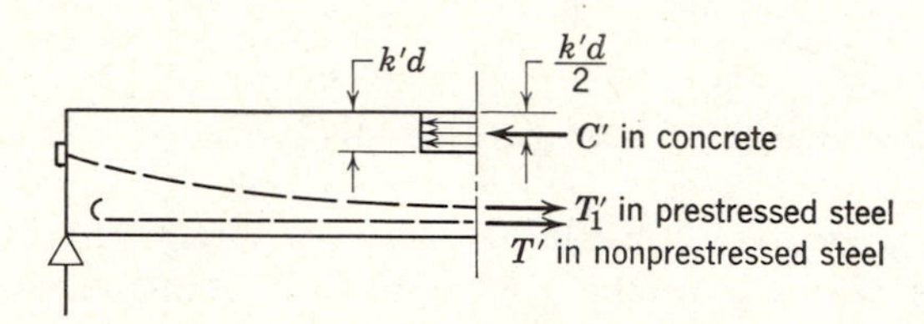

(b) Nonprestressed Steel in Tension Flange

Fig. 9-10. Ultimate design of nonprestressed steel.

assuming that the ultimate center of compression in concrete coincides with the c.g.s. A more accurate computation would require the determination of the actual center of compression at the ultimate moment. More importantly, a proper factor of safety has to be chosen against the ultimate load capacity.

The ultimate design of nonprestressed steel in the tension flange can be formulated as follows. Referring to Fig. 9-10(b), the total tension in the steel, both prestressed and nonprestressed, can be estimated. Corresponding to that total tension, the depth of compression in the concrete can be figured,

$$k'd = \frac{T' + T_1'}{k_1 f_c' b}$$

With the neutral axis thus located for the ultimate load, the ultimate tension in the steels can be obtained from diagrams and curves such as Fig. 9-8 and 9-9. Then $k'd$ can be revised, if necessary. The lever arms for the tensile forces, a' and a_1', are easily obtained and the ultimate moment computed,

$$M' = T'a' + T_1'a_1'$$

Then the allowable moment can be obtained from M' by using a proper factor of safety.

This procedure will be illustrated in the following example.

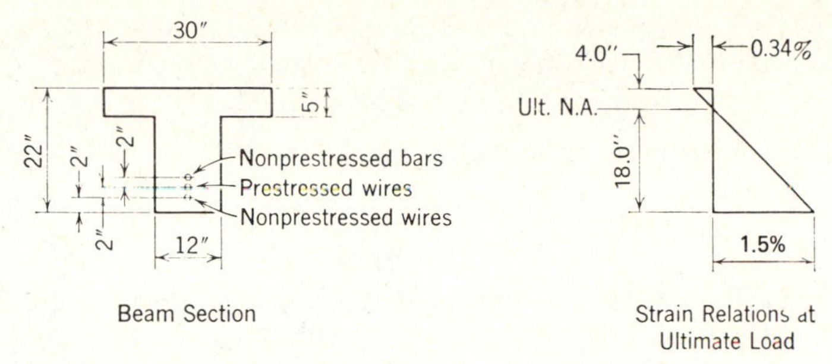

Fig. 9-11. Example 9-1.

EXAMPLE 9-1

A prestressed concrete beam has a T section as shown, Fig. 9-11. It is prestressed with high-tensile wires (A_{ps} = 1.06 sq in.) and additionally reinforced with nonprestressed wires (A_s = 0.47 sq in.) and nonprestressed mild steel bars (A_s = 1.32 sq in.). The c.g.s. of each type of steel is shown in the figure. $f_c' = 5000$ psi; for wires, $f_{pu} = 250$ ksi, $f_{se} = 125$ ksi; for mild steel bars, $f_y = 40$ ksi. Estimate the ultimate flexural strength of the section, assuming no failure in shear or bond. ($f_c' = 34$ N/mm^2, $f_{pu} = 1724$ N/mm^2, $f_{se} = 862$ N/mm^2, and $f_y = 276$ N/mm^2)

Solution Assuming that at rupture the prestressed wires will be stressed to 250 ksi (1724 N/mm^2), the nonprestressed wires stressed to 230 ksi (1,586 N/mm^2), and the mild steel to 40 ksi (276 N/mm^2), then the total tensile force at rupture will be

$$250 \times 1.06 = 265 \text{ k (1179 kN)}$$
$$230 \times 0.47 = 108 \text{ k (480 kN)}$$
$$40 \times 1.32 = 53 \text{ k (236 kN)}$$
$$\text{Total} = \overline{426 \text{ k}} \text{ (1895 kN)}$$

Assuming the average stress in concrete to be $0.85f_c' = 4250$ psi (29.3 N/mm^2), the ultimate depth of compression will be

$$k'd = \frac{426{,}000}{30 \times 4250}$$
$$= 3.4 \text{ in. (86 mm)}$$

Thus, $NA \cong 4$ in. (101.6 mm) from top of beam. For a concrete ultimate strain of 0.0034 or 0.34% (note that ACI Code would suggest use of 0.0030 or 0.30%), the ultimate strain in steel can be computed by a simple diagram as in the figure, thus,

$$\left(\frac{0.34}{4.0}\right) \times 18.0 = 1.5\%$$

corresponding to which the stresses in both the prestressed and the nonprestressed wires can be taken as 250 ksi and 240 ksi (1.724 N/mm^2 and 1655 N/mm^2), respectively (see Fig. 9-8).

Further revision of $k'd$ is deemed unnecessary; hence the resisting moments of the various steels, taken to the middepth of $k'd$, can be listed as below.

$$250 \times 1.06 \times 16.3 = 4320 \text{ k-in. } (488 \text{ kN-m})$$

$$240 \times 0.47 \times 18.3 = 2064 \text{ k-in. } (233 \text{ kN-m})$$

$$40 \times 1.32 \times 14.3 = 750 \text{ k-in. } (85 \text{ kN-m})$$

$$\text{Total} = \overline{7134 \text{ k-in.}} = 595 \text{ k-ft } (806 \text{ kN-m})$$

which is considered a rather close estimate of the ultimate resisting moment of the section. For design under the ACI Code we would multiply this nominal ultimate moment by a strength reduction factor, $\phi = 0.9$. The design ultimate moment should be determined by applying proper load factors to the service loads. In addition, the stresses in the concrete and the amount of deflection under the unfactored service load should be investigated.

9-5 Nonprestressed Reinforcements for Transfer Strength

Section 5-10 discusses the stress conditions "at transfer," when there exists little external moment or when a negative moment is applied to adversely increase the eccentricity of prestress. This section will present the design of reinforcement to improve the behavior of such beams and to increase their ultimate strength.

To improve the behavior of these beams immediately after cracking, nonprestressed reinforcement can be added to limit the cracks. This is usually done by providing sufficient steel to replace the tensile force represented in the elastic stress block. Thus, referring to example 5-13, Fig. 5-38, the total tension is given by

$$930 \times 12 \times 5.2/2 = 29{,}000 \text{ lb } (129 \text{ kN})$$

Using an allowable stress of 20,000 psi (138 N/mm^2), the required A_s is

$$29{,}000/20{,}000 = 1.45 \text{ sq in. } (935 \text{ mm}^2)$$

This design approach is often improper, because the total tensile force in the stress block is quite sensitive to the change in the magnitude of the external moment. Since it is frequently impossible to accurately predict the external moment, the computed tension force may have little meaning. Furthermore, it is not easy to determine the actual stress in the reinforcement across the cracks, and using an arbitrary value such as 20,000 psi (138 N/mm^2) may not be satisfactory. However, when judiciously applied, this method can work well for the limitation and control of cracking.

The ultimate strength method is considered to be a more rational approach to the design of these reinforcements for resisting adverse moments. Using this method, the maximum adverse moment that can be expected is computed or estimated, using a suitable margin of safety in its estimation. Then the reinforcement is designed to resist this moment.

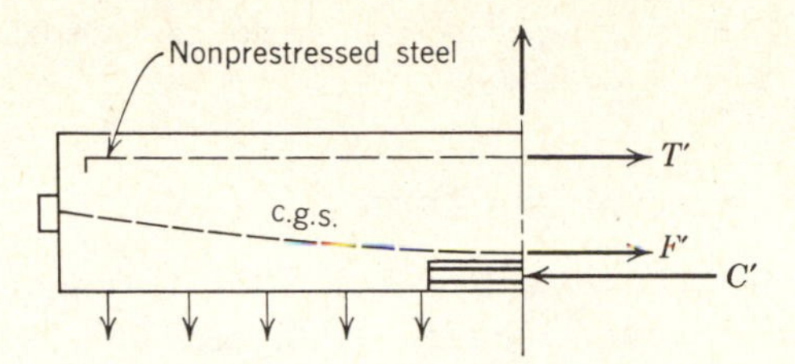

Fig. 9-12. Freebody of half beam reinforced for adverse moments.

Under a high prestress, acting on the bottom flange, it is possible that a compression failure in the concrete could occur previous to the yielding of the top tensile steel (see Chapter 5, reference 9). Since such failures are relatively rare in practice, we will consider only tension failure resulting from the top steel reaching its yield point.

A freebody of one-half of a simple beam being picked up at midspan is shown in Fig. 9-12, which indicates the weight of the beam acting on the cantilever producing a negative moment at midspan. For the purpose of design, this moment can be multiplied by factor of safety to obtain the ultimate negative moment M'. Other forces acting on the section are

T' = tension force in the nonprestressed top steel, assumed to be $A_s f_y$
F' = prestressing force under the ultimate moment, which is less than the effective prestress F_e owing to the compression in the concrete
C' = ultimate compressive force in concrete $= F' + T'$

As a rough approximation, we can neglect C' and F', and design the section like a conventionally reinforced concrete beam, thus,

$$M' = T' jd = A_s f_y jd$$

Hence,

$$A_s = \frac{M'}{f_y jd} \tag{9-2}$$

For a more accurate answer, the magnitude and location of both C' and F' should be taken into account, considering the strain relations between the concrete and the steels (Chapter 5, reference 9).

EXAMPLE 9-2

Design the nonprestressed steel required for the top of the beam in example 5-15 to resist an ultimate moment of the beam's own weight plus 100% impact.

Solution Using approximate formula 9-2, and assuming $j = 0.87$, $f_y = 40{,}000$ psi (276 N/mm^2), we have, required steel area,

$$A_s = \frac{M'}{f_y jd} = \frac{120{,}000 \times 12}{40{,}000 \times 0.87 \times 22}$$
$$= 1.85 \text{ sq in. } (1194 \text{ mm}^2)$$

The effect of the tendons may either increase or decrease the above value of A_s and should be considered in many instances.

9-6 Tendons Stressed at Low Level

In order to meet the requirement of high ultimate strength, it is sometimes necessary to use an unusually large tendon area. If the tendons are fully prestressed, the concrete will be subjected to high stresses, and excessive camber will occur. Thus it becomes necessary to stress the steel to a low level of stress (example 9-3) or to stress only a portion of the steel and leave the remainder unstressed. The first alternative, when carried too far, could end up in a high percentage of loss of prestress, while the second alternative may not enable the unstressed steel to develop its ultimate stress. Both methods, however, are useful for certain conditions.

EXAMPLE 9-3

A pretensioned beam, with a section as shown, Fig. 9-13, is to be designed to resist a high ultimate moment of 1600 k-ft. Compute the amount of steel and determine its level of prestress. $M_G = 200$ k-ft. $I = 73{,}800$ in.^{4}, $A = 680$ in.^{2} (Ultimate moment = 2170 kN–m, $M_G = 271$ kN–m, $I = 30.7 \times 10^9$ mm^4, and $A = 439 \times 10^3$ mm^2)

Solution Locate the c.g.s. at 5 in. (127 mm) above the bottom fiber, and assume a lever arm of $a = 28$ in. (711 mm). We have for steel stressed to 230 ksi (1,586 N/mm^2),

$$A_s = \frac{1600}{230} \times \frac{12}{28} = 2.98 \text{ sq in. } (1{,}923 \text{ mm}^2)$$

Suppose it is desired to limit the top fiber tension to 200 psi (1.38 N/mm^2) and the bottom fiber compression to 2400 psi (16.55 N/mm^2) at transfer; we can compute the maximum total prestress F:

$$f = \frac{F}{A} \pm \frac{Fec}{I} + \frac{Mc}{I}$$

For top fiber,

$$200 = \frac{-F}{680} + \frac{F \times 19.6 \times 11.4}{73{,}800} - \frac{200 \times 12{,}000 \times 11.4}{73{,}800}$$

$$200 = -0.00147F + 0.00303F - 372$$

$$F = \frac{572}{0.00156} = 367{,}000 \text{ lb } (1632\ kN)$$

For bottom fiber,

$$-2400 = \frac{-F}{680} - \frac{F \times 19.6 \times 24.6}{73{,}800} + \frac{200 \times 12{,}000 \times 24.6}{73{,}800}$$

$$= -0.00147F - 0.00655F + 800$$

$$F = \frac{3200}{0.00802} = 399{,}000 \text{ lb } (1775 \text{ kN})$$

Top fiber controls, $F = 367$ k (1632 kN).

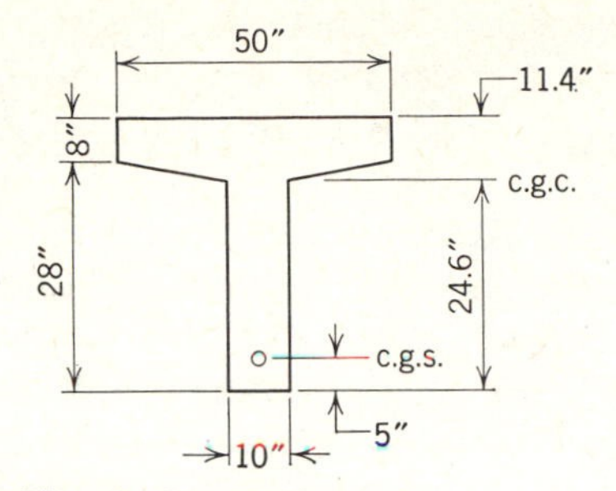

Fig. 9-13. Example 9-3.

For $A_s = 2.98$ in.^{2} (1923 mm^2),

$$f_s = \frac{F}{A_s} = \frac{367}{2.98} = 123 \text{ ksi } (848 \text{ N/mm}^2)$$

Hence it is only permissible to stress the steel to 123 ksi (848 N/mm^2) at transfer, instead of the usual 175 ksi (1207 N/mm^2). If the loss of prestress is 35 ksi (241 N/mm^2), the percentage loss will be 35/123 = 28.4%.

9-7 Combination of Prestressed and Reinforced Concrete

While a combination of prestressed and reinforced concrete is evidenced in the use of nonprestressed reinforcement, the flexural strength is essentially supplied by the tendons, with the nonprestressed steel playing only a minor role. For certain types of construction, a full combination of prestressed and reinforced concrete could be the best design, making use of the advantages of both. Reinforced concrete has the advantage of simplicity in construction, monolithic in behavior, no camber, less creep, and resonably high ultimate strength. Prestressed concrete utilizes high-strength steel economically, produces a favorable distribution of stress under certain conditions of loading, and controls deflection and cracking.

Certain structural elements and systems would favor pure reinforced concrete; others would favor pure prestressed concrete; still others, partially prestressed concrete. But some will be best designed with a combination of reinforced and prestressed concrete having the nonprestressed steel carrying perhaps 50% or more of the total ultimate load.

One occasion for the economical use of this combination would be the case of high live load to dead load ratio, when prestressing alone could produce excessive camber. Another case would be high added dead load requiring prestressing in stages which may be cumbersome. A third case would be the requirement of high ultimate strength to resist heavy blast loadings.

There is also reason to believe that use of nonprestressed steel in conjunction with unbonded tendons will result in economy and in developing a stress in the tendons at ultimate perhaps higher than that allowed by the ACI Code for unbonded beams. At least a minimum amount of unbonded reinforcement must be provided with unbonded tendons to distribute cracking. Tests show such beams behave well.[4]

For precast columns, prestressing will help control cracking during transportation and erection; it will contribute to the bending strength. Nonprestressed steel will increase both the axial load and the flexural capacity. Hence a combination may be the best solution for certain cases. The use of nonprestressed reinforcement for joineries and continuity is, of course, frequently a simple and economical solution.

There are new problems involved in this combination. First of all, shrinkage and creep due to prestressing will put the nonprestressed steel into compression. The nonprestressed steel does not act until the concrete cracks, and does not contribute toward the precracking strength. Hence if cracking could result in a primary or a secondary failure, nonprestressed steel may not be of any help. Another problem is the possibility of corrosion of the prestressing steel if the member cracks too early or too often. However, it is clear that this field has not been explored and may prove to be a fertile one, especially when different grades of steel and types of tendons are considered in combination.

References

1. P. W. Abeles, "The Use of High Strength Steel in Ordinary Reinforced and Prestressed Concrete Beams," *Preliminary Publications, Fourth Congress, Int. Assn. for Bridge and Structural Engg.*, 1952. Also supplement to above, *Final Report, Fourth Congress*, 1953.
2. P. W. Abeles, "Static and Fatigue Tests on Partially Prestressed Concrete Constructions," *J. Am. Conc. Inst.*, December 1954 (*Proc.*, Vol. 50), pp. 361–376.
3. N. H. Burns, "Moment-Curvature Relationships for Partially Prestressed Concrete Beams," *J. Prestressed Conc. Inst.*, Vol. 9 No. 1, February 1964, pp. 52–63.
4. N. H. Burns, and D. W. Pierce, "Strength and Behavior of Prestressed Concrete Beams with Unbonded Tendons," *J. Prestressed Conc. Inst.*, Vol. 12 No. 5, October 1967, pp. 15–29.
5. P. W. Abeles, "The Practical Application of Prestressing", IABSE Final Report, 1968.
6. T. Y. Lin, "Partial Prestressing Design Philosophies", FIP Notes 69, July-August 1977, pp. 5–9.
7. Fritz Leonhardt, "Recommendations for the Degree of Prestressing in Prestressed Concrete Structures", FIP Notes 69, July-August, 1977 pp. 9–14.
8. A. Naaman, "Serviceability Based Design of Partially Prestressed Beams," *J. Prestressed Conc. Inst.*, Vol. 24 No. 2, March/April 1979, pp. 64–89.

10

CONTINUOUS BEAMS

10-1 Continuity, Pros and Cons

A simple comparison between the strength of simply supported and a continuous beam will demonstrate the basic economy inherent in continuous construction of prestressed concrete. Consider a simple prestressed beam loaded with a uniformly distributed load of intensity w', Fig. 10-1(a). The total w' that can be ultimately carried by the beam is determined by the ultimate moment capacity of the midspan section. If the ultimate tension developed in the tendon is T', acting with a lever arm a', then the ultimate resisting moment at midspan is $T'a'$. With one half of the span as a freebody, Fig. 10-1(b), and taking moment about the left support, we have

$$\frac{w'L^2}{8} = T'a'$$

$$w' = \frac{8T'a'}{L^2} \tag{10-1}$$

The moment diagram produced by the load w' is shown in Fig. 10-1(c). It should be noted that the ultimate load w' carried by the beam is controlled by the capacity of the midspan section and cannot be increased by any change in the end eccentricities of the c.g.s.

Now consider the intermediate span of continuous beam, Fig. 10-2(a), with the same section, same span, and same prestressing steel as the simple beam of Fig. 10-1(a). Again with one half of the span as a freebody, Fig. 10-2(b), and taking moment about the left support, we have

$$\frac{w_c'L^2}{8} = 2T'a'$$

$$w_c' = \frac{16T'a'}{L^2} \tag{10-2}$$

noting that there are two resisting moments, one at midspan and another over the support. Hence the load-carrying capacity is definitely affected by the position of c.g.s. over the intermediate support. The moment diagram produced by the load w' is now plotted in Fig. 10-2(c).

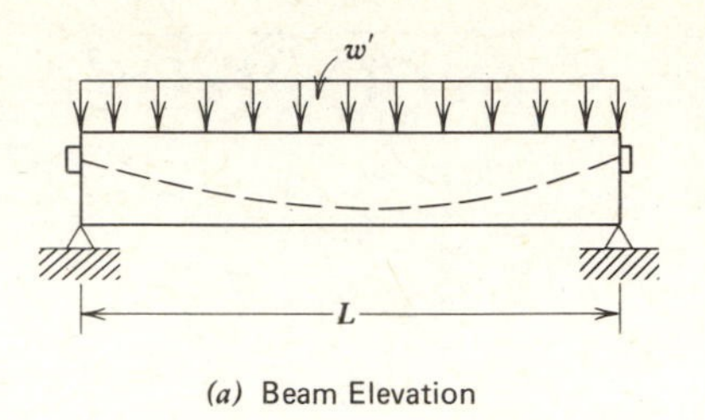

(a) Beam Elevation

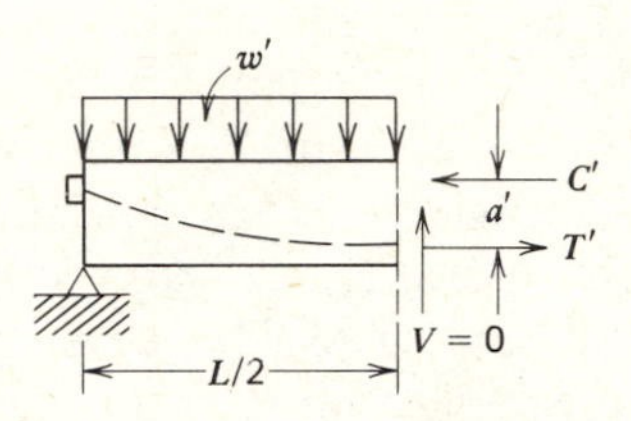

(b) Freebody of Half Span

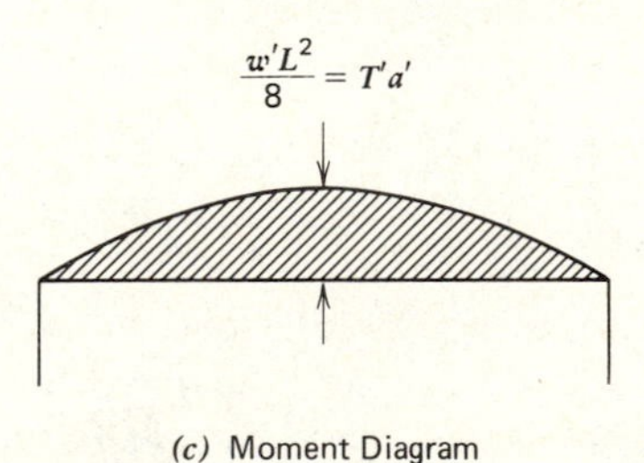

(c) Moment Diagram

Fig. 10-1. Load-carrying capacity of a simple beam.

Comparing Fig. 10-1(c) with Fig. 10-2(c), or equation 10-1 with equation 10-2, it is readily seen that $w_c' = 2w'$. This means that twice the load on the simple span can be carried by the continuous span for the same amount of concrete and steel. This represents a very significant basic economy that should be realized by engineers designing prestressed concrete structures. Because of this strength inherent in continuous construction, it is possible to employ smaller concrete sections for the same load and span, thus reducing the dead weight of the structure and attaining all the resulting economies.

Although it is generally conceded that continuity is economical in reinforced concrete, it is seldom known that, from certain points of view, even greater economy can be attained in prestressed construction. In reinforced concrete, the negative steel often laps with the positive steel bars, and both sets of bars are extended for additional anchorage, thus canceling some of the economy of

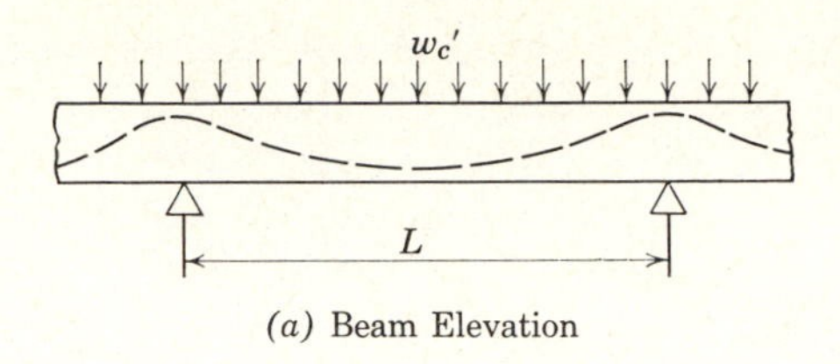

(a) Beam Elevation

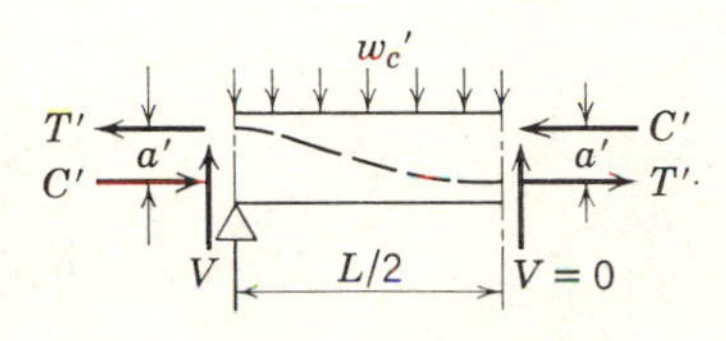

(b) Freebody of Half Span

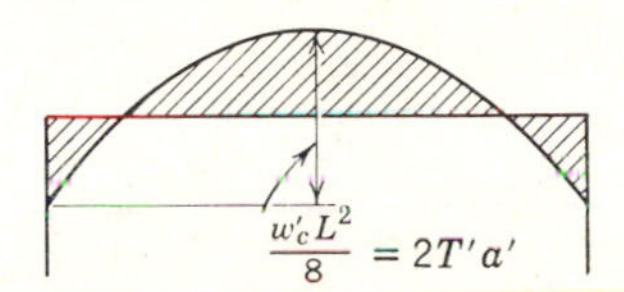

(c) Moment Diagram

Fig. 10-2. Load-carrying capacity of a continuous beam.

continuity. In prestressed concrete, the same cable for the $+M$ is bent over the other side to resist the $-M$, with no loss of overlapping. In addition, continuity in prestressed concrete saves end anchorages otherwise required over the intermediate supports, thus resulting in further economy and convenience.

The above discussion refers to the ultimate capacity of continuous beams, but the same general principles hold true within the elastic range. For both the elastic and the plastic ranges, with one half of the beam a freebody, there are two resisting moments in a continuous beam, but only one in a simple beam. Within the elastic range, however, the two resisting moments acting on the beam may not have equal capacity (allowable moment). Then one of the moments will control the design stresses, and the elastic resisting capacity of the continuous may not be twice that of the simple beam.

Economical design of continuous prestressed beams can be achieved in several ways. Owing to the variation of moment along the beam, the concrete section and the amount of steel are often varied accordingly. The peaks of the negative moments can be reinforced with nonprestressed steel, thus reducing the amount of prestressing steel. Advantage can be taken of the redundant reactions to obtain favorable lines of pressure in the concrete, which will be discussed in sections 10-4 and 10-5. Designs can be based on the ultimate strength of such beams, applying the principles of limit design. Some of these, however, are more delicate problems which should be handled with care.

It is perhaps unnecessary to add that, as is true with other continuous structures, the deflections will be less than comparable simple spans. Hence, for continuous spans, smaller depth is sufficient not only for strength but also for rigidity.

Like any type of construction, there are advantages and also shortcomings, which, under certain conditions, could outweigh the advantages. The choice of a particular type of design must be made after considering all the factors involved in the job. Disadvantages inherent in continuous prestressed concrete beams can be enumerated as follows.

1. Frictional loss in continuous tendons. This can be serious if there are many reversed curves, if the curves possess large deflection angles, or if the tendons are excessively long. Such loss can be minimized by using relatively straight cables in undulating or haunched beams. The usual methods of overtensioning, with stressing from both ends, can also be used to reduce frictional losses, as discussed in Chapter 4.
2. Shortening of long continuous beams under prestress. This may produce excessive lateral force and moments in the supporting columns, if they are rigidly connected to the beams during prestressing. Provisions are usually made to permit movement at the beam bearings or rocking of the columns.
3. Secondary stresses. Secondary stresses due to prestressing, creep and shrinkage effects, temperature changes, and settlements of supports could be serious for continuous structures unless they are controlled or allowed for in the design. One interesting point in continuous prestressed structures is that these secondary stresses can often be utilized to good advantage so that they will add to the economy of the structure.
4. Concurrence of maximum moment and shear over supports. It is believed that the concurrence of maximum moment and shear at the same section may decrease the ultimate capacity of a beam. This happens over the supports of most continuous beams. Hence care must be taken to reinforce such points properly for both shear and moment if high ultimate strength is desired. The elastic strength, however, is not affected by such concurrence.
5. Reversal of moments. If live loads are much heavier than dead load, and if partial loadings on the spans are considered, continuous beams can be subjected to serious reversal of moments. This can sometimes be overcome by proper design, such as extensive use of nonprestressed steel in combination with prestressed concrete.

6. Moment peaks. Peaks of maximum negative moments may sometimes control the number of tendons required for the entire length of the beam. These peaks, however, can be strengthened by employing deeper sections or by adding prestressed and nonprestressed reinforcements over the portions where they are needed. Moment redistribution at ultimate will keep this from being a serious problem, provided we use cross sections with sufficient ductility for the peak support moment to be shifted to the adjoining sections near midspan which have smaller moment.
7. Difficulty in achieving continuity for precast elements. It is easy and natural to obtain continuity for cast-in-place construction, but continuity for precast elements cannot always be achieved without special effort. On account of difficulties in handling precast continuous beams, they are often precast as simple elements, to be made continuous after they are erected in place.
8. Difficulty in designing. It is more difficult to design continuous than simple structures. But, with the development of simpler methods, the design of continuous prestressed concrete beams can be made into a more or less routine procedure applying basic principles for continuous structures familiar to most engineers. These methods will be presented in the following sections. The use of the load-balancing concept is a much more useful design approach for continuous beams than for simple beams as will be demonstrated in the following chapter.

10-2 Layouts for Continuous Beams

Several methods for providing continuity in prestressed concrete construction have been applied in practice.[1] These methods permit various layouts to be adopted, some of which are shown in Fig. 10-4 and 10-5 and will be described below. There are other methods and layouts that are perhaps less frequently used. Still other arrangements are being developed. But it is considered sufficient to present the more common methods, leaving the variability as well as the desirability of each to the judgment of the designer, who should of course take into account the particular conditions surrounding each structure when selecting his layout.

Continuous beams may be divided into two classes: fully continuous beams and partially continuous beams. For full continuity, all the tendons are prestressed in place and are generally continuous from one end to the other although some can be anchored at intermediate points if found desirable. The concrete may be either poured in place or made of blocks assembled on falsework. The tendons may be encased in the concrete during pouring, threaded through preformed holes, or placed outside the webs. They may be either bonded or unbonded, depending on the requirements of the structure.

Fig. 10-3. Precast members made continuous by coupling the tendons with high-tensile rods (button head wire system).

Precast elements can also be made fully continuous by coupling the tendons together with a high-tensile rod and then stressing one or both of the tendons. A typical case is shown in Fig. 10-3. Some other typical layouts for full continuity are shown in Fig. 10-4:

(a) In (*a*) is shown a straight beam with curved tendons, which follows in general the tensile side of the beam. This layout is often used for slabs or short-span beams, where simple formwork is more important than the saving of steel and concrete. The main objections here are the heavy fricitional loss and the difficulty of threading the tendons through when they are continuous over several spans.

(b) For longer spans and heavier loads, it will be more economical to haunch

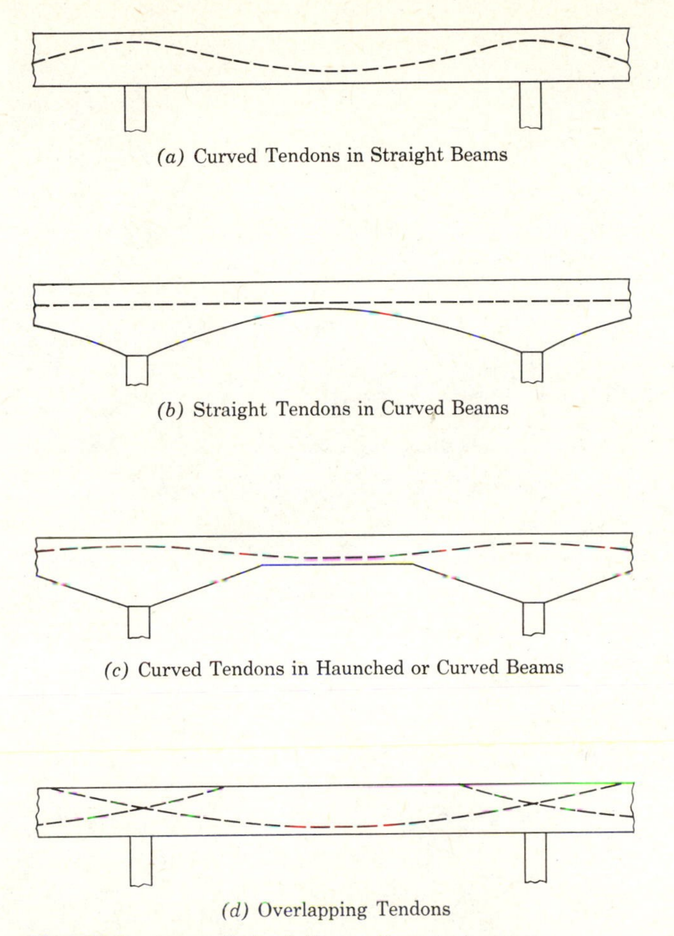

(a) Curved Tendons in Straight Beams

(b) Straight Tendons in Curved Beams

(c) Curved Tendons in Haunched or Curved Beams

(d) Overlapping Tendons

Fig. 10-4. Layouts for fully continuous beams.

or curve the beams, as in (*b*). This will not only save concrete and steel but also permit the use of straight tendons, likewise positioned on the tensile side of the beam. However, it is often diffcult to get the optimum eccentricities all along the beam if the tendons are to remain entirely straight.

(c) The best layout is often obtained with a compromise of the above two arrangements, using curved beams and slightly curved tendons at the same time, as in (*c*). This would permit optimum depth of beam as well as ideal position of steel at all points, while avoiding excessive frictional loss.

(d) Cables protruding at intermediate points, as in (*d*), offer a possibility of varying prestressing force along the beam. The arrangement shown here has no reversed curves in the tendons, so that heavier cables and rods can be more easily threaded through and stressed with less frictional loss. This

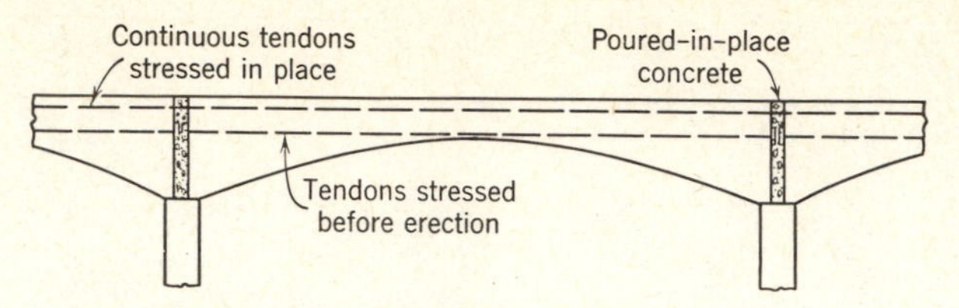

(a) Continuous Tendons Stressed after Erection

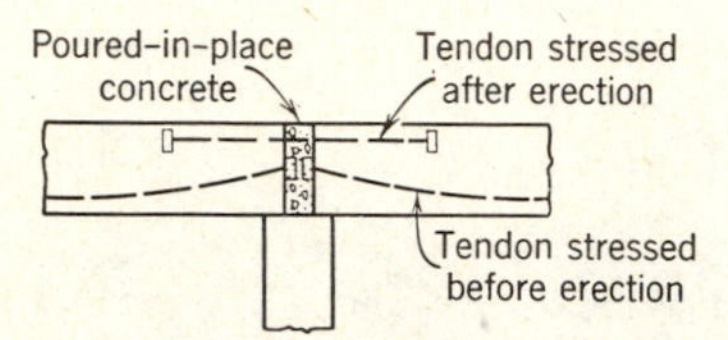

(b) Short Tendons Stressed over Supports

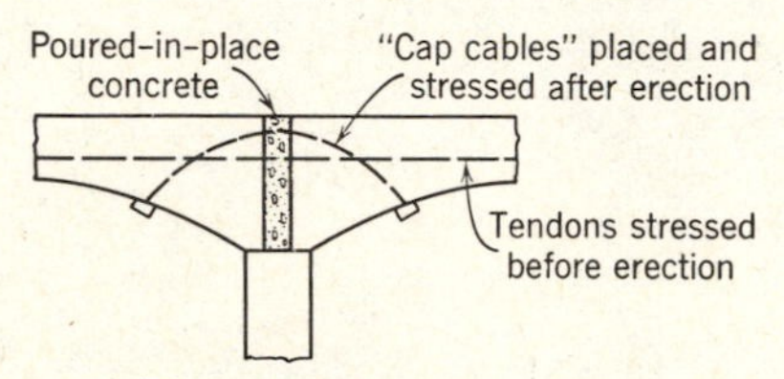

(c) Cap Cables over Supports

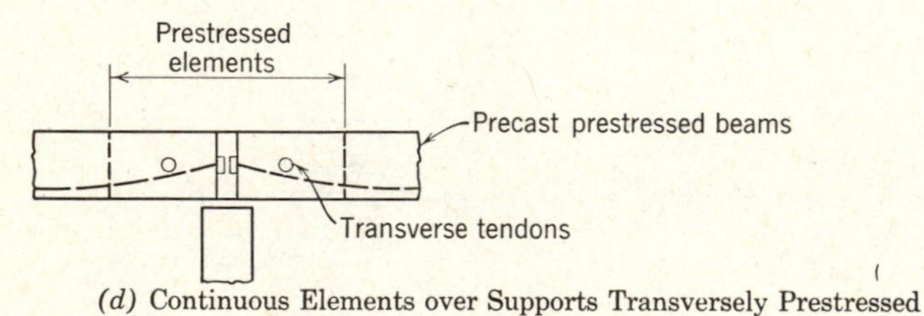

(d) Continuous Elements over Supports Transversely Prestressed

Temporary anchorage
Coupler
Tendons being prestressed
Jack
Beam with tendons prestressed
Poured-in-place concrete
Next beam

(e) Couplers over Supports

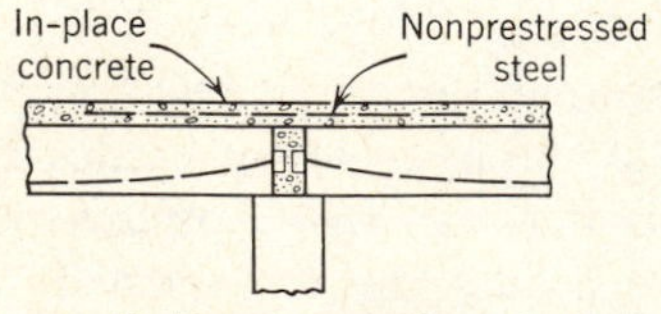

(f) Nonprestressed Steel over Supports

Fig. 10-5. Layouts for partially continuous beams.

arrangement requires more end anchorages to accomplish the posttensioning of a multiple span beam.

For partial continuity, each span can be precast as a simple beam, using a sufficient amount of prestressed steel for handling and erection. Generally, no falsework is required for erection. Concrete at the support section is cast-in-place after the precast beams are erected. Several possible details are shown in Fig. 10-5, for such partially continuous beams. Sometimes, additional elements, prestressed longitudinally or transversely, are inserted to provide continuity over the support, Fig. 10-5(*b*), (*c*), (*d*).

(a) In (*a*) are shown continuous prestressed tendons placed in conduits or grooves left in the structure. After erection, concrete is poured between the beams over the supports. When the concrete hardens, the continuous cables can be stressed to provide continuity. The construction is relatively simple, but economy in steel cannot be easily attained since the same cable area is provided throughout the entire length, whether needed or not.

(b) In (*b*) is shown a layout similar to (*a*), but with continuous tendons placed over the supports only. This saves steel but requires more anchorages than the first layout. Moreover, the anchorages are located at intermediate points and are more difficult to tension.

(c) Another method of supplying continuity over the supports is to add the so-called cap cables, as in (*c*). These tendons, usually made of wires or strands, can be conveniently stressed from the soffit of the beam, but they possess an appreciable curvature and hence corresponding frictional loss. It is not possible to thread big rods through the holes unless the profile is made into a circular curve and the bars are prebent to a definite curvature.

(d) Still another way to provide continuity is to insert tensile elements over the supports, as in (*d*), and to attach them to the precast beams by transverse prestressing, which supplies a sort of bolt action clamping the elements together. These tying elements can be made of reinforced- or prestressed-concrete planks, and can be either precast or poured-in-place. Sometimes, the precast elements themselves can be cantilevered at the ends so that they overlap over the supports and transverse prestress is applied to hold them together, thus making them continuous under live load.

(e) Especially applicable to high-tensile bars, but also to other forms of tendons, is the use of couplers as a means of obtaining continuity, (*e*). This permits the stressing of tendons one span at a time, thus minimizing the frictional loss encountered when prestressing tendons running through several continuous spans. Suppose we erect the beams in (*e*) successively

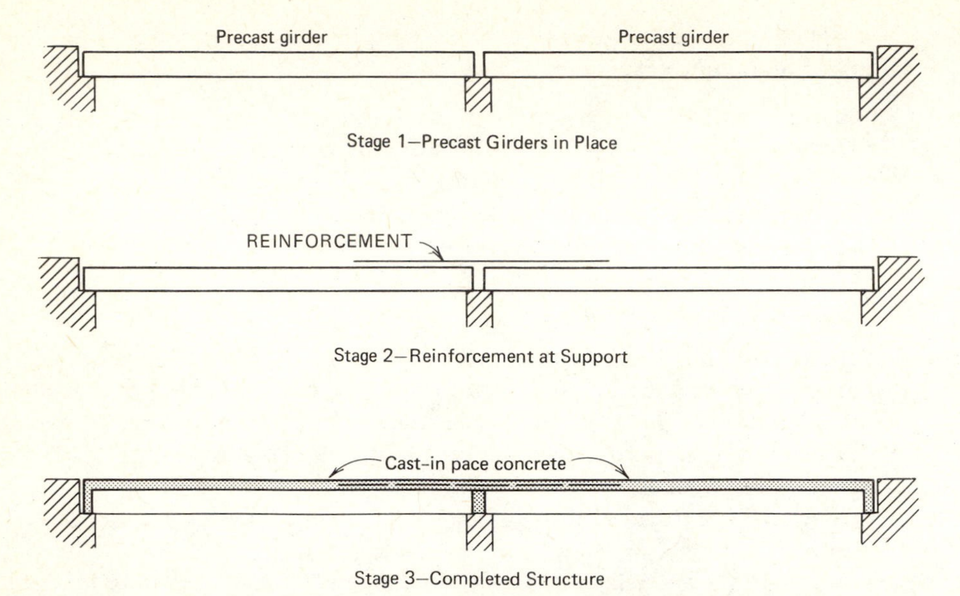

Fig. 10-5(*g*). Sequence of construction for highway bridge using detail (*g*) and detail of Figure 10-6.

from left to right. After one beam is fully prestressed, the next beam is erected and its unstressed bar is connected to the stressed bar of the previous beam by a coupler. Then a jack is applied to the right end for tensioning. This method is also applicable to cast-in-place beams, provided the sequence of construction permits the insertion of jacks.

(f) Precast elements can be conveniently made continuous for live load by placing nonprestressed steel over the support (*f*).[2,3] This is especially true for composite construction where a topping will be concreted in place. If dead load continuity is desired propping of the precast elements will be required prior to casting the composite slab. The sequence of construction for a bridge structure using this detail is shown in Fig. 10.5(*g*).

A partial continuity detail similar in general concept to that of Fig. 10-5(*f*) is shown in Fig. 10-6. This detail makes use of prestressed, precast slender rods, Fig. 10-6(*a*), placed in the composite slab at support section, similar to the detail of Fig. 10-5(*f*) with reinforcing bars, Fig. 10-5(*g*). The precast rods will have high stress locked into the strand from the pretensioned construction of the simple elements. To prevent end splitting, short lengths of spiral reinforcing must be included at the ends of the rods, and research has shown that the strand should be cut off at a distance of about 50 diameters beyond the end of the rod. This allows the protruding strand to be developed beyond the end of the precast element when the cast-in-place slab is added. It is possible to design the support

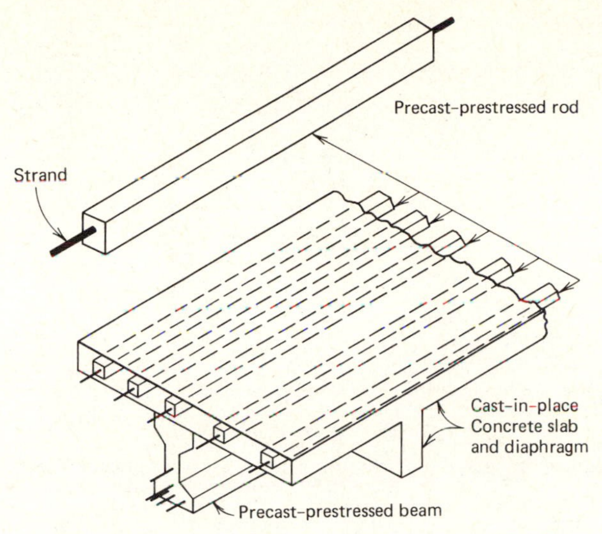

(*a*) Connection Detail with Precast Prestressed Rod Reinforcement

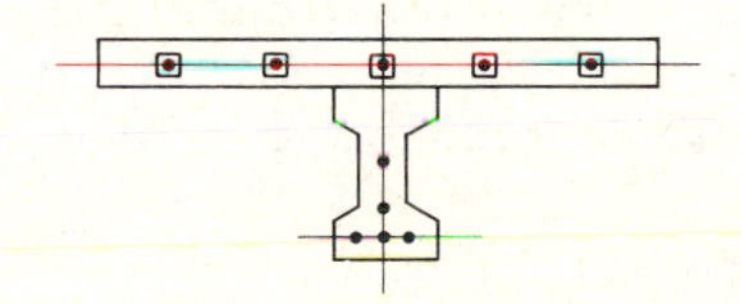

(*b*) Cross Section with Rod Reinforcement in Slab

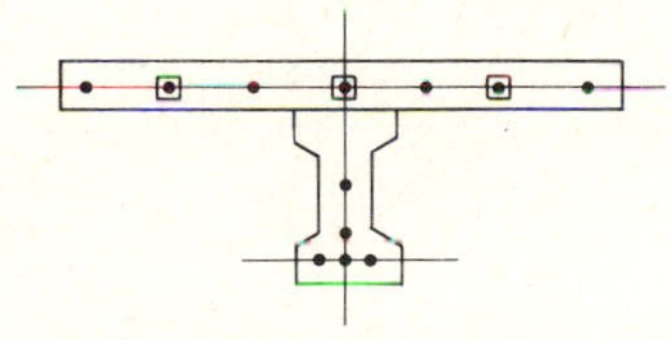

(*c*) Cross Section with Combination of Bars and Rods in Slab

Fig. 10-6. Composite structure continuity detail using precast-prestressed rods (reference 4).

continuity detail for a given required flexural strength using these precast, prestressed elements in the composite slab, Fig. 10-6(*b*), or a combination of these elements alternating with unstressed deformed bars, Fig. 10-6(*c*). Of course, one might develop the same ultimate moment capacity at support as using only bars with the detail of Fig. 10-5(*f*).

Tests of double cantilever type specimens were made to compare performance of the different details utilizing precast rods,[4] Fig. 10-7. Similar flexural strength

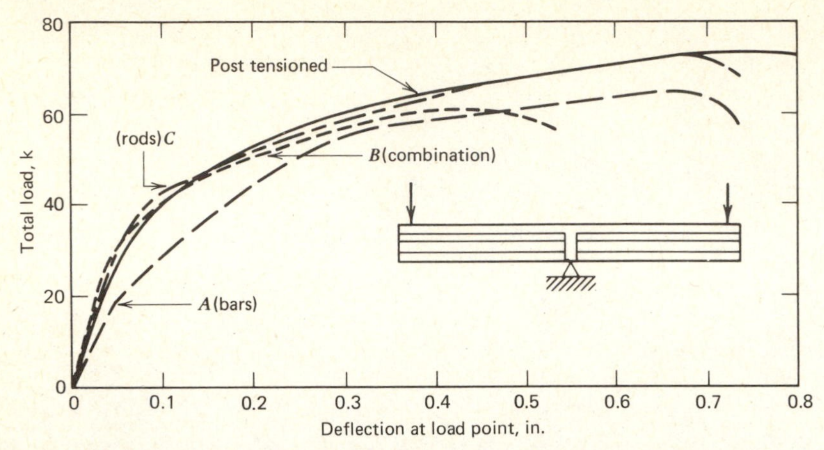

Fig. 10-7. Load-deflection curves for double cantilever specimens (reference 4).

can be attained by each of the combinations as shown. Use of the precast elements has the advantage of stiffening the support detail, gaining some of the benefits of prestressing but avoiding a posttensioning operation on the site. The concrete surrounding the strand in the precast elements is precompressed while the composite slab concrete is unstressed. Cracking is thus delayed by the rods which are bonded to the surrounding concrete, resulting in the stiffer performance as indicated in the tests. Also, the presence of high initial tension in the strand provides a positive force across cracks in the composite slab. Thus cracks which have developed under applied loading close almost completely when the load is removed. This technique of prestressing the strand in the precast element allows use of the high strength strand as reinforcement without the development of wide cracks as would be the case if the same material were used without initial stress. Other research[5] has shown similar performance of this type detail, and it has been quite sucessfully used in a bridge construction in Canada.[6]

10-3 Analysis, Elastic Theory

Tests on continuous prestressed-concrete beams have shown that the elastic theory can be applied with accuracy within the working range. Since there is little or no tensile stress in the beam under working loads, there are no cracks, and the beam behaves as a homogeneous elastic material, more so than an ordinary reinforced-concrete beam which usually is cracked in certain portions. By making proper allowance for shrinkage and creep, the elastic theory can be applied for all practical purposes to compute the deflections, strains, and stresses up to cracking. This is true for the effect of prestress as well as of dead and live loads.

The method of analysis presented here is based on the classical elastic theory. Fundamentally, the theory is applicable to all statically indeterminate structures of prestressed concrete, provided that consideration is given to the axial shortening effect in frames and similar structures. However, for simplicity, only fully continuous beams are referred to in the following, although most of the discussions apply to rigid frames, slabs, and partially continuous beams as well.

The analysis and design of continuous prestressed concrete structures are considered by most engineers to present a rather difficult problem. This is an erroneous impression. Undoubtedly, the design is more complicated than that of continuous reinforced-concrete structures or of statically determinate prestressed structures. But the basic theories involved are the same, and the additional complications are limited in nature. Hence a person who understands the analysis of statically indeterminate structures and the design of simple prestressed concrete beams can learn the design of continuous prestressed concrete beams with little difficulty.

Several methods are available for the analysis of prestressed continuous beams.[7, 8, 9] All of them are similar to those followed in the analysis of any statically indeterminate structures; they are all based on the same assumptions and yield the same answers. Hence only one method, which the author considers the simplest, will be discussed in this chapter. The method will be based on no involved mathematics, and only the following principles will be utilized.

1. Moment and shear diagrams for ordinary continuous beams.
2. Moment distribution method.[10, 11]
3. Location of line of pressure in a prestressed-concrete beam.

Before starting on this method of analysis, let us examine first the difference between a continuous prestressed beam and a simple one. Owing to the application of external loads, the moments in a bonded continuous prestressed beam are computed by the elastic theory, like any other type of statically indeterminate structure. In an unbonded beam, the effect of change in prestress due to beam curvature should be added, although the magnitude is usually small and can be neglected. Owing to the application of prestress, the moments in a continuous beam are directly affected by the prestress and indirectly by the support reactions induced by the bending of the beam. In a simple beam, or any other statically determinate beam, no support reactions can be induced by prestressing.

Consider a prestressed simple beam, Fig. 10-8(a). No matter how much the beam is prestressed, only the internal stresses will be affected by prestressing. The external reactions, being determined by statics, will depend on the dead and live load (including the weight of the beam), but are not affected by the prestress. Without load on the beam, no matter how we prestress the beam internally, the external reactions will be zero, hence the external moment will be

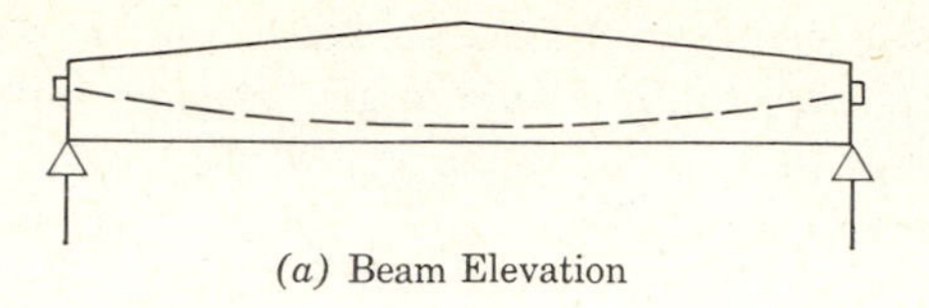

(a) Beam Elevation

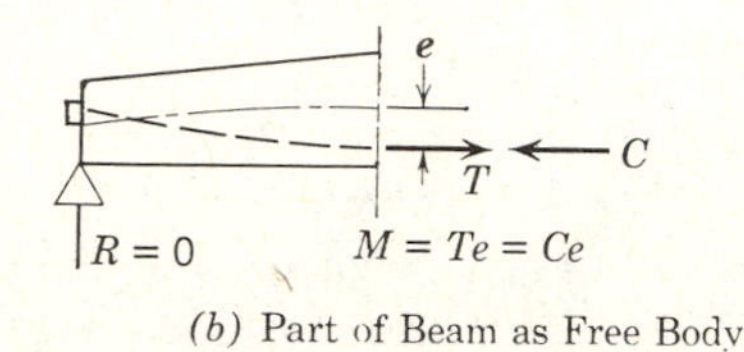

(b) Part of Beam as Free Body

Fig. 10-8. Moment in concrete due to prestressing in a simple beam.

zero. With no external moment on the beam, the internal resisting moment must be zero, hence the *C*-line (which is the line of pressure in the concrete) must coincide with the *T*-line in the steel (which is the c.g.s. line), as in (*b*). The *C*-line in the concrete being known, the moment in the concrete at any section can be determined by $M = Te = Ce$.

Next, let us consider a continuous prestressed-concrete beam, Fig. 10-9(*a*). When the beam is prestressed, it bends and deflects. The bending of the beam can be such that the beam will tend to deflect itself away from some of the supports, as in (*b*). If the beam is refrained from deflection at these supports, (*c*), reactions must be exerted on the beam to hold it there. Thus reactions are induced when a continuous beam is prestressed (unless, by intent or by chance, the prestress has no tendency to deflect the beam from any of its supports). These induced reactions produce moments in the beam, (*d*). To resist these moments, the *C*-line must be at a distance *a* from the *T*-line, (*e*), such that the internal resisting moment equals the external moment *M* caused by the reactions, that is,

$$a = \frac{M}{T}$$

Now, let us compare the simple beam with the continuous beam under the action of prestress, neglecting the weight of the beam and all other external loads. In the simple beam, the *C*-line coincides with the *T*-line. In the continuous beam, the *C*-line usually deviates from the *T*-line. In the simple beam, the stress distribution in the concrete at any section is given by the location of the *T*-line. In the continuous beam, it is given by the location of the *C*-line which does not coincide with the *T*-line. The difference between the two beams lies in the presence of external reactions and moments in the continuous beam, produced as a result of prestressing. Since the external moment is solely

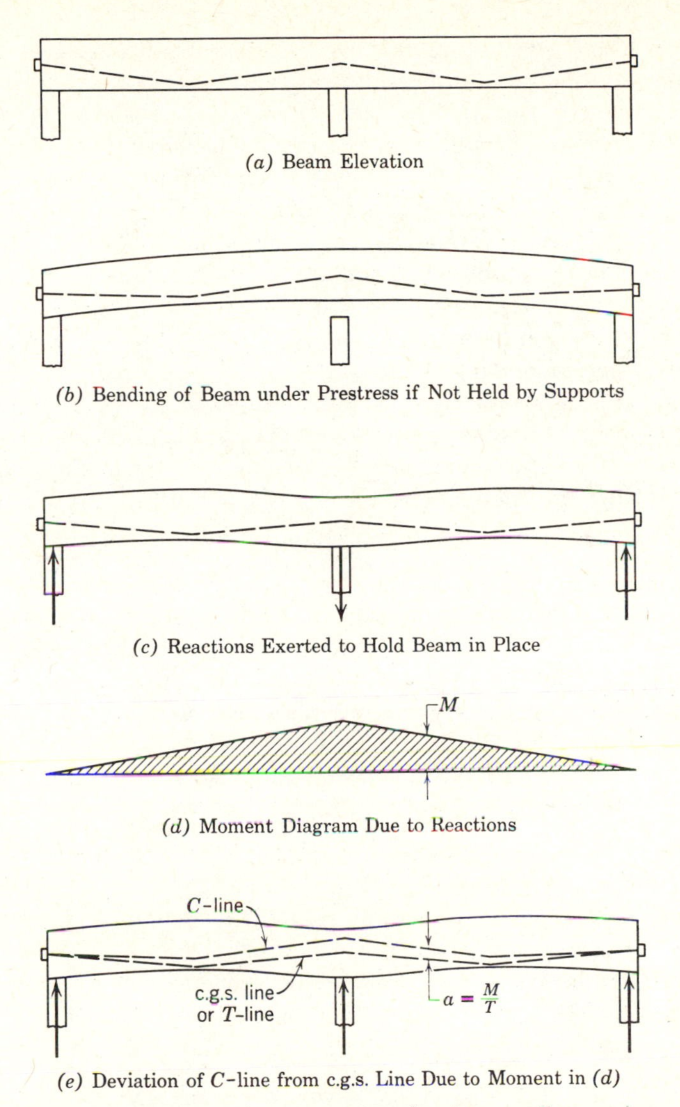

(a) Beam Elevation

(b) Bending of Beam under Prestress if Not Held by Supports

(c) Reactions Exerted to Hold Beam in Place

(d) Moment Diagram Due to Reactions

(e) Deviation of C-line from c.g.s. Line Due to Moment in (d)

Fig. 10-9. Moment in concrete due to prestressing in a continuous beam.

produced by the reactions, and since the reactions are only applied at the supports, the variation of moment between any two consecutive supports is a linear one. If T remains constant between the supports, then the deviation a, being directly proportional to M, also has to vary linearly, Fig. 10-9.

From another point of view, the difference between a simple and a continuous beam under prestress can be represented by the existence of "secondary moments." Once these moments over the supports are determined, they can be interpolated for any point along the beam. These moments are called secondary because they are by-products of prestressing and because they do not exist in a statically determinate beam. The term "secondary" is somewhat misleading, since sometimes the moments are not secondary in magnitude but play a most important part in the stresses and strength of the beam.

From this same point of view, the moment in the concrete given by the eccentricity of the prestress is designated as the primary moment, such as would exist if the beam were simple. With known primary moment acting on a continuous beam, the secondary moments caused by the induced reactions can be computed. The resulting moment due to prestress, then, is the algebraic sum of the primary and secondary moments.

The following gives a procedure for computing directly the resulting moments in the concrete sections over the supports, based on the moment distribution method. Once the resulting moments are obtained, the secondary moments can be computed from the relation

$$\text{Secondary moment} + \text{Primary moment} = \text{Resulting moment}$$

It is also possible to consider some reactions as redundant and solve for the values required to produce zero deflections at the supports. This can be done by the classical method of redundant reactions, and sometimes may be simple for a single redundancy. But this method and others will not be discussed here.

Before going any further, it would be well to summarize first the assumptions made for our method of design and analysis. These are the usual assumptions made for continuous prestressed-concrete beams, and their effects on the computed values are known to be negligible in most cases.

1. The eccentricities of the prestressing cables are small compared to the length of the members.
2. Frictional loss of prestress is negligible (where frictional loss is appreciable, it should be taken into account).
3. The same tendons run through the entire length of the member (varying steel areas can be included with some modifications, which will be evident to the designer once he learns the basic procedure presented herein).

As a result of the above assumptions, analysis can be made on the following bases.

1. The axial component of the prestress is constant for the member and is equal to the prestressing force F.
2. The primary moment M_1 at any section in the concrete is given by

   $$M_1 = Fe_1$$

   where e_1 is the eccentricity of the c.g.s. with respect to c.g.c.

On these bases, the procedure of analysis can be formulated as follows.

First treat the entire beam as if it had no supports. Plot the moment diagram for the concrete produced by the eccentricity of prestress. Compute the loading on the beam corresponding to that moment diagram; this is the loading produced by the steel on the concrete. Now, with this loading acting on the continuous beam as it is actually supported, compute the resulting moment by moment distribution or other similar method. Referring to Fig. 10-10 the various steps will be further outlined.

1. Plot the primary moment diagram for the entire continuous beam, as produced only by prestress eccentricity, as if there were no supports to the

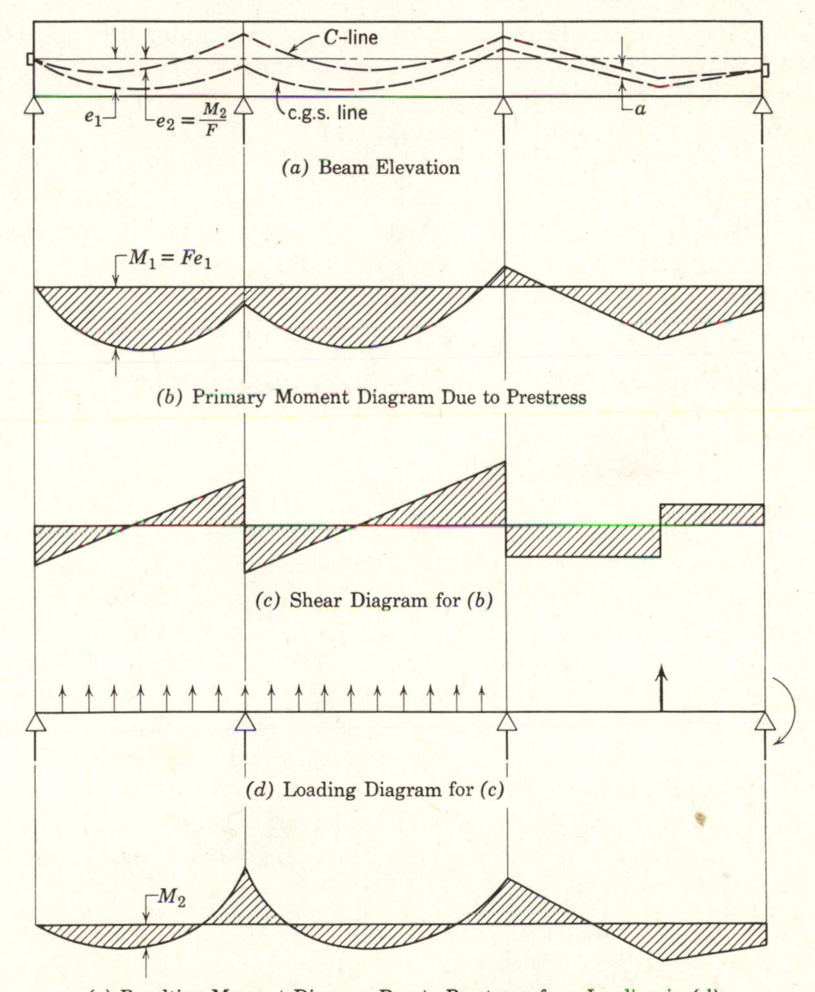

Fig. 10-10. Computation of moments due to prestress in continuous beam.

beam. This is simply given by the eccentricity curve plotted to some suitable scale, as in (b), since $M_1 = Fe_1$, and F is a constant.

2. From the above moment diagram, plot the shear diagram corresponding to it, (c). This can be done either graphically or algebraically.
3. From the above shear diagram plot the loading diagram corresponding to it, (d). This can also be done either graphically or algebraically.
4. Now, for the loading obtained above acting on the continuous beam with the actual supports, and including any singular moments such as might occur at the ends of the beam due to the eccentricity of c.g.s., compute the resulting moments M_2 by moment distribution, (e).
5. The C-line in (a) is now obtained by linearly transforming the c.g.s. line so that it will have new eccentricities e_2 over the supports corresponding to the resulting moments M_2, thus,

$$e_2 = M_2 / F$$

Since the C-line deviates linearly from the c.g.s. line, it will have the same intrinsic shape as the c.g.s. line, and can be easily plotted. The secondary moment is represented by the deviation between the C-line and the c.g.s. line. If desired, it can be computed by the simple relation

$$\text{Secondary moment} = M_2 - M_1$$

and the deviation a of the C-line from the c.g.s. line, Fig. 10-10(a), is given by

$$a = \frac{M_2 - M_1}{F}$$

Note that the above procedure involves only principles familiar to the engineer except perhaps the plotting of loading and shear diagram from given moment diagrams. While engineers can plot shear from loading diagrams and moment from shear diagrams, which is essentially a process of integration, most are not familiar with the reverse of the process, plotting shear from moment diagrams and loading from shear diagrams, which is essentially a process of differentiation. However, with a little experience, the art can be easily mastered. In fact, very often it is not necessary to plot the primary moment and shear diagrams since the loading diagram can be obtained directly from the known tendon layout and prestress force.

In order to facilitate the plotting of loading diagrams directly from the tendon layout and prestress (yielding the known moment diagrams), the following hints are given for reference, Fig. 10.11.

1. At the end of the tendons, the force F from the tendons on the concrete can be resolved into three components:
   a. An axial force, $F \cos \theta_1 = F$(since $\cos \theta = 1$), acting at the end of the anchorage. This usually has no effect on the bending moment in a

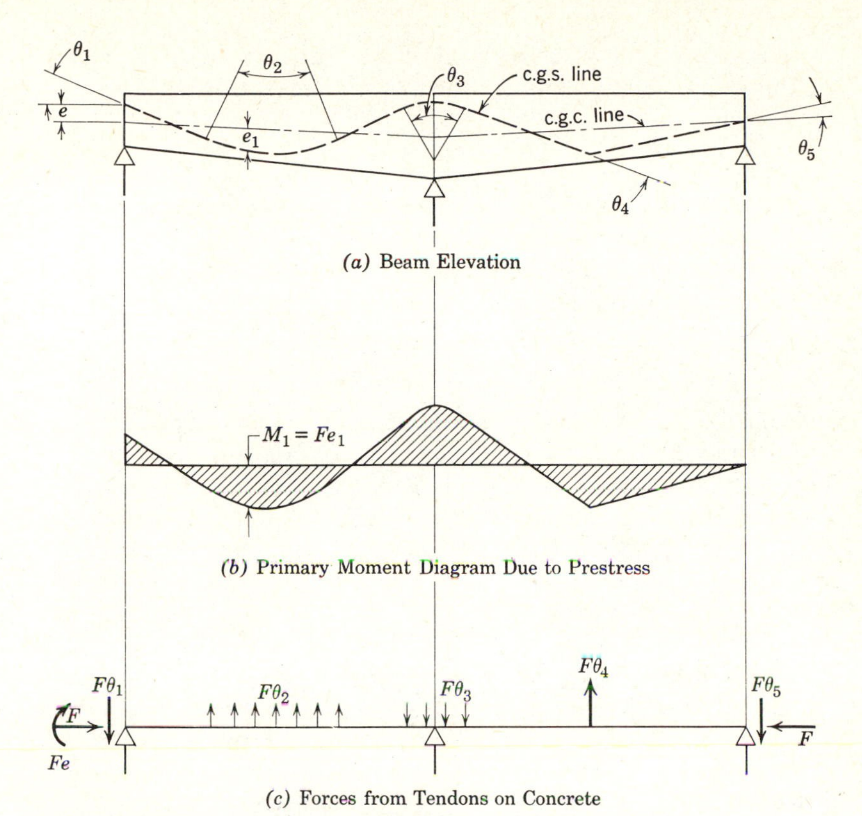

Fig. 10-11. Obtaining loading diagrams due to prestress.

continuous beam but may produce moments in a rigid frame, owing to the axial shortening effect.

b. A transverse force, $F \sin \theta_1 = F\theta_1 = F \tan \theta$, applied at the support and balanced by the vertical reaction from the support directly beneath. This again produces no moment in a continuous beam, unless it is applied away from the support. Its effect in a rigid frame will be small.

c. A moment, $F \cos \theta_1 e = Fe$, acting at the end of the beam. This will produce moments along the entire length of the continuous beam, and it must be included when following the moment distribution procedure.

2. Along the span of a member, where the c.g.s. or the c.g.c. line of the member bends and curves, transverse loads are applied to the concrete. Two common cases can be considered:

   a. When the moment diagram takes the shape of a parabolic or a circular curve (note: owing to the assumption of flat curvature, parabolic and circular curves are considered to have the same effect in producing transverse loads), a uniformly distributed load is applied to the concrete along the length of the curve. The total force for each curve is given by the change in slope between the two end tangents; thus the total force at θ_2 is given by

$$W = F \sin \theta_2 = F\theta_2$$

For practical purposes, the load W can be considered as uniformly distributed along the length of the curve.

b. When the moment diagram changes direction sharply, the force can be considered as concentrated at one point; the amount, at θ_4, for example, is

$$F \sin \theta_4 = F\theta_4$$

3. Over the interior supports, where the moment diagram changes direction, a load is applied directly over these supports. Again two cases can be considered:
   a. If the moment diagram curves gradually over the support, again a uniformly distributed load is applied, as shown for θ_3. This will affect moments in the beam, and the load must be considered in performing the moment distribution.
   b. If the moment diagram is bent abruptly over the supports, a concentrated load is applied thereon. Such a concentrated load is directly reacted by the support underneath and produces no moments on the beam. It can be neglected in performing the moment distribution.

Having computed the loads on the concrete, Fig. 10-11(c), we can proceed to determine the bending moments in the concrete, as for any continuous beam. This can be done by any method, but only the moment distribution method will be followed here. The application of moment distribution, which is also based on the elastic theory, necessarily involves other assumptions, such as the validity of Hooke's law, the principle of superposition, and linear variation of strain along the depth of a beam. Although these assumptions may not be exactly correct, the method has been considered accurate enough for reinforced concrete; because of the absence of cracks in prestressed concrete under working loads, the method can be applied with great precision and is considered sufficiently accurate for purpose of design.

The application of the above method, together with moment distribution, will be illustrated by example 10-1. Two points should be noted in the example. First, the example assumes F is constant while the effect of change in prestress in the tendons due to beam curvature should be considered (although the effect is generally small, as previously mentioned). Next, a beam with a straight c.g.c. line is illustrated. Should the beam posses a curved or bent axis, it is only necessary to plot the primary moment diagram by measuring the c.g.s. eccentricity from the curved or bent c.g.c. line instead of from a straight base line. Gross section is used for I, neglecting any refinement for transformed or net section which might be more exact for bonded or unbonded tendons, respectively. This effect can be shown to be negligable for usual design problems.

EXAMPLE 10-1

A continuous prestressed-concrete beam with bonded tendons is shown in Fig. 10-12(*a*). The c.g.s. has an eccentricity at *A*, is bent sharply at *D* and *B*, and has a parabolic curve for the span *BC*. Locate the line of pressure (the *C*-line) in the concrete due to prestress alone, not considering the dead load of the beam. Consider a prestress of 250 k. (1112 kN)

Solution The primary moment diagram for the concrete is shown in (*b*). The corresponding shear diagram is computed and shown in (*c*), from which the loading diagram is drawn in (*d*). For the loading in (*d*) acting on the continuous beam, the fixed-end moments are computed: Span *AB* at *A*, in addition to 50 k-ft (68 kN-m) singular moment, we have

$$\frac{20\times 20^2\times 30}{50^2} = +96 \text{ k-ft } (+130 \text{ kN-m})$$

at *B*,

$$\frac{20\times 30^2\times 20}{50^2} = -144 \text{ k-ft } (-195 \text{ kN-m})$$

Span *BC*,

$$\frac{0.88\times 50^2}{12} = \pm 183 \text{ k-ft } (\pm 248 \text{ kN-m}) \text{ at } B \text{ and } C$$

Moment distribution is performed in (*e*). The exterior end moments are first distributed, -96 and $+183$ k-ft (-130 and $+248$ kN-m) being obtained. Together with the eccentric moment of -50 k-ft (68 kN-m) at *A*, these are carried over to *B*, obtaining -73 and $+92$ k-ft (-99 and $+125$ kN-m). Now the total unbalanced moment of $+58$ k-ft ($+79$ kN-m) at *B* is distributed, obtaining -29 k-ft (-39 kN-m) for each span. The resulting moment is 246 k-ft (334 kN-m) at *B*. The eccentricity of the line of pressure at *B* is, then,

$$246/250 = 0.98 \text{ ft } (0.299 \text{ m})$$

The line of pressure for the entire beam can be computed by plotting its moment diagram and dividing the ordinates by the value of the prestress. But this is not necessary; since the line of pressure deviates only linearly from the c.g.s. line, it is only necessary to move the c.g.s. line linearly so that it will pass through the points located over the supports, (*f*). (This is known as linear transformation and will be discussed more fully in the next section.) Thus the line of pressure at *D* will be translated upward by the amount of

$$(0.98-0.4)30/50 = 0.35 \text{ ft } (0.107 \text{ m})$$

and is now located at $0.80-0.35=0.45$ ft (0.137 m) below the c.g.c. line. At midspan of *BC*, the line of pressure will be translated upward by the amount of $(0.98-0.4)25/50=$ 0.29 ft (0.088 m) and is now located at 0.61 ft (0.186 m) below the c.g.c. line.

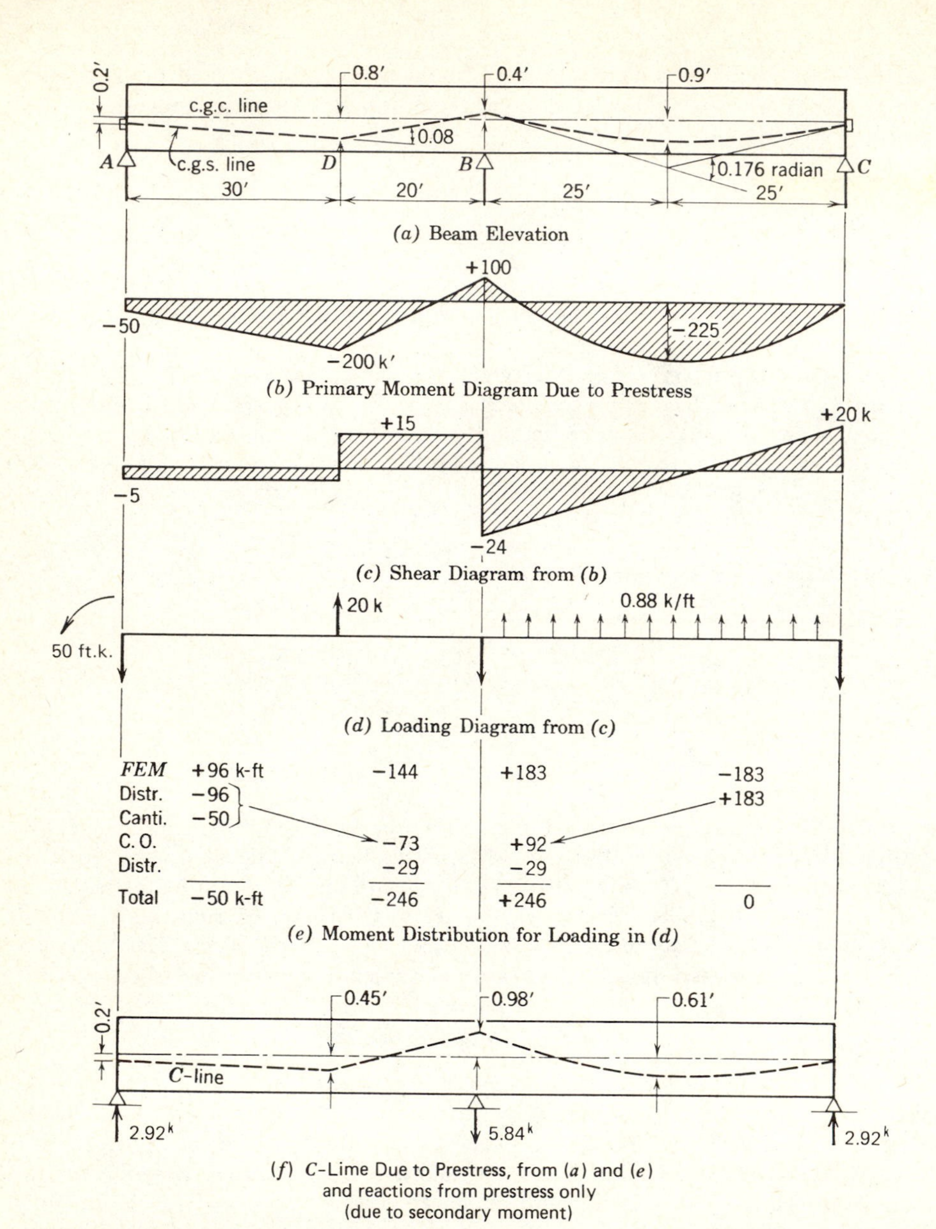

Fig. 10-12. Example 10-1.

As an exercise, the reader may plot the entire moment diagram for the continuous beam and divide it by the amount of prestress to obtain the line of pressure. Of course, it should check exactly with the line here obtained.

If desired, the secondary moment over the center support can be computed as

$$M_2 - M_1 = 246 - 100$$
$$= 146 \text{ k-ft } (198 \text{ kN-m})$$

For the beam of example 10-1. we might also consider the reactions which accompany the loadings from prestress alone. Figure 10-13(a) shows the "reactions" due to primary moment. The loads were previously obtained in Fig. 10-12, replacing the tendon with loads acting on the concrete free body. Note that at A we have an internal force of 5 k (22 kN) acting downward from primary moment as shown in Fig. 10-13(a). This agrees exactly with the -5 k (-22 kN) shear at A shown on the primary shear diagram of Fig. 10-12(c). The solution of moment due to prestress in example 10-1 indicated that the secondary moment was 146 ft-k (198 kN-m) at B, resulting in the free body diagram shown in Fig.

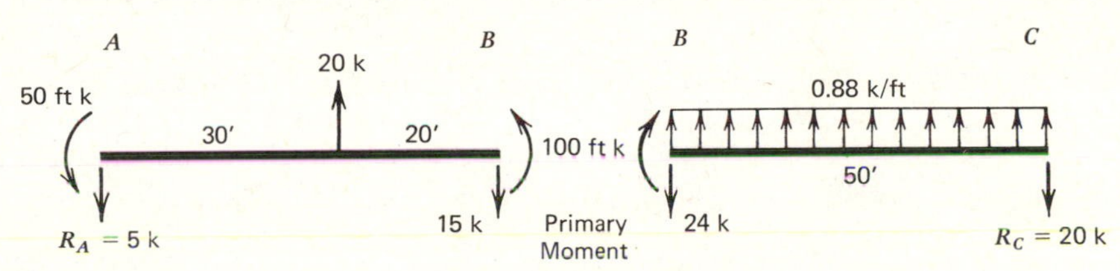

(a) Loads and "reactions" Due To Primary Moment
(*Internal forces* required for equilibrium)

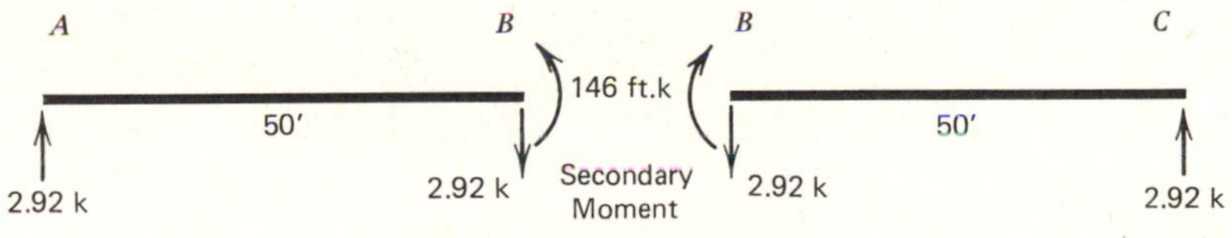

(b) Reactions Due To Secondary Moment
(*External forces* due to beam deformations)

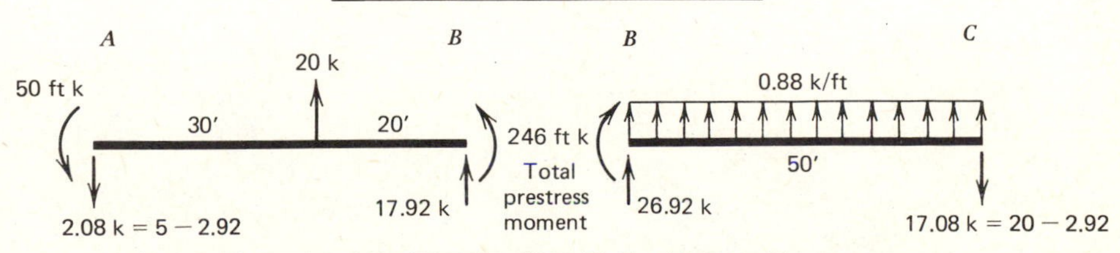

(c) Loads and Net "Reactions" Due To Total Prestress Moment
(*Net forces* required for equilibrium)

Fig. 10-13. Beam of Example 10-1, reactions.

10-13(b). An external reaction of 2.92 k (13 kN) at A must exist for the equilibrium of forces from this secondary moment. If these two cases are combined in Fig. 10-13(c), we obtain the total moment from prestress of 246 ft-k (334 knN-m) at B. This agrees with the resulting moment from example 10-1. Solving the internal force at A from the free-body diagram of the beam with prestress moment = 246 ft-k (334 kN-m) at B and the loads from prestress acting in each span, Fig. 10-13(c), the "reaction" at A is now 2.08 k (down). This net force at A required for equilibrium of the free body agrees with the summation of 5 k (down) from primary moment plus 2.92 k (up) from secondary moment obtained in (a) and (b).

The external reactions induced by prestressing this beam are those arising from secondary moment, Fig. 10-13(b). Figure 10-14 summarizes the results of our solution for the beam of example 10-1:(a) the primary moments would cause the beam to lift from support B, and (b) restraining the beam by allowing no deflection at B results in secondary moments with the external reactions shown due to prestress only.

The above procedure outlines the method for locating the line of pressure due to prestressing in a continuous beam. It is seen that the induced reactions produce moments in a continuous beam, which shift the C-line away from the c.g.s. line. Now, when external loads are applied on the beam, additional moments will be produced, and the C-line will again be shifted. Two methods of computation are possible.

1. Moments in the continuous beam due to the external loads (including the weight of the beam) are computed by the usual elastic theory, using methods

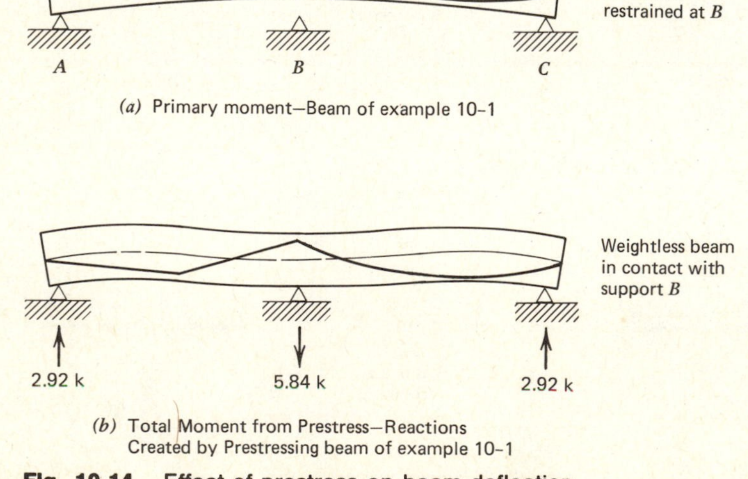

(a) Primary moment—Beam of example 10-1

(b) Total Moment from Prestress—Reactions Created by Prestressing beam of example 10-1

Fig. 10-14. Effect of prestress on beam deflection.

such as moment distribution. These moments are added to the prestressing moments previously calculated, thus yielding the final moments in the beam. This can also be performed by shifting the *C*-line from that obtained for prestressing only. The amount of shifting equals the moments due to external loads divided by the prestress. For simplicity, the effective prestress and the gross concrete area can be used for all computations. This method is generally preferred when there is more than one condition of loading.

2. When there is only one condition of loading to be investigated, it may be easier to consider the effects of prestress and external loading together. Since the effect of prestressing can be reduced to a system of forces acting on the beam, it is only necessary to add these forces to the external loads to obtain the total loads on the beam. One moment distribution will then be sufficient for the two sets of forces. The load balancing concept in Chapter 11 utilizes this method.

EXAMPLE 10-2

For the prestressed beam in example 10-1, a uniform load of 1.2 k/ft is applied to the entire length of the two spans (including the weight of the beam itself). Locate the line of pressure in the concrete due to the combined action of the prestress and the external loads. Compute the stresses in concrete at section *B*, if the concrete section is as shown in Fig. 10-15, with $I = 39{,}700$ in.^{4} and $A_c = 288$ sq in. ($w = 17.5$ kN/m, $I = 16.5 \times 10^9$ mm^4, and $A_c = 186 \times 10^3$ mm^2)

Solution Two methods are possible. Since the moments and line of pressure due to prestress have been obtained in example 10-1, we shall now follow the first method, obtaining only the effect of external loads. By a simple moment distribution, see (*b*), the moment diagram can be plotted as in (*c*). (Note that for this particular case the moment diagram can be plotted by using ready-made tables from handbooks if desired.) Dividing this moment by the prestress of 250 k (1112 kN) the shifting of *C*-line due to this external loading is given in (*d*). Adding (*d*) to Fig. 10-12(*f*) of the last example, the final location of the resulting line of pressure for both prestress and external load is given in (*e*).

The resulting moment in the concrete at section *B* is $250 \times 0.52 = -130$ k-ft (-176 kN-m) and the stresses are

Top fiber:

$$\frac{-250}{288} + \frac{130 \times 12 \times 18}{39{,}700} = -0.867 + 0.707 = -0.160 \text{ ksi } (-1.10 \text{ N/mm}^2)$$

Bottom fiber:

$$-0.867 - 0.707 = -1.574 \text{ ksi } (-10.85 \text{ N/mm}^2)$$

The resulting moment in the concrete at section *B* could easily be obtained by directly combining moment from prestress with that due to applied loading. Thus, at section *B*

$$\text{Resulting moment} = 246 - 375 = -129 \text{ k-ft } (-175 \text{ kN-m})$$

At other points along the span the moment from prestress may be found as *F* times distance to *C*-line, Fig. 10-12(*f*), and this moment is combined algebraically with the moment from applied loading, Fig. 10-15(*c*).

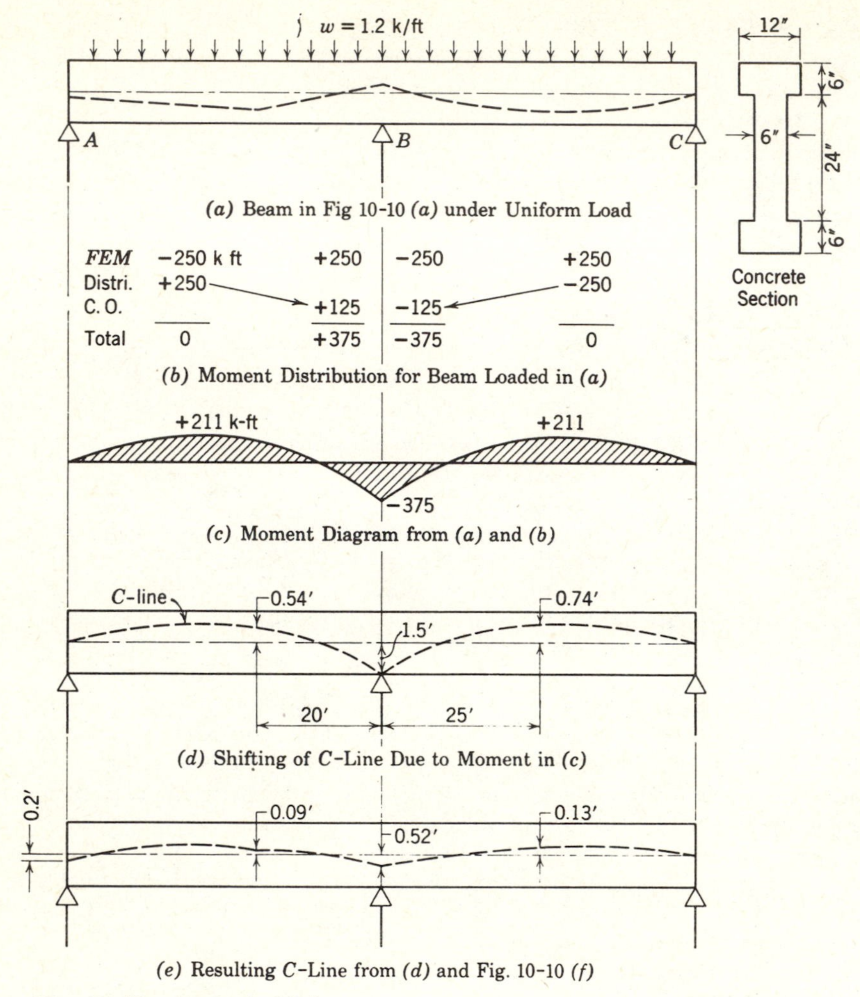

Fig. 10-15. Example 10-2.

Note: The reader can try to combine the forces in Fig. 10-12(d) with those in Fig. 10-15(a) and perform one moment distribution to obtain the line of pressure. Obviously you should get the same results.

10-4 Linear Transformation and Concordancy of Cables

The previous section explains the analysis of prestressed continuous beams; the design of such beams is a more complicated problem. In analysis, the concrete section, the steel, and the location of the steel are already known or assumed. It is only necessary to compute the stresses for the given loading conditions. This is not true in design, which is essentially a trial-and-error process in an effort to reach the best proportions. The designer must be well acquainted with the method of analysis before he can perform efficiently in design. In order to design well, we must be conversant with some of the mechanics of continuous prestressed beams.

In this connection, two terms will be explained first: linear transformation and concordancy of cables.[12] After a thorough study of this section, the designer should be able to perform linear transformation with ease and skill and to obtain either concordant or nonconcordant cables to satisfy the most desirable conditions. First of all, let us define linear transformation:

When the position of c.g.s. line or of a C-line is moved over the interior supports of a continuous beam without changing the intrinsic shape (i.e., the curvature and bends) of the line within each individual span, the line is said to be linearly transformed. Linear transformation of a c.g.s. line is illustrated in Fig. 10-16(*a*).

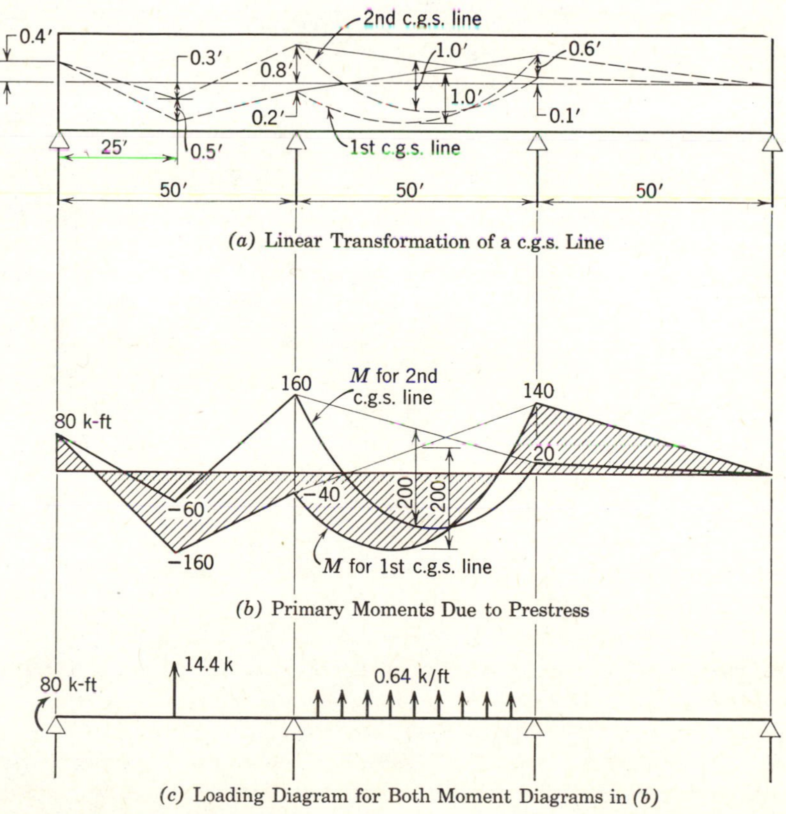

(*a*) Linear Transformation of a c.g.s. Line

(*b*) Primary Moments Due to Prestress

(*c*) Loading Diagram for Both Moment Diagrams in (*b*)

Fig. 10-16. Example 10-3.

In further explaination of the above definition, attention is called to the following points. First, the position of the line is moved only over the interior supports whenever desired, but not at the ends of a beam. Strictly speaking, a line can still be termed linearly transformed if it is moved at the ends in addition. However, for purpose of design, linear transformation without involving movement at the ends is much more useful; hence we will define it as such for the sake of convenience. Second, by linear transformation, the intrinsic shape of the line within each span remains unchanged; only the amount of bending of the line over the interior supports is changed.

It may be well to remind the reader that one use of linear transformation was described in the previous section, where we stated: "The *C*-line resulting from prestressing a continuous beam is a linearly transformed line from the c.g.s. line," which was explained by the fact that the secondary moment that produces the deviation between the two lines varies linearly between any two consecutive supports. Now, another interesting theorem concerning linear transformation is that *in a continuous beam, any c.g.s. line can be linearly transformed without changing the position of the resulting C-line*. This means that the linear transformation of c.g.s. line does not affect the stresses in the concrete, since the *C*-line remains unchanged. Thus the two c.g.s. lines in Fig. 10-16(a) will produce the same *C*-line, and hence the same stresses in the concrete, despite their apparently divergent locations.

The proof of this theorem is as follows. c.g.s. lines having the same intrinsic shape within each individual span will produce primary moment diagrams also having same intrinsic shapes in each individual span. For moment diagrams with the same intrinsic shapes (i.e., the same curvatures and bends), the corresponding loading diagrams along the span are the same, since load is given by the curvature (or second derivative) of the moment. Since the loads are the same, the resulting moments must be the same, which means that the *C*-lines will have the same position. It must be noted here that, though the resulting moments are the same, the primary moments differ; hence the secondary moments will necessarily differ, since

$$\text{Secondary moment} = \text{Resulting moment} - \text{Primary moment}$$

Any bending of the c.g.s. line over the supports will produce transverse forces acting on the beam which are directly counteracted by reactions from the supports. Hence such bending will not affect the moment along the beam. Since the moment is not affected, the *C*-line is not affected. Thus, linear transformation involving bending of the c.g.s. line over the interior supports will not change the location of the *C*-line. On the other hand, any movement of the c.g.s. line at the ends of the beam changes the magnitude of the applied end moments, which do affect the moments along all spans of continuous beam and change the location of the *C*-line on all spans. Hence linear transformation cannot involve the movement of the c.g.s. line over the ends of the beam or over the exterior support of a cantilever, but it can involve movement of the c.g.s. line over the interior supports.

The above theorem, permitting the linear transformation of the c.g.s. line without changing the *C*-line, offers many possible adjustments in the location of the c.g.s. line which cannot be easily accomplished without that knowledge. Some of these possibilities are evident from the above; others will be discussed later. The validity of this theorem has been proved experimentally within the elastic range, although such a logical theory hardly needs any experimental proof. The effect of linear transformation of the c.g.s. line on the ultimate strength of continuous beams will be discussed in section 10-6.

EXAMPLE 10-3

The first c.g.s. line in the beam, Fig. 10-16(a) is linearly transformed to the second position. Show that the *C*-line in the concrete is the same for both positions. Assume prestress $F=200$ k. (890 kN)

Solution The two moment diagrams are shown in (b). The loading diagrams corresponding to those moment diagrams are exactly the same and are both shown in (c). Hence the line of pressure must also be the same for both c.g.s. lines. The only forces which are different are those directly over the intermediate and end supports. Since they do not produce any moments in the beam, they do not affect our calculations and are not shown in the figure.

Having defined "linear transformation," let us now define "concordant cable,": *A concordant cable in a continuous beam is a c.g.s. line which produces a C-line coincident with the c.g.s. line.*

In other words, a concordant cable produces no secondary moments. Thus, every cable in a statically determinate structure is concordant, because no external reaction is induced, and there is no secondary moment in the structure. For a continuous beam, on the other hand, external reactions will usually be induced by prestressing. These reactions will produce secondary moments in the beam, and the *C*-line will shift away from the c.g.s. line. When this happens, the cable is termed nonconcordant. When, by chance or by purpose, reactions are induced in a continuous beam by prestressing, then there will be no secondary moments and the cable is a concordant one. When a concordant cable is prestressed, it will tend to produce no deflection of the beam over the supports, and hence no reactions will be induced (not considering the weight of the beam). The essential differences between a concordant and a nonconcordant cable are shown in Fig. 10-17.

Besides the fact that a concordant cable line is easier for analysis, there is seldom a necessity for using a concordant one. There were, at first, some doubts as to the behavior of a nonconcordant cable, whether its *C*-line would change with time or with the elastic and plastic properties of concrete. A little thinking

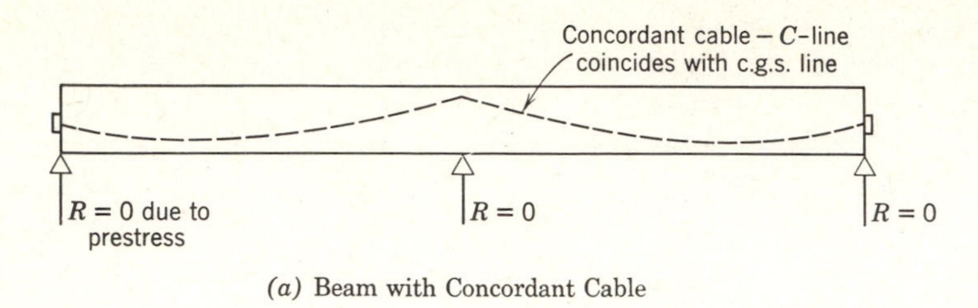

(a) Beam with Concordant Cable

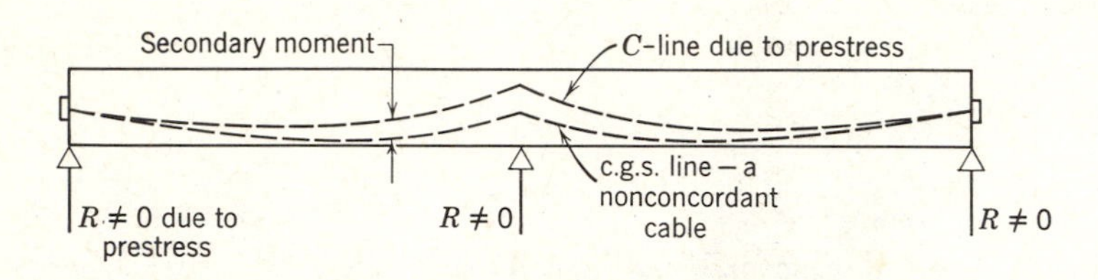

(b) Beam with Nonconcordant Cable

Fig. 10-17. Properties of nonconcordant cables.

on the subject would clear these doubts. If E_c of the entire beam changes uniformly, there will be no change in the secondary moments, since the secondary moments due to prestressing are computed independent of the E_c value, as, for example, by moment distribution. If the E_c of one portion of concrete changes at a different rate from that of another portion, slight changes in the secondary moments might result, but such effects are generally considered negligible, since the elastic theory assuming uniform E_c for a beam is believed to be sufficiently accurate for the analysis of both reinforced and prestressed concrete.

While no significant reason can be given for preferring a concordant cable, there is even less justification for locating a nonconcordant cable for the sake of nonconcordancy. The real choice of a good c.g.s. location depends on the production of a desirable *C*-line and the satisfaction of other practical requirements, but not on the concordancy or nonconcordancy of the cable. A concordant cable, being somewhat easier to compute, is slightly preferred, other things being equal.

A convenient procedure in design is to obtain a concordant cable that gives good positions of the c.g.s. in resisting the external moment. If that location falls outside the beam it can be linearly transformed to give a more practical location without changing its *C*-line. According to this procedure, the finding of locations for concordant cables becomes a useful means to an end.

Several methods have been proposed for obtaining concordant cables, but the authors advocate the following method, utilizing only one basic theorem as follows. *Every real moment diagram for a continuous beam on nonsettling supports, produced by any combination of external loadings, whether transverse loads or moments, plotted to any scale, is one location for a concordant cable in that beam.* The application of this theorem is illustrated in example 10-4.

EXAMPLE 10-4

For continuous prestressed-concrete beam loaded as shown in Fig. 10-18(a), obtain some desirable locations for concordant cables to support that loading.

Solution Note that every moment diagram plotted to any scale is a concordant cable. If we plot the continuous beam moment diagram for the given loading, we obtain (b). Two concordant cable locations are shown in (c) and (d), both proportional to the moment diagram in (b), and hence both are concordant. (f) gives another location of a concordant cable, which is proportional to the moment diagram for loading in (e). Many similar concordant cables can be found by drawing all kinds of moment diagrams. The most desirable concordant cable will be governed by practical requirements of the particular problem as well as by the ability of the cable to resist the applied loads. For example, the location in (c) gives larger resisting arms for the steel but may overstress the concrete if the weight of the beam is light, in which case (d) may be a better location. (f) does not suit this particular loading as well but gives a symmetrical layout and may carry other loadings, such as the beam's own weight, more efficiently.

The above theorem can be easily proved. Since any moment diagram due to the loads on a continuous beam is computed on the basis of no deflection over the supports, and since any c.g.s. line following that diagram will produce a similar moment diagram, that c.g.s. line will also produce no deflection over the supports; hence it will induce no reactions and is a concordant cable. The theorem applies not only when the beam is under a constant prestress. If the amount of prestress varies along the beam, the application of the theorem can be modified to suit. In fact, the theorem can be extended to include beams on elastic supports.

Based on this general theorem, many corollaries can be derived which will help the designer in selecting proper position for concordant cables. After the designer has mastered the theorem and its corollaries discussed herein, his work of obtaining a concordant cable is reduced to that of finding a proper moment diagram, which is a familiar operation with most engineers. Some corollaries will be stated.

1. The reverse of the theorem is also true: The eccentricity of any concordant cable measured from the c.g.c. is a moment diagram for some system of loading on the continuous beam plotted to some scale.
2. Any C-line is a concordant cable, since it is obtained by computing the moments due to a system of loads on the continuous beam.
3. Superposing two or more concordant cables will result in another concordant cable. Superposing a concordant and a nonconcordant cable will result in a nonconcordant one.

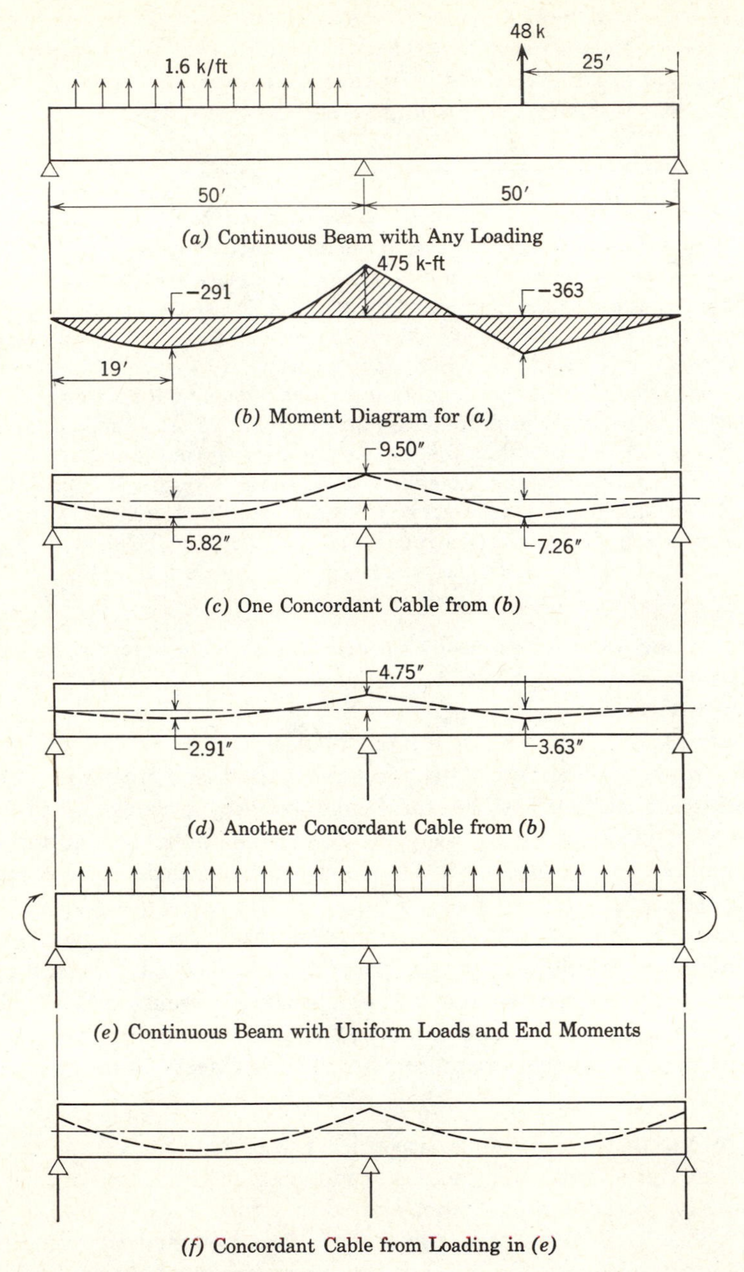

Fig. 10-18. Example 10-4.

4. When a sudden change in direction is desired, a concentrated load is applied. When a gradual change is desired, a uniform load is applied. One moment diagram can thus be modified into another by the addition of loads. Hence one concordant cable can be easily modified into another.
5. In order to obtain a concordant cable from another by linear transformation involving the moving of eccentricities over the ends of a beam, the following procedure can be used. Apply an end moment on the continuous beam; compute the moment diagram due to that moment. When one end is moved by a given amount, the entire cable must be transformed linearly in proportion to that moment diagram. If the movement of the eccentricities at both ends is desired, apply end moments at both ends proportional to the respective amount of movement, and shift the entire cable in proportion to the moment diagram so obtained. This will yield another concordant cable, as is illustrated in example 10-5.

Much ingenuity can be exercised in the location of concordant cables, but it should be left to the skill of the designer after he understands the basic theorem and its main corollaries. When applied to rigid frames, the effect of sidesway and rib shortening should be additionally considered. For varying prestress along the beam, the moment diagram should be divided by the corresponding prestress at each point in order to obtain the location of a concordant cable. Or the tendons may be treated separately. If each individual tendon or group of tendons forms a concordant cable, then, when acting together, they also form a concordant cable.

Like all statically indeterminate structures, it is sometimes desirable to purposely adjust the elevations of the supports in order to produce favorable moments in the beams. The moments so produced are of a different nature from the secondary moments caused by prestressing. It was previously mentioned that the secondary moments due to prestressing would not change with the value of E_c. Moments due to adjusted support elevations, on the other hand, do change with the value of E_c, because the moments induced by a given displacement are a function of E_c. Hence such moments will change with time as creep takes place and E_c changes. Therefore, when attempts are made to produce moments by support displacements, either the possible change in E_c must be allowed for, or the displacements must be adjusted from time to time. Thus the economy of such a manipulation, though feasible for certain large structures, may be doubtful for small and even medium-sized ones.

EXAMPLE 10-5

Obtain a new concordant cable, with its intrinsic shape the same as that of Fig. 10-18(c), but with the right end of the cable 4 in. (254 mm) above the c.g.c.

Solution Apply a unit moment at the right end of beam: by the method of moment distribution, plot the moment diagram as in Fig. 10-19(a). The concordant cable in Fig.

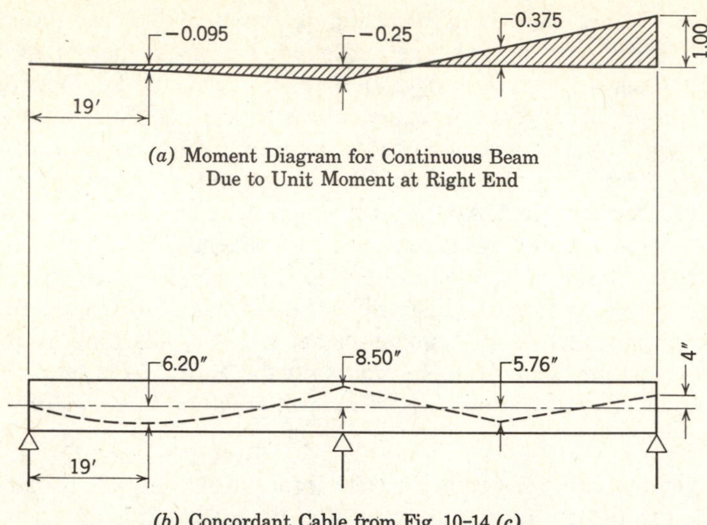

(*a*) Moment Diagram for Continuous Beam Due to Unit Moment at Right End

(*b*) Concordant Cable from Fig. 10-14 (*c*) with 4-in. Displacement at Right End

Fig. 10-19. Example 10-5.

10-18(*c*) can now be linearly transformed in proportion to the moment diagram Fig. 10-19(*a*), giving a new concordant cable as in Fig. 10-19(*b*). Note that the same moment diagram (*a*) can be used to shift the end eccentricity any other amount, not only for the 4 in. (254 mm) illustrated here. Also, owing to the symmetry of the beam, (*a*) can be similarly used for moving the end eccentricity at the left. A combination of two moment diagrams due to a moment at each end will permit the simultaneous shifting of both end eccentricities.

An infinite number of concordant cables can be obtained by rotating one concordant cable about the points of inflection, because such rotation simply represents the addition of one concordant cable to another, and should result in a concordant one. The points of inflection in these moment diagrams are called "nodal points" by some European authors.

10-5 Cable Location

Here, again, by cable location is meant the location of the centroid of the tendons, that is, the c.g.s. line. After the c.g.s. line is determined, the location of the individual position of the various tendons is an easier problem which will not be discussed here.

Designing a continuous prestressed-concrete beam, like that of any other continuous structure, is essentially a procedure of trial and error. Knowledge regarding the analysis of such structures, together with a systematic approach to the solution, will aid greatly in arriving at desired results. The following steps are recommended for designing continuous prestressed beam.

Step 1. Assume section of members for dead-load computation.

Step 2. Compute maximum and minimum moments at critical points for various combinations of dead, live, and other external loads. Fig. 10-20(a). Compute the amount of prestress required for these moments and the corresponding depth of concrete. Modify section of members, and repeat steps 1 and 2 if necessary.

Step 3. Plot top and bottom kern lines for the members, Fig. 10-20(b). From the bottom kern line, plot

$$a_{min} = \frac{M_{min}}{F} \quad \text{also} \quad a_G = \frac{M_G}{F_0}$$

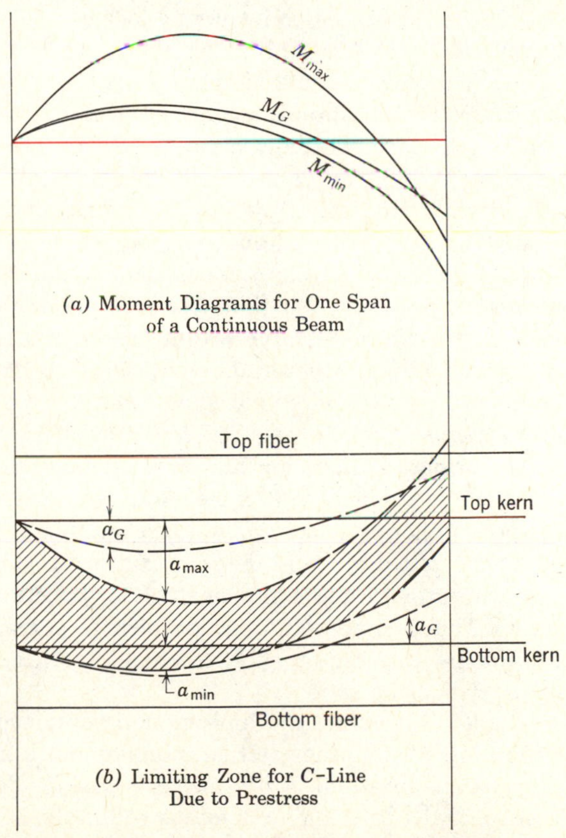

(a) Moment Diagrams for One Span of a Continuous Beam

(b) Limiting Zone for C–Line Due to Prestress

Fig. 10-20. Obtaining limiting zone for the *C*-line due to prestress.

where M_{min} = the algebraically smallest moment. The distances a_{min} and a_G should be plotted upward for $-M$ and downward for $+M$.

From the top kern, plot

$$a_{max} = \frac{M_{max}}{F} \quad \text{also} \quad a_G = \frac{M_g}{F_0}$$

again upward for $-M$ and downward for $+M$.

The shaded area, between the limit of these four lines, obtained by a_{max} and a_{min} and a_G, represents the zone in which the line of pressure must lie if no tension is permitted. As in previous discussions on cable location for simple and cantilever beams, when the zone is too wide, an excess of prestress or of concrete section or of girder load is generally indicated. If the limiting line from one kern crosses a limiting line from another kern, an inadequacy is evident. An ideal layout is obtained when there exists a narrow limiting zone within the beam where the centroid of the cables can be conveniently located.

Step 4. Select a trial cable location within the above zone. Note that, if the cable follows the shape of some moment diagram, it will be a concordant cable. If this trial location is a concordant cable, it is a satisfactory solution. If it is a nonconcordant cable, the *C*-line can be determined by moment distribution as described in the previous section. If the *C*-line still lies within the limiting zone, then two locations are possible: either the trial location giving a nonconcordant cable, or a new location following the *C*-line, thus giving a concordant cable. If this *C*-line lies outside the zone, new cable locations can be tried. An attempt should be made to get a concordant cable within the zone. It is generally best to try concordant cables, because they coincide with their *C*-lines and give a more direct solution. Note that, after having obtained one concordant cable, it is much easier to derive from it other concordant cables by the general theorem that any moment diagram for the continuous beam is a concordant cable. Adding to the first concordant cable any form of moment diagram will give another concordant cable. The shape of the added moment diagram can be obtained by applying couples or concentrated and uniform loads anywhere along the continuous beam. Thus it is not a difficult problem to add another moment diagram to the first concordant cable to obtain another concordant one which will lie within the zone.

Step 5. The concordant cable within the limiting zone obtained in step 4 is a good location for resisting the external moment, but it may or may not be a good practical location. For example, it may be desirable to shift the c.g.s. line in order to bring the tendons within the boundaries of the beam. To achieve this without shifting the *C*-line, the concordant cable can be linearly transformed as desired. After this linear transformation, the cable becomes nonconcordant. This procedure is illustrated in example 10-6.

EXAMPLE 10-6

A pedestrian bridge of prestressed-concrete slab (the Harkness Avenue Bridge in San Francisco, California, of the California Division of Highways) has a three-span symmetrical continuous layout as shown, Fig. 10-21(a). The bridge is 9 ft 4 in. (2.85 m) wide with

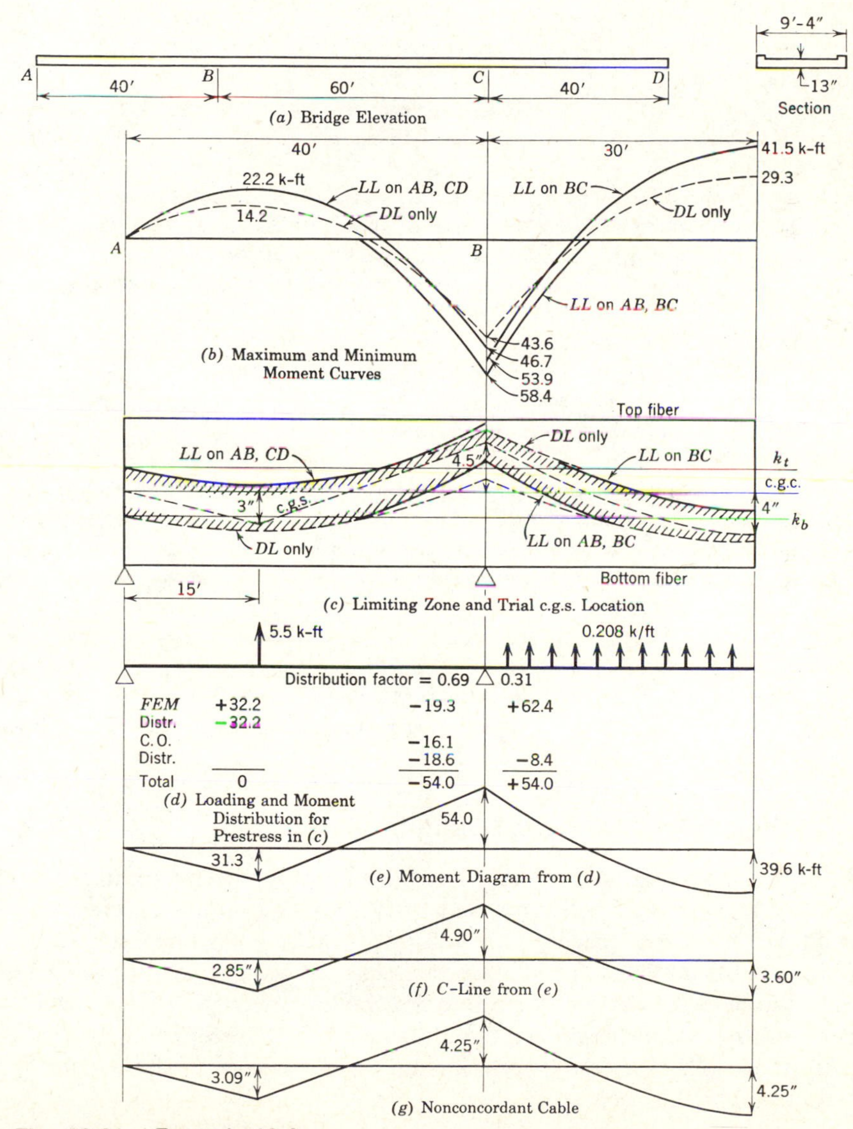

Fig. 10-21. Example 10-6.

a uniform thickness of 13 in. (330 mm) (neglecting curbs). The total effective prestressing force is 1,230,000 lb (5,741 kN) after deducting a loss of 15%. Design live load is 50 psf (2.39 kN/m^2). Choose a suitable location for the cable, allowing no tension in the concrete, $f_c' = 5000$ psi. (34 N/mm^2)

Solution Following the procedure described above and considering 1-ft (0.30 m) width of slab:

Step 1. The section is already chosen, and the dead load is 162 psf or 162 plf (7.76 kN/m^2 or 2.36 kN/m) for 1-ft (0.30 m) width.

Step 2. The amount of prestress is already chosen; it is 1,230,000/9.33 = 132 k per ft (1926 kN per meter) width of slab for the effective prestress, or 156 k (2276 kN per meter) for the initial prestress.

Step 3. Kern lines for a rectangular section are located at the third points. The maximum and minimum diagrams together with the girder moment diagrams are shown in Fig. 10-21(*b*) for one half of the structure. These diagrams are divided by the respective prestress, F for those with live loads, and F_0 for dead load only. The a values thus obtained are plotted from the kern lines as shown in Fig. 10-21(*c*), giving the limits for the zone within which the *C*-line due to prestressing must lie.

Step 4. Using the moment diagrams as guides, select a trial c.g.s. location within the zone as shown in Fig. 10-21(*c*). For the purpose of illustration, assume the c.g.s. line to possess the following characteristics.

1. Passing through the c.g.c. (mid-depth of slab) at end supports.
2. One sharp bend for each side span.
3. One sharp bend over each intermediate support.
4. A parabolic curve for the center span.

For this c.g.s. location, the corresponding loading on the concrete is shown in (*d*), moment distribution for which gives a moment diagram as in (*e*). Dividing the moment diagram by the prestress yields a *C*-line as shown in (*f*) which is very close to the trial location and is still within the limiting zone. Hence this *C*-line is a location for a satisfactory concordant cable.

Step 5. A more practical location for the c.g.s. is shown in (*g*), affording better protection for the steel. This is obtained by linearly transforming the concordant cable into a nonconcordant one. This nonconcordant cable will yield the same *C*-line as the concordant one and hence will serve the same purpose as far as stresses are concerned.

10-6 Cracking and Ultimate Strength

Tests have shown that the elastic theory can be applied to continuous prestressed-concrete beams with great accuracy[13] as long as the concrete has not cracked. Occasionally, structures are subjected to overloads beyond the point of cracking or are designed for some permissible cracking under working loads. Then it will be neccessary to determine the cracking strength of such structures. In addition, knowledge regarding the ultimate strength of these beams is also of interest in providing criterion for designing.[14] Such strengths will be investigated and discussed in this section.

Since a prestressed structure is nearly a homogeneous material before cracking, the elastic theory can be applied to the calculation of strength up to that point. Even when some cracks have occurred, a prestressed structure is no less homogeneous than a reinforced-concrete structure under working loads. In fact, there is every reason to believe that the elastic theory would be more applicable to prestressed concrete at the start of cracking than reinforced concrete under working loads, since reinforced concrete usually starts to crack at about one-third the working load.

What should be the tensile stress in continuous prestressed-concrete beams at the point of cracking? Some engineers believe that the cracking tensile strength is higher than the modulus of rupture measured from plain concrete specimens. Experiments have shown, however, that the modulus of rupture is a reasonably accurate measure of the start of cracking in continuous prestressed beams. It must be realized that only hair cracks are produced when the modulus of rupture is reached. These cracks, at first, will not be easily visible to the unaided eye but can be detected by strain gages or microscopic examinations.

Before the start of actual cracking, some plastic deformation is usually exhibited in the concrete. Such deformation occurs only in limited regions and does not affect the general behavior of the structure as an elastic body. Hence the validity of the elastic theory can still be counted on, up to (and perhaps, slightly beyond) the cracking of concrete.

Accurate determination of the ultimate strength of a continuous prestressed-concrete beam involves may difficulties. However, for design purposes, the ultimate strength can be estimated on the basis of the limit design theory,[15] if the plastic hinges are formed at critical points of maximum moment. This is true for underreinforced sections, which deform extensively before final rupture. For overreinforced sections, which may suddenly fail in the compressive zone of concrete before any appreciable rotation, a perfect plastic action cannot be expected. The action then is partly elastic and partly plastic. The ACI code procedure sets the amount of moment redistribution at ultimate as a function of ω, which is an index of ductility for the cross section.

The ACI Code allows some moment redistribution in the design of continuous members, provided the cross section has a reinforcement index $(\omega+\omega_p-\omega')$ which is equal to or less than 0.20, where $\omega=\rho f_y/f'_c$, $\omega_p=\rho_p f_{ps}/f'_c$, and $\omega'=\rho' f_y/f'_c$. While the reinforcement index may be as much as 0.30 for design under the ACI Code (analogous to $\frac{3}{4}\rho$ for balanced design of reinforced concrete members), there is concern that members with this much reinforcement will not be sufficiently underreinforced, since it may enable only a very small degree of plastic hinging. Sudden crushing of the concrete with brittle failure may be expected for members with $(\omega+\omega_p-\omega')\geqslant 0.30$. The limit of $(\omega+\omega/_p-\omega')\leqslant 0.20$ for considering moment redistribution is analogeous to the $\frac{1}{2}\rho$ balanced limit for reinforced concrete. Such a section is certain to be underreinforced, and some plastic behavior can be expected before failure. The percentage by which such members can have the support moment increased or decreased in connection with inelastic moment redistribution is given by the following expression from the ACI Code:

$$20\left[1-\frac{(\omega+\omega_p-\omega')}{0.30}\right]$$

The secondary moment acting should be considered in checking ultimate strength of continuous members.[16] When secondary moment results from induced reactions caused by prestressing, this moment must be combined algebraically with moment from applied loads to check strength under the ACI Code (318-77). It is quite possible for a positive secondary moment to exist at a support. In this case more negative moment from applied loading will be required to impose any inelastic action on the cross section at the support than would be the case for a concordant tendon layout. Of course, if the secondary moment at the support were negative, the negative support moment would demand inelastic action of the support cross section sooner than would be the case if the tendon were concordant (no secondary moment). Secondary moments do exist in most cases and they influence the elastic moments at all sections along the structure over the full range of loading up to failure. As shown in reference 12 the neglect of the secondary moment can be unsafe in some cases.

Most continuous beams are designed for limiting allowable stresses at service load and then checked for strength. The ACI Code Commentary suggests that strength be checked by the following sequence:

(a) Determine elastic moments under factored loading pattern $(1.4D+1.7L)$

(b) Compute what maximum percent redistribution (if any) is allowed from $20[1-(\omega+\omega_p-\omega')/30]$. Adjust support moment by increasing or decreasing its value by this percentage (or less).

(c) Combine the elastic secondary moment (if any) with the adjusted elastic moment diagram from (a) and (b). Check to see that the moment does not exceed M_u for the cross section at the support or at any other section of peak moment.

The following example 10-7 illustrates an approximate strength analysis for the three-span structure analyzed in example 10-6 for service load. This is an overreinforced section with no capability for redistribution of moment as evaluated in step (b) of the ACI procedure described above. This structure will be redesigned in Chapter 11 with a more ductile cross section, and its strength will be reevaluated. Example 10-8 at the end of this chapter will show the procedure for strength evaluation of an underreinforced section.

EXAMPLE 10-7

For the continuous prestressed slab in example 10-6, with the c.g.s. located as finally chosen, Fig. 10-17(g), compute the ultimate load-carrying capacity for uniform load on all the spans. For effective prestress of 120 ksi, steel area per foot width of slab is $132/120 = 1.1$ sq in. f_{pu} for the steel wire is given as 240 ksi. $f_c' = 5000$ psi. Assume tendons are grouted after stressing is completed (bonded beam) ($f_{se} = 827$ N/mm^2, $f_{pu} = 1655$ N/mm^2, and $f_c' = 34$ N/mm^2).

Solution The probable plastic hinge locations are over the intermediate supports, at center of middle span, and near the 0.4 points E and G of the outside spans. These sections are shown in Fig. 10-22(a). Let us first determine the ultimate moment capacities of these sections. A little calculation will show that the ultimate strength of steel cannot be developed for these sections.

Use the ACI Code equation for bonded members to estimate the tendon stress at ultimate moment. The effective steel stress, $f_{se} = 120$ ksi (827 N/mm^2) for this design from example 10-6. Since this is $0.5f_{pu}$, the formula is valid to use here. For this problem, we will first estimate actual strength, M', although for design we must use $\phi = 0.9$ to check the M_u at support section B and C.

$$\rho_p = \frac{A_s}{bd} = \frac{1.10}{12 \times 10.75} = 0.008527$$

$$f_{ps} = f_{pn}\left(1 - 0.5\rho_p \frac{f_{pn}}{f_c'}\right)$$

$$f_{ps} = 240\left(1 - 0.5 \times 0.008527 \times \frac{240}{5}\right)$$

$$f_{ps} = 191 \text{ ksi } (1317 \text{ N/mm}^2)$$

$$\omega_p = \rho_p \frac{f_{ps}}{f_c'} = 0.008527 \times \frac{191}{5} = 0.326 > 0.30$$

Note that this slightly exceeds the limiting value allowed currently by ACI Code based on prestressed steel only. The presence of compression steel at support could make this a satisfactory design. Here, our concern is to estimate the ductility which might be expected. This $\omega = 0.326$ shows that the section is overreinforced, and we must not expect ductility of the cross section to redistribute moment. The ACI Code allows redistribution only if $(\omega_p - \omega) \leqslant 0.20$. The maximum moment the section can carry at B and C is

$$M' = A_{ps} f_{ps}\left(d - \frac{a}{2}\right)$$

$$a = \frac{A_{ps} f_{ps}}{0.85 f_c' b} = \frac{1.1 \times 191}{0.85 \times 5 \times 12} = 4.12 \text{ in. } (105 \text{ mm})$$

$$M' = 1.1 \times 191\left(10.75 - \frac{4.12}{2}\right) = 1826 \text{ in.-k}$$

$$M' = 152 \text{ ft-k } (206 \text{ kN-m})$$

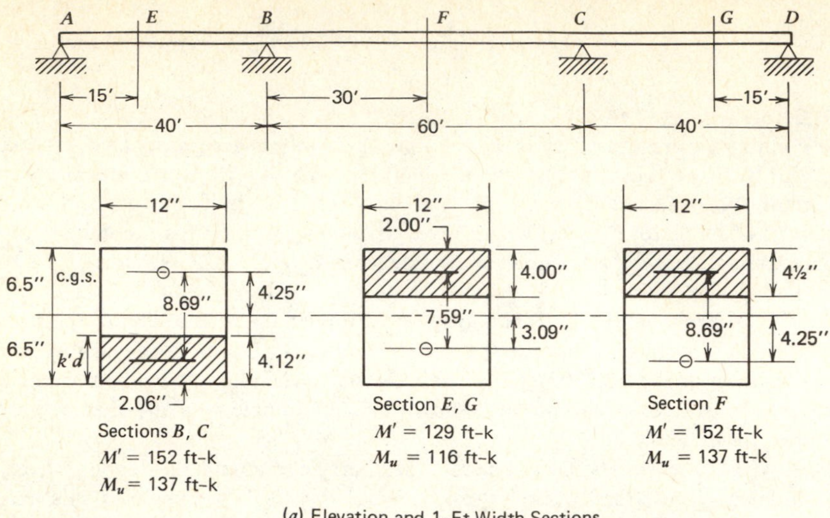

(*a*) Elevation and 1-Ft Width Sections

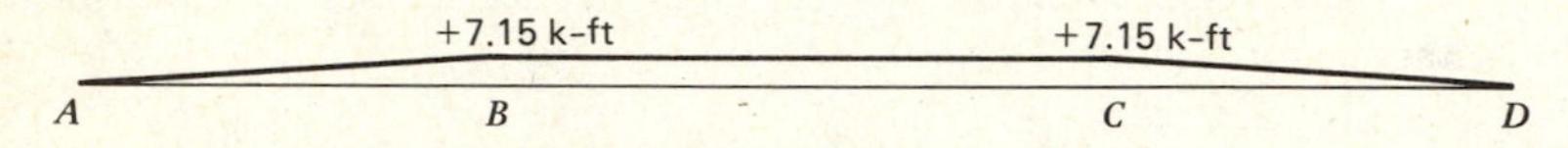

(*b*) Secondary Moment Due to Prestressing (tendon layout Fig. 10-21 g and $F = 132$ k)

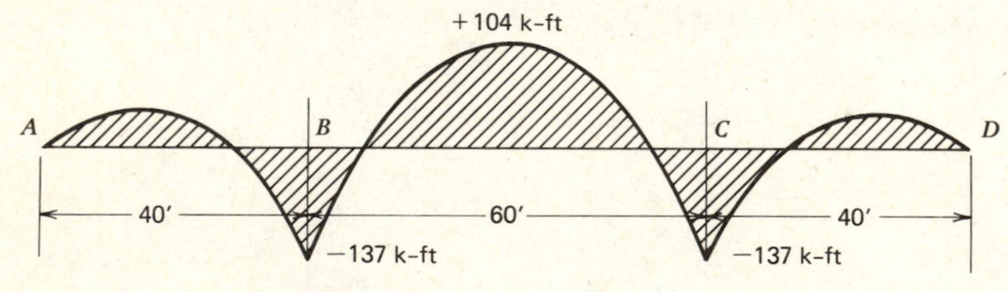

(*c*) Ultimate Moments All Spans Loaded $w_u = 0.536$ k/ft

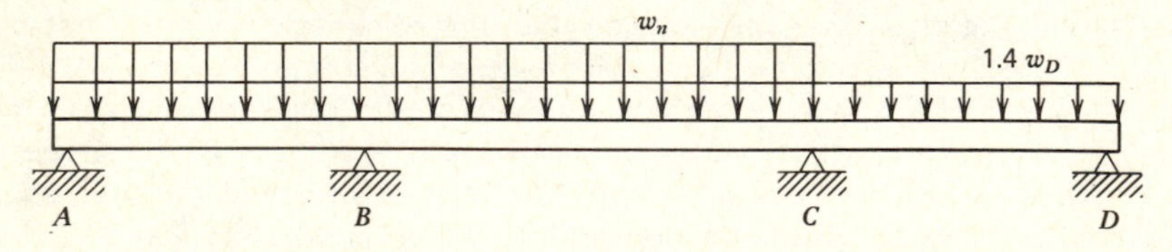

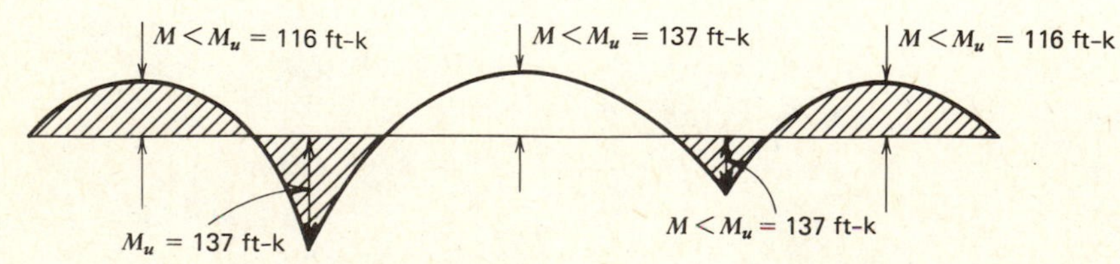

(*d*) Ultimate Moments, Partial Loading

Fig. 10-22. Example 10-7.

According to the elastic theory, maximum moment is given as follows for this series of spans with all spans loaded:

$$-M' = 0.0747wL^2$$
$$152 = 0.0747w \times 60^2$$
$$w' = 0.565 \text{ k/ft } (8.24 \text{ kN/m})$$

The ACI Code recognizes the secondary moment as a factor in calculating ultimate strength of continuous members. We can combine the secondary moment as follows:

The secondary moment at B and C from example 10-6 is

$$M_{sec} = (4.90 - 4.25)(132) = 85.8 \text{ in-k} = 7.15 \text{ ft-k } (9.70 \text{ kN-m})$$

Since this is $+M$ and the applied load moment at support is $-M$, we can allow additional load which will produce 7.15 ft-k moment at support

$$\text{Additional } w = \frac{7.15}{0.0747 \times 60^2} = 0.0266 \text{ k/ft } (0.39 \text{ kN/m})$$

$$\text{Maximum } w' = 0.565 + 0.0266 = 0.592 \text{ k/ft. } (8.63 \text{ kN/m})$$

Based on the analysis with $\phi = 1.0$, we can estimate the factor of safety for combined dead and live load as follows: (service dead plus live load = 0.212 k/ft.)

$$\text{F.S.} = \frac{0.592}{0.212} = 2.79$$

Revising M_u for $\phi = 0.9$ following the ACI Code may be easily done here:

$$M_u = 0.9M' = (0.9)(152) = 137 \text{ ft-k } (186 \text{ kN-m})$$

which gives the following ultimate load and moments as shown in Fig. 10-22(c)

$$w_u = \frac{137}{0.0747 \times 60^2} = 0.509 \text{ k/ft } (7.43 \text{ kN/m})$$

$$\text{Limit } w_u = 0.509 + 0.0266 = 0.536 \text{ k/ft } (7.82 \text{ kN/m})$$

$$\text{F.S.} = \frac{0.536}{0.212} = 2.53$$

With partial loading of spans AB and BC, Fig. 10-22(d), the elastic moment is slightly more at support B, and this worst condition is

$$-M = 0.0765wL^2$$
$$137 = 0.0765 \times w \times 60^2$$
$$w_u = 0.497 \text{ k/ft } (7.25 \text{ kN/m})$$

considering secondary moment benefit

$$\text{Limit } w_u = 0.497 + 0.026 = 0.523 \text{ k/ft } (7.63 \text{ kN/m})$$

$$\text{F.S.} = \frac{0.523}{0.212} = 2.47$$

By inspection this structure can carry more than the required $(1.4D + 1.7L)$ ultimate load required by ACI Code. As far as maximum $-M$ at support is concerned, loading with live load on exterior spans AB and CD would produce maximum $+M$ in the exterior spans near midspan. A similar check on strength, including moment redistribution and secondary moment, will show that this structure also exceeds ACI Code requirements for carrying $(1.4D+1.7L)$ with this pattern of loading.

Tests have been run to prove the validity of the theory of linear transformation in the ultimate range, which can be stated as follows. *Linear transformation of the c.g.s. line does not change the ultimate load-carrying capacity of a continuous beam.* Theoretical proof of this statement has also been made but will not be attempted here. It should be mentioned, however, that the theory is valid in the ultimate range only under the following two conditions.

1. The steel must be sufficiently far from the compressive side of concrete so as not to produce sudden compression failures in the concrete. In other words, the plastic hinges must remain plastic, and the sections must remain under-reinforced.
2. The location of plastic hinges must not be changed as a result of linear transformation. For uniformly distributed loads and curved cables, linear transformation may change the location of the plastic hinge near midspan and thus modify the ultimate load-carrying capacity. In general, however, such change in location does not affect the strength seriously.

When nonprestressed steel is added to the critical points of a continuous beam, the ultimate moment capacity of these sections is increased. The amount of increase can be figured by the method presented in example 9-1. When the nonprestressed steel consists of deformed bars, which is quite common, the additional force offered by this steel may usually be computed at its f_y. ACI Code calls for minimum amounts of bonded reinforcement to be supplied at the points of peak moment for beams with unbonded tendons. Proper addition of bonded nonprestressed steel also helps to distribute the cracks and to increase the shear strength as well as the fatigue strength of critical sections. Design of continuous beams by the ACI Code sometimes involves plastic moment redistribution and this is a simple procedure as is illustrated below in example 10-8. However, the elastic theory will still be needed to compute the stresses at transfer, deflections under working loads, cracking loads, and possibly crack width, and secondary moments.

How the secondary moments changes as the beam goes from elastic into the ultimate range has not been fully clarified (see reference 16) if full moment redistribution does not take place. However, it is generally agreed that the elastic secondary moment can be used for strength computation in any case. The ACI Code requires the consideration of secondary moment found on the basis of the effective prestress force F_e. This is illustrated in example 10-8.

EXAMPLE 10-8

A three-span continuous beam has the peak negative moment from elastic analysis of 300 ft-k (407 kN-m) as shown in Fig. 10-23(a). Assume the cross section at B has $(\omega+\omega_p-\omega')=0.15$ and analysis shows the M_u for the support B section is 260 ft-k (353 kN-m). The secondary moment is +15 ft-k (20 kN-m) over span BC with linear variation to zero at A and D as shown in Fig. 10-23(b). Is the section satisfactory at B following ACI Code?

Solution Refer to Fig. 10-23(a) for moments along the span. We can adjust the peak support moment as follows:

$$20\left(1-\frac{0.15}{0.30}\right)=10\%$$

We can adjust the 300-ft-k elastic moment, reducing it by the allowable 10%, thus

$$M_B=-300+(0.10)(300)=-270 \text{ ft-k } (-366 \text{ kN-m})$$

This moment still exceeds the $M_u=260$ ft-k (353 kN-m) given as strength at support B for this member, but we must combine the +15 ft-k secondary moment before checking

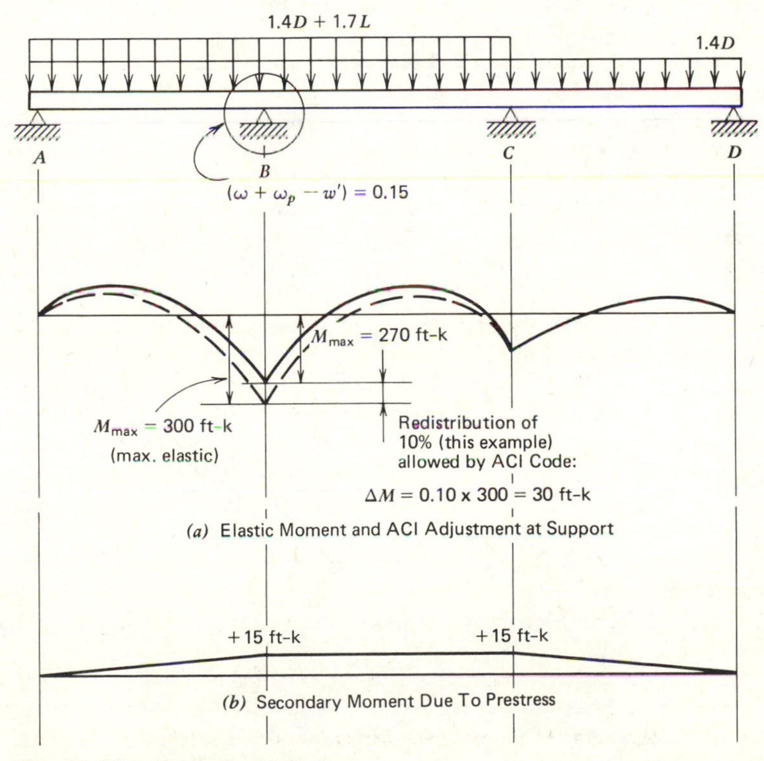

Fig. 10-23. Example 10-8.

strength required.

$$M_B = -270 + 15 = \underset{(M_u \text{ Required})}{-255 \text{ ft-k} \quad (-346\ kN\text{-}m)} < \underset{(M_u \textit{ Provided})}{-260\ ft\text{-}k \quad (-353 \text{ kN-m})}$$

The section is satisfactory for strength following ACI Code.

Readers interested in continuous prestressed-concrete structures are urged to study the concept of load balancing described in Chapter 11. The method presented therein will appear to be much simpler. However, both the elastic behavior and the ultimate strength described in this chapter are essential to the proper analysis and design of continuous prestressed-concrete structures and should be mastered by the designer. An example of the ACI Code procedure for checking strength of an underreinforced continuous slab will be presented in the following chapter to show the procedure.

References

1. *Symposium on Prestressed Concrete Continuous and Framed Structures*, Cement and Concrete Assn., London, 1951.
2. F. S. Rostasy, "Connections in Precast Concrete Structures—Continuity in Double-T Floor Construction," *J. Prestressed Conc. Inst.*, August 1962.
3. P. H. Karr, L. B. Kriz, and E. Hognestad, "Precast-Prestressed Concrete Bridges (1. Pilot Tests of Continuous Girders)," *Journal of the PCA Research and Development Laboratories*, Vol. 2 No. 2, May 1960, pp. 21–37.
4. N. Burns, "Development of Continuity between Precast Prestressed Concrete Beams," *J. Prestressed Conc. Inst.*, Vol. 11 No. 3, June 1966, pp. 23–36.
5. A. F. Shaikh and D. E. Branson, "Non-Tensioned Steel in Prestressed Concrete Beams," *J. Prestressed Conc. Inst.*, February 1970.
6. "Canadian Bridge Uses Prestressed Concrete Reinforcing Elements," Bridge Report, Prestressed Concrete Institute, 1968.
7. A. L. Parme and G. H. Paris, "Designing for Continuity in Prestressed Concrete Structures," *J. Am. Conc. Inst.*, September 1951 (*Proc.*, Vol. 47), pp. 45–64.
8. R. B. B. Moorman, "Equivalent Load Method for Analyzing Prestressed Concrete Structures," *J. Am. Conc. Inst.*, January 1952 (*Proc.*, Vol. 48), pp. 405–416.
9. E. I. Fiesenheiser, "Rapid Design of Continuous Prestressed Members," *J. Am. Conc. Inst.*, April 1954 (*Proc.*, Vol. 50), pp. 669–676.
10. H. Cross and N. D. Morgan, *Continuous Frames of Reinforced Concrete*, John Wiley & Sons, New York, 1932.
11. T. Y. Lin, "A Direct Method of Moment Distribution," *Trans. Am. Soc. C.E.*, 1937, pp. 561–605.
12. Y. Guyon, "Statically Indeterminate Structures in the Elastic and Plastic States," *General Report*, First Int. Congress, Int. Federation of Prestressing, London, 1953.
13. T. Y. Lin, "Strengths of Continuous Prestressed Concrete Beams under Static and Repeated Loads," *J. Am. Conc. Inst.*, June 1955 (*Proc.*, Vol. 51).

14. N. M. Hawkins, M. A. Sozen, and C. P. Siess, "Strength and Behavior of Two-span Continuous Prestressed Concrete Beams," *Structural Research Series No*. 225, Univ. of Illinois, September 1961.
15. J. A. Van den Broek, *Theory of Limit Design*, John Wiley & Sons, New York, 1948. (out of print).
16. T. Y. Lin and K. Thornton, "Secondary Moment and Moment Redistribution in Continuous Prestressed Concrete Beams," *J. Prestressed Conc. Inst*., January/February 1972, Vol. 17 No. 1, pp. 8–20.

11

LOAD-BALANCING METHOD

11-1 Stress-Concept, Strength-Concept, and Balanced-Load Concept

In section 1-2, three basic concepts for prestressed concrete are discussed. Briefly, the first concept is to treat prestressed concrete as an elastic material so that it can be designed and analyzed with respect to its elastic stresses. This "stress-concept" forms the basis for methods and formulas presented in the major portion of Chapters 5 through 10. The second concept considers prestressed concrete similar to reinforced concrete, and deals with its ultimate strength. It might be termed as the "strength-concept" and is also discussed liberally in most of the above-mentioned chapters.

The third concept sees prestressed concrete as primarily an attempt to balance a portion of the load on the structure.[1,2] It was only mentioned in section 1-2 and will now be expounded upon in this chapter. While this "balanced-load-concept" often represents the simplest approach to prestressed design and analysis, its advantage over the two other concepts is not significant for statically determinate structures. When dealing with statically indeterminate structural systems, this balanced-load concept offers tremendous advantages both in calculating and in visualizing. It is noticeably simpler than the methods explained in Chapter 10 and is highly recommended. Although preliminary designs and often final analyses for indeterminate structures can be achieved far more easily using this method, it is urged that engineers also learn the other concepts, including the material presented in Chapter 10 on continuous beams.

To understand this balanced-load concept relative to the other two concepts, let us first examine the life history of a prestressed member under flexure, Fig. 11-1. While this figure is intended to describe the load-deflection relationship of a member, such as a simple beam, it also applies to a section of a member. There are several critical points in the life history, which follow.

1. The point of *no deflection* which indicates a rectangular stress block across the section.
2. The point of *no tension* which indicates a triangular stress block with zero stress at the bottom fiber of a simple beam.
3. The point of *cracking* which occurs when the extreme fiber is stressed to the modulus of rupture.

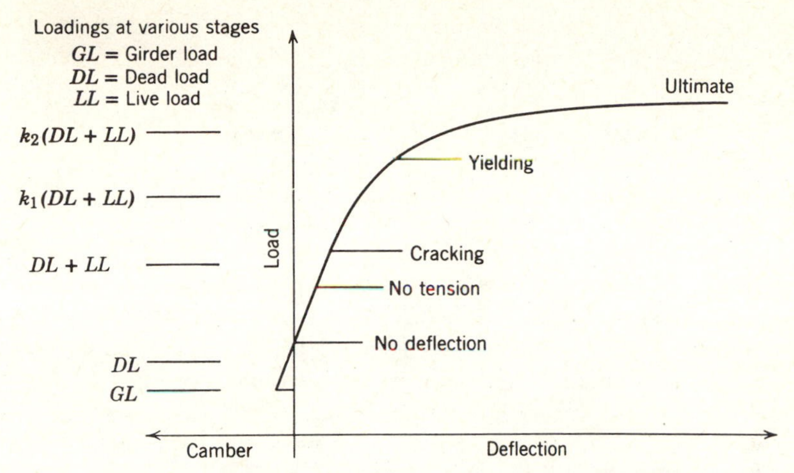

Fig. 11-1. Life history of prestressed member under flexure.

4. The point of *yielding* at which the steel is stressed beyond its yield point so that complete recovery will not be obtained.
5. The *ultimate load* which represents the maximum load carried by the member at failure.

In Fig. 11-1 the various loading conditions to which a beam is subjected to are indicated as follows.

1. Girder load, GL
2. Total dead load, DL
3. The working load, made up of dead plus live load, $DL+LL$
4. A factor of safety applied to the working load to obtain the minimum yield point load, $k_1(DL+LL)$
5. Another factor of safety k_2 applied to obtain the minimum ultimate load, $k_2(DL+LL)$.

Design by stress-concept actually consists of matching the $(DL+LL)$ with the point of some allowable tension on the beam (or possibly "no tension"). Design by strength-concept consists of matching the $k_2(DL+LL)$ with the "ultimate strength" of the beam. Design by balanced-load concept consists of matching the $DL+k_3LL$ (where k_3 is zero or some value much less than 1) with the point of no deflection. In some design situations, balancing less than full dead load may be the best solution, while in others some of the live load should be balanced.

It is clear that, depending on the relative values of the three stages of loadings as compared to the relative values of the three stages of the beam behavior (see table below), designs based on the three approaches could yield the same proportions or widely varying ones.

Applied Loadings	Stages of Beam Behavior
$DL+k_3LL$	No deflection
$DL+LL$	No tension or allowable tension
$k_2(DL+LL)$	Ultimate

It is also noted that, regardless of what concept is followed in the design, it is common practice to check for the behavior of the beam at the other stages. For example, if the stress concept is used for design, the ultimate strength requirements and the deflection of the beam under dead load are usually computed in addition. We are not generally concerned with the other stages, such as the $k_1(DL+LL)$ loading, the cracking and the yielding stages, although sometimes they could be more important and deserve careful consideration.

If the load deflection of a beam or the moment-curvature relationship (Chapter 5, section 5-7) of a section is of a definite shape, it is then possible to determine all critical points whenever one point is known. Actually, on account of the difference in the shape of the section, the amount and location of prestressed and nonprestressed steel, as well as different stress-strain relationships of both the concrete and the steel, these load-deflection or moment-curvature relationships may possess divergent forms. Thus it is often necessary to determine more than one critical point in order to be sure that the beam will behave properly under various loading conditions.

Which concept is the best to follow will depend on the circumstances. Generally it is desirable to choose the one which will control the proportioning of the member. If it is not certain that the other requirements will be met automatically, analysis for these other critical stages will be made, and modifications of the design may be effected. Since the balanced-load point is often representative of the behavior during the greater portion of the life span of the structure, it will deserve more consideration than either the elastic stresses or the ultimate strength.

Another consideration in the choice of the proper concept is the simplicity of analysis and design. It is believed that the balanced-load concept offers by far the simpler approach for statically indeterminate structures, especially for a preliminary design. It also gives a better picture of the structural behavior and thus enables a more intelligent approach to design and layout, as will be shown in the following sections.

A further advantage of the balanced-load approach is the convenience in the computation of deflections. Since the loading under which there will be no deflection anywhere along the beam is already known, the net deflection produced under any other condition of loading is simply computed by treating the loading differential acting on an elastic beam. Thus if the effective prestress balances the sustained loading, the beam will remain perfectly level regardless of the modulus of elasticity or the flexural creep of concrete.

11-2 Simple Beams and Cantilevers

While the load-balancing approach is not usually the best method for designing a simple beam, it can be well introduced with this simple case. Figure 11-2 illustrates how to balance a concentrated load by sharply bending the c.g.s., at midspan, thus creating an upward component,

$$V = 2F\sin\theta$$

If this V exactly balances a concentrated load P applied also at midspan, the beam is not subjected to any transverse load (neglecting the weight of the beam). At the ends, the vertical component of prestress $F\sin\theta$ is transmitted directly into the supports, while the horizontal component $F\cos\theta$ creates a uniform compression along the entire beam. Thus the stresses in the beam at any section (except for local stress concentrations) are simply given by

$$f = \frac{F\cos\theta}{A_c}$$

$$= \frac{F}{A_c}$$

for small values of θ. Any loading in addition to P will now cause bending in an elastic homogeneous beam (up to point of cracking), and the additional stresses can be simply computed by

$$f = \frac{Mc}{I}$$

where M = the moment produced by load in addition to P.

Similarly, Fig. 11-3 illustrates the balancing of a uniformly distributed load by means of a parabolic cable whose upward component w_b (lb/ft) is given by

$$w_b = \frac{8Fh}{L^2}$$

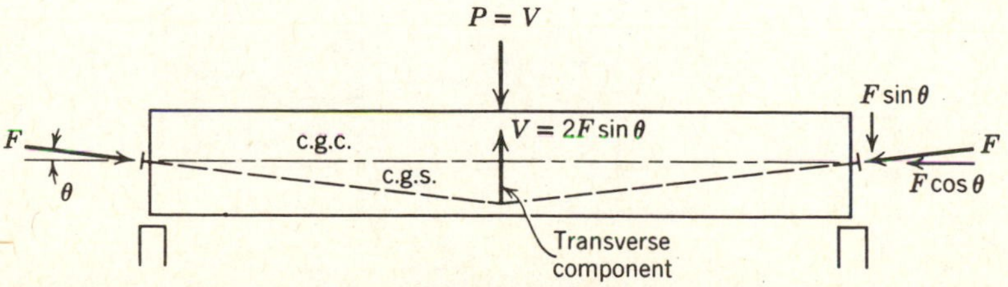

Fig. 11-2. Balancing of a concentrated load.

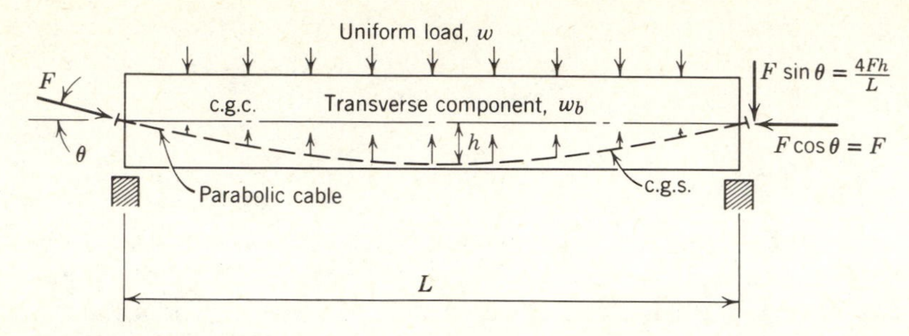

Fig. 11-3. Balancing of a uniform load.

If the externally applied load w (including the weight of the beam) is exactly balanced by the component w_b there is no bending in the beam. The beam is again under a uniform compression with stress,

$$f = \frac{F}{A_c}$$

Should the external load be different from w_b, it is only necessary to analyze the moment M produced by the load differential and compute the corresponding stresses by the formula,

$$f = \frac{Mc}{I}$$

This procedure has already been illustrated in example 1-4 of Chapter 1.

Now consider a cantilever beam, Fig. 11-4. The conditions for load balancing become slightly more complicated, because any vertical component at the cantilever end C will upset the balance, unless there is an externally applied load at that tip. To balance a uniformly distributed load w, the tangent to the c.g.s. at C will have to be horizontal. Then the parabola for the cantilever portion can best be located by computing

$$h = \frac{wL^2}{2F}$$

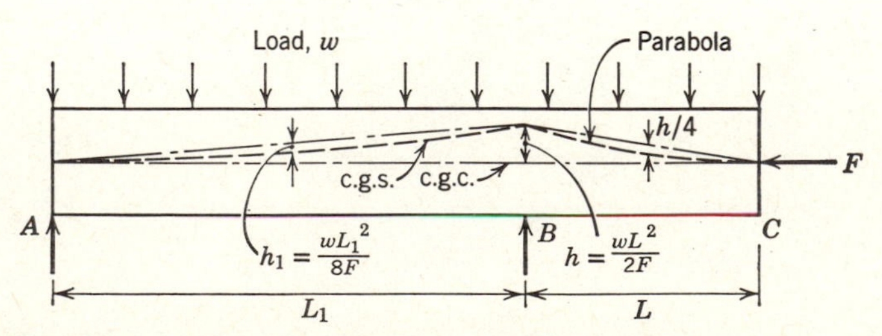

Fig. 11-4. Load balancing for a cantilever beam.

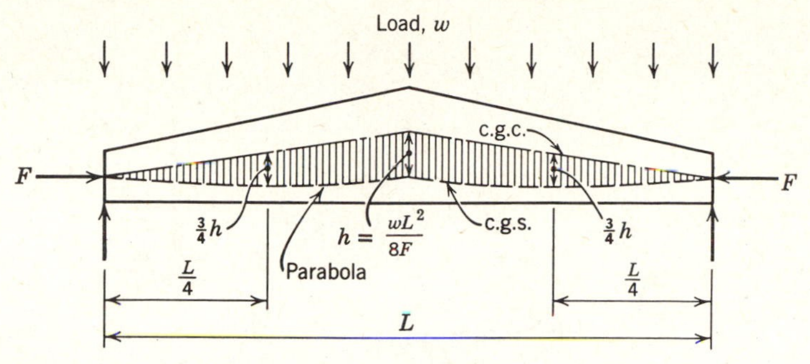

Fig. 11-5. Beam with curved c.g.c.

and the parabola for the anchor arm by

$$h_1 = \frac{wL_1^2}{8F}$$

It soon becomes apparent that the load-balancing approach may not always be the simplest, and the same result may be achieved by locating the c.g.s. line, such that the area between the c.g.s. and the c.g.c. will be proportional to the external moment diagram.

In order to compute the transverse component of prestress for beams with curved c.g.c. (Fig. 11-5), it is necessary to use the area between the c.g.s. and the c.g.c, rather than just the curvature of the cable itself. This becomes evident when the eccentric moment created by the horizontal end component of the prestress is taken into consideration.

When exact load balancing is required, the c.g.s. line should always be located at the c.g.c. for the ends of the beam. For practical reasons, it may not be desirable to do so, as for a pretensioned beam of short span (Fig. 11-6), where exact load balancing is not obtained, and other methods of analysis may be simpler.

EXAMPLE 11-1

A double cantilever beam is to be designed so that its prestress will exactly balance the total uniform load of 1.6 k/ft (23.3 kN/m) normally carried on the beam, Fig. 11-7. Design the beam using the least amount of prestress, assuming that the c.g.s. must have a

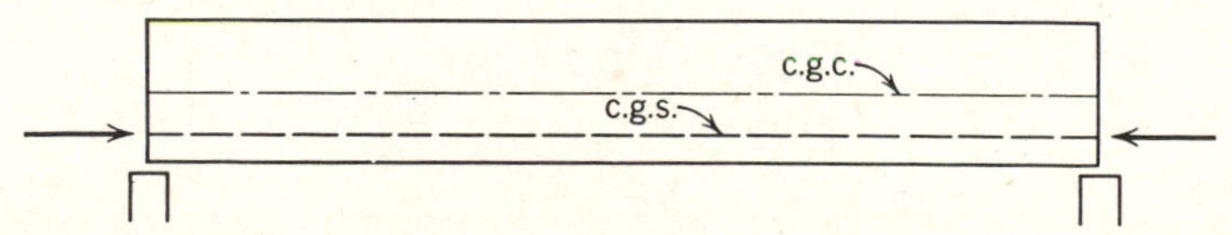

Fig. 11-6. Beam with straight cable.

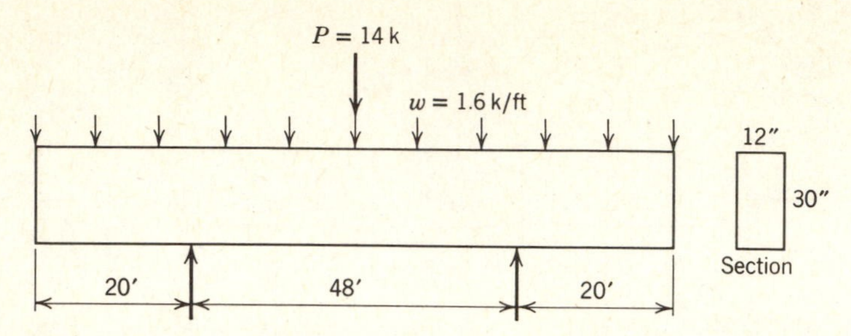

Fig. 11-7. Example 11-1.

concrete protection of at least 3 in. (76.2 mm). If a concentrated load of $P=14$ k (62 kN) is added at midspan, compute the maximum top and bottom fiber stresses.

Solution In order to balance the load in the cantilever, the c.g.s. at the tip must be located at the c.g.c with a horizontal tangent. To use the least amount of prestress, the eccentricity over the support should be a maximum, that is, $h=12$ in. or 1 ft (0.305 m). The prestress required is

$$F=\frac{wL^2}{2h}$$
$$=\frac{1.6\times 20^2}{2\times 1}$$
$$=320 \text{ k } (1423 \text{ kN})$$

In order to balance the same load on the center span, using the same prestress, $F=320$ k (1423 kN), the sag for the parabola must be

$$h_1=\frac{wL_1^2}{8F}$$
$$=\frac{1.6\times 48^2}{8\times 320}$$
$$=1.44 \text{ ft } or \text{ } 17.3 \text{ in. } (0.439 \text{ m})$$

Hence the c.g.s. is located as shown, Fig. 11-8.

Under the combined action of the uniform load and the prestress, the beam will have no

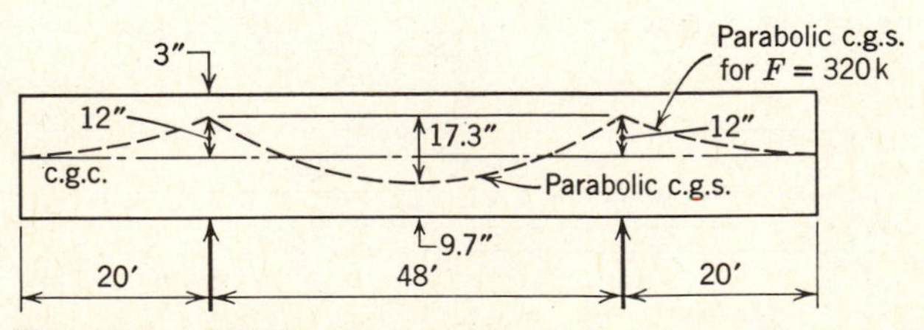

Fig. 11-8. Solution for example 11-1.

deflection anywhere and is under uniform compressive stress of

$$f = \frac{F}{A_c} = \frac{320{,}000}{360} = -889 \text{ psi } (-6.13 \text{ N/mm}^2)$$

Owing to the $P = 14$ k (62 kN), the moment M at midspan is,

$$M = \frac{PL}{4} = \frac{14 \times 48}{4} = 168 \text{ k-ft } (228 \text{ kN-m})$$

and the extreme fiber stresses are

$$f = \frac{Mc}{I} = \frac{6M}{bd^2} = \frac{6 \times 168 \times 12{,}000}{12 \times 30^2}$$

$$= \pm 1120 \text{ psi } (\pm 7.72 \text{ N/mm}^2)$$

The resulting stresses at midspan are

$$f_{top} = -889 - 1120 = -2009 \text{ psi } (-13.85 \text{ N/mm}^2) \text{ compression}$$

$$f_{bot} = -889 + 1120 = +231 \text{ psi } (+1.59 \text{ N/mm}^2) \text{ tension}$$

Note that the actual cable placement may not possess the sharp bend shown over the supports, and the effect of any deviation from the theoretical position must be investigated accordingly. Also note that $F = 320$ k (1423 kN) is the effective prestress, so that under the initial prestress there will be a slight camber at midspan and either a camber or a deflection at the tips which can be computed.

For better stress conditions under the load P and better deflection control if this load normally exists for long periods of time, it would be desirable to relocate the c.g.s. so that it would have more sag at midspan. Then a balanced condition would not exist under the uniform load w acting alone.

11-3 Continuous Beams

Several methods are available for designing a prestressed continuous beam for balanced load. One method is to compute the actual moment diagram produced by the loading and plot the c.g.s. with an ordinate $y = M/F$ measured from the c.g.c. This involves the plotting of moment diagrams for continous beams as a start, and it yields only a concordant cable, which is generally not the most economical one. The application of the balanced-load concept to prestressed continuous beams not only greatly simplifies their design and analysis, but also gives an approach to the use of nonconcordant cables, as well as concordant ones.

A continuous beam under the balanced action between the transverse component of the prestress and the applied external load has a uniform stress f across any section of the beam. This is given by

$$f = \frac{F}{A_c}$$

For any change from the balanced-load condition, ordinary elastic analysis (such as moment distribution) can be applied to the load differential to obtain the moment M at any section. The resulting stresses, in addition to the $f=(F/A_c)$ stress, can be computed from the familiar formula,

$$f=\frac{My}{I}$$

This means that, after load balancing, the analysis of prestressed continuous beams is reduced to the analysis of a nonprestressed continuous beam. Furthermore, since such analysis will be applied to only the unbalanced portion of the load, any inaccuracies in the method of analysis become a relatively insignificant factor, and approximate methods may often prove sufficient.

While analysis by this balanced-load approach is easier for continuous beams, the stress values obtained are no different than those found by the method described in Chapter 10. Design by this method, however, gives a different visualization of the problem and may yield different layouts and proportions. In order to cultivate the ability to design, it will be necessary to first learn the method of analysis. A simple case is illustrated in example 11-2.

EXAMPLE 11-2

For the symmetrical continuous beam in Fig. 11-9, prestressed with $F=320$ k (1423 kN) along a parabolic cable as shown, compute the extreme fiber stresses over the center support for $DL+LL=1.6$ k/ft (23.3 kN/m). Use the balanced-load method.

Solution The upward transverse component of prestress is

$$w_b=\frac{8Fh}{L^2}=\frac{8\times 320\times 1}{50^2}=1.03 \text{ k/ft } (15.0 \text{ kN/m})$$

For an applied downward load of 1.03 k/ft (15.0 kN/m), the beam is balanced under a uniform stress of

$$f=\frac{320{,}000}{360}=-889 \text{ psi } (-6.13 \text{ N/mm}^2)$$

For applied load $w=1.6$ k/ft (23.3 kN/m), the unbalanced downward load is,

$$1.60-1.03=0.57 \text{ k/ft } (8.3 \text{ kN/m})$$

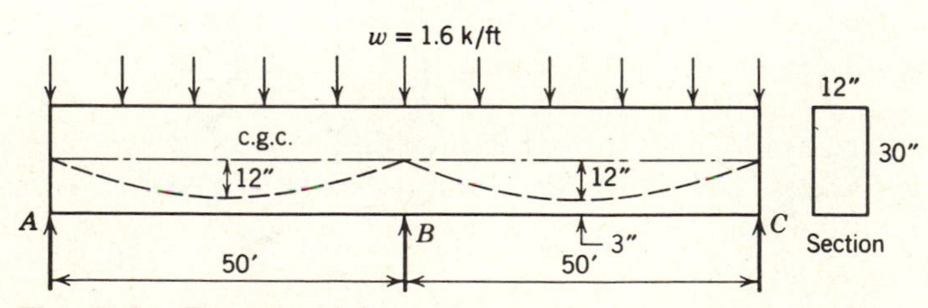

Fig. 11-9. Example 11-2.

which produces a negative moment over the center support,

$$M=\frac{wL^2}{8}=\frac{0.57\times 50^2}{8}$$
$$=-178\text{ k-ft }(-241\text{ kN-m})$$

and fiber stresses,

$$f=\frac{Mc}{I}=\frac{6\times 178\times 12{,}000}{12\times 30^2}$$
$$=\pm 1187\text{ psi }(\pm 8.18\text{ N/mm}^2)$$

The resulting stresses are

$$f_{\text{top}}=-889+1187=+298\text{ psi }(+2.05\text{ N/mm}^2)\text{ tension}$$

$$f_{\text{bot}}=-889-1187=-2076\text{ psi }(-14.31\text{ N/mm}^2)\text{ compression}$$

Note that we have analyzed the stresses without referring to the concordancy and linear transformation of cables as discussed in Chapter 10. Those who have studied Chapter 10 will notice that this is a nonconcordant cable, and reactions are induced by prestressing. It will be easy to compute the total reactions by the balanced-load approach. Thus, under the action of 1.03 k/ft (15.0 kN/m) applied load (including beam weight), the reactions are simply the vertical components of the cables:

Exterior support, reaction $R_A=1.03\times 25=25.8$ k (115 kN)

Interior support, reaction $R_B=2\times 25.8=51.6$ k (230 kN)

Using the principlies of "primary" and "secondary" moment, we can easily verify the reactions found above by load balancing. The tendon profile of Fig. 11-9 shows that there is zero eccentricity of c.g.s. from c.g.c. at support B. Thus, no "primary" moment is present at B, and moment from prestress is "secondary." Moment due to prestress only at this section is simply the moment created by the $w_b=1.03$ k/ft (15.0 kN/m) (upward) balanced load.

$$M_b=\frac{(1.03)(50)^2}{8}=322\text{ ft-k }(437\text{ kN-m})$$

This "secondary" moment causes a reaction at A

$$R_A=\frac{332}{50}=6.44\text{ k }(28.6\text{ kN})\text{ (upward)}$$

The reaction due to $w_b=1.03$ k/ft (15.0 kN/m) applied to the two-span continuous beam is

$$R_A=\left(\frac{3}{8}\right)(1.03)(50)=19.31\text{ k }(85.9\text{ kN})\text{ (upward)}$$

Thus, under the balanced load the reaction at A is

$$R_A=6.44+19.31=25.75\text{ k }(115\text{ kN})$$

(same as 25.8 k from load balancing)

Under the action of the additional 0.57 k/ft (8.3 kN/m) load, the reactions can be computed by using ordinary continuous beam formulas:

$$\text{Exterior support, } R_A = \tfrac{3}{8}wL = \tfrac{3}{8} \times 0.57 \times 50 = 10.6 \text{ k } (47.1 \text{ kN})$$

$$\text{Interior support, } R_B = 2 \times \tfrac{5}{8}wL = 2 \times \tfrac{5}{8} \times 0.57 \times 50 = 35.6 \text{ k } (158.3 \text{ kN})$$

Hence the total reactions are due to 1.60 k/ft and the effect prestress,

$$\text{Exterior support, } R_A = 25.8 + 10.6 = 36.4 \text{ k } (162 \text{ kN})$$

$$\text{Interior support, } R_B = 51.6 + 35.6 = 87.2 \text{ k } (388 \text{ kN})$$

The simple symmetrical cable profile in example 11-2 is chosen for the convenience of illustrating the method of analysis; it does not represent the most economical location. But once the principle of load-balancing is well understood, it is possible to design the beam economically and in a straightforward manner, as will be shown in example 11-3.

EXAMPLE 11-3

For the continuous beam in example 11-2, determine the prestress F required to balance a uniform load of 1.03 k/ft (15.0 kN/m), using the most economical location of cable. Assume a concrete protection of at least 3 in. (76.2 mm) for the c.g.s. Compute the stresses at support B and the reactions for the effect of prestress together with an external load of 1.6 k/ft (23.3 kN/m).

Solution The most economical cable location is one with the maximum sag so that the least amount of prestress will be required to balance the load. As shown in Fig. 11-10, a 3-in. (76.2 mm) protection is given to the c.g.s. over the center support and at midspan. [A theoretical parabola based on these clearances will have slightly less than 3 in. (76.2 mm) at a point about 20 ft (6.10 m) in from the exterior support.] The c.g.s. at the beam ends should coincide with the c.g.c. and cannot be raised, not only because such raising will destroy the load balancing, but it will not help to increase the efficiency of the cable, since unfavorable end moments will be introduced.

The cable now has a sag of 18 in. (457.2 mm) and the prestress F required to balance the load of 1.03 k/ft (15.0 kN/m) is

$$F = \frac{wL^2}{8h} = \frac{1.03 \times 50^2}{8 \times 1.5}$$

$$= 214 \text{ k } (952 \text{ kN})$$

The fiber stress under this balanced load condition is now

$$f = \frac{F}{A_c} = \frac{214{,}000}{360} = -593 \text{ psi } (-4.09 \text{ N/mm}^2)$$

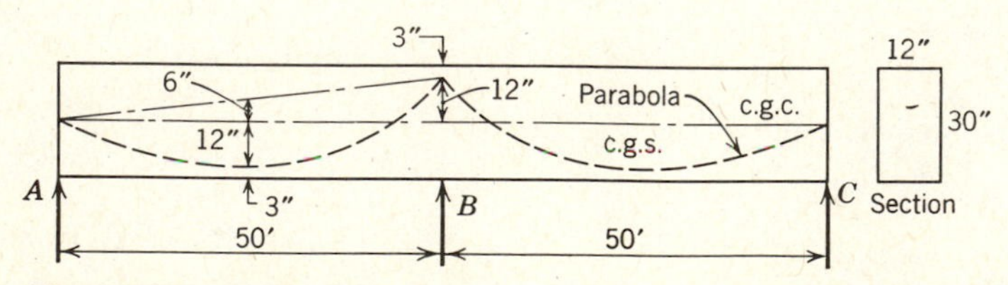

Fig. 11-10. Example 11-3.

Owing to the additional load of 0.57 k/ft (8.3 kN/m), the stresses produced over center support are, as in example 11-2,

$$f = \pm 1187 \text{ psi } (8.18 \text{ N/mm}^2)$$

And the resulting stresses over center support are

$$f_{top} = -593 + 1187 = +594 \text{ psi } (+4.10 \text{ N/mm}^2) \text{ tension}$$

$$f_{bot} = -593 - 1187 = -1780 \text{ psi } (-12.27 \text{ N/mm}^2) \text{ compression}$$

The reactions due to 1.03 k/ft (15.0 kN/m) can be computed from the vertical components of the cable and are, very closely,

$$\text{Exterior support, } R_A = 1.03 \times 25 - \tfrac{1}{50}(214) = 25.8 - 4.3 = 21.5 \text{ k } (95.6 \text{ kN})$$

$$\text{Interior support, } R_B = 51.6 + 2 \times 4.3 = 60.2 \text{ k } (267.8 \text{ kN})$$

The reaction at A can be verified in this example, as in example 11-2, by using the principles of primary and secondary moment from Chapter 10. We have the same moment of 322 ft-k (437 kN–m) due to prestress only since the same $w_b = 1.03$ k/ft (15.0 kN/m) is developed in this beam. In this case the "primary" portion of this moment is due to $F = 214$ k (952 kN) at $e_b = 1.0$ ft (0.30 m).

$$\text{Total prestress moment} = 322 \text{ ft-k } (437 \text{ kN–m})$$

$$\text{"Primary" moment} = 214 \times 1 = 214 \text{ ft-k } (290 \text{ kN–m})$$

Thus, we know the "secondary" moment at B is

$$\text{"Secondary" moment} = 322 - 214 = 108 \text{ ft-k } (147 \text{ kN–m})$$

This "secondary" moment causes a reaction at A

$$R_A = \frac{108}{50} = 2.16 \text{ k } (9.6 \text{ kN}) \text{ (upward)}$$

With the balanced load of 1.03 k-ft (15.0 kN/m) applied to the beam, the total reaction is that from prestress (secondary moment) plus $R_A = 19.31$ k (85.9 kN) from the balanced load (upward as found previously). The total reaction at A is thus

$$R_A = 2.16 + 19.31 = 21.47 \text{ k } (95.5 \text{ kN}) \text{ (same as 21.5 k from load balancing)}$$

Under the action of 0.57 k/ft (8.3 kN/m) load, the reactions are, as obtained for example 11-2,

$$\text{Exterior support, } R_A = 10.6 \text{ k } (47.1 \text{ kN})$$

$$\text{Interior support, } R_B = 35.6 \text{ k } (158.3 \text{ kN})$$

Hence the total reactions due to 1.6 k/ft (23.3 kN/m) load and the effect of $F = 214$ k (952 kN) are

$$\text{Exterior support, } R_A = 21.5 + 10.6 = 32.1 \text{ k } (143 \text{ kN})$$

$$\text{Interior support, } R_B = 60.2 + 35.6 = 95.8 \text{ k } (426 \text{ kN})$$

Although these examples are limited to uniform loads, the same principle can be applied to concentrated loads as illustrated in example 11-1. When the cable transverse component is higher than the externally applied load, the load differential would be upward instead of downward, and computation can be made accordingly.

Since the cables in the two previous examples are nonconcordant, it was worthwhile to compute the reactions at A by following the method of linear transformation explained in Chapter 10. Evidently, more work is required even for the analysis. But what is more important is the clarity of design afforded by this load balancing method. For actual design, the load to be balanced by prestress should be chosen after individual study and with the guidance of experience.

The linear transformation of c.g.s. lines can be easily explained by the balanced load-concept. Since the transverse force applied from the cable on the concrete remains unchanged by linear transformation, it is evident that the elastic behavior of the beam remains the same. On the other hand, since the vertical components from the cable do change directly over the supports by linear transformation of c.g.s. lines, the reactions will be accordingly modified. This phenomenon has been explained in Chapter 10 and is again explained here by a new approach.

This method can be conveniently applied to continuous beams of multiple and unequal spans, including the use of unequal prestress in the spans. For example, the beam in Fig. 11-11 carrying various loads can have the loads in each span balanced by choosing a proper profile for the c.g.s. The cantilever span at the left has the c.g.s. tangent to the c.g.c. at the tip. The middle span has a sag h_{3a} to balance the concentrated load P and another sag h_{3b} to balance the uniform load w_3. The longer span at the right end can have higher prestress by adding more cables. The total prestress in each section is computed by using the corresponding F, w, L, and h for the section span in question. Either the F is known and the h is computed or, if h is predetermined, then F is computed.

If this is done, the entire beam is balanced due to the effect of prestress and the given loading. Then the stress at any section is computed simply by

$$f = \frac{F}{A_c}$$

where F and A_c are the prestress and the concrete area for that section.

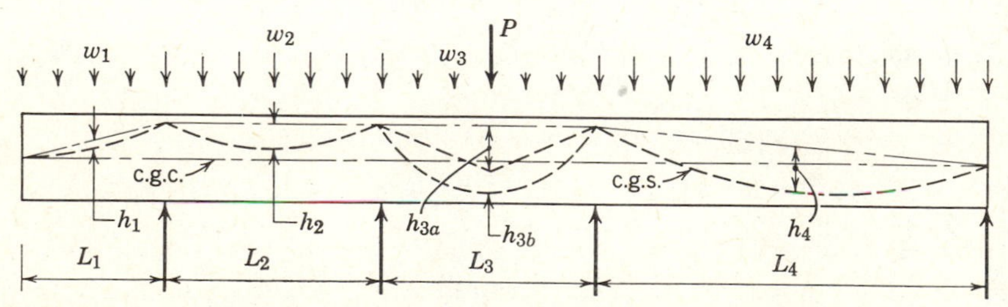

Fig. 11-11. Load balancing for irregular spans and loadings.

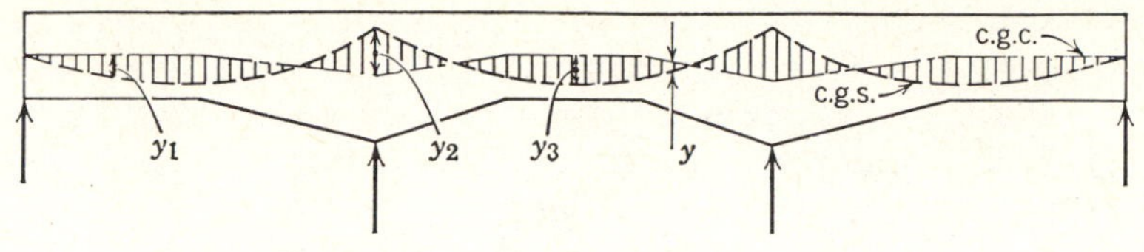

(a) Actual c.g.s. Location for Balanced Load.

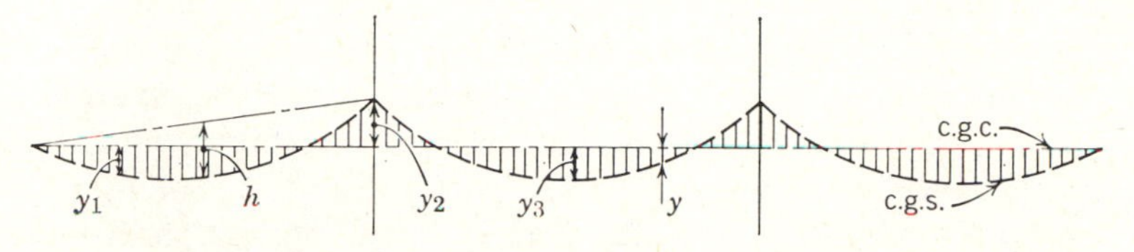

(b) Graphical Solution to Obtain c.g.s. Location

Fig. 11-12. Balanced-load design for beam with curved c.g.c.

The proper use of this method for design depends greatly on the choice of the proper loading to be balanced by the prestress. A simple uniform load for the balanced design may not be the most desirable, or the most economical. With some experience, excellent design can be obtained with such simplicity that the design of continuous prestressed-concrete structures is no longer a difficult problem.

For beams with curved or bent c.g.c., the determination of c.g.s. for a balanced load design can be made using a semi-graphical method as in Fig. 11-12. First estimate from (a) the controlling eccentricites (probably y_1 and y_2) as dictated by physical dimensions and prestress requirements. Transfer these eccentricities to (b) and join them with parabolas or other c.g.s. profiles. The sag h and the prestress F required for load balancing can now be computed. Transpose all the y values from (b) back into (a), obtaining the actual cable layout. A few trials and errors might be required to obtain the most desirable location, but no great difficulty is involved. Again, adjustments in the profile might have to be made to suit certain physical requirements. These include rounding off the peaks of the profile over the interior supports, moving the c.g.s. at the ends away from the c.g.c. etc. The effect of such variations should be calculated or at least estimated.

11-4 Rigid Frames

The analysis and design of prestressed-concrete rigid frames can be made by following the method explained in Chapter 10 for continuous beams. This consists of taking freebodies of the concrete separated from the tendons and of determining the linear transformation of the tendons. Although the method is not too difficult, the designer can get lost in the mechanics of analysis so that he cannot easily visualize the problem and reach an economical design. When axial shortening and induced stresses are considered, or when precast elements are joined to form a frame, the design of rigid frames becomes much more complicated than that of continuous beams.

The balanced-load approach, on the other hand, can be applied to rigid frames quite easily, since it quickly leads to a condition of uniform stress distribution for all members of a rigid frame. Since the relatively unfamiliar effect of prestressing has been eliminated by load balancing, it is only necessary to analyze an ordinary elastic rigid frame subjected to the additional loading or to the effect of axial shortening. Since engineers are already familiar with such analysis, and since we are only concerned with the added loading or with the effect of axial shortening, without the effects of bending, the problem becomes easily controllable.

Consider a simple case of a one-story single rigid frame, Fig. 11-13. If it is desired to balance the uniform load w so that there will be no bending at all, a parabolic c.g.s. can be designed and located such that

$$F_1 = \frac{wL^2}{8h_1}$$

If the span is short and the columns long, it will be better to locate the c.g.s. of F_1 without any eccentricity with respect to the c.g.c. Then it will not be necessary to prestress the column in order to balance the loading. On the other hand, if the span is long and the columns short, it will often be economical to place the ends of F_1 as high as possible, so as to obtain a higher value for h_1 and thus a smaller value of F_1, as shown in Fig. 11-13. In this case, the prestress F_1 produces an eccentric end moment F_1e_1 which must be balanced by another

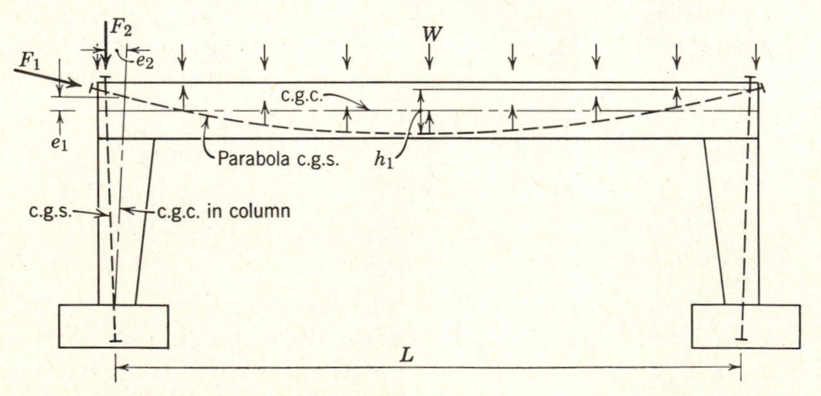

Fig. 11-13. Prestressed rigid frame.

cable in the column placed with an eccentricity e_2 such that

$$F_2e_2 = F_1e_1$$

When the frame is so designed, all sections in the frame are subjected to uniform stress distribution (except at points of stress concentration) with stress in the beam equal to F_1/A_1, and stress in the column equal to $(F_2+V_2)/A_2$ where V_2 is the vertical load on the column. A_1 and A_2 are the cross-sectional areas of the beam and the column, respectively.

Fig. 11-14. 120-ft precast girders, pretensioned to balance own weight and posttensioned to balance in-place slabs, form rigid frames with posttensioned columns—Telecomputing Facilities, Chatsworth, California (T. Y. Lin International, Consulting Engineers).

The elastic shortening of the beam under prestress, as well as its shrinkage and creep, will tend to move the columns inward, thus producing bending in both the beam and the columns, even though the external loadings are balanced. Such stresses should be computed and if necessary counteracted by relocating the c.g.s. or by putting additional prestress in both the beam and the columns, or they can be resisted by nonprestressed steel.

Flexural stresses produced by bending of columns resulting from creep shortening in the beams can be computed on the basis of a reduced modulus of elasticity E_{cr}, which takes into account the flexural creep in the columns. Otherwise, the computed flexural stresses in the columns would be unreasonably high.

While the above explains the design of a cast-in-place rigid frame, a slight modification will make it applicable to precast rigid frames. It is generally desirable to first balance the weight of the beam. Then, when the beam is placed in position and more load is superimposed, additional prestress is applied to balance the load. An interesting example is shown in Fig. 11-14, where a 120-ft (36.6 m) precast T-beam was pretensioned to balance its own weight, and then formed into a rigid frame by posttensioning the beam to balance the weight of the concrete slab poured on top. High tensile bars were used to prestress the column and to supply the end moment F_2e_2 as indicated in Fig. 11-13.

Some multistory rigid frames may require prestressing only for the beams. Balancing the load on the beams will take away the major part of the moment in the columns. Then if the value of e_1 is small, reinforcing the columns may be sufficient to carry the bending stresses.

11-5 Actual Tendon Layouts for Design

In posttensioning of actual structures, the idealized simple parabolic curves used in the load-balancing design prodecure (presented in previous sections of this chapter) may be only an approximate approach. Actual tendon layout made up of parabolic segments forming a smooth reversed curve will often produce moments from prestressing very close to a broken series of simple parabolas. We will first consider how an actual layout may be modeled in our calculations, then we will compare moments from companion idealized and actual layouts.

Let us compare the idealized tendon layout used for example 11-3 with an actual layout as shown in Fig. 11-15. The idealized layout was established with control points shown by the heavy dots in Fig. 11-15(a) and the $F = 214$ k (952 kN) was established on the basis of load balancing. Such a layout cannot be used exactly due to the cusp at B which would require the tendon to be bent through a sharp angle. The actual layout of Fig. 11-15(b) overcomes this problem, while holding the control points and $F = 214$ k (952 kN) as previously

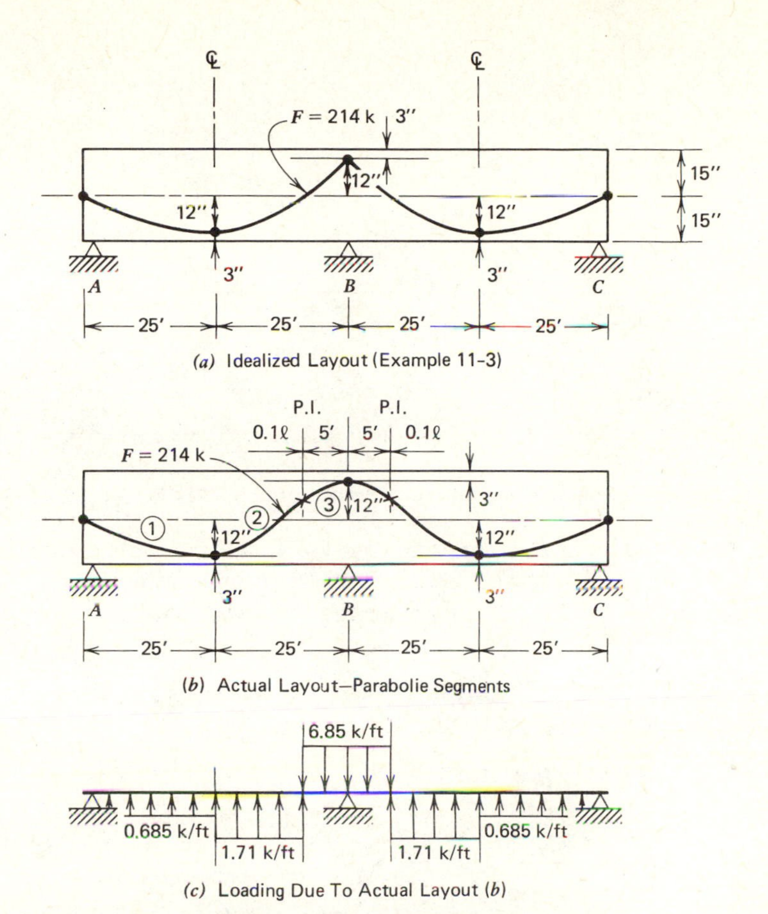

(a) Idealized Layout (Example 11-3)

(b) Actual Layout—Parabolie Segments

(c) Loading Due To Actual Layout (b)

Fig. 11-15. Idealized and actual layouts, examples 11-3 and 11-4.

established. At midspan, the actual layout has two parabolic segments joined at the control point, and each segment has a horizontal tangent. From this point to support B we use two reversed parabolic segments which join with a common slope at the point of inflection ($0.1 \times$ span away from B). A detail of this part of the span is shown in Fig. 11-16.

In Fig. 11-16 an important property of the two intersecting parabolas joining at the P.I. is shown. The two parabolas join with a common slope at the point X, which lies on a chord joining the two control points shown as heavy dots. The point of inflection will typically be at about $\frac{1}{10}$ of span as indicated, but it can generally be moved within the range $1/8$ to $1/12$ of span. A typical tendon can accommodate this much curvature according to actual practice, with $0.1 \times$ span being a common assumption for distance from high point, such as B, to point of inflection, X, where curvature reverses.

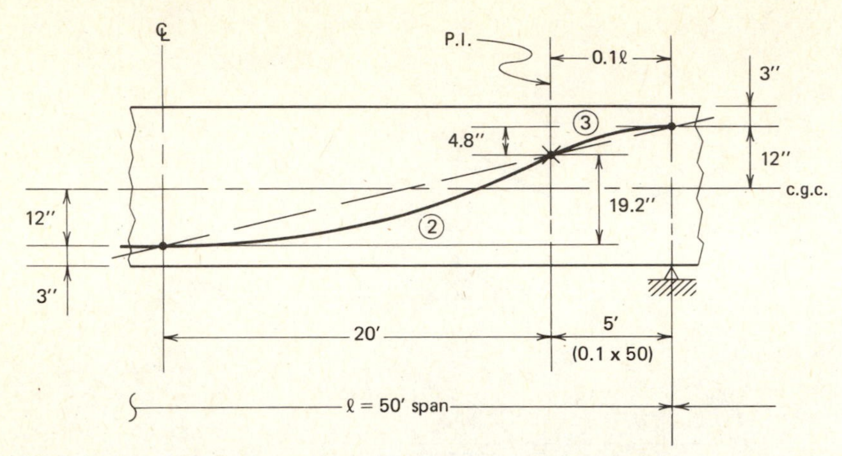

Fig. 11-16. Detail of actual tendon layout from Fig. 11-15(*b*).

Each of the three parabolic segments is a half-parabola, similar to the idealized parabolas previously used in load balancing. We can use the same simple relationship to get the loadings for the three segments as follows: (refer to Figs. 11-15 and 11-16)

(1) $$w_b = \frac{8Fh}{L^2} = \frac{(8)(214)(1.0)}{(2\times 25)^2} = 0.685 \text{ k/ft } (10.0 \text{ kN/m})$$

(2) $$w_b = \frac{(8)(214)\left(\frac{19.2}{12}\right)}{(2\times 20)^2} = 1.71 \text{ k/ft } (25.0 \text{ kN/m})$$

(3) $$w_b = \frac{(8)(214)\left(\frac{4.8}{12}\right)}{(2\times 5)^2} = 6.85 \text{ k/ft } (100.0 \text{ kN/m})$$

These loads along the span, Fig. 11-15(*c*), are used for the actual layout to obtain moment due to prestress at *B*. Design aids are readily available to allow determination of fixed-end moments for partial span loadings such as those shown in Fig. 11-15(*c*). Any available analysis solution could be used, and it is rather common for design offices to have small computers which could easily handle this type problem.

Comparing results obtained for this actual layout with the moment produced by the idealized layout, Fig. 11-15(*a*), we note the following:

Idealized layout: $M_B = +322$ ft-k (437 kN–m)

(Uniform $w_b = 1.03$ k/ft)

Actual layout: $M_B = +298$ ft-k (404 kN–m)

(see w_b in Fig. 11-15(*c*))

The actual loading is found to produce moment at B, which is only 7.5% different from the much simpler idealized layout which would commonly be used with load balancing as described in earlier sections of this chapter. This result is representative of the accuracy one might expect for similar problems where idealized layout control points (Fig. 11-15) are held in describing the actual layout.

EXAMPLE 11-4

(*a*) For the continuous beam of Example 11-3 having the actual tendon layout of Fig. 11-15(*b*) and a prestress force $F=214$ k, find the stresses at support B due to prestress along with 1.6 k/ft external load. Compare results with the stresses solved in example 11-3 with idealized tendon.

(*b*) Also find the stresses at midspan AB with actual tendon layout. For rectangular section $b=12$ in. and $h=30$ in.: $A=360$ in.^{2}, $I_X=27{,}000$ in.^{4}, and $S_x=1800$ in.^{3} ($F=952$ kN, $w=23.3$ kN/m, $A=232\times10^3$ mm^2, $I_x=11.2\times10^9$ mm^4, and $S_x=29.5\times10^6$ mm^3).

Solution (*a*) We have already shown the actual tendon profile and resulting loads along the span in Fig. 11-15. It was also shown that the moment at B due to prestress $F=214$ k (952 kN) with this layout was $M_p=+298$ ft-k (404 kN-m). The moment at B from applied load of 1.6 k/ft (23.3 kN/m) is:

$$M_B=-\frac{(1.6)(50)^2}{8}=-500 \text{ ft-k } (-678 \text{ kN-m})$$

Stresses at B are $f=\dfrac{F}{A}\pm\dfrac{M_p}{S_x}\pm\dfrac{M_B}{S_x}$

$$f_{top}=-\frac{214}{360}-\frac{298\times12}{1800}+\frac{500\times12}{1800}=+0.752 \text{ ksi } (+5.19 \text{ N/mm}^2) \text{ (tension)}$$

$$f_{bot}=-\frac{214}{360}+\frac{298\times12}{1800}-\frac{500\times12}{1800}=-1.941 \text{ ksi } (-13.38 \text{ N/mm}^2) \text{ (compression)}$$

Note that the stresses exceed the values from Example 11-3 due to the fact that the moment from prestress with actual layout was 298 ft-k (404 kN–m) while that from the idealized layout was 322 ft-k (437 kN–m). If these stresses are found to be excessive, F could be increased slightly to balance a little more of the applied load. The moment due to prestress would change directly with F if the actual layout remains the same.

(*b*) Primary moment at $B-214\times1.0=214$ ft-k (290 kN-m)

Moment due to Prestreess at $B=298$ ft-k (404 kN-m)

Thus the secondary moment at $B=298-214=+84$ ft-k (114 kN-m)

Shift of C-line at $B=\dfrac{84\times12}{214}=4.71$ in. (120 mm) as shown in Fig. 11-17.

Shift of C-line at midspan $AB=\dfrac{4.71}{2}=2.36$ in. (60 mm), and

distance to C-line$=12-2.36=9.64$ in. (245 mm) as shown in Fig. 11-17.

Thus moment at midspan AB from prestress$=9.64\times\dfrac{214}{12}=172$ ft-k (233 kN-m)

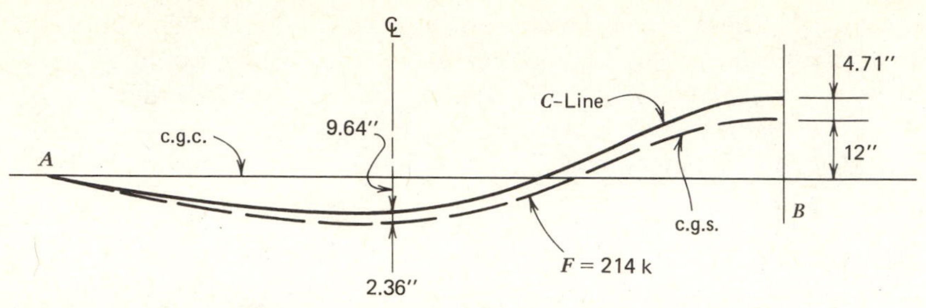

Fig. 11-17. *C*-line and c.g.s. line, example 11-4.

We can show that moment from applied load is:

$$M = \frac{wl^2}{16} = \frac{(1.6)(50)^2}{16} = 250 \text{ ft-k } (339 \text{ kN-m})$$

Thus the stresses at midspan AB are:

$$f_{\text{top}} = -\frac{214}{360} + \frac{172 \times 12}{1800} - \frac{250 \times 12}{1800} = -1.114 \text{ ksi } (-7.68 \text{ N/mm}^2) \text{ (comp.)}$$

$$f_{\text{bot}} = -\frac{214}{360} - \frac{172 \times 12}{1800} + \frac{250 \times 12}{1800} = -0.074 \text{ ksi } (-0.51 \text{ N/mm}^2) \text{ (comp.)}$$

This shows the support B section has higher stresses than midspan AB. Somewhat higher positive moment from loads actually exists at a section away from midspan (maximum moment at $\frac{3}{8}$ span distance from A), and this section could easily be checked in the same way that stresses at midspan AB were checked above.

11-6 Two-Dimensional Load Balancing

The principles of two-dimensional load balancing will be briefly explained in this section, including their application to a two-way slab and a grid system. The detailed design of prestressed slabs, and particularly flat slabs, will be presented in Chapter 12.

Two-dimensional load balancing differs from linear load balancing for beams and columns in that the transverse component of the tendons in one direction either adds to or subtracts from that component in the other direction. Thus the prestress design in the two directions or dimensions are closely related, one to the other. However, the basic principle of load balancing still holds, and the main aim of the design is to balance a given loading so that the entire structure (whether a slab or a grid) will possess uniform stress distribution in each direction and will not have deflection or camber under this loading. Any deviation from this balanced loading will then be analyzed as loads acting on an elastic slab without further considering the transverse component of prestress.

As a simple example of two-dimensional load balancing, let us consider a two-way slab simply supported on four walls Fig. 11-18. The cables in both

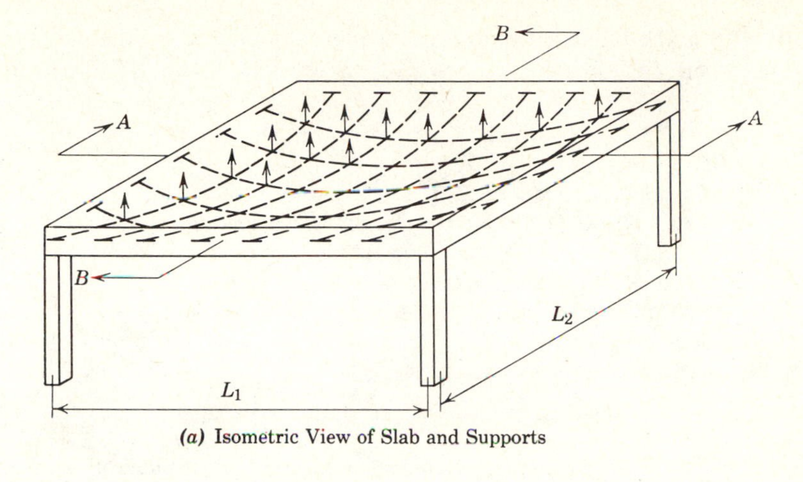

(a) Isometric View of Slab and Supports

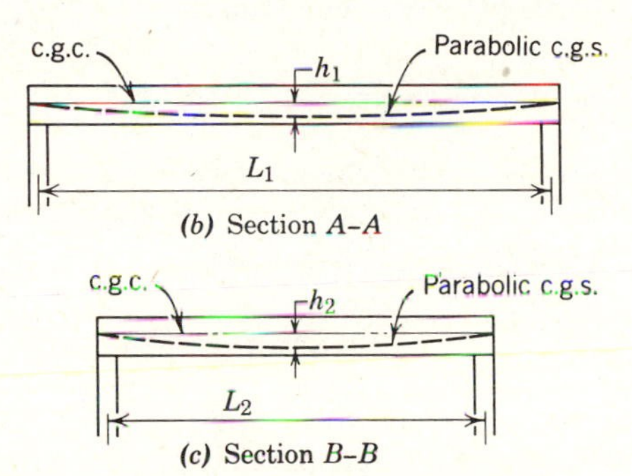

(b) Section A–A

(c) Section B–B

Fig. 11-18. Load balancing for two-way slabs.

directions exert an upward force on the slab, and if the sum of the upward components balances the downward load w, then we have a balanced design. Thus, if F_1 and F_2 are the prestressing forces in the two directions per foot width of slab, we have

$$\frac{8F_1h_1}{L_1^2} + \frac{8F_2h_2}{L_2^2} = w$$

Note that many combinations of F_1 and F_2 will satisfy the above equation. While the most economical design is to carry the load only in the short direction (or to carry $0.50w$ in each direction in case of a square panel), practical considerations might suggest different distributions. For example, if both directions are properly prestressed, it is possible to obtain a crack-free slab.

Under the action of F_1, F_2, and the load w, the entire slab has a uniform stress distribution in each direction equal to F_1/t and F_2/t, respectively. Any change in loading from the balanced amount of w can be analyzed by the elastic theory for slabs, such as Timoshenko[3] or O'Rourke. Example 11-5 illustrates the method.

EXAMPLE 11-5

An 8-in. slab supported on four walls, Fig. 11-19, is to be posttensioned in two directions. Design live load is 100 psf (4.79 kN/m^2). Compute the amount of prestress, assuming that a minimum of 200 psi (1.38 N/mm^2) compression is desired in the concrete in each direction for the purpose of getting a watertight roof slab.

Solution Since it is more economical to carry the load in the short direction, a minimum amount of prestress will be used in the long direction. At 200 psi (1.38 N/mm^2) compression in concrete, the prestress is,

$$200 \times 8 \times 12 = 19.2 \text{ k/ft (280 kN/m) of slab}$$

A parabolic c.g.s. with this prestress will supply a uniformly distributed upward force, as computed below, for an eccentricity of 2.75 in. (70 mm),

$$w = \frac{8Fe}{L^2} = \frac{8 \times 19{,}200 \times 2.75/12}{40^2} = 21.9 \text{ psf } (1.05 \text{ kN/m}^2)$$

Since the weight of slab is 100 psf (4.79 kN/m^2), it will be necessary to supply another upward force of $100 - 21.9 = 78.1$ psf (3.74 kN/m) in order to balance the dead load. This will require a prestress in the 30-ft (9.14 m) direction of

$$F = \frac{wL^2}{8e} = \frac{78.1 \times 30^2}{8 \times 2.75/12} = 38.4 \text{ k/ft (560 kN/m)}$$

and assuming an eccentricity of 2.75 in. (70 mm) (note that this eccentricity cannot be exactly maintained for cables in both directions where they cross each other at center

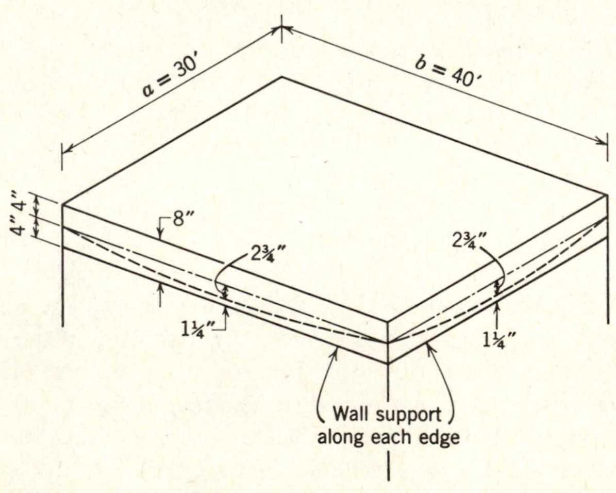

Fig. 11-19. Example 11-5.

portion of slab), which will give a uniform compression in the concrete of

$$\frac{38{,}400}{8\times 12}=400 \text{ psi } (2.76 \text{ N/mm}^2)$$

Thus, under the action of dead load alone, the slab will be under uniform stress of 200 psi (1.38 N/mm²) in the 40-ft (12.2 m) direction and 400 psi (2.76 N/mm²) in the 30-ft (9.14 m) direction.

The effect of live load can now be investigated. Referring to Timoshenko's treatise,[3] for span ratio $q=\frac{40}{30}=1.33$, we have $B=0.0713$ and $B_1=0.0505$. Thus, for $w=100$ psf (4.79 kN/m²) and $a=30$ ft (9.14 m),

$$M=Bwa^2=0.0713\times 100\times 30^2=6420 \text{ ft-lb/ft } (28.56 \text{ kN}-\text{m/m}) \text{ of slab}$$

Hence, the concrete fiber stresses are

$$f=\frac{F}{A}\pm\frac{Mc}{I}$$

$$=-400\pm\frac{6420\times 12\times 6}{12\times 8^2}$$

$$=-400\pm 602$$

$$=-1002 \text{ psi } (-6.91 \text{ N/mm}^2) \text{ compression top fiber,}$$

$$+202 \text{ psi } (+1.39 \text{ N/mm}^2) \text{ tension bottom fiber}$$

Stresses in the 40-ft direction can be similarly computed to be −626 psi (−4.32 N/mm²) compression for top fiber and +226 psi (+1.56 N/mm²) tension for bottom fiber.

If uniform stress distribution and zero deflection is not essential for a structure, balanced-load design may not be the most economical approach. For example, a cable placed along the middle strips will evidently be more effective than one along the wall. If more cables are located along the middle strips than along the walls, a stronger design might be obtained than the balanced-load design suggested above. If this is done, the slab will not be level under a uniform load or under its own weight. However, it will have no deflection under a varying load intensity which is everywhere equal and opposite to the upward component of the prestress.

The same principle of load balancing for the slab just described will apply to a grid system, if at each intersection the load is balanced by the upward component of prestress. Figure 11-20 shows a grid system (also called a waffle slab) with cables running in two directions X and Y, producing vertical components V_x and V_y. Near the center of the panel, both V_x and V_y act upward. Near the center of column line A–A, V_x acting upward is partly counteracted by V_y acting downward. Over a column capital, both V_x and V_y act downward but are balanced by another cable in the cap supplying a component V_c. Thus a prestressed grid system can be designed for a balanced-load condition with simple application of statics.

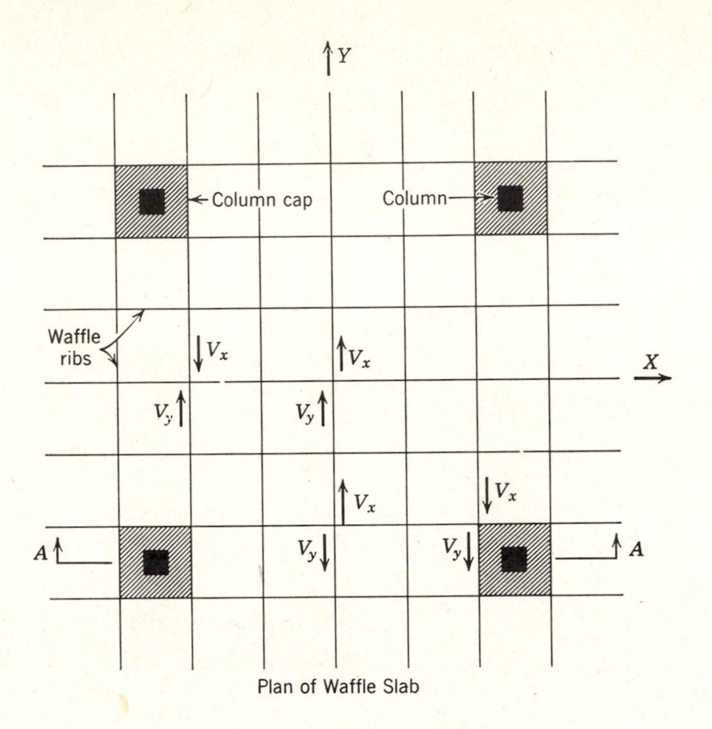

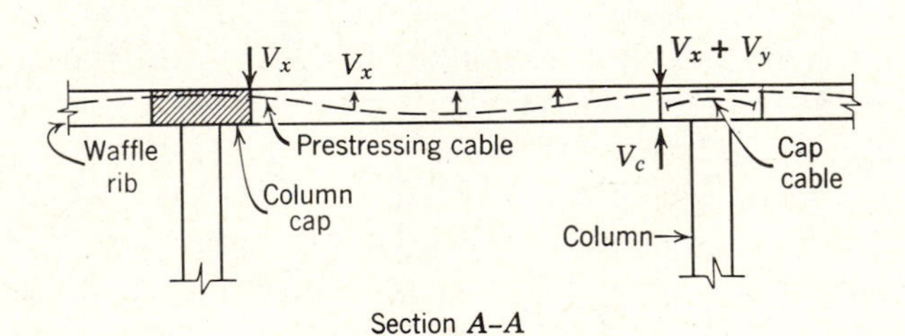

Fig. 11-20. Load balancing in a grid system.

If the dead load on the grid system is 100 psf and the live load 60 psf, and the design is made to balance a load of 110 psf, then it is only necessary to analyze the elastic slab for 10 psf upward load when no live load is on and for 50 psf downward when full live load is on. This approach will enable us to design prestressed grid systems of various layouts, since the elastic analysis needs to be made only for a small fraction of the total load and can be solved by a suitable approximate method.

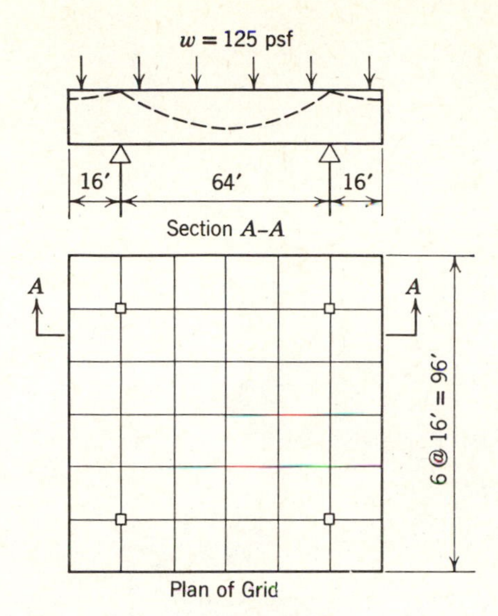

Fig. 11-21. Example 11-6.

EXAMPLE 11-6

A square grid system with 7 beams in each direction, Fig. 11-21, is to be designed for no deflection under a uniform load of 125 psf (6.0 kN/m^2) (which includes its own weight). Compute the vertical components of the tendons for a balanced-load condition.

Solution A solution is given in Fig. 11-22, indicating the vertical component of tendons in both directions, with a net upward force from the tendons at each interior panel point of $125 \times 16^2 = 32$ k (142 kN), each exterior panel point of 16 k (71 kN), and at each corner 8 k (35.5 kN). Only half of the force from the tendons along the center lines (on the X and Y axes) are listed in order that the diagrams will represent exactly a quadrant of the grid.

Although a balanced-load design is achieved by this arrangement, the design may not be the best, or necessarily the most economical, since the four beams passing the columns will carry very heavy loads and may have to be unusually large. Should the spans be longer than shown, some special arrangement around the columns might have to be made similar to the capital shown in Fig. 11-20.

For the purpose of illustration a tendon layout to furnish the proper vertical components is made for the beam along the column line, Fig. 11-22(c). It is noted that the c.g.s. must start from the c.g.c. at the ends of the beam in order to produce a balanced-load design. On account of the high location of c.g.c. for a waffle slab, the drape of c.g.s. for the cantilever is limited and a large amount of prestress might be required. For the center portion, a much greater drape is possible, and the prestress required is correspondingly less. Although a broken

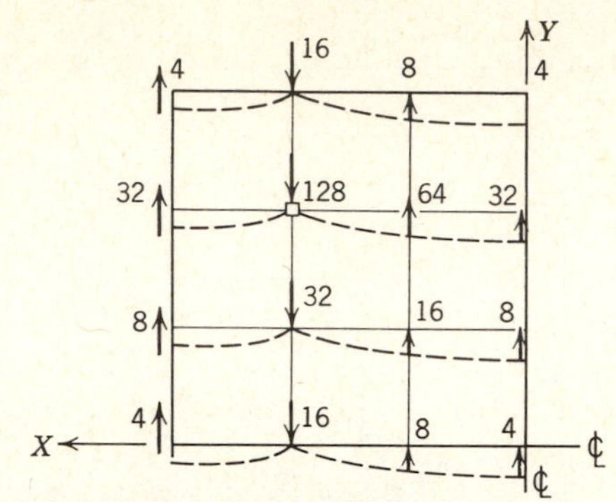

(a) Quarter Plan Showing Vertical Component of Tendons in *X* Direction

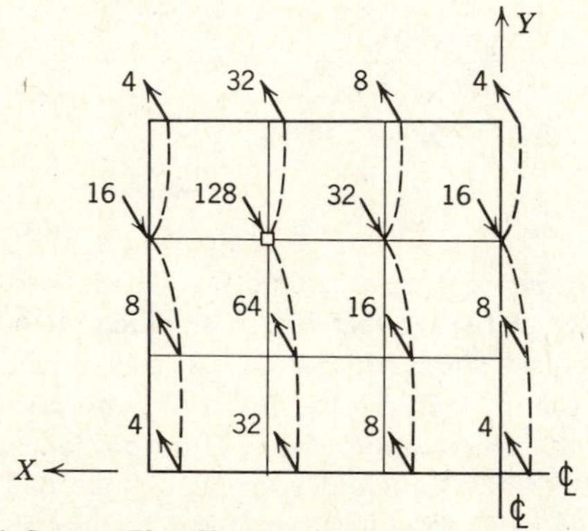

(b) Quarter Plan Showing Vertical Component of Tendons in *Y* Direction

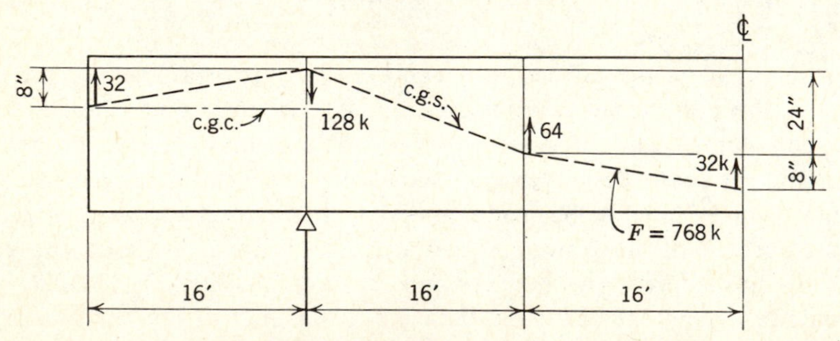

(c) Cable Profile for Half of Beam along Column Line

Fig. 11-22. Balanced solution for example 11-6.

line is shown here for the profile, a smooth curve passing through the controlling points would be a better solution, not only for obvious practical reasons, but also to balance the loads along the beam between the intersections.

Note that while other methods could also be used to solve this particular example, the balancing-load approach reveals design possibilities not so clearly indicated by others. The method can very well be used for the layout of cables in a complicated system of waffle or flat slabs. This can be accomplished either manually, using simple arithmetic, or with the help of electronic computers to solve a system of simultaneous equations. Generally, numerous sets of cable profiles can be used for each balanced-load condition, but the most economical arrangement has to be determined for each case, taking into account practical as well as theoretical requirements.

11-7 Three-Dimensional Load Balancing

One example of three-dimensional load balancing is illustrated in Figs. 11-23 and 11-24, which show two pedestrian bridges connecting two buildings across a wide street. For architectural reasons, these bridges with a maximum span of 102 ft (31 m) are limited to a structural depth of only 27 in. (686 mm), Fig. 11-24(*c*). Furthermore, the main supporting piers, Fig. 11-24(*a*), had to be located on the outside away from the center line of the bridges. To achieve a balanced-load condition for dead load (which included 800 lb of marbles per ft of each bridge), the bridge was prestressed in three directions, *X*, *Y*, and *Z*.

Fig. 11-23. Pedestrian bridge joining two Union Oil Buildings, Los Angeles, California, has 27-in. depth spanning 102 ft (Architect Periera and Luckman; T. Y. Lin International Consulting Engineers).

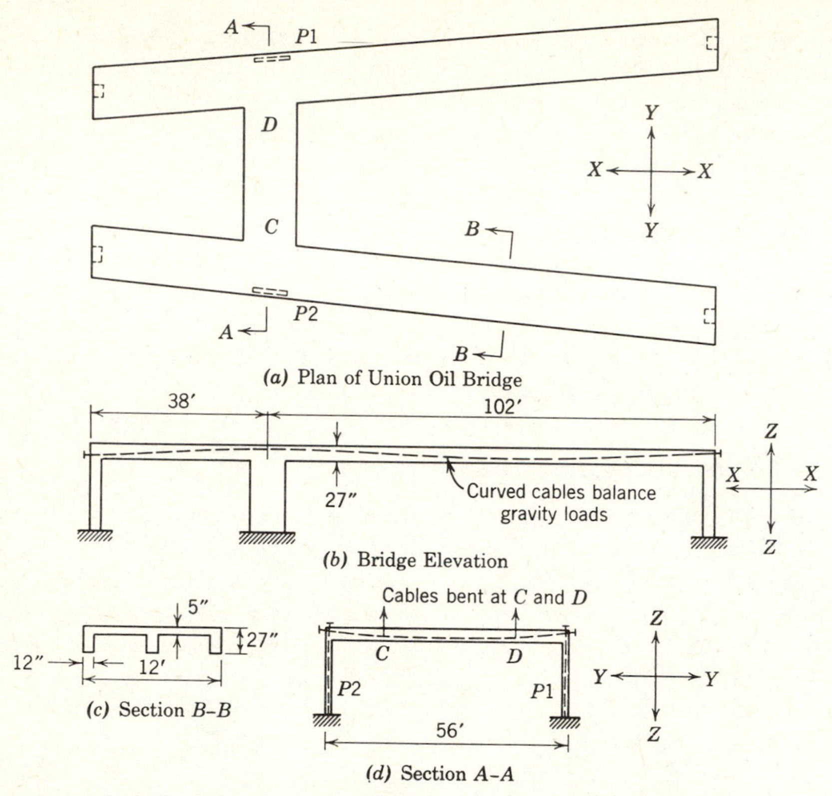

(*a*) Plan of Union Oil Bridge

(*b*) Bridge Elevation

(*c*) Section *B–B*

(*d*) Section *A–A*

Fig. 11-24. Plan, elevation, and sections of the Union Oil Bridge, Los Angeles.

Along the *X* direction, each bridge is a two-span continuous beam, Fig. 11-24(*b*). Parabolic cables are posttensioned in the 12 in.×27 in. (305 mm×686 mm) ribs to balance the dead load. Along the *Y* direction, bent cables are posttensioned in the crossover (*a* and *d*), supplying two upward concentrated forces at *C* and *D* to serve as invisible piers. Since these cables are located high at the ends, they produce end moments which are balanced by tendons in piers, *P*1 and *P*2, in the *Z* direction.

While these intricate load-balancing techniques can be used to enable various structural layouts, and while the basic principles are relatively simple, a word of warning should be injected here. In all designs of this type, careful consideration should be given to the effects of shrinkage and creep in concrete; to the problems of stresses and camber during construction; to the amount of deflection vibration under applied loads; to the detailing of tendons, anchorages, supports, and auxiliary steel; to construction sequence and supervision; and to the economics of design and layout. All of these, however, are beyond the scope of this chapter, and indeed much of them are beyond the scope of this treatise.

After the engineers have mastered the basic principles, it is up to them to learn the techniques and the application by careful study as well as actual experience.

A different application of three-dimensional load balancing lies in thin shells and folded plates.[4] It should first be pointed out that complete load balancing in all directions cannot be easily achieved for such structures. In theory, we do not know enough of shell action so as to be able to balance the loads. In practice, it may not be possible or economical to prestress the shells in all directions. However, sometimes the load-balancing approach can be used to advantage.

Consider a cylindrical shell, Figs. 11-25 and 11-26. Cables can be posttensioned along the shell surface so that the vertical component will balance the gravity load. This is given by the condition that the prestressing force is

$$F = \frac{wL^2}{8h}$$

where $w = wt$ of the shell per linear ft along the Y axis. For a long shell behaving in accordance with the beam theory, a cable with a parabolic vertical projection

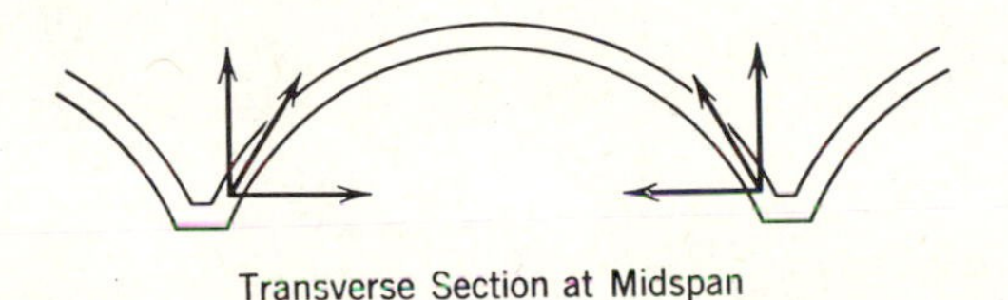

Transverse Section at Midspan

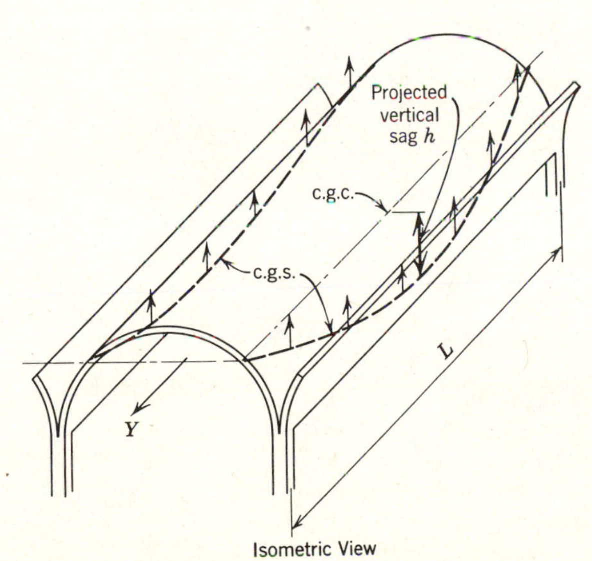

Isometric View

Fig. 11-25. Cylindrical shell prestressed for load balancing.

Fig. 11-26. Four-inch pneumatically placed concrete shell spans 120 ft for High School Gymnasium, Mira Costa, California. (Architects Flewelling and Moody; Structural Engineer Carl B. Johnson; T. Y. Lin International Consulting Engineers). (*a*) Layout of cables. (*b*) Completed building.

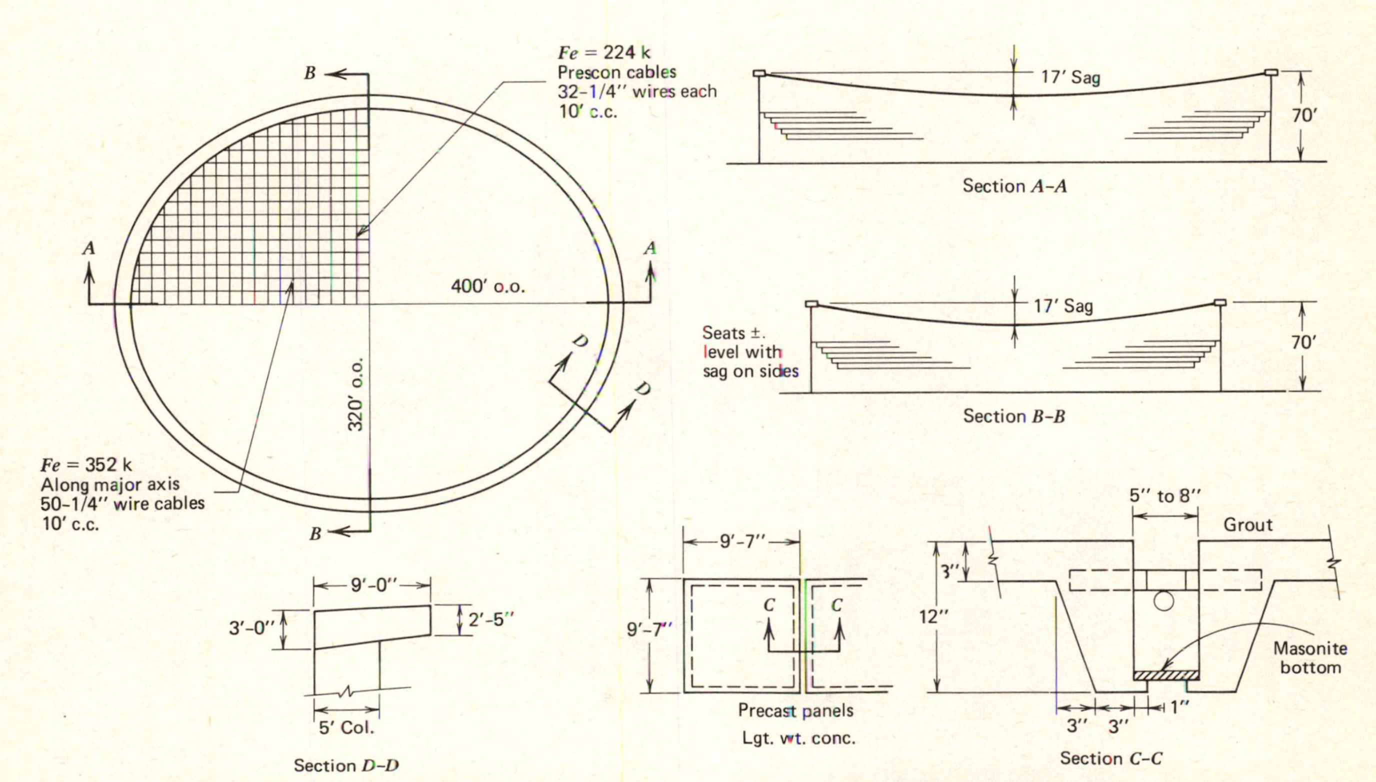

Fig. 11-27. Roof design, Oklahoma City Arena (Architects Jack Scott & Associates; Coston, Frankfurt, Short, Inc.; T. Y. Lin International, Consulting Engineers).

will counteract the beam action. There will then be no deflection along the length of the shell, and the stress distribution will be essentially uniform. Bending stresses will still exist in the transverse direction of the shell, but they can be analyzed as strips of arches supported at different points. For the unbalanced load (generally the live load or a portion of it), ordinary shell analysis can be applied to determine the stresses.

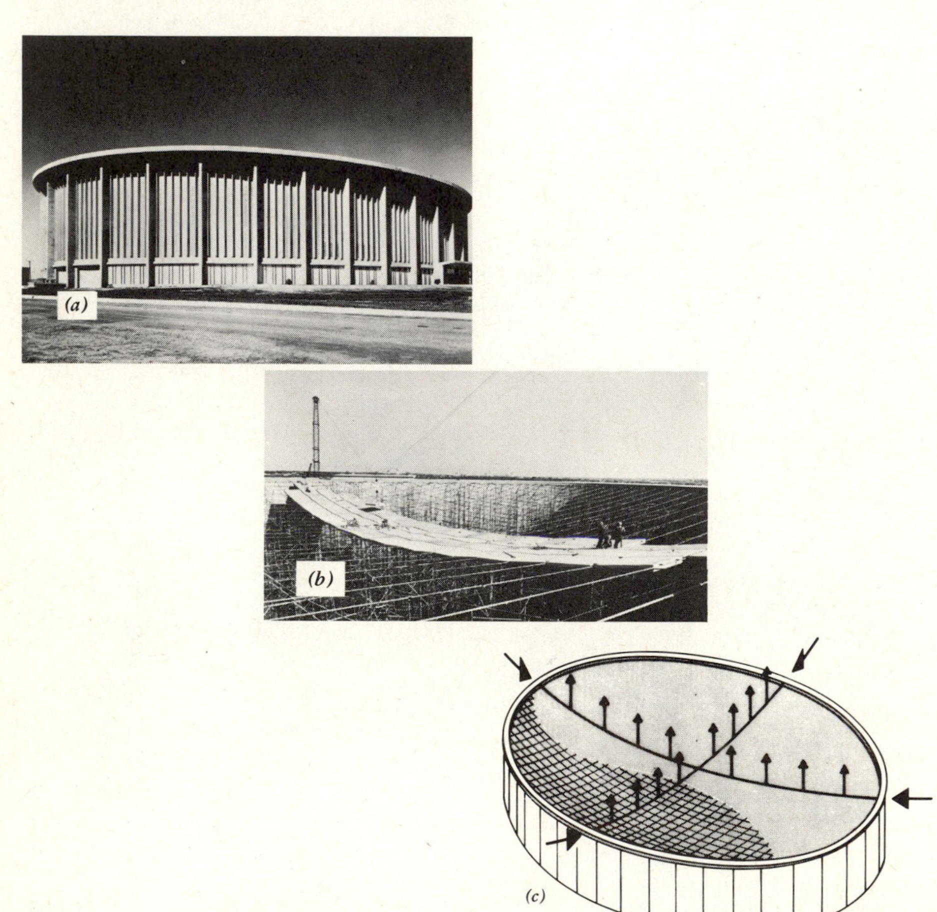

Fig. 11-28. (*a*) Oklahoma State Fair Coliseum. 320×400 ft (97.6×122 m) elliptical shell dish roof, with compression ring on columns and precast double tee walls (T. Y. Lin International, Consulting Engineers). (*b*) Construction of Oklahoma State Fair Coliseum Roof with precast waffles of 3 in. (76 mm) thick concrete, posttensioned between the waffles (T. Y. Lin International, Consulting Engineers). (*c*) Load-balancing theory to Oklahoma State Fair Coliseum (T. Y. Lin International, Consulting Engineers).

If the deflections of the shells are minimized by load balancing, their secondary stresses are also reduced.

A very dramatic example of three-dimensional load balancing is illustrated by the 320×400 ft (97.6 m $\times 122$ m) elliptical cable-supported roof over the Oklahoma State Fair Coliseum in Oklahoma City,[6] Fig. 11-27. Precast concrete panels 9 ft 7 in. (2.92 m) square are supported on cables at 10 ft (3.05 m) spacing forming the two-way grid shown in Figs. 11-27 and 11-28. Simultaneously with the cable stringing came installation of steel scaffold towers which were adjusted to correct cable elevation with screw jacks. Cables were supported temporarily during construction at alternate intersections on a 20 ft. (6.10 m) square grid as shown in Fig. 11-28(*b*). To avoid bending moments that might crack the compression ring, a four-stage, load-balancing sequence of prestressing was used on this project. To minimize the ring bending, pairs of mutually perpendicular cables were stressed simultaneously. In the first three stages of stressing, the prestress offset the sag under dead load by an upward deflection that just balanced it as illustrated in Fig. 11-28(*c*). In the fourth stage, a final stressing operation balanced the 56 psf (2.68 kN/m^2) dead load plus 12 psf (0.575 kN/m^2) live load. The grid of continuous strips between panels was grouted one day before the third stressing stage, and the roof was a rigid monolithic concrete membrane following this stage of stressing. The final stressing stage came eight days later, completing the load-balancing of dead load plus live load.

11-8 Criteria for Load Balancing; Accuracy of the Method

Using this concept of load balancing, an important question is: What should be the loading to be balanced by the prestress? The answer to this question may not be simple. As a starting point, it is often assumed that the dead load of the structure or element be completely balanced by the effective prestress. This would mean that a slight amount of camber may exist under the initial prestress. In the course of time, when all the losses of prestress have taken place, the structure or element would come back to a level position.

Although it seems logical to balance all the dead load, such balancing may require too much prestress. Since a certain amount of deflection is always permitted for a nonprestressed structure under dead load, it is reasonable to also permit a limited amount of deflection if it would not become objectionable. However, there is a greater tendency in prestressed structures to increase their deflections as a result of creep and shrinkage. Hence the deflections should be limited to a smaller value at the beginning.

When the live load to be carried by the structure is high compared to its dead load, it may be necessary to balance some of the live load as well as the dead load. One interesting approach is to balance the dead load plus half the live load, $(DL + \frac{1}{2} LL)$. If this is done, the structure will be subjected to no bending when one half of the live load is acting. Then, it is only necessary to design for one-half live load acting up when no live load exists, and for one-half live load acting down when full live load is on the structure. This idea of balancing dead load plus one-half live load, while theoretically interesting, could result in excessive camber if the live load consists essentially of transient load. If the live load represents actual sustained loading such as encountered in warehouses, excessive camber may not occur.

When attempting to evaluate the amount of live load to be balanced by prestressing, it is necessary to consider the real live load and not the specified design live load. If the specified design live load is higher than the actual live load, only a small amount of the live load or even no live load at all should be balanced. On the other hand, if the actual live load could be much higher than the design live load, especially if the live loading would be sustained, it would be desirable to balance a greater portion of the live load. The engineer should exercise judgment when choosing the proper amount of loading to be balanced by prestressing. This should be done while keeping in mind the satisfaction of other requirements such as elastic stress limitations, crack control, and ultimate strength.

A balanced load design can be achieved with considerable accuracy because both the gravity load and the prestressing force can often be predicted with precision. However, variations may be encountered so that the actual loading and the actual prestress may not be as expected. For a relatively stiff member, errors in estimating the weight and the prestress will usually be negligible. For a slender member, even slight variations may result in considerable errors in the estimation of load balancing, and either camber of deflection may result.

As is well known, the modulus of elasticity of concrete and the creep characteristics cannot be predetermined with accuracy. Fortunately, neither the modulus nor the flexural creep would enter into the picture if the sustained load is exactly balanced by the prestressing component. In other words, since there is no transverse load on the member, there will be no bending regardless of the value of the modulus or the creep coefficient.

Depending on the accuracy desired in the control of camber and deflection, the amount of loading to be balanced must be chosen. If the limits of error can be estimated and if the significance of deflection or camber control can be assessed, it will not be difficult to design the member so as to possess the desired behavior.

References

1. T. Y. Lin, "A New Concept for Prestressed Concrete," *Construction Review*, Sydney, Australia pp. 21–31; (Reprinted in *J. Prestressed Conc. Inst.*, December 1961, pp.

36–52.) Also T. Y. Lin, "Revolution in Concrete," *Architectural Forum Part I*, May 1961, pp. 121–127; *Part II*, June 1961, pp. 116–121.

2. T. Y. Lin, "Load Balancing Method for Design and Analysis of Prestressed Concrete Structures," *J. Am. Conc. Inst.*, June 1963.
3. S. Timoshenko, "Theory of Plates and Shells," McGraw-Hill, 1940, p. 133.
4. T. Y. Lin, "Prestressed Concrete—Slabs and Shells—Design and Research in the United States," *Civil Engineering*, October 1958, pp. 75–77.
5. Henry M. Layne and T. Y. Lin, "Prestressed Concrete Shell for Grandstand Roofs," *J. Am. Conc. Inst.*, November 1959, pp. 409–422.
6. W. Nashert, "Elliptical Dish Sets Record for Cable Supported Roof," *Eng. News-Rec.*, March 18, 1965.

12

SLABS

12-1 Introduction; One-Way Slabs

The flexural design of a simple one-way slab is usually made similar to a beam, as discussed in Chapters 5 and 6. Most continuous slabs are designed like continuous beams (Chapter 10 and 11). Balanced-load design for two-way slabs and waffle slabs is described in section 11-5. The design for punching shear in prestressed slabs is discussed in section 12-5, but this chapter will be devoted primarily to the flexural design of flat slabs and certain other special features.

A one-way slab has main reinforcement only along the length of the slab. All its supports extend the full width of the slab, Fig. 12-1, there being no isolated point-supports or supports running along the length of the slab. Occasionally, the supports may be interrupted or stopped before they reach the entire width, in which case the remaining portion should be designed for a different condition of support.

The usual procedure to design a one-way prestressed slab is to consider a typical 1-ft width of slab, and treat it as if it were a beam, as is done for a reinforced-concrete one-way slab. Whether the slab is simple, cantilever, or continuous, it is designed like a beam, with identical supports and hinges. Hence all the analysis and design of beams discussed in the previous chapters can be directly applied to slabs without any amplification. For example, the theory of linear transformation and of concordant cables for beams is also valid for one-way slabs.

Although the main prestressing steel runs only along the length of the slab, transverse steel, either prestressed or not, may be added to take care of shrinkage and to distribute any concentration of loads. The design of transverse steel, in both reinforced and prestressed structures, has always been a controversial issue, although the analysis of transverse stresses produced by concentrated loads has been solved both theoretically and experimentally for certain simple cases.[1] The main difficulties in design are the choice of a proper amount of concentration, the combination of such concentrations, and the employment of correct allowable stresses in the design. In addition, the actual structure might have to be simplified in order to suit the conditions of analysis.

If the transverse reinforcement is prestressed, two additional questions are raised. One question is whether Poisson's ratio effect will have a significant influence on the loss of prestress in a slab prestressed in two directions. If we

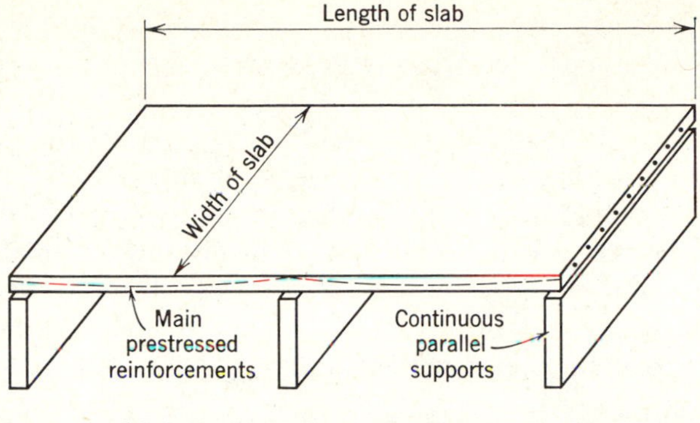

(a) A Two-Span Continuous Prestressed Slab

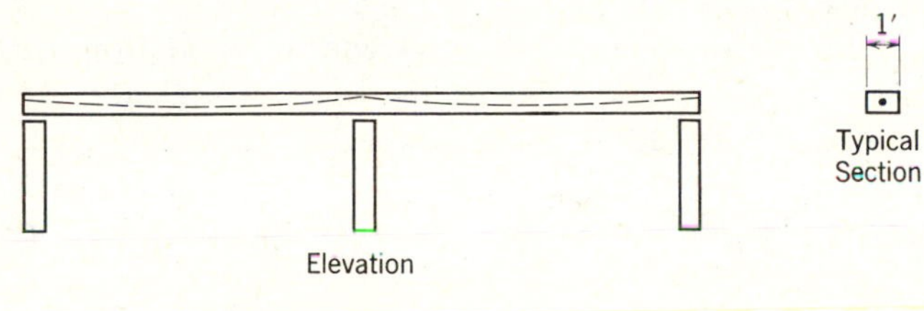

(b) Analysis for Slab in (a) Based on an Equivalent Beam 1 Ft Wide

Fig. 12-1. One-way slabs.

assume that the concrete has a Poisson's ratio of 0.15 and that the slab is subject to the same prestress in both directions, corresponding to a loss of prestress due to elastic shortening and creep in concrete of 8%, the decrease in loss due to Poisson's ratio effect will be $0.15 \times 8\% = 1.2\%$, which is not too significant as far as practical design is concerned.

Another question is whether such biaxial prestressing will change the basic strength and strain characteristics of concrete, so that they can no longer be predicted by the application of the elastic theory in conjunction with the properties obtained from ordinary test specimens. Regarding this question, there were two schools of thought. Some engineers believed that concrete as a material was more of a complex "solid-liquid" rather than a simple elastic solid, so that the laws of thermodynamics, rather than the laws of elasticity of solid, should be applied.[2] This belief was partly substantiated by several field tests in France, notably the testing of the prestressed runway at Orly. These tests brought forth the possibility that, for a slab prestressed in two directions, the cracking strength might be much higher than the value given by the elastic theory. It was proposed that, in a statically indeterminate system, there could be no relation between the appearance of cracks and the existence of tension equal to the modulus of rupture.

The second school of thought directly contradicts the first. It is believed that, when concrete is prestressed in two directions, its basic behavior and properties have not been changed. Hence the elastic theory, together with the tested values for modulus of rupture, can be applied to statically indeterminate structures, including slabs prestressed in two directions. This was shown by several laboratory tests conducted at the university of Ghent. In one test on a two-way prestressed slab, the elastic theory was shown to be quite accurate for predicting the cracking load.[3]

In another series of tests on continuous beams, the same conclusion was drawn (see reference 13, Chapter 10). More recently, tests on simple and continuous slabs prestressed in two directions[4,5,6,7] all indicated the validity of the elastic theory, and it is now generally agreed that the cracking strength is, at least, approximated by the modulus of rupture of concrete.

For narrow one-way slabs the transverse reinforcements are usually nonprestressed, because short prestressing is neither economical nor accurate. When the width is small compared to the span, any concentrated load is assumed to be carried by the entire width of the slab, and little transverse reinforcing is required for load distribution. Generally, the nonprestressed reinforcement required for shrinkage is sufficient also for load distribution.

Tests of one-way slabs, continuous over three spans, with unbonded tendons[8,9,10] have verified the effectiveness of bonded bars in regions of peak moments to provide design strength and assure good distribution of cracking. These tests of half-scale model slabs indicated that the first localized cracking at the point of maximum moment was at (or slightly above) the load level corresponding to modulus of rupture stress in the concrete. For the continuous slab, first cracking was very localized, and the overall load-deflection curve, Fig. 12-2, did not show a sudden change in slope at this load level as might be expected in a simple beam. Design of the prototype assumed load-balancing of slightly less than full dead load, and the slab shown in Fig. 12-2 had a tension stress of $6\sqrt{f_c'}$ (as calculated by the beam method) at the support under the loading for maximum moment. The performance at service load was that predicted by elastic analysis, and the strength was in excess of $(1.4D+1.7L)$ loading assumed in design. ACI rules for moment redistribution in continuous prestressed structures were followed in the design and analysis of the slab. The tests[10] indicated these to be excellent guide lines for design of one-way, posttensioned slabs.

For the two one-way slabs referred to above, loading was uniformly distributed as would be expected in many buildings. With this type loading, the whole slab is uniformly loaded, and the transverse moment is not a problem. (This

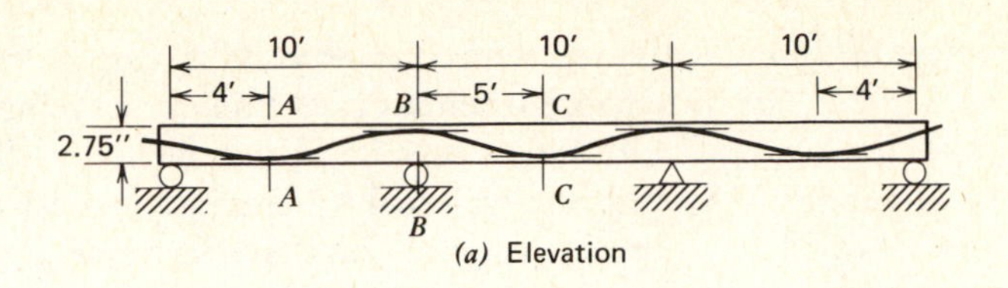

(a) Elevation

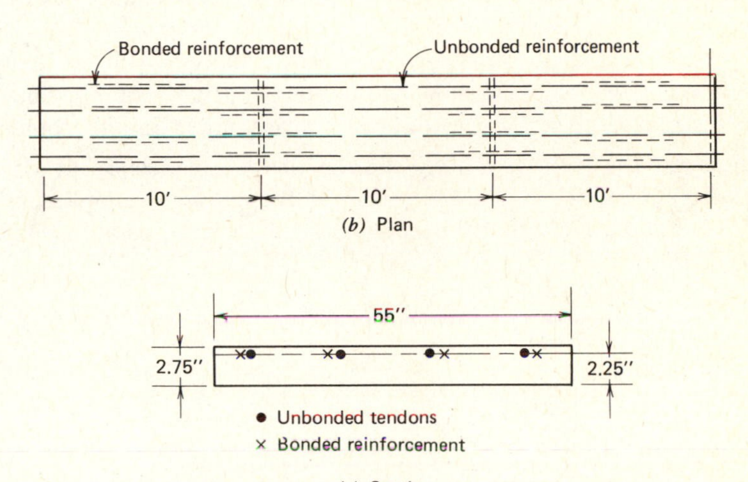

(b) Plan

(c) Section
(As Shown at B–B, Inverted at A–A and CC)

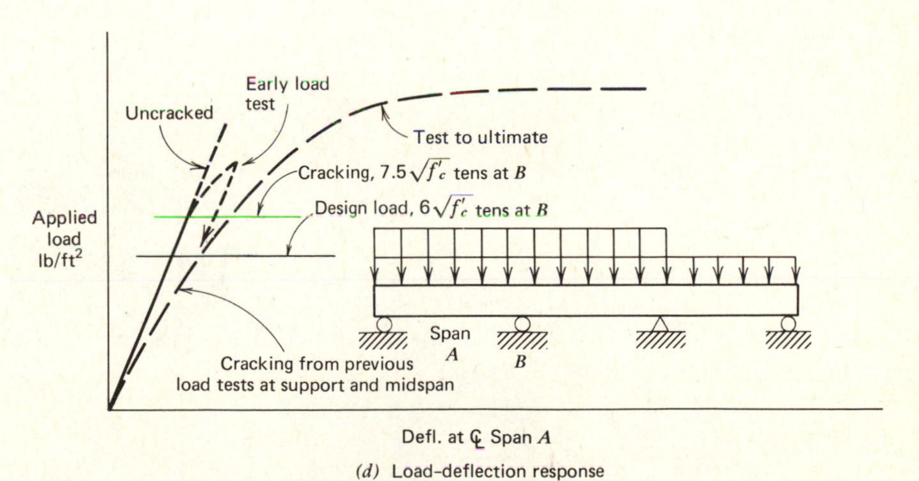

(d) Load–deflection response

Fig. 12-2. Test of one-way slab with unbonded tendons and bonded bars.

design problem is discussed later with one-way slabs carrying concentrated loading, Fig. 12-3.) The provision of minimum transverse reinforcement for shrinkage and temperature will satisfy design requirements in the transverse direction for a uniformly loaded one-way slab. A minimum prestress of 100 psi (0.69 N/mm^2) provided by posttensioning in the transverse direction as an alternative to the use of unstressed reinforcing bars for shrinkage and temperature requirements has been used in some structures with satisfactory performance.

Deflection in one-way slabs follows elastic analysis with the deflection at load levels above first cracking showing gradual departure from the linear response of the uncracked slab,[8,10] Fig. 12-2(d). It was noted in the companion slab[9,10] which was designed for $9\sqrt{f_c'}$ tension that the deflection at service load was closely estimated by elastic analysis, even though very minor cracks were present at one section. Note that the application of several loads in different patterns

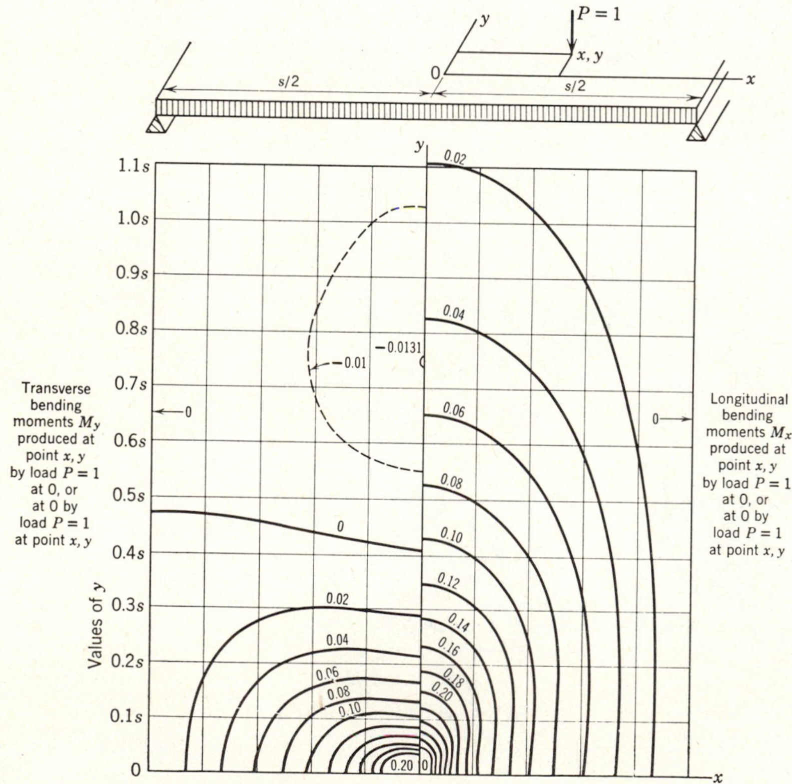

Fig. 12-3. Transverse and longitudinal moments in one-way slab due to concentrated load.

producing numerous cracks in both positive and negative moment regions results in a reduced stiffness as shown in the test to ultimate of the one-way slab, Fig. 12-2(*d*). Such loading is not typical of most actual structures which experience few applications of overload beyond that shown in the early load test of Fig. 12-2(*d*).

For one-way slabs with width greater than about 50% of the span, the deflection of the different slices of the slab may vary considerably under concentrated loads. This indicates heavy transverse bending, which must be resisted by reinforcements whether prestressed or nonprestressed. If nonprestressed, the bending moments can be calculated by the usual elastic theory and the proper amount of steel provided as for any reinforced-concrete design. If economic or other considerations justify the use of transverse prestressing, the moments can also be calculated by the elastic theory and the transverse prestress designed by the ordinary procedure for designing prestressed beam sections. This is believed to be a safe procedure and, if properly applied, should give reasonable results.

In order to convey some idea of the moments in a wide slab, the longitudinal and transverse moments in a one-way slab under concentrated load[1] are given in Fig. 12-3. These are based on the elasticity theory as applied to an infinitely wide slab with a Poisson's ratio of 0.15. In such slabs, there also exist torsional moments, but the magnitude is small, and they are not often considered in design.

After the transverse moments to be resisted by prestressing have been computed, the determination of the amount of prestress is a relatively simple matter, it being remembered that, if no tension is allowed, the resisting moment is given by the prestress times its lever arm measured to the opposite kern point. This simple procedure permits the design of prestressed transverse reinforcements to be made with the same ease as nonprestressed reinforcements. The method is illustrated in example 12-1.

Concentric transverse prestressing is often preferred for slabs, although it is not as economical as eccentric prestressing. As can be seen from Fig. 12-3, the transverse positive moments are higher than the negative moments. Hence the steel should be positioned farther from the top kern than from the bottom kern, so as to possess a greater lever arm for resisting the positive moments. But such eccentric prestressing will tend to bend the slab transversely, which may be objectionable.

EXAMPLE 12-1

The Bacon Street Highway Bridge in San Francisco has a simple span of 60 ft (18.3 m) and a width of 100 ft (30.5 m), Fig 12-4. It is prestressed with an effective prestress of 196,000 lb per ft (2860 kN/m) of width along the 60-ft (18.3 m) span. Compute the amount of concentric transverse prestress required per foot of span. Design for a concentrated load of 16,000 lb (71 kN) and no tensile stress in the concrete.

$$M = 16{,}000 \times 0.20$$
$$= 3200 \text{ ft-lb/ft width}$$

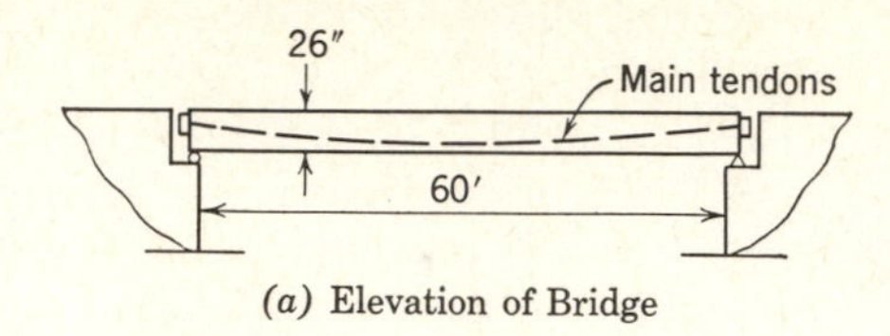

(*a*) Elevation of Bridge

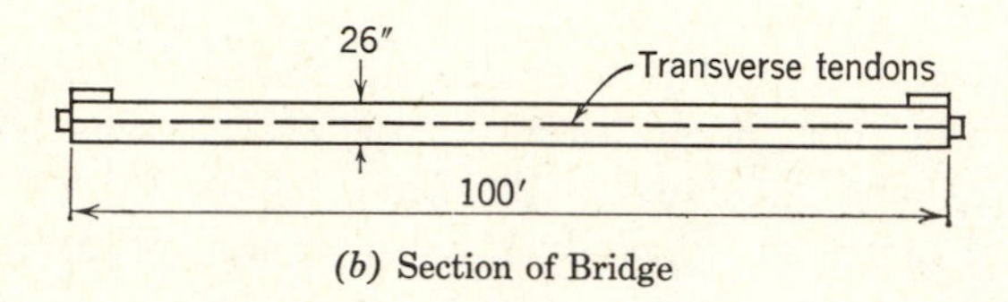

(*b*) Section of Bridge

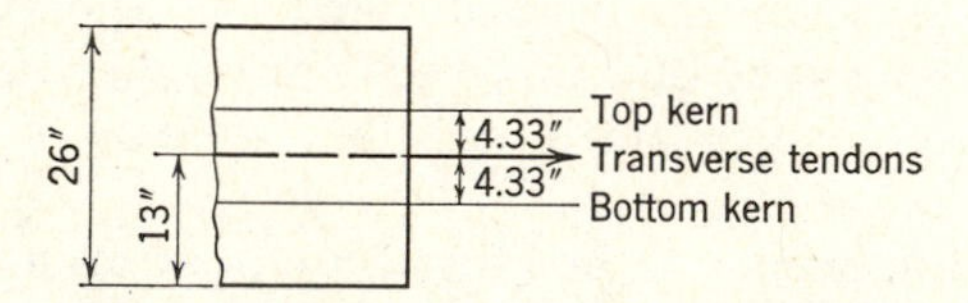

(*c*) Lever Arm for Resisting Moment

Fig. 12-4. Example 12-1.

Solution Since the width of the bridge is greater than its span, it is close enough to use Fig. 12-3 for computing the transverse moments. For a maximum coefficient of 0.20,

$$M = 16{,}000 \times 0.20$$
$$= 3200 \text{ ft-lb/ft } (14.2 \text{ kN} - \text{m/m}) \text{ width}$$

In order to resist that moment without producing tension in the concrete, the most economical position for the transverse tendons is at the lower kern point. However, if the tendons are located at the lower kern point, the entire slab will be subjected to a negative transverse moment, resulting in a convex bending across the width of the bridge. Moreover, the slab may be subjected to some negative moments under concentrated load, and it would be better to place the cable within the kern. Since the amount of prestress required is small, it will be convenient to locate them through the middepth of the slab; then the lever arm to either kern is 4.33 in. (110 mm), Fig. 12-4(*c*). To resist the above moment, the amount of prestress required per foot is

$$\frac{3200 \times 12}{4.33} = 8860 \text{ lb } (39.4 \text{ kN})$$

Actual stress measurements on the slab of the Bacon Street Bridge indicated the accuracy of the elastic theory for load distribution and transverse moments.[11] The load distribution is apparently little affected by the presence of the transverse prestress.

12-2 Two-Way and Simple Flat Slabs

Though a one-way slab may be prestressed in two directions, it is not a two-way slab, because the transverse prestressing only serves to strengthen the concrete locally but is not intended for carrying any portion of the load to the supports. A two-way prestressed slab is one whose prestressing steels in two perpendicular directions both serve to transfer the load to its supports. Thus a two-way slab rests on continuous supports in the form of beams or walls running in two perpendicular directions. When a slab is supported by a network of columns, either with or without capitals, it can properly be called a prestressed flat slab, using that term as in reinforced-concrete construction. A prestressed flat slab can be designed using the method of load balancing as explained for the grid system in section 11-6, or the beam method as will be explained in this chapter. The two methods will yield similar resisting moment in each direction, but the distribution of tendons will be quite different. The beam method with the normal distribution of tendons between column and middle strips will not yield a balanced design for uniform loads, although slabs so designed have been found to be fairly level from a practical point of view.

Before considering the "Recommendations for Design of Prestressed Flat Slabs" proposed by the ACI-ASCE Committee of Prestressed Concrete, we will consider design approaches used prior to the 1970s. One basis for the design of prestressed flat slabs is to use moment coefficients for the design of reinforced-concrete flat slabs which are available from building codes and handbooks. This method is not commonly used since load balancing has been a great simplification in the structural analysis with prestressed concrete frames. When applied to prestressed concrete, the procedure of design can be discussed in two parts: the acting moments due to loads, and the resisting moments provided by the prestressing steel. As far as the load moment is concerned, there is no major difference between reinforced and prestressed two-way slabs. Within the working load, they both behave according to the elastic theory, with the prestressed slabs following it more closely. Although, near the ultimate load, they behave less nearly alike, there is reason to believe that the moment coefficients for reinforced concrete, based essentially on elastic analysis, can be used for prestressed concrete without serious adjustments. This does not mean that we are satisfied with these coefficients. In fact we are not satisfied with them even as applied to reinforced-concrete slabs themselves. However, we might with some discretion apply these coefficients to prestressed concrete.

The second part of the problem is to provide the resisting moments. As usual, the resisting moments in prestressed concrete are supplied by the steel acting with a lever arm up to around the kern point. For continuous spans, the resisting couple, instead of being measured from the steel, should be measured from the *C*-line produced by prestress (within elastic or service load level), the determination of which is a more complicated problem, although it can be solved either by the theory of elasticity or the use of model tests. In this connection, a thorough understanding of the principles discussed in Chapter 10 is essential. Instead of the elastic theory, the application of ultimate design together with proper choice of load factors may also result in satisfactory proportions. It must be remembered that, for prestressed slabs, the initial condition at transfer could be a critical situation that must be examined for overstress in concrete. Flat slabs of prestressed concrete supported by a network of columns have already found wide application in this country. This is especially true in connection with lift slabs, where the slabs are cast on the ground and lifted along the columns to their proper height.[12] An interesting example is the 13-story apartment building shown in Fig. 1-8, where all the slabs were cast on the ground, posttensioned and then lifted into position.[13] Here, an 8-in. (203 mm) flat slab of lightweight concrete spans a typical bay of 28 ft (8.53 m) and a maximum of 32 ft (9.75 m). By proper application of the laws of statics and the theory of elasticity, plus a thorough knowledge of prestressed concrete, flat slabs can be designed with satisfactory results.

The state-of-the-art in design of posttensioned flat plate structures was summarized in the "Tentative Recommendations for Prestressed Concrete Flat Plates" in 1974.[14] Revisions in the ACI Code in 1976 provide guidelines for the amount and location of the minimum bonded reinforcement for flat plates with unbonded tendons. Thus, the usual practice for design and detailing of the posttensioning of flat plates is rather well established based on these references.

Let us first consider a flat slab supported on four columns, Fig. 12-5. This is a statically determinate system as far as the reactions are concerned. The total moments across any section, such as *A–A* or *B–B*, for example, can be readily determined from statics. But the distribution of the total moments along the length of the section is a problem in elasticity. Such distribution can be obtained theoretically by the theory of elasticity, or it can be measured experimentally by means of models. (Note that a load balancing method similar to example 11-4 could be a simpler approach.) In general, the total moment along the columns strip *B–B* will be greater than those along the middle strip *A–A*, but the distribution will not be the same. Maximum moment will be found in the column strip with the moment more nearly uniform across line *AA* than line *BB*.

The magnitude of the slab moments at each point having been determined, the next step is to provide enough steel to resist the moments. An ideal arrangement would be to provide in both directions exactly the required amount of steel and eccentricity at each point. But this may not be possible in practice, and a reasonably satisfactory solution can be obtained by a good estimation of

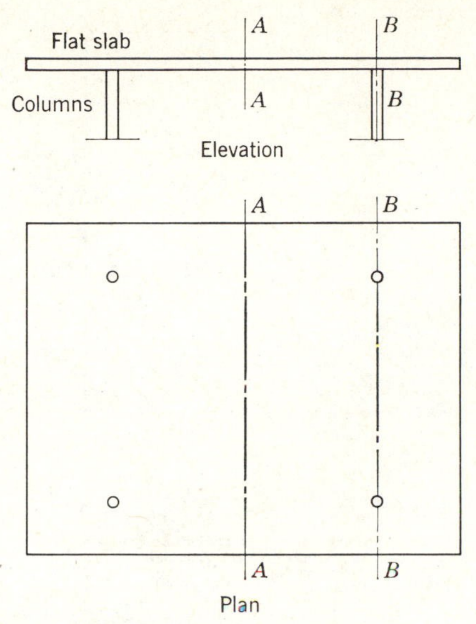

Fig. 12-5. A simple flat slab.

the distribution of the moments. So long as the total resisting moment equals the external moment, any slight error in distribution is not of serious consequence, since the transverse rigidity of the slab can be depended on to a certain extent to transfer the resistance across the slab. Again, stresses in the concrete should also be investigated for the initial condition at transfer of prestress.

An example is given in the following, illustrating the computation of steel area for a simple flat slab prestressed in two directions.
The complete design of such a slab would involve the following:

1. Locating the cable profiles
2. Spacing the cables.
3. Checking stresses in concrete both at transfer and under working loads.
4. Computation for deflections at various stages, including the effect of plastic flow.
5. Computation of cracking and ultimate loads.
6. Design for end anchorage details.

The reader is referred to other parts of this treatise where these are discussed.

EXAMPLE 12-2

A simple flat slab 40 ft by 30 ft is supported by four columns as shown, Fig. 12-6(a). The 6-in. concrete slab weighs 75 psf and carries a roof live load of 20 psf; $f_c' = 4000$ psi; $\frac{1}{4}$-in. wires grouped in 4 wires per unit are to be used for prestressing in two directions. The cables are greased and wrapped to prevent bond to the concrete. Ultimate strength of the

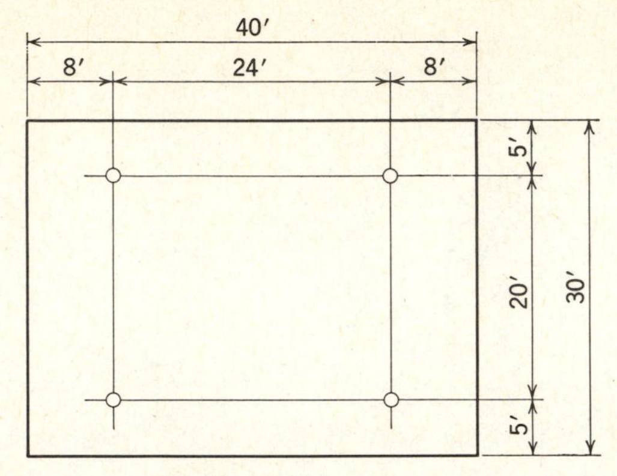

(a) Plan of Slab

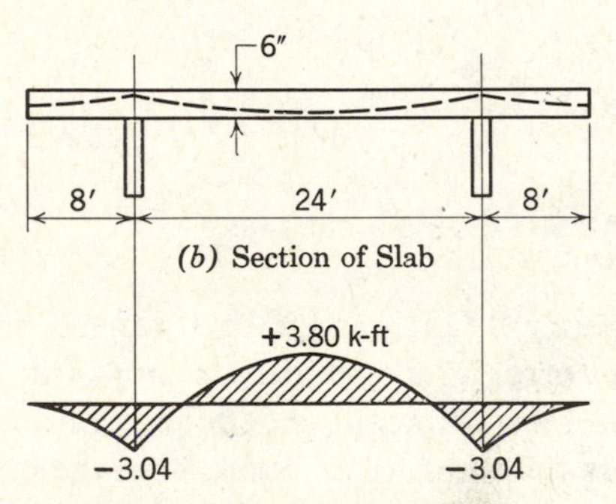

(b) Section of Slab

(c) Moment Diagram for 1-Ft Width

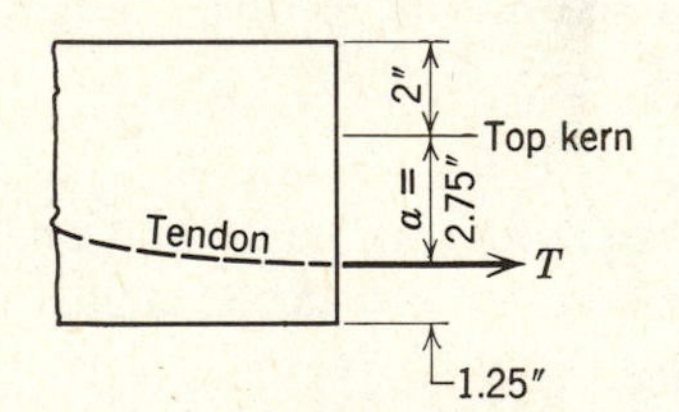

(d) Lever Arm *a* for Prestress at Midspan

Fig. 12-6. Example 12-2.

wires is 250 ksi, with an initial prestress of 150 ksi and an effective prestress of 125 ksi. Minimum clear coverage for the cables is to be $\frac{3}{4}$ in., which is equivalent to $1\frac{1}{4}$-in. protection measured to the center line of the cables. Compute the required number of 40-ft cables per slab. (Flat slab = 12.2 m × 9.14 m, $w_d = 3.59$ kN/m^2, $w_l = 0.96$ kN/m^2, $f_c' = 34$ N/mm^2, $f_{pu} = 1{,}724$ N/mm^2, $f_{se} = 862$ N/mm^2, and initial prestress = 1034 N/mm^2.)

Solution

$$DL = 75 \text{ psf } (3.59 \text{ kN/m}^2)$$
$$LL = 20 \text{ psf } (0.96 \text{ kN/m}^2)$$
$$\text{Total load} = \overline{95} \text{ psf } (4.55 \text{ kN/m}^2)$$

For the 40-ft direction, the average cantilever moment is

$$-wL^2/2 = (95 \times 8^2)/2$$
$$= 3.04 \text{ k-ft/ft } (13.52 \text{ kN}-\text{m/m}) \text{ of width}$$

and the average positive moment at midspan is

$$wL^2/8 - 3.04 = (95 \times 24^2)/8 - 3.04$$
$$= 6.84 - 3.04$$
$$= 3.80 \text{ k-ft/ft } (16.90 \text{ kN}-\text{m/m}) \text{ of width}$$

For the entire width of 30 ft, the moment is

$$3.80 \times 30 = 114 \text{ k-ft } (155 \text{ kN}-\text{m})$$

The resisting moment is furnished by the steel with a lever arm of 2.75 in. (70 mm) measured to the top kern point, allowing no tension in concrete, Fig. 12-6(d). Hence the total prestress required is, as controlled by the $+M$,

$$(114 \times 12)/2.75 = 497 \text{ k } (2211 \text{ kN})$$

Each cable has 4 wires with $A = 0.05$ sq in. (32.3 mm^2); hence A_s per cable is 0.20 sq in. (129 mm). For an effective prestress of 125 ksi (862 N/mm^2), each cable has a total prestress of $0.20 \times 125 = 25$ k (111.2 kN). The total number of cables required is

$$497/25 = 19.9 \qquad \text{Use 20 cables.}$$

(Place 55% in column strip and 45% in middle strip.) Bonded reinforcement must be provided along with the tendons in the immediate column region. If we supply approximately $0.0015 \times$ column strip concrete area,

$$A_s \cong 0.0015 \times 10 \times 12 \times 6 = 1.08 \text{ in.}^2 \text{ } (397 \text{ mm}^2)$$

Use six No. 4 bars each direction placed within a region extending 1.5 in. $\times$ 6 in. = 9 in. (229 mm) either side of column.

12-3 Continuous Flat Slabs—Design Approach and Test Results

Continuous flat slabs are frequently constructed of prestressed concrete, sometimes combined with the lifting process. A prestressed slab is lighter than a reinforced one, and its flexibility lends itself to the lifting process. Dead-load deflection in the slab can be largely balanced by the camber produced by prestress. Then there are the saving in formwork and other conveniences inherent in flat-slab construction.

The design of a continuous prestressed flat slab is based on a knowledge of the design of simple flat slabs as outlined in the previous section. As a result of continuity, two additional problems should be discussed: the negative moments over the interior supports due to loads, and the effect of prestressing a statically indeterminate structure, including the problems of linear transformation and cable concordancy as discussed for continuous beams.

For simple flat slabs, the total moment across any section is definitely known, because all the reactions are statically determinate. The reactions for continuous slabs, however, are statically indeterminate, and hence the total moment across a section cannot be computed from statics alone.

Since prestressed concrete can be treated as a homogeneous and elastic material in the analysis of moments, the theory of elasticity can be depended on to yield reasonably accurate results before the cracking of concrete. But to apply a rigid elastic theory to a continuous slab is a very tedious operation which would consume a great deal of time even for a simple case. Hence some easier procedure must be devised for its design.

While the method of load balancing for a grid system would be found convenient for certain cases, the beam method presented in this chapter probably offers the simplest solution, especially when combined with the concept of load balancing for continuous beams.

According to most building codes, reinforced-concrete flat slabs can be designed as continuous beams. So far as moment due to external load is concerned, there is as much justification for applying such a method to prestressed flat slabs. This beam method is illustrated in Fig. 12-7, which assumes continuous knife-edge supports along one direction when analysis is being made for moment in the other direction.

A continuous slab having been transformed into a continuous beam, the problem is greatly simplified. The effect of prestressing such a slab can then be computed as for continuous beams. On this assumption, then, it is possible to apply the method of linear transformation just as is done for a continuous beam.

If concordant cables are desired for a flat slab, a real moment surface should be computed by the elastic theory; then any set of cables producing eccentric moments proportional to the moment surface is a set of concordant cables. Although this is theoretically interesting, it will not give a better layout than the beam method.

By analyzing a continuous slab as a continuous beam, the total moment across any section due to loading and the average position of the *C*-line under prestressing can be obtained. But the distribution of the total moment and the variation of the position of *C*-line along the width of slab still remain to be determined. Approximations have been used, for example, assuming 45% of the total moment to be carried by the middle strip and 55% by the column strip for a simple flat slab of uniform thickness supported by 4 columns.[4] For the interior span of a slab, continuous in both directions, a better approximation seems to be 25% by the middle strip and 75% by the column strip.[5] This can also be partly

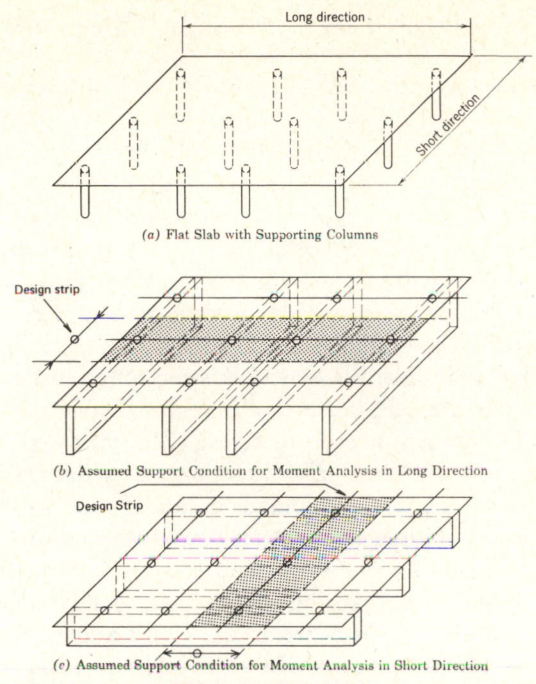

Fig. 12-7. Moment analysis of continuous flat slabs.

explained by the balanced-load concept since at midspan of a middle strip, the cables from both directions act upward; while at midspan of a column strip, one set of cables act up with the other set acting down.

The Prestressed Concrete Committee of ACI has recommended[14] the beam method described above, using the equivalent frame of the ACI Code to consider relative stiffness of beams and columns in the moment distribution analysis. Moment in the slab at the column line is calculated for a design strip as shown in Fig. 12-7. Total moment at support is known to be much higher at the column (column strips) than in the outside regions (half middle strips) of these design strips in each direction. Rather than deal with this variation by assigning moments with coefficients, the ACI Committee recommendation[14] calls for distribution of tendons as follows:

Tendon distribution The ultimate strength of a flat plate is controlled primarily by the total amount of tendons in each direction. However, tests,[15,16] indicate that tendons passing through columns or directly around column edges contribute more to load carrying capacity than tendons remote from the columns. For this reason, it is recommended that some tendons should be placed through the columns or at least around their edges. In lift slab construction, some tendons should be placed over the lifting collars.[15,16]

For panels with length-width ratios not exceeding 1.33, the following approximate distribution may be used:

Simple spans; 55 to 60% of the tendons in the column strip with the remainder in the middle strip.

Continuous spans; 65 to 75% of the tendons in the column strip with the remainder in the middle strip.[5]

The total number of tendons is obtained by use of load-balancing to simplify design of each strip as a one-way continuous beam, with some flexibility in the assignment of tendons within guidelines suggested above. Experience with design has shown this to be a simple procedure, and tests[6,7,18] have shown excellent behavior of slabs designed in this manner.

Figure 12-8 shows the typical layout of posttensioned unbonded tendons and bonded reinforcement for a $\frac{1}{3}$-scale model test slab (slab I) with 70% of the tendons in the column strip in each direction. The prototype for this slab was designed for 30 ft (9.14 m) spans each direction and a thickness of 8.25 in. (210 mm) (span/depth ratio of 44) with tendons balancing 100% of slab dead load in each span. Additional loading of 20 lb/ft^2 (0.96 kN/m^2) for partitions (dead load) plus 50 lb/ft^2 (2.40 kN/m^2) live load was considered for design at both service load and ultimate. The average F/A stress from the unbonded posttensioned tendons was 325 psi (2.24 N/mm^2) for slab I and bonded reinforcement, Fig. 12-8, provided 0.15% of column strip slab area as recommended by reference 14. The ACI Code now requires this same amount of bonded reinforcement as minimum in two-way slabs, but it is expressed in a more general manner, as mentioned later. Behavior of slab I was excellent at both service load and ultimate.[7]

The measured unbonded tendon stresses at ultimate for slab I were 2 to 13% below the predicted ultimate stress using the ACI Code equation. Recommendations of reference 14 for design of flat slabs with unbonded tendons were verified by the test. It was noted that the shear strength using v_{cw} as the shear stress at failure was a safe assumption. The equivalent frame analysis was recommended for design of prestressed flat plates in the report of this test of slab I.

The ACI Code was revised in 1976 to require the minimum bonded reinforcement which had proven adequate for crack control in the slab I test described above. Many slabs have rectangular proportions rather than square, thus the provisions are stated in terms of the dimensions in the span direction and the transverse direction as follows:

In negative moment areas at column supports, the bonded reinforcement, A_s, in each direction shall be

$$A_s = 0.00075hl \qquad \text{ACI (18-7)}$$

where l is the length of the span in the direction parallel to that of the reinforcement

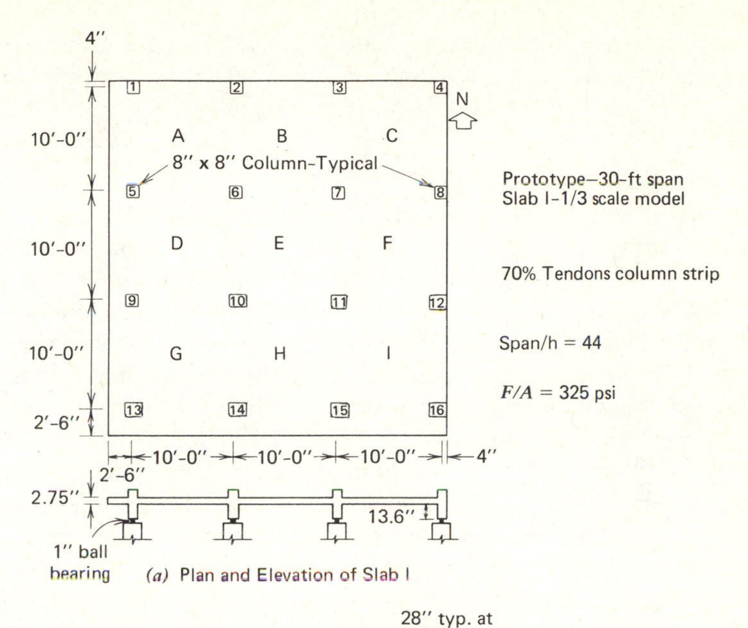

(a) Plan and Elevation of Slab I

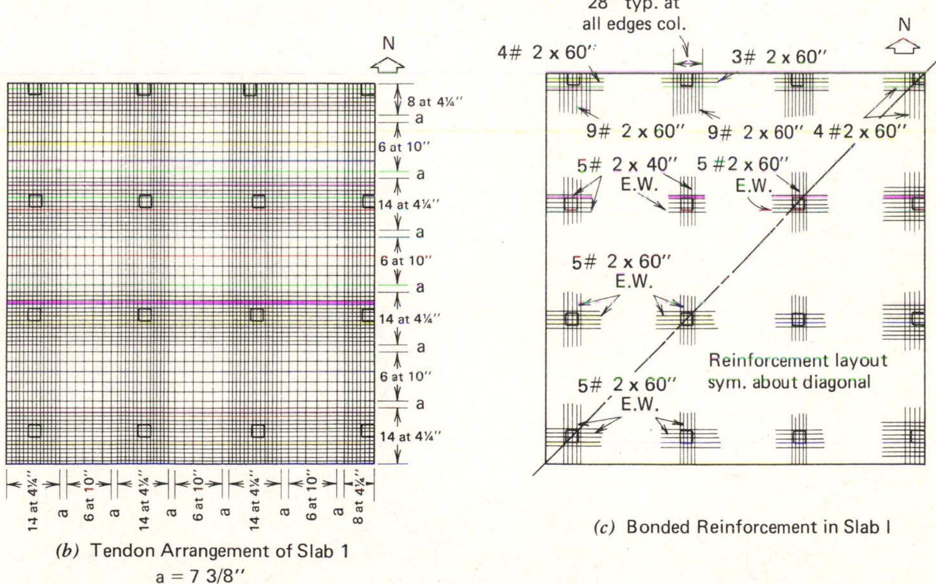

(b) Tendon Arrangement of Slab 1
a = 7 3/8″

(c) Bonded Reinforcement in Slab I

Fig. 12-8. Unbonded tendons and bonded reinforcement for continuous Slab I, reference 7.

being determined. The bonded reinforcement shall be distributed within a slab width between lines that are $1.5h$ outside opposite column faces, shall be spaced not greater than 12 in. (305 mm), and not less than four bars or wires shall be provided in each direction.

The tests of single-column[17] specimens which were companion tests to slab I indicated that the increase of the bonded slab reinforcement at the column from zero (control test) to 0.15% of column strip (minimum requirement), to 60% more than the minimum requirement resulted in better crack control and improved shear strength at ultimate. All three tests had similar secondary punching shear failures following the development of primary flexural failure. All three of the single-column tests showed a radial cracking pattern at the column, but the cracks were fewer and wider in the control specimen which had only unbonded tendons. The cracking at ultimate for slab I is shown in Fig. 12-9 with the radial cracking at columns identical to that in the single-column test which had bonded reinforcement equal 0.15% of column strip. Observing these cracks it becomes clear that the bonded reinforcement of Fig. 12-8(c) (following present ACI Code requirements) is properly placed for crack control in the slab in the region of peak shear and moment at the column. The development of flexural failure (yield lines) with large deflection was observed in the load-deflection curve for the failure test of slab I.

A second nine-panel slab test[6] was a half-scale model of a prototype with 20-ft (6.10 m) spans in each direction, Fig. 12-10. Slab II was a model for a

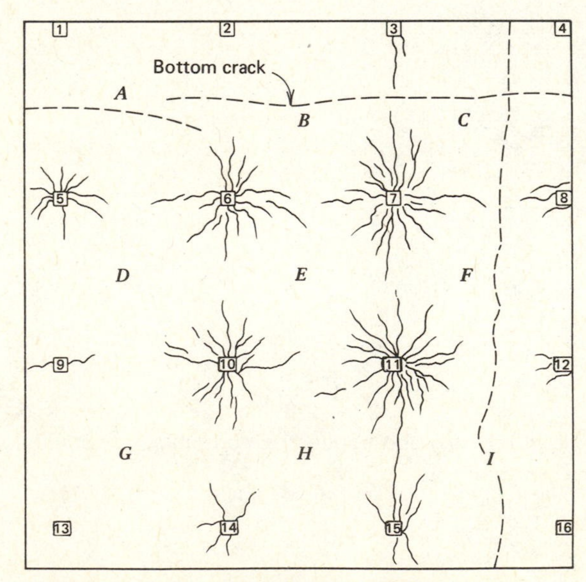

Fig. 12-9. Crack pattern for Slab I after Test 110 (failure).

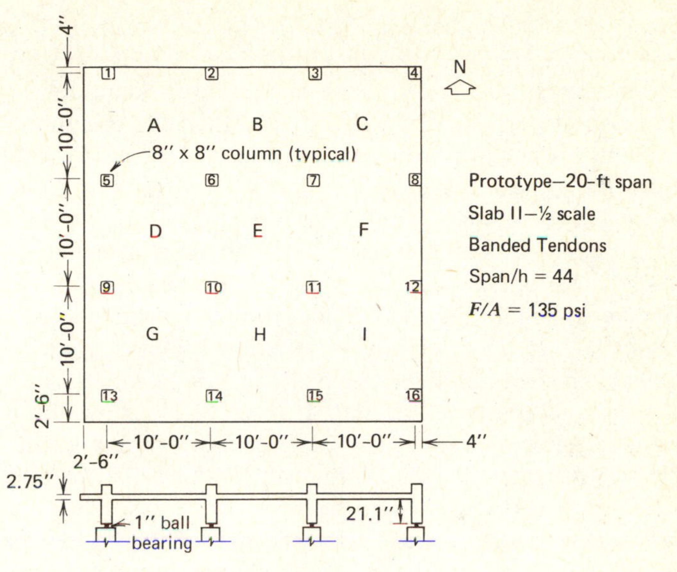

(a) Plan and Elevation of Slab II

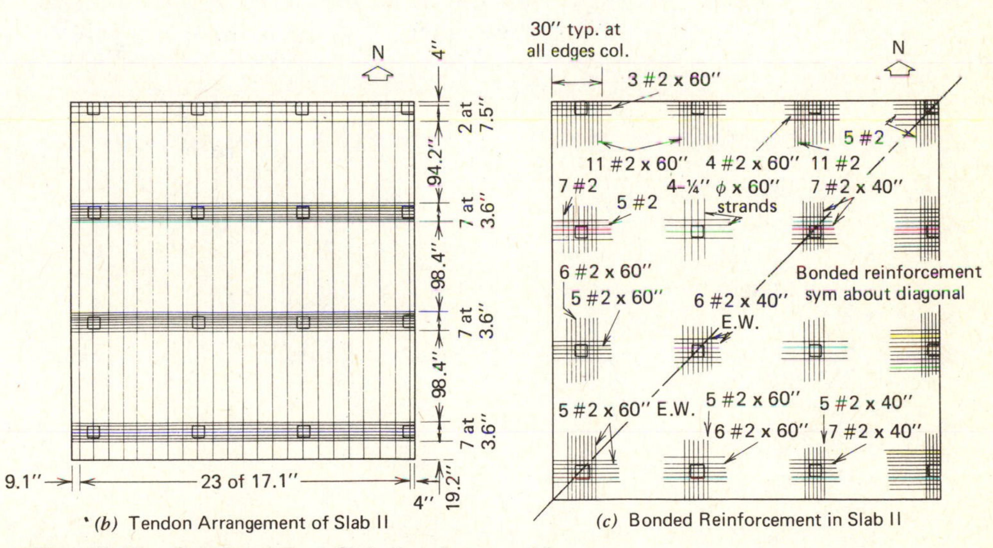

(b) Tendon Arrangement of Slab II

(c) Bonded Reinforcement in Slab II

Fig. 12-10. Details of Test Slab II, reference 14.

prototype with smaller spans and balancing of slightly less than full slab dead load in design, resulting in smaller average prestress ($F/A = 135$ psi $= 0.93$ N/mm^2) than slab I. This is perhaps a lower-bound level of prestress for design and slab II purposely had slightly less than the minimum F/A of 150 psi (1.03 N/mm^2) which was recommended in reference 14. Another major departure from the typical design at the time of the test was to use the banded tendon arrangement shown in Fig. 12-10(*b*). The only weakness observed in the performance of slab II was that the ultimate strength of exterior panels *A*, *B*, and *C* was almost exactly ($1.4D + 1.7L$) while the interior panels had more reserve strength, Fig. 12-11. Location of bonded reinforcement outside the immediate column vicinity of exterior columns contributed to the (6%) deficiency in strength of exterior panels.

The crack patterns for slab II, Fig. 12-11, show the importance of placing bonded reinforcement in the region within $1.5t$ either side of columns. Note the cracking at exterior columns for slab II which did not occur in slab I because of inadequate restraint of the stub columns in the slab I model. Since slab II was less conservatively designed, the strength of the slab at the exterior column was more important for strength of the exterior panels than was the case with slab I. Bonded reinforcement for slab II was distributed over a width of 30 in. (762 mm) as shown in Fig. 12-10(*c*). This exceeds the $1.5t$ width either side of column

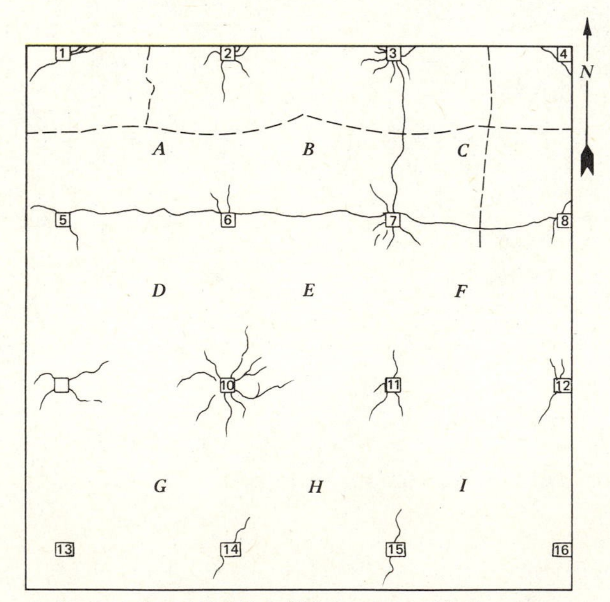

Panels *A*, *B*, and *C*, Failure 6% below factored load
Panels *D*, *E*, *F*, *G*, *H*, and *I*, Failure 13% above factored load

Fig. 12-11. Crack pattern of Slab II after Test 207 (failure).

(16.25 in. = 413 mm) which the ACI Code now uses as a limit. It was concluded from the slab II test[6] that the exterior moment at the column should be based on only the tendons and bonded reinforcement in the immediate column region. Present ACI Code now reflects this design recommendation.

A four-panel model, slab III, was tested to further verify performance of a slab designed with the banded tendon arrangement. This half-scale model represented a prototype designed for two 20-ft. (6.10 m) spans in each direction with no edge overhangs. The tendons and bonded reinforcement, Fig. 12-12, were generally similar to slab II except that the average F/A was 180 psi (1.24

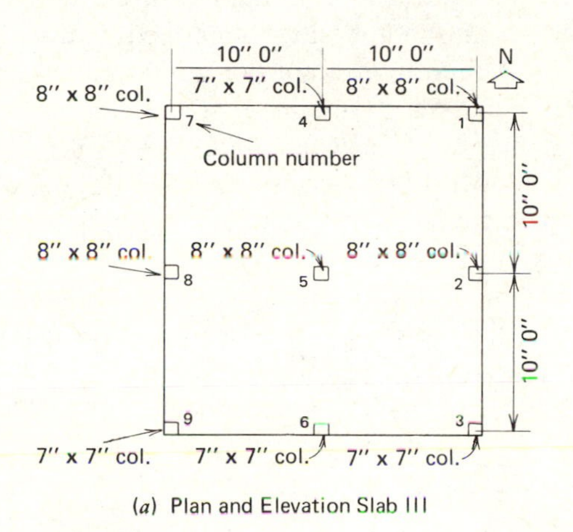

(*a*) Plan and Elevation Slab III

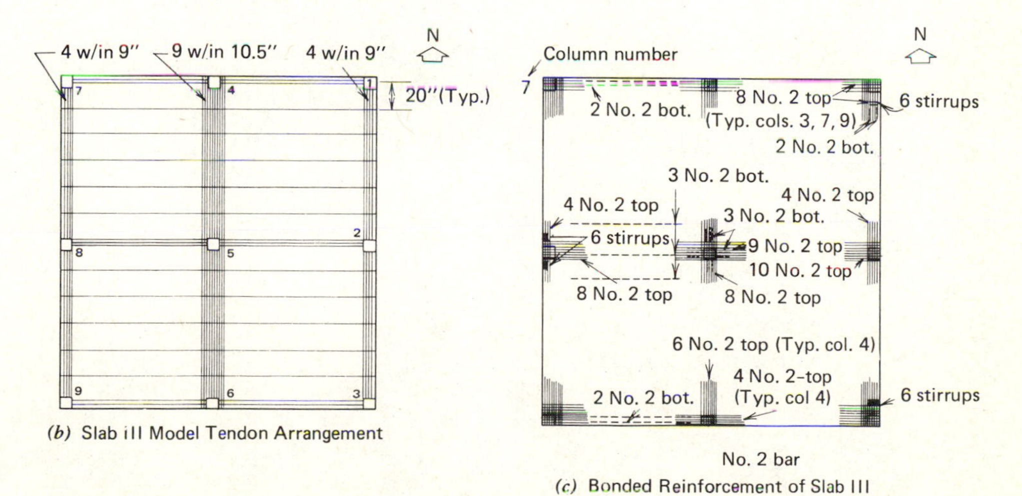

(*b*) Slab III Model Tendon Arrangement

(*c*) Bonded Reinforcement of Slab III

Fig. 12-12. Details of Test Slab III, reference 16.

N/mm^2) and all unbonded tendons and bonded bars counted for moment at exterior columns were within lines 1.5 to either side of faces of column. The performance at both service load and at ultimate for slab III was excellent. Two or three tendons were carried through the column rather than a single tendon in slab II, Fig. 12-10(*b*), in the uniformly distributed direction. Figure 12-13 shows the cracking at ultimate for slab III.

The provisions of the ACI Code may be used with the beam design approach outlined above in each direction. The tests indicate the validity of this procedure for design, even if tendon distribution varies; 70% in column strip and 30% in middle strip each direction or a modified "banded" tendon arrangement gave slab performance which was excellent. The minimum amount and the location of bonded bars provided along with the unbonded tendons in column regions is adequate to control cracking although more than the minimum amount may be used if required for strength. Strength of all three of the test slabs described above was closely estimated by the yield line analysis[20] for flexure. Shear strength may be estimated following the recommendation of the ACI-ASCE Committee on Prestressed Concrete in reference 14. The v_{cw} equation for shear strength from the ACI Code may be used, reflecting the contribution of average F/A to shear strength (see section 12-5).

Example 12-3 shows the approach to design of a two-way prestressed lift slab. This slab is supported at column locations but there is no continuity between the slab and the column. The design approach in this example uses a continuous

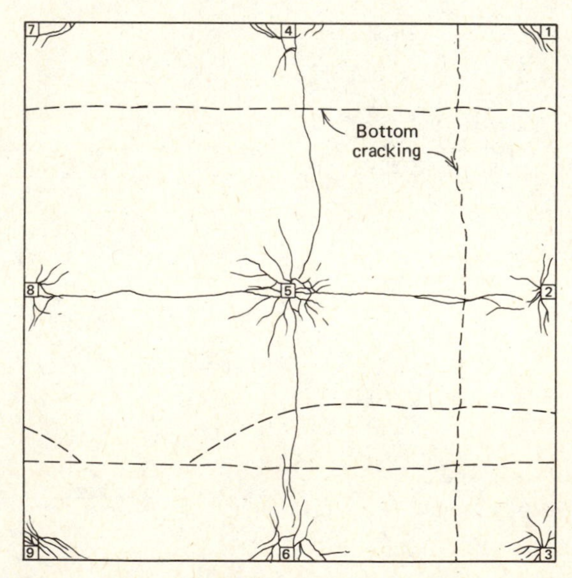

Fig. 12-13. Cracking pattern—Test 309 for Slab III (failure).

beam method of analysis as discussed in Chapter 10 and 11. The design strip is taken to be the width of slab to midspan each direction from the column line as shown in Fig. 12-7. The basic design follows that of a continuous beam. The total prestress force and number of tendons in the design strip is found, then the distribution of the tendons is specified by the structural engineer with practical considerations of tendon layout in mind. The test slabs described above have shown the perhaps surprisingly wide variation in tendon distribution which may be used with satisfactory slab performance. Many flat slabs are now being built with the banded tendon arrangement much like slabs II and III previously described. The test results have been presented in more complete fashion here to provide guidance in making this basic design decision concerning tendon distribution. More flexibility in the tendon distribution is allowed than one might suspect, but the detailing of the placement of the boned reinforcement in the column region is now more explicitly limited ($1.5h$ each side of column) by the ACI Code primarily as a result of these tests.

A very complete prestressed slab example problem is given in Chapter 16 as design example 16-4. The reader should follow this example for monolithic construction where columns are continuous with the slab. The design example 16-4 makes use of load balancing in the design process and the equivalent frame analysis is used in obtaining design moments. The equivalent frame analysis for slabs following the ACI Code is described in all current texts on reinforced concrete design and in the ACI Code Commentary. Refer to reference 24 for the details of the analysis if it is not familiar; the research results justify use of this same method of analysis for reinforced concrete or prestressed concrete slabs. Note that the provisions of the ACI Code (Chapter 13) which relate to the distribution of reinforcing bars in reinforced concrete slabs do not apply. We would use the tendon distribution and reinforcement distribution as discussed in this chapter and in the ACI Code provisions for prestressed concrete (Chapter 18). Reference 14 (ACI-ASCE Commentary 423) also gives recommendations for the design of prestressed flat plates which are very helpful with respect to tendon placement.

EXAMPLE 12-3

A two-way prestressed lift slab has a plan as shown, Fig. 12-14(a). The $7\frac{1}{2}$-in. concrete slab weighs 94 psf and carries a live load of 75 psf; $f_c' = 4000$ psi; $\frac{1}{4}$-in. wires grouped in 6 wires per unit are to be used for prestressing with ultimate strength of 250 ksi; $f_0 = 150$ ksi; $f_{se} = 125$ ksi. Minimum coverage for the cables is $1\frac{1}{4}$ in. measured to the center line. (a) Allowing no tension in the concrete, choose the location for the cables and compute the number of 64-ft long cables required for the slab. (b) Use the beam method for analysis. ($w_d = 4.50$ kN/m^2, $w_l = 3.59$ kN/m^2, $f_c' = 28$ N/mm^2, $f_{pu} = 1724$ N/mm^2, $f_0 = 1034$ N/mm^2, and $f_{se} = 862$ N/mm^2).

Solution (a) Following the method explained in Chapter 10 using the theory of linear transformation. Assume a cable layout with the maximum possible eccentricities

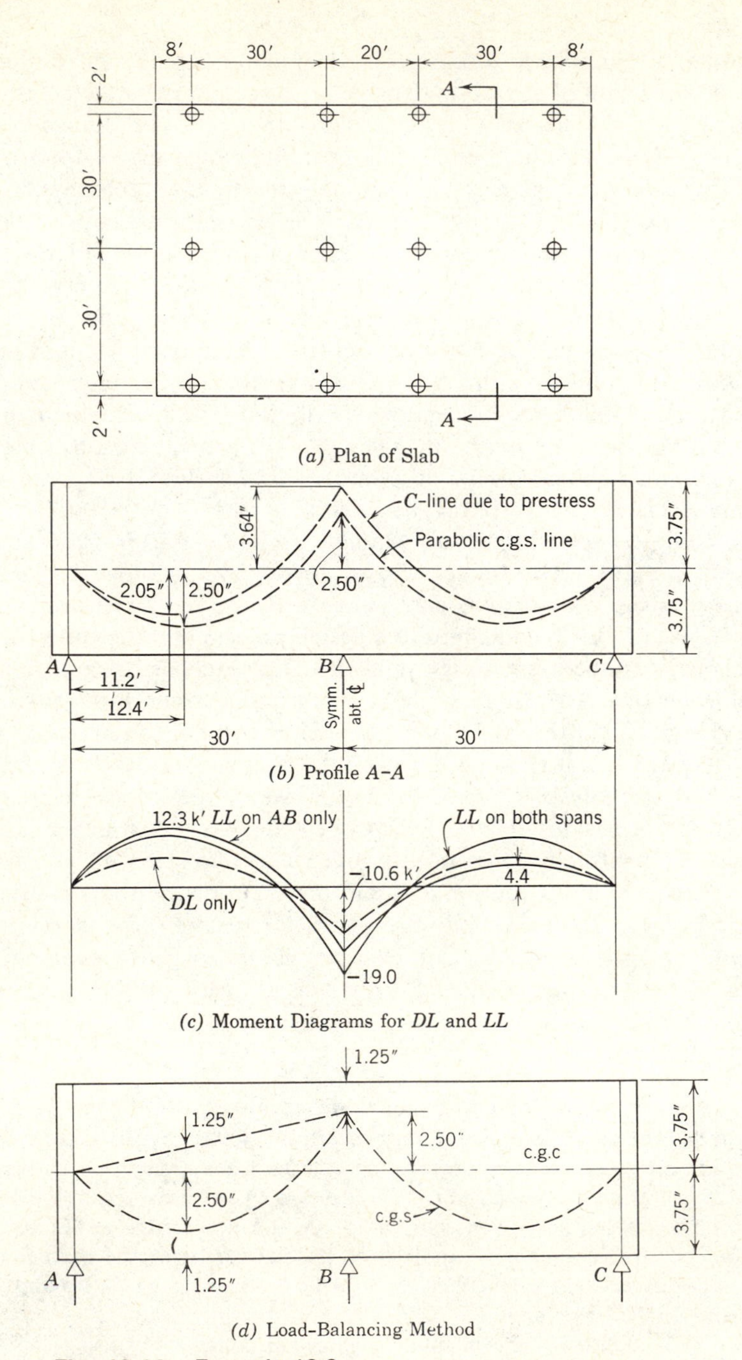

Fig. 12-14. Example 12-3.

for both the positive and the negative moments, as shown in Fig. 12-14(b). This will result in maximum curvature for the cables, and hence the maximum upward force from the cables on the slab. A parabolic cable is used, with an eccentricity of 2.50 in. (63.5 mm) (corresponding to a concrete protection of $1\frac{1}{4}$ in.) at the points of maximum positive and negative moments. Note that the maximum $-M$ occurs at 12.4 ft (3.78 m) from the exterior supports, which is the lowest point for the parabolic curve. This trial location is not likely a concordant cable but offers the maximum lever arm for the steel at critical points.

In order to obtain the C-line under prestressing for this cable, we can proceed as outlined in section 10-4, for continuous beams. But, for a simple problem like this one, it is not necessary to go through all the steps outlined for the procedure. The C-line here can be obtained by inspection after the principles discussed in the previous sections have been mastered. It is noted first that the C-line is a curve linearly transformed from the curve of the cable. Then it is seen that, for a parabolic cable on a beam with straight axis, the force from the cable on the slab is a uniformly distributed load. Neglecting the minor effect of the 2-ft (0.61 m) cantilevers, the moment diagram for a uniform load on two equal spans is well known, having a value of $wL^2/8$ over the center support and $9wL^2/128$ at 11.2 ft (3.41 m) (the $\frac{3}{8}$ point) from the exterior supports. This moment diagram, when plotted to proper scale, gives the eccentricity of the C-line produced by prestress. Thus the trial parabolic cable is linearly transformed to obtain the C-line. A little geometry will show that the position of the cable over the center support is moved upward by 1.14 in. (29.0 mm) to obtain the C-line, and at the $\frac{3}{8}$ point is moved up by $\frac{3}{8}\times 1.14=0.43$ in. (10.9 mm), leaving an eccentricity of 2.05 in. (52.1 mm).

In other words, by the theory of prestressed continuous beams, a cable located as shown in (b) will produce the same effect as though it were located through the computed C-line. Since it is not practicable to locate the cable through the above C-line (too near the top surface over the center support), it is just as well to place it along the trial line, resulting in the same effect. We will see later that this line of pressure seems to lie in a very desirable location.

Next, let us compute the maximum and minimum moment diagrams, (c). The greatest $+M$ is obtained with live load on its own span only; the greatest $-M$ is obtained with live load on both spans. The smallest $+M$ is obtained with live load on the other span, but the smallest $-M$ with dead load only. For a final location of the cables, the graphical solution explained in section 10-5 should be used. But just to obtain the number of cables, we will compute for the critical points only, that is, the moment over the center support and near the $\frac{3}{8}$ points from the exterior supports.

First, let us design for the total moments. Over the center support, the lever arm available for the resisting couple is measured to the bottom kern point, 1.25 in. (31.75 mm) below the middepth, or $3.64+1.25=4.89$ in. (124.2 mm). Hence the effective prestress required is

$$(19.0\times 12)/4.89=46.6 \text{ k/ft } (680 \text{ kN/m}) \text{ of width}$$

For the $\frac{3}{8}$ points, the lever arm available is $2.05+1.25=3.30$ in. (83.8 mm), and the effective prestress required is

$$(12.3\times 12)/3.30=44.7 \text{ k/ft } (652 \text{ kN/m}) \text{ of width}$$

Hence the moment over the center support controls the design, and a total prestress for the entire slab should be

$$96 \text{ ft} \times 46.6 = 4480 \text{ k } (19.927 \text{ kN})$$

A 6-wire unit of $\frac{1}{4}$-in. (6.35 mm) wires will have an effective prestress of $6 \times 0.049 \times 125 =$ 36.8 k (164 kN); hence the total number of units required is

$$4480/36.8 = 122$$

Note that this number is not too excessive for the $+M$, which would require 117 units, indicating that this is a well-balanced layout.

Now we have to check whether the line of pressure would fall outside the kern under the action of prestress and the minimum moments. If we permit some tension in the concrete, the C-line might be allowed to fall slightly outside the kern. The same two critical points as above are chosen for investigation. Over the center support, the minimum DL moment is 10.6 k-ft per ft. (47.1 kN-m/m). The initial prestress of the 122 cables will be

$$F_0 = \frac{122 \times 6 \times 0.049 \times 150}{96} = 56.0 \text{ k/ft } (249 \text{ kN/m}) \text{ of width}$$

$$\frac{M_G}{F_0} = \frac{10.6 \times 12}{56.0} = 2.27 \text{ in. } (57.7 \text{ mm})$$

which means that the dead-load moment will bring the C-line from 3.64 in. (92.5 mm) down to $3.64 - 2.27 = 1.37$ in. (34.8 mm) above the mid-depth. Since the top kern is only 1.25 in. (31.8 mm) above the mid-depth, the C-line under dead load only will be $1.37 - 1.25 = 0.12$ in. (3.0 mm) outside the kern, and some tension will exist in the bottom fiber over the center support under the initial prestress, but the value is evidently small and will be reduced as soon as loss of prestress takes place. Hence, this is considered satisfactory.

Now, near the $\frac{3}{8}$ points, the minimum moment occurs when live load exists on the other span only, a total moment of 4.4 k-ft. Corresponding to the prestress of 56.0 k/ft (249 kN/m), this moment will move the C-line upward by the amount of

$$4.4 \times 12/56.0 = 0.94 \text{ in. } (23.9 \text{ mm})$$

This will place the C-line $2.05 - 0.94 = 1.11$ in. (28.2 mm) below the middepth, which is within the kern, and no tension will exist.

Thus, 122 cables with critical points located as above can be considered sufficient. To make a complete design of the slab, many related problems, such as those mentioned in section 12-2 for simple flat slabs, must yet be considered. In addition, the sharp bend over the center support should be smoothed out. Also note that at the intersection of the two sets of cables in the two directions the maximum lever arm for resisting moment cannot be obtained for both sets.

(b). Following the load-balancing method in Chapter 11 is a much more common approach for design as indicated above in discussing present state-of-the-art. We can start off by assuming that for optimum behavior it will be desirable to balance the 94 psf (4.50 kN/m^2) of dead load plus 15 psf (0.72 kN/m^2) of the live load, or a total of 109 psf (5.22 kN/m^2). Referring to Fig. 12-14(d), the cable sag h is very nearly 3.75 in. (95.25 mm). Hence the effective prestress required is

$$F = \frac{wL^2}{8h} = \frac{109 \times 30^2 \times 12}{8 \times 3.75} = 39.4 \text{ k/ft } (575 \text{ kN/m})$$

and the slab is under uniform prestress of

$$f_{av} = \frac{39,400}{12 \times 7.5} = -437 \text{ psi } (-3.01 \text{ N/mm}^2)$$

for its dead load plus 15 psf (0.72 kN/m^2) live load.

To check the stresses under full live load, we compute the effect of $75 - 15 = 60$ psf (2.87 kN/m^2) additional live load, which will produce a maximum moment over the center support of

$$-M = \frac{wL^2}{8} = \frac{60 \times 30^2}{8} = 6750 \text{ lb-ft } (9.15 \text{ kN-m})$$

And the maximum bending stresses, according to the beam theory, are

$$f = \frac{Mc}{I} = \frac{6M}{bd^2} = \frac{6 \times 6750 \times 12}{12 \times 7.5^2} = 720 \text{ psi } (4.96 \text{ N/mm}^2)$$

The resultant maximum fibers stresses are

$$f_{top} = -437 + 720 = +283 \ (+1.95 \text{ N/mm}^2) \text{ psi tension}$$

$$f_{bot} = -437 - 720 = -1157 \ (-7.98 \text{ N/mm}^2) \text{ psi compression}$$

These stresses are not considered excessive, and $F = 39.4$ k/ft (575 kN/m) is satisfactory. This solution is clearly much easier than (*a*). Note that if no tension is used as the criterion and $\frac{3}{8}L$ is used as the controlling point, the required prestress would be $F = 46.6$ k/ft (680 kN/m) as obtained in solution (*a*).

Deflections of flat slabs can be obtained by the theory of elasticity, but the time consumed for such an analysis would be enormous. When only approximate results are desired, it is possible to treat strips of the slab as beams and compute the accumulated deflection.[21] For example, the center deflection of a slab is the sum of two deflections, one due to a continuous beam along the columns, another due to a perpendicular continuous beam along the middle, Fig. 12-15. If the moments along these two strips are known, the deflections

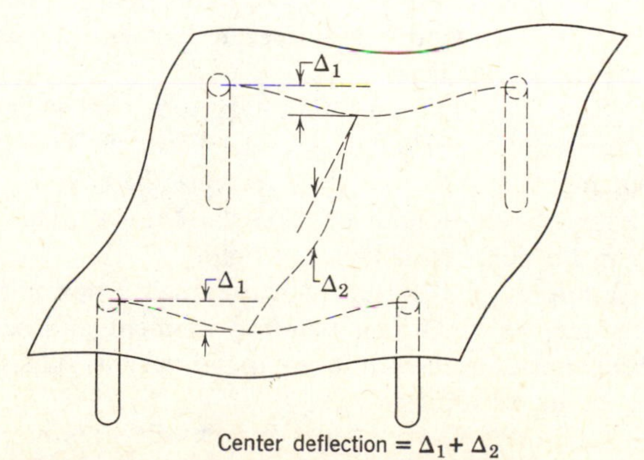

Fig. 12-15. Estimating slab deflections.

produced by both prestress and external loads can be computed with precision. The effect of plastic flow has to be considered separately and added to the initial deflections.

12-4 Flat Slabs, Some Practical Remarks on Design

Haunched Slabs. Most prestressed-concrete flat slabs have been built of uniform thickness. This was because they were sometimes used in conjunction with lifting. In order to be cast conveniently on the ground with no formwork underneath, it is desirable to employ a flat soffit. Even for cast-in-place slabs the forming is most economical with uniform thickness. If the spans are long, and if the slabs are to be cast in place, it may sometimes be economical to design haunched slabs or slabs with drop panels similar to reinforced-concrete construction.

Hollow Slabs or Waffle Slabs. If the spans are long, it is often economical to keep the dead load within limits. This is done by hollowing the slab, or by using waffle slabs. The sections are thus either I or T in shape and should be designed accordingly. For area over the columns, these slabs are often made solid in order to carry the heavy shear and the negative moments. Lift slabs of the waffle type should be handled with care, both in designing and during lifting. They are thicker and therefore stiffer than equivalent solid slabs and are subjected to higher stresses.

Lift Collars. Collars for lifting the slabs are of various designs usually made by welding angles or channel-shaped structural steel sections. For details of collars, refer to current brochures from the companies which market the lifting equipment.

Partition Walls for Lift Slabs. After the prestressed slabs are lifted in position, partition walls may sometimes be constructed beneath them. These partitions actually serve as bearing walls to some extent. The existence of such walls generally strengthens the slab and reduces its deflections. When located at odd positions, however, they may tend to increase the moments at certain points, and cracking of the slabs may result.

Long Slabs. When the continuous slabs are too long in one direction, say much over 100 ft, special problems may arise. First, the friction in the cables may increase appreciably and thus tend to decrease the effective prestress. Next, there may be excessive shortening of the slab under prestress which may produce bending in the columns if they are rigid.

Cantilevers. Cantilevering the slabs beyond their exterior row of columns often helps to reduce the maximum bending moment and saves prestressing steel. But the deflections of such cantilevers under various stages of loading may be excessive and should be studied.

Tendon Spacing. Tendon spacing for flat slabs will often be rather large, Fig. 12-16, enabling easy placing of concrete. Reference 14 gives guidelines for

Fig. 12-16. Tendons (unbonded) and bonded reinforcement at the column for a posttensioned flat plate. Note banded tendons one way with distributed tendons in the other direction. (Seneca Construction Systems, Canoga Park, California)

tendon spacing. The tendons in a banded slab may be bundled side by side with two to four strands in a group.

Average Prestress. Average prestress is defined as the amount of prestressing force divided by the cross-sectional area of slab concrete. A minimum average prestress is required, if it is desired to eliminate or minimize cracks in the slab. Experience indicates that this minimum value is about 150 psi (1.03 N/mm^2), probably ranging between 100 and 200 psi (0.69 and 1.38 N/mm^2). Too high an average prestress would induce excessive creep, and should be avoided. No definite rule can be given, although 500–600 psi (3.45–4.14 N/mm^2) is considered fairly high for flat slabs.

Shear Wall Location. Stiff vertical elements, such as walls and shafts, rigidly connected to the prestress slabs should not be located so as to restrain their shortening. As a result of creep, prestressed slabs may tear themselves away from the supports or produce cracks in them.

Slabs Posttensioned In Place. For multistory buildings, it is frequently quite economical to posttension the slabs in place. For in-place posttensioning, proper provision should be made for the shrinkage and creep of the slabs. Fortunately, most of the floors tend to shorten together and thus will not create a serious problem.

12-5 Flat Slabs, Shear Strength

The two-way action of the flat slab is recognized in design, and the shear is checked at a critical section $d/2$ away from and around the face of column. Tests[4,5,15,16,17] show that the level of prestress influences the shear strength of a flat slab with higher average F/A yielding more shear strength. Failure typically involves a final punching shear at the column with a rather flat angle of the failure plane as a result of the prestress in the slab.

The ACI-ASCE Committee recommendation[14] for shear stress at ultimate is shown in Fig. 12-17. The value $4\sqrt{f_c'}$ for ultimate shear stress in reinforced concrete slabs and footings with two-way action (ACI Code) is shown by tests of prestressed slabs to be conservative for design. Data from tests support the use of the shear stress shown by the dashed line in Fig. 12-17:

$$v_{cw} = 3.5\sqrt{f_c'} + 0.3f_{pc} \qquad \text{(equation 11-13) ACI}$$

The shear at an interior column, Fig. 12-18, is maximum when the surrounding panels are loaded with live load as shown. There will be only a small moment transfer with this loading, and the shear will be slightly nonuniform around the critical section $d/2$ away from face of column A. The check on shear

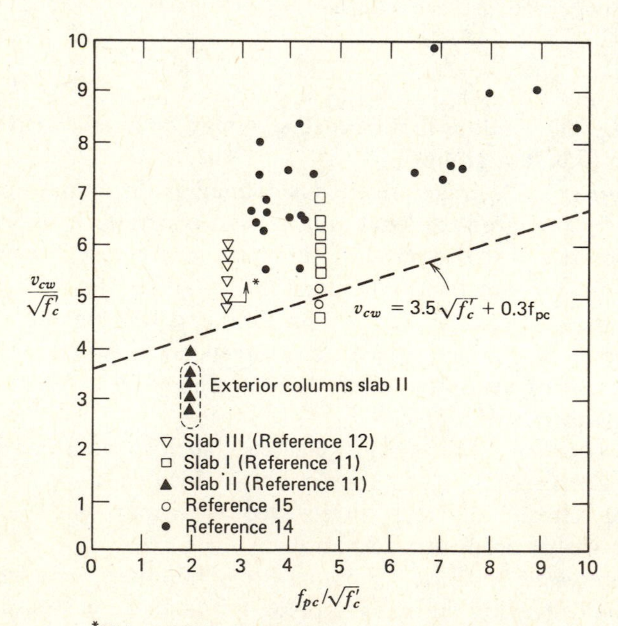

*Did not fail in punching shear: maximum stress attained

Fig. 12-17. Shear Test Data Versus equation 11-13 of ACI 318-77.

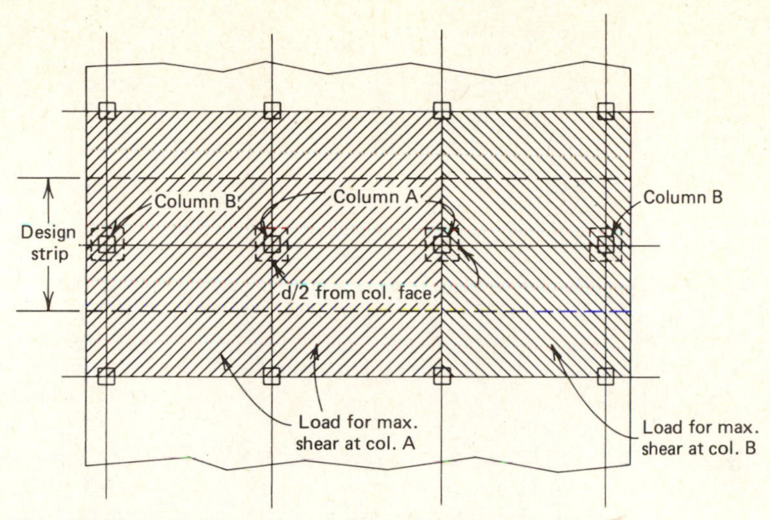

Fig. 12-18. Loading for maximum shear at columns.

strength involves a comparison between v_{cw} and the ultimate shear stress, v_u, along the line where it is maximum as shown in Fig. 12-19. This can be written in equation form as

$$\phi v_u = \frac{V_u}{b_0 d} + \frac{\alpha M_t c_3}{J_c} \tag{12-1}$$

where V_u = factored shear force at section
v_u = shear stress at design (factored) loads
b_0 = perimeter of shear section at $d/2$ from face of column as defined by ACI Code
d = distance from centroid of tendon to compression face in direction of moment transfer, but need not be less than $0.8h$, where h is member thickness
ϕ = capacity reduction factor for shear, 0.85
α = fraction of moment transferred by shear

$$\alpha = 1 - \frac{1}{1 + 2/3\left(\dfrac{c_1 + d}{c_2 + d}\right)^{1/2}}$$

c_1 = support dimension in the direction of moment transfer
c_2 = support dimension perpendicular to c_1
c_3 = distance from centroid of critical shear section to extreme fiber in direction of moment transfer
M_t = net moment to be transferred to column
J_c = polar moment of inertia of critical section

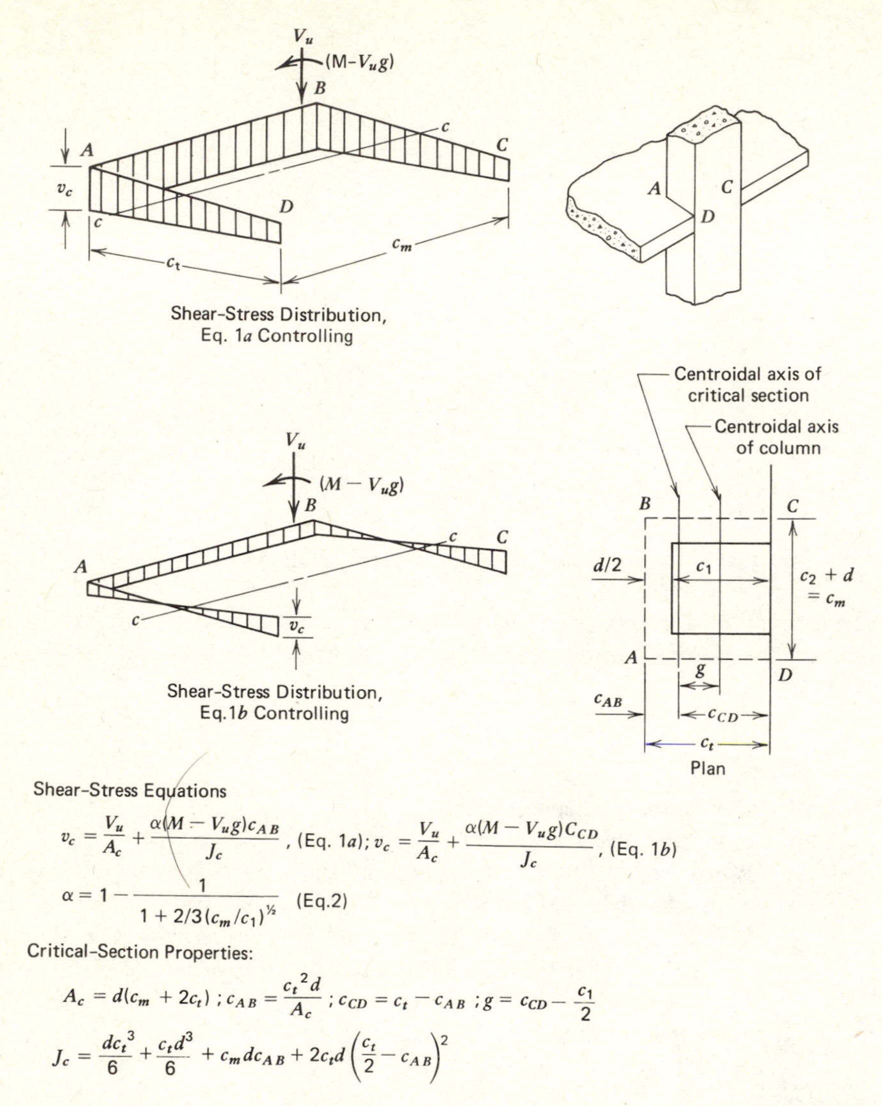

Fig. 12-19. Moment-shear interaction-relationships for edge column connections.

The ACI Code Commentary includes equations which allow the calculation of the shear due to moment transfer. Sketches similar to Fig. 12-19 are also given to help the designer visualize that there is a combined maximum shear stress which must be considered. The edge columns experience higher unbalance moment and thus more contribution from the second term of equation 12-1.

For exterior columns such as column B in Fig. 12-18 the second term of equation 12-1 is frequently more important than for interior columns. Figure

12-19 illustrates the possible shear stress distributions which would develop. Part of the moment at this column is transferred by the eccentricity of the centroid of the shear section, g, with respect to the column centroid, creating a moment $V_u g$. Thus the moment, M_t, to be transferred by shear and torsion is the unbalanced moment minus $V_u g$, the moment from shear eccentricity, Fig. 12-19. Design Example 16-4, Chapter 16, shows the calculations for shear at an exterior column of this type where the shear from the moment transfer is quite significant.

Research on the shear strength of flat slabs in both reinforced and prestressed concrete is still needed in order to fully understand the moment-shear-torsion interaction. As stated in reference 23, recent tests[6,19,22] confirm the applicability of the above approach for calculation of the shear capacity of connections transferring moment as well as shear.

References

1. H. M. Westergaard, "Computation of Stresses in Bridge Slabs Due to Wheel Loads," *Public Roads*, March 1930 (also *Public Roads*, March 1926, paper by Kelley giving test results).
2. E. Freyssinet, "The Deformation of Concrete," *Magazine of Concrete Research*, December 1951.
3. G. L. Rogers, "Validity of Certain Assumptions in the Mechanics of Prestressed Concrete," *J. Am. Conc. Inst.*, December 1953 (*Proc.*, Vol. 49), pp. 317–330.
4. A. C. Scordelis, K. S. Pister, and T. Y. Lin, "Strength of a Concrete Slab Prestressed in Two Directions," *J. Am. Conc. Inst.*, September 1956 (*Proc.*, Vol. 53), pp. 241–256.
5. T. Y. Lin, A. C. Scordelis, and R. Itaya, "Behavior of a Continuous Slab Prestressed in Two Directions," *J. Am. Conc. Inst.*, Vol. 31, No. 6, December 1959 (*Proc.* Vol. 56), pp. 441–459.
6. R. Hemakom, "Strength and Behavior of Post-Tensioned Flat Plates with Unbonded Tendons," Ph.D. dissertation, The University of Texas at Austin, December 1975, 272 pp.
7. N. H. Burns, and R. Hemakom, "Test of Scale Model Post-Tensioned Flat Plate," *J. Struct, Div. ASCE*, Vol. 103, No. ST6, June 1977, pp. 1237–1255.
8. F. A. Charney, "Strength and Behavior of a Partially Prestressed Concrete Slab with Unbonded Tendons," M. Sc. thesis, The University of Texas at Austin, May 1976, 180 pp.
9. W. R. Vines, "Strength and Behavior of Concrete Slab with Unbonded Tendons," M. Sc. thesis, The University of Texas at Austin, May 1976, 179 pp.
10. N. H. Burns, F. A. Charney, and W. R. Vines, "Tests of One-Way Post-Tensioned Slabs with Unbonded Tendons," *J. Prestressed Conc. Inst.*, Vol. 23, No. 5, September/October 1978.
11. A. C. Scordelis, W. Samarzich, and D. Pirtz, "Load Distribution on Prestressed Concrete Slab Bridge," *J. Prestressed Conc. Inst.*, June 1960.

12. Charles Peterson and A. H. Brownfield, "Our Experience with Prestressed Lift-Slabs," *Proc. World Conference on Prestressed Concrete*, San Francisco, 1957.
13. H. Korner, "A 13-story Building by Lift Slab and Slip Form," *Civil Engineering*, September 1960, pp. 62–64.
14. "Tentative Recommendations for Prestressed Concrete Flat Plates," reported by ACI-ASCE Commentary 423, *J. Am. Conc. Inst.*, Vol. 71, No. 2, February 1974, pp. 61–71.
15. Louis L. Gerber, and Ned. H. Burns, "Ultimate Strength Tests of Post-Tensioned Flat Plates," *J. Prestressed Conc. Inst.*, Vol. 16, No. 6, November–December 1971, pp. 40–58.
16. T. Y. Lin, A. C. Scordelis, and H. R. May, "Shearing Strength of Prestressed Concrete Lift Slabs," Document Section, State of California, Sacramento, October 1957. Also, summarized in "Shearing Strength of Prestressed Lift Slabs," *J. Am. Conc. Inst.*, (*Proc.* Vol. 55), No. 4, October 1958, pp. 485–506.
17. S. W. Smith, and N. H. Burns, "Post-tensioned Flat Plate to Column Connection," *J. Prestressed Conc. Inst.*, Vol. 19, No. 3, May-June 1974, pp. 74–91.
18. J. C. Winter, "Flexural Behavior of a Post-Tensioned Flat Plate with Unbonded Tendons," M. Sc. thesis, The University of Texas at Austin, December 1978, 154 pp.
19. G. M. Kosut, "Shear Strength of a Post-Tensioned Concrete Flat Plate at the Column Connections," M. Sc. thesis, The University of Texas at Austin, January 1977, 104 pp.
20. E. Hognestad, "Yield-Line Theory for the Flexural Strength of Reinforced Concrete Slabs," *J. Am. Conc. Inst.*, March 1953 (*Proc.*, Vol. 49), pp. 637–658.
21. E. K. Rice and F. Kulka, "Design of Prestressed Lift-Slabs for Deflection Control," *J. Am. Conc. Inst.*, February 1960, pp. 681–693.
22. N. M. Hawkins, and N. Trongtham, "Moment Transfer Between unbonded Post-Tensioned Concrete Slabs and Columns," Progress Report to the Post Tensioning Institute and Reinforced Concrete Research Council on Project #39, Structures and Mechanics Division, Department of Civil Engineering, University of Washington, Seattle, November 1976.
23. "Design of Post-Tensioned Slabs," Post-Tensioning Institute, 1977, 52 pp.
24. P. M. Ferguson, *Reinforced Concrete Fundamentals*, 4th ed., Wiley, New York, 1979.

13

TENSION MEMBERS; CIRCULAR PRESTRESSING

13-1 Tension Members, Elastic Design

Prestressed tension members combine the strength of high-tensile steel with the rigidity of concrete and provide a unique resistance to tension consistent with small deformations that cannot be obtained by either steel or concrete acting alone. The rigidity of prestressed concrete serves well, especially for long tension members such as tie rods for arches or staybacks for wharves and retaining walls. When prestressed, concrete is given strength to resist any local bending and at the same time steel is stiffened and protected. Numerous prestressed concrete tanks and pressure vessels have been constructed both in this country and abroad. With better understanding, wider application should follow.

The basic behavior of prestressed tension members can be explained from three points of view.

1. The member can be considered as essentially made of concrete which is put under uniform compression so that it can carry tension produced by internal pressure or external loads. If the concrete has not cracked, it is able to carry a total tensile force equal to the total effective precompression plus the tensile capacity of the concrete itself.
2. The member can be considered as essentially made of high-tensile steel which is preelongated to reduce its deflection under load. From this viewpoint, the ultimate strength of the member is dependent upon the tensile strength of the steel, but the usable strength is often limited by excessive elongation of the steel following the cracking of the concrete.
3. The member can be considered as a combined steel and concrete member whose strains and stresses before cracking can be evaluated, assuming elastic behavior and taking into account the effect of shrinkage and creep.

Each of the three points of view furnishes some basic concepts from which the engineer can visualize his design, but the third viewpoint is most convenient for analysis by the elastic theory and will be explained first.

If the total initial prestress is F_0 and the total effective prestress F, then the stresses in the concrete will be

$$f_c = \frac{F_0}{A_c}$$

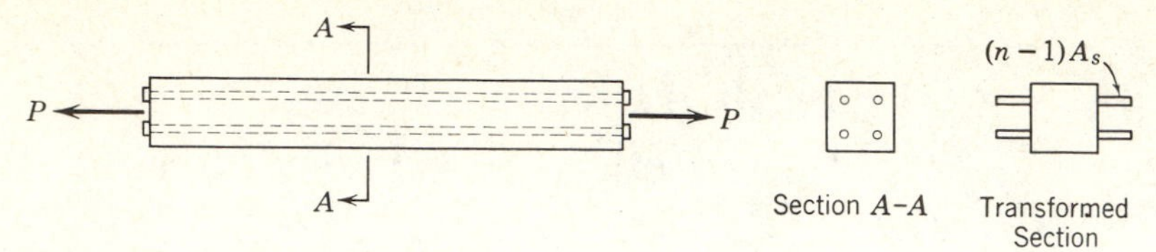

Fig. 13-1. Prestressed-concrete tension member.

for the initial prestress and

$$f_c = F/A_c$$

for the effective prestress.

Because of a load P applied externally, Fig. 13-1, both the steel and the concrete will elongate the same amount. Hence the usual transformed-section method for reinforced concrete can be applied here. Thus the cross section of the member can be transformed into an equivalent area of concrete equal to

$$A_t = nA_s + A_c \tag{13-1}$$

If the gross area of concrete A_g is used, the transformed area can be expressed as

$$A_t = nA_s + A_g - A_s = A_g + (n-1)A_s \tag{13-2}$$

This formula is valid only when the section is grouted. Otherwise the hole in the concrete will be greater than A_s and formula 13-1 can be more conveniently applied, with A_c referring to the net concrete area.

The stresses produced by P will be, for concrete,

$$f_c = \frac{P}{A_t}$$

and for steel,

$$f_s = \frac{nP}{A_t}$$

The value of $n = E_s/E_c$ should be chosen for the proper stress and duration of loading, taking into account the effect of creep if necessary.

Thus the resultant stresses due to the effective prestress plus the external load are, for concrete,

$$f_c = \frac{F}{A_c} + \frac{P}{A_t} \tag{13-3}$$

and for steel,

$$f_s = f_e + \frac{nP}{A_t} \tag{13-4}$$

If it is desired to determine the load P which will produce zero stress in the concrete, it is only necessary to put $f_c=0$ in equation 13-3, thus

$$\frac{F}{A_c}+\frac{P}{A_t}=0$$

$$P=-F\frac{A_t}{A_c}=-F(1+np) \tag{13-5}$$

It is seen that, with no stress in the concrete, the load carried by the member is somewhat greater than the effective prestress F. This is because the stress in the steel has been somewhat increased under the action of the external load P.

It is important to investigate the strains in a prestressed-concrete member, both due to prestressing and due to external loads. Under the initial prestress F_0, the stress in the concrete being F_0/A_c, the corresponding instantaneous unit strain will be

$$\delta=\frac{F_0}{EA_c}$$

which will reduce to F/EA_c after the losses have taken place.

Under the action of external load P, the instantaneous strain is given by

$$\delta=\frac{P}{EA_t}$$

In all cases, the value of E must be chosen with regard to the level of stress and the age of concrete, and the effect of creep must be considered.

Let us first compare the magnitude of strains in a prestressed-concrete member with those in an ordinary steel member. For a structural steel member stressed to 20,000 psi (137.9 N/mm^2), corresponding to a value of $E_s=30{,}000{,}000$ psi (207 kN/mm^2), the unit elongation is

$$\delta=\frac{20{,}000}{30{,}000{,}000}=0.00067$$

For a prestressed-concrete member, with the stresses in concrete changing from -1000 psi (-6.9 N/mm^2) to 0, for an E_c of 4,000,000 psi (27.6 kN/mm^2), the unit strain is

$$\delta=\frac{1000}{4{,}000{,}000}=0.00025$$

which is less than half of the strain in structural steel.

High-tensile steel, by itself, cannot be used for long tension members where elongation must be limited. In order to be stressed to its working strength of 125,000 psi (862 N/mm^2), the unit elongation will be

$$\delta=\frac{125{,}000}{30{,}000{,}000}=0.00417$$

which is more than 6 times that of structural steel and 16 times that of prestressed concrete in the above example.

Strains in prestressed-concrete members are influenced by several factors. If the precompression in the concrete remains over a period of time, the shortening of that member due to creep could be considerable. Such creep strain, however, would be gradually recovered (though not completely) under the application of an external tension (see Chapter 2, reference 3). Since a greater portion of the creep may be eventually recovered, the lengthening of the member under sustained external load may be greater than is indicated by the elastic calculations.

On the other hand, there are ways to limit further the elongation of prestressed-concrete tension members. One obvious method is to increase the cross-sectional area of concrete. For example, if the concrete area is doubled, the stress range will be halved, and so will the strain. There is, of course, an economical limit to this method, since the area of concrete cannot be indefinitely increased. Another way to control the elongation is to time the application of prestress to the application of the external dead load. If this is carefully done, the elongation due to dead load can really be reduced to a minimum, although practical considerations may not permit such an ideal sequence of application of forces. Still another way is to use concrete possessing high modulus of elasticity. It is known that, for high strength concrete (say $f_c' > 5000$ psi $= 34$ N/mm^2) over two or three years old, the instantaneous modulus of elasticity would be as high as 6,000,000 psi (41 kN/mm^2).

EXAMPLE 13-1

A straight concrete member 150 ft long is prestressed with a high-tensile steel strand through the centroid of the section. The strand is anchored to the concrete with end anchorages but separated from it by bond-breaking agents along the length. $A_c = 80$ in.^{2}. $A_{ps} = 0.80$ in.^{2}. $f_c' = 4000$ psi, $f_{pu} = 250{,}000$ psi, $f_0 = 150{,}000$ psi, $f_{se} = 127{,}500$ psi, $E_c = 4{,}000{,}000$ psi, $E_s = 30{,}000{,}000$ psi

(*a*) Compute the allowable external load on the member, allowing no tension in the concrete. (*b*) Compute the shortening of concrete due to prestress, assuming a creep coefficient of 1.5. (*c*) Compute the lengthening of the member due to the external load obtained in (*a*), neglecting creep. (*d*) If the member were designed of structural steel with an allowable stress of 20,000 psi, compute the lengthening under the load. (*e*) Compute the lengthening if the strand is used alone by itself with an allowable stress of 127,500 psi. (Span $= 45.7$ m, $A_c = 51.6 \times 10^3$ mm^2, $A_{ps} = 516$ mm^2, $f_c' = 28$ N/mm^2), $f_{pu} = 1{,}724$ N/mm^2, $f_0 = 1{,}034$ N/mm^2, $f_{se} = 879$ N/mm^2, $E_c = 27.6$ kN/mm^2, and $E_s = 207$ kN/mm^2.)

Solution (*a*) From formula 13-5, allowable external load is

$$P = F(1 + np)$$
$$= 127{,}500 \times 0.80(1 + 7.5 \times 0.80/80)$$
$$= 110{,}000 \text{ lb } (489 \text{ kN})$$

(*b*) Under the initial prestress, the shortening of concrete will be

$$\frac{F_0 L}{E_c A_c} = \frac{150{,}000 \times 0.80 \times 150 \times 12}{4{,}000{,}000 \times 80}$$

$$= 0.675 \text{ in. } (17.15 \text{ mm})$$

If the effective prestress is considered, the shortening will be

$$0.675 \times \frac{127{,}500}{150{,}000} = 0.573 \text{ in. } (14.55 \text{ mm})$$

If the creep coefficient is based on the effective prestress, the total elastic and creep shortening will be

$$0.573 \times 1.5 = 0.860 \text{ in. } (21.84 \text{ mm})$$

(*c*) Under the external load of 110 k (489 kN), for a transformed area of $A_t = 80 + 7.5 \times 0.80 = 86$ in.^{2} (55.5×10^3 mm^2), again using $E_c = 4{,}000{,}000$ psi (27.6 kN/mm^2), the lengthening of the member will be

$$\frac{110{,}000 \times 150 \times 12}{4{,}000{,}000 \times 86} = 0.575 \text{ in. } (14.61 \text{ mm})$$

this checks closely with the shortening of the concrete computed in (*b*).

(*d*) For a structural steel stressed to 20,000 psi (137.9 N/mm^2), the elongation will be

$$\frac{20{,}000 \times 150 \times 12}{30{,}000{,}000} = 1.20 \text{ in. } (30.48 \text{ mm})$$

(*e*) For high-tensile steel stressed to 127,500 psi (879 N/mm^2), the elongation will be

$$\frac{127{,}500 \times 150 \times 12}{30{,}000{,}000} = 7.65 \text{ in. } (194.31 \text{ mm})$$

13-2 Tension Members, Cracking and Ultimate Strengths

The previous section discusses the computation of stresses in a prestressed-concrete tension member, up to zero compression in the concrete. The design of such a member may or may not be made on this basis, depending on the probable amount of overloading to which the member may be subjected. In order to get a sufficient factor of safety, it may be necessary to design the member so that, under working loads, there will always be some residual compression in the concrete. This will become evident after a study of the cracking and ultimate strengths of the member. Tension members are one of the typical cases in prestressed concrete where design by the allowable stress method may err very much on the dangerous side and may not yield consistent results.

Generally speaking, prestressed-concrete tension members have a very low reserve strength above the point of zero stress. If the member is not cast as one piece, for example, if it is made up of blocks, cracking may coincide with zero stress. Then any additional load on the member will be carried by the steel alone. Since the prestressing steel has a relatively small area of cross section, excessive elongation will immediately start at the cracking of concrete, and failure of other parts of the structure may result. For such a member, then, it is evident that a considerable amount of residual compression is necessary in order to ensure safety, the amount being governed by the magnitude of the probable overloads.

If the member is cast as one piece, and if shrinkage and other cracks have not occurred, it will be able to take some tension before cracking. The direct tensile strength of concrete is variable and generally ranges from 0.06 to $0.10f_c'$. Thus, for a concrete of 4000 psi (28 N/mm^2), the tensile strength may be from 240 to 400 psi (1.65 to 2.76 N/mm^2), which may provide a good margin of safety if the strength exists and has not been destroyed. But, once the concrete has cracked, the margin of safety is gone. In fact, failure of the entire structure may result as soon as the concrete cracks, because at this moment the tensile load carried by the concrete in tension is suddenly transferred to the steel. Thus there may be a sudden elongation of steel which may have serious effects, even though the ultimate strength of the steel is far from being reached.

The above discussion must not be construed to mean that such tension members are unsafe. They are just as safe as any other type of tension members and perhaps safer if properly designed. When heavy overloads are possible, they should not be designed on the basis of allowable stresses, but rather on the basis of the cracking or ultimate strength, with proper load factors.

Load factors should vary with the type of structure. Their choice will depend on the possibilities of overloading. If dead load predominates in a member, as in most buildings and long-span bridges, the over all load factor required will be smaller. For liquid storage tanks and certain pressure vessels, both the possibility and the magnitude of overloading are small, and a very low load factor is employed.

EXAMPLE 13-2

For the tension member in example 13-1, what working load can it carry, using a factor of safety of 2.0 against the cracking of concrete, assuming the direct tensile strength in concrete to be $0.08f_c' = 320$ psi (2.21 N/mm^2)? Compute the residual compression in concrete under that working load.

Solution From formula 13-3, for $f_c = 320$ psi (2.21 N/mm^2),

$$\frac{F}{A_c} + \frac{P}{A_t} = 320$$

From example 13-1,

$$F = -102{,}000 \text{ lb } (-454 \text{ kN})$$

$$A_c = 80 \text{ sq in. } (51.6 \times 10^3 \text{ mm}^2)$$

$$A_t = 86 \text{ sq in. } (55.5 \times 10^3 \text{ mm}^2)$$

Substituting,

$$\frac{-102{,}000}{80} + \frac{P}{86} = 320$$

$$P = 137{,}000 \text{ lb } (609 \text{ kN})$$

which is the cracking load.

For a factor of safety of 2.0, the working load will be

$$137{,}000/2 = 68{,}500 \text{ lb } (305 \text{ kN})$$

Though the load factor of 2.0 is not always necessary, the great difference between this answer and the last one of 110 k (489 kN) should be noticed.

The residual compression can be computed using the same formula, for $P = 68{,}500$ lb (305 kN),

$$f_c = \frac{F}{A_c} + \frac{P}{A_t} = \frac{-102{,}000}{80} + \frac{68{,}500}{86}$$

$$= -1275 + 795 = -480 \text{ psi } (-3.31 \text{ N/mm}^2)$$

13-3 Circular Prestressing

The term "circular prestressing" is employed to denote the prestressing of circular structures such as pipes and tanks where the prestressing wires are wound in circles. In contrast to this term, "linear prestessing" is used to include all other types of prestressing, where the cables may be either straight or curved, but not wound in circles around a circular structure. In most prestressed circular structures, prestress is applied both circumferentially and longitudinally, the circumferential prestress being circular and the longitudinal prestress actually linear. For convenience, both types of prestress as they are applied to circular structures will be discussed in this chapter.

The basic theories of circular prestressing are the same as those for linear prestressing; hence practically all the general principles presented in the previous chapters can be applied to circular structures as well, although such application necessarily involves certain details not discussed for linear structures. The practice of circular prestressing differs from linear prestressing in that the techniques of applying the prestress and of anchoring the tendons are often different.

In this chapter, the discussion will be centered on the design of circular liquid containers and pressure vessels. Most of these principles are applicable also to the design of pipes, which will not be discussed in detail. Instead, some citations on prestressed pipes are given to which readers can refer if they are interested in the subject.[1]

Prestressed-concrete pipes in this country can be divided into two types: those with and those without steel cylinders. The construction of those with steel cylinders is now a standardized procedure, as evidenced by the specifications[2]

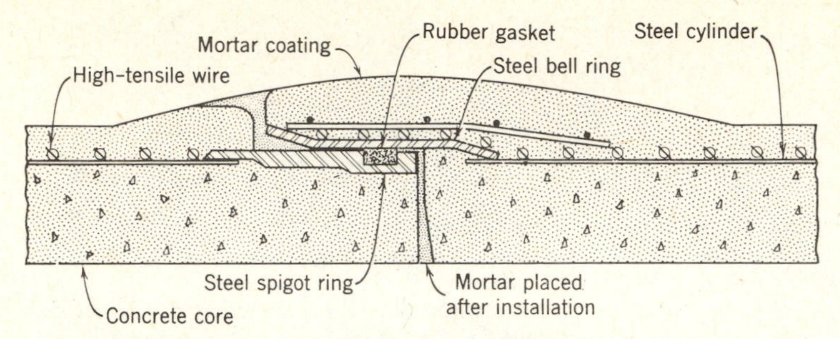

Fig. 13-2. Longitudinal section through joint of prestressed-concrete cylinder pipe.

approved by the A.W.W.A. in 1952. These specifications cover the manufacture of such water pipes ranging in size from 16 to 54 in. (406–1372 mm) and designed for static loads from 100 to 600 ft (30.5–183 m) of water. A typical longitudinal section of the pipe through the joint is shown in Fig. 13-2. The pipe consists of a continuously welded sheet-steel cylinder with steel joint rings welded to its ends, the cylinder being lined on the inside with dense concrete of suitable thickness. After proper curing of concrete, high-tensile wire is wound around the outside of the steel cylinder at a specified prestress and securely fastened to it at its ends. Then a coating of mortar or concrete is deposited over the cylinder and wire for protection. A self-centering joint with rubber gasket as the sealing element is designed so as to be watertight under all conditions of service.

Pipes without steel cylinders are manufactured by simply winding prestressed wires around a concrete core and covering the wires with air-applied mortar. Longitudinal prestress is sometimes provided by pretensioning longitudinal wires against the inner steel form.[3] In another method, helical wire wrapping is applied in a basket-weave pattern so as to produce a longitudinal component of prestress.[4] The concrete core is often cast by the Rocla roller compaction method, by which a rotating mold rolls and places the dense concrete to form a thin-walled pipe. The Rocla firm in Australia applies pretensioning techniques to concrete pipes by embedding circular reinforcement in the pipe concrete which immediately after having been placed is subjected to high pressure applied to the inside of the pipe.[5] When the concrete hardens, the steel remains stretched; then the inside pressure is relieved, and the concrete becomes compressed.

In this country prestressed-concrete tanks were first constructed by the preload method, using their wire winding machines. Up to 1951, about 700 large tanks with a total capacity of more than 500 million gallons and 300 spherical shell roofs in spans up to 205 ft (62.5 m) had been built of prestressed concrete in North America, using almost exclusively the Preload method of prestressing.[6] The Preload procedure consists of the following process. First, the walls for the tanks are built of either concrete or pneumatic mortar, mortar being generally used if the walls are less than 6 in. (152 mm) thick. Often, the walls are poured in alternate vertical slices keyed together. After the concrete wall have attained sufficient strength they are prestressed circumferentially by a self-propelled machine, which winds the wire around the walls in a continuous operation, stressing it and spacing it at the same time. Under favorable conditions, the machine can place the wire up to 7 miles an hour and can complete the horizontal prestressing of an average million-gallon tank in about two days.

After the circumferential prestressing is completed for each layer, a coat of pneumatic mortar is placed around the tank for protection. Two or more layers of prestressing are used for large tanks. Vertical prestressing for the tanks can be applied using any system of linear prestressing, whichever may be the most economical.

Since the 1950's, several methods of linear prestressing have been applied to tanks and pressure vessels with the tendons in equal lengths of portions of a circle. The tendons are threaded through preformed ducts and are stressed from both ends and anchored against pylons spaced uniformly around the tank. By staggering the end anchors in adjacent tendons, the frictional loss of prestress is nearly equalized around the circle. This type of circular prestressing has an advantage over the wire-winding process, whose gunite coating is not prestressed and may not give as good a protection for the tendons. It is gaining popularity and has dominated the field of pressure vessels.

13-4 Circumferential Prestressing

Circumferential prestress is designed to resist hoop tension produced by liquid pressure. Hence, essentially, each horizontal slice of the wall forms a ring subject to uniform internal pressure. In several senses, such a ring can be regarded as a prestressed-concrete member under tension, and much of the discussion in sections 13-1 and 13-2 on tension members can be applied to the design of circumferential prestressing as well.

Consider one half of a thin horizontal slice of a tank as a freebody, Fig. 13-3(a). Under the action of prestess F_0 in the steel, the total compression C in the concrete is equal to F_0. The location of the line of pressure or the C-line in the concrete does not usually coincide with the c.g.s. line. In a circular ring under circular prestress, the C-line always coincides with the c.g.c. line. This is because a closed ring is a statically indeterminate structure, and the theory of linear transformation explained in Chapter 10 for continuous beams is applicable to such a ring. A cable through the c.g.c. is a concordant cable; any other cable parallel to it is simply that line, linearly transformed, whose line of pressure will still remain through the c.g.c. This phenomenon can also be explained by the simple fact that the effect of circular prestress is to produce an

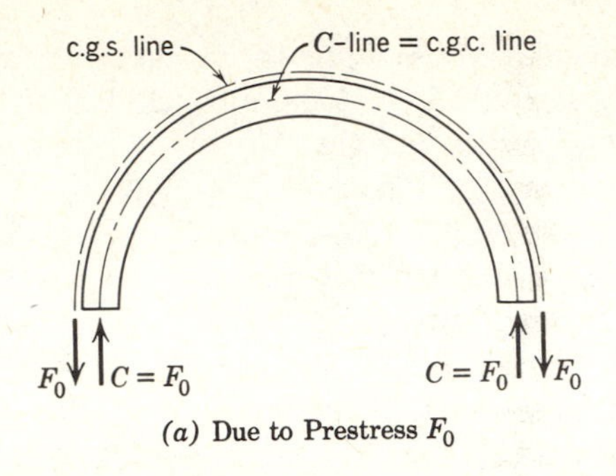

(a) Due to Prestress F_0

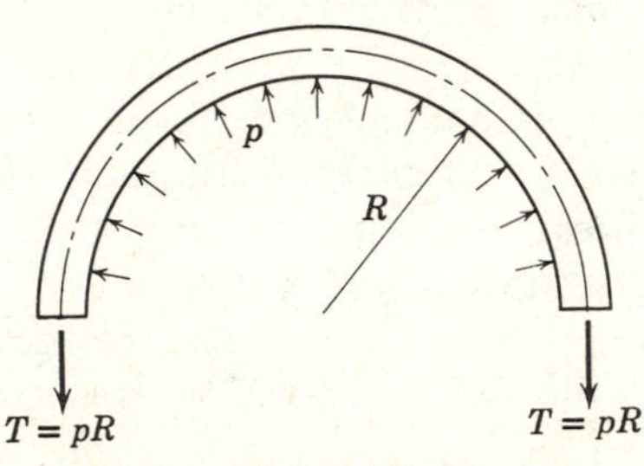

(b) Due to Internal Pressure p

Fig. 13-3. Forces in a horizontal slice of tank (half slice as freebody).

initial hoop compression on the concrete, which is always axial irrespective of the point of application of the prestress. Hence, owing to circular prestress, the stress in the concrete is always axial and is given by the formula

$$f_c = -\frac{F_0}{A_c}$$

which reduces to

$$f_c = -F/A_c$$

after the losses in prestress have taken place.

With the application of internal liquid pressure, Fig. 13-3(*b*), the steel and concrete act together, and the stresses can be obtained by the usual elastic theory. Using the method of transformed section, we have

$$f_c = pR/A_t$$

where p = internal pressure intensity, R = internal radius of the tank, A_t = transformed area $= A_c + (n-1)A_s$.

The resultant stress in the concrete under the effective prestress F and the internal pressure p is

$$f_c = -\frac{F}{A_c} + \frac{pR}{A_t} \tag{13-6}$$

In order to be exact, the value of n has to be chosen correctly, considering the level of stress and the effect of creep. In practice, slight variation in the value of n may not affect the stresses very much, and an approximate value will usually suffice. If a coating of concrete or mortar is added after the application of prestress, then the area A_c under prestress may be the core area while the A_c sustaining the liquid pressure may include the additional coating. Such refinements in calculation may or may not be necessary, depending on the circumstances.

The criteria for designing prestressed tanks vary. The practice in this country has been to provide a slight residual compression in the concrete under the working pressure. This is accomplished by the following procedure of design.

Assume that the hoop tension produced by internal pressure is entirely carried by the effective prestress in the steel; we have

$$F = A_s f_s = pR \tag{13-7}$$

thus the total steel area required is

$$A_s = \frac{pR}{f_s} \tag{13-8}$$

The total initial prestress is then

$$F_0 = A_s f_0 \tag{13-9}$$

For an allowable compressive stress f_c in concrete, the concrete area required to resist the initial prestress F_0 is

$$A_c = -\frac{F_0}{f_c} \tag{13-10}$$

From this value of required A_c, the thickness for the tank can be determined.

Corresponding to the adopted value of A_c, the stresses in the concrete and steel under the internal pressure p can be obtained by

$$\text{Stress in concrete} = -\frac{F}{A_c} + \frac{pR}{A_t} \tag{13-11}$$

$$\text{Stress in steel} = f_s + nf_c \tag{13-12}$$

Since F is equal and opposite to pR, and A_t is always greater than A_c, it can be seen from equation 13-11 that there will be some residual compression in the concrete under the working pressure. This residual compression serves as a margin of safety in addition to whatever tension may be taken by the concrete.

Since the serviceability of a tank is impaired as soon as the concrete begins to crack, it is of utmost importance that an adequate margin of safety be provided against cracking. Where overflow pipes are installed for tanks so that there cannot exist any excessive pressure, a smaller margin of safety is required. Thus the English *First Report on Prestressed Concrete* (Institution of Structural Engineers, London, England, 1951) recommends a factor of safety of 1.25 against cracking. For pipes that may be subjected to much higher pressure than the working value, a greater factor of safety is necessary. For the design of prestressed concrete pipes with steel cylinders, the A.W.W.A. species that the concrete core should be sufficiently compressed to withstand an internal hydrostatic pressure equal to at least 1.25 times the designed pressure without tensile stress being induced in the core. In addition, the pressure producing elastic limit stresses in the steel cylinder and wire is sometimes required to be 2.25 times the normal operating pressure.[1]

The conventional method of design equating the effective prestress to the hoop tension may or may not provide the necessary factor of safety. If a factor of safety of m against cracking is required, the following procedure of design may be adopted.

Assuming f_t = tensile strength in concrete at cracking (which averages about $0.08f_c'$ but may be zero if the concrete has previously cracked or if precast blocks are used), we may write

$$-\frac{F}{A_c}+\frac{mpR}{A_t}=f_t \qquad (13\text{-}13)$$

At the same time, in order to limit the maximum compression in concrete to f_c, we have

$$A_c=-F_0/f_c$$

Substituting this value of A_c into equation 13-13, and noting that $A_t=A_c+nA_s$, $F=f_sA_s$, and $F_0=f_0A_s$, we have

$$-\frac{f_sA_sf_c}{f_0A_s}+\frac{mpR}{(f_0A_s/f_c)+nA_s}=f_t \qquad (13\text{-}14)$$

Solving for A_s, we have

$$A_s=\frac{mpR}{[f_s-(f_t/f_c)f_0](1-nf_c/f_0)} \qquad (13\text{-}15)$$

After A_s is obtained, F_0 and A_c can be computed using equations 13-9 and 13-10, and the stresses in the concrete and steel can be evaluated by equations 13-11 and 13-12.

Recommendations have been prepared by the FIP Commission on Concrete Pressure and Storage Vessels on the design of prestressed concrete oil storage tanks.[7] These will also serve as guidelines for other liquid storage tanks.

One of the important items in the design of tanks is the evaluation of the losses of prestress. Although the details of the sources of loss are discussed in Chapter 4, the usual amount of loss occurring and allowed for in prestressed tanks will be mentioned here. Extensive experiments have been made to measure the amount of losses in prestressed tanks.[8] The average loss of prestess seems to be about 25,000 psi (172 N/mm^2), resulting chiefly from the shrinkage and creep of concrete. An allowance of 35,000 psi (241 N/mm^2) is considered quite conservative, although, under extremely adverse conditions, losses up to 40,000 psi (276 N/mm^2) might take place.

Analyzing the principal sources of these losses, it might be estimated that concrete under a constant load of about 600 psi (4.14 N/mm^2) may attain a total elastic and creep deformation of about 0.0006. Since the concrete is under low compression when the tank is full, the amount of creep strain may be much smaller if the tank is kept filled most of the time. The amount of shrinkage will depend chiefly upon the moisture content in the concrete. Although the worst possible shrinkage strain can be as much as 0.0010, there have been tanks whose concrete expanded instead of contracted, thus resulting in a gain of prestress instead of a loss. For example, if a tank is prestressed after the concrete has aged for several months under dry climatic conditions, expansion will take place when it is filled with water.

The following may be taken as a safe average value.

Elastic and creep strain in concrete	=0.0005
Shrinkage	=0.0005
Total loss	=0.0010

which amounts to about 28,000 psi (193 N/mm^2), taking E_s as 28,000,000 psi (193 kN/mm^2). If accurate values are desired, the possible losses must be considered for each individual tank and duly allowed for.

EXAMPLE 13-3

Determine the area of steel wire required per foot of height of a prestressed-concrete water tank 60 ft in inside diameter to resist 20 ft of water pressure. Compute the thickness of concrete required. $f_c'=3000$ psi, $f_c=750$ psi, $n=10$, $f_0=150{,}000$ psi, $f_s=120{,}000$ psi. Neglect the mortar coating in the calculations. Design both steel and concrete on the following two bases:

1. Assuming all hoop tension carried by the effective prestress.
2. For a load factor of 1.25, producing zero stress in concrete ($f_c'=21$ N/mm^2, $f_c=5.17$ N/mm^2, $f_0=1034$ N/mm^2, and $f_s=827$ N/mm^2).

Solution (*a*) Pressure of 20 ft of water

$$p=20\times 62.4=1248 \text{ psf } (59.78 \text{ kN/m}^2)$$

Using equations 13-8 and 13-10,

$$A_s=pR/f_s = \frac{1248\times 30}{120{,}000} = 0.312 \text{ sq in. } (201 \text{ mm}^2)$$

$$A_c=-F_0/f_c = \frac{-0.312\times 150{,}000}{-750} = 62.5 \text{ sq in. } (40.3\times 10^3 \text{ mm}^2)$$

For a height of 12 in. (305 mm), the thickness required is 62.5/12=5.2 in. (132 mm). Suppose that a thickness of 5.5 in. (140 mm), is adopted; then, under the action of the internal pressure, equation 13-11 gives

$$f_c = -\frac{F}{A_c} + \frac{pR}{A_t}$$
$$= -\frac{0.312 \times 120{,}000}{5.5 \times 12} + \frac{1248 \times 30}{66 + 10 \times 0.312}$$
$$= -567 + 541$$
$$= -26 \text{ psi } (-0.179 \text{ N/mm}^2)$$

Note here that, the thicker the concrete, the smaller will be the residual compression under load, unless the amount of wire is proportionately increased.

(*b*) Using equation 13-15,

$$A_s = \frac{mpR}{[f_s - (f_t/f_c)f_0](1 - nf_c/f_0)}$$
$$= \frac{1.25 \times 1248 \times 30}{[120{,}000 + 0](1 - 10 \times -750/150{,}000)}$$
$$= 0.372 \text{ in.}^2 \ (240 \text{ mm}^2)$$
$$A_c = F_0/f_c$$
$$= 0.372 \times 150{,}000/750$$
$$= 74.4 \text{ in.}^2 \ (48.0 \times 10^3 \text{mm}^2)$$

Thickness required = 74.4/12 = 6.2 in. (157 mm). If a thickness of 6.5 in. (165 mm) is adopted, the resulting stress in the concrete under full water pressure will be

$$f_c = \frac{F}{A_c} + \frac{pR}{A_t}$$
$$= \frac{-0.372 \times 120{,}000}{6.5 \times 12} + \frac{1248 \times 30}{78 + 10 \times 0.372}$$
$$= -573 + 458$$
$$= -115 \text{ psi } (-0.793 \text{ N/mm}^2)$$

which provides a margin of saftey of 25% up to zero compression in concrete.

Note that designing by this second method gives heavier sections for both concrete and steel. The design can be economized if some tension in the concrete is allowed at 25% overload.

13-5 Vertical Prestressing in Tanks

The design of prestressed-concrete structures is based on a knowledge of the behavior of nonprestressed structures plus an understanding of the effect of prestressing. This is as true for the design of tanks as for beams and slabs. Before analyzing the stresses in a prestressed tank, let us consider an ordinary

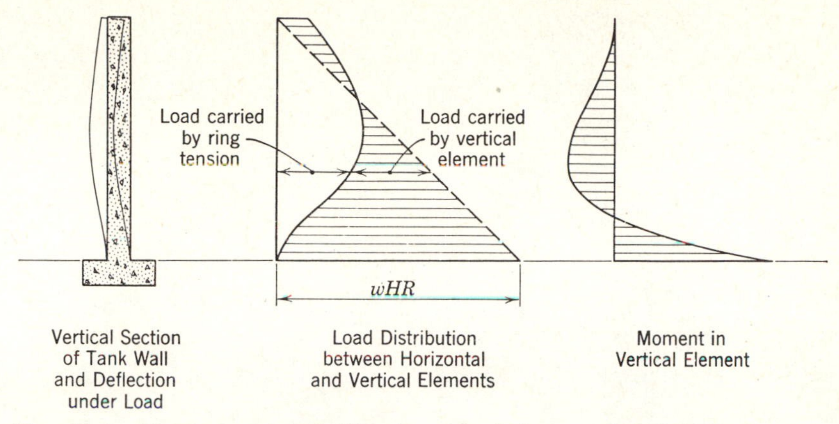

Fig. 13-4. Moment and deflection in vertical element of tank wall.

reinforced-concrete tank under the action of internal liquid pressure. It is well known that, whereas the horizontal elements of the tank are subject to hoop tension, the vertical elements are under bending, Fig. 13-4. The amount and variation of bending in the vertical elements will depend on several factors.

1. The condition of support at the bottom of the wall, whether fixed, hinged, free to slide, or restrained by friction.
2. The condition of support at the top of the wall, whether fully or partially restrained or free to move.
3. The variation of concrete thickness along the height of the wall.
4. The variation of pressure along the depth, whether triangular or trapezoidal.
5. The ratio of the height of the tank to its diameter.

Theoretical solution for several of these combinations are given by Timoshenko[9] and numerical values, convenient for application, are tabulated in some pamphlets.[10] European books give solutions for additional cases, such as walls of varying thickness, and the results are plotted in some publications.[11] Readers interested in the problem are referred to these and to the bibliographies listed in them. To give an idea of such distribution of loads among the horizontal and vertical elements, two graphs are presented in Fig. 13-5. It is evident from these graphs that the active pressure on the horizontal elements is not a direct function of depth, but often decreases with it, while the vertical elements may carry a considerable amount of load, especially if the structure is squattier than usual.

For prestressed-concrete tanks, an additional problem is introduced: the effect of prestressing, both circumferential and vertical. Since horizontal pressure will produce vertical moments in the walls, it is evident that circumferential pre stressing will also induce such moments. These vertical moments caused by

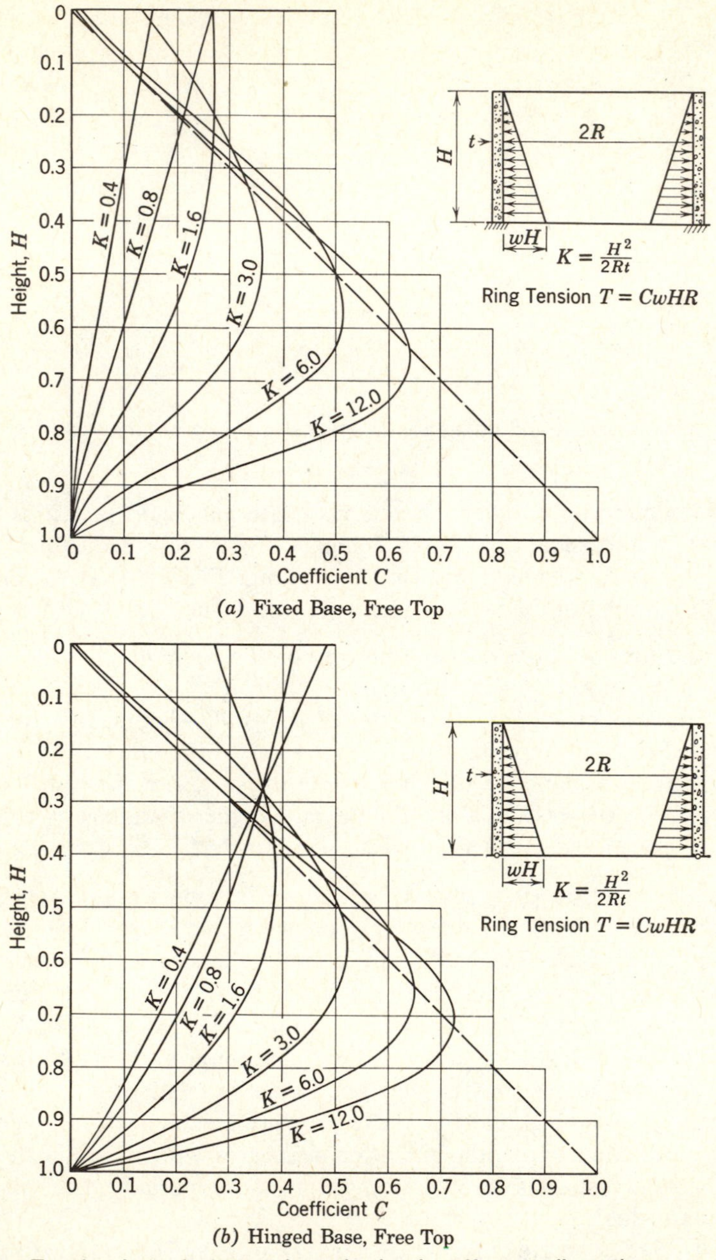

(a) Fixed Base, Free Top

(b) Hinged Base, Free Top

Fig. 13-5. Tension in tank rings, triangular load, uniform wall section.

circumferential prestressing will exist by themselves when the tank is empty and will act jointly with the moments produced by liquid pressure when the tank is filled. To reinforce the wall against these moments, vertical prestressing may be applied. If vertical prestress is concentrically applied to the concrete, only direct compressive stress is produced and the solution is simple. If the vertical tendons are bent or curved, the vertical prestress produces radial components which, in turn, influence the circumferential prestress. Hence the analysis can become quite complicated.

Let us investigate the effect of circumferential prestressing on the vertical moments. If the circumferential prestress varies triangularly from zero at the top to a maximum at the bottom, its effect is equal but opposite to the application of an equivalent liquid pressure. If the circumferential prestress is constant throughout the entire height of the wall, it is the same as the application of an equivalent gaseous pressure. For both cases, tables are available for the computation of vertical moments.[10] To obtain the optimum results, the circumferential prestress along the depth of the wall should be varied to suit the variation of the active pressure on the horizontal elements. However, the effect of such circumferential prestressing on vertical moments cannot be readily determined.

Vertical prestressing should be designed to stand the stresses produced by various possible combinations of the following forces.

1. The vertical weight of the roof and the walls themselves.
2. The vertical moments produced by internal liquid pressure.
3. The vertical moments produced by the applied circumferential prestress.

In addition to the above, stresses may be produced as a result of differential temperature between the inner and outer faces of the wall, and by shrinkage of the concrete walls unless they are entirely free to slide on the foundation. These forces cannot be easily evaluated and hence are often neglected or provided for indirectly in an overall factor of safety.

It must be noted that the maximum stresses in the concrete usually exist when the tank is empty, because then the circumferential prestress would have its full effect. When the tank is filled, the liquid pressure tends to counterbalance the effect of circumferential prestress and the vertical moments are smaller. Since it is convenient to use the same amount of vertical prestress throughout the entire height of the wall, the amount will be controlled by the point of maximum moment. By properly locating the vertical tendons to resist such moment, a most economical design can be obtained. However, efforts are seldom made to do so, and the amount of prestress as well as the location of the tendons is generally determined empirically rather than by any logical method of design.

EXAMPLE 13-4

A 1-ft vertical element of a water tank is shown in Fig. 13-6. It carries 1500 lb (6.67 kN) of weight from the roof. At a point 20 ft (6.1 m) below the top, the vertical moments are:

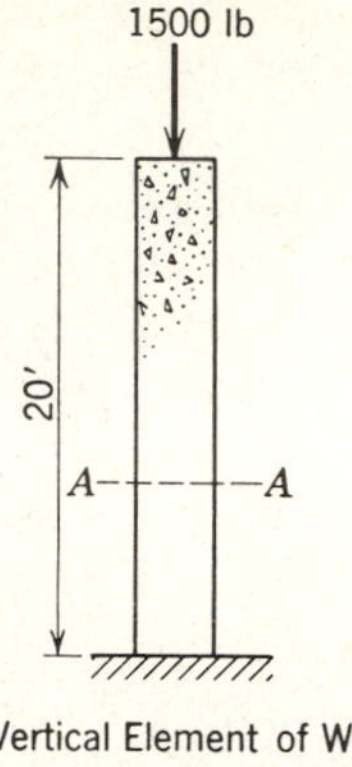

Vertical Element of Wall

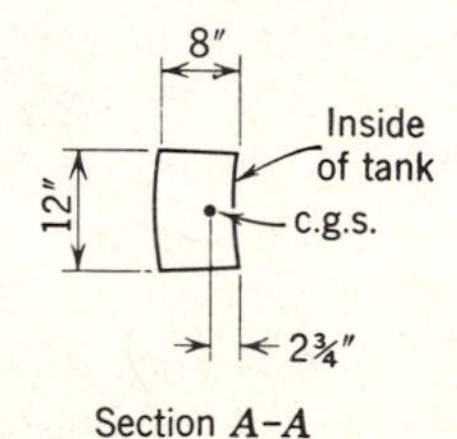

Section A–A

Fig. 13-6. Example 13-4.

for initial circumferential prestress, $M = 3200$ ft-lb (4.34 kN-m) (tension on the inside fibers), which reduces to 2500 ft-lb (3.39 kN-m) eventually. For full liquid pressure, $M = 2400$ ft-lb (3.25 kN-m) (tension on the outside face). The vertical prestressing wire is located $2\frac{3}{4}$ in. (70 mm) from the inside face and exerts an initial prestress of 11,000 lb/ft (160.5 kN/m), which reduces to 8000 lb/ft (116.7 kN/m) eventually. Compute stresses in the extreme vertical fibers of the concrete under the initial and final conditions, considering both an empty and a full tank.

Solution The stresses for both the inside and outside fibers under both initial and final conditions are computed and listed as in the table. It is seen from the table that a slight tension of 36 to 41 psi (0.248 to 0.283 N/mm^2) exists on the inside vertical fibers when the tank is empty. Otherwise, compressive stresses are obtained throughout. (See Table 13-1.)

13-6 Dome Ring Prestressing

It is beyond the scope of this book to discuss the design of domes. Only the general principles and practice of dome prestressing especially as applied to tank roofs will be mentioned here. Readers interested in the subject are referred to other publications for additional details.[12] Generally speaking, for domes with diameter greater than 100 ft (30.5 m), the economy of prestressing should be

Table 13-1 Computation for Stresses in Concrete (Example 13-4)

Conditions		Initial		Final	
Fiber		**Inside**	**Outside**	**Inside**	**Outside**
A. Weight of roof	$\frac{1500}{8\times12}=$	−16	−16	−16	−16
B. Weight of wall	$\frac{20\times150}{144}=$	−21	−21	−21	−21
C. Axial component of vertical prestress	$\frac{11{,}000}{8\times12}=$	−115	−115		
	$\frac{8000}{8\times12}=$			−83	−83
D. Eccentricity of vertical prestress	$\frac{6M}{bd^2}=\frac{6\times11{,}000\times1.25}{12\times8^2}=$	−107	+107		
	$\frac{6\times8000\times1.25}{12\times8^2}=$			−78	+78
E. Vertical moment due to circumferential prestress	$\frac{6M}{bd^2}=\frac{6\times3200\times12}{12\times8^2}=$	+300	−300		
	$\frac{6\times2500\times12}{12\times8^2}=$			+234	−234
Total for tank empty		+41	−345	+36	−276
F. Vertical moment due to liquid pressure	$\frac{6M}{bd^2}=\frac{6\times2400\times12}{12\times8^2}=$	−225	+225	−225	+225
Total for tank full		−184	−120	−198	−51

seriously considered. Domes for tanks up to 230 ft (70 m) in diameter have been constructed.

The dome roof itself is made of concrete or pneumatic mortar with thickness varying from 2 to 6 in. (50.8–152.4 mm). For domes of large diameter, variable thicknesses may be employed and thicknesses greater than 6 in. (152.4 mm) are used for the lower portion. Before concreting the dome, some erection bars are prestressed around the base of the dome. After the hardening of the shell concrete, wires are prestressed around it, Fig. 13-7. During this operation, the dome shell rises from its forms as it is compressed, thus simplifying the careful procedure of decentering required for nonprestressed domes.

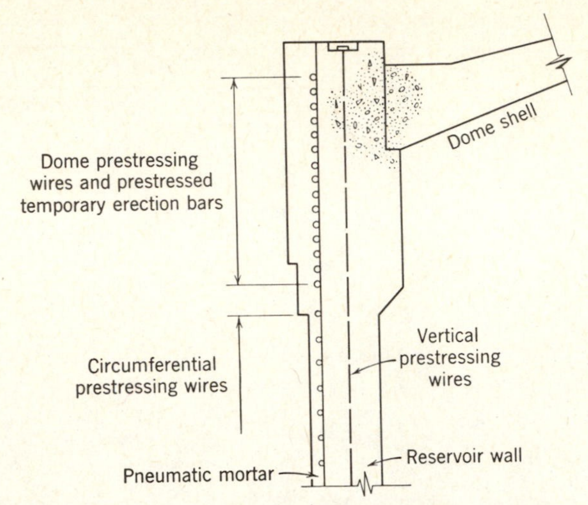

Fig. 13-7. Typical section of dome ring for tanks.

Methods and formulas, though available for the analysis of dome stresses under uniform loads, are applicable only to points on the domes removed from the discontinuous edge. The computation of stresses in the edge ring becomes a very complicated problem if the edge ring is prestressed. However, for purposes of design, a conventional method is available. It consists of prestressing the ring to induce sufficient compressive stresses to counteract the tensile stresses set up in the ring under the maximum live and dead loads. With this prestress, it is usually possible to raise the dome from its false work, since only the dead load is actually acting on the dome.

Consider a spherical dome carrying loads symmetrical about the axis of rotation, that is, load with intensity constant along any given latitude, Fig. 13-8. If the total load is W, the vertical reaction per foot of length along the edge member will be

$$V = \frac{W}{2\pi R \sin\theta}$$

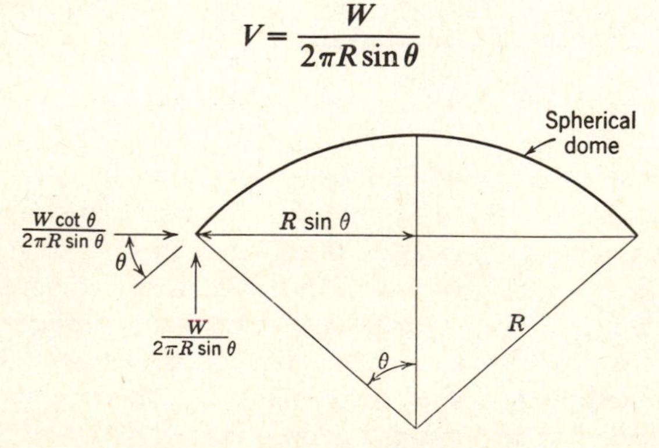

Fig. 13-8. Design for prestress in edge ring of dome.

Since a dome is not supposed to carry any appreciable moment, the resultant reaction along the edge must be tangent to the surface. Hence the horizontal reaction per foot of length must be

$$H = V\cot\theta = \frac{W\cot\theta}{2\pi R\sin\theta}$$

Assuming this horizontal reaction to be entirely supplied by the prestressing force F acting in hoop tension, then,

$$\begin{aligned} F &= HR\sin\theta \\ &= \frac{W}{2\pi}\cot\theta \end{aligned} \tag{13-16}$$

The effective prestressing force F having been determined, the cross-sectional area of the ring concrete can be designed by

$$A_c = \frac{F_0}{f_c} \tag{13-17}$$

where F_0 = the initial prestressing force, and f_c = the allowable compressive stress in concrete.

It is desirable to keep f_c at a relatively low value, say about $0.2f_c'$ and not greater than 800 psi (5.52 N/mm^2). This is necessary in order to minimize excessive strain in the edge ring which might in turn produce high stresses in the shell. It must be further observed that this procedure of design is satisfactory only when there is no possibility of heavy overloads, because the prestressed edge ring does not possess a high factor of safety against overloads, although the factor of safety is sufficient for ordinary roof loading.

One of the world's largest concrete roofs ever built[13] is the circular dome cover for the Assembly Hall of the University of Illinois, Urbana, Fig. 13-9. The folded plate dome has a diameter of 400 ft (122 m). It is of lightweight concrete

Fig. 13-9. University of Illinois Stadium, Urbana, Ill., has 400-ft dome with posttensioned ring to deflect the membrane forces inward, forming a bowl. (Architect Harrison and Abramovitz; Structural Engineer Ammann and Whitney).

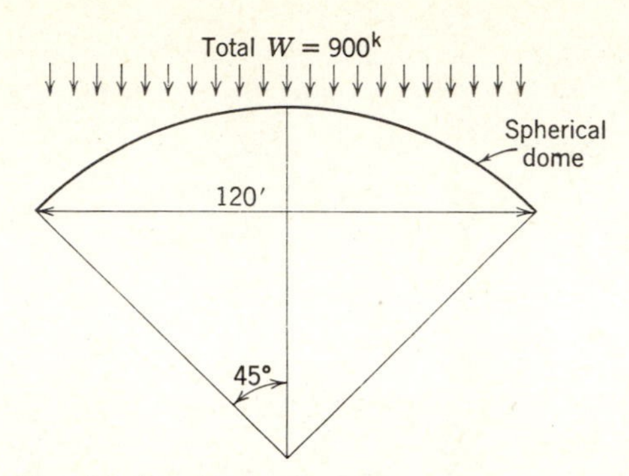

Fig. 13-10. Example 13-5.

and weighs 10,700 k (47,000 kN). The prestessing consisted of 2503 circles of 0.236-in. (6.0 mm) wires with a total steel area of 85 sq in. (54.8×10^3 mm^2), stressed to about 150,000 psi (1034 N/mm^2), producing a stess in the edge beam of about 1000 psi (6.9 N/mm^2) at transfer.

EXAMPLE 13-5

A spherical dome, Fig. 13-10, carries a total live and dead load of 900 k. Design the prestress in the edge ring and the cross-sectional area of concrete required for the edge ring. Loss of prestress = 20%. $f_c = 600$ psi ($W = 4000$ kN, $f_c = 4.14$ N/mm^2).

Solution From equation 13-16, for $W = 900$ k (4000 kN) and $\theta = 45°$, we have

$$\begin{aligned} F &= (W/2\pi)\cot\theta \\ &= 900/2\pi \\ &= 143 \text{ k } (636 \text{ kN}) \end{aligned}$$

which will result in zero tension in the dome ring under full live and dead load. From equation 13-17, area of concrete required, for $F_0 = 143/0.8 = 179$ k (796 kN), is

$$\begin{aligned} A_c &= F_0/f_c \\ &= 179/0.6 \\ &= 298 \text{ in.}^2 \ (192 \times 10^3 \text{ mm}^2) \end{aligned}$$

13-7 Nuclear Plant Containment Vessels

Prestressed concrete pressure vessels for nuclear reactors and/or their containment were initiated in Europe in the early 1960's. Their introduction into the United States was pioneered by the testing of two reactor vessel models by General Atomic, with T. Y. Lin International as consultants, 1963 (see Fig. 13-11). The $\frac{1}{4}$-scale models had 12-ft (3.66 m) I.D. and 2-ft (0.61 m) thick walls to resist internal design pressure of 600 psi (4.14 N/mm^2). After being subjected to pressures over 1200 psi (8.28 N/mm^2), at which the concrete walls seriously

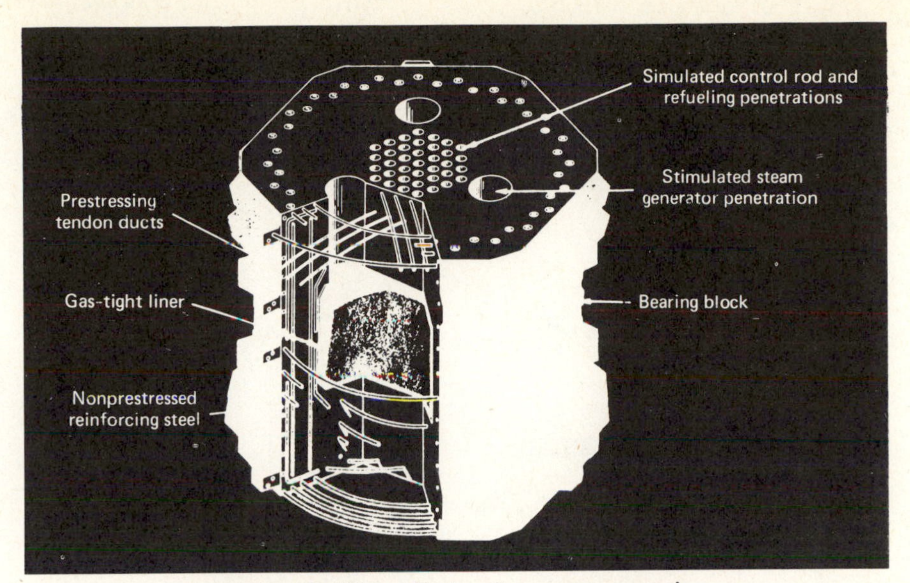

Fig. 13-11(*a*). Parts of pressure vessel.

Fig. 13-11(*b*). Scale model for tests.

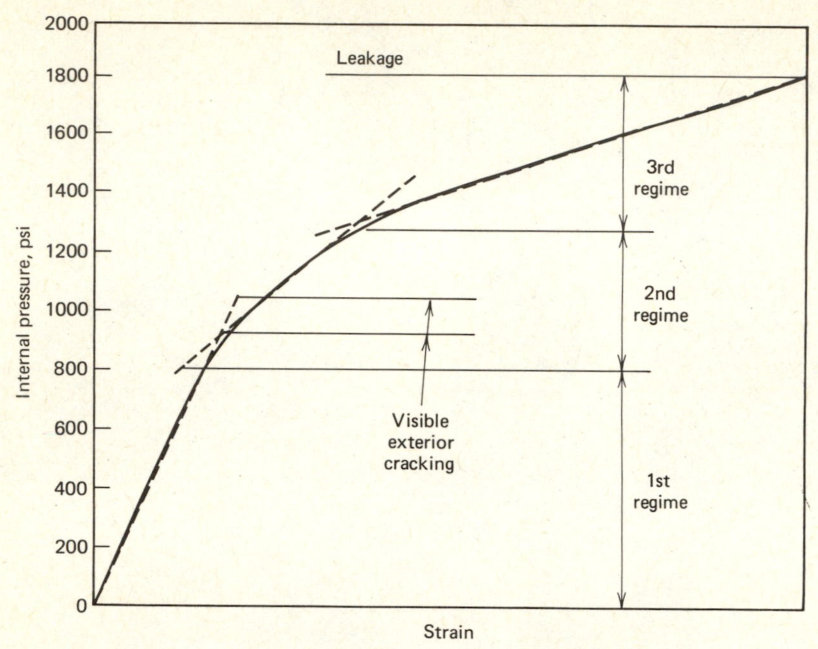

Fig. 13-11(*c*). Model test results.

cracked, the reduction of internal pressure was followed by the apparent closing of the cracks. The vessels finally failed at about 1800 psi (12.41 N/mm^2) internal pressure in a ductile manner, which demonstrated the extra safety inherent in prestressed vessels.

From 1965 to 1978, over 60 prestressed concrete nuclear containment vessels (as distinguished from reactor vessels), have been built or designed in the United States alone, Fig. 13-12. These are typically about 100 to 140 ft (30.5–42.7 m) in diameter, 150 to 210 ft (45.7–64.0 m) high, with walls 3 to 4 ft (0.91–1.22 m) thick, resisting an internal pressure around 60 psi (0.414 N/mm^2). They have been more or less standardized since the ACI, together with ASME, set up the ASME Boiler and Pressure Vessel Code. Section III, Division 2, on "Concrete Nuclear Vessels and Containment," 1977. The International Federation for Prestressing has prepared a Report on PCRV[14] which describes the background material and gives guidance for the design and construction of such structures.

The development and details of these containment vessels are very well described by D. W. Halligan.[15] His paper presents improvements made from 1966 to 1976 and indicates the general acceptance of such vessels in the United States and throughout the world.

The preliminary design of these vessels can be accomplished rather simply, using half cylinder as a freebody, as shown in Fig. 13-3. Normally, the design

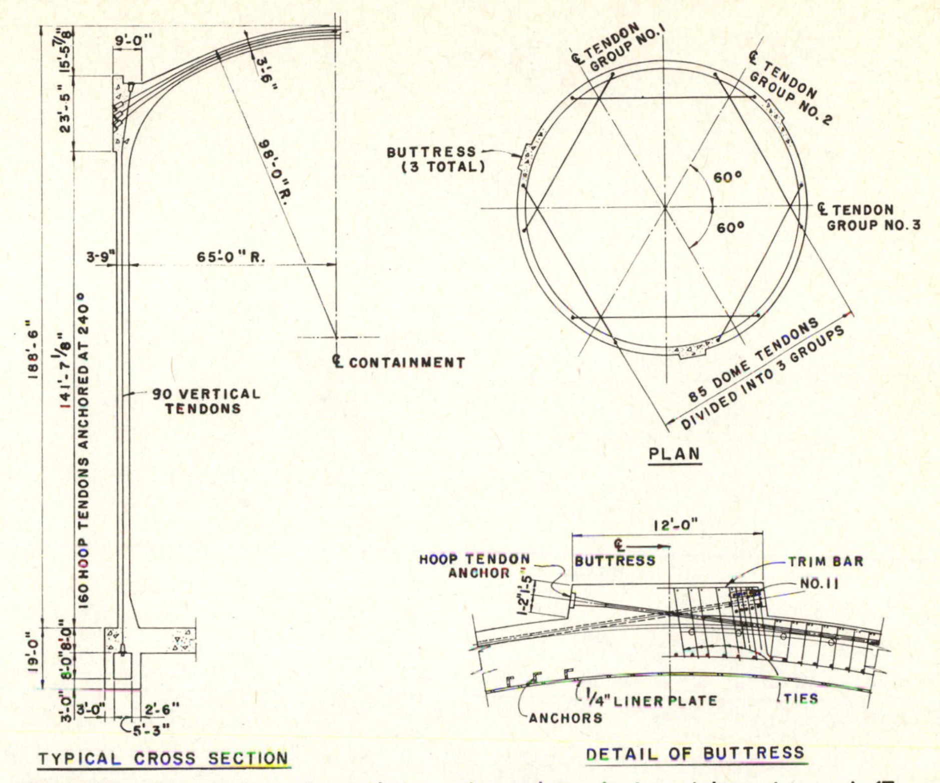

Fig. 13-12. A typical prestressed concrete nuclear plant containment vessel. (From *PCI Journal*, September–October 1976.)

internal pressure will be balanced by the applied effective prestress with a safety factor of 1.15. The effect of penetrations in a nuclear containment vessel, of which there are many, are carefully analyzed and reinforced accordingly.

Circular posttensioning is achieved by looping the tendons around the cylinder and anchoring them against the buttresses. By providing enough butresses to anchor the tendons, frictional loss due to curvature can be controlled. These circumferential tendons are usually tensioned from both ends, and since frictional losses even out themselves on account of buttress location around the circumference, total loss at any point is not too large and 360° hoop tendons can sometimes be installed.

By carefully positioning the ducts, the wobble factor K can be maintained at 0.0003 per foot. Pregreasing the tendons can cut the coefficient of friction down to 0.12. For grouted tendons, water-soluble oil will need to be injected first and then washed out after stressing. Around the dome of the vessel, triangular tendon arrangements and others have been used, Fig. 13-12. By combining the vertical tendons with the dome tendons into continuous tendons, the heavy ring

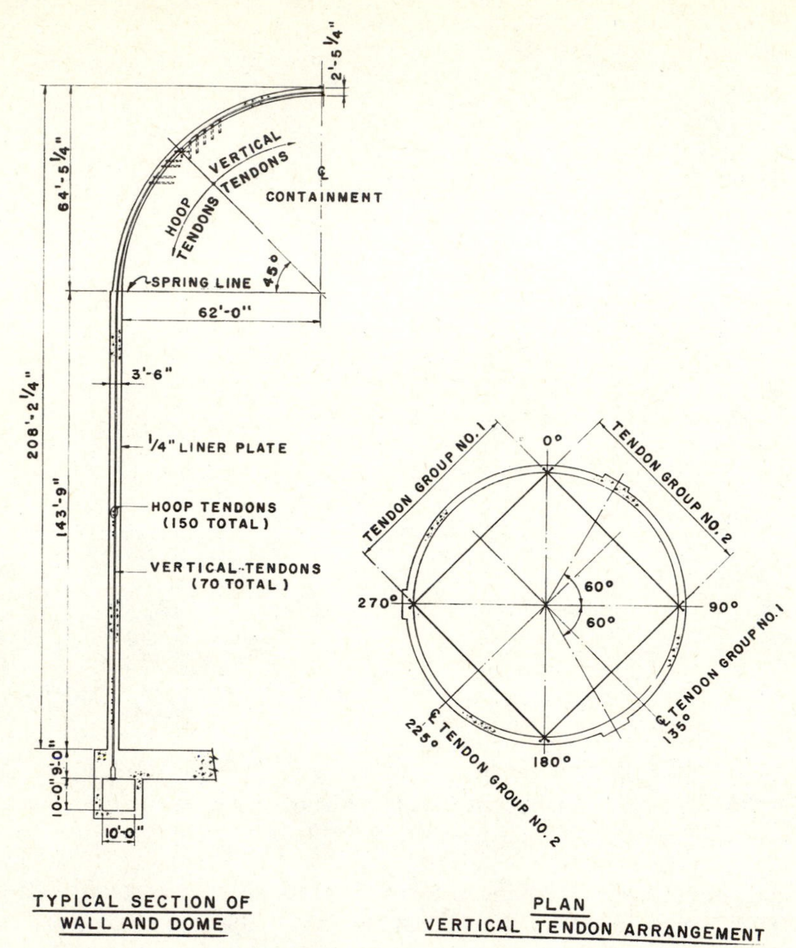

Fig. 13-13. Containment vessel without ring girder, using inverted-U vertical tendons. (From *PCI Journal*, September–October 1976.)

girder otherwise required around the top of the cylinder can be eliminated. This made it possible to install inverted-U vertical tendons, Fig. 13-13.

The use of prestressed concrete nuclear "reactors" have been so far been limited to only one project in the United States, namely, the Fort St. Vrain reactor in Colorado, designed for an internal pressure of 600 psi (4.14 N/mm^2). But it is envisioned that much higher internal pressure can be resisted by prestressed thick-walled vessels with economy and safety, as presented in a report prepared for the Department of Energy, Washington, D. C.[16]

References

1. H. F. Kennison, "Design of Prestressed Concrete Cylinder Pipe," *J. Am. Water Works Assn.*, November 1950, p. 1049.
2. *Standard Specifications for Reinforced Concrete Water Pipe—Steel Cylinder Type, Prestressed*, Am. Water Works Assn., 1952.
3. R. M. Doull, "Prestressed Pipe without Steel Cylinders," *Eng. News-Rec.*, June 24, 1948, p. 68.
4. "New Casting and Prestressing Techniques for Ultra-Strong Concrete Pipe," *Eng. News-Rec.*, October 6, 1949, p. 24.
5. "Prestressing Concrete Pipe," *Concrete*, September 1947, p. 38.
6. C. Dobell, "Prestressed Concrete Tanks," *Proc. First U.S. Conference on Prestressed Concrete*, 1951.
7. "Recommendations for the Design of Prestressed Concrete Oil Storage Tanks," FIP, CACA, Wexham Springs, Slough SL3 6PL, England, 1977.
8. J. M. Grom, "Design of Prestressed Tanks," *Proc. Am. Soc. C.E.*, October 1950, (separate No. 37).
9. S. Timoshenko, *Theory of Plates and Shells*, McGraw-Hill Book Co., New York, 1940.
10. "Circular Concrete Tanks without Prestressing," *Bull. Portland Cement Assn.*
11. W. S. Gray, *Reinforced Concrete Reservoirs and Tanks*, Concrete Publications Ltd., London, 1954.
12. "Design of Circular Domes," *Bull. Portland Cement Assn.*
13. "A 400-ft Prestressed Saucer," *Eng. News-Rec.*, June 1, 1961 pp. 32–36; also May 24, 1962, pp. 44–46.
14. "The Design and Construction of Prestressed Concrete Reactor Vessels," FIP, England, 1977.
15. D. W. Halligan, "Prestressed Concrete Nuclear Plant Containment Structures," *J. Prestressed Conc. Inst.*, September–October, 1976, pp. 158–175.
16. James O'Hara, Richard Howell, T. Y. Lin, Philip Chow, De Ngo, "Prestressed Concrete Pressure Vessels for Coal Conversion," Department of Energy, Washington D.C., 1978.

14

COMPRESSION MEMBERS; PILES

14-1 Column Action Due to Prestress

The question is often brought up whether a concrete member under prestress will have a tendency to buckle like an ordinary column under compression. The answer is that, if the prestressing element is in direct contact with concrete all along its length, there will be no "column action" in the member due to prestress.

Consider an ordinary column under an external load, Fig. 14-1(a). When the column deflects, additional moment in a section A–A is created by the deflection, because the external load now acts with a different eccentricity on that section. This additional moment is the cause of column action. Now consider a member internally prestressed but not externally loaded, (b); so long as the steel and concrete deflect together, there is no change in the eccentricity of the prestress on the concrete, no matter how the member is deflected. Hence there is no change in moment due to any deflection of the member and no column action. When an external load is applied to a prestressed-concrete column, any deflection of the column will change the moment, and column action will result.

Another way to look at the problem is to separate the steel from the concrete and treat them as two free bodies, Fig. 14-2. Considering the concrete alone, it is a column under direct compression, and any slight bending of the column will result in an eccentricity on a section such as A—A, and hence in a tendency to buckle. But, considering the steel as a freebody, there will exist an equal eccentricity with an equal but opposite force, producing a tendency to straighten itself out. The tendency to straighten is exactly equal and opposite to the tendency to buckle, and hence the resulting effect is zero. This is not true, of course, when the member is externally prestressed, say against the abutments, because there will be no balancing effect from the prestressing element, and column action will result.

If the steel and concrete are not in direct contact along the entire length, the problem will be different, Fig. 14-3. The concrete under compression will have a tendency to deflect laterally. That deflection will not at first bring the steel to deflect together with it; hence the eccentricity of prestress on the concrete is actually changed, thus resulting in column action. After a certain amount of

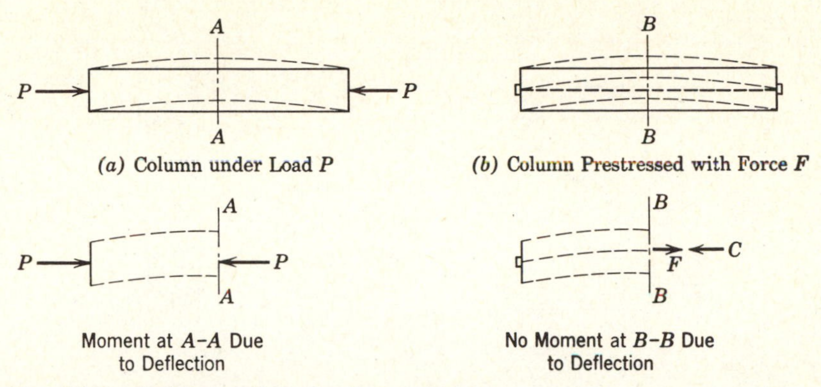

Fig. 14-1. Column action due to prestress.

deflection, the steel is brought into contact with the concrete and the two will begin to deflect together. Hence the column action is limited to the differential deflection of the two materials.

If the steel is in contact with the concrete at several points, say at E and F, but not along the entire length, Fig. 14-4, then the column action is limited to the length between the points of contact. If such length is short, column action will not be serious.

Next, consider an isolated or statically determinate curved member subject to internal prestress, Fig. 14-5(a). If the prestress is concentric at all sections (the c.g.s. line coinciding with the c.g.c. line), then the concrete is behaving like an arch subject to axial force with the exception that the applied force from the steel will move with the deflection of the concrete and will always remain concentric. Hence there is no tendency to buckle as in an ordinary arch under external loads, whose line of pressure is determined by the loads and may not

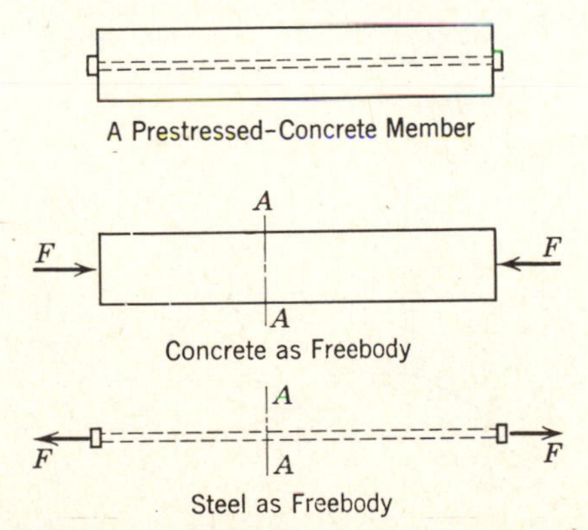

Fig. 14-2. Balancing action of concrete and steel.

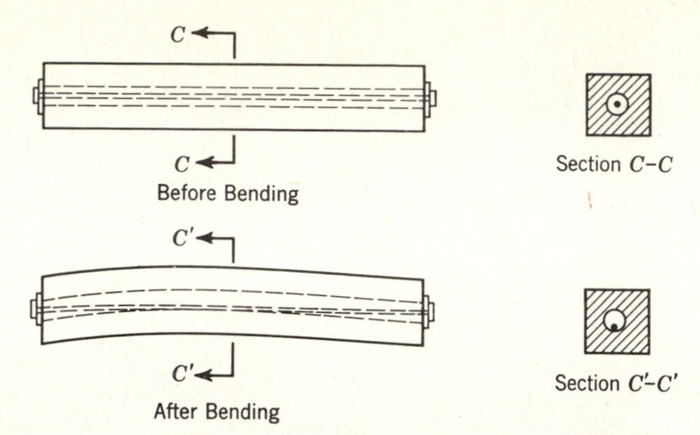

Fig. 14-3. Steel and concrete in contact after bending.

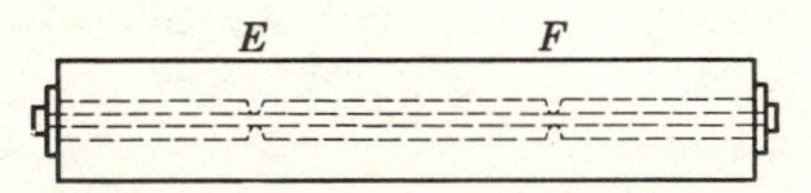

Fig. 14-4. Steel and concrete in contact at several points.

shift together with the deflection of the arch. As an extreme example, even if the member has a reverse curve, Fig. 14-5(*b*), the application of concentric prestress will not tend to straighten the member although the local effects produced by the transverse pressure of the curved tendon must always be designed for. Also note the above statement is not true for a statically indeterminate member since it is subject to secondary reactions and moments.

If the prestress is eccentric, as on sections *G* and *H*, Fig. 14-6, the compression in the concrete is still equal and opposite to the tension in the steel.

Any deflection of the member will still displace both of them together, and there will be no column action due to prestress. The effect of an eccentric prestress on the concrete, however, will produce deflection of the member. If the

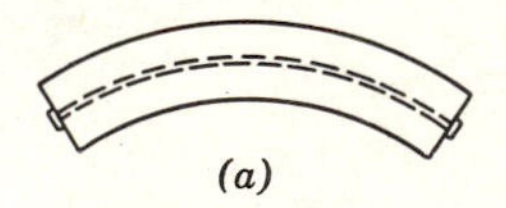

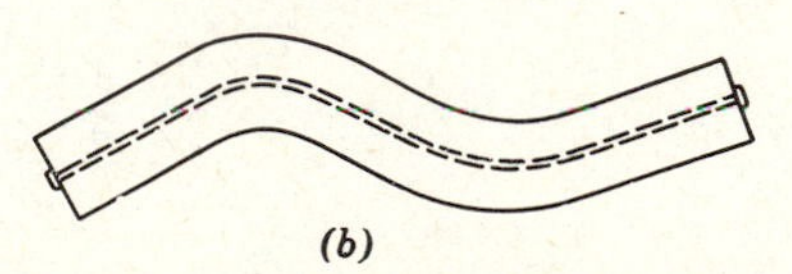

Fig. 14-5. Bent members under concentric prestress.

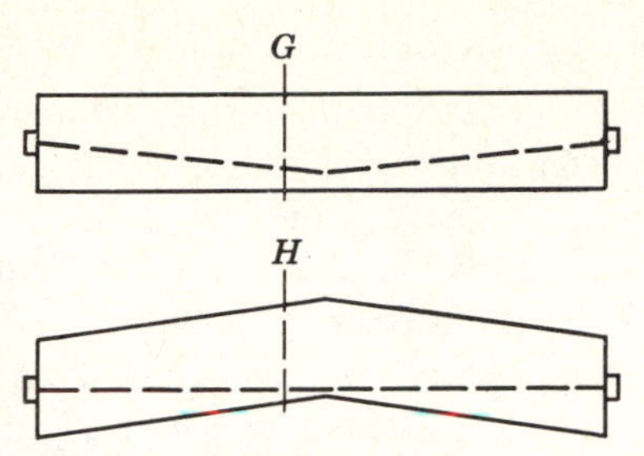

Fig. 14-6. Eccentric prestress and column action.

deflection is appreciable, the deflected axis of the member should be used in computing column effects due to external loads.

As far as column action is concerned, it is immaterial whether there is any frictional loss along the length of the prestressing tendon, because the tension in the steel is always balanced by the compression in the concrete at any section, whatever frictional losses may occur. Hence, whether there is frictional loss or not, there will be no column action due to prestress.

14-2 Compression Members

A prestressed-concrete compression member is one that carries external compressive load. A member that is simply compressed by its prestress is not a compression member. As explained in the previous section, a prestressed member is not under column action due to its own prestress, but it is subject to column action under an external compressive load just like a column of any other material.

It is seldom that a prestressed-concrete member is utilized to stand compression and is prestressed for compression's sake. Evidently, concrete can carry compressive load better without being precompressed by steel. And it is difficult to conceive of steel wires as adding any appreciable strength to a member carrying axial compression. However, many compression members, besides carrying direct compressive loads, are subject to transverse loads as well. Bending due to these transverse loads may more than offset the axial compressive stress at certain points, so as to produce some resulting tension in the concrete. Then it will be advisable to reinforce such columns for possible tension. In other words, some compression members are actually flexural members, and all the advantages of prestressing a beam would apply to the prestressing of those members.

Consider an industrial building of one story, for example; the columns or bearing walls may carry only light vertical loads. But they may be subject to bending during handling and erection if they are precast, or they may carry lateral force such as that due to wind and earthquake after the completion of the building. Similar conditions may exist in bridges. Then it is often feasible to precompress the member so that it can stand a certain amount of bending.

One beneficial effect of prestressing a compression member is the reduction of its deflection under transverse loads. One such pylon, 100 ft (30.5 m) high, was prestressed to resist an earthquake load of 2450 k (10,900 kN) applied horizontally along the pylon.[1] Since the deflection of an uncracked section is about 40% that of a cracked section, a prestressed pylon could be about 2.5 times as stiff as an ordinary reinforced one. In this instance, the reduction of deflection at the top of the pylon minimizes the relative movements between the building floors and saves a tremendous amount of steel otherwise required to reinforce other parts of the building.

Very long precast-prestressed concrete columns have been used in multistory structures. One structure of this type was an office building on the campus of the University of California, Davis, which had 90-ft (27.5 m) long precast-prestressed columns as shown in Fig. 14.7. The prestressing of these columns allowed handling of the 90 ft (27.5 m) long pieces without cracking. Intermediate temporary bracing was needed during construction to support these slender elements until the beams framing in at each floor level were in place. These bracing points can be seen in the photograph of Fig. 14.7. Safety during the construction stage is an important consideration when such long columns are being erected.

Fig. 14-7. Erection of prestressed (pretensioned) concrete 90-ft columns. Office building University of California—Davis (T. Y. Lin International, Consulting Engineers).

Within the working range, the stresses in a prestressed compression member due to both prestress and external loads can be analyzed by the usual elastic theory. But the design of the member is another question, because the empirical methods for designing reinforced-concrete columns cannot be directly applied to prestressed ones. The stresses ordinarily allowed for reinforced concrete are not applicable to prestressed concrete, partly because the stresses due to internal prestressing are of different nature from those due to external loads, which have column action, while internal prestressing has not. For proper design of prestressed-concrete members, one must go into basic theories of columns and prestress and choose a proper standard for the safety of the structure in each particular case.

If a section of a column is under an effective prestress F with an eccentricity e, and loaded by a concentric load P plus an external moment M, the extreme fiber stresses at that section can be computed by the following formula.

$$f_c = \frac{F}{A_c} \pm \frac{Fec}{I_c} + \frac{P}{A_t} \pm \frac{Mc}{I_t} \tag{14-1}$$

If the column is a slender one, the deflection of the member due to both the prestress eccentricity and the external load may significantly affect the magnitude of the external moment M and must be included in it, as will be shown in the next section.

An approximate investigation of the effect of axial prestressing on the ultimate strength of columns can be made. Under the action of an external compressive load, the column will shorten and the prestress in the steel will be decreased. If, at the ultimate load, the unit compressive strain in the concrete is of the order of 0.0030, then the pretensioned strain in the steel will be decreased by that same amount, and the remaining prestress at the moment of failure will be less than the original effective prestress.

If the effective prestress is 120,000 psi (827 N/mm^2), the remaining prestress will be only

$$\begin{aligned} f_s &= f_{se} - 0.0030E_s \\ &= 120{,}000 - 0.0030 \times 30{,}000{,}000 \\ &= 30{,}000 \text{ psi } (207 \text{ N/mm}^2) \end{aligned}$$

In other words, the major part of the prestress may be lost at the ultimate compressive strength of the concrete. This means that the ultimate load-carrying capacity of the column is not much decreased by prestressing. On the other hand, if the column fails on the tensile side as the result of bending or buckling, the steel on that side can be stressed to near its ultimate strength.

The ultimate strength of concentrically prestressed slender columns under axial loads has been investigated both theoretically and experimentally at various universities.[2, 3, 4, 5] The general conclusion is that the axial prestressing of a slender column has no effect on the superimposed axial load which will cause that column to buckle. Use of high-strength materials results in cross sections which are smaller for prestressed concrete columns than might be required for ordinary reinforced concrete. The possibility of strength reduction due to slenderness must always be investigated based on elastic theory taking column slenderness and end conditions of column into account. If the prestressing exceeds the difference between the buckling stress and the ultimate strength of the concrete, the column will fail in compression before it will buckle. When the super-imposed load is not axial, prestressing could increase both the cracking and the ultimate strength as will be explained in section 14-3.

The buckling of the compressive flange of prestressed beams is subject to the same reasoning. There is no danger of flange buckling produced by internal prestress in a beam. For external loads, a tendency to buckle in the flange is governed by the usual theory of elasticity, so long as there are no cracks in the concrete. After cracking or near the ultimate load, little is known about the buckling of the compressive flange in prestressed beams.

EXAMPLE 14-1

A concrete column 16 in. by 16 in. in cross section and 18 ft high, Fig. 14-8, is pretensioned with eight $\frac{3}{8}$-in. wires, which are end-anchored to the concrete. The effective prestress is 100,000 psi in the steel. For a concentric compressive load of 80 k and a horizontal load of 8 k at the midheight of the column, compute the maximum and minimum stresses in the column, assuming it to be hinged at the ends. Investigate the secondary moments in the column due to deflection. Discuss the safety of the column under such loads and also during handling. Assume that $n=7$, $f_c'=4000$ psi, $f_{pu}=200{,}000$ psi, $E_c=4{,}000{,}000$ psi. (Column cross section = 406 mm by 406 mm, height = 5.5 m,

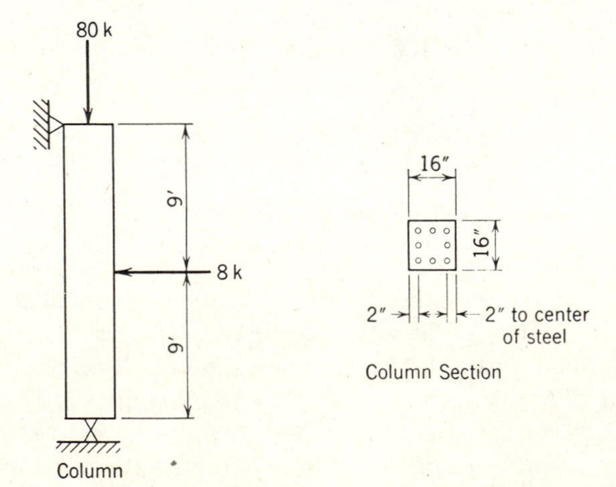

Fig. 14-8. Example 14-1.

vertical load=356 kN, horizontal load=35.6 kN, f_{se}=689.5 N/m, f_c'=28 N/mm^2, f_{pu}=1,379 N/mm^2, and E_c=27.6 kN/mm^2).

Solution Stress in the concrete due to prestress is

$$\frac{F}{A_c} = \frac{8\times 0.11\times(-100{,}000)}{256-(8\times 0.11)}$$

$$= -344 \text{ psi } (-2.37 \text{ N/mm}^2)$$

Stress due to the axial load of 80 k (356 kN), disregarding deflection of column, is

$$\frac{P}{A_t} = \frac{-80{,}000}{256+(7-1)8\times 0.11}$$

$$= -80{,}000/261$$

$$= -306 \text{ psi } (-2.11 \text{ N/mm}^2)$$

The maximum bending moment occurs at the midheight of column, and is

$$8\times 18/4 = 36 \text{ k-ft } (48.8 \text{ kN-m})$$

The I_t of the transformed section is

$$\frac{16^4}{12} + 6\times 0.11\times(7-1)\times 6^2 = 5460+142$$

$$= 5602 \text{ in.}^4 \; (2.33\times 10^9 \text{ mm}^4)$$

The extreme fiber stresses are

$$\frac{Mc}{I_t} = \frac{36{,}000\times 12\times 8}{5602}$$

$$= \pm 616 \text{ psi } (\pm 4.25 \text{ N/mm}^2)$$

The maximum and minimum stresses are hence

$$-344-306-616 = -1266 \text{ psi } (-8.73 \text{ N/mm}^2) \text{ compression}$$

$$-344-306+616 = -34 \text{ psi } (-0.23 \text{ N/mm}^2) \text{ compression}$$

The maximum deflection of column due to the horizontal load is

$$\frac{PL^3}{48E_cI_t} = \frac{8000\times 18^3\times 12^3}{48\times 4{,}000{,}000\times 5602}$$

$$= 0.075 \text{ in. } (1.91 \text{ mm})$$

This will increase the moment due to axial load by the amount of 80,000×0.075=6000 in.-lb=0.5 k-ft (0.68 kN/m). This moment will produce more deflection and further increase the eccentric moment in the column, but the magnitude is seen to be quite small and may be neglected. Hence the above-computed stesses can be considered sufficiently correct. The maximum compressive stess of 1266 psi (−8.73 N/mm^2) would appear high for a reinforced-concrete column but is not excessive for a prestressed member which is more a beam than a column in this example.

The safety of the column can be determined only if we know the ultimate strength of the column under such combined axial and transverse loads and also if we know the possibilities of overloading, that is, to what extent the axial or the horizontal loads may be increased, and whether eccentricity of the applied axial load may be possible.

For the purpose of investigation, let us assume that both the horizontal and the axial load are increased by 50% while, in addition, there will be an eccentricity of 2 in. for the axial load. Then the stresses will be

Due to axial load, $1.5\times306=-459$ psi. (-3.16 N/mm^2)

Due to eccentricity of axial load, $1.5\times80\times2$ in. $=240$ k-in. $=20$ k-ft (27.1 kN-m), which will produce stresses of

$$616\times20/36=\pm342 \text{ psi } (\pm2.36 \text{ N/mm}^2)$$

Due to horizontal load, $1.5\times616=\pm924$ psi. (±6.37 N.mm^2)

Resulting stress:

$$-344-459-342-924=-2069 \text{ psi } (-14.27 \text{ N/mm}^2)$$

$$-344-459+342+924=+463 \text{ psi } (+3.19 \text{ N/mm}^2)$$

Note that the compressive stress of 2069 psi (14.27 N/mm^2) is only about $0.52f_c'$ while the tensile stress is below the modulus of rupture of about $7.5\sqrt{f_c'}=530$ psi. Hence the column would not have cracked, and the midheight deflection can still be computed by the elastic theory to be not more than 0.2 in. (5.1 mm), which is not a significant value. Thus it can be concluded that the column is safe.

For investigating handling stresses, let us assume that the column is picked up at the midheight.

The moment produced will be

$$\frac{wL^2}{2}=\frac{256\times(150/144)\times9^2}{2}$$

$$=10.8 \text{ k-ft } (14.64 \text{ kN-m})$$

which will produce a maximum tensile stress of

$$\frac{Mc}{I_t}=\frac{10.8\times12{,}000\times8}{5602}=+185 \text{ psi } (+1.28 \text{ N/mm}^2)$$

This is much less than the precompression of 344 psi (2.37 N/mm^2), and the column is safe during handling.

14-3 Columns under Eccentric Load

Precast bearing walls and columns can be prestressed to improve their elastic behavior and handling characteristics, and to increase their resistance to lateral forces both in the elastic and the ultimate ranges. They cannot be designed following rules of thumb applied to reinforced concrete walls and columns. But they can be properly designed on basic principles of mechanics and properties of materials. The behavior and strength of prestressed columns under eccentric loading, Fig. 14-9, can be predicted with fair precision, although they have been confirmed only by a limited number of tests.[3, 5, 6] The degree of accuracy will

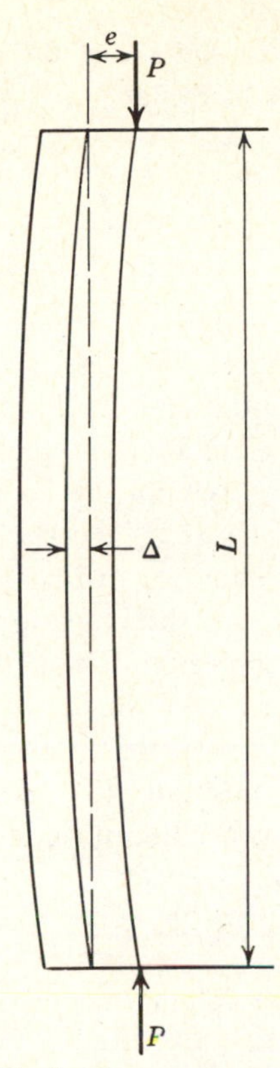

Fig. 14-9. Column under eccentric load.

depend on the choice of values for the modulus of elasticity, the modulus of rupture, and the compressive strength of concrete. Before cracking, the stresses and deflections can be calculated assuming the column to behave elastically. The stress at any section is the sum of the stresses due to prestress, direct axial load, moment due to the eccentricity, and the moment due to the deflection.

$$f_{\substack{\max \\ \min}} = -\frac{F}{A_t} - \frac{P}{A_t} \pm \frac{Pec}{I_t} \pm \frac{P\Delta c}{I_t} \tag{14-2}$$

where F = effective total prestress including all losses except elastic shortening of concrete due to superimposed load

P = superimposed load

e = eccentricity of load from the centroid of the section

c = distance to the extreme fiber from the centroid of the section

A_t = area of transformed section

I_t = moment of inertia of transformed section

Δ = deflection of column at the section

Critical stresses occur at the midheight of the column, where the deflection is given by the well-known secant formula:

$$\Delta = e\left(\sec\sqrt{\frac{PL^2}{4E_cI_t}} - 1\right) \tag{14-3}$$

By the elastic theory, cracking can be assumed to occur when the fiber stress reaches the modulus of rupture. Beyond cracking, the elastic theory is no longer accurate. An estimation of the ultimate load can be made by the elastic theory, assuming it to be the load at which the extreme fiber stress reaches the compressive strength of the concrete. Since such approximate analysis can be way off (by perhaps some 10% even for ordinary cases), it is desirable to apply plastic analysis, taking into account the cracking of concrete under tension, the plasticity of concrete under compression, and the plasticity of steel. Furthermore, while the elastic analysis mentioned above will generally err on the conservative side, it is conceivable that, under unusual conditions, erratic conclusions could be reached unless plastic analysis is applied.

Elastic analysis for a prestressed column under eccentric load is illustrated in example 14-2.

EXAMPLE 14-2

A pretensioned concrete pin-ended column has elevation and section as shown, Fig. 14-10. The effective prestess in the six $\frac{3}{8}$-in. 7-wire strands ($A_{ps}=0.08$ in.^{2} each) is 150,000 psi or 12,000 lb per strand. $E_s=30{,}000{,}000$ psi. Concrete has cylinder strength of 5700 psi, modulus of rupture of 600 psi, and $E_c=4{,}000{,}000$ psi. It is loaded by load P with an eccentricity of 1.5 in. along the weak direction. Compute the cracking and the ultimate value of P using the elastic theory, assuming noncracked section ($A_{ps}=51.6$ mm^2, $F_{se}=1034$ N/mm^2, $E_s=207$ kN/mm^2, $f_c'=39.3$ N/mm^2, $f_r=4.14$ N/mm^2, $E_c=27.6$ kN/mm^2, and $e=38.1$ mm).

Solution (a) Compute properties of the section,

$$n=\frac{30{,}000{,}000}{4{,}000{,}000}=7.5$$

$$A_t=8\times 12+0.48\times(7.5-1)=96+3=99 \text{ sq in. } (63.9\times 10^3 \text{ mm}^2)$$

$$I_t=\frac{12\times 8^3}{12}+0.48\times(7.5-1)\times 2.5^2$$

$$=512+19=531 \text{ in.}^4\ (221\times 10^6 \text{ mm}^4)$$

(b) Compute deflection of the column at midheight by formula 14-3 for various values of P, say $P=50$, 100, 120, 140, and 160 k (222.4, 444.8, 533.8, 622.7 and 711.7 kN).

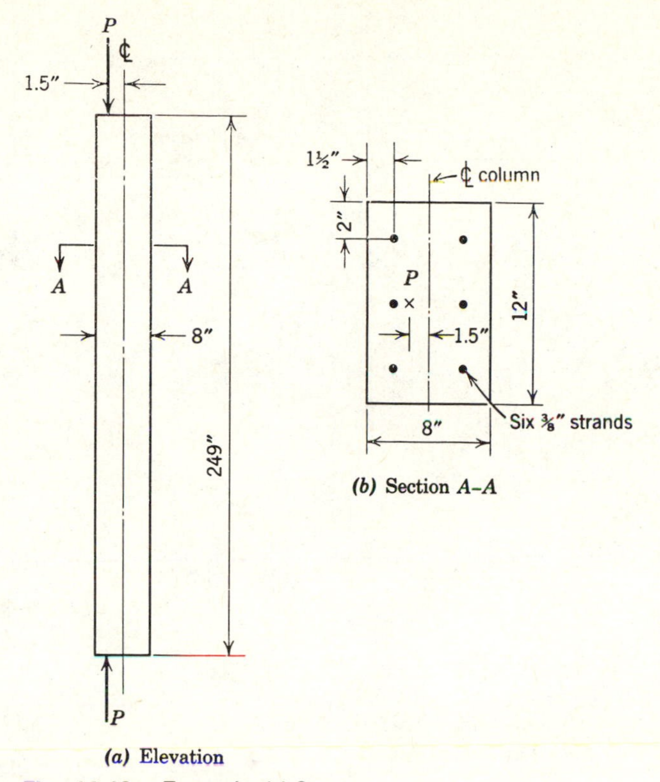

Fig. 14-10. Example 14-2.

Only calculation for $P=120$ k (533.8 kN) will be given here:

$$\Delta = e\left(\sec\sqrt{\frac{PL^2}{4E_cI_t}}-1\right)$$

$$= 1.5\left(\sec\sqrt{\frac{120{,}000\times 249^2}{4\times 4{,}000{,}000\times 531}}-1\right)$$

$$= 1.5(\sec 0.935-1)$$

$$= 1.5(\sec 53^\circ.7-1)$$

$$= 1.5(1.688-1)$$

$$= 1.03 \text{ in. } (26.2 \text{ mm})$$

(*c*) Compute stresses in concrete at midspan section by formula 14-2.

$$f_{\substack{\max\\ \min}} = -\frac{F}{A_t}-\frac{P}{A_t}\pm\frac{Pec}{I_t}\pm\frac{P\Delta c}{I_t}$$

Assuming $F=72$ k (320 kN) is not reduced by the presence of P, and using $P=120$ k

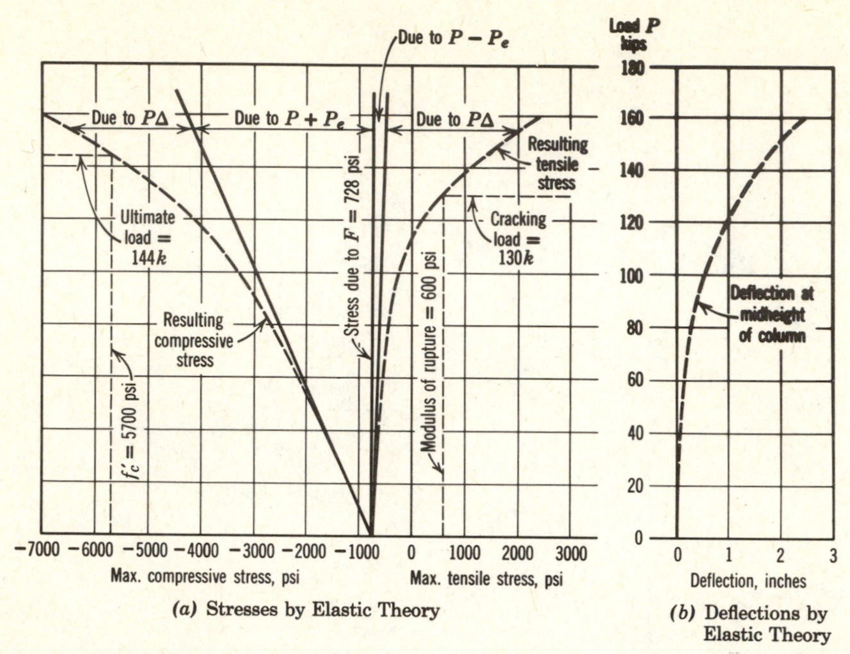

Fig. 14-11. Example 14-2.

(533.8 kN) with $\Delta = 1.03$ in. (26.2 mm), we have

$$f_{\substack{max \\ min}} = -\frac{72{,}000}{99} - \frac{120{,}000}{99} \pm \frac{120{,}000 \times 1.5 \times 4}{531} \pm \frac{120{,}000 \times 1.03 \times 4}{531}$$

$$= -728 - 1210 \pm 1355 \pm 930$$

$$f_{max} = -4223 \text{ psi } (-29.1 \text{ N/mm}^2) \text{ compression}$$

$$f_{min} = +347 \text{ psi } (+2.39 \text{ N/mm}^2) \text{ tension}$$

Elastic stresses and deflections for various loads P are calculated and plotted in Fig. 14-11, assuming noncracked sections. From the graph, it can be seen that the cracking load located at f_{min} = modulus of rupture of 600 psi (4.14 N/mm^2) corresponds to P = 130 k (578.2 kN), and the ultimate load located at f_{max} = 5700 psi (39.3 N/mm^2) gives P = 144 k (640.5 kN) at ultimate. (Although the actual tested values were 130 k = 578.2 kN at cracking and 155 k = 689.4kN at ultimate, it should be mentioned that any correctness of this ultimate-load analysis by the elastic theory is only coincidental. Should the eccentricity be greater and more cracking occur, the assumption of noncracked section could lead to rather nonconservative estimates.)

The ultimate strength of a section under combined axial load and moment can be estimated by the following relationship. Thus, in Fig. 14-12, static equilibrium of the section requires that for $\Sigma V = 0$,

$$P_n = C - T_1 - T_2$$

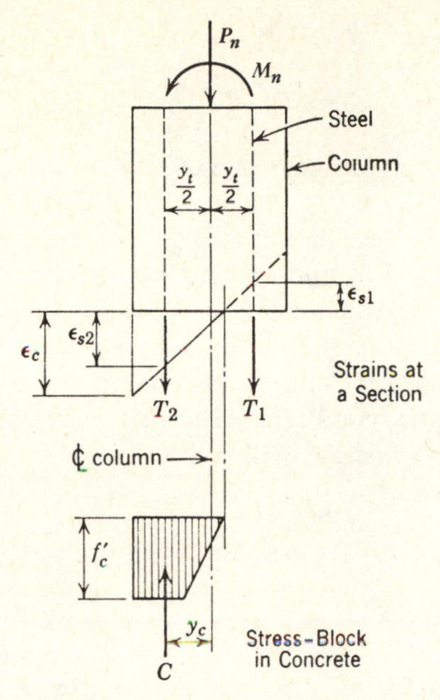

Fig. 14-12(*a*). Ultimate strength under combined axial load and moment.

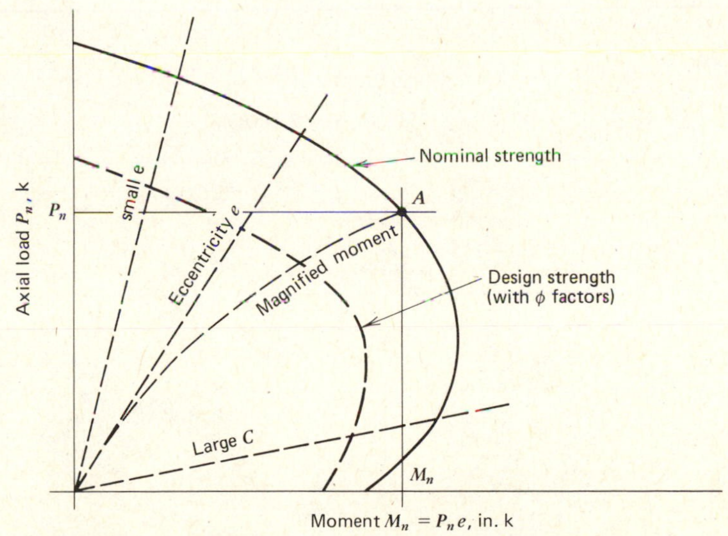

Fig. 14-12(*b*). Interaction diagram for eccentric load.

and for $\Sigma M = 0$,

$$M_n = (T_1 - T_2)\frac{y_t}{2} + Cy_c$$

By assuming a location for the neutral axis at ultimate load, setting ε_c as the ultimate strain in concrete and f_c' as the ultimate stess of concrete, and by assigning ultimate stress distribution curves for concrete, it is possible to compute the combination of P_n and M_n that results in this ultimate failure.

For slender columns, the value of M just computed should include the effect of deflection, which can be computed by a numerical procedure provided the load-moment curvature relationship of the column section is known. Ultimate load analysis taking into account the effect of cracking and the plasticity of concrete and steel is explained in references 5 and 6.

EXAMPLE 14-3

For the column section shown in example 14-2, if the ultimate neutral axis were located at 3 in. from one edge as shown in Fig. 14-13, compute the combined P_n and M_n producing that failure.

Solution Assume $\varepsilon_c = 0.0030$, by proportion, changes in strain for the steel are shown as $+0.0009$ for T_1 and -0.0021 for T_2, which gives stresses in steel (for $E_s = 30{,}000{,}000$

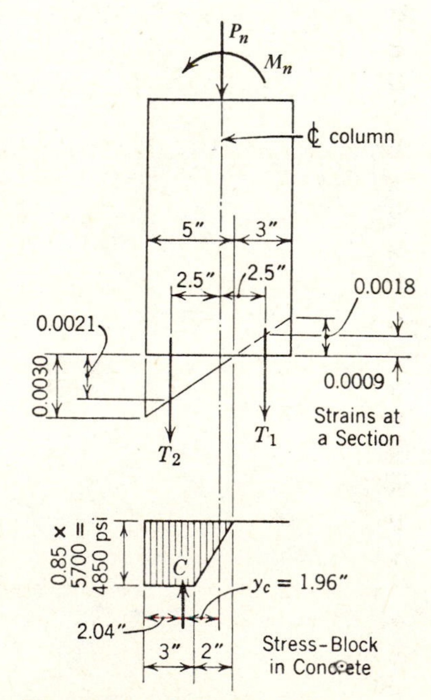

Fig. 14-13. Example 14-3.

and $A_s = 3 \times 0.08 = 0.24$ sq in.),

$$T_1 = 150{,}000 + 0.0009 \times 30{,}000{,}000 = 177{,}000 \text{ psi} \times 0.24 = 42.5 \text{ k (189 kN)}$$

$$T_2 = 150{,}000 - 0.0021 \times 30{,}000{,}000 = 87{,}000 \text{ psi} \times 0.24 = 20.8 \text{ k (92.5 kN)}$$

Using trapezoidal stress distribution for concrete with ultimate strength at $0.85 \times 5700 = 4850$ psi (33.4 N/mm^2), we have

$$C = 4850 \times 12 \times (3 + 2/2) = 232{,}000 \text{ lb (1032 kN)}$$

acting at 2.04 in. (51.8 mm) from the edge. Hence,

$$P_n = C - T_1 - T_2 = 232 - 42.5 - 20.8 = 169 \text{ k (752 kN)}$$

$$M_n = (T_1 - T_2)2.5 + C(1.96)$$

$$= 21.7 \times 2.5 + 169 \times 1.96$$

$$= 54 + 331 = 385 \text{ k-in. (43.5 kN-m)}$$

$$e = \frac{M}{P} = \frac{385}{169} = 2.28 \text{ in. (57.9 mm)}$$

which indicates that a load of 169 k (752 kN) with a total eccentricity of 2.28 in. (57.9 mm) (including the column deflection if any) will produce failure of the column, when the ultimate neutral axis is assumed located at 3 in. (67.2 mm) from the edge. By locating the neutral axis at various places, other combinations of P_n and M_n to produce failure can be obtained. Figure 14-12(b) shows the form of the resulting interaction diagram.

This example indicates how simple it is to compute the ultimate P_n and M_n for a given location of ultimate neutral axis. On the other hand, it would be a major mathematical problem to determine the neutral axis, if either P_n or M_n or both are given.

Thus the resisting capacity of a given section can be best obtained by computing the interaction curve for ultimate P_n and M_n for various assumed locations of ultimate neutral axis. This method can be extended to include the presence of nonprestressed steel in prestressed columns and is indeed a simple approach to the solution of combined ultimate strength of columns.

The ultimate strength of prestressed columns may be controlled by the tensile strength of steel or by the compressive strength of concrete, depending on the amount of reinforcement, the eccentricity of loading, etc. For slender columns, deflections may appreciably affect the ultimate load. When the eccentricity of the load is small, column deflections may be predicted by the elastic theory, assuming noncracked sections. When moment predominates, it would be necessary to estimate the deflections on the basis of cracked sections. If the loading on a column is sustained, creep and shrinkage could become a major problem and should be carefully considered.

Point A on Fig. 14-12(b) illustrates the failure point on the interaction diagram when the moment magnification from column deflection is taken into account. For design we may use the ACI Code method for prestressed columns with reasonable accuracy including inelastic effects. Even though this approach was developed for reinforced concrete columns, it is considered the best method for prestressed columns until further data are available.

14-4 Piles*

Since piles are subjected to tensile stresses during transportation, driving, and under certain service conditions, the desirability of prestressing is evident. Posttensioned concrete piles, essentially of the Raymond cylinder type, have been produced since 1949.[7] About 1953, pretensioned concrete piles were developed; they are now readily available because of the establishment of hundreds of pretensioning plants throughout the country, and indeed all over the world.[8]

In common with many construction materials and techniques, pretensioned concrete piles were developed by the industry rather than by the profession. A process of trial and error, rather than a rational approach, was employed during their development. At present, enough experience has been accumulated to permit safe and economical utilization of these piles.[9]

Discussion follows under five headings: design, details, manufacture, driving, and special applications.

Design. Experience seems to indicate that a prestress of about 700 psi (4.83 N/mm^2) in the piles will insure safety during handling and driving under normal conditions. While the amount of prestress required will vary with the size and shape of the pile, the hammer blow, and the cushioning effects, as well as the soil conditions, it is obviously impractical to vary the prestress in each pile. Of course, higher or lower values than 700 psi (4.83 N/mm^2) may be desirable for special cases.

The bearing capacity of concrete piles is seldom if ever controlled by their strength under direct compression, but it is convenient to express the bearing capacity in terms of the compressive strength or stresses. Strictly speaking, if the bearing capacity were limited by the compressive stress, there would be no need for prestressing. Therefore, current formulas are empirical in nature. For reference, they are outlined as follows.

The design load on such piles is often based on the ultimate strength, using an arbitrary factor of safety of about 4. Such a high factor of safety is hardly necessary so far as the service load is concerned, but it is believed that, for piles so designed, the compressive stresses during driving will seldom be critical, and it should be possible to attain the desired bearing value without damaging the pile.

*Material in this section was taken from the article "Pretensioned Concrete Piles—present knowledge summarized," by T. Y. Lin and W. J. Talbot, *Civil Engineering*, May 1961, pp. 53—58 with revisions following reference 9 recommendations (1973) and other more recent references.

If the cylinder strength of the concrete is f_c', the ultimate strength of the concrete in a pile can be safely assumed as $0.85f_c'$. At ultimate load, the amount of prestress remaining in the tendons is approximately 60% of the effective prestress. Thus, if a 6000-psi (41 N/mm^2) concrete pile is prestressed to an effective prestress of 700 psi (4.83 N/mm^2), the ultimate strength can be computed by the formula,

$$N' = (0.85 \times 6000 - 0.60 \times 700)A_c = 4680A_c$$

where A_c is the cross-sectional area of the concrete pile in sq in. Using a factor of safety of 4, the design load, N, is one fourth of this, or $1170A_c$.

Standards set up by the Prestressed Concrete Institute[9] state that the maximum compressive load, N, on prestressed concrete piles (in addition to the effective prestress) shall not exceed the following:

$$N = (0.33f_c' - 0.27f_{pc})A_c$$

where $\frac{h'}{r} < 50$

h' = effective unsupported length of pile, taken as the actual length of pile when both ends are hinged

r = radius of gyration of transformed section of pile

$f_{pc} = \frac{F}{A_c}$ (effective prestress after losses)

A_c = area of concrete

The buckling load of a slender pretensioned concrete pile can be computed by Euler's formula,

$$N_{cr} = \frac{\pi^2 EI}{L^2}$$

where N_{cr} = critical buckling load, lb

E = modulus of elasticity for concrete, psi

I = moment of inertia of concrete pile section, in.^{4}

L = length of pile, in.

The above formula assumes hinged supports for both ends and can be modified for other end conditions. The value of E should be chosen to fit the duration of loading—that is, a higher value should be used for dynamic load and a reduced value for sustained load. Since the possibility of an increase in actual load is remote, a factor of safety of 2 is considered sufficient. Thus the allowable load is often set at

$$N = \frac{N_{cr}}{2} = \frac{\pi^2 EI}{2L^2}$$

If no tensile stress is allowed, a high factor of safety is obtained for concentrically prestressed members subjected to bending. It is therefore often permissible to allow tensile stresses in the concrete under design moments.

For piles with $h'/r > 50$ the design procedure is based on ultimate strength of the pile under combinations of axial loading and moment following the PCI Recommendations.[9]

The ultimate moment, in inch-pounds, for a pretensioned pile can be computed by the methods in section 14-3 or approximated by

$$M_{\text{ult}} = CA_{ps}f_{ps}d$$

where C = a coefficient depending on shape of pile section, etc., varying usually from 0.32 to 0.38
A_{ps} = total area of prestressing steel, sq in.
f_{ps} = ultimate strength of steel, psi
d = diameter or size of pile, in.

The design moment, based on the ultimate, should have a factor of safety of 2, while a factor of 1.5 to 1.7 will be sufficient for earthquake and wind loads.

According to the elastic theory, the existence of direct external loads delays the cracking of the concrete pile and thereby increases the moment-carrying capacity. On the other hand, the ultimate moment capacity is reduced by the presence of direct external loads. Hence, when the design is for combined moment and direct loads, the moment capacity of the pile should be checked by the ultimate-load theories, as described in section 14-3. Reference 9 gives interaction diagrams for standard piles as design aids, and design examples which illustrate their use.

Details and Connections. Some typical pretensioned pile sections are given in Fig. 14-14. These piles are usually pretensioned with 7-wire strands, $\frac{1}{2}$ in. (12.7 mm) in diameter. The spacing of sprial steel also was established by experience. Its design has not been rationalized, but it is generally agreed that steel of No. 5 gage about 0.2 in. (5.1 mm) in diameter, at a 6 in. (152.4 mm) pitch, will suffice for the middle part of the pile, while a 3-in. (76.2 mm) pitch is used for the end portions. Four or five tight turns at about a 1-in. (25.4 mm) pitch are usually extended to within 1 in. (25.4 mm) from the end.

Where the pile tops are encased in a heavy footing, very often no connection other than sufficient embedment is required. In this case, the pile can be either driven to grade or cut off to the desired level with ordinary concrete chipping hammers.

If additional anchorage is required, one or more of the following types of connections can be used.

The prestressing strand can be extended from the pile head and used as reinforcement. Actual tests have shown that an embedment of about 50 diameters is adequate to develop the full strength of a strand. The strands may be

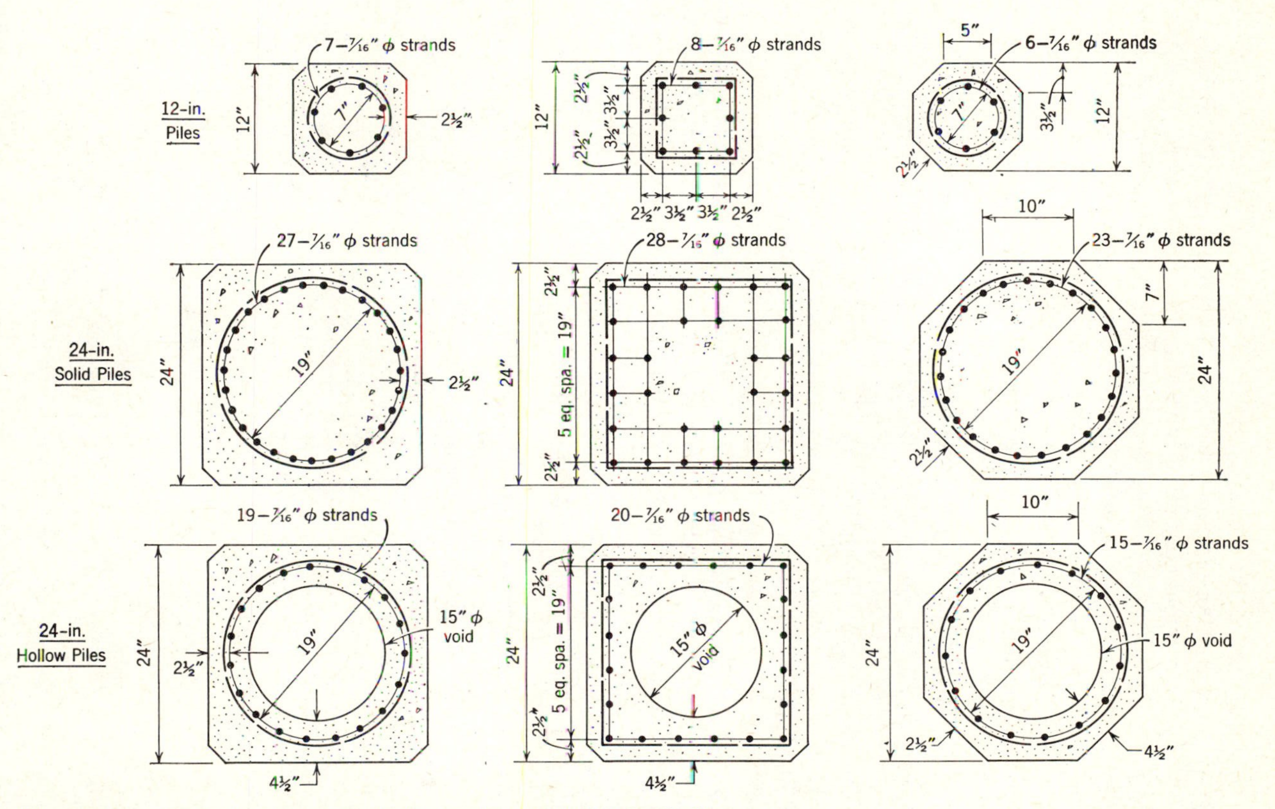

Fig. 14-14. Typical pretensioned pile sections (AASHTO-PCI Standard).

exposed by stripping off the concrete after driving. If the piles are to be driven to grade, the strand extension must be allowed for at the time of casting.

If the pile reinforcement consists of more than 12 or 14 strands, threading the projecting steel through the driving head becomes costly and time consuming. Since most precast piles are ordered slightly on the long side, it may be simpler to cut off the pile after driving and expose the reinforcing for anchorage. It should be noted that the full working strength of the strands cannot be developed as unprestressed reinforcement, owing to the excessive elongation that could be produced.

To posttension a dozen or more small strands through the cap or footing is difficult and costly so that this method has not been used to any extent for pretensioned piles. In the posttensioned pile, particularly where alloy bars of large diameter are used for reinforcement, a tensioned connection is easier to make and has been used in Europe, where fixity or full anchorage is required at the pile head.

Probably the most versatile and widely used connection of pile to cap is the simple reinforcing-bar anchorage similar to that used for many years in conventionally reinforced piles, Fig. 14-15. The reinforcing can be cast in the head of the prestressed pile where the piles can be driven reasonably close to final elevation, but a special driving head or follower must be used. Bars can be

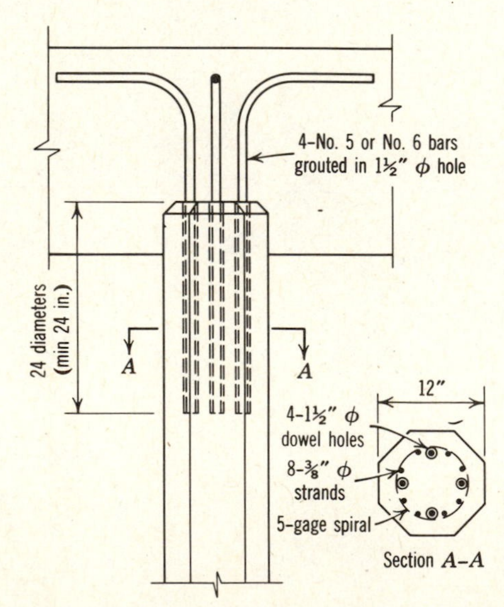

Fig. 14-15. Holes drilled or formed ¾ in. or 1 in. larger than dowels, by lightweight tubing (without bond breaker or lubricant), provide adequate bond for reinforcing when carefully grouted. Section A-A shows method of ensuring load transfer for piles with 3/8-in. strand.

grouted into either precast holes or holes drilled after driving. If precast holes are to be used, the pile must be driven within 1 or 2 ft (0.30 or 0.61 m) of final grade. With greater variations, the field-drilled holes may prove to be cheaper.

Pull-out tests have indicated that special grouting mixtures are not required to develop the full bond strength. A good plastic sand-and-cement grout with an ultimate strength equal to that of the pile, properly worked into place and cured, will be satisfactory. A water-reducing plasticizing admixture is desirable to reduce shrinkage.

For a pile with a hollow core, the connection can be made by using a poured-in-place concrete plug with an embedded reinforcing-steel cage placed inside the core and extended into the cap or footing, Fig. 14-16. Where welded connections are desired, they are provided by grouting a plate or other steel section into the pile with extended steel reinforcing bars attached. The connection is completed by welding to the superstructure.

Splicing of precast piles can be rather difficult. The splice or extension should be at least as durable as the prestressed concrete section; it should have equivalent load-carrying capacity, bending, and shear strength. The splice must be economical and must allow driving to be resumed within a reasonable time.

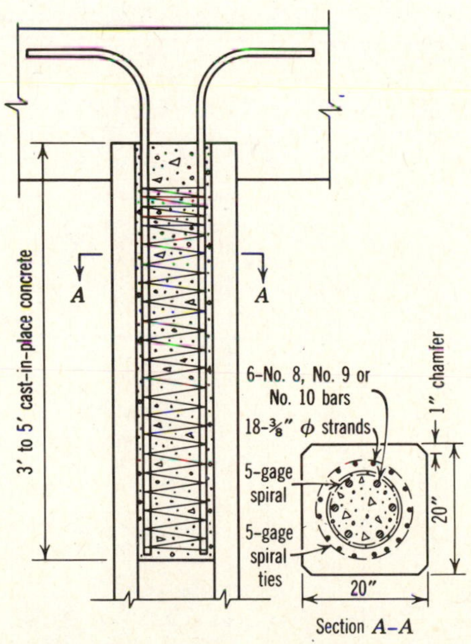

Fig. 14-16. Connection of a hollow pile to footing or bent, using steel dowels. The poured-in-place plug can be put at any depth by suspending an expendable form at the proper depth.

Several methods have been developed and successfully used for splicing prestressed concrete piles. A very complete review of the various types of splices which have been used and their performance is given in reference 10.

A splice developed in Connecticut consists of two pretensioned sections joined by means of reinforcing steel dowels and a plasticized cement. The lower section is first driven and the head drilled to receive reinforcing bars cast in the upper section. The jointing compound is placed in a molten state and fills the dowel holes and the space between the pile sections. Because of the high strength and fast set of the cementing compound, the spliced section can be driven after a very short curing time. On the Pacific Coast, similar experimental splices have been made using a slower setting epoxy compound, Fig. 14-17. Reference 11 reports extensive test results on this type of splice with very satisfactory performance.

A splice can be made by welding the two sections in the field using steel pipe sleeves or anchored plates embedded in the prestressed sections at the time of casting. An epoxy compound can be used between the concrete sections for load transfer.

Where full moment-carrying splices are not necessary, simpler connections can be used. In Honolulu, long piles composed of three prestressed sections were spliced by driving the upper section into a snugly fitting steel sleeve that had previously been driven over the lower section. A center dowel was used to align the two sections.

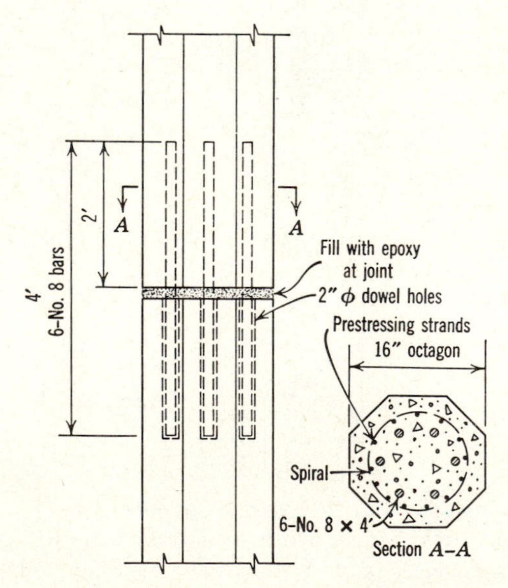

Fig. 14-17. Pile splice detail. Dowels are secured with epoxy.

Manufacture. Pretensioned piles are usually manufactured by the long-line method, Fig. 14-18; multiple sections are cast in a single stressing line up to perhaps 600 ft (183 m) in length. Pretensioning is used for piles up to about 30 in. (762 mm) in cross section, although it is possible to pretension larger sections if it is economical to set up the heavier tensioning bed and auxiliary equipment.

The pretensioning bed should be strong enough to resist the maximum stressing force to be applied, rigid enough to prevent excessive deflections, and accurately level so that the strands and the final product will be in true alignment. The strands can be stressed as a group by large-capacity jacks, or individually by single-strand jacks.

Forms for the piles should be preferably of steel, sufficiently rigid to eliminate distortion. Joints between sections, or at end forms, should be accurately fitted so that offsets or openings are eliminated. Thermal movements, particularly with high-temperature curing, can produce cracking at points where large offsets or fins restrict movement. Forms should be so constructed as to permit movement of the pile without damage during release of the prestressing force. The form at the pile ends should be perfectly square with the pile axis and reasonably plane. To form the core for hollow piles, either paraffin-treated cardboard tubes or inflatable rubber mandrels are used.

Most pretensioned piles made in the United States have a square or octagonal cross section. The square section is perhaps somewhat easier to form and pour; the octagonal section is often preferred because of its smaller area and weight for a given least dimension. The octagonal pile requires a slightly more complicated form process, but generally requires less complicated spiral ties. Both fixed and collapsible-type steel forms are used for either type of pile.

Low-slump concrete with a cement content of $6\frac{1}{2}$ to 8 sacks per cu yd is generally used to provide a 28-day strength of 4500 to 7000 psi (31 to 48 N/mm^2). A water-reducing admixture of some kind is usually added. Calcium chloride should not be used. Most piling specifications require Type II low-alkali cement, particularly where piles are used in sea water, although cements of Types I and III are occasionally permitted. Aggregates usually are $\frac{3}{4}$ to 1 in. (19.1–25.4 mm) in maximum size.

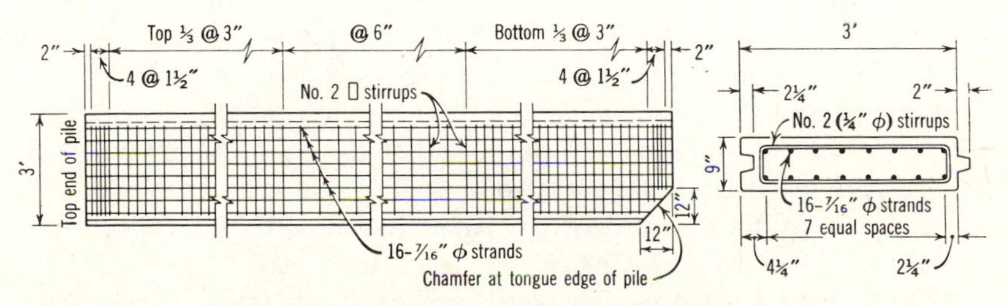

Fig. 14-18. A typical prestressed concrete sheetpile.

When the concrete is placed, particular care should be taken to insure that the head and tip are well vibrated as these areas usually have close spiral or tie spacing and may have dowels or tubes for preformed holes at the head. In the case of piles with a hollow core formed with paper tubes or inflated rubber tubes, it is necessary to provide external hold-downs at intervals close enough to prevent flotation or deflection of the tube. Attempts to secure the tube by means of the encircling strands usually result in excessive flotation or deflection on all but very short piles.

Since economy in manufacture requires a rapid turnover of piles on the casting bed, most producers of pretensioned piles employ accelerated curing. Low-pressure steam, radiant heat, and hot air are all used for curing. Curing is usually done at temperatures ranging from 130 to 165°F for 10 to 18 hours. In hot, arid areas, additional water or membrane curing may be required for an additional 7 to 10 days. In most cases air curing will develop the required ultimate strengths before 28 days.

Since pretensioned piles employ concentric prestressing seldom exceeding 1000 psi (6.90 N/mm^2), the primary consideration is bond rather than compression or tension in the concrete at stress transfer. Concrete strengths for stress release are generally set at 3500 psi (24 N/mm^2) as a minimum for strands. Minimum transfer strengths are usually adequate for handling of the section so that piles can be removed from beds immediately after stress transfer.

Driving. Prestressed concrete piles have proven their ability to take an unbelievable amount of punishment without structural damage. They are very strong in bending but are not indestructible. Experience has shown that if the criteria of no tension in the concrete is used for handling and transporting, a sufficient factor of safety against cracking is available to take care of impact and shock loads for all but extreme cases. When no extra loads are expected, tensile stresses may be permitted during handling.

As previously mentioned the pile head must be truly perpendicular to the pile axis and reasonably plane. Irregular or inclined heads tend to concentrate the driving blow and may spalling or cracking at the head. The wire strands should be burned flush with the pile head. Projecting wire stubs, even when covered by a cushion block, have been known to cause spalling at the head. A chamfered edge at the head also helps to prevent spalling. If jetting is required, the internal-type jet has proved superior to the external type in preventing wandering of the pile tip and consequent eccentricity during driving.

One of the most important details in connection with the driving operation is the cushion block. Generally the best performance has been obtained with 4 to 8 in. (101.6–203.2 mm) of laminated softwood, such as Douglas fir, placed directly between the pile head and the driving helmet. Hardwood blocks have proven unsatisfactory. Plywood cushion blocks are sometimes used, but in more than a few cases these have caused trouble until replaced with thicker laminations of softwood. Cushion blocks should be replaced frequently; once they are fully compressed, they cease to perform their function.

The type of cushion block directly affects the magnitude of the driving stresses. In severe cases, the use of a reduced hammer blow during the soft driving stage may also help.

In one instance, what was originally thought to be tensile cracking was directly traced to torsional stresses induced by the twisting of the pile head in the driving helmet. The solution to the problem in this case was to round the pile head so as to eliminate restraint at the driving head. The pile head should not be completely restrained against rotation.

It has been found that pile hammers with an energy within the limits shown below are usually adequate for driving in the moderate to hard driving range. These values are listed for reference rather than for absolute guidance:

Pile Size, in.	Ft-lb of Energy
10	8000–15,000
12	15,000–19,000
14	15,000–24,000
16–18	24,000–32,000
20–21	24,000–36,000
24 and over	32,000–38,000

Large-diameter cylinder piles have been driven with hammers having a rated energy up to 50,000 ft-lb (67.8 kN-m).

Properly designed prestressed concrete piles can be used economically for almost all types of foundations, piers, wharves, mooring and fender dolphins, marine ways, bulkheads, sea walls, and other structures.

When very long piles are used, the problem of handling and transporting may be a critical one. Hollow-core piles can be used to reduce the handling weight; 20-in. (508 mm) octagonal hollow-core piles up to 132 ft (40.2 m) long were driven for a waterfront structure in San Francisco.

Special Applications. Where batter piles are used in marine structures in deep water, critical bending stresses may occur during setting and driving when the pile may be cantilevered far below the pile driver leads. A pile with higher prestress or one of greater section modulus may be required.

When the pile tip is to be seated in rock or other hard material, a steel H-pile section may be embedded in the prestressed concrete pile. For example, hollow-core piles 26 in. (660 mm) square, with 14-in. (356 mm) steel H-pile tips, were driven to rock for the foundation of the Petaluma Creek Bridge near San Francisco. The design load on these piles was 200 tons.

Prestressed sheetpiles, Fig. 14-18, are being used for marine installations where corrosion resistance is important. Since the sheetpile is primarily a

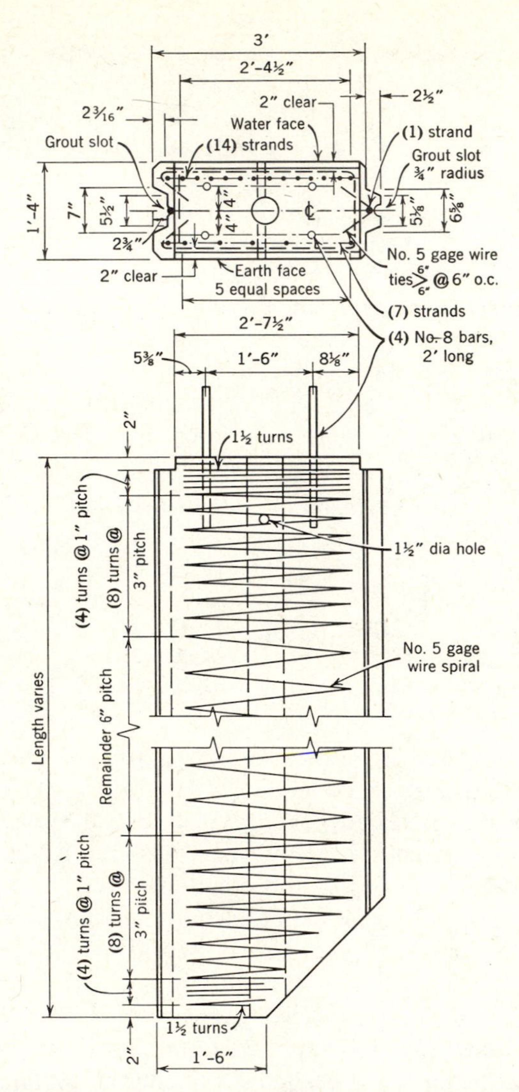

Fig. 14-19. Sheetpile with eccentric prestressing for a bulkhead wall in Long Beach, California. Slight eccentricity does not affect the driving characteristics of the pile.

bending member, the effective prestress may vary depending on design moments. The efficiency of a sheetpile in bending can be substantially increased by using eccentric prestressing, Fig. 14-19. Where moment reversals may occur, this is not feasible and somewhat greater concentric prestressing should be used.

The use of prestressed-concrete soldier piles for bulkheads, sea walls, and retaining walls is another application as a bending member. These piles can be

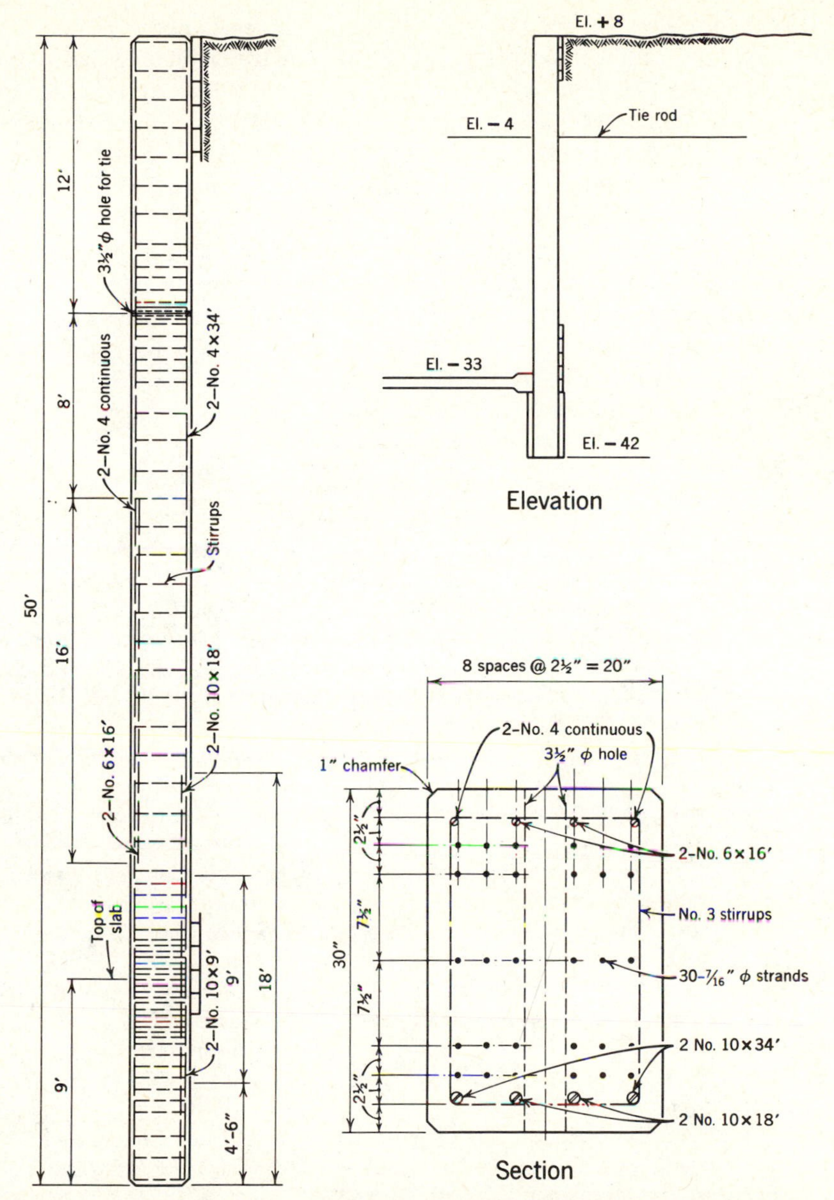

Fig. 14-20. Soldier piles with reinforced and prestressed steel, to take care of peak moment in different parts of the pile.

made in the form of an H-section with timber lagging or concrete planks placed in slots or grooves on the sides of the piles, or as square or rectangular members with lagging placed behind the pile. On a graving dock built in Alameda, California, 20×30-in. (508×762 mm) prestressed soldier piles are being used as a retaining wall for a cut 40 ft (12.2 m) deep, Fig. 14-20. Mild-steel reinforcing is used for additional strength at the moment peaks, the main reinforcing being provided by the prestressing steel.

While sufficient experience has been accumulated for the proper design, production, driving, and utilization of pretensioned concrete piles, it must be admitted that very little has been done to rationalize the procedures or to analyze the results. If economical applications have already been obtained by sheer experience, it is only natural to expect even better performance with a more refined and scientific approach.

References

1. W. H. Ellison and T. Y. Lin, "Parking Garage Built for $5.28 per sq. ft," *Civil Engineering*, June 1955, pp. 37–40.
2. P. Zia and F. L. Moreadith, "Ultimate Load Capacity of Prestressed Concrete Columns," *J. Am. Conc. Inst.*, Vol. 63, No. 7, July 1966, pp. 767–788.
3. P. Zia and E. C. Guillermo, "Combined Bending and Axial load in Prestressed Concrete Column," *J. Prestressed Conc. Inst.*, Vol. 12, No. 3, June 1967, pp. 52–59.
4. S. Aroni, "The Strength of Slender Prestressed Concrete Columns," *J. Prestressed Conc. Inst.*, Vol. 13, No. 2, April 1968, pp. 19–33.
5. N. D. Nathan, "Slenderness of Prestressed Concrete Beam-Columns," *J. Prestressed Conc. Inst.*, Vol. 17, No. 6, November–December 1972, pp. 45–57.
6. T. Y. Lin and R. Itaya, "A prestressed Concrete Column under Eccentric Loading," *J. Prestressed Conc. Inst.*, December 1957.
7. A. S. Hall, *Buckling of Prestressed Columns*, Cement and Concrete Association, Sydney, Australia, 1961.
8. N. H. E. Weller, "Prestressed Concrete Piles," *J. Prestressed Conc. Inst.*, Vol. 7, No. 5, October 1962, pp. 46–55.
9. ACI committee 543, "Recommendations for Design, Manufacture, and Installation of Concrete Piles," *J. Am. Conc. Inst.*, Vol. 70, No. 8, August 1973, pp. 509–644.
10. R. N. Bruce and D. C. Hebert, "Splicing Concrete Piles: Part 1—Review and Performance of Splices," *J. Prestressed Conc. Inst.*, Vol. 19, No. 5, September–October 1974, pp. 70–97.
11. R. N. Bruce and D. C. Hebert, "Splicing of Precast Prestressed Concrete Piles: Part 2—Test and Analysis of Cement-Dowel Splice," *J. Prestressed Conc. Inst.*, Vol. 19, No. 6, November–December 1974, pp. 40–66.

15

ECONOMICS; STRUCTURAL TYPES AND LAYOUTS

15-1 Economics

When prestressed concrete was first used in United States in the early 1950's, the problem of the relative economy of this type of construction as compared to others was a controversial issue. Some zealous advocates held an optimistic outlook on its saving in materials; other conservative engineers overestimated the additional labor involved and condemned its popular adoption in this country. That period is now concluded. With the numerous prestressed-concrete structures built all over the country, its economy is no longer in doubt. Like any new promising type of construction, it will continue to grow as more engineers and builders master its technique. But, like any other type of construction, it has its own limitations of economy and feasibility so that it will suit certain conditions and not others.

The time is also past when one or two specific instances of the relative economy of prestressed concrete as against other types could be cited as positive proof either for or against its adoption. Basic quantity data are now known for prestressed concrete, and the unit price for prestressing is stabilized.

In prestressed concrete we have materials that are much stronger than those for ordinary reinforced concrete. At the same time, these materials cost more and require more labor and better technique for placement. Speaking in general, the working stress in prestressing steel is 5 to 7 times as high as structural or reinforcing steel, and its unit price in place is 2 to 4 times as much. Concrete is some 50% stronger than reinforced concrete, and costs about 10% more, not including formwork which may cost from 0% to 100% more than that for reinforced concrete. Between the various possible combinations of strength and cost of these materials, it can be readily seen that the net result can be either for or against the use of prestressed concrete.

From an economic point of view, conditions favoring prestressed construction can be listed as follows:

1. Long spans, where the ratio of dead to live load is large, so that saving in weight of structure becomes a significant item in economy. A minimum dead-to-live-load ratio is necessary in order to permit the placement of steel near the tensile fiber, thus giving it the greatest possible lever arm for

resisting moment. For long members, the relative cost of anchorages is also lowered.

2. Heavy loads, where large quantities of materials are involved so that saving in materials becomes worthwhile.
3. Multiple units, where forms can be reused and labor mechanized so that the additional cost of labor and forms can be minimized.
4. Precasting units, where work can be centralized so as to reduce the additional cost of labor and to obtain better control of the products.
5. Pretensioning units, where the cost of anchorage, sheathing, and grouting can be saved.

There are other conditions which, for certain locale, are not favorable to the economy of prestressed concrete but which are bound to improve as time goes on. These are:

1. The availability of builders experienced with the work of prestressing.
2. The availability of equipments for posttensioning and of plants for pretensioning.
3. The availability of engineers experienced with the design of prestressed concrete.
4. The improvement of codes for prestressed concrete design and construction.

While the problem of designing prestressed-concrete structures economically is discussed in every chapter of this treatise, it may be well to summarize the major issues involved so that the designer can grasp the vital points of economy and not lose himself in a maze of minor details.

The first and foremost decision to be made is whether the structure can be economically designed for prestressed concrete or whether it might be better to employ some other type of construction, be it timber, steel, or reinforced concrete. There are, of course, other problems besides economy, such as aesthetic or functional requirements, which might force the choice one way or another, but by far the majority of structures will be decided on an economic basis. There are structures where prestressed concrete would be most suitable, but there are also those where it simply cannot compete with other types. Therefore the first motto for the designer is to employ prestressed concrete only where it belongs and not to use it as a cure-for-all.

The choice for prestressed-concrete construction must be considered together with the possible change in layout of the entire structure. Since most engineers are more familiar with other types of construction, span lengths and proportions are usually laid out with those in mind, the designer not realizing that, with prestressed concrete, radical changes may be possible and often desirable. For example, longer spans, smaller depth, thinner members, bold cantilevers, precasting procedures, reuse of forms not contemplated for reinforced concrete, may be embodied in the layout.

Once the use of prestressed concrete has been decided upon, the next important decision concerns the right type of prestressed construction. Should the members be pretensioned or posttensioned? Should they be precast or cast-in-place, or should composite construction be adopted? Is bonded reinforcement necessary for the job, or would the unbonded type suffice for the service? These questions must be answered first, although often not until some preliminary work has been done on the design.

Engineers seldom realize an important prerequisite in design, namely, that the design loadings for structures must be chosen with care and judgment. Too often the loadings are specified for the structures by certain code requirements, and the engineer simply takes them for granted. Such specified loadings can be either too heavy or too light. Although they may have proved to be satisfactory for other types of construction whose methods of design have been empirically devised to suit such loadings, they may not be directly applicable to a new type of construction like prestressed concrete. If such loadings are adopted the designer should exercise care in employing a suitable set of allowable stresses and load factors so as to attain a proper but not excessive degree of safety in the structure.

Having settled these essential premises, the engineer can now proceed to design the members in detail. The first item to be settled, then, is usually the depth or thickness of the member. Although this must be determined for each particular structure, speaking in general, the economical depth of beams and thickness of slabs would be about 70% to 80% that of reinforced concrete. The depth having been chosen, the concrete section can be designed, taking into account the cost of formwork, ease in concrete placing, as well as stress and deflection requirements. At the same time, the strength of concrete for the job can be decided on. Often it is not the 28-day strength but the strength at transfer that controls the economy, since it is frequently necessary to obtain high strength fast in order to permit prestress transfer at an early age and thus speed up the production and construction.

Although prestressing steel is an item peculiar to prestressed work, its choice ranks least in importance when compared to the major considerations mentioned above. When the layout of a member is fixed, there will result a certain amount of prestress required, which is often beyond the control of the designer. Sometimes, the adoption of a particular system of prestressing may modify the sectional dimensions of the members, but such modifications are minor. Often it is best to design a structure so that several systems may fit into it with very little modification, and the problem of selecting an economic system is left to the competition of the bidders. There are, however, some exceptional cases when one system is definitely superior to the others, and then the engineer should design on that basis.

One of the vital problems in the design of prestress steel is the positioning of the steel so as to give it a maximum resisting lever arm. In order that such a position of steel will not overstress the concrete at transfer, it is sometimes necessary to add superimposed weight to the member prior to prestressing, or to tension the steel in stages. Another economic method is to place nonprestressed reinforcement at strategic points in order to increase the ultimate strength or to improve local behavior. Means such as those must be kept in mind and applied when conditions warrant.

15-2 Building Types and Layouts

Prestressed concrete can be economically applied to many types of buildings, whether of long or short spans, multistory apartments or single-story industrial buildings. Each structure must be studied by itself, with respect to its particular functions, environment, facilities, and other requirements. One should always take into consideration related factors, such as architectural, mechanical, occupancy, and operational economies.

Some of the prevalent structural systems and layouts for prestressed concrete buildings will be mentioned below.

Prestressed Slabs. In-place, posttensioned slabs are widely constructed in the United States to the extent of some 4,000,000 m^2 in 1974 alone. They employ almost exclusively unbonded tendons,[1] and have proven to be economical for 20- to 35-ft (6.1–10.7 m) spans. As an example of its economy, a 40-story office building in Singapore, Fig. 15-1, has typical floors 8-in. (203 mm) thick flat slabs spanning 30 ft (9.14 m). When compared to its original conventional design with beam-and-slab construction at 20-in. (508 mm) depth, a saving of structural depth of 1-ft (0.30 m) per floor was effected, resulting in 40-ft (9.14 m) height reduction for the high-rise building, resulting in great economy.

For slabs with span from 30 to about 45 ft (9.14 to about 13.7 m), haunches and dropped panels can be added around the columns. Waffle slabs can be made to span about 100 ft (30.5 m) without excessive concrete or steel, if suitable structural depth can be obtained.

The shrinkage, creep, and elastic shortening of concrete must be allowed or compensated for when prestressed in place. This can be accomplished by arranging the vertical elements to permit movement in the direction of the shortening, Fig. 15-2. Either the elements are made slender, or sliding joints and hinges can be provided. For a multistory building, only the shortening of the second floor relative to the ground floor needs to be considered, while the relative shrinkage and creep between the various floors are usually small.

Pretensioned hollow slabs of various sizes and designs are produced in the United States. They are about 8-in. (203 mm) thick with 5-in. (127 mm) round or oval holes.

Pretensioned solid slabs over continuous spans are occasionally employed. Such long and thin slabs over several spans reduce the number of joints and

Fig. 15-1. The United Industrial Commercial Development (UICD) Building near completion—Singapore (Special Structural Consultant T. Y. Lin International, Consulting Engineers). The United Industrial Commercial Development (UICD) 40-story office building, 490 ft high, is one of the world's tallest in prestressed concrete. It consists of seven floors in the 600-ft-long podium and a tower block 121 × 110 ft rising therefrom. The typical floor is made up of a posttensioned flat slab 8 in. thick spanning 30 ft, with no interior beams. The typical floor, overhanging on four sides up to 15 ft, is supported by cantilever beams bracketing outward from the columns. The height of the building was reduced by a total of 40 ft, resulting in great economy.

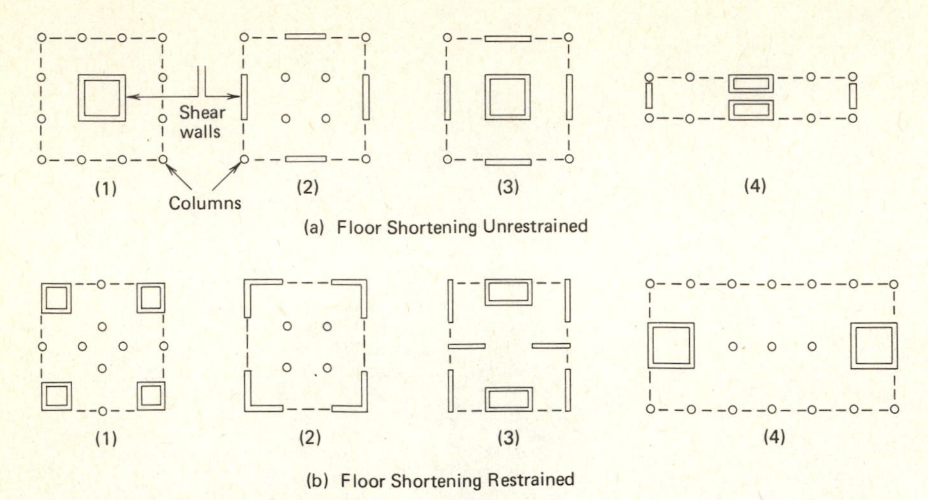

Fig. 15-2. Shear wall location—floor plans.

may prove economical for short spans (10–20 ft, i.e., 3.0–6.1 m) when they can be transported and erected without special equipment.

Another type of construction utilizes precast pretensioned slabs as the form on which the top concrete can be poured in place. This type of composite precast construction is economical when the live load is heavy or when an in-place topping is required.

The lift-slab type of construction resulted from an effort to reduce the cost of concrete construction by eliminating the expense of soffit forming and shoring. These slabs are cast and prestressed at ground level and then lifted into position by hydraulic jacks mounted on top of the columns. The prestressing of the slab reduces the slab's thickness, controls the deflection, eliminates cracks in the slab, and cuts down the cost.

When a flat slab with long spans in one direction is required, a dropped beam and slab type is used. For roof parking on factories and commercial buildings, this type of framing can be made to yield watertight roofs without conventional roofing or membrane, since the concrete is prestressed in two directions.

Precast Floor and Roof Panels. Various types of precast prestressed panels are developed in different countries. In the United States, the double-tee and single-tee types have found wide acceptance. Their economy lies in the utilization of the top slab to carry the major portion of the compressive force, in combination with the narrow stems, which house the prestressing tendons, acting in tension.

Figures 15-3 and 15-4 give the quantities of concrete and steel required for double-tee sections for different spans and live loads. Simply by increasing the

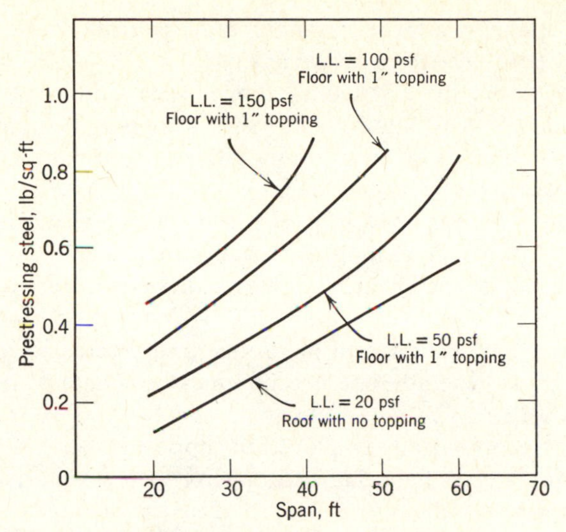

Fig. 15-3. Quantity of prestressing steel for double-tee panels.

depth of the stems, the double-tee sections can be made to span greater distance, say 50 to 60 ft (15.2–18.3 m), with little additional material.

The single-tee sections provide greater fire protection and more flexibility by concentrating two stems in one. The forms for the single-tee can be easily stripped and adjusted to accommodate variations in stem depth and thickness.

The channel section has also been successfully employed for buildings. When spans are short, say below 20 ft (6.1 m), reinforced concrete channels are often

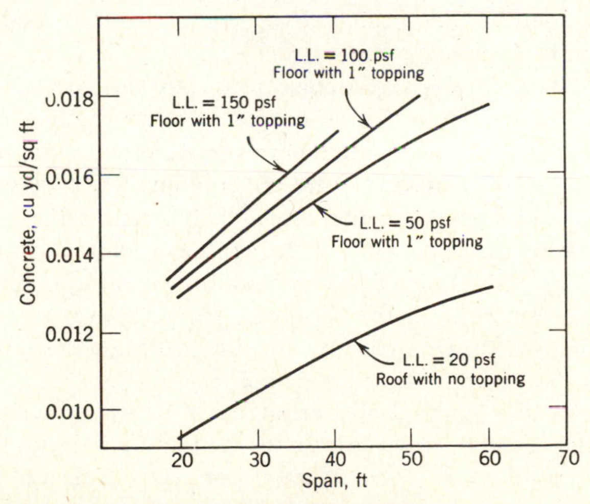

Fig. 15-4. Quantity of concrete for double-tee panels.

used. For long spans, prestressed sections are preferred, both for their lighter weight and for better deflection characteristics.

When used for floors, these panels are topped with a thin layer of cast-in-place concrete 1 to 3-in. (25.4–67.2 mm) thick. Such a topping is employed to smooth out any uneven camber, to encase conduits in the floor, and to provide a diaphragm for earthquake or wind loads.

Beams. Beams posttensioned in place are economical when carrying heavy loads over long spans. Precasting the beams in a plant or at the site will sometimes be found to be cheaper, if they can be transferred and lifted into position.

Joists of rectangular sections can be precast and erected to receive a poured-in-place slab. This is generally not as economical as single or double-tee panels discussed previously.

I-beams can be economically employed to support in-place floor slabs. One of the interesting features is the relative simplicity with which a number of large holes for ducts and conducts can be provided in the prestressed beam web, without damaging its shear strength, if properly designed.

Continuity for composite beams can be obtained in several ways. Often, mild steel bars are embedded in the concrete over the supports to carry the negative moments. This type of combined action, with mild steel reinforcing bars carrying the negative moments and the prestressing wires carrying the positive moments, is best designed by the plastic theory, with their deflections checked by the elastic theory. The usual values of allowable stresses may not serve as a proper guide. Fig. 10-5 shows several schemes for developing continuity.

Continuity for beams can also be achieved by posttensioning over the columns. Many methods have been employed, including the lapping of cables previously embedded in the beams. Excellent control of deflection can be obtained if properly designed and constructed.

Columns and Wall Panels. Although it is seldom necessary to prestress cast-in-place building columns, there are cases when such a design may be desirable and economical. If a column carries only direct compression, there is no need for prestressing. However, there are columns which serve as anchors—for example, of a cantilever anchor arm—then prestressing gives a definite hold-down and limits the deflections of the cantilever. Some columns carry high lateral loads and are therefore subject to bending, resulting in tensile stresses in certain portions. Prestressing is then desirable for both strength and rigidity.

Precast columns which are subject to bending during transportation can also be economically built of prestressed concrete, Fig. 14-7.

Wall panels can be economically prestressed to permit easier handling, to minimize cracking, as well as to increase the bending resistance.

Trusses. Pretensioned trusses of shorter spans have been used in Europe for buildings. While there is a great saving in materials, their formwork is costly and cannot be justified unless there are sufficient repetitions of the same design.

They may be economical for medium and long spans, say from 100 ft (30.5 m) up.

Tied arches can also be made of precast elements posttensioned together. The bottom chord is made of a posttensioned tie, while the vertical hangers are also of posttensioned concrete. Precast panels may be placed either on the top or the bottom chord. Thus a long span roof can be constructed with no interior columns.

Vierendeel trusses can also be posttensioned economically for long span roofs, particularly when the live load is light, and the top chord can be curved.

Conclusions. There are many different types of prestressed concrete construction available for buildings.[2,3] While it cannot be always cheaper than other materials, its adaptability will be greatly enhanced by the knowledge and experience of engineers and builders. Standardization of certain products will help to reduce the cost; but individual attention must always be given to each case, taking into account the local and specific requirements.

Finally, no building should be judged and designed solely on the basis of first cost. The cost of maintenance, of insurance, and of operation must be simultaneously considered. The value of the building, its earning power, its rental income, its appearance, its adaptability and durability all play an important part which should affect the decision of the engineer.

15-3 Connections for Buildings

In the design of precast-prestressed or in-place posttensioned buildings one should pay special attention to the detailing of the connections. In addition to the usual problems encountered in detailing a precast building, prestressing introduces additional ones. Prestressed elements tend to shrink and creep more than ordinary reinforced ones. Whereas reinforced members have numerous shrinkage cracks spread throughout their length, prestressed members concentrate their shrinkage and creep at the connections which are usually the weaker links. Prestressed members are more slender and subject to greater rotation at the ends. Under temperature variations, they also camber and deflect more than conventional designs. Hence the connections should be designed to permit both longitudinal and rotational movements. The Prestressed Concrete Institute has prepared a manual for such connection details.[4] It includes numerous designs for column base, beam-to-column, beam-to-girder, slab-to-beam, slab-to-wall, and other details. It is an excellent treatise, to which a designer should refer. Most of the following details are reproduced therefrom.

Fig. 15-5 shows some sample connections between different elements.

Fig. 15-6 shows details, including welding for hard connections and neoprene for soft connections—the latter to permit movement as often needed to allow for shrinkage, creep, and temperature effects.

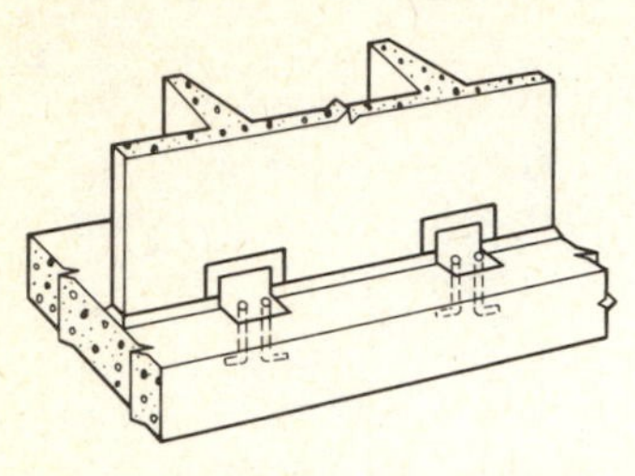

(a) Wall-Foundation Connection

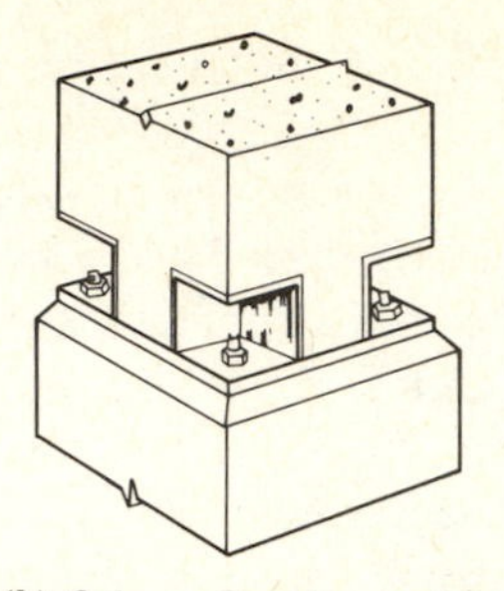

(b) Column-Base Connection

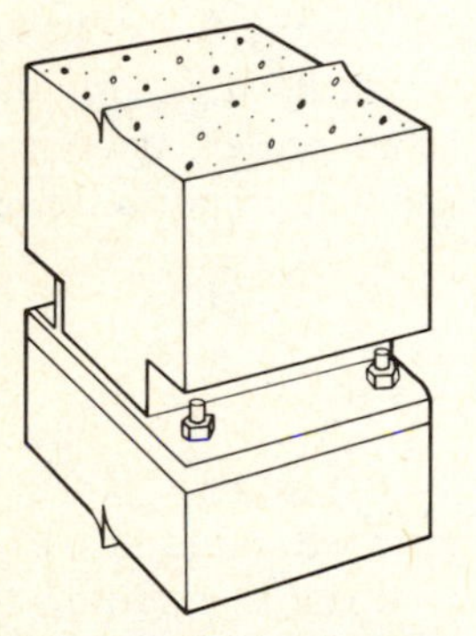

(c) Column-to-Column Connection

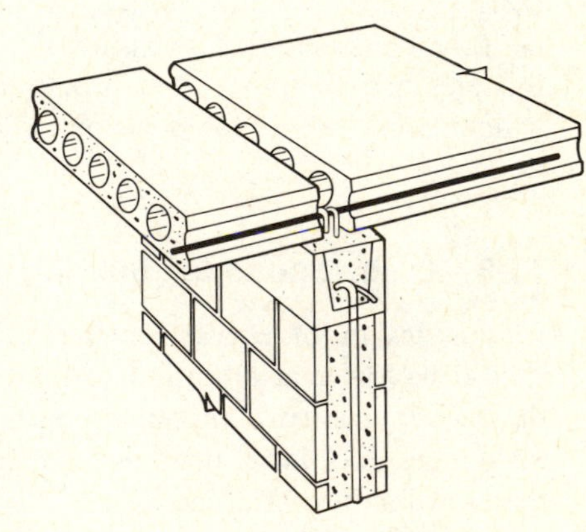

(d) Slab-Wall Connection

Fig. 15-5. Typical precast element connections.

As examples of reinforcement within members to transmit forces at connections, Fig. 15-7 suggests some approaches.

For in-place posttensioned slabs, it is often necessary to allow for elastic shortening of the slabs during posttensioning. Figure 15-8 shows an arrangement which will also take care of shrinkage and creep effect previous to the grouting of the joints.

To develop moment resistance for continuity, as well as to provide resistance to seismic forces, a number of joint details have been worked out be the FIP Commission on Seismic Studies.[5] Readers are advised to look up these for their reference.

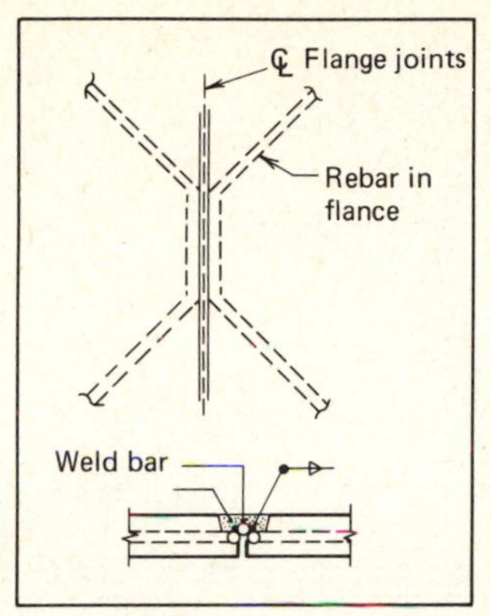

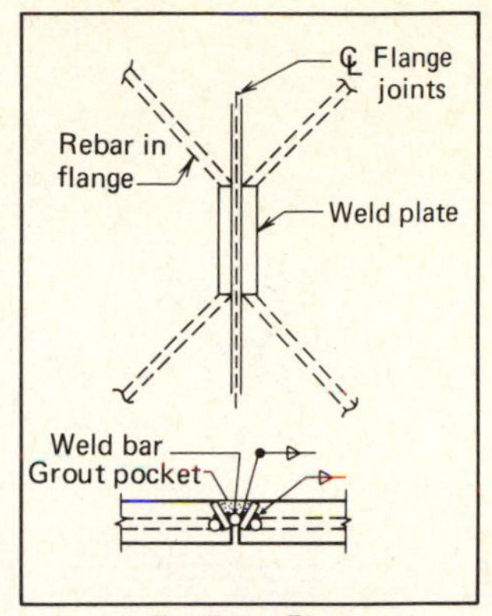

(*a*) Shear Connectors Between Tee or Double Tee Beam Flanges

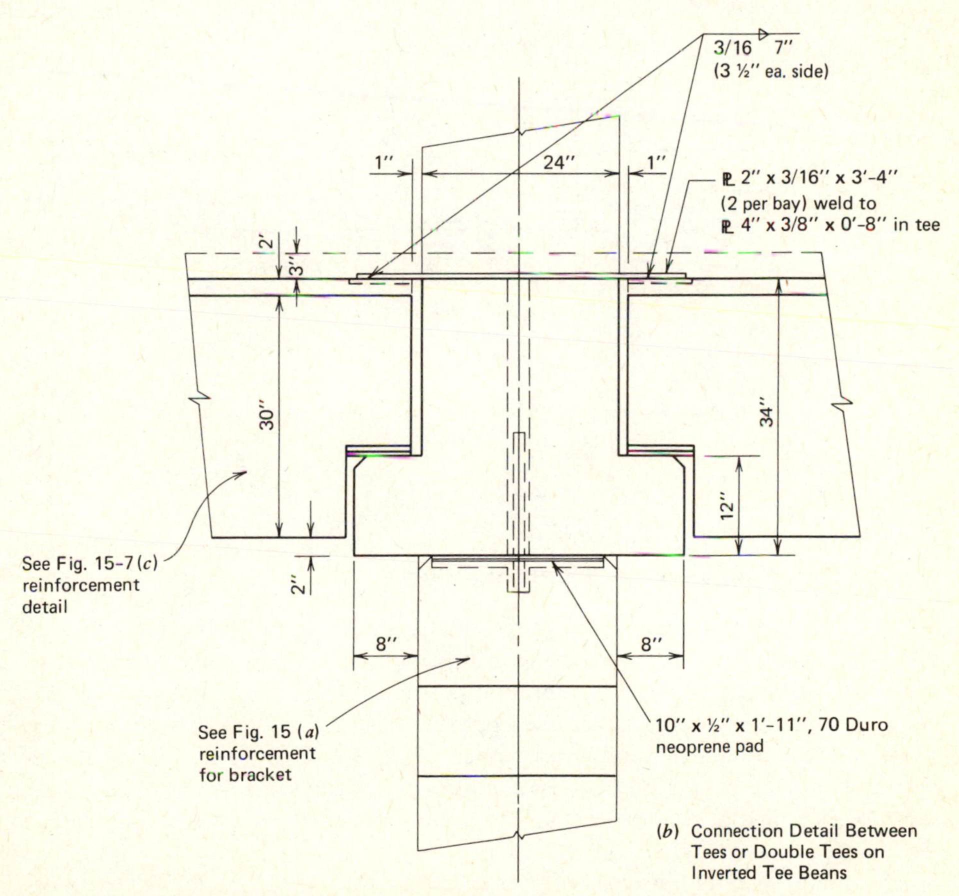

(*b*) Connection Detail Between Tees or Double Tees on Inverted Tee Beans

Fig. 15-6. Connection details for precast elements.

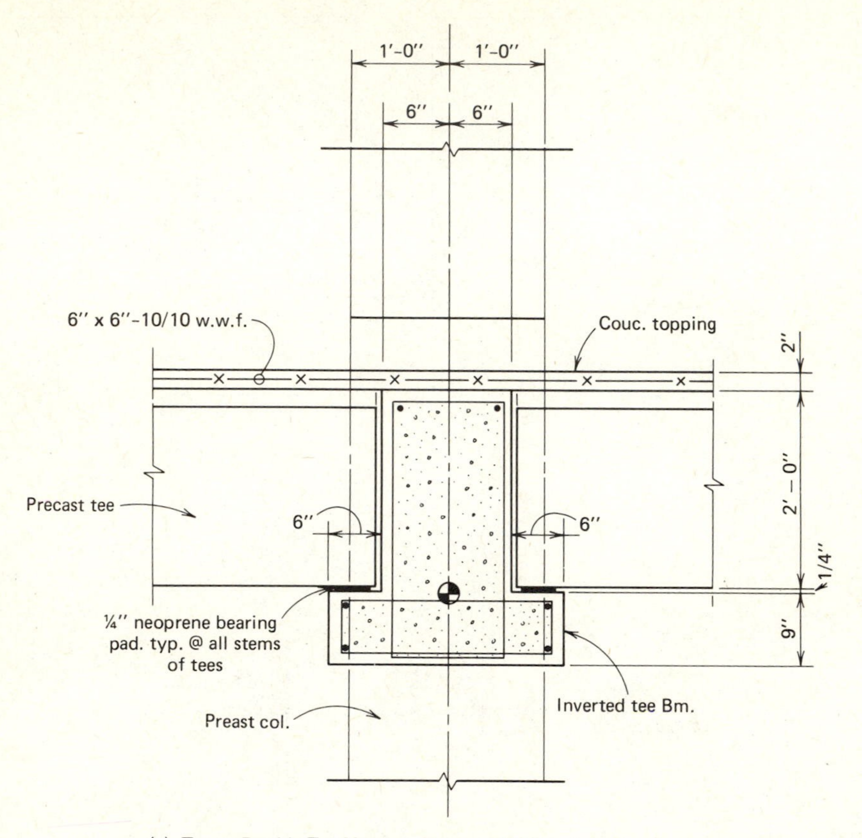

(*c*) Tee or Double Tee Members Beams—With In-place Topping

Fig. 15-6. (*continued*)

15-4 Thin Shells for Roof Construction

The prestressing of thin shell roofs opens up a completely new field for long-span construction, in which, with ingenuity, great economy can be developed. It is possible to use the minimum practical shell thickness, say 3 in. (67.2 mm), to span 200-ft (61 m) simple spans or 100-ft (30.5 m) cantilevers, completely eliminating structural interferences, Fig. 1-4, Fig. 11-25. The economy results from the balancing of gravity forces by prestressing leaving only direct compression in the concrete shell. The additional cost for prestressing steel is only a minor item (Fig. 15-9).

The major part of the cost of a prestressed thin shell lies in its formwork. Means must be devised to reduce its cost if economy in construction is to be

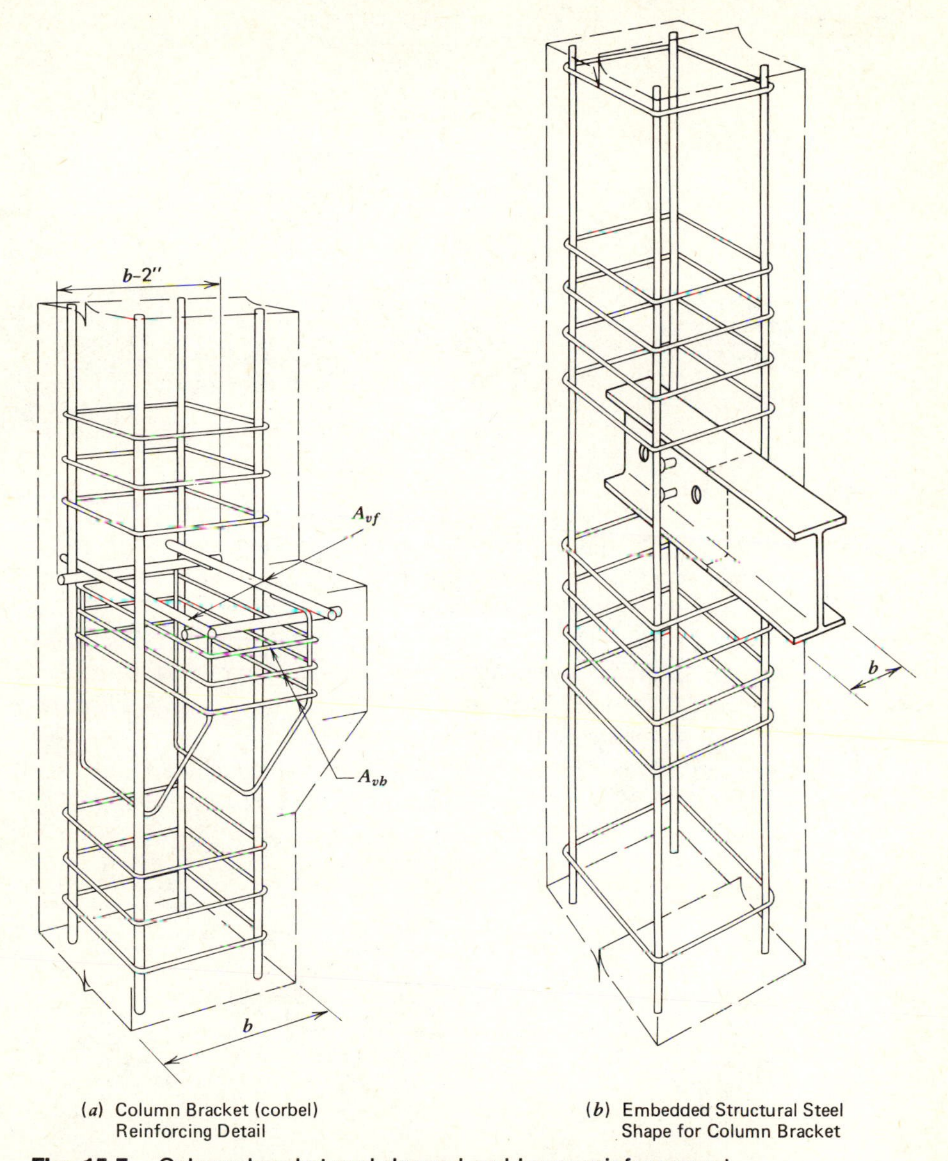

(*a*) Column Bracket (corbel) Reinforcing Detail

(*b*) Embedded Structural Steel Shape for Column Bracket

Fig. 15-7. Column bracket and dapped-end beam reinforcement.

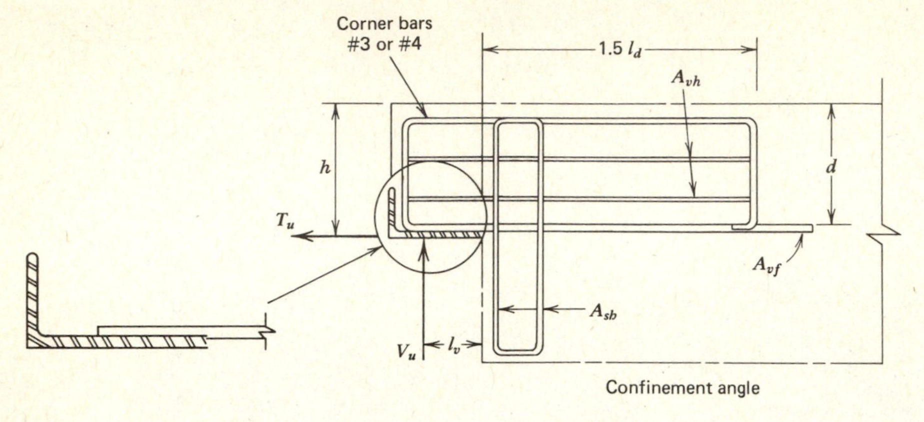

(c) Draped End Beam Reinforcement

Fig. 15-7. *(continued)*

obtained. This can be achieved in three ways. First, the shells can be precast on the ground so that a set of forms can be used many dozens of times. A long shell can be precast in parts and joined by posttensioning.

Secondly, traveling formwork can be used. This can be done for cylindrical or cloverleaf shells. Temporary shores can be installed at intervals to support the edge beams before the final posttensioning is applied. Domes and disks have been built of precast elements posttensioned together.[6]

The Oklahoma State Fair Arena, Oklahoma City,[7] has an elliptical roof shell of 400 ft by 320 ft (122 m by 97.5 m), Figs. 11-27 and 11-28. It was built of

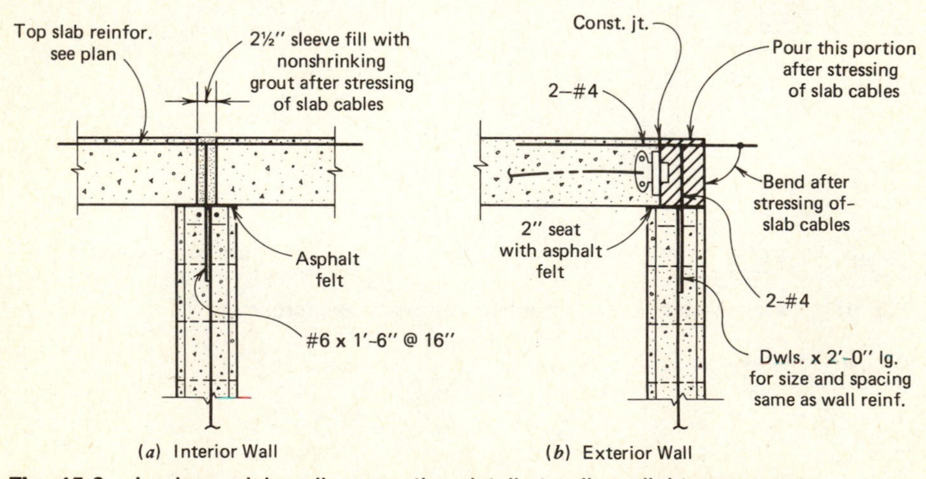

(a) Interior Wall (b) Exterior Wall

Fig. 15-8. **In-place, slab-wall connection details to allow slight movement.**

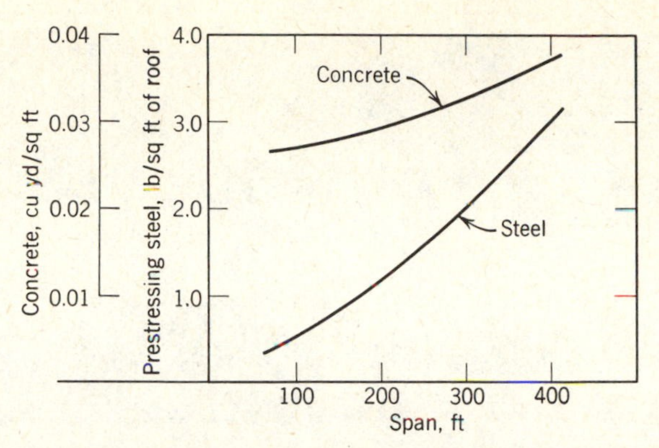

Fig. 15-9. Quantities of steel and concrete for prestressed cylindrical shell or folded-plate roofs.

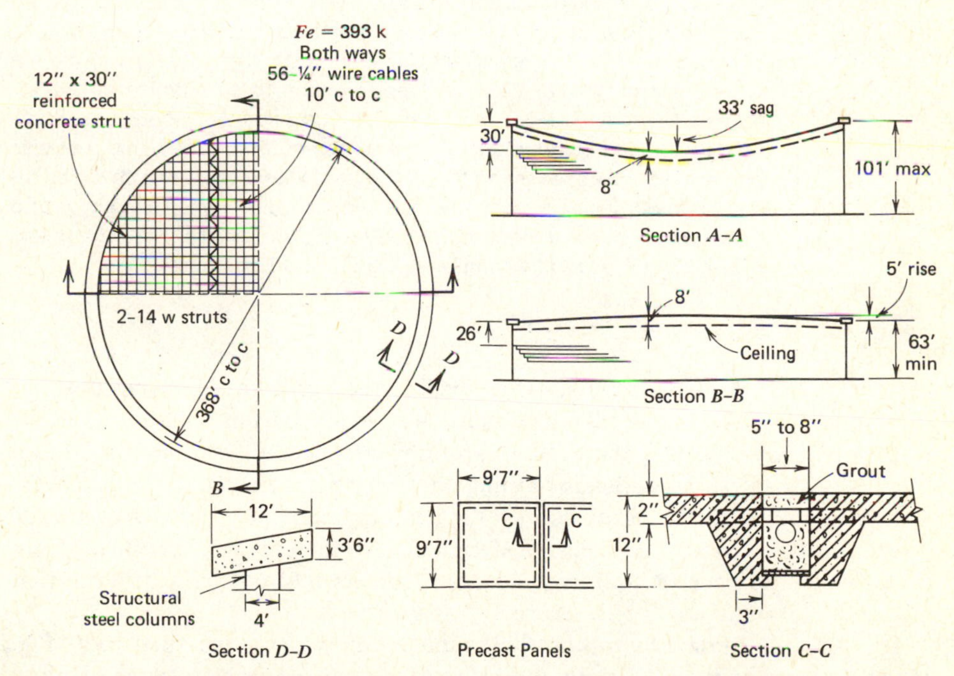

Fig. 15-10. Roof for Arizona State Fairgrounds Coliseum at Phoenix is a circular hyperbolic paraboloid with precast panels supported by posttensioned cables.

Fig. 15-11(a). Arizona State Fair Coliseum—Saddle Dome. This structure features an unusual saddle dome of hyperbolic paraboloid circular shape with a diameter of 380 ft. This shape is ideal to fit the seating around an oblong arena. It minimizes enclosed volume to economize air conditioning. It provides easy roof drainage and excellent acoustic properties. Using load-balancing method and computer analysis, this roof was built in 1965. By precasting the entire roof with a standard shape of 10 ft×10 ft waffles and posttensioning in two directions, simplicity and economy in construction was achieved.[7] (See details, Fig. 15-10 and Fig. 1-23.) (T. Y. Lin International, Consulting Engineers).

precast waffles 10 ft (3.05 m) square which formed grid lines housing posttensioning cables. One standard waffle panel was made into an elliptic parabolic surface by simply varying the width of grouting between adjacent panels. Thus a total of 72 cables, 32 and 40, running in two perpendicular directions were posttensioned in four stages to control the bending moment in the elliptical edge beam during erection. The shell was designed with the upward component from the cables to balance the full live and dead load on the roof. It is believed that this shell still holds the world record for its type.

The Arizona State Fairground Coliseum[7] is another unique shell roof, Fig. 15-10. It has a hyperbolic paraboloid surface on a circular plane, with a diameter of 380 ft (116 m), Fig. 15-11. It employed a similar design and construction technique like the previous elliptical roof. In this case, the cables

Fig. 5-11(b). Erection of precast panels (See details Fig. 15-10). Fig. 5-11 (continued) Arizona State Fair Coliseum.

sagging in one direction produced an upward vertical component, while humping cables in a perpendicular direction push downward. The sagging cables were sufficient to balance both the downward prestress and the external live and dead loads.

Thin cantilever shells of the hyperbolic paraboloid type can be economically built using posttensioned construction. An outstanding example is the Ponce Coliseum in Puerto Rico[8], Fig. 1-11. This 4-in. (102 mm) shell cantilevers 138 ft (42 m) with an edge beam whose depth decreases from 56 in. (1422 mm) near the support to 18 in. (457 mm) at the tip, Fig. 15-12. This shell was posttensioned along the diametrices of the H-P shell itself and was additionally posttensioned along the edge beams. A careful computer analysis brought out the fact that the edge beam carried only about one-sixth of its own weight in bending, while the remaining five-sixths was carried by the interaction between the edge beam and the shell. Similarly, the posttensioning forces applied along the edge beam was shared among the beam itself and the shell-beam interaction approximately in the ratio of 1:2. All of this illustrates that the application of posttensioning and precasting to thin shells is perhaps only beginning. Note particularly that posttensioning minimizes the deflection of the shell structure so that its secondary stresses can be greatly diminished.

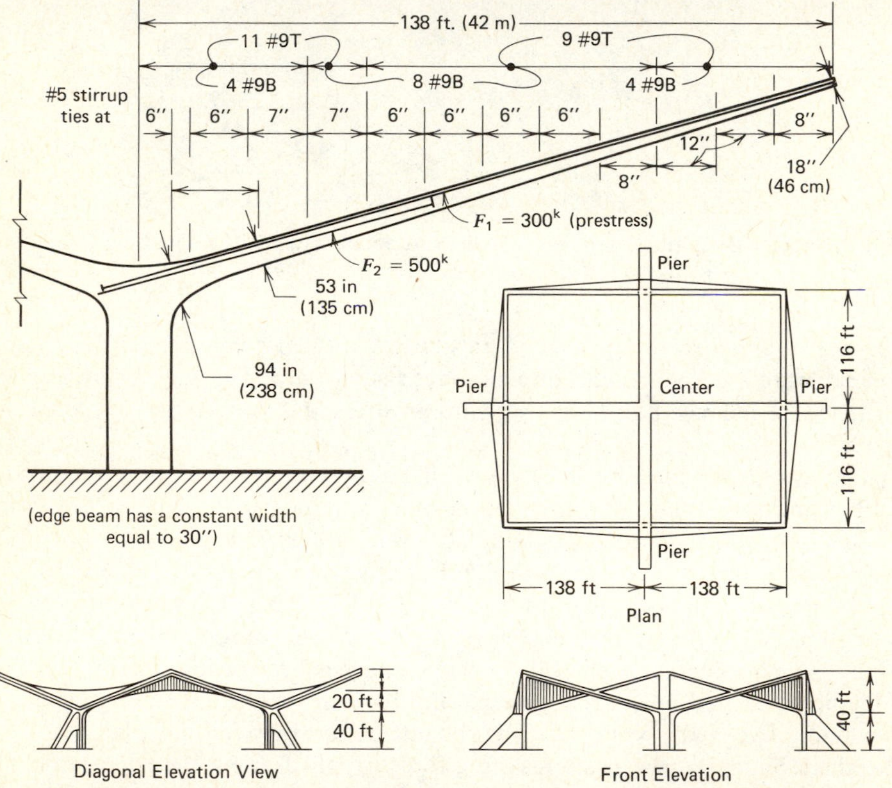

Fig. 15-12. Ponce Coliseum, Puerto Rico. This 10,000-seat multipurpose coliseum in Ponce, Puerto Rico, represents great progress in the design and construction of shell roofs. An area of 276 ft×232 ft was covered with a 4-in. hyperbolic paraboloid concrete shell supported on four columns only. The entire roof was posttensioned in place, with edge beams cantilevering 138 ft. (Refer to reference 8.) (See also Fig. 1-11, Chapter 1.) (T. Y. Lin International, Consulting Engineers).

15-5 Bridge Types and Layouts

The development of prestressed concrete design and construction, mostly pioneered in Europe, is spreading fast into the United States and all throughout the world.[9]

From 1950 to 1964, 85% of the total number of bridges built in West Germany were concrete, leaving only 15% to steel. The ratio by length of bridges was 75% for concrete and 26% for steel. It is also known that by far the greater part of the concrete bridges built in Germany since 1950 were of prestressed concrete, and not conventionally reinforced concrete. However, for this period, only 22% of bridges over a 330 ft (100 m) span has been built using concrete, versus 78% using steel. It was further estimated that for 2185 bridges built in Germany from 1974 to 1977, 90% are concrete and 10% are steel. Most of the prestressed portion was due to the increase in long-span bridges.

In the United States, prestressed concrete bridges occupy 27% of the total bridge construction in 1973, reinforced concrete 14% and steel bridges 59%. In the state of California, however, prestressed concrete bridges occupied 61% of constructed bridge area in 1972. While common spans for prestressed concrete were from 40 to 100 ft (12.2–30.5 m), they are getting into much longer spans starting with the 1970's.

Bridge sections in the United States vary approximately in the following order: I-sections, box sections, T-sections, hollow core, and channel sections. While there are inherent advantages to each section, its availability and local economic factors often determine the adoption of one or the other.

Precast bridge of short spans are generally simply supported, but they are often made continuous by placing reinforcement over the piers. For span length exceeding some 120 ft (36.6 m), continuous spans are definitely more economical. While truss bridges, arch bridges, rigid frame bridges, cantilever bridges, and suspension bridges have all been built of prestressed concrete, it seems clear that, at the present time, two types of bridge construction dominate the prestressed field: (1) segmental cantilever construction and (2) cable-stayed construction. For both types, either precasting or in-place slip-forming can be used to reduce forming costs to a minimum, while free cantilever construction extending from piers eliminates falsework requirements.

Cantilever Segmental vs. Cable-Stayed Bridges. Cantilever segmental construction with tendons buried in the top flange presents a neat elevation, either with a straight soffit or a curved one, varying according to the moment diagram. Note that the span of the Rio Higuamo Bridge is 623 ft (190 m), Fig. 15-13. The deck structure also adapts itself well to any width of roadway using cantilever segmental construction. It possesses a certain simplicity and straight-forwardness in erection and connections, but the amount of steel and concrete becomes excessive for extremely long spans.

Cable-stayed bridges require less material but their erection and connections are more complicated and they are limited to through bridges. The structural

Fig. 15-13. Double cantilever segmental construction using traveling wagon for in-place concreting. Rio Higuamo Bridge, Dominican Republic, main span 623 ft (190 m), 1972.

depth below the roadway is less than required for the cantilever type. Figure 15-14 shows the Brotonne cable stayed bridge in France with its record setting 1050 ft (320 m) span. This bridge and the Pasco-Kennewick cable stayed bridge in the United States (Fig. 1-31) with its 951-ft (290 m) main span illustrate the slender roadway. As mentioned before, both bridge types are built in one of the following ways:

1. Precast concrete segments, see Fig. 1-31(*b*).
2. In-place concreting, using horizontal slipforming, see Fig. 15-13(*a*).

The Rio Colorado Bridge was erected like a conventional suspension bridge, Fig. 15-15. In this case the cables are below the deck for the main span of 492 ft (150 m).

On account of its simplicity, segmental cantilever construction has taken over bridge spans from about 50 up to 240 m (164–787 ft), with the most economical spans generally limited to between 60 and 200 m (197 and 656 ft). When span exceeds some 200 m (656 ft), cable-stayed construction will likely be more

Fig. 15-14. Brotonne cable-stayed bridge near Roven, France, with 1050-ft (320 m) main span is the longest concrete span (Freyssinet International).

economical. If a through bridge is justified, cable-stayed type may be the better design even for some shorter spans.

Precast vs. Slip-forming. A comparison of precast segmental construction against in-place horizontal slip-forming would be interesting. It may be noted that precast segmental construction originated and prospered in France, whereas the carriage-type construction using in-place concreting was developed and widely applied in Germany. While each bridge site must be considered by itself, some general statements can be advanced as follows.

1. The amount of steel and concrete used for both the precast and the in-place construction is about the same. Therefore, the decision will depend upon the cost of equipment and labor for fabrication, transportation and erection.
2. Precast segmental construction would require a precasting yard which must be equipped for fabricating and handling. Then the segment must be transported to the site, and erection equipment provided thereon. To economically write off such plant and equipment normally requires multiple reuse.
3. In-place segmental concreting requires only horizontal slip-forming carriages. No expansive fabricating, transporting and erecting equipment are needed. On the other hand, the reuse of these vehicles may be limited by critical path schedule. Should many sets be needed, the cost can be too high.

Fig. 15-15. Upside down suspension bridge utilizes cables below the deck to support precast concrete slabs forming a platform for superstructure erection. Rio Colorado Bridge, Costa Rica, main span 492 ft (150 m), 1973 (T. Y. International, Consulting Engineers).

Cable-stayed vs. Suspension Bridges. Cable-stayed bridge are much more economical than conventional suspension bridges, for the following reasons.

1. The cost of stay cables is less than 50% of the cable cost for suspension bridges, due to a more direct load transmission and the higher allowable stress currently used for stay cables. The higher stress is partly justified by the relative rigidity of cable-stayed bridges.

2. Cable anchorages for suspension bridges are dispensed with in cable-stayed bridges and very compact sockets for gripping many wires have been developed.
3. The erection of cable-stayed bridges by free cantilevering is often easier than the cable-spinning process and truss erection for suspension spans.
4. The cost of stiffening truss and deck construction for suspension bridge of medium spans, say between 300 and 600 m (984 and 1968 ft), can exceed the cost of box sections for cable-stayed bridges.

Concrete vs. Steel Deck Girder for Cable-stayed Bridges. Can cable-stayed bridges be built more economically using steel orthotropic or concrete box sections? As far as the deck girder itself is concerned, concrete is almost always more economical than steel. However, more cables are needed to carry the heavier concrete deck—perhaps twice as much. For the concrete deck, the towers and foundations will also cost more. At the span range of 600-m (1968 ft), the added cost of cables and towers will just about offset the saving in the cost of the deck.

Since towers carry compression, concrete towers should cost less than steel towers. This is particularly true when using concrete deck girder, because the significant load increase over steel deck girder can be economically resisted by concrete towers acting in compression. Although 600 m (1968 ft) was mentioned as the limit for choosing concrete versus steel girders, it will be interesting to note the Ruck-a-Chucky Bridge, Fig. 1-32, which has a curved span of 400 m (1312 ft) hung with cables from the adjoining mountain slopes. Two complete designs for this bridge, one with a steel deck girder and another with a concrete deck girder, came out with identical estimated costs. The concrete deck girder itself costs $2.5 million less, but it requires more cables and anchorages costing $2.5 million more. On the surface, this would indicate the critical span between steel and concrete to be 400 m (1312 ft). But in this particular bridge, more cables are needed than a bridge of similar span. In fact, this bridge study would confirm the general conclusion that the span can be well over 600 m (1968 ft) before a steel deck will be cheaper.

Why have the majority of cable-stayed bridges now built incorporated steel girders rather than prestressed concrete? This, in the authors' opinion, was due to the lack of experience and understanding on the part of engineers in the design and erection of cable-stayed concrete girders as compared to steel. Steel is a material which takes both tension and compression in a straightforward manner, while concrete cannot take tension by itself. If the available knowledge and technology in prestressed concrete is properly applied to cable-stayed bridges, concrete construction can be competitive with steel up to about 600 m (1968 ft) and perhaps beyond.

Conclusion. A new era of medium and long-span concrete bridges has begun with widespread construction of double cantilever and cable-stayed structures.[10,11] The cantilever deck bridges possess simplicity in form and in construction, while the cable-stayed through bridges utilize minimum material for long spans. The use of precast segments, traveling wagons, and other methods will make these bridges competitive for a wide span range to meet various environmental requirements.

Of course, there are other types of concrete bridges, such as rigid frames, inclined pier arches, stressed ribbons, and inverted suspension bridges, each of which may fit special situations. But cantilever segmental and cable-stayed bridges will dominate the field for some time to come. Innovations, extensions, and optimizations for their design and construction are only beginning. As engineers from all countries are getting acquainted with these methods and details, we will see great advances ahead. The Ruck-a-Chucky hanging-arc bridge which spans 1312 ft (400 m) over the American River in California represents a most striking example of the potential for cable-stayed bridge construction for a unique situation where the stays were anchored to the rock walls of the canyon, Fig. 1-32.

15-6 Railroad Ties and Pavements

Railroad Ties. In the United States the first recorded use of concrete ties was in 1893 when 200 were installed by the Reading Company in Germantown, Pa.[12] An extensive series of laboratory investigation and field performance testing on prestressed-concrete railroad ties has been carried out (1961) by the American Association of American Railroads.[13,14] A preliminary study to replace one wood tie with one concrete tie, both spaced at 20 in. (508 mm), indicated satisfactory performance from a technical standpoint but was not desirable from an economic point of view. By increasing the concrete tie spacing to 30 in. (762 mm), it was found that the rail stresses would be increased by only 10%, that the tie loading would be increased from 40% of the axle load to 50%, and that the pressure of the 12-in. (305 mm) concrete tie on the ballast would be the same as that under the 9-in. (229 mm) wide-wood ties.

Consequently, experimental ties were designed and subjected to both static and repeated loads in the laboratory. Four $\frac{7}{16}$-in. (11.11 mm) 7-wire strands were used to apply a total initial prestress of 81,000 lb (360 kN) on the tie. An important feature of this tie design is its wedge shape for the central 3-ft (0.91 m) portion, reducing the direct bearing surface on the ballast to only 2 in. (50.8 mm). This practically eliminates "center binding," which results when the ballast supporting the tie under the rails is pounded and shifted until the tie is supported only at its midlength. It is also noted that the bottom surface of the tie under the rails is slightly concave, thus helping to hold the supporting ballast in place.

While these test ties made with lime-rock aggregates were not satisfactory, those made with granite were able to meet the design criteria of a static bending moment of 150,000 in.-lb (16.95 kN–m) at center line of rail and to sustain without failure 2,000,000 cycles of repeated load, producing a bending moment of 200,000 in.-lb (22.6 kN–m) at center line of rail. Suitable rail fastenings, such as rail clips, insulating pads, anchor bolts, and thimbles for electrical insulation were also developed.

There has been renewed interest in the use of prestressed concrete ties in the United States in the 1970's with a major research effort at the Portland Cement Association Laboratories. The extensive use of pretensioned concrete ties by railroads all over the world indicates that the concrete tie has become an important constituent of the modern railway track structure.[15] The growing use of concrete ties is justified by the greater consistency of the product quality, suitability of concrete ties for use with continuously welded rails, expected long service life, and reduced maintenance requirements.

The long-line method of manufacturing pretensioned prestressed concrete ties is used in most tie plants in North America, and the prestressed monoblock tie is today the most widely used type. It is estimated that by 1972 over 100 million prestressed monoblock ties were laid in tracks of railroads all over the world.[15] Figure 15-16 shows the tie configuration now in use for most prestressed ties. Use of high-strength concrete is absolutely necessary in conjunction with indented wires (or indented wire strands) with not over $\frac{3}{8}$ in. diameter for good performance of these ties in service.

Major installations where prestressed concrete ties with 24 in. or 26 in. spacings have served well in test sections are[15,16] the Alaska Railroad (1973), 100 miles north of Anchorage, Alaska; Chessie System (1974) near Lorraine, Va.; Santa Fe (1974) near Streator, Ill.; Norfolk and Western (1974) near Roanoke, Va.; and the Facility for Accelerated Service Testing of the U. S. Department of Transportation (1976) at Pueblo, Colo. All the ties are reported to be performing well, and the pad displacement between rail and tie has been minimized with improved details for the attachments.[15,16]

Airfield and Highway Pavements. Little has been done in the United States in the development of prestressed-concrete highway pavements. It is generally believed that, although cheaper in maintenance, a prestressed slab of 4 or 5 in. (102 or 127 mm) cannot compete in initial cost with the commonly used 8-in. (203 mm) pavement of plain concrete. This situation, however, may change if suitable design for prestressed pavements can be evolved and if economical methods of construction can be developed.

The foremost problem confronting the design of prestressed pavements is the loss of prestress through friction between the slab and the ground. Another problem is the devising of a reliable and economical expansion joint, although it is required only every few hundred feet. Construction costs could be lowered if

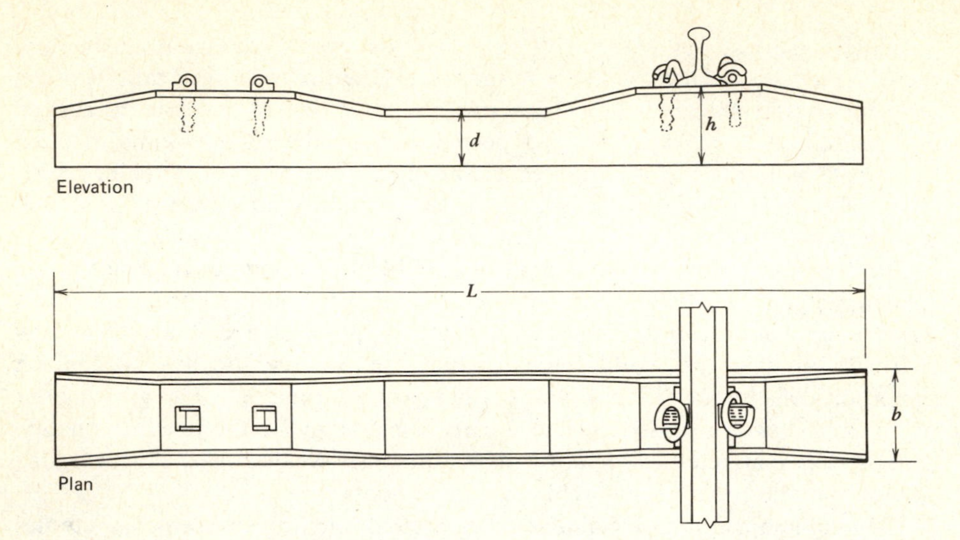

L 8′-3 to 9′-0″ (2.515 to 2.743 m)
b 10½ to 12″ (267 to 305 mm)
h 9 to 10″ (229 to 254 mm)
d 6 to 7½″ (152 to 191 mm)

Prestressing force
110 to 150 k (489 to 667 kN)

(typically 8-3/8 in. diameter prestressing strands)

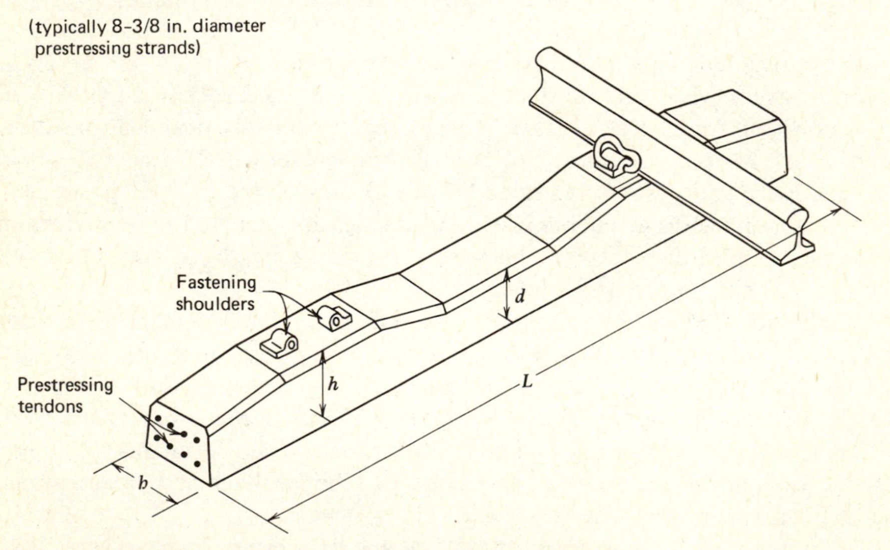

Fig. 15-16. Tie configuration and typical dimensions.[15]

cheap methods for prestressing the steel can be devised. It does appear that chemical prestressing by expansive cements may eventually provide a solution, since the labor for stressing operations is eliminated, and the friction developed during slab expansion could help to precompress the concrete.

For airfield pavements, an 8-in. (203 mm) prestressed slab could economically replace a 24-in. (610 mm) plain concrete one, if maintenance is taken into consideration. Prestressed-concrete pavements possess several advantages compared to a conventional type,[17,18] cracks are practically eliminated, the number of transverse joints are reduced, a more watertight pavement and a smoother surface are provided, and a longer life is expected. All these advantages point to further development of prestressed pavements.

Power Masts. While prestressed concrete poles and masts have found wide acceptance in European and Mediterranean countries, their adoption in the United States has so far been limited. Prestressed concrete masts are lighter than those of conventionally reinforced concrete, they are more durable than timber, and require less maintenance than steel. When produced on a massive scale, they will likely compete favorably with other conventional types.[19] An experimental tower for high voltage transmission up to 750,000 volts has been constructed by the General Electric Company.[20] This tower 100-ft (30.5 m) high is made of two columns of hollow box section 18×36 in. (457×914 mm), with 5-in. (127 mm) wall thickness. These columns are cross-braced by 12×12 in. (305×305 mm) solid sections.

References

1. "Tentative Recommendations for Prestressed Concrete Flat Slabs," ACI-ASCE Committee on Prestressed Concrete, *J. Am. Conc. Inst.*, Vol. 71, No. 2, February, 1974, pp. 74–91.
2. Alfred A. Yee, "Prestressed Concrete for Buildings," *J. Prestressed Conc. Inst.*, September–October 1976, pp. 112–156.
3. F. Kulka, T. Y. Lin, and Y. C. Yang, "Prestressed Concrete Building Construction Using Precast Wall Panels," *J. Prestressed Conc. Inst.*, Vol. 20, No. 1, January–February 1975. pp. 62–73.
4. "PCI Manuel on Design of Connections for Precast Prestressed Concrete," Prestressed Concrete Institute, 20 N. Wacker Dr., Chicago, Ill., 60606, 1974.
5. "Connections for Seismic Resistance," FIP Report, 1978.
6. Keith Thornton, "Posttensioned precast Thin Shell Concrete Dome for Garden State Art Center, New Jersey," paper SP28-9, Proceedings ACI Concrete Thin Shells Symposium, 1969; "Garden State Arts Center is Appropriately Artistic," *Eng. News-Rec.*, March 14, 1968.
7. T. Y. Lin and Ben Young, "Two Large Shells of Post-tensioned Concrete," *Civil Engineering*, July 1965, ASCE, pp. 55–59.
8. T. Y. Lin, F. Kulka, and K. Lo, "Giant Prestressed HP Shell for Ponce Coliseum," *J. Prestressed Conc. Inst.*, Vol. 18, No. 5, September/October 1973, pp. 60–72.

9. T. Y. Lin and Ben C. Gerwick, Jr., "Design of Long-span Concrete Bridges with Special Reference to Prestressing, Precasting, Erection, Structural Behavior, and Economics," Concrete Bridge Design Symposium, ACI, 1968, pp. 693–704.
10. T. Y. Lin and F. Kulka, "Fifty-year Advancement in Concrete Bridge Construction," *ASCE Journal of the Construction Division*, Vol. 101, No. C03, September 1975 pp. 491–510.
11. John J. Kozak and Thomas J. Bezouska, "Twenty Five Years of Progress in Prestressed Concrete Bridges," *J. Prestressed Conc. Inst.*, Vol. 21, No. 5, September–October 1976, pp. 90–110.
12. J. W. Weber, "Concrete Crossties in the United States," *J. Prestressed Con. Inst.*, Vol. 14, No. 1, February 1969, pp. 46–61.
13. "Prestressed Concrete Tie Investigation," Association of American Railroads, Engineering Research Division, *Report No. ER*-20, November 1961.
14. G. M. Magee and E. J. Ruble, "Service Tests of Prestressed Concrete Ties," *Railway Track and Structures*, September 1960.
15. A. N. Hanna, "State-of-the-Art of Prestressed Concrete Ties for North American Railroads," *J. Prestressed Conc. Inst.*, Vol. 24, No. 5, September–October 1979, pp. 32–61.
16. W. J. Venuti, "Concrete Railroad Ties in North America," *Concrete International*, Vol. 2, No. 1, January 1980, pp. 25–32.
17. F. M. Mellinger, "A Summary of Prestressed Concrete Pavement Practices," *Journal of the Air Transport Division* (*Proc. ASCE*, Vol. 87, No. AT2), August 1961.
18. J. P. McIntyre and F. M. Mellenger, "Prestressed Concrete Taxiway Biggs Air Force Base, Texas, U.S.A.," *Fourth Congress International Federation for Prestressing*, Rome, May 1962; C. F. Renz and P. L. Melville, "Experience with Prestressed Concrete Airfield Pavements in the United States," *J. Prestressed Conc. Inst.*, March 1961; Y. Osawa. "Strength of Prestressed Concrete Pavements," *Journal of the Structural Division* (*Proc. ASCE*, Vol. 88, No. 575) October 1962.
19. Robert J. D. Finfrock, "Prestressed Concrete Masts for Power Lines," *Proc. World Conference on Prestressed Concrete*, San Francisco, 1957.
20. M. Schupack, "Design on an Extra-high Voltage Transmission Tower 100 Feet High," *J. Prestressed Conc. Inst.*, February 1962.

16

DESIGN EXAMPLES

In order to illustrate the principles of analysis and design of prestressed concrete members discussed in previous chapters, four example problems are given in this chapter. The design examples are more complete than previous short example problems which illustrated specific points covered as material was presented in previous chapters. While these design situations are realistic and cover a range of types of flexural members, it is realized that they present only a small portion of the possible range of applicability of prestressed concrete to meet structural design requirements.

16-1 Design Example, Posttensioned, Multispan Continuous Bridge

This example is a redesign of the Harkness Avenue Pedestrian Bridge in San Francisco, California, which was initially discussed in Chapter 10, Examples 10-6 and 10-7. In the following revised design, the load-balancing approach to design of a continuous structure is illustrated. This redesign will show how the state-of-the-art was advanced to greatly simplify the design of continuous structures since the initial design was done about 20 years ago. Also, the economy in design which is accomplished by allowing higher tensile stresses in the concrete at service load, without sacrifice of adequate strength, is illustrated by comparison with the original design.

Figure 16-1 shows the three-span structure which consists of a one-way solid slab 13 in. (330 mm) thick carrying pedestrian traffic. As in Example 10-6, we will take the design dead load to be the weight of the 13-in. thick slab of normal weight concrete (162 lb./ft.^{2} = 7.76 kN/m^2) and the live load to be pedestrian loading (50 lb./ft.^{2} = 2.40 kN/m^2). The moment for a 1-ft (0.30 m) wide strip of the structure from the loading conditions for maximum and minimum moments are shown in Fig. 16-1(*b*). Concrete strength remains $f_c' = 5000$ psi (34 N/mm^2), and the grouted tendons will consist of $\frac{1}{4}$ in. (6.35 mm) diameter wires with $f_{pu} = 240$ ksi (1655 N/mm^2).

Solution.

Tendon Layout. The tendon layout will initially be assumed to consist of idealized parabolic tendons with 2.25 in. (57.2 mm) minimum cover from extreme fiber, top or bottom, to centroid of tendon (c.g.s.). For obtaining a trial layout, we will assume that the idealized tendon balances the same load in each

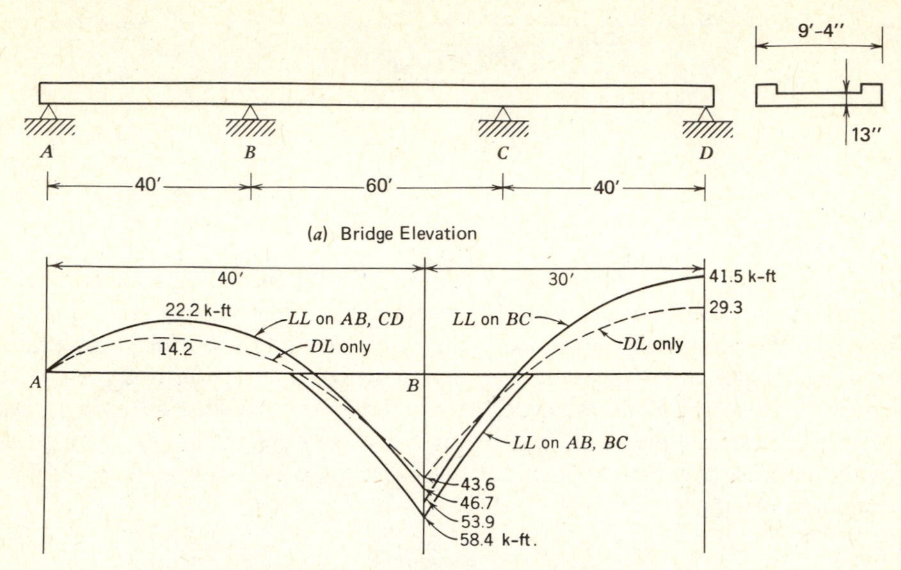

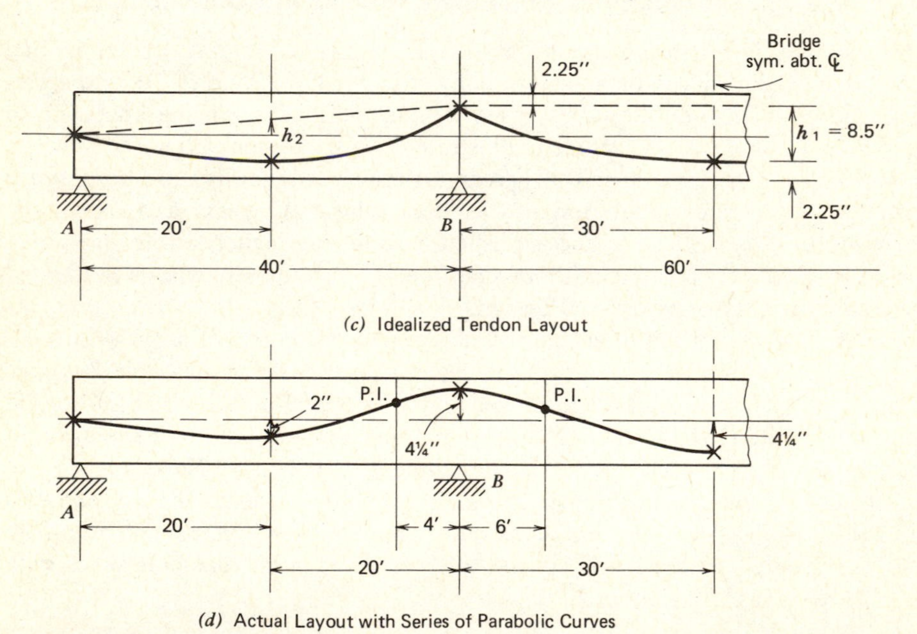

Fig. 16-1. Design Example 16-1.

span, and we will initially try balancing the full-slab dead load of 162 lb/ft (2.36 kN/m).

For the 60-ft (18.3 m) span *BC*, we can obtain *F* to balance this load with $h_1 = 8.5/12 = 0.708$ ft (0.216 m):

$$F = \frac{wL^2}{8h_1} = \frac{(0.162)(60)^2}{(8)(0.708)} = 103 \text{ k (458 kN)}$$

Balancing the same load (162 lb/ft) with $F = 103$ k in the outer 40 ft (12.2 m) spans, we find:

$$h_2 = \frac{wL^2}{8F} = \frac{(0.162)(40)^2}{(8)(103)} = 0.315 \text{ ft} = 3.78 \text{ in. (0.096 m} = 96 \text{ mm)}$$

At midspan *AB* (Fig. 16-1) we measure this h_2 for the parabolic drape from the chord shown as a dashed line and find the tendon eccentricity:

$$\text{at } \mathbb{C}\!\!\!\text{L} \text{ span } AB\text{: } e = 3.78 - \frac{4.25}{2} = 1.66 \text{ in. (42.2 mm)}$$

Try $e = 2.0$ in. (50.8 mm) as practical dimension.

$$\text{Cover provided} = 6.50 - 2.0 = 4.50 \text{ in. (114.3 mm)} > 2\tfrac{1}{4} \text{ in. (57.2 mm) minimum}$$

With this quick check on *F* and key control points along the layout indicated by *X* on Fig. 16-1(*c*), we may establish the actual trial layout of Fig. 16-1(*d*). The points of inflection are taken at $0.1L$ from support forming the smooth continuous curve which we would actually use for posttensioning with $F = 103$ k (458 kN). This layout along with resulting loads created by prestressing are shown in detail in Fig. 16-2.

Moments Due to Prestress. The loads which the tendon applies to the concrete are computed in Fig. 16-2 using the relationship $w = 8Fh/l_0^2$ where l_0 for each of the segments is twice the length of individual half parabolic segments making up the actual layout. We can use usual indeterminate structural analysis to solve the moment due to prestress at supports *B* and *C* for the continuous structure subjected to these loads. Convenient expressions have been developed which make it easy to obtain the fixed-end moments at *B* and *C* directly from the tendon layout and prestress force. With moment distribution the moments due to prestress can then be found by balancing these fixed-end moments. Of course, the same result is obtained by either method; and this analysis gives us the total moment due to prestress only. For this structure the moment at *B* is 486 in.-k = 40.5 ft-k (54.9 kN-m) due to prestressing the continuous 13-in. (330 mm)-thick slab with the tendon layout of Fig. 16-1(*d*).

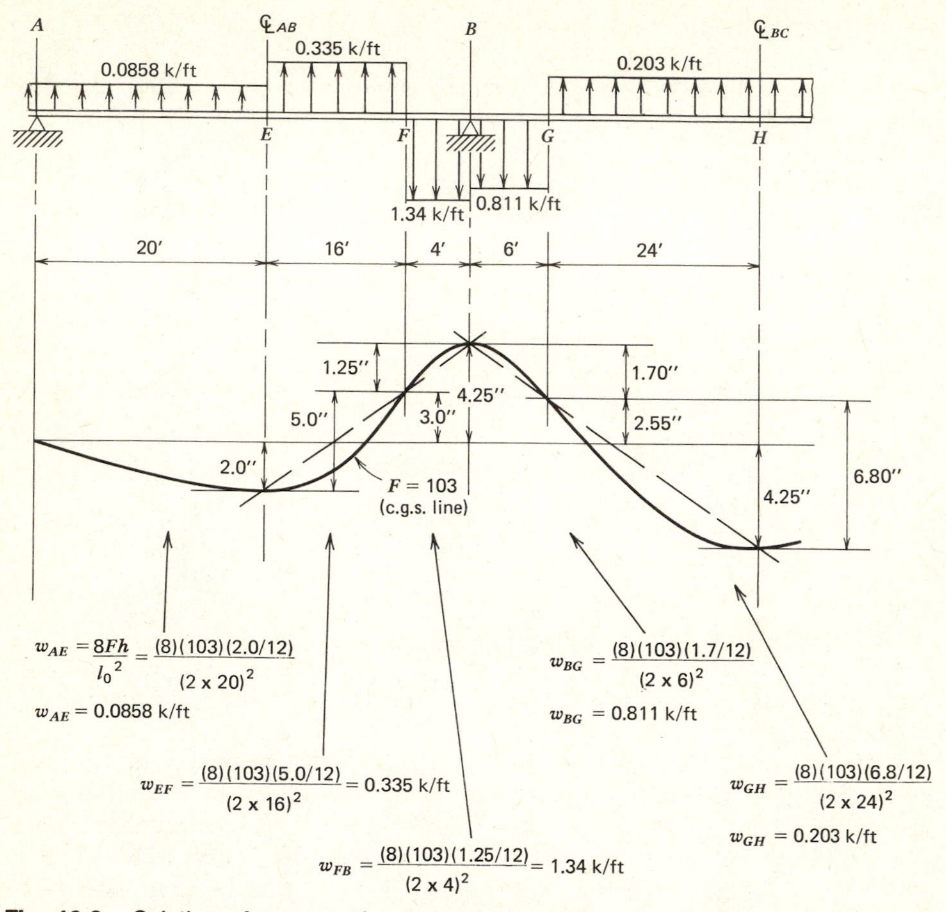

Fig. 16-2. Solution of moment due to prestress. From analysis of structure with this loading, we obtain $M_B = 486$ in-k $= 40.5$ ft-k due to posttensioning only.

The distance from centroid of section (c.g.c.) to the C-line will be located at B, and we can find whether posttensioning causes secondary moment.

$$\text{At } B \text{ distance to } C\text{-line} = \frac{\text{moment from prestress}}{F}$$

$$= \frac{486}{103} = 4.72 \text{ in. (120 mm)}$$

At B distance to c.g.s. of tendon $= 4.25$ in. (108 mm) < 4.72 in. (120 mm)

Thus this secondary moment results from posttensioning with now-concordant tendon.

$$M_{\text{sec at } B} = (103)(4.72 - 4.25) = 48.4 \text{ in.-k} = 4.03 \text{ ft-k (5.46 kN-m)}$$

Stresses Due to Prestress Combined with Worst Service Loadings. Combining the c.g.s. layout from Fig. 16-2 with the known C-line location at B, we can establish the C-line location along the structure as shown in Fig. 16-3. The moment from prestress follows the shape of the C-line. This C-line profile is easily established with respect to the known c.g.s. line using linear transformation as shown for key locations in Fig. 16-3. Combining moments from prestress with those from loading, Fig. 16-1(b), we can now find the net moment at key points along the span. The stress from this net moment is combined with the F/A stress to check extreme fiber stresses in the concrete.

Check of the slab stresses at B gives:

$$\text{Top}: f_B = -\frac{F}{A} + \frac{M_{\text{net}}}{S} = -\frac{103}{156} + \frac{(17.9)(12)}{338}$$

$$= -0.660 + 0.636 = -0.024 \text{ ksi } (-0.165 \text{ N/mm}^2) \text{ (comp.)}$$

$$\text{Bottom}: f_B = -\frac{F}{A} - \frac{M_{\text{net}}}{S} = -0.660 - 0.636 = -1.296 \text{ ksi } (-8.936 \text{ N/mm}^2) \text{ (comp.)}$$

Note: Since the net moments at other sections shown in Fig. 16-3 are of lesser magnitude than at B, the stress distribution will be less critical than that at B as shown with maximum $f_c = 1.296$ ksi (-8.936 N/mm^2).

If the concrete strength is specified as $f_c' = 5000$ psi (34 N/mm^2) we would allow $f_c = 0.45 \times 5000/7000 = 2.25$ ksi (15.51 N/mm^2) $>$ 1.296 ksi (8.936 N/mm^2). We might use f_c' less than 5000 psi (34 N/mm^2) without overstress (for 4000 psi = 28 N/mm^2 concrete, $f_c = 0.45 \times 4000/1000 = 1.80$ ksi = 12.41 N/mm^2 $>$ 1.292 ksi = 8.936 N/mm^2). Stresses in solid slabs tend to be of lower magnitude than we might find for beams and the average $F/A = 660$ psi (4.56 N/mm^2) for this slab is higher than would be expected in slabs for many typical building situations.

To illustrate the use of the C-line from Fig. 16-3 in getting net moments and resulting stresses at other points away from the support, we will check the distribution of stresses over the depth of the slab at E and H. As shown in Fig. 16-3, the moment due to prestress is found as the product of prestress force and distance to the C-line. The C-line is linearly transformed from the c.g.s. layout which was previously established (Fig. 16-1).

At point H (midspan BC) we find the C-line shifts upward 0.47 in. (11.9 mm) from the c.g.s. line at interior supports B and C. Thus the distance to the C-line at any point within span BC is adjusted by this shift of 0.47 in. (11.9 mm) and at point H the distance to the C-line is $4.25 - 0.47 = 3.78$ in. (96.0 mm) as shown in Fig. 16-3. The resulting moment from prestress is of magnitude $103 \times 3.78 = 389$ in.-k (44.0 kN-m), and with the signs used for our bending moments this is negative moment (-389 in.-k $= -32.4$ ft-k $= 44.0$ kN-m). The worst loading condition for moment at point H is a pattern loading with dead load on all

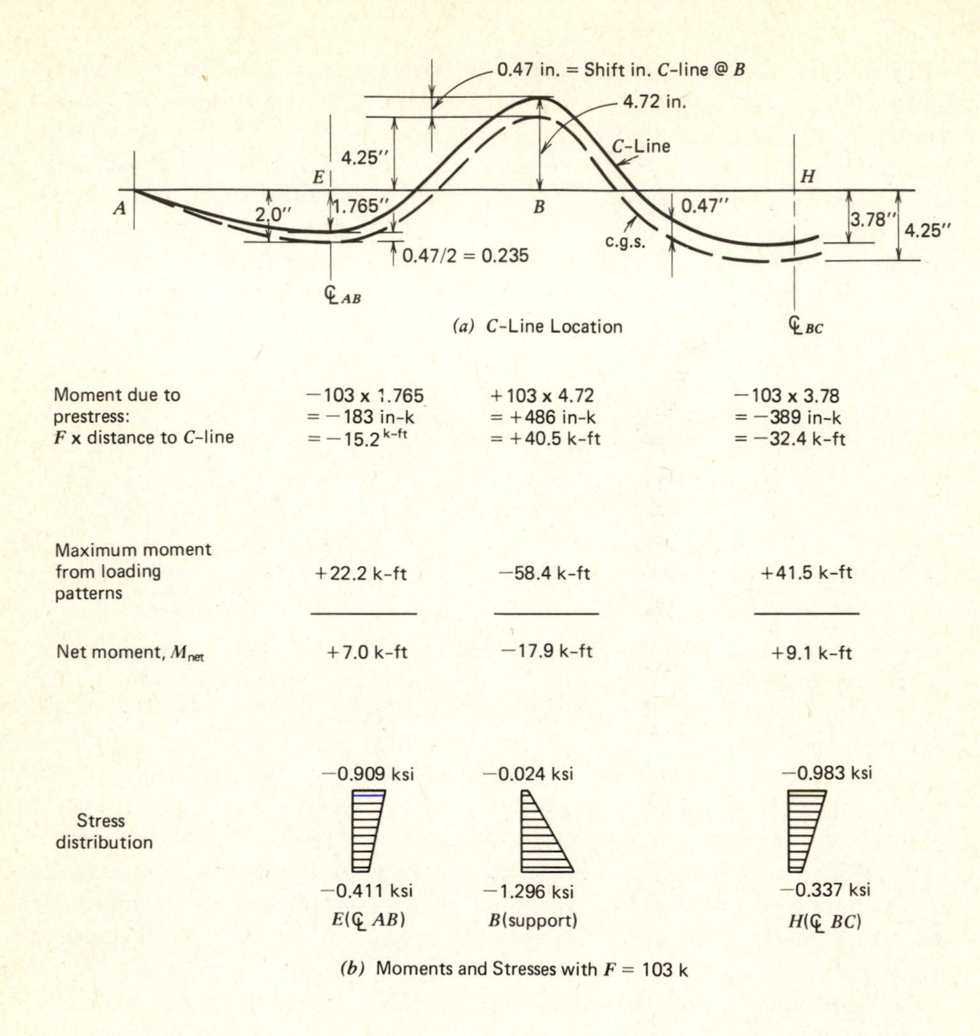

(a) C-Line Location

(b) Moments and Stresses with $F = 103$ k

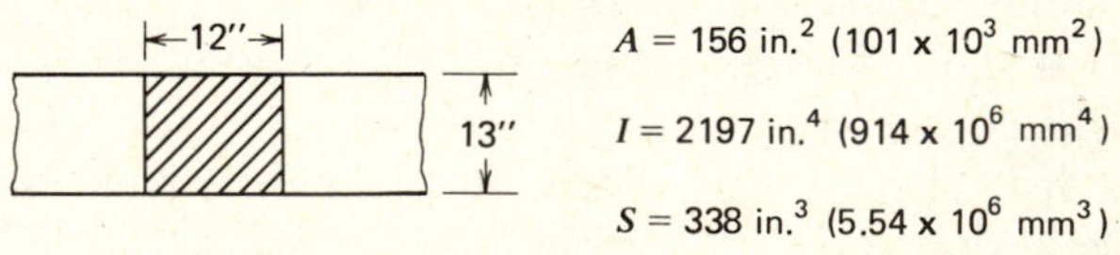

$A = 156$ in.^{2} (101×10^3 mm^2)

$I = 2197$ in.^{4} (914×10^6 mm^4)

$S = 338$ in.^{3} (5.54×10^6 mm^3)

(c) Section Properties for 1′ Wide Strip Analyzed

Fig. 16-3. Moments and stresses from analysis.

spans and live load only on span BC, resulting in moment of $+41.5$ ft-k ($+56.3$ kN-m) at H. Thus the net moment at H is $+41.5-32.4=9.1$ ft-k (12.3 kN-m) and the concrete stresses at H are:

$$\text{Top: } f_H = -\frac{103}{156} - \frac{(9.1)(12)}{338} = -0.660-0.323 = -0.983 \text{ ksi } (-6.78 \text{ N/mm}^2) \text{ (comp.)}$$

$$\text{Bottom: } f_H = -\frac{103}{156} + \frac{(9.1)(12)}{338} = -0.660+0.323$$

$$= -0.337 \text{ ksi } (-2.32 \text{ N/mm}^2) \text{ (comp.)}$$

At point E the shift of the C-line away from the c.g.s. line is half that at support B ($0.47/2=0.235$ in. $=5.97$ mm) as shown in Fig. 16-3. Thus the distance to the C-line at E is $2.0-0.235=1.765$ in. (44.83 mm) and the moment due to prestress is of magnitude $103\times 1.765=182$ in.-k (20.6 kN-m). This is negative moment of -182 in.-k $=-15.2$ ft-k (20.6 kN-m) at E (midspan AB) due to prestress. For the worst loading condition (live load on spans AB and CD with dead load on all spans) we find the maximum positive moment to be $+22.2$ ft-k ($+30.1$ kN-m). While this maximum positive moment from loading is actually slightly to the left of midspan, Fig. 16-1(b) it will be on the safe side to combine it with the moment due to prestress to obtain the net moment at E as $+22.2-15.2=+7.0$ ft-k (9.5 kN-m). The resulting concrete stresses at E are

$$\text{Top: } f_E = -\frac{103}{156} - \frac{(7.0)(12)}{338} = -0.660-0.249 = -0.909 \text{ ksi } (-6.27 \text{ N/mm}^2) \text{ (comp.)}$$

$$\text{Bottom: } f_E = -\frac{103}{156} + \frac{(7.0)(12)}{338} = -0.660+0.249$$

$$= -0.411 \text{ ksi } (-2.83 \text{ N/mm}^2) \text{ (comp.)}$$

Comparing the tendon force $F=103$ k/ft (1503 kN/m) obtained in this design by the load-balancing approach with the $F=132$ k/ft. (1926 kN/m) used in example 10-6 shows about 22% less prestress, which would be more economical than the original design. The load-balancing design procedure has the advantage of providing deflection control by balancing the dead load (based on the idealized layout). The span/depth ratio of the 13 in. (330 mm) solid slab is about 55 for the center span and about 37 for the outer spans. The table of approximate limits for span/depth ratios at the end of Chapter 8 shows that we are probably close to the limiting slenderness for one-way solid continuous slabs and the load balancing of the slab dead load is instrumental in preventing deflection problems.

Examination of the stress at B with $F=103$ k $=458$ kN (balanced 100% of slab DL) shows that we might allow slightly more design live load without getting too much tension stress in the concrete. Alternatively, if the design loads used are kept unchanged, we might balance less than 100% of slab dead load. For example, we might balance 85% of dead load $=(0.85)(162)=138$ k/ft (6.86 kN/m^2), giving us

$$F=\frac{wL^2}{8h_1}=\frac{(0.138)(60)^2}{(8)(0.708)}=88.0\text{ k }(391\text{ kN})$$

The moment due to prestress would change in direct proportion to F if we maintain the same layout shown in Figs. 16-2 and 16-3. Thus we may revise the moment at B due to this prestress force:

$$\text{Moment at } B \text{ from prestress}=\frac{88}{103}\times 486=415\text{ in.-k}$$

$$=34.6\text{ ft-k}=46.9\text{ kN-m (or we may use } F\times\text{dist. to the } C\text{-line}$$

$$=88\times 4.72=415\text{ in.-k}=34.6\text{ ft-k}=46.9\text{ kN-m)}$$

Thus the resulting net moment at B becomes

$$M_{\text{net}}=58.4-34.6=23.8\text{ ft.-k }(32.3\text{ kN-m})$$

The stress distribution at B would become (with $F=88$ k $=391$ kN)

$$\text{Top: } f_B=-\frac{88}{156}+\frac{(23.8)(12)}{338}=-0.564+0.845$$

$$=\underset{\text{(tension)}}{+0.281\text{ ksi}}\ (+1.937\text{ N/mm}^2)\text{ less than } 6\sqrt{f_c'}=\underset{\text{(o.k.)}}{0.424\text{ ksi}}\ (2.923\text{ N/mm}^2)$$

$$\text{Bottom: } f_B=-\frac{88}{156}-\frac{(23.8)(12)}{338}=-0.564-0.845$$

$$=\underset{\text{(comp.)}}{-1.409\text{ ksi}}\ (-9.715\text{ N/mm}^2)\text{ less than } 0.45f_c'=\underset{\text{(o.k.)}}{2.25\text{ ksi}}\ (15.51\text{ N/mm}^2)$$

Thus with $f_c'=5000$ psi (34 N/mm^2) and $F=88$ k/ft (1284 kN/m) we don't get any overstress at B where the largest net moment occurs. By inspection we note that all other sections will have less severe stresses than at B and will not be rechecked with $F=88$ k (391 kN) rather than $F=103$ k (458 kN) (see Fig. 16-3).

Check on Strength for Factored Loading. For this problem we will assume that $F\cong 103$ k (458 kN) is supplied by tendons at 12 in. (304.8 mm) spacing which consist of several $\frac{1}{4}$-in. (6.35 mm) diameter wires with $f_{pu}=240$ ksi (1655 N/mm^2). Using the allowable steel stresses of the ACI Code, we can show that 14 wires ($A_{ps}=0.687$ in.$^2=443$ mm^2) would allow an initial prestress force of 115 k (512 kN) at the time of stressing the tendon (0.7 f_{pu}) and after losses this would be reduced to about 92 k (409 kN). This is within the range of prestress force which we checked above (88 k/ft to 103 k/ft i.e., 1284 kN/m to 1503 kN/m) without exceeding allowable stresses in the concrete at service load with the worst pattern of loading. Also, it can be found from the manufacturer's literature that the diameter of a duct for the grouted tendon consisting of $14-\frac{1}{4}$ in. diameter wires will be approximately 2 in. (50.8 mm). The minimum 1 in. (25.4 mm) cover required by the ACI Code (Section 7.7.3.1) for prestressed slabs exposed to the weather would be satisfied with our allowance of $2\frac{1}{4}$ in. from the c.g.s. centerline to the outside face of concrete (see Fig. 16-4).

$$A_{ps} = 0.687 \text{ in. } (443 \text{ mm}^2)$$

$$\rho_p = \frac{A_{ps}}{bd} = \frac{0.687}{(12)(10.75)} = 0.00533$$

$$\omega_p = \frac{\rho_p f_{ps}}{f'_c} = \frac{0.00533 \times 209}{5.0} = 0.223 < 0.30$$

$$f_{ps} = f_{pu}\left(1 - 0.5\rho_p \frac{f_{pu}}{f'_c}\right)$$

$$= 240\left(1 - 0.5 \times 0.00533 \times \frac{240}{5.0}\right)$$

$$f_{ps} = 209 \text{ ksi } (1441 \text{ N/mm}^2)$$

The ω_p satisfies ACI Code requirements, assuring ductility of the cross section. Note that $\omega_p = 0.223$ will not qualify for moment redistribution in checking strength under the ACI Code. We might also note that we are on the safe side in

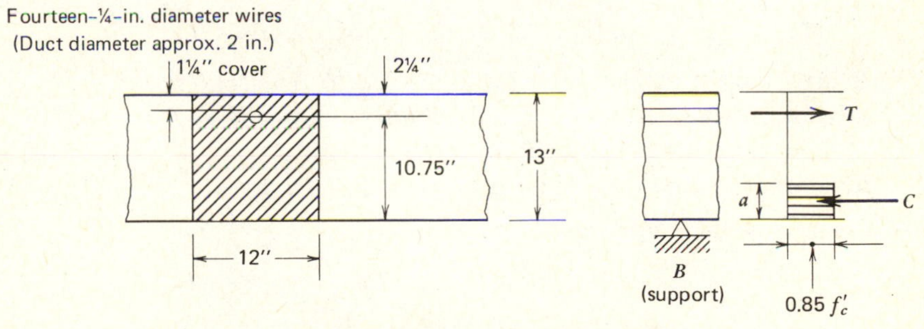

Fig. 16-4. Cross section at *B* for check on strength.

design to neglect the redistribution, but actual behavior of an under reinforced section with $\omega_p = 0.223$ would exhibit some rearrangement of elastic moments (and thus carry some additional load) before failure.

Using the moment couple of Fig. 16-4 we will evaluate the ultimate moment capability of the cross section at B.

$$T = A_{ps} f_{ps} = (0.687)(209) = 144 \text{ k } (640.5 \text{ kN})$$

$$C = T = 144 \text{ k} = 0.85 f_c' ab$$

With $b = 12$ in. (304.8 mm) and $f_c' = 5000$ psi $= 5.0$ ksi (34 N/mm^2)

$$a = \frac{144}{(0.85)(5.0)(12)} = 2.82 \text{ in. } (71.6 \text{ mm})$$

$$M_n = -(144)\left(10.75 - \frac{2.82}{2}\right) = -1345 \text{ in.-k}$$

$$= -122 \text{ ft-k } (-152 \text{ kN-m})$$

For flexure the ACI Code specifies the strength reduction factor $\phi = 0.9$, thus we may count the following moment capacity for this cross section at ultimate:

$$M_u = -(0.9)(112) = -101 \text{ ft.-k } (137 \text{ kN-m}) \text{ (provided)}$$

Now we must evaluate the maximum moment at B with $1.4D + 1.7L$ as shown in Fig. 16-5. The maximum moment from factored loading gives the following

$$M_B \text{ at ultimate} = (1.4)(-43.6) + (1.7)(-18.4)$$

$$M_u = -86.2 \text{ ft-k } (-117 \text{ kN-m})$$

where (moments at B)

$$M_D = -43.6 \text{ ft-k } (-59.1 \text{ kN-m}) - \text{dead load all spans}$$

$$M_L = -18.4 \text{ ft-k } (25.0 \text{ kN-m}) - \text{live load spans } AB \text{ and } BC$$

Combining the secondary moment at B due to prestress as required by the ACI Code before evaluating strength [secondary moment at B $103 \times 0.47 = 48.4$ in.-k $= +4.0$ ft-k $= 5.4$ kN-m based on the prestress force of 103 k and the shift of C-line above the c.g.s. line at B from Fig. 16-3(a)] gives

$$\text{Required } M_u = -86.2 + 4.0$$

$$= -82.2 \text{ ft-k } (-111.5 \text{ kN-m}) < -101 \text{ ft-k } (-137 \text{ kN-m})$$

The strength at B is more than adequate for this worst loading condition without moment redistribution.

The check on stresses summarized in Fig. 16-3 indicates clearly that the worst loading condition for this structure involves the maximum moment at B (Fig. 16-5) and we will not carry the strength analysis further for this example. It can be shown that with factored live load in span BC and factored dead load on all

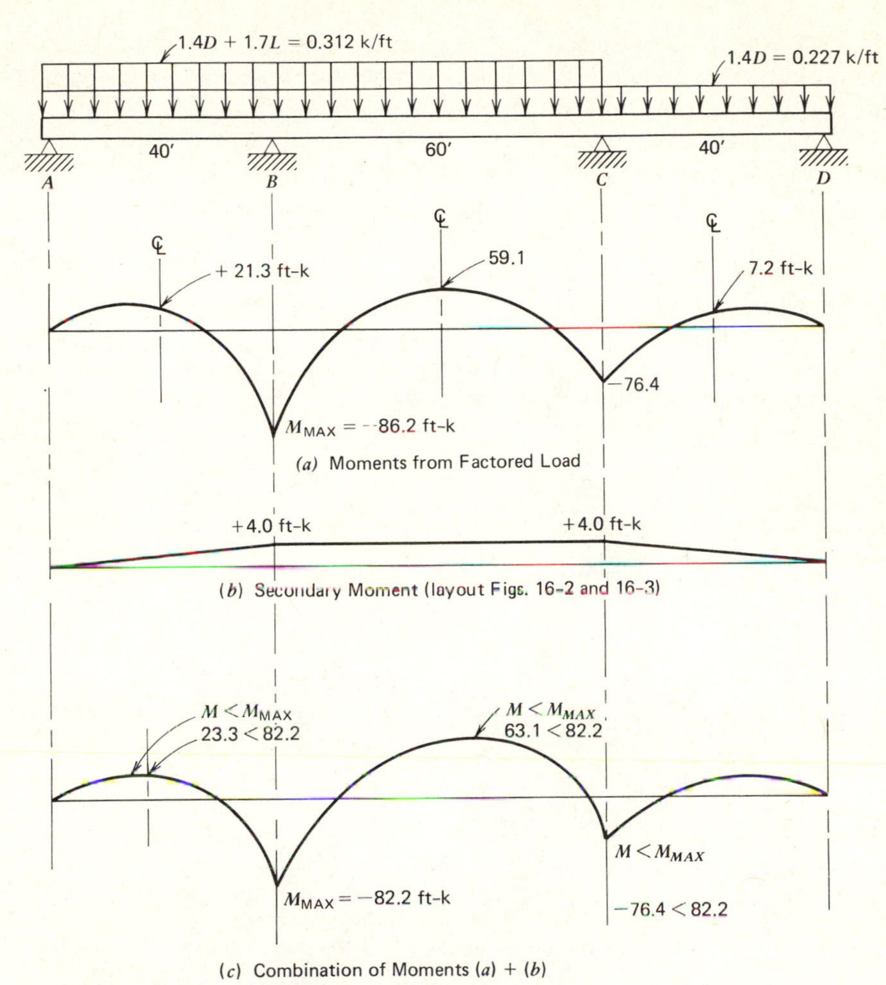

Fig. 16-5. Maximum moment at *B* for strength design.

spans combined with secondary moment, the maximum positive moment at *H* (midspan *BC*) is less than that found above (82.2 ft-k = 111.5 kN-m) for maximum negative moment. The cross-section is identical at *B* and *H* (inverted) and we obviously have ample strength for this loading. Similarly, the maximum positive moment from partial loading of the outside spans is much less than the strength of the cross section at *E*.

Shear strength of this solid slab is more than adequate without stirrups. The maximum shear of 9.5 k (42.3 kN) accompanies the loading of Fig. 16-5. Thus, $V_{u/\phi} = 9.5/0.85 = 11.2$ k (49.8 kN) while the minimum $V_n = 1.7\sqrt{f_c'}\ bd = 15.5$ k

(68.9 kN) exceeds this without analysis for V_{ci} and V_{cw}. Some local reinforcement would be provided in the vicinity of the anchorages but there is ample space for bearing plates (of size approximately 6 in. × 8 in. i.e., 152 mm × 203 mm) at midheight of the 13-in. (330 mm)-thick slab at the ends.

We may note that the strength check as well as the check on stresses would allow the live load to be more than the 50 lb/ft^2 (2.40 kN/m^2) assumed in the design of the Harkness Avenue pedestrian bridge. Some building codes might specify a slightly higher design live loading, which could be accommodated by the structure as revised. The economy in design and simplicity of analysis for the structure using load-balancing principles should again be noted. We obviously have less prestress force than the original design (example 10-6) leading to less tendon steel, but we still have more than ample strength.

16-2 Design Example, Pretensioned Roof Beam

This example illustrates the design of a tapered roof beam, precast and pretensioned. The design was used in a warehouse, San Diego, California, Fig. 16-6.

Fig. 16-6. Erecting a 66-ft pretensioned roof beam (Southwest Structural Concrete Corp., San Diego, California).

One of the beams was tested to failure and showed a factor of safety of 1.3 (girder + $D + L$) against cracking and a factor of 2.2 (girder + $D + L$) at ultimate.

This precast pretensioned roof beam is to span a clear distance of 64 ft (19.5 m) with an additional length of 1 ft (0.30 m) over each support. The beam carries a superimposed dead and live load of 520 plf (7.59 kN/m), in addition to its own weight. The top chord of the beam is to be tapered so that the depth of beam is 40 in. (1016 mm) at midspan and 16 in. (406 mm) at the ends, the bottom of the beam to remain straight. Use concrete with $f'_c = 4500$ psi (31 N/mm^2) at 28 days and $f'_{ci} = 4000$ psi (28 N/mm^2) at transfer. Steel for pretensioning will have an ultimate strength of 200,000 psi (1379 N/mm^2) ($\frac{3}{8}$-in. (9.525 mm) wires with Dorland anchorages for pretensioning). Allowable stresses are as follows.

	At Transfer	Under Working Load
Concrete		
Compressive fiber	$f_b = 0.60 f'_{ci} = 2400$	$f_t = 0.45 f'_c = 2025$ psi
Tensile fiber	$f'_t = 0.06 f'_{ci} = 240$	$f'_b = 0$
Principal tension without web steel		$f''_t \simeq 4\sqrt{f'_c}\ /2 \simeq 135$ psi
Steel:		
Initial prestress	$f_i = 130{,}000$ psi	
Prestress just after transfer	$f_0 = 120{,}000$ psi	
Effective prestress		$f_{se} = 110{,}000$ psi

(*a*) Make a preliminary design for the beam.

(*b*) Design the beam for flexure, and locate the profile for the c.g.s.

(*c*) Check the beam for shear, and provide necessary web reinforcement.

Solution.

(a) Preliminary Design. For this tapered beam carrying a uniform load, the controlling section for flexure is not at midspan but is nearly at the third point. For $w_S = 520$ plf (7.59 kN/m),

$$M_S = \frac{w_S L^2}{8} = \frac{520 \times 65^2}{8} = 274 \text{ k-ft } (372 \text{ kN-m})$$

The weight of the beam is estimated by an empirical formula

$$A_c = \frac{5M}{hf_c} = \frac{5 \times 274 \times 12}{40 \times 2.025} = 203 \text{ in.}^2\ (131 \times 10^3 \text{ mm}^2)$$

Hence, w_G = about 220 plf (3.21 kN/m). The total load on the beam is 520 + 220 = 740 plf (10.8 kN/m). Since the ratio of M_G/M_T is not too high, critical stresses may be found both at transfer and under the working load; hence an I-shape is considered economical. The moments at midspan and third points are, for span of 65 ft (19.8 m),

	Midspan	Third Points
$M_S = 520 \times 65^2/8 = 274$ k-ft $\times 8/9 =$		244
$M_G = 220 \times 65^2/8 = 116$ k-ft $\times 8/9 =$		103
M_T	$= 390$ k-ft (529 kN-m)	347 k-ft (471 kN-m)

The depth of beam at the third points is nearly 32 in. (813 mm). Using formulas 6-1 and 6-4, we have

$$F = \frac{M_T}{0.65h} = \frac{347 \times 12}{0.65 \times 32} = 200 \text{ k (890 kN)}$$

and

$$F = \frac{M_L}{0.50h} = \frac{244 \times 12}{0.50 \times 32} = 183 \text{ k (814 kN)}$$

Corresponding to $F = 200$ k (890 kN) and $f_c = 2025$ psi (13.96 N/mm^2), we have, from formula 6-3

$$A_c = F/0.50 f_c$$

$$= 200/(0.50 \times 2.025) = 198 \text{ sq in. } (128 \times 10^3 \text{ mm}^2)$$

A trial midspan section is sketched as in Fig. 16-7, with $A_c = 208$ sq in. (134×10^6 mm^2) at the third point. In order to maintain a uniform flange section throughout the entire length of the beam, a trial layout for the beam is now sketched as shown. The end 3 ft (0.91 m) of the beam is made rectangular in section in order to distribute the stress at anchorage and to provide sufficient area for the end shear.

For a section 3 ft (0.91 m) from the end of beam, the shear is

$$V_T = 740 \times 30 = 22{,}200 \text{ lb (98.7 kN)}$$

Neglecting the shear taken by steel, the maximum unit shear stress in concrete can be approximated by

$$v = 1.2 \frac{V}{A_{\text{web}}}$$

$$= 1.2 \frac{22{,}200}{4 \times 18.2} = 366 \text{ psi } (2.52 \text{ N/mm}^2)$$

for a web 4 in. (102 mm) thick and a depth of beam = 18.2 in. (462 mm). Corresponding to a stress f_c at c.g.c. of about 600 psi (4.14 N/mm^2), this would indicate a principal tension of

$$f_t'' = \sqrt{366^2 + 300^2} - 300 = 173 \text{ psi } (1.19 \text{ N/mm}^2)$$

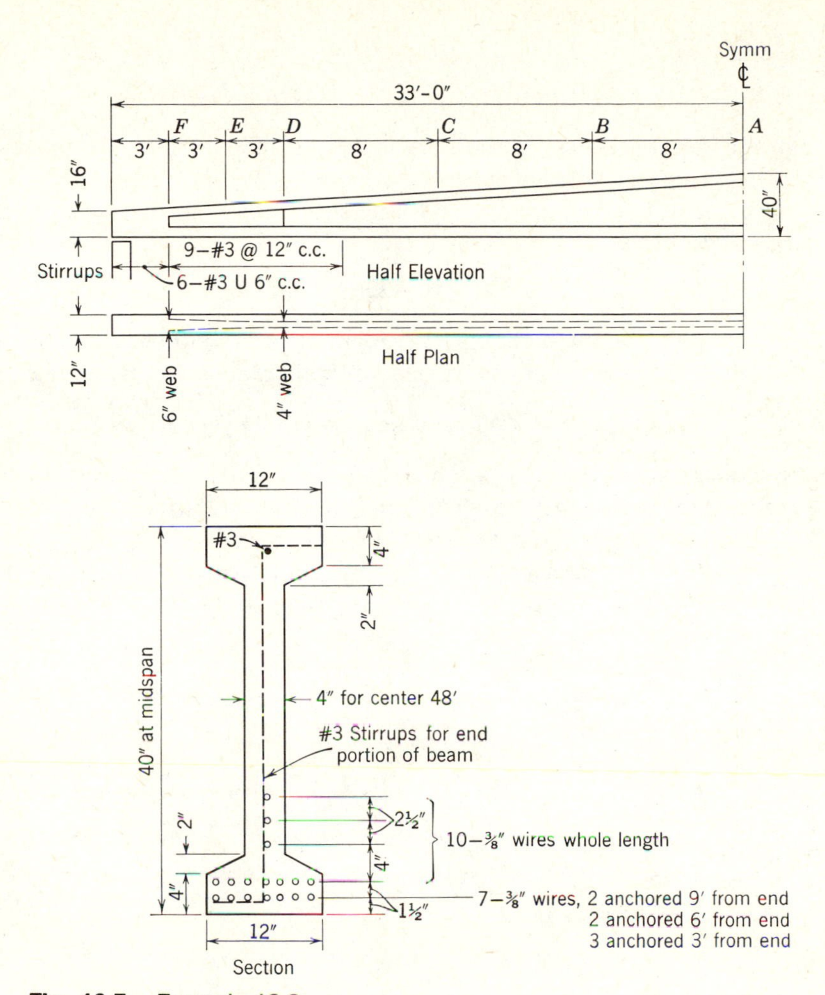

Fig. 16-7. Example 16-2.

which is somewhat above the approximate value of 135 psi (0.93 N/mm^2) used at service load for this design. (This value corresponds to 50% of the $4\sqrt{f_c'}$ principal tensile stress permitted at factored load, assuming that web cracking controls shear strength.) Hence it seems desirable to thicken the web near the support. A gradual increase from 4 in. (102 mm) to 6 in. (152 mm) within a distance of 6 ft (1.83 m) is thus adopted as shown. This also increases the A_c near the ends so as to keep the flexural stress within limits.

(b) Design for Flexure. To provide for effective prestress of about 200 k (890 kN), seventeen $\frac{3}{8}$-in. (9.525 mm) wires will be used, furnishing total

effective prestress

$$F = 17 \times 0.11 \times 110 = 206 \text{ k } (916 \text{ kN})$$

Although the initial prestress in the steel is 130 ksi (896 N/mm^2), immediately after transfer, some relaxation in steel and the elastic shortening of concrete would have already taken place, and the steel stress at that time can be considered to be 120 ksi (827 N/mm^2). Hence,

$$F_0 = 206 \times 120/110 = 225 \text{ k } (1000 \text{ kN})$$

Now, again considering the section at 3 ft (0.91 m) from end of beam, the A_c is 169 sq in. (109×10^3 mm^2), and the maximum compression can be as high as

$$(225/169) \times 2 = 2660 \text{ psi} = 2.66 \text{ ksi } (18.34 \text{ N/mm}^2)$$

which is somewhat too high. It would be desirable to cut off bond between concrete and wires before they reach the ends. After some trial, it is decided to cut off 3 wires at *F*, 2 at *E*, and another 2 at *D*, Fig. 16-7.

In order to determine the location of the c.g.s., a procedure similar to that of example 8-4 is followed. Since both A_c and A_{ps} vary along the beam, it will be convenient to tabulate the computations as shown.

Section	*F*		*E*		*D*		*C*	*B*	*A*
x, ft from center of support	2.5		5.5		8.5		16.5	24.5	32.5
h, in.	18.2		20.3		22.5		28.4	34.2	40.0
Web thickness, in.	12	6	5		4		4	4	4
A_c, in.^{2}	218	169	171		170		193	217	240
I, in.^{4}		5700	7700		10,000		18,800	30,500	46,000
r^2, in.^{2}		34	45		60		98	141	192
$k_t = k_b$, in.	3.0	3.8	4.4		5.4		6.9	8.2	9.6
No. of $\frac{3}{8}''$ wires	10	13	13	15	15	17	17	17	17
F_0 at 120 ksi	132	172	172	198	198	225	225	225	225
F at 110 ksi	121	158	158	182	182	206	206	206	206
M_T, k-ft	59		121		179		294	368	390
M_G, k-ft	17		36		53		87	109	116
M_T/F, in.	5.9	4.5	9.2	8.0	11.8	10.4	17.1	21.4	22.7
M_G/F_0, in.	1.5	1.2	2.5	2.2	3.2	2.8	4.6	5.8	6.2
$f_t' A_c k_b / F_0$, in. ($f_t' = 0.24$ ksi)	1.2	0.9	1.1	1.0	1.1	1.0	1.4	1.9	2.5
e_b, in.	1.7	0.9	2.2	1.4	2.3	1.4	2.1	2.2	1.8
$f_b = \frac{F_0}{A_c}\left(\frac{h}{c_t} - \frac{e_b}{k_t}\right) + 0.24$, ksi	1.11	2.03	1.74	2.19	2.08	2.54	2.29	2.04	1.94
e_t, in.	1.5	3.9	1.1	2.7	0.9	2.5	0.4	0.4	3.4
$f_t = \frac{F}{A_c}\left(\frac{h}{c_b} - \frac{e_t}{k_b}\right)$, ksi,	0.83	0.91	1.62	1.48	1.96	1.86	2.07	1.85	1.42

Values from the table are plotted in Fig. 16-8, which shows a half profile of the beam. First, the kern points are plotted from the c.g.c.; then, from the respective kern lines, the values of M_T/F and $M_G/F_0 + f_t' A_c k_b / F_0$ are plotted to

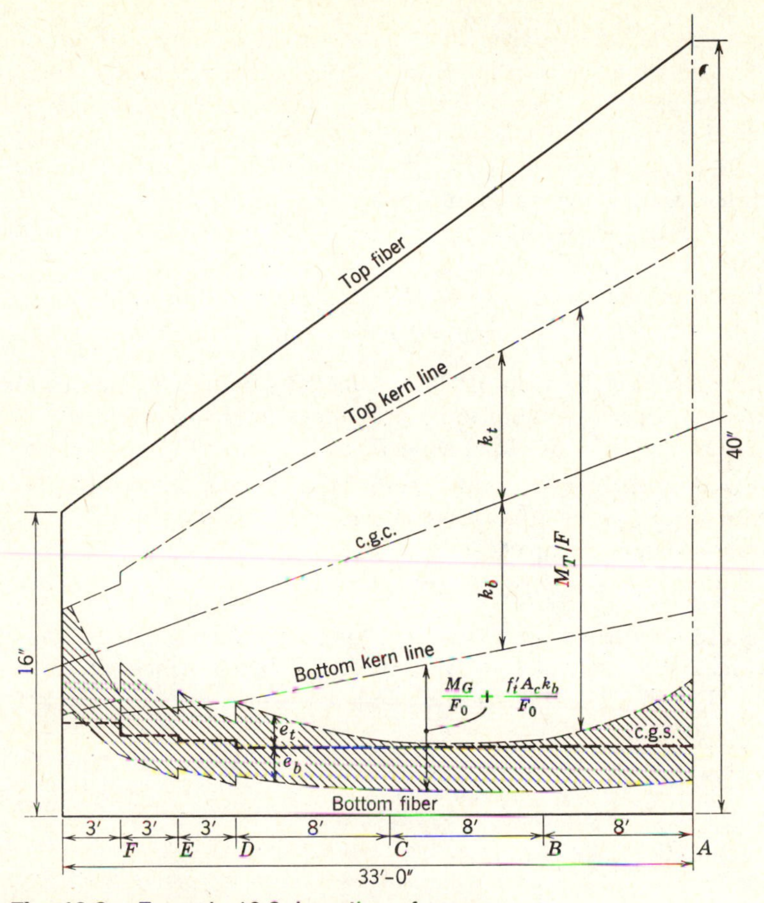

Fig. 16-8. Example 16-2. Location of c.g.s.

obtain the limiting zone for c.g.s. If the c.g.s. is located within this zone, there will be no tension in the bottom fiber under working load and the tension in the top fiber will be no greater than 240 psi (1.65 N/mm^2) at transfer.

The actual location of c.g.s. is given by the heavy dotted line in Fig. 16-8, with the wires placed and cut off as indicated in Fig. 16-7. For this profile of c.g.s., the bottom compressive stress at transfer is

$$f_b = \frac{F_0}{A_c}\left(\frac{h}{c_t} - \frac{e_b}{k_t}\right) + 0.24$$

and the top fiber compressive stress under working load is

$$f_t = \frac{F}{A_c}\left(\frac{h}{c_b} - \frac{e_t}{k_b}\right)$$

where e_b and e_t are the distances of c.g.s. within the limiting zone measured from the bottom and top limits, respectively. These values of f_b and f_t are computed and listed in the table. It will be observed that the greatest compressive stress occurs at D at transfer; its magnitude of 2.54 ksi (17.5 N/mm^2) is slightly exceeds that of the allowable of 2.40 ksi (16.55 N/mm^2) but is not considered serious. The greatest compressive stress under working load is 2.07 ksi (14.27 N/mm^2), very close to the allowable of 2.03 ksi (14.00 N/mm^2), and is considered satisfactory.

The percentage of reinforcement at D is checked in Fig. 16-9:

$$\omega_p = 0.273 < 0.30$$

which indicates that the beam is close to the ACI limit and compression failure in concrete may occur without too much elongation in the wires.

Actual test showed a live load deflection of $7\frac{1}{2}$ in. (190.5 mm) at rupture.

(c) Check Shear. After a little investigation, it is shown that the critical section for shear is a 3 ft (0.91 m) from the end of beam, where we have a 6-in. (152 mm) web and a shear of

$$30 \times 740 = 22{,}200 \text{ lb } (98.7 \text{ kN})$$

which is also very nearly the shear perpendicular to the c.g.c. line. The prestressing steel makes an angle of 1/33 with the c.g.c. line and hence will carry some of the shear. The 10 wires at this section with a prestress of 121 k (538 kN) carry a

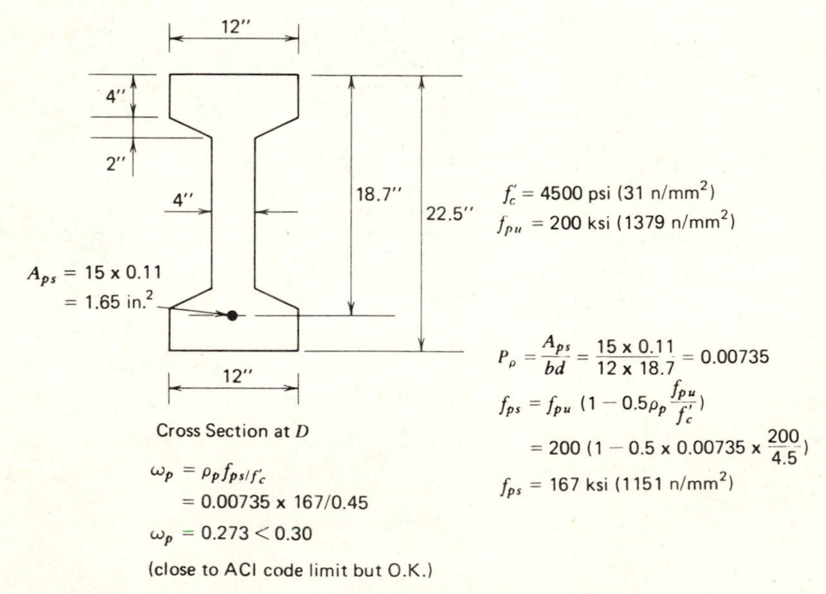

Fig. 16-9. Check of reinforcement ratio at D.

shear of

$$121{,}000/33 = 4000 \text{ lb } (17.8 \text{ kN})$$

leaving $V_c = 22{,}200 - 4000 = 18{,}200$ lb (81.0 kN) to be carried by the concrete. The maximum shearing stress at the c.g.c. is

$$v = \frac{V_c Q}{Ib} = \frac{18{,}200(48 \times 7.1 + 6 \times 4.4 + 30.6 \times 2.55)}{5700 \times 6}$$

$$= 240 \text{ psi } (1.65 \text{ N/mm}^2)$$

The compressive stress at this point is $121{,}000/169 = 716$ psi (4.94 N/mm²). Hence the principal tension is

$$f_t'' = \sqrt{240^2 + 358^2} - 358 = 72 \text{ psi } (0.50 \text{ N/mm}^2)$$

Although this is well within the limit of 135 psi (0.93 N/mm²) used at service load (only about 25% of ACI allowable principal tensile stress at the factored load level), it is desirable to provide some web steel to increase the shear resistance of the beam. For the end 3 ft (0.91 m), $\frac{3}{8}$-in (9.525 mm) U-stirrups at 6-in. (152 mm) spacing are used to distribute the load, Fig. 16-7. For the next 9 ft (2.74 m), single stirrups at 12-in. (305 mm) spacing are employed. No stirrups are used for the remainder of the beam.

16-3 Design Example, Posttensioned Bridge Girder

This example illustrates the design of a precast posttensioned girder for composite action in a highway bridge. It can be easily modified for pretensioning as may be desired.

Precast girders of a highway bridge are to be posttensioned, grouted, then lifted to the bridge site to be connected together by concrete poured in place, Fig. 16-10. The two-lane bridge is to carry H20-S16-44 loading, and the girders are spaced 6 ft (1.83 m) on centers. Overall length of girder is 96 ft (29.3 m), with 95 ft (29.0 m) between centers of supports. Following the AASHTO Specifications for Highway Bridges, generally when applicable, design an interior girder as follows:

(*a*) Design the midspan section, indicating the required amount of prestressing steel.
(*b*) Design the end section, showing the mild-steel stirrups.
(*c*) Design a longitudinal layout of the girder showing the profile for c.g.s. and the intermediate and end diaphragms.
(*d*) Investigate the factor of safety of the girder at cracking and ultimate strengths.
(*e*) Compute the deflection of the girder at transfer and under the working load.

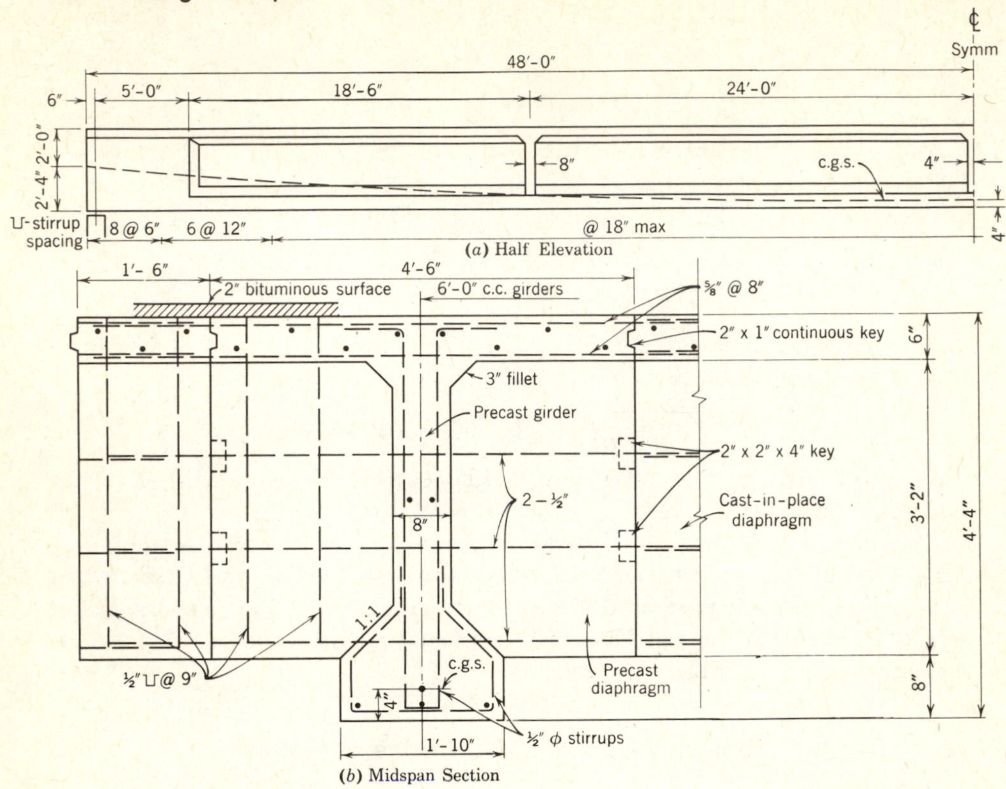

Fig. 16-10. Example 16-3. Girder layout.

(*f*) Detail the midspan and the end sections using the Freyssinet system. Compute the loss of prestress due to friction and the initial prestress required at the jack.

Strength of concrete is to be 4500 psi (31 N/mm^2) at 28 days and 4000 psi (28 N/mm^2) at transfer.

The high-tensile steel used is to have a minimum ultimate tensile strength of 250,000 psi (1724 N/mm^2) and a minimum yield point of 200,000 psi (1379 N/mm^2) at 0.2% plastic set. The steel stress at transfer will be 160,000 psi (1103 N/mm^2), and the effective prestress used in this solution is 135,000 psi (931 N/mm^2). $E_s = 28{,}000{,}000$ psi (193 kN/mm^2). $E_c = 4{,}000{,}000$ psi (27.6 kN/mm^2). Use intermediate-grade reinforcing bars for the mild-steel reinforcement.

Solution. (*a*) From Appendix A of the AASHTO Specification for Highway Bridges, the maximum moment for one lane of H20-S16-44 loading on a span of 95 ft (29.0 m) is found to be 1433 k-ft (1943 kN-m) AASHTO. The proportion of lane load carried by an interior stringer is approximated by the formula $S/10$ for interior stringers. Note that this is more conservative than required following the AASHTO Standard Specification where $S/(2)(5.5)=S/11$ could be used. For spacing $S=6$ ft (1.83 m), the moment per girder is

$$LLM = 1433 \times 6/10 = 860 \text{ k-ft } (1.166 \text{ kN-m})$$

Impact on highway bridges is given by the formula

$$I = 50/(L+125)$$
$$= 50/(95+125)$$
$$= 0.227$$

Impact moment $IM = 860 \times 0.227 = 195$ k-ft (264 kN-m)

$$LLM + IM = 860 + 195 = 1055 \text{ k-ft } (1430 \text{ kN-m})$$

Note the above value of maximum moment actually does not occur at midspan, but for all practical purposes it can be assumed to occur there.

After some preliminary design and trials, a section is assumed as in Fig. 16-10 from which

Bituminous paving 2 in. at 150 pcf = 25 psf × 6 ft = 150 plf (2.19 kN/m)

In-place concrete slab and diaphragms = 133 plf (1.94 kN/m)

Total added DL = 283 plf (4.13 kN/m)

Added DL moment $= (283 \times 95^2)/8 = 319$ k-ft (433 kN-m)
Weight of girder and diaphragms = 940 plf (13.71 kN/m)
Girder load moment $M_G = (940 \times 95^2)/8 = 1060$ k-ft (1437 kN-m)
Total moment $M_T = 1055 + 319 + 1060 = 2434$ k-ft (3300 kN-m)
The c.g.c. of the section, using the gross area, is as follows.

$$6 \times 54 = 324 \times 3 = 972$$
$$3 \times 3 = 9 \times 7 = 63$$
$$38 \times 8 = 304 \times 25 = 7600$$
$$7 \times 7 = 49 \times 41.67 = 2040$$
$$22 \times 8 = \underline{176} \times 48 = \underline{8450}$$
$$862 \qquad 19{,}125 \div 862 = 22.5 \text{ in. } (571.5 \text{ mm}) = c_t$$
$$52 - 22.5 = 29.5 \text{ in. } (749.3 \text{ mm}) = c_b$$

The moment of inertia of the concrete section about the c.g.c. is

$$
\begin{aligned}
324(6^2/12+19.5^2) &= 124{,}000\\
9(3^2/18+15.5^2) &= 2{,}200\\
304(38^2/12+2.5^2) &= 38{,}600\\
49(7^2/18+19.17^2) &= 18{,}000\\
176(8^2/12+25.5^2) &= \underline{115{,}000}
\end{aligned}
$$

$$297{,}800 \div 862 = 345 = r^2$$

$$k_t = r^2/c_b = 345/29.5 = 11.7 \text{ in. } (297 \text{ mm})$$

$$k_b = r^2/c_t = 345/22.5 = 15.4 \text{ in. } (391 \text{ mm})$$

Owing to the relatively large ratio of M_G/M_T, it is evident that the c.g.s. can be located as low as practicable without producing any tension in the top fiber. Assuming that the c.g.s. is located 4 in. (102 mm) above the bottom fiber, the total arm for the internal resisting moment is

$$a = 11.7 + 29.5 - 4 = 37.2 \text{ in. } (945 \text{ mm})$$

The total effective prestress required is

$$F = M_T/a = (2434 \times 12)/37.2 = 786 \text{ k } (3496 \text{ kN})$$

For a loss of prestress at 15%, the initial prestress required will be

$$F_0 = F/0.85 = 786/0.85 = 925 \text{ k } (4114 \text{ kN})$$

To limit the top fibers to a maximum stress of 1.8 ksi (12.41 N/mm²), we must have

$$A_c = \frac{Fh}{f_t c_b} = \frac{786 \times 52}{1.8 \times 29.5} = 770 \text{ in.}^2 \ (497 \times 10^3 \text{ mm}^2)$$

To limit the bottom fibers to a maximum of 2.2 ksi (15.17 N/mm²), we must have

$$
\begin{aligned}
A_c &= \frac{F_0}{f_b}\left(1 + \frac{e-(M_G/F_0)}{k_t}\right)\\
&= \frac{925}{2.2}\left(1 + \frac{25.5 - 1060 \times 12/925}{11.7}\right)\\
&= 841 \text{ in.}^2 \ (543 \times 10^3 \text{ mm}^2)
\end{aligned}
$$

The actual gross area furnished is 862 sq in. (556×10^3 mm²), which seems to be just about sufficient for the required area of 841 in.² (543×10^3 mm²). The top fibers will not be stressed to the allowable value, but the width and thickness of the top flange are governed by the slab requirements, which will not be discussed here. Also note that the in-place concrete will further reduce the stress in the tip fibers, and may be included in the computation if desired. Hence the section is considered satisfactory and is adopted without further changes. Note that it generally takes two or three trials to arrive at this adopted section rather than just one trial as illustrated here.

To supply the effective prestress of 786 k (3,496 kN) at an allowable stress of 135 ksi (931 N/mm^2), steel area required will be

$$786/135 = 5.83 \text{ in.}^2 \ (3761 \text{ mm}^2)$$

(*b*) Shearing stresses can be checked for two sections, one at the support and another 5 ft (1.52 m) from the support where the web is 8 in. (203 mm) thick. At the support, the web is 22 in. (559 mm) thick; shear is evidently not controlling. The shear at 5 ft (1.52 m) from support is

LL shear, from AASHTO Specifications	= 61.3 k/lane
$61.3 \times 6/10$	= 36.8 k/girder
Impact shear = 0.227×36.8	= 8.4 k
Bituminous paving and in-place concrete 0.283×42.5	= 12.0 k
Girder own wt 0.940×42.5	= 40.0 k
Total shear	97.2 k (432 kN)

Shear V_p carried by the tendons at end of span, assuming a parabolic rise of $h = 2$ ft (0.61 m) on a length of $L = 96$ ft (29.3 m), is given by

$$\begin{aligned} V_p &= 4Fh/L \\ &= (4 \times 786 \times 2)/96 \\ &= 65.5 \text{ k (291 kN)} \end{aligned}$$

At 5.5 ft (1.68 m) from end of girder,

$$\begin{aligned} V_p &= (42.5/48)65.5 \\ &= 58.0 \text{ k (258 kN)} \end{aligned}$$

Hence V_c by concrete is

$$97.2 - 58.0 = 39.2 \text{ k (174 kN)}$$

Maximum shearing stress in concrete occurs at c.g.c. and is given by

$$v = V_c Q/Ib$$

Since Q, the statical moment of the area above the c.g.c. about it, is

$$(324\times 19.5)+(9\times 15.5)+(8\times 16.5^2/2)=7550,$$

$$v=\frac{39{,}200\times 7550}{297{,}800\times 8}=124 \text{ psi } (0.855 \text{ N/mm}^2)$$

The compressive fiber stress at c.g.c. is given by F/A_c

$$\begin{aligned} f_c &= F/A_c \\ &= 786{,}000/862 \\ &= 912 \text{ psi } (6.288 \text{ N/mm}^2) \end{aligned}$$

The principal tensile stress is

$$\begin{aligned} f_t'' &= \sqrt{v^2+(f_c/2)^2} - f_c/2 \\ &= \sqrt{124^2+456^2} - 456 \\ &= 20 \text{ psi } (0.138 \text{ N/mm}^2) \end{aligned}$$

The moment is relatively small at this section of maximum shear; hence the fiber stress is nearly uniform throughout the depth of the section, and the maximum principal tensile stress at the c.g.c. represents rather closely the greatest tensile stress. This maximum principal tension does not approach the $4\sqrt{f_c'}$ stress associated with web cracking, which assures us that there will be no web shear cracking in the girder at working load.

To investigate the ultimate strength for shear we will use the following factored load shear for this example:

$$\begin{aligned} V_u &= 1.5D+2.5(L+I) \\ &= 1.5\times 52.0+2.5\times 45.2 \\ &= 191 \text{ k } (850 \text{ kN}) \end{aligned}$$

Shear carried by the tendons can be conservatively approximated by $V_p=58.0$ k (258 kN). Using the recently revised AASHTO Specifications (very similar to the ACI Code):

$$\begin{aligned} V_{cw} &= \left(3.5\sqrt{f_c'}+0.3f_{pc}\right)b_w d+V_p \\ &= \left[3.5\sqrt{4500}+(0.3)(912)\right](8.0)(41.6)+58{,}000 \end{aligned}$$

using $d=0.8h=(0.8)(52)=41.6$ in. (1057 mm)

$$V_{cw}=113{,}400+58{,}000=171{,}400 \text{ lb } (762 \text{ kN})=V_c$$

Thus, the concrete and vertical component of tendon carries 171.4 k (762 kN). Stirrups must carry the $(V_u/\phi - V_c)$ excess which is

$$V_s = \frac{191}{0.85} - 171.4 = 53.3\ k\ (237\ kN)$$

$$V_s = \frac{A_v f_y d}{s} = 53.3\ \text{k}\ (237\ \text{kN})$$

Assuming No. 4 stirrups with $f_y = 40$ ksi (276 N/mm^2)

$$V_s = \frac{(0.4)(40)(41.6)}{s} = 53.3$$

$$s = 12.5\ \text{in.}\ (317.5\ \text{mm})$$

Use 12 in. (304.8 mm) spacing for stirrups near end (Fig. 16-10).

In much of the span the shear is lower and we may find the minimum $A_v = 100 b_w s/f_y$ will govern under AASHTO. This gives

$$0.40 = \frac{(100)(8)(s)}{40{,}000}$$

$$s = 20\ \text{in.}\ (508\ \text{mm})$$

Use 18 in. (457.2 mm) spacing for most of span as shown in Fig. 16-10.

Similar computation can be made for other points along the girder. So far as shear V_{cw} is concerned, more stirrups are required near the ends than along the middle portion of the girder, but the reverse is true when considering the effect of combined moment and shear based on V_{ci}. Hence judgment should be exercised in the actual spacing of the stirrups.

For the end section, stirrups are required to distribute the anchorage stresses. Since the anchorages are to be fairly uniformly distributed, the computed tensile stresses in the anchorage zone will be low and analysis is not required. Nominal stirrups, however, are provided as shown on Fig. 16-10.

(*c*) A half elevation of the girder is shown in Fig. 16-10. The midspan section is adopted for the entire girder, except the 5 ft (1.52 m) near the ends where a uniform web thickness equal to the bottom flange width of 22 in. (559 mm) is used in order to accommodate the end anchorages, to permit the curving up of some tendons, and to distribute the prestress. Three intermediate diaphragms are placed along the length of the span. Sometimes transverse prestressing is employed to bind the girders together. But, for this design transverse dowels are provided in these diaphragms to be joined together by in-place concrete. In either case, the theoretical calculation for the steel in these diaphragms can be a complicated problem. But the amount of steel is not excessive for these girders; some nominal reinforcement is employed as shown.

The most common location of c.g.s. for a posttensioned simple beam is a parabola with c.g.s. near the c.g.c. at the ends. Such a profile will give ample moment resistance along the entire beam. If the c.g.s. is above the c.g.c. at the ends, the tendons will carry greater shear but lose some of the reserve moment resistance. If the c.g.s. is below the c.g.c. at the ends, the tendons will carry less shear, but the positive prestressing moment at the ends will tend to decrease the principal tension. Also note that the c.g.c.'s for the midspan and the end sections actually differ slightly. For this design, the c.g.s. will be placed a little below the c.g.c. of the end section.

(*d*) The cracking moment is computed as follows. The resisting moment up to zero stress in the bottom fiber is given by

$$Fa = 786 \times 37.2/12 = 2434 \text{ k-ft } (3300 \text{ kN-m})$$

Assuming the modulus of rupture at bottom fiber to be $7.5\sqrt{f'_c} = 503$ psi (3.47 N/mm^2), the additional resisting moment from zero stress to 503 psi (3.47 N/mm^2) is

$$M = fI/c_b = \frac{503 \times 297{,}800}{12{,}000 \times 29.5}$$

$$= 423 \text{ k-ft } (574 \text{ kN-m})$$

Total resisting moment at cracking is

$$M_{cr} = 2434 + 423 = 2857 \text{ k-ft } (3{,}874 \text{ kN-m})$$

Overall factor of safety against cracking is

$$2857/2434 = 1.17$$

Factor of safety for live load and impact is

$$\frac{M_{cr} - (M_G + DLM)}{LLM + IM} = \frac{2857 - (1060 + 319)}{1055} = 1.4$$

which indicates that the girder will begin to crack only when the live load plus impact is increased by as much as 40%.

The ultimate resisting moment will be computed as follows.

$$\rho_p = \frac{A_{ps}}{bd} = 5.83/54 \times 48 = 0.00225$$

$$f_{ps} = f_{pu}\left(1 - 0.5\rho_p \frac{f_{pu}}{f'_c}\right) = 250\left(1 - 0.5 \times 0.00225 \frac{250}{4.5}\right)$$

$$f_{ps} = 234 \text{ ksi } (1613 \text{ N/mm}^2)$$

The tension force at ultimate is

$$A_{ps} f_{ps} = 5.83 \times 234 = 1364 \text{ k } (6067 \text{ kN})$$

Thus the depth of the equivalent stress block can be found as follows assuming $0.85f_c' = 3.82$ ksi (26.34 N/mm²)

$$\text{Area of compression} = \frac{13644}{3.82} = 357 \text{ in.}^2\ (230 \times 10^3 \text{ mm}^2)$$

$$\text{Area top flange} = 6 \times 54 = 324 \text{ in.}^2\ (209 \times 10^3 \text{ mm}^2)$$

$$\text{Compression web area} = 33 \text{ in.}^2\ (21 \times 10^3 \text{ mm}^2)$$

The rectangular stress block extends down 9 in. (229 mm) to the base of the 3 in. (76 mm) fillet (Fig. 16-10). The centroid of this compressive area is located at

$$(324 \times 3 + 9 \times 7 + 24 \times 7.5)/357 = 3.4 \text{ in. } (86 \text{ mm})$$

below the top fiber, and the ultimate lever arm is

$$52 - 4 - 3.4 = 44.6 \text{ in. } (1{,}133 \text{ mm})$$

Hence the ultimate moment capacity of the cross section is

$$M_u = 1364 \times 44.6/12 = 5070 \text{ k-ft } (6{,}875 \text{ kN-m})$$

However, the factored load moment used for this example is

$$1.5D + 2.5(L + I) = 1.5 \times 1379 + 2.5 \times 1055$$
$$= 4706 \text{ k-ft } (6381 \text{ kN-m}) < 5070 \text{ k-ft } (6875 \text{ kN-m})$$

Hence the ultimate strength is considered sufficient.

(*e*) Referring to Fig. 16-11, it is seen that the deflection due to the initial prestress of 925 k (4114 kN) can be computed as due to a uniform moment of $M_1 = 116$ k-ft (157 kN-m) for the whole length of the beam plus a parabolic moment with $M_2 = 1850$ k-ft (2509 kN-m) at midspan. Downward deflection due to M_G is given by a parabolic moment with 1060 k-ft (1437 kN-m) at midspan. Thus the instantaneous upward deflection due to prestress is given by

$$\frac{M_1L^2}{8EI} + \frac{5M_2L^2}{48EI} = \left(\frac{116}{8} + \frac{5 \times 1850}{48}\right)\frac{96^2 \times 12^2 \times 12{,}000}{4{,}000{,}000 \times 297{,}800}$$
$$= 2.78 \text{ in. } (70.6 \text{ mm})$$

To simplify the numerical work, gross I for the concrete is used for all computations. Owing to the loss of prestress, this deflection will reduce to

$$0.85 \times 2.78 = 2.36 \text{ in. } (59.9 \text{ mm})$$

Downward deflection due to M_G is

$$\frac{5M_GL^2}{48EI} = \left(\frac{5 \times 1060}{48}\right)\frac{96^2 \times 12^2 \times 12{,}000}{4{,}000{,}000 \times 297{,}800}$$
$$= 1.47 \text{ in. } (37.3 \text{ mm})$$

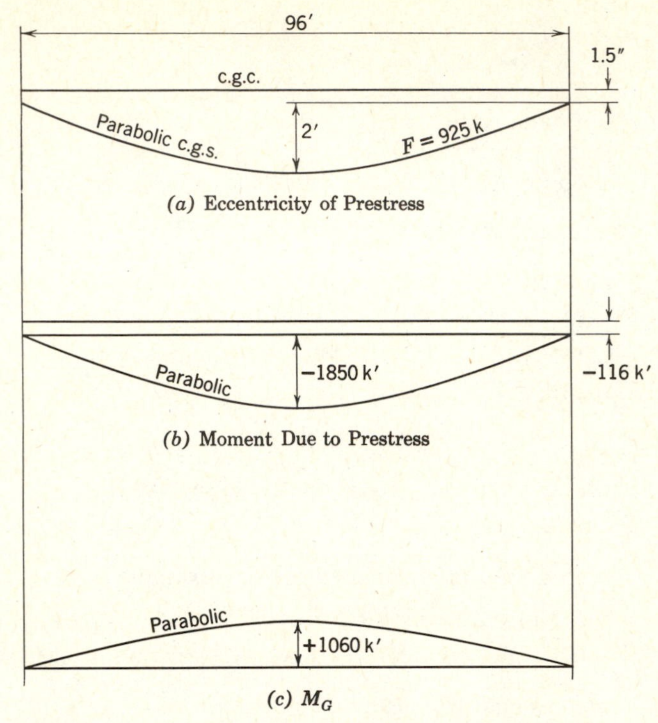

Fig. 16-11. Example 16-3. Deflection computation.

Hence the immediate upward deflection of the girder at transfer will be

$$2.78 - 1.47 = 1.31 \text{ in. } (33.3 \text{ mm})$$

The added dead load will produce a downward deflection of

$$(319/1060) \times 1.47 = 0.44 \text{ in. } (11.2 \text{ mm})$$

Not considering the composite action of in-place concrete, the eventual upward deflection will be decreased by the effect of loss of prestress, but increased by the creep effect of concrete. Assuming creep coefficient of 2 for the period considered, we have the resultant upward deflection after loss of prestress

$$(2.36 - 1.47 - 0.44)2 = 0.90 \text{ in. } (22.9 \text{ mm})$$

The instantaneous downward deflection due to the design live load and impact assuming parabolic moment diagram, is

$$(1055/1060)1.47 = 1.46 \text{ in. } (37.1 \text{ mm})$$

(*f*) The above design will be applicable to most prestressing systems now used in this country, although minor modifications may be desirable for certain cases. For purpose of illustration, detailed arrangement of the tendons will be as

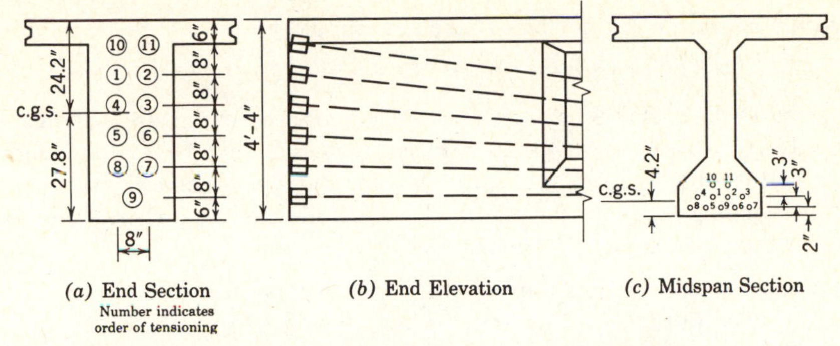

(*a*) End Section
(*b*) End Elevation
(*c*) Midspan Section

Fig. 16-12. Example 16-3. Cable location for Freyssinet system.

shown in Fig. 16-12. Using cables of eighteen 0.196-in. (4.98 mm) wires, 11 tendons are required. They will supply a steel area of 5.97 in.^{2} (3,852 mm^2) which is more than the 5.83 in.^{2} (3,761 mm^2) required in part (*a*) of this solution. The midspan and end sections are drawn showing the arrangement of tendons to give the required locations of c.g.s. Curving of the tendons in both horizontal and vertical planes is necessary to conform with the required location of c.g.s. It is noted that some deviation from the required parabola is permissible, because it will not affect the strength of the girder. A recommended order for tensioning the cables is indicated as shown.

Loss for the anchorage slip in the cones is assumed to average 0.2 in. (5.08 mm), which, if averaged throughout the entire length of 96 ft (29.3 m), indicates a loss of prestress equal to

$$[0.2/(96\times 12)]\times 28{,}000{,}000 = 4900 \text{ psi } (33.8 \text{ N/mm}^2)$$

To estimate the frictional loss, let us assume a coefficient of friction = 0.35 for wire cables in metal sheathing and a $K = 0.0010$ per ft for wobble effect. The average change in direction for the cables is given by $8\times 2/96 = 0.167$ radian, for a parabolic rise of 2 ft (0.61 m) in a span of 96 ft (29.3 m). Hence the maximum frictional loss at the far end, if tensioned only from one end, can be computed as

$$\begin{aligned} u\alpha + KL &= 0.35\times 0.167 + 0.0010\times 96 \\ &= 0.058 + 0.096 \\ &= 15.4\% \end{aligned}$$

The controlling point is the midspan which has a loss equal to half of 15.4% = 7.7%. 7.7% of 160,000 = 12,300 psi (84.8 N/mm^2). If the tendons are overtensioned by 12,300 psi (84.8 N/mm^2) at the anchorages, the anchorage loss of 4900 psi (33.8 N/mm^2) will automatically be balanced. Using the Freyssinet jack, there is an additional loss at the jack of about 8,000 psi (55.2 N/mm^2).

Hence the maximum initial stress at the jack should be

$$160{,}000 + 12{,}300 + 8000 = 180{,}300 \text{ psi } (1243 \text{ N/mm}^2)$$

According to AASHTO Specifications this temporary jacking stress should not exceed $0.80 f_{pu} = 0.80 \times 250{,}000 = 200{,}000$ psi (1379 N/mm^2). In fact, the tendons can be more highly stressed so as to reduce the required steel area if desired.

Note: It is suggested that the reader try to rework this example using some of the following possible modifications:

1. For comparison in detailing, try the Freyssinet system, but using tendons with either 120.276 in. (7.01 mm) wires or with $12\frac{1}{2}$-in. (12.7 mm) 7-wire-strands.
2. For comparison of posttensioning systems, try the BBRV, or the Prescon system with $\frac{1}{4}$-in. diameter buttonhead wires instead of the Freyssinet system.
3. For comparison of different girder sections and the use of pretensioning instead of posttensioning, try the Type IV I-girder spaced at 6-ft (1.83 m) centers with a 6-in. (152 mm) poured in-place slab. A section of this girder together with its properties are shown in Fig. 16-13. The composite section properties are computed for slab concrete with $f'_c = 3000$ psi (21 N/mm^2)

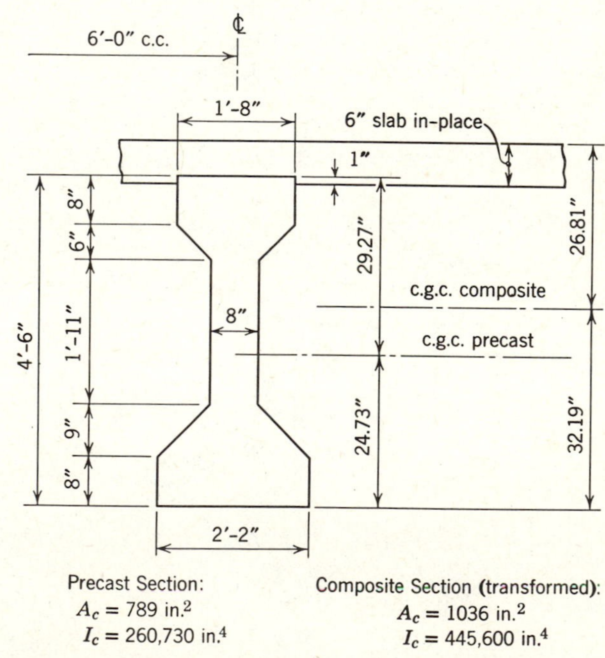

Fig. 16-13. Alternate section for example 16-3. (Type IV I-beam).

and girder concrete with $f'_c = 7000$ psi (34 N/mm^2), using a transformed section considering 0.6 of the slab area. Also note that for these pretensioned sections, no end blocks are used with the girder, but additional mild steel is added as described in section 7-8. Seven-wire strands of $\frac{1}{2}$-in. (12.1 mm) size are commonly used in the United States for girders of this size. To approximate a parabola, try 2-point harping for these pretensioned strands.

16-4 Design Example, Flat Plate Apartments

Posttension with Unbonded Tendons.* Design typical transverse strip as described in Fig. 16-14. Use normal weight concrete with $f'_c = 4000$ psi (28 N/mm^2) in slabs and columns. Assume a set of loads to be balanced by parabolic tendons. Analyze an equivalent frame subjected to the net downward loads, according to the principles of ACI Code, section 13.7. Check flexural stresses at critical sections and revise load balancing tendon forces as required to obtain net flexural tension stresses in accordance with ACI Code.

When final forces are determined obtain frame moments for factored dead and live loads. Calculate secondary moments induced in the frame by posttensioning forces, and combine with factored load moments to obtain design ultimate moments. Provide minimum mild steel reinforcement in accordance with ACI Code, section 18.9. Check ultimate flexural capacity and increase mild steel if required by strength criteria. Investigate shear strength, including shear due to vertical load and due to moment transfer by torsion; compare total to allowable value calculated from ACI Code equation 11-13.

Solution.

(1) Service Load Design. Since, for this example, spans and loads are low, and slab thickness is a little generous, it is likely that minimum values will govern many aspects of the design for flexure. Therefore, a force corresponding to an average compression stress of 150 psi (1.03 N/mm^2), with maximum parabolic tendon profile, will be used for the initial estimate of balanced load.

$$\text{Then } F_e = 0.150 \times 6.5 \times 12 = 11.7 \text{ k/ft (170.7 kN/m)}$$

Assuming $\frac{1}{2}$-in. (12.7 mm) diameter, 270 ksi (1862 N/mm^2) strand tendons, and 30 ksi (207 N/mm^2) long-term losses, effective force per tendon is $0.153 \times (0.7 \times 270 - 30) = 24.33$ k (108 kN).

For a 20-ft bay, $20 \times 11.7/24.33 = 9.6$ tendons, say 10.

$$\text{Then } F_e = 10 \times \frac{24.33}{20} = 12.16 \text{ k/ft (177.4 kN/m)}$$

$$F/A = 12.16/78 = 0.155 \text{ ksi } (1.07 \text{ N/mm}^2)$$

*The author helped develop this design example from the *Post-Tensioning Manual*, 1976, of the Post-Tensioning Institute. It is used with PTI permission.

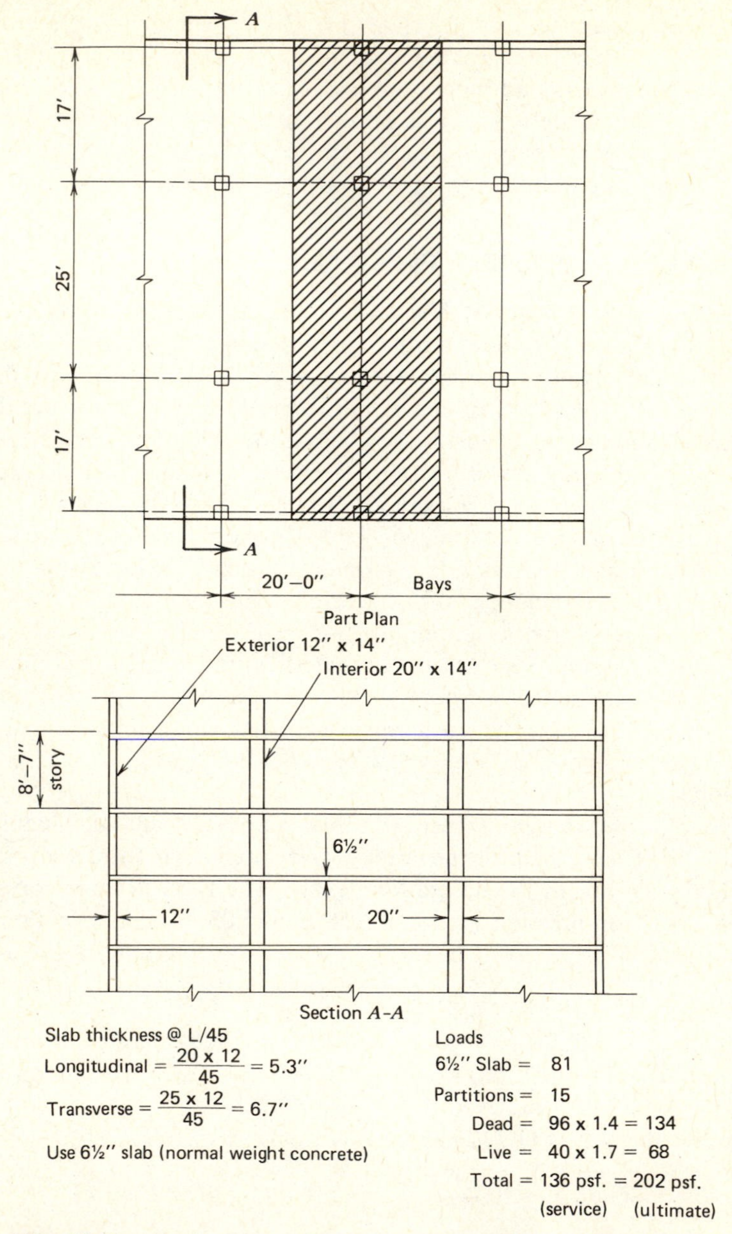

Slab thickness @ L/45

$$\text{Longitudinal} = \frac{20 \times 12}{45} = 5.3''$$

$$\text{Transverse} = \frac{25 \times 12}{45} = 6.7''$$

Use 6½″ slab (normal weight concrete)

Loads

6½″ Slab =	81		
Partitions =	15		
Dead =	96 x 1.4 =	134	
Live =	40 x 1.7 =	68	
Total =	136 psf. =	202 psf.	
	(service)	(ultimate)	

Fig. 16-14. Design Example 16-4, prestressed flat plate.

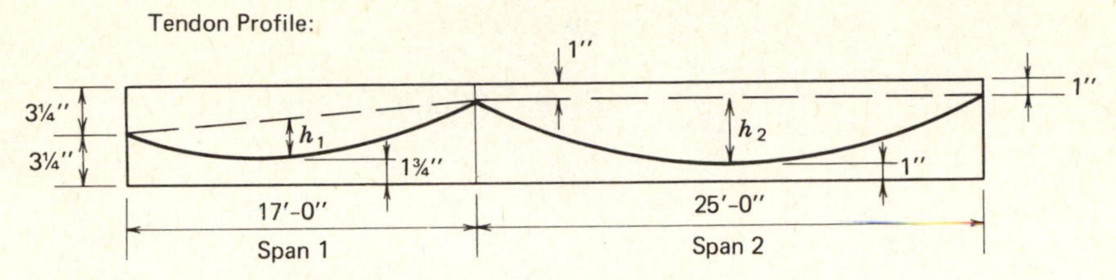

Fig. 16-15(a). Tendon profile, Design Example 16-4.

For spans 1 and 3 (Fig. 16-15a):

$$h_1 = \left(3\tfrac{1}{4} + 5\tfrac{1}{2}\right)/2 - 1\tfrac{3}{4} = 2.625 \text{ in. } (66.675 \text{ mm})$$

$$w_{\text{bal}} = \frac{8Fh}{L^2} = 8 \times 12.16 \times 2.625/12 \times 17^2$$

$$= \frac{12.16 \times 2.625}{1.5 \times 289} = 0.074 \text{ ksf } (3.54 \text{ kN/m}^2)$$

net load causing bending =

$$w_{\text{net}} = 0.136 - 0.074 = 0.062 \text{ ksf } (2.97 \text{ kN/m}^2)$$

For span 2:

$$h_2 = 6\tfrac{1}{2} - 1 - 1 = 4\tfrac{1}{2} \text{ in.}$$

$$w_{\text{bal}} = \frac{8 \times 12.16 \times 4.5}{12 \times 25^2} = 0.058 \text{ ksf } (2.78 \text{ kN/m}^2)$$

$$w_{\text{net}} = 0.078 \text{ ksf } (3.74 \text{ N/mm}^2)$$

Equivalent Frame Properties
(see ACI Code, section 13.7)

(a) *Equivalent Columns*
The basic stiffness of columns, including the effects of "infinite" stiffness at joints may be calculated by classical methods or by simplified methods which are in close agreement.

A simple approximation for K_c is shown by Rice and Hoffman in "Structural Design Guide to the ACI Building Code" (reference 2):

$$K_c = \frac{4EI}{L - 2h'} \qquad \text{where } h' = \text{slab thickness} = 6.5 \text{ in. } (165 \text{ mm})$$

The approximate formulas give results within 5% of the "exact" values, and considering the nature of assumptions necessary for design of the highly complex two-way flat plate, these formulas are completely adequate. Refer to Rice and Hoffman,[2] for a comparison of approximate and classical methods.

Exterior column-14 in. × 12 in. (356 mm × 305 mm)

$$I=\frac{14\times 12^3}{12}=2016 \text{ in.}^4\ (839\times 10^6 \text{ mm}^2)$$

$$\frac{E_{\text{col}}}{E_{\text{slab}}}=1.0$$

$$K_c=\frac{4\times 1.0\times 2016}{103-2\times 6.5}=90\times 2=180 \quad \text{joint total}$$

Torsional stiffness of slab in column line, K_t, is calculated as follows:

$$C=\left(1-0.63\frac{x}{y}\right)\frac{x^3y}{3}=\left(1-0.63\frac{6.5}{12}\right)\frac{(6.5)^3\times 12.0}{3}=724$$

$$K_t=\frac{\Sigma 9\times C\times E}{L_2\times(1-c_2/L_2)^3}$$

$$=\frac{9\times 724\times 1}{20\times 12(1.0-1.17/20)^3}+\frac{9\times 724\times 1}{20\times 12(1.0-1.17/20)^3}=65$$

Equivalent column stiffness is then obtained:

$$\frac{1}{K_{ec}}=\frac{1}{K_t}+\frac{1}{K_c}$$

$$K_{ec}=\left(\frac{1}{65}+\frac{1}{180}\right)^{-1}=48$$

Interior column = 14 in. × 20 in. (356 mm × 508 mm)

$$I=\frac{14\times(20)^3}{12}=9333 \text{ in.}^4\ (3.88\times 10^9 \text{ mm}^2)$$

$$K_c=\frac{4\times 1\times 9333}{103-2\times 6.5}=415\times 2=830 \text{ joint total}$$

$$C=\left(1-0.63\times\frac{6.5}{20}\right)\times\frac{(6.5)^3\times 20}{3}=1456$$

$$K_t=\frac{9\times 1456}{240(1-1.17/20)^3}+\frac{9\times 1456}{240(1-1.17/20)^3}=130$$

$$K_{ec}=\left(\frac{1}{830}+\frac{1}{130}\right)^{-1}=112$$

(*b*) Slab stiffness (see Rice and Hoffman,[2] width of slab-beam = 20/2 + 20/2 = 20 ft (6.1 m).

$$K_s=\frac{4EI}{L_1-c_1/2}$$

where L_1 is center-line span
C_1 is column depth

At exterior column

$$K_s=\frac{4\times 1\times 20\times (6.5)^3}{12\times 17-12/2}=111$$

At interior column spans 1 and 3

$$K_s=\frac{4\times 1\times 20\times (6.5)^3}{12\times 17-20/2}=110$$

Use single value of 111 for both ends of span since there is so little difference.

At interior span 2

$$K_s=\frac{4\times 1\times 20\times (6.5)^3}{12\times 25-20/2}=76$$

(*c*) Distribution factors for analysis by moment distribution at exterior joint slab distribution factor $=111/(111+48)=0.70$; at interior joints for spans 1&3 $=111/(111+76+111)=0.37$; and span $2=76/299=0.25$.

Moment Distribution—Net Loads

Since the nonprismatic section causes only very small effects on fixed end moments and carryover factors, fixed end moments will be calculated from $wl^2/12$ and carry over factors taken as $\frac{1}{2}$. Span 1&3 net load $FEM=0.062\times 289/12=1.49$ ft-k (2.02 kN-m)

Span 2 $FEM=0.078\times 625/12=4.06$ ft-k (5.51 kN-m)

Moment Distribution

0.70	**0.37**	**0.35 Symm.**
−1.49	−1.49	−4.06*FEM*
+1.04	−0.95	+0.64 dist
+0.48	−0.52	−0.32 C.O.
−0.34	+0.07	−0.05 dist
−0.31	−2.89	−3.79

(*a*) Check net tensile stresses at face of column:

Moment at column face is center-line moment $+\,Vc/3$

$$-M_{\max}=-3.79+\frac{20}{3\times 12}(12.5\times 0.078)$$

$$=-3.79+0.54=-3.25 \text{ ft-k } (-4.41 \text{ kN-m})$$

$$S=bt^2/6=12\times 6.5\times \frac{6.5}{6}=84.5 \text{ in.}^3\ (1.38\times 10^6 \text{ mm}^3)$$

$$=84.5 \text{ in.}^3\ \frac{\text{ft}}{12 \text{ in.}}=7.04 \text{ in.}^2 \text{ ft}$$

$$\text{then } f_t=\frac{3.25}{7.04}-0.155=0.462-0.155=0.307 \text{ ksi}$$

$$=+307 \text{ psi OK since } 6\sqrt{f_c'}=380 \text{ psi } (2.62 \text{ N/mm}^2)$$

(*b*) Check midspan tensile stress:

$$+M_{\max}=6.09-3.79=+2.30\text{ ft-k }(3.12\text{ kN-m})$$

$$f_t=\frac{2.30}{7.04}-0.155=0.175\text{ ksi }(1.21\text{ N/mm}^2)$$

$$f_t>2\sqrt{f_c'}\text{ i.e., 126 psi }(0.87\text{ N/mm}^2)$$

By requirements of ACI Code, section 18.9, when tensile stress of $2\sqrt{f_c'}$ is exceeded, the entire tensile force must be replaced by mild reinforcing at a stress of $f_y/2$ [Fig. 16-15(*b*)].

$$f_c=-0.330-0.155=-0.485<0.45\times4000$$

then $y=6.5\left(\dfrac{0.175}{(0.175+0.485)}\right)=1.72$ in. (43.7 mm)

$$T=0.175\times1.72\times\frac{12}{2}=1.8\text{ k/ft }(26.3\text{ kN/m})$$

$$A_s=\frac{1.8}{60/2}=0.06\text{ in.}^2/\text{ft }(127\text{ mm}^2/\text{m})$$

Say six No. 4 bars at 40 in. (1016 mm) on center bottom of midspan 2, $A_s=0.06$ in.^{2}/ft. (127 mm^2/m)

This completes the service load portion of the design, but the ultimate strength in flexure and shear must be verified to complete the design.

(2) Ultimate Flexural Capacity

(*a*) *Calculation of Design Moments*

Design moments for statically indeterminate posttensioned members are determined by combining frame moments due to factored dead and live loads with secondary moments induced into the frame by the tendons. The load balancing approach directly includes both primary and secondary effects, so that for

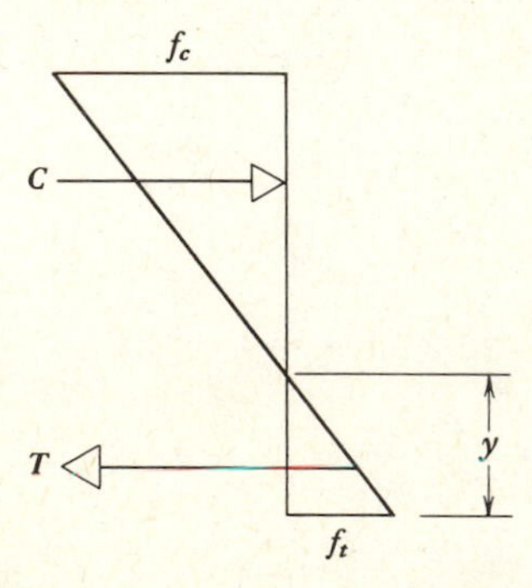

Fig. 16-15(*b*). Stress distribution at midspan, Design Example 16-4.

service conditions only "net loads" need be considered. At ultimate, the balanced load moments are used to determine secondary moments by subtracting the primary moment, which is simply $F \times e$, at each support. For multistory buildings where typical vertical load design is combined with varying moments due to lateral loading, an efficient design approach would be to analyze the equivalent frame under each case of dead, live, balanced, and lateral loads, and combine these cases for each design condition with appropriate load factors. For this example the balanced load moments are determined by moment distribution as follows:

$$\text{Span 1\&3 balanced load } FEM = \frac{0.074 \times 289}{12} = -1.78$$

$$\text{Span 2 } FEM = 0.058 \times \frac{625}{12} = -3.02$$

Moment Distribution Balanced Loads

0.7	0.37	0.25 Symm.
−1.78	−1.78	−3.02 *FEM*
+1.25	−0.46	+0.31 dist.
+0.23	−0.63	−0.16 C.O.
−0.16	+0.17	−0.12 dist.
−0.46	−2.70	−2.99

Since load balancing accounts for both primary and secondary moment directly, secondary moments can be found from the following relationship:

$$M_{bal} = M_1 + M_2 \text{ then } M_2 = M_{bal} - M_1$$

The primary moment, M_1 is $F \times e$ at each support. Thus at exterior columns, the secondary moment, M_2 is

$$M_2 = 0.46 - 12.16 \text{ k} \times \frac{0 \text{ in.}}{12 \text{ in./ft}} = 0.46 \text{ ft-k } (0.62 \text{ kN-m})$$

At interior column, span 1&3

$$M_2 = 2.70 - 12.16 \times \frac{(3.25 - 1.0)}{12} = 0.42 \text{ ft-k } (0.57 \text{ kN-m})$$

Span 2

$$M_2 = 2.99 - 12.16 \times \frac{2.25}{12} = 0.71 \text{ ft.k } (0.96 \text{ kN-m})$$

These moments are shown in Fig. 16-15(c) with balanced load moments.

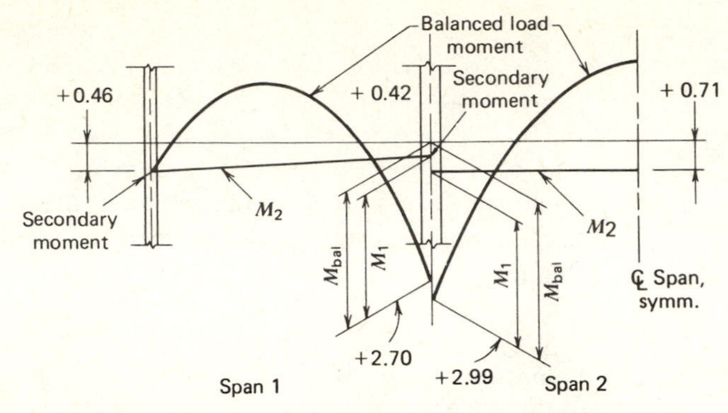

Fig. 16-15(*c*). Moments from balanced loads, Design Example 16-4.

Factored Load Moments

Span 1&3:

1.4 dead + 1.7 live *FEM*

$$=0.202\times\frac{289}{12}=4.86 \text{ ft-k } (6.59 \text{ kN-m})$$

Span 2:

$$FEM=0.202\times\frac{625}{12}=10.52 \text{ ft-k } (14.27 \text{ kN-m})$$

Moment Distribution Factored Loads

0.7	0.37	0.25 Symm.
−4.86	−4.86	−10.52 *FEM*
+3.40	−2.09	+1.42 dist.
+1.05	−1.70	−0.71 C.O.
−0.74	+0.37	−0.25 dist.
−1.15	−8.28	−10.06

Combine factored load and secondary moments to obtain design moments:

	Span 1		Span 2
Factored load moments	−1.15	−8.28	−10.06
Secondary moments	+0.46	+0.42	+0.71
Design moments at column	−0.69	−7.86	−9.35
Moment reduction to face, Vc/3	+0.43	+1.19	+1.40
Design moments at critical section	−0.26	−6.67	−7.95

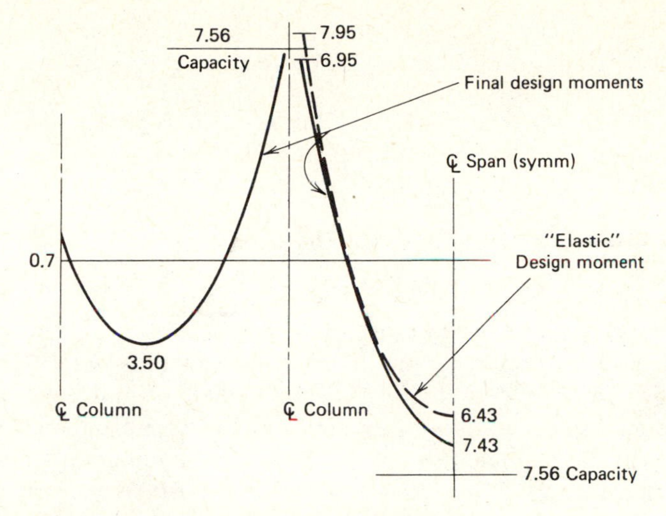

Fig. 16-16. Factored load design moments, Design Example 16-4.

Calculate design midspan moments:
Span 1

$$V_{\text{exterior}} = \frac{0.202\times 17}{2} - \frac{(8.28-1.15)}{17} = 1.72-0.42=1.30\ \text{k/ft}\ (18.97\ \text{kN/m})$$

$$V_{\text{center}} = 01.72+0.42=2.14\ \text{k/ft}\ (31.22\ \text{kN/mm})$$

Point of zero shear and maximum moment:

$$x=\frac{1.30}{0.202}=6.45\ \mathit{ft}\ (1.97\ \mathit{m})\ \text{from exterior column}$$

End span positive moment =

$$1.30\times 6.45 = +8.38$$

$$-0.202\frac{(6.45)^2}{2} = -4.20$$

$$\text{End moment} = \underline{-1.15}$$

$$+3.03$$

$$\text{Secondary moment } M_2 = \underline{\ 0.45\ }$$

$$+M\ \text{max} = +3.50\ \text{ft k/ft}\ (15.57\ \text{kN-m/m})$$

Span 2

$$V = \frac{0.202 \times 25}{2} = 2.52 \text{ k/ft } (36.77 \text{ kN/m})$$

$$+\text{Moment} = 0.202 \times 25^2/8 = 15.78 \text{ ft-k/ft}$$

$$\text{End moment} = \frac{-10.06}{+5.72}$$

$$\text{Secondary moment } M_2 = \underline{+0.71}$$

$$+M \text{ max.} = 6.43 \text{ ft-k/ft } (28.60 \text{ kN-m/m})$$

(*b*) *Calculation of Flexural Strength*
Check interior support section. Section 18.9 of the ACI Code requires a minimum amount of mild steel at the immediate column zone regardless of service load stress conditions, or strength, unless more than the minimum is required for ultimate flexural capacity. This minimum amount is to help insure the integrity of the punching zone so that full shear capacity can be developed, and is determined by the following expression:

$$A_s = 0.00075h \times L$$

The initial check of flexural strength will be made considering this steel.

$$A_s = 0.00075 \times 6.5 \times (17+25)/2 \times 12 = 1.22 \text{ in.}^2 \ (787 \text{ mm}^2)$$

Say six No. 4 Bars × 9 ft (2.74 m). Space at maximum 6 in. (152 mm) on center so that bars are placed within a width of column plus $1\frac{1}{2}$ slab thickness each side of column. Then for average one-foot strip.

$$A_s = 6 \times \frac{0.2}{20} = 0.06 \text{ in.}^2/\text{ft } (127 \text{ mm}^2/\text{m})$$

Calculate stress in tendon at ultimate, use ACI Code equation 18-4:

$$f_{ps} = f_{se} + \frac{f'_c}{100\rho_p} + 10 \text{ ksi}$$

since we have $10 - \frac{1}{2}$ in. (12.7 mm) diameter tendons in a 20-ft (6.1 m) bay, each with area = 0.153 in.^{2} (98.7mm^2):

$$\rho_p = \frac{A_{ps}}{bd} = \frac{10 \times 0.153}{20 \times 12 \times 5.5} = 0.00116$$

$$f_{se} = 0.7 \times 270 - 30 \text{ ksi losses} = 159 \text{ ksi } (1096 \text{ N/m})$$

$$f_{pe} = 159 + 10 + \frac{4.000}{0.00116 \times 100} = 169 + 34.5$$

$$= 203.5 \text{ ksi } (1403 \text{ N/mm}^2)$$

$$F_{su} = \frac{203.5 \times 0.153 \times 10}{20} = 15.57 \text{ k/ft}$$

$$F_u = 60 \times 0.06 = 3.60$$

$$F = \text{total force} = 19.17 \text{ k/ft } (279.7 \text{ kN/m})$$

$$\text{Depth of compression block } a = \frac{F}{0.85 b f_c'} = \frac{19.17}{0.85 \times 12 \times 4} = 0.47 \text{ in.}$$

since bars and tendons are in the same layer:

$$\left(d - \frac{a}{2}\right) = \frac{(5.5 - 0.47/2)}{12} = 0.438 \text{ ft } (134 \text{ mm})$$

Moment capacity at column centerline =

$$M_u = 0.9 \times 0.438 \times 19.17 = 7.56 \text{ ft-k/ft } (33.63 \text{ kN-m})$$

This value is less than the required capacity of 7.95 ft-k/ft (35.36 kN-m/m).

Calculate available capacity at midspan and allowable inelastic moment redistribution at column. See the ACI Code, section 18.10.

Allowable redistribution:

$$= 20\% \times \left(1 - \frac{\omega_p}{0.30}\right)$$

$$\Sigma\omega = \frac{19.12}{5.5 \times 12 \times 4} = 0.072 < 0.20$$

$$R = 20\left(1 - \frac{0.072}{0.30}\right) = 15.2\%$$

$$M_R = 0.152 \times 7.56 = 1.15 \text{ ft-k } (1.56 \text{ kN-m})$$

Since the midspan of span 2 requires six No. 4 bars from service load consideration, the flexural strength is also 7.56 ft-k (10.25 kN-m). The required moment capacity is 6.43 ft-k (8.72 kN-m), which leaves 1.13 ft-k (1.53 kN-m) available to accommodate moment redistributed from the support section. If 1.0 ft-k (1.36 kN-m) is redistributed:

$$-M = -7.95 + 1.00 = -6.95 < 7.56 \text{ OK}$$

$$+M = +6.43 + 1.00 = +7.43 < 7.56 \text{ OK}$$

Thus minimum rebar and tendons are adequate for strength. Midspan sections of spans 1 and 3 have more than adequate capacity by comparison with span 2.

The flexural capacity at exterior columns is governed by moment transfer requirements. Since moment transfer also involves shear stresses, the two aspects will be treated under the heading of shear.

(3) Ultimate Shear Capacity

(a) *Shear at Exterior Column*

Vertical shear at exterior column calculated above at 1.30 k/ft (18.97 kN/m). Total shear at exterior column = 20 × 1.30 = 26.0 k (115.6 kN).

Assume exterior skin is masonry and glass averaging 0.4 klf (5.84 kN/m).

$$\text{Slab shear} = 26.0$$
$$V_u = 1.4 \times 0.4 \times 20 = 11.2$$
$$\text{Total shear} = 37.2 \text{ k } (165.5 \text{ kN})$$

(b) Moment Transfer
At exterior columns, a portion of the total bay moment is transferred to the column by the eccentricity of the critical section relative to the column center. In order to evaluate these cases the properties of the critical section must first be calculated.

Shear Section Properties

The critical shear section is taken at $d/2$ from the face of the column as per Fig. 12-19. Referring to Fig. 12-19,

$$\text{Assume } d = 0.8 \times 6.5 = 5.2 \text{ in. } (132 \text{ mm})$$
$$c_1 = 12 \text{ in. } (305 \text{ mm})$$
$$c_2 = 14 \text{ in. } (356 \text{ mm})$$
$$c_m = 14 + 5.2 = 19.2 \text{ in. } (488 \text{ mm})$$
$$c_t = 12 + \frac{5.2}{2} = 14.6 \text{ in } (371 \text{ mm})$$
$$A_c = 5.2(19.2 + 2 \times 14.6) = 252 \text{ in.}^2 \ (163 \times 10^3 \text{ mm}^2)$$
$$c_{AB} = 14.6^2 \times 5.2/252 = 4.40 \text{ in. } (112 \text{ mm})$$
$$c_{CD} = 14.6 - 4.40 = 10.2 \text{ in. } (259 \text{ mm})$$
$$g = 10.2 - 12/2 = 4.2$$

$$\alpha = 1 - \frac{1}{1 + 2/3(c_m/c_t)^{1/2}} = 1 - \frac{1}{1 + 2/3(19.2/14.6)^{1/2}} = 0.43$$

$$J_c = \frac{dc_t^3}{6} + \frac{c_t d^3}{6} + c_m d c_{AB}^2 + 2c_t d\left(\frac{c_t}{2} - c_{AB}\right)^2$$

$$J_c = \frac{5.2 \times 14.6^3}{6} + \frac{14.6 \times 5.2^3}{6} + 19.2 \times 5.2 \times 4.40^2 + 2 \times 14.6 \times 5.2\left(\frac{14.6}{2} - 4.40\right)^2$$

$$J_c = 2697 + 342 + 1937 + 1277 = 6249 \text{ in.}^4 \ (2.60 \times 10^9 \text{ mm}^4)$$

Total bay moment at column centerline:

$$M_u = 20 \times (-0.69) = -13.8 \text{ ft.k } (-18.7 \text{ kN-m})$$

Moment transferred by eccentricity of shear reaction:

$$Vg = 26.0 \times \frac{4.2}{12} = 9.10 \text{ ft-k } (12.3 \text{ kN-m})$$

Net moment to be transferred $= M_t = 13.8 - 9.10 = 4.70$ ft-k (6.37 kN-m)
Amount to be transferred by shear $= \alpha M_t = 0.43 \times 4.70 = 2.02$ ft-k (2.74 kN-m)

$$v_c = \frac{37{,}200}{252 \times 0.85} + \frac{2{,}020 \times 12 \times 4.40}{6249 \times 0.85}$$

$$v_c = 173 + 20 = 193 \text{ psi } (1.33 \text{ N/mm}^2)$$

Shear stress allowable according to Equation 11-13 of the ACI Code:

$$v_c = 3.5\sqrt{4000} + 0.3 \times 155 \text{ (ignoring } V_p/A_c)$$

$$v_c = 221 + 46 = 267 \text{ psi } (1.84 \text{ N/mm}^2)$$

Allowable shear stress exceeds calculated stress.

Check flexural moment transfer.

Although the flexural moment to be transferred is small, for illustrative purposes, calculate the capacity of the section of width equal to the width of the column plus $1\frac{1}{2}$ slab thicknesses each side. Assume that of the 10 tendons required for the 20-ft (6.1 m) bay width, 2 are anchored within the column cage and are bundled together across the building. This minimum amount and location of posttensioning force should be noted on the structural drawings. Besides providing flexural capacity, this prestress will act directly on the critical section for shear and enhance the shear strength. As previously shown, a minimum amount of mild steel is required at all columns. For this joint the area of rebar is:

$$A_s = 0.00075 \times 6.5 \times 12 \times 17 = 1.0 \text{ in.}^2 \text{ } (645 \text{ mm}^2)$$

Say five No. 4 bars $\times$ 5 ft (1.52 m) including standard hook

$$A_s = 1.0 \text{ in.}^2 \text{ } (645 \text{ mm}^2)$$

Calculate stress in strand tendon:

$$b = 14 + 3\times 6.5 = 33.5 \text{ in. } (851 \text{ mm})$$

$$f_{pe} = \frac{33.5\times 3.25\times 4}{100\times 0.153\times 2} + 169 = 183 \text{ ksi } (1262 \text{ N/mm}^2)$$

$$F_p = \frac{183}{159}\times 24.33\times 2 = 56.0 \text{ k } (249 \text{ kN})$$

$$F_y = 60\times 5\times 0.2 = 60.0 \text{ k } (267 \text{ kN})$$

$$T_u = 116.0 \text{ k } (516 \text{ kN})$$

$$a = \frac{116}{0.85\times 4\times 33.5} = 1.02 \text{ in } (25.9 \text{ mm})$$

$$\text{Tendon } j_u d = \frac{(3.25 - 1.02/2)}{12} = 0.23 \text{ ft } (70.1 \text{ mm})$$

$$\text{Rebar } j_u d = (5.5 - 1.02/2)/12 = 0.42 \text{ ft } (128.0 \text{ mm})$$

$$\phi M_u = 0.9\times(0.23\times 56 + 0.42\times 60) = 34.3 \text{ ft-k } (46.5 \text{ kN-m})$$

Much greater than moment transfer requirement.

(c) Shear at Interior Column

Direct shear left and right of interior columns is calculated in (2) above.
Total direct shear = (2.14 + 2.52)20 = 93.20 k (414.6 kN)
Moment transfer $M_t = 20(9.35 - 7.86) = 29.80$ ft. kips (40.41 kN-m)
Shear section properties (See Fig. 12-19)

$$d = 6.5 - 1.0 = 5.5 \text{ in. } (140 \text{ mm})$$

$$d + c_1 = 5.5 + 20 = 25.5 \text{ in. } (648 \text{ mm})$$

$$\text{at torsion faces, } d = 0.8\times 6.5 = 5.2 \text{ in. } (132 \text{ mm})$$

$$d + c_2 = 5.2 + 14 = 19.2 \text{ in. } (488 \text{ mm})$$

$$b_0 d = 2\times(25.5\times 5.2 + 19.2\times 5.5)$$

$$= 476 \text{ in.}^2 \ (307\times 10^3 \text{ mm}^2)$$

Polar moment of inertia:

$$J = 2\left(5.2\times(25.5)^3/12 + 25.5\times\frac{(5.2)^3}{12} + 19.2\times 5.5\times\left(\frac{25.5}{2}\right)^2\right)$$

$$= 49{,}300 \text{ in.}^4 \ (20.5\times 10^9 \text{ mm}^4)$$

$$\frac{J}{12\times c} = \frac{49{,}300}{12\times 25.5/2} = 322.2 \text{ in.}^2\text{-ft. } (63.4\times 10^6 \text{ mm})$$

Portion of moment to be transferred by torsional shear:

$$\alpha = 1.0 - \frac{1.0}{1.0 + 2/3\sqrt{\frac{25.5}{19.2}}} = 43.3\%$$

$$M_{vt} = 0.433 \times 29.80 = 12.90 \text{ ft-k } (17.49 \text{ kN-m})$$

$$M_{vf} = 29.80 - 12.90 = 16.90 \text{ ft-k } (22.92 \text{ kN-m})$$

Shear stresses:

$$\text{Direct shear} = v_u = \frac{93.20}{476} = 0.196 \text{ ksi } (1.35 \text{ N/mm})$$

$$\text{Torsional shear} = v_t = \frac{12.90}{322.2} = 0.040$$

$$\text{Total shear} = 0.236 \text{ ksi } (1.63 \text{ N/mm}^2)$$

$$\text{divided by } \phi = 0.85 = 0.276 \text{ ksi } (1.90 \text{ N/mm})$$

Calculated allowable shear stress by ACI Code equation 11-13:

$$v_{cw} = 3.5\sqrt{f_c'} + 0.3F/A + V_p/A_c$$

$$v_{cw} = 3.5\sqrt{4000} + 0.3 \times 0.155 = 0.268 \text{ ksi } (1.85 \text{ N/mm}^2)$$

$0.268 < 0.276$, but V_p component of tendon force crossing the critical section will obviously make up the slight deficiency. Moment transfer by flexure:

$$M_{vf} = 16.90 \text{ ft-k } (22.92 \text{ kN-m})$$

$$b = 14 + 3 \times 6.5 = 33.5 \text{ in. } (851 \text{ mm})$$

Say $F_p = 56.0$ k (249 kN) as per exterior column.

$$A_s = 0.00075 \times 6.5 \times 21.0 \times 12 = 1.22 \text{ in.}^2 \ (787 \text{ mm}^2)$$

Use six No. 4 bars $A_s = 1.20$ in.^{2} (774 mm^2)

$$F_y = 1.20 \times 60 = 72.0 \text{ k } (320 \text{ kN})$$

$$T_u = 56.0 + 72.0 = 128.0 \text{ k } (569 \text{ kN})$$

$$\text{depth of compression block, } a = \frac{128}{0.85 \times 4 \times 33.5} = 1.12 \text{ in. } (28.4 \text{ mm})$$

$$\left(d - \frac{a}{2}\right) = \frac{[5.5 - (1.12/2)]}{12} = 0.41 \text{ ft } (125.0 \text{ mm})$$

$$\phi M_u = 0.9 \times 0.41 \times 128 = 47.2 \text{ ft-k } (64.0 \text{ kN-m})$$

Moment transfer capacity in flexure of 47.2 ft-k (64.0 kN-m) is much greater than the required flexural moment transfer of 16.90 ft-k (22.9 kN-m).

It is strongly emphasized that tendons in each direction should pass directly through the column reinforcing cage. In addition to shear contribution at design loads, these tendons provide a residual shear capacity after complete failure of the punching section by acting in pure tension or "catenary" action. Shear reinforcement may be designed using the ACI Code, section 11.11.

(4) Deflection. Calculate live load deflection of a 1-ft (0.30 m) strip in 25-ft (7.62 m) span by area moment procedure.

$$\text{Live load} = 0.040 \text{ ksf}, \ (1.92 \text{ kN/m}^2)$$

$$I = 12 \times 6.5^3/12 = 275 \text{ in.}^4 \ (114 \times 10^6 \text{ mm}^4)$$

$$E = 3.8 \times 10^3 \text{ ksi} \ (26.2 \text{ kN/mm}^2)$$

$$EI = 1045 \times 10^3 \text{ k-in.}^2$$

$$\text{Support moments} = \frac{0.040}{0.202} \times 10.06 = 1.99 \text{ ft-k} \ (2.70 \text{ kN-m})$$

Deflection at midspan can be solved for these properties = +0.081 in (+2.06 mm) approximate live load deflection in 20-ft (6.1 m) spans interior span is similar to fixed end beam

$$= \frac{wL^4}{384EI} = \frac{0.04 \text{ x } 240^4}{384 \times 1{,}045{,}000 \times 12} = 0.027 \text{ in. } (0.68 \text{ mm})$$

total deflection at center of panel = 0.108 in. (2.74 mm).
Calculated live load deflection = 25 × 12/0.108 = span 2778.

The net dead load in this span is 0.038 ksf (1.82 kn/m), so by comparison with the live load deflection calculated above, the long-term dead load deflection, assuming a creep factor of 2, is approximately 0.24 in. (6.10 mm) and the total long-term dead plus instantaneous live load deflection at the center of the panel would be less than 0.36 in. (9.14 mm). Permissible long-term plus short-term deflection = span/838. The calculated deflections of this structure are satisfactory.

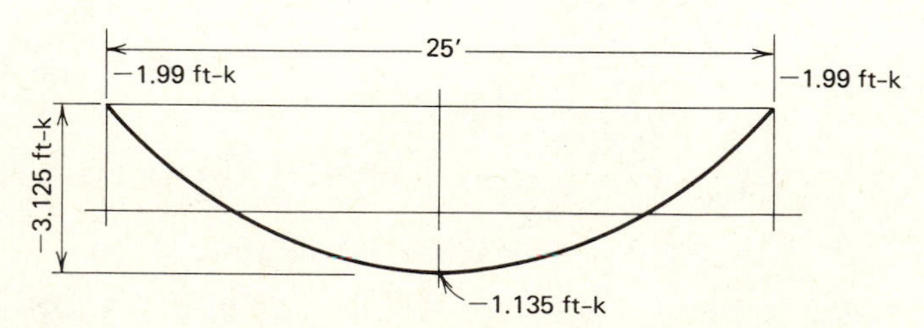

Fig. 16-17. Moments for deflection calculation, Design Example 16-4.

(5) Distribution of Tendons. The Commentary to Section 18.12 of the ACI Code includes the following discussion of tendon distribution and spacing:

"Tests indicate that the strength of prestressed slabs is controlled by total tendon strength rather than by tendon distribution. While it is important that some tendons pass directly over columns, distribution elsewhere is not critical and any rational method which satisfies statics may be used. It is suggested that maximum spacing of tendons in column strips should not exceed 4 times the slab thickness and that the maximum spacing in middle strips or for uniformly distributed tendons should not exceed 6 times the slab thickness."

In accordance with the above discussion, the 10 tendons per 20 ft (6.1 m) bay in this design will be distributed in a group of 2 tendons directly through the column with the remaing 8 tendons spaced at 2 ft 3 in. (0.686 m) centers (about four times the slab thickness). Tendons in the direction perpendicular to the tendons designed in this example to be placed in a narrow band through and immediately adjacent to the columns.

References

1. *Post-Tensioning Manual*, Post-Tensioning Institute, Phoenix, Arizona, 1976.
2. P. F. Rice and E. S. Hoffman, *Structural Design Guide to the ACI Building Code*, Van Nostrand Reinhold Company, New York, 1972, 437 pp.

APPENDIX A

DEFINITIONS, NOTATIONS, ABBREVIATIONS

Definitions

1. *Pretensioning and posttensioning.* Any method of prestressing concrete members in which the reinforcement is tensioned before (after) the concrete is placed.
2. *Full and partial prestressing.* Degree of prestress applied to concrete in which no tension (some tension) is permitted in the concrete under the working loads.
3. *Circular and linear prestressing.* Circular prestressing refers to prestressing in round members like tanks and pipes; prestressing in all other members is termed linear.
4. *Transfer.* The transferring of prestress to the concrete. For pretensioned members, transfer takes place at the release of prestress from the bulkheads; for posttensioned members it takes place after the completion of the tensioning process.
5. *Bonded and unbonded reinforcement.* Reinforcement bonded (not bonded) throughout its length to the surrounding concrete.
6. *Anchored and non-end-anchored reinforcement.* Reinforcement anchored at its ends (not anchored) by means of mechanical devices capable of transmitting the tensioning force to the concrete.
7. *Prestressed and nonprestressed reinforcement.* Reinforcement in prestressed -concrete members, which are elongated (not elongated) with respect to the surrounding concrete.
8. *Tendons.* Another name for prestressed reinforcement, whether wires, bars, or strands.
9. *Cables.* A group of tendons, or the c.g.s. of all the tendons.
10. *Concordant and nonconcordant cables.* Cables or c.g.s. lines which produce a *C*-line or line of pressure coincident (noncoincident) with the c.g.s. line itself.
11. *Linear transformation.* Moving the position of a c.g.s. line over the interior supports of a continuous beam without changing the intrinsic shape of the line within each individual span.
12. *Girder load, working load, service load, cracking load, and ultimate load.* GIRDER LOAD: The weight of the beam or girder itself pluS whatever weight

is on it at the time of transfer. WORKING LOAD OR SERVICE LOAD: The normally maximum total load which the structure is specified or expected to carry. CRACKING LOAD: The total load required to initiate cracks in a prestressed-concrete member. ULTIMATE LOAD: The total load which a member or structure can carry up to total rupture.

13. *Load factor.* Constant used to increase working or service load to the ultimate load required in strength design. Table 1-2 gives some of the values.
14. *Creep.* Time-dependent inelastic deformation of concrete or steel resulting solely from the presence of stress and a function thereof.
15. *Shrinkage of concrete.* Contraction of concrete due to drying and chemical changes, dependent on time but not directly dependent on stresses induced by external loading.
16. *Relaxation.* Time-dependent inelastic deformation of steel resulting from sustained stress and a function thereof.

Notations

Greek Letters

Δ = deflection of beams.
Δ_a = total deformation of anchorage.
Δ_s = total strain in steel.
δ = unit strain.
δ_i = initial unit strain in concrete, due to elastic shortening.
δ_t = final unit strain in concrete, including the effect of creep but not of shrinkage.
δ_s = unit strain in steel.
μ = coefficient of friction.
θ or α = change in angle of tendons; angles in general.
α, β, γ = torsional constants.
$\rho = A_s/bd$.
= ratio of nonprestressed tension reinforcement.
$\rho' = A'_s/bd$.
$\rho_p = A_{ps}/bd$.
= ratio of prestressed reinforcement.
ϕ = capacity reduction factor.
$\omega = \rho f_y/f'_c$.
$\omega' = \rho' f_y/f'_c$.
$\omega_p = \rho_p f_{ps}/f'_c$.
$\omega_w, \omega_{pw}, \omega'_w$ = reinforcement indices for flanged sections computed as for ω, ω_p, and ω' except that b shall be the web width, and the steel area shall be that required to develop the compressive strength of the web only.

English Letters

A = cross-sectional area in general.

A_c = net cross-sectional area of concrete; or area of precast portion.

A_g = gross cross-sectional area of concrete.

A_t = gross cross-sectional area of concrete, including steel transformed by ratio n.

C = center of compressive force, center of pressure, or center of thrust; or carry-over moment, in moment distribution.

C' = C at ultimate load.

C_c = coefficient of creep = δ_t/δ_i.

c = distance from c.g.c. to extreme fiber.

c_b, c_t = c for bottom (top) fibers; $c_{b1}, c_{t1}, c_{b2}, c_{t2}$ for compressive portion at transfer or under working load, respectively; c_b', c_t' for composite sections.

c.g.s. = center of gravity of steel area.

c.g.c. = center of gravity (centroid) of concrete section; c.g.c.′ for composite section.

e = eccentricity in general.

$e_1 e_2, e_b, e_t$ = various eccentricities as defined locally in text.

F = total effective prestress after deducting losses.

F_a = average prestress in steel for a given length.

F_1, F_2 = total prestress at points 1 and 2, respectively.

F_i = total initial prestress before transfer.

F_0 = total prestress, just after transfer.

FEM = fixed-end moment, in moment distribution.

f = unit stress in general.

f_1, f_2 = unit stresses at stages or points 1 and 2, respectively.

f_r = modulus of rupture of concrete.

f_a = average unit stress in steel for a given length.

f_c = unit stress in concrete.

f_c' = ultimate unit stress in concrete, generally at 28 days old.

f_{ci}' = ultimate unit stress in concrete, at time of transfer.

f_i = initial unit prestress in steel before transfer.

f_0 = unit prestress in steel, just after transfer.

f_e = effective unit prestress in steel after deducting losses.

Δf_s = change in f_s.

f_t, f_b = fiber stress at top (bottom) fibers.

f_t', f_b' = tensile fiber stress at top (bottom) fibers.

f_c'' = principal compressive stress.

f_t'' = principal tensile stress.

k = coefficient for depth of compressive area kd in a beam section; or as defined locally.

$k' = k$ at ultimate load.

k_1 = ratio of average stress in ultimate compression area of beam to f'_c.

k_t, k_b = kern distances from c.g.c. for top (bottom) $= r^2/c_b(r^2/c_t)$; $k_{t1}, k_{b1}, k_{t2}, k_{b2}$ for compressive portion at transfer or under working load, respectively; k'_t, k'_b for composite section.

M_u = moment at factored load.

M_n, M' = ultimate moment capacity $\geqslant M_u/\phi$.

M_A, M_B = moment at point $A(B)$.

M_C = moment acting on composite section.

M_{cr} = cracking moment.

M_L = moment due to total live load only.

M_G = moment due to girder load, including any load on the beam or girder at time of transfer.

M_P = moment on precast portion of composite section.

M_S = moment due to superimposed load.

M_T = moment due to total load.

M_1, M_2 = primary (resulting) moments in a continuous beam.

m = load factor or factor of safety.

m_b, m_t = ratio of section moduli of precast portion to composite section for bottom (top) fiber.

n = modular ratio E_s/E_c.

T = total tension in prestressed steel or center of total tension.

$T' = T$ at ultimate load.

$T_1 = T$ in nonprestressed steel.

$T'_1 = T'$ in nonprestressed steel.

V_u = total shear in beam at factored load.

V_c = total shear carried by concrete. (V_{ci} or V_{cw} lesser value).

V_s = total shear carried by steel.

V_n = nominal ultimate shear strength $\geqslant V_u/\phi$.

w = load or weight per length.

$w' = w$ at ultimate load.

$w_c, w'_c = w$ and w' for continuous beams, respectively.

w_G = girder load, plf.

w_S = superimposed load, plf.

y = perpendicular distance from c.g.c. line to said fiber.

Refer to the Appendix of the ACI Code for all the equations of the code in parallel U.S. customary and SI metric units. The following conversion factors will be helpful, along with the other equations, as engineers begin making this conversion.

CONVERSION FACTORS—U.S. CUSTOMARY UNITS TO SI METRIC UNITS

OVERALL GEOMETRY

Spans	1 ft = 0.3048 m
Displacements	1 in. = 25.4 mm
Surface area	1 ft^2 = 0.0929 m^2
Volume	1 ft^3 = 0.0283 m^3
	1 yd^3 = 0.765 m^3

STRUCTURAL PROPERTIES

Cross-sectional dimensions	1 in. = 25.4 mm
Area	1 in.^{2} = 645.2 mm^2
Section modulus	1 in.^{3} = 16.39×10^3 mm^3
Moment of inertia	1 in.^{4} = 0.4162×10^6 mm^4

MATERIAL PROPERTIES

Density	1 lb/ft^3 = 16.03 kg/m^3
Modulus and stress	1 lb/in.^{2} = 0.006895 N/mm^2
	1 k/in.^{2} = 6.895 N/mm^2

LOADINGS

Concentrated loads	1 lb = 4.448 N
	1 k = 4.448 kN
Density	1 lb/ft^3 = 0.1571 kN/m^3
Linear loads	1 k/ft = 14.59 kN/m
Surface loads	1 lb/ft^2 = 0.0479 kN/m^2
	1 k/ft^2 = 47.9 kN/m^2

STRESSES AND MOMENTS

Stress	1 lb/in.^{2} = 0.006895 N/mm^2
	1 k/in.^{2} = 6.895 N/mm^2
Moment or torque	1 ft-lb = 1.356 N-m
	1 ft-k = 1.356 kN-m

Metric equivalent of limiting values

U.S. CUSTOMARY	SI METRIC
$3\sqrt{f_c'}$	$0.25\sqrt{f_c'}$
$6\sqrt{f_c'}$	$0.50\sqrt{f_c'}$
$7.5\sqrt{f_c'}$	$0.62\sqrt{f_c'}$

APPENDIX B

DATA FOR SOME PRESTRESSING SYSTEMS

POSTTENSIONING COMPANIES—ADDRESSES

CONCRETE CONSTRUCTION SUPPLY, INC.
3609 Dividend Drive
Garland, Texas 75042

CONTINENTAL CONCRETE STRUCTURES
4487 "C" Park Drive
Norcross, GA 30091

CONTINENTAL STRUCTURES, INC.
20815 Belshaw Avenue
Carson, California 90746

DYCKERHOFF AND WIDMANN, INC.
Dywidag Prestressed Concrete
529 Fifth Avenue
New York, NY 10017
[Other offices:
Baltimore, Maryland
Greenville, South Carolina
Lemont, Illinois
Lincoln Park, New Jersey
Richardson, Texas
San Diego, California]

INRYCO, INC.
Post-Tensioning Division
P. O. Box 1056
Melrose Park, IL 60160

THE PRESCON CORPORATION
8918 Tesoro Drive
P. O. Box 17450
San Antonio, TX 78217
[Other offices in or near:
Atlanta, GA; Baltimore, MD;
Columbus, OH; Denver CO;
Houston, TX; Los Angeles, CA;
Memphis, TN; New York, NY;
San Francisco, CA]

PRESCON/FREYSSINET INTERNATIONAL
66, Route de la Reine
92100 Boulogne, Billancourt
France

TITAN PRESTRESSING CORPORATION, LTD
#209-1035-64 Avenue S.E.
Calgary, Alberta T2H 2J7
Canada

VSL CORPORATION
101 Albright Way
P. O. Box 459
Los Gatos, CA 95030
[Other offices in or near:
Atlanta, GA; Chicago, IL;
Dallas, TX; Denver, CO;
Honolulu, HI; Houston, TX;
Los Angeles, CA; Miami, FL;
Washington, DC]

WESTERN CONCRETE STRUCTURES CO., INC.
9113 South Hamilton Avenue
P. O. Box 440
Gardena, CA 90247
[Other offices in or near: New
Orleans, LA; Atlanta, GA]

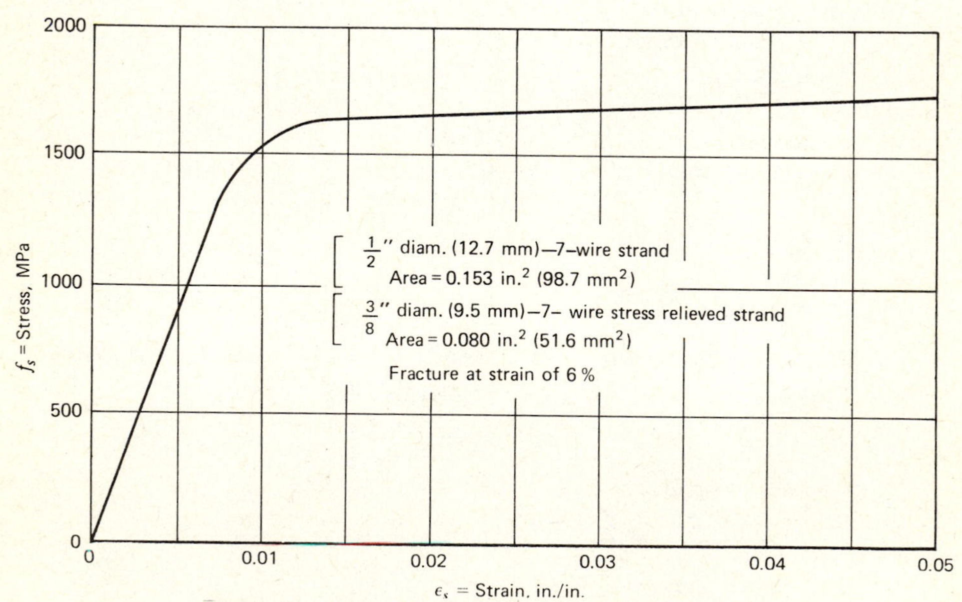

Fig. B-1. Stress strain curve for prestressing steel.

Properties of Prestressing Steel

	U.S. Customary			Metric		
Type	Nominal Diameter (in.)	Nominal Area (in.^{2})	Nominal Weight (lb/ft)	Nominal Diameter (mm)	Nominal Area (mm^2)	Nominal Mass (kg/m)
Seven-wire strand (Grade 250)	$\frac{1}{4}$(0.250)	0.036	0.12	6.350	23.2	0.179
	$\frac{5}{16}$(0.313)	0.058	0.20	7.950	37.4	0.298
	$\frac{3}{8}$(0.375)	0.080	0.27	9.525	51.6	0.402
	$\frac{7}{16}$(0.438)	0.108	0.37	11.125	69.7	0.551
	$\frac{1}{2}$(0.500)	0.144	0.49	12.700	92.9	0.729
	(0.600)	0.216	0.74	15.240	139.4	1.101
Seven-wire strand (Grade 270)	$\frac{3}{8}$(0.375)	0.085	0.29	9.525	54.8	0.432
	$\frac{7}{16}$(0.438)	0.115	0.40	11.125	74.2	0.595
	$\frac{1}{2}$(0.500)	0.153	0.53	12.700	98.7	0.789
	(0.600)	0.215	0.74	15.240	138.7	1.101
Prestressing wire	0.192	0.029	0.098	4.877	18.7	0.146
	0.196	0.030	0.10	4.978	19.4	0.149
	0.250	0.049	0.17	6.350	31.6	0.253
	0.276	0.060	0.20	7.010	38.7	0.298

	U.S. Customary			Metric		
Type	Nominal Diameter (in.)	Nominal Area (in.^{2})	Nominal Weight (lb/ft)	Nominal Diameter (mm)	Nominal Area (mm^2)	Nominal Mass (kg/m)
Prestressing bars (smooth)	$\frac{3}{4}$	0.44	1.50	19.050	283.9	2.232
	$\frac{7}{8}$	0.60	2.04	22.225	387.1	3.036
	1	0.78	2.67	25.400	503.2	3.973
	$1\frac{1}{8}$	0.99	3.38	28.575	638.7	5.030
	$1\frac{1}{4}$	1.23	4.17	31.750	793.5	6.206
	$1\frac{3}{8}$	1.48	5.05	34.925	954.8	7.515
Prestressing bars (deformed)	$\frac{5}{8}$	0.28	0.98	15.875	180.6	1.458
	$\frac{3}{4}$	0.42	1.49	19.050	271.0	2.217
	1	0.85	3.01	25.400	548.4	4.480
	$1\frac{1}{4}$	1.25	4.39	31.750	806.5	6.535
	$1\frac{3}{8}$	1.58	5.56	34.925	1006.5	8.274

Properties of Reinforcing Steel (unstressed)

Reinforcing Bars

	U.S. Customary			Metric		
Bar Size	Nominal Diameter (in.)	Nominal Area (in.^{2})	Nominal Weight (lb/ft)	Nominal Diameter (mm)	Nominal Area (mm^2)	Nominal Mass (kg/m)
3	0.375	0.11	0.376	9.525	71	0.560
4	0.500	0.20	0.668	12.700	129	0.994
5	0.625	0.31	1.043	15.875	200	1.552
6	0.750	0.44	1.502	19.050	284	2.235
7	0.875	0.60	2.044	22.225	387	3.042
8	1.000	0.79	2.670	25.400	510	3.973
9	1.128	1.00	3.400	28.651	645	5.060
10	1.270	1.27	4.303	32.258	819	6.404
11	1.410	1.56	5.313	35.814	1,006	7.907
14	1.693	2.25	7.650	43.002	1,452	11.385
18	2.257	4.00	13.600	57.328	2,581	20.240

Wire Reinforcement

W and D Size		U.S. Customary			Metric		
Smooth	Deformed	Nominal Diameter (in.)	Nominal Area (in.^{2})	Nominal Weight (lb/ft)	Nominal Diameter (mm)	Nominal Area (mm^2)	Nominal Mass (kg/m)
W31	D31	0.628	0.310	1.054	15.951	200.0	1.569
W30	D30	0.618	0.300	1.020	15.697	193.6	1.518
W28	D28	0.597	0.280	0.952	15.164	180.7	1.417
W26	D26	0.575	0.260	0.934	14.605	167.7	1.390
W24	D24	0.553	0.240	0.816	14.046	154.8	1.214
W22	D22	0.529	0.220	0.748	13.437	141.9	1.113
W20	D20	0.504	0.200	0.680	12.802	129.0	1.012
W18	D18	0.478	0.180	0.612	12.141	116.1	0.911
W16	D16	0.451	0.160	0.544	11.455	103.2	0.810
W14	D14	0.422	0.140	0.476	10.719	90.3	0.708
W12	D12	0.390	0.120	0.408	9.906	77.4	0.607
W11	D11	0.374	0.110	0.374	9.500	71.0	0.557
W10.5		0.366	0.105	0.357	9.296	67.7	0.531
W10	D10	0.356	0.100	0.340	9.042	64.5	0.506
W9.5		0.348	0.095	0.323	8.839	61.3	0.481
W9	D9	0.338	0.090	0.306	8.585	58.1	0.455
W8.5		0.329	0.085	0.289	8.357	54.8	0.430
W8	D8	0.319	0.080	0.272	8.103	51.6	0.405
W7.5		0.309	0.075	0.255	7.849	48.4	0 380
W7	D7	0.298	0.070	0.238	7.569	45.2	0.354
W6.5		0.288	0.065	0.221	7.315	41.9	0.329
W6	D6	0.276	0.060	0.204	7.010	38.7	0.304

DYWIDAG POSTTENSIONING SYSTEM DETAILS*

Available in $\frac{5}{8}$ in., 1 in., $1\frac{1}{4}$ in., and $1\frac{3}{8}$ in. nominal diameter, Dywidag Threadbars are hot rolled and proof stressed alloy steel conforming to ASTM A-722–75.

The Dywidag Threadbar prestressing steel has a continuous rolled-in pattern of threadlike deformations along its entire length. More durable than machined threads, the deformations allow anchorages and couplers to thread onto the threadbar at any point.

Available in mill lengths to 60 ft, threadbars may be cut to specified lengths before shipment to the job site. Or where circumstances warrant, the threadbars may be shipped to the job site in mill lengths for field cutting with a portable friction or band saw. Threadbars may be coupled for ease of handling or to extend a previously stressed bar.

Prestressing steel properties

Nominal Threadbar Diameter (in.)	Ultimate Stress (f_{pu}-ksi)	Cross-Section Area (A_{ps}-in.^{2})	Ultimate Strength ($f_{pu}A_{ps}$)	Prestressing Force–(k) 0.80 $f_{pu}A_{ps}$	0.70 $f_{pu}A_{ps}$	0.60 $f_{pu}A_{ps}$	Weight[b] (lb/ft)	Minimum Elastic Bending Radius (ft)
$\frac{5}{8}$	157	0.28	43.5	34.8	30.5	26.1	0.98	26
1	150	0.85	127.5	102.0	89.3	76.5	3.01	52
1	160[a]	0.85	136.0	108.8	95.2	81.6	3.01	49
$1\frac{1}{4}$	150	1.25	187.5	150.0	131.3	112.5	4.47	64
$1\frac{1}{4}$	160[a]	1.25	200.0	160.0	140.0	120.0	4.47	60
$1\frac{3}{8}$	150	1.58	237.0	189.6	165.9	142.2	5.56	72

[a] Check on availability before specifying.
[b] Shipping weight may vary.

Anchorage Details				
Threadbar diameter (in.)	$\frac{5}{8}$	1	$1\frac{1}{4}$	$1\frac{3}{8}$
Bell anchor size (in.)	$3\frac{1}{4}\varnothing \times 1\frac{1}{2}$	$5\frac{1}{2}\varnothing \times 2\frac{5}{8}$	$6\frac{3}{4}\varnothing \times 2\frac{5}{8}$	$7\frac{3}{4}\varnothing \times 3\frac{1}{8}$
Anchor plate size[a] (in.)	$3\times 3\times \frac{3}{4}$	$5\times 5\frac{1}{2}\times 1\frac{1}{4}$	$6\times 7\times 1\frac{1}{2}$	$7\times 7\frac{1}{2}\times 1\frac{3}{4}$
	$2\times 5\times 1$	$4\times 6\frac{1}{2}\times 1\frac{1}{4}$	$5\times 8\times 1\frac{1}{2}$	$5\times 9\frac{1}{2}\times 1\frac{3}{4}$
Nut extension (in.)	1	$1\frac{7}{8}$	$2\frac{1}{2}$	$2\frac{3}{4}$
Min. bar protrusion[b] (in.)	$2\frac{1}{2}$	3	$3\frac{1}{2}$	4

*From *Post-Tensioning Manual*, 1981. Used with permission of the Post-Tensioning Institute and Dyckerhoff and Widmann, Inc.

Coupler Details

Threadbar diameter (in.)	5/8	1	1 1/4	1 3/8
Length (in.)	3 1/2	5 1/2	6 3/4	8 5/8
Diameter (in.)	1 1/8	2	2 3/8	2 5/8

Sheathing Details

Threadbar diameter (in.)	5/8	1	1 1/4	1 3/8
Threadbar sheathing O.D. (in.)	1	1 1/2	1 3/4	2
Threadbar sheathing I.D. (in.)	3/4	1 1/4	1 1/2	1 3/4
Coupler sheathing O.D. (in.)	1 3/4	2 3/4	3 1/4	3 3/4
Coupler sheathing I.D. (in.)	1 3/8	2 3/8	2 7/8	3 3/8

Pocket Former Details

Threadbar diameter (in.)	5/8	1	1 1/4	1 3/8
Length (in.)	4 3/4	7	8	8 5/8
Maximum diameter (in.)	3 1/8	5 1/8	6 1/2	6 1/2

[a] Other plate sizes available on special order.
[b] To accommodate stressing.

Tendon Assembly with Bell Anchorage

The components of Dywidag Threadbar System are manufactured in the United States exclusively by Dyckerhoff & Widmann, Inc.

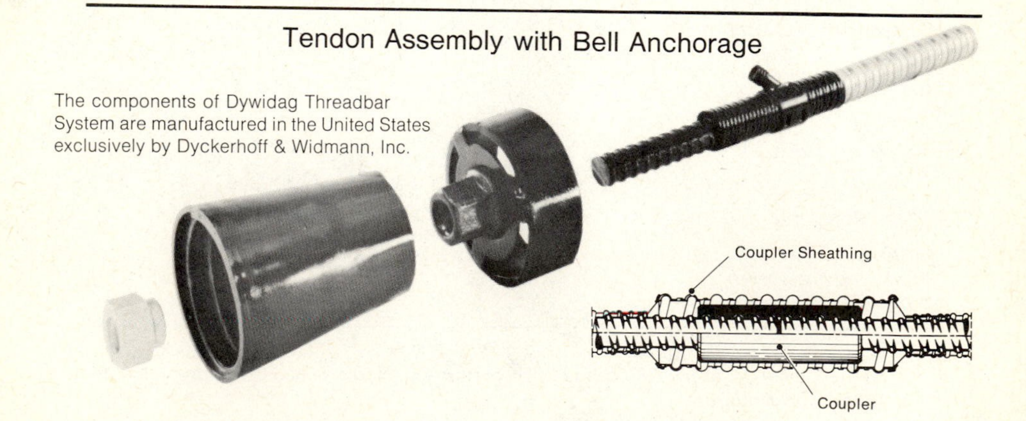

BBRV TENDONS AND ANCHORS*

BBRV posttensioning was developed in 1949 by four Swiss engineers, Birkenmeier, Brandestini, Ros and Vogt, whose initials give the system its name.

A BBRV tendon consists of several (or many) parallel lengths of $\frac{1}{4}$ in. high-strength wire with each end of each wire terminating in a cold-formed buttonhead, after the wire passes separately through a machined anchorage fixture. The system provides for the simultaneous stressing of all the wires in a tendon and the buttonheads allow development of ultimate tendon forces.

BBRV tendons are generally used in bonded installations, and the tendons completely shop fabricated to exact job-site specifications.

In building applications tendon sizes up to 52 wires are most commonly used, while sizes up to 170 wires are normally required for bridge construction, large rock anchors and in containment structures of nuclear power plants.

Stressing Anchors for BBRV Tendons

Type BG **(see p. 604)**

Anchor designation		18 BG	24 BG	30 BG	38 BG	46 BG	52 BG
No. of wires (max.)		18	24	30	38	46	52
Bearing plate	A	$7\frac{1}{2}\times7\frac{1}{2}$	$8\frac{1}{2}\times8\frac{1}{2}$	$9\frac{1}{2}\times9\frac{1}{2}$	$10\frac{1}{2}\times10\frac{1}{2}$	$11\frac{1}{2}\times11\frac{1}{2}$	$12\frac{1}{4}\times12\frac{1}{4}$
size (in.)	B	$6\times9\frac{1}{2}$	$7\times10\frac{1}{2}$	$8\times11\frac{1}{2}$	$9\times12\frac{1}{4}$	$10\times13\frac{1}{2}$	$11\times13\frac{1}{2}$
Trumpet O.D. (in.)		$3\frac{3}{4}$	$4\frac{1}{2}$	5	$5\frac{1}{4}$	$5\frac{1}{2}$	$5\frac{3}{4}$
Conduit O.D. (in.)		$1\frac{5}{8}$	$1\frac{7}{8}$	2	$2\frac{3}{8}$	$2\frac{5}{8}$	$2\frac{3}{4}$
Effective prestressing force at $0.6f_{pu}$ (k)		127.3	169.7	212.1	268.7	325.2	367.6

Recessed anchor used where stressing pocket depth is limited. Type BM also available.

Type MG **(see p. 604)**

Anchor designation		8 MG	12 MG	18 MG	24 MG	30 MG	38 MG	46 MG
No. of wires (max.)		8	12	18	24	30	38	46
Bearing plate	A	$3\frac{1}{2}\times8\frac{1}{4}$	$4\times8\frac{1}{2}$	5×10	$6\times11\frac{1}{2}$	$7\times12\frac{1}{2}$	$8\times13\frac{3}{4}$	$9\times14\frac{3}{4}$
size (in.)	B	$3\frac{1}{2}\times10$	$5\times7\frac{3}{4}$	$6\times9\frac{3}{4}$	$7\times11\frac{1}{4}$	$8\times12\frac{1}{4}$	9×14	10×15
Trumpet O.D. (in.)		$1\frac{1}{2}$	3	3	4	$4\frac{1}{2}$	5	$5\frac{1}{2}$
Conduit O.D. (in.)		$1\frac{1}{4}$	$1\frac{3}{8}$	$1\frac{5}{8}$	$1\frac{7}{8}$	2	$2\frac{3}{8}$	$2\frac{5}{8}$
Effective prestressing force at $0.6f_{pu}$ (k)		56.6	84.8	127.3	169.7	212.1	268.7	325.2

Nonrecessed type, preferred where there is no limitation on depth of stressing pocket.

*Used with permission of Inryco, Inc.

Tendons with $\frac{1}{4}$ in. Wires—Constants
(Prescon, BBRV, Western Concrete Systems)

Wires per Tendon	Tendon Area (sq in.)	Tendon Weight (lb/ft)	P_e 144 ksi (k)	P_i 168 ksi (k)	P_{max} 192 ksi (k)	P_{ult} 240 ksi (k)
			$(0.6f_{pu})$	$(0.7f_{pu})$	$(0.8f_{pu})$	(f_{pu})
4	0.1964	0.667	28.3	33.0	37.7	47.1
5	0.2455	0.833	35.4	41.2	47.1	58.9
6	0.2946	1.000	42.4	49.5	56.6	70.7
7	0.3437	1.167	49.5	57.7	66.0	82.5
8	0.3928	1.333	56.6	66.0	75.4	94.3
9	0.4419	1.500	63.6	74.2	84.8	106.1
10	0.4910	1.667	70.7	82.5	94.3	117.8
12	0.5892	2.000	84.8	99.0	113.1	141.4
14	0.6874	2.333	99.0	115.5	132.0	165.0
16	0.7856	2.667	113.1	132.0	150.8	188.5
18	0.8838	3.000	127.3	148.5	169.7	212.1
20	0.9820	3.333	141.4	165.0	188.5	235.7
24	1.1784	4.000	169.7	198.0	226.2	282.8
30	1.4730	5.000	212.1	247.5	282.8	353.5
38	1.8658	6.333	268.7	313.5	358.2	447.8

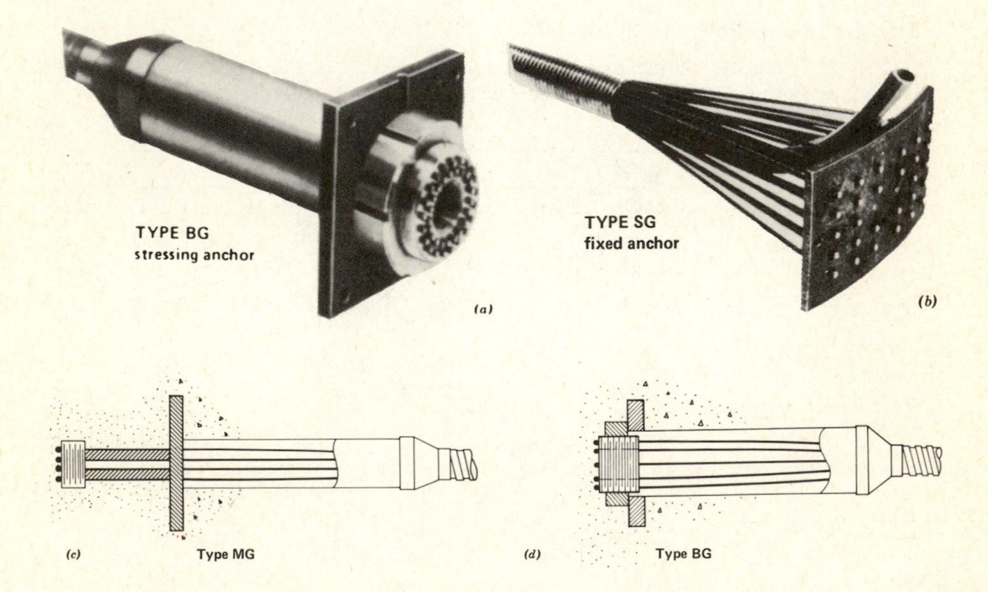

The Freyssinet K-Range system*

General All tendons in the K system can either be premade and pulled into the sheath or the strands pushed one by one into the sheath, before or after concreting, to suit the construction sequence.

Duct friction Duct diameters are chosen from experience to give the best compromise between the practical problems of tendon-threading and grout injection and the economical factors of tendon concentration and material costs.

The following constants are recommended for design purposes where normal steel sheath is used:

For curvature: $\mu = 0{\cdot}26$

For unintentional variation (wobble): $K = 0.0010/\text{ft}$.

The above figures based on the FIP Recommendations are for clean strands and sheath free of internal surface rusting. However, where specially coated sheath or water soluble emulsions on the cable are used, lower figures may be adopted and figures can be recommended for individual cases.

Special operations All K range tendons may at the time of stressing be:

1. Retensioned as many times as necessary to achieve the force of extension required.
2. Tensioned in as many movements as is necessary for structural considerations.
3. Detensioned—this operation is carried out on each strand individually using a single-strand type of jack which needs to be made available. This operation should only be requested in an emergency.

The 0.5-in. K monogroup range

Anchorage	No. of	Internal Sheath Diameter (min)	Prestressing Force (k) % Min. Breaking Strength			System and
Type	Strands	(in.)	70%	80%	100%	Jack Type
1M5	1	1	28·9	33·0	41·3	Monostrand
7K5	2		57·8	66·0	82·6	K100
	3		86·7	99·0	124	
	4		116	132	165	
	5		145	165	207	
	6		174	198	248	
	7	$2\frac{1}{8}$	202	231	289	

*Used with permission of Freyssinet International.

The 0.5-in. K monogroup range

Anchorage Type	No. of Strands	Internal Sheath Diameter (m) (in.)	Prestressing Force (k) % Min. Breaking Strength 70%	80%	100%	System and Jack Type
12K5	8		231	264	330	K200
	9		260	297	372	
	10		289	330	413	
	11		318	363	454	
	12	$2\frac{1}{2}$	347	396	496	
19K5	13		376	429	537	K350
	14		405	462	578	
	15		434	495	620	
	16		463	528	661	
	17		491	561	702	
	18		520	594	743	
	19	$3\frac{3}{8}$	549	627	785	
27K5	20		578	660	826	K500
	21		607	693	867	
	22		636	726	909	
	23		665	759	950	
	24		694	792	991	
	25		729	825	1032	
	26		752	858	1074	
	27	$3\frac{3}{4}$	781	891	1115	
37K5	28		810	924	1156	K700
	29		838	957	1198	
	30		867	990	1239	
	31		896	1023	1280	
	32		925	1056	1322	
	33		954	1089	1363	
	34		982	1122	1404	
	35		1012	1155	1033	
	36		1041	1188	1074	
	37	$4\frac{1}{2}$	1070	1221	1528	
55K5	38 to 55	$5\frac{1}{8}$	1590	1817	2271	K1000

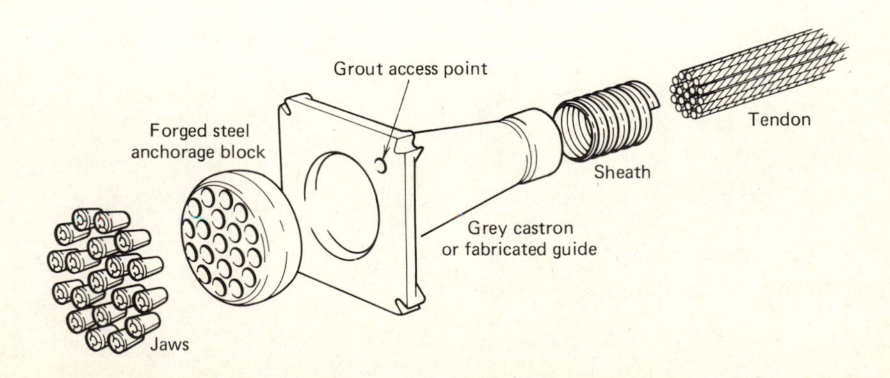

Inches

	1M5	1M6	7K5 4K6	12K5 7K6	12K6	19K5	27K5 19K6	37K5 27K6	55K5 37K6
a	7·75	7·75	7·75	10·5	11·75	12·75	14·75	17·75	20·75
b	3·5	3·5	4·5	5·0	6·75	7·75	9·25	10·75	12·75
c	3·5	3·5	7·0	9·75	11·75	13·75	9·75	21·75	25·5
d	3·5	3·5	4·0	4·75	5·0	5·0	5·5	6·0	6·25
e	—	—	8·75	10·5	14·0	15·0	17·75	21·5	25·0
f	—	—	4·25	5·0	6·75	7·0	8·25	10·0	11·75
g	—	—	27·5	34·5	39·25	39·25	49·25	59·0	69·0
h	—	—	4·5	6·0	7·0	7·75	9·25	10·75	12·75
k	—	—	13·0	15·0	17·75	17·75	22·5	27·25	31·5
l	—	—	10·5	11·75	14·75	14·75	19·0	22·75	26·25

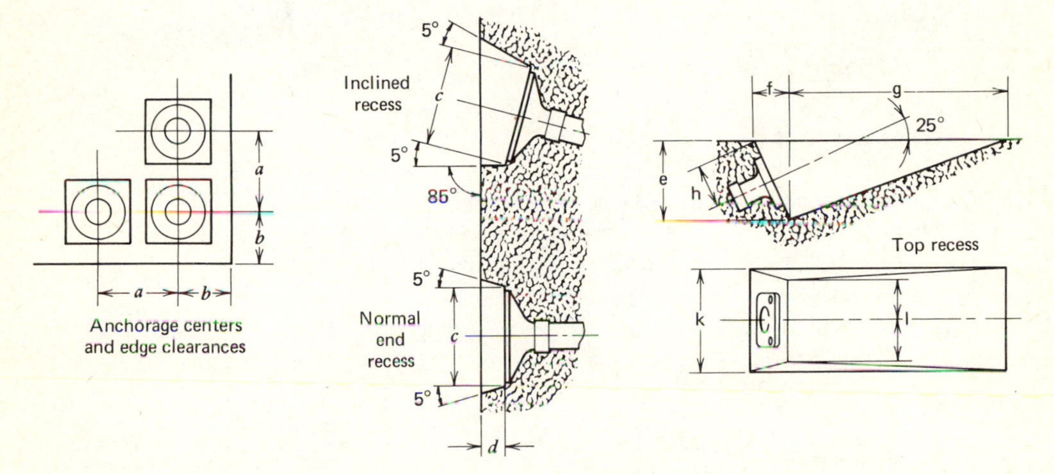

The K Range jacks The K-Range of Jacks are center-hole rams of the hydraulic double-acting type with fixed cylinder and moving piston.

The attachment of the strand to the jack is by specially designed wide-angle, multiuse Jaws, which are self-releasing on completion of jacking.

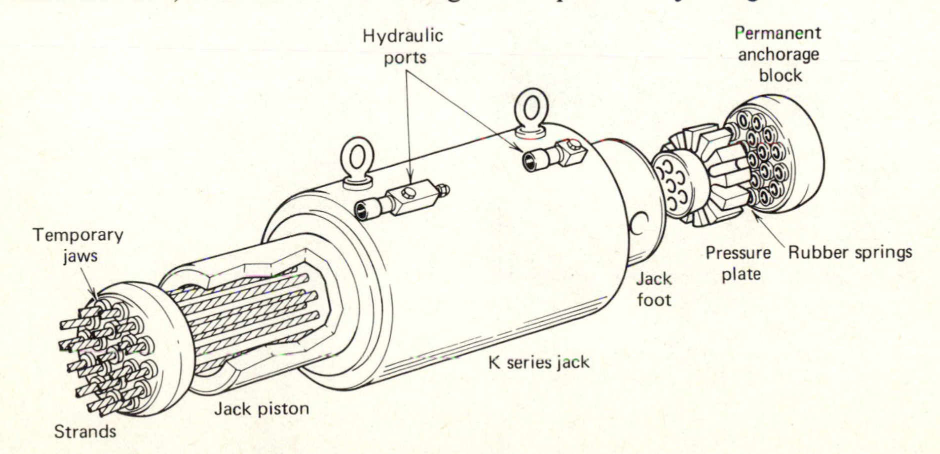

Multistrand Systems*

The 270-k 7-wire strand for prestressed concrete with an ultimate stress (f_s') of 270,000 psi is produced and tested in accordance with the requirements of ASTM A-416. Physical properties of the $\frac{1}{2}$ in. dia. 270K, uncoated and stress-relieved strand are as follows:

Ultimate strength	41,300 lb
Yield strength (at 1% extension)	35,100 lb
Approx. modulus of elasticity	27,000,000 psi[a]
Min. elongation at rupture	3.5% in 24 in.

Anchorages

Stressing Anchorage VSL Type E The stressing anchorage, a VSL type E, consists of an anchor head, VSL grippers, bearing plate and sleeve.

This anchorage can also be used as an accessible fixed (dead-end) anchorage when an immediate introduction of the full prestressing force is required at the end of the member to be prestressed.

Bearing plates shown are standard and are based on a bearing stress of 3000 psi at working force. Compressive strength of concrete at the time of initial prestress $f_{ci}' = 3500$ psi.

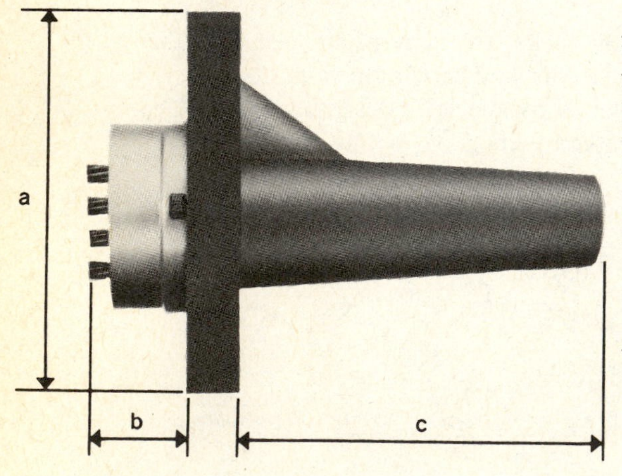

Unit	*a*	*b*	*c*
E5-3	$5\frac{1}{4} \times 5\frac{1}{4}$	$3\frac{1}{2}$	4
E5-4	$6\frac{1}{8} \times 6\frac{1}{8}$	$3\frac{1}{2}$	$4\frac{1}{2}$
E5-7	8×8	$3\frac{1}{2}$	8
E5-12	$10\frac{1}{2} \times 10\frac{1}{2}$	$3\frac{1}{2}$	12
E5-19	$13\frac{1}{4} \times 13\frac{1}{4}$	4	23
E5-22	$14\frac{3}{8} \times 14\frac{3}{8}$	$4\frac{1}{2}$	26
E5-31	17×17	5	34
E5-55	23×23	8	42

Dimensions in inches.

*Used with permission of VSL Corporation.

[a] These figures may vary slightly for different manufacturers.

$\frac{1}{2}$ in. dia. 270 ksi strands

Unit	No. of Strands	Steel Area (sq in.)	Weight (lb/ft)	Sheath Diameter in Inches: Flexible Metal Tubing I.D.	Sheath Diameter in Inches: Rigid Thin Wall Tubing O.D.	Max. Temp. Force $0.8 f_{pu}$ (k)	Initial Force $0.7 f_{pu}$ (k)
E5-3	2	0.306	1.050	$1\frac{1}{4}$	$1\frac{1}{2}$	66.1	57.8
	3	0.459	1.575	$1\frac{1}{2}$	$1\frac{3}{4}$	99.1	86.7
E5-4	4	0.612	2.100	$1\frac{5}{8}$	$1\frac{3}{4}$	132.2	115.6
E5-7	5	0.765	2.625	$1\frac{3}{4}$	$2\frac{1}{16}$	165.2	144.5
	6	0.918	3.150	$1\frac{7}{8}$	$2\frac{1}{16}$	198.2	173.5
	7	1.071	3.675	2	$2\frac{1}{4}$	231.3	202.4
E5-12	8	1.224	4.200	2	$2\frac{1}{4}$	264.3	231.3
	9	1.377	4.725	$2\frac{1}{8}$	$2\frac{7}{16}$	297.4	260.2
	10	1.530	5.250	$2\frac{1}{4}$	$2\frac{7}{16}$	330.4	289.1
	11	1.683	5.775	$2\frac{3}{8}$	$2\frac{7}{16}$	363.4	318.0
	12	1.836	6.300	$2\frac{1}{2}$	$2\frac{13}{16}$	396.5	346.9
E5-19	13	1.989	6.825	$2\frac{5}{8}$	3	429.5	375.8
	14	2.142	7.350	$2\frac{5}{8}$	3	462.6	404.7
	15	2.295	7.875	$2\frac{3}{4}$	$3\frac{3}{16}$	495.6	433.6
	16	2.448	8.400	$2\frac{7}{8}$	$3\frac{3}{16}$	528.6	462.6
	17	2.601	8.925	3	$3\frac{9}{16}$	561.7	491.5
	18	2.754	9.450	3	$3\frac{9}{16}$	594.7	520.4
	19	2.907	9.975	$3\frac{1}{8}$	$3\frac{9}{16}$	627.8	549.3
E5-22	20	3.060	10.500	$3\frac{1}{4}$	$3\frac{3}{4}$	660.8	578.2
	21	3.213	11.025	$3\frac{1}{4}$	$3\frac{3}{4}$	693.8	607.1
	22	3.366	11.550	$3\frac{3}{8}$	$3\frac{3}{4}$	726.9	636.0
E5-31	23	3.519	12.075	$3\frac{1}{2}$	$3\frac{15}{16}$	759.9	664.9
	24	3.672	12.600	$3\frac{1}{2}$	$3\frac{15}{16}$	793.0	693.8
	25	3.825	13.125	$3\frac{5}{8}$	$3\frac{15}{16}$	826.0	722.7
	26	3.978	13.650	$3\frac{5}{8}$	$3\frac{15}{16}$	859.0	751.7
	27	4.131	14.175	$3\frac{3}{4}$	$4\frac{5}{16}$	892.1	780.6
	28	4.284	14.700	$3\frac{7}{8}$	$4\frac{5}{16}$	925.1	809.5
	29	4.437	15.225	$3\frac{7}{8}$	$4\frac{5}{16}$	958.2	838.4
	30	4.590	15.750	4	$4\frac{1}{2}$	991.2	867.3
	31	4.743	16.275	4	$4\frac{1}{2}$	1024.2	896.2
E5-55	55	8.415	28.875	$5\frac{1}{2}$	6	1817.6	1590.4

APPENDIX C

CONSTANTS FOR BEAM SECTIONS

Table C-1 Constants for T-Sections

Section	b'/b	t/h	A^a	$c_b{}^b$	$c_t{}^b$	I^c	$r^{2\,d}$	$k_t{}^b$	$k_b{}^b$
1-*a*	0.1	0.1	$0.19bh$	$0.714h$	$0.286h$	$0.0179bh^3$	$0.0945h^5$	$0.132h$	$0.333h$
1-*b*	0.1	0.2	0.28	0.756	0.244	0.0192	0.0688	0.0910	0.282
1-*c*	0.1	0.3	0.37	0.755	0.245	0.0193	0.0520	0.0689	0.212
1-*d*	0.1	0.4	0.46	0.735	0.265	0.0202	0.0439	0.0597	0.165
1-*e*	0.2	0.1	0.28	0.629	0.371	0.0283	0.1010	0.161	0.272
1-*f*	0.2	0.2	0.36	0.678	0.322	0.0315	0.0875	0.129	0.272
1-*g*	0.2	0.3	0.44	0.691	0.309	0.0319	0.0725	0.105	0.234
20*h*	0.2	0.4	0.52	0.684	0.316	0.0316	0.0616	0.090	0.195
1-*i*	0.3	0.1	0.37	0.585	0.415	0.0365	0.0985	0.169	0.237
1-*j*	0.3	0.2	0.44	0.626	0.374	0.0408	0.0928	0.148	0.248
1-*k*	0.3	0.3	0.51	0.645	0.355	0.0417	0.0819	0.127	0.231
1-*l*	0.3	0.4	0.58	0.645	0.355	0.0417	0.0720	0.112	0.203
1-*m*	0.4	0.1	0.46	0.559	0.441	0.0440	0.0954	0.171	0.216
1-*n*	0.4	0.2	0.52	0.592	0.408	0.0486	0.0935	0.158	0.229
1-*o*	0.4	0.3	0.58	0.609	0.391	0.0499	0.0860	0.141	0.220
1-*p*	0.4	0.4	0.64	0.612	0.388	0.0502	0.0785	0.128	0.205
1-*q*	1.0	1.0	1.00	0.500	0.500	0.0833	0.0833	0.167	0.167

[a] Given as a function of bh.
[b] Given as a function of h.
[c] Given as a function of bh^3.
[d] Given as a function of h^2.

TABLE 1

Constants for T-Sections

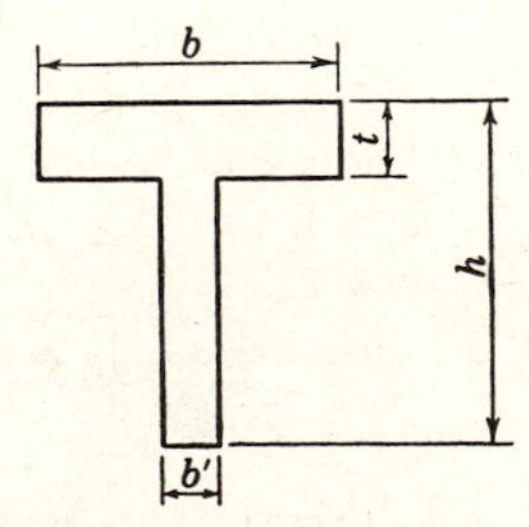

Cross section for Table C-1 constants.

Table C-2 Constants for I-Sections

Section	b'/b	t/h	A[a]	c_b[b]	c_t[b]	I[c]	r^2[d]	k_t[b]	k_b[b]
2-*a*	0.1	0.1	0.21bh	0.650h	0.350h	0.0260bh^3	0.1236h^2	0.190h	0.354h
2-*b*	0.1	0.2	0.32	0.675	0.325	0.0345	0.1080	0.160	0.332
2-*c*	0.1	0.3	0.43	0.672	0.328	0.0387	0.0900	0.134	0.274
2-*d*	0.2	0.1	0.29	0.610	0.390	0.0316	0.1090	0.179	0.280
2-*e*	0.2	0.2	0.38	0.647	0.353	0.0378	0.0994	0.153	0.282
2-*f*	0.2	0.3	0.47	0.655	0.345	0.0402	0.0856	0.131	0.248

[a]Given as a function of bh.
[b]Given as a function of h.
[c]Given as a function of bh^3.
[d]Given as a function of h^2.

Cross section for Table C-2 constants.

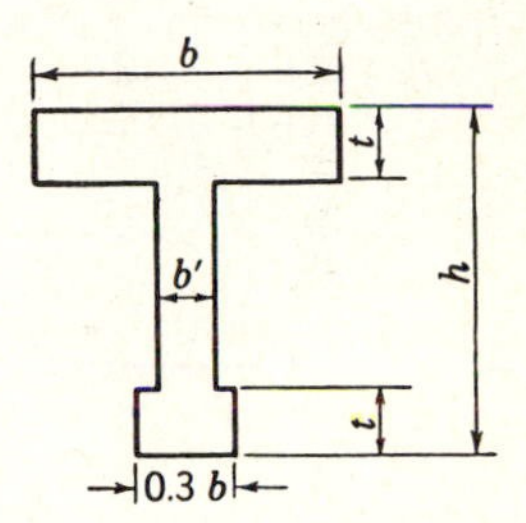

Table C-3 Constants for I-Sections

Section	b'/b	t/h	A[a]	c_b[b]	c_t[b]	I[c]	r^2[d]	k_t[b]	k_b[b]
3-*a*	0.1	0.1	0.23bh	0.597h	0.403h	0.0326bh^3	0.1420h^2	0.238h	0.352h
3-*b*	0.1	0.2	0.36	0.611	0.389	0.0464	0.1288	0.210	0.331
3-*c*	0.1	0.3	0.49	0.606	0.394	0.0535	0.1090	0.180	0.274
3-*d*	0.2	0.1	0.31	0.572	0.428	0.0373	0.1204	0.210	0.282
3-*e*	0.2	0.2	0.42	0.595	0.405	0.0488	0.1160	0.195	0.286
3-*f*	0.2	0.3	0.53	0.599	0.401	0.0540	0.1020	0.170	0.254
3-*g*	0.3	0.1	0.39	0.557	0.430	0.0443	0.1103	0.198	0.250
3-*h*	0.3	0.2	0.48	0.582	0.418	0.0510	0.1065	0.183	0.255
3-*i*	0.3	0.3	0.57	0.592	0.408	0.0553	0.0970	0.164	0.238

[a]Given as a function of bh.
[b]Given as a function of h.
[c]Given as a function of bh^3.
[d]Given as a function of h^2.

Cross section for Table C-3 constants.

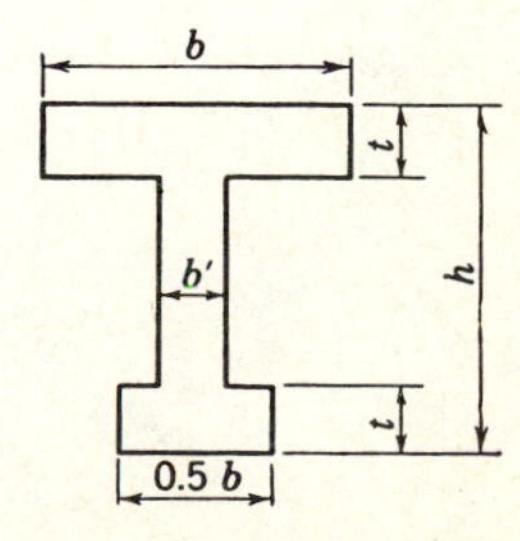

Table C-4 Constants for I-Sections

Section	b'/b	t/h	A[a]	c_b[b]	c_t[b]	I[c]	r^2[d]	k_t[b]	k_b[b]
4-*a*	0.1	0.1	$0.25bh$	0.554	$0.446h$	$0.0381bh^3$	$0.1525h^2$	$0.276h$	$0.342h$
4-*b*	0.1	0.2	0.40	0.560	0.440	0.0560	0.1391	0.248	0.316
4-*c*	0.1	0.3	0.55	0.557	0.443	0.0651	0.1182	0.212	0.267
4-*d*	0.2	0.1	0.33	0.540	0.460	0.0425	0.1290	0.239	0.280
4-*e*	0.2	0.2	0.46	0.552	0.448	0.0578	0.1258	0.228	0.281
4-*f*	0.2	0.3	0.59	0.553	0.447	0.0657	0.1113	0.202	0.249
4-*g*	0.3	0.1	0.41	0.534	0.466	0.0467	0.1140	0.214	0.244
4-*h*	0.3	0.2	0.52	0.546	0.454	0.0598	0.1150	0.210	0.254
4-*i*	0.3	0.3	0.63	0.550	0.450	0.0663	0.1051	0.191	0.234

[a]Given as a function of bh.
[b]Given as a function of h.
[c]Given as a function of bh^3.
[b]Given as a function of h^2.

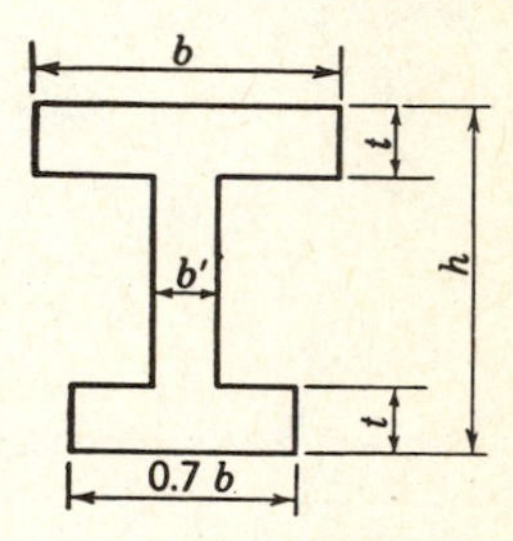

Cross section for Table C-4 constants.

Table C-5 Constants for I-Sections

Section	b'/b	t/h	A[a]	c_b[b]	c_t[b]	I[c]	r^2[d]	k_t[b]	k_b[b]
5-*a*	0.1	0.1	$0.21bh$	$0.350h$	$0.650h$	$0.0260bh^3$	$0.1236h^2$	$0.354h$	$0.190h$
5-*b*	0.1	0.2	0.32	0.325	0.675	0.0345	0.1080	0.332	0.160
5-*c*	0.1	0.3	0.43	0.328	0.672	0.0387	0.0900	0.274	0.134
5-*d*	0.2	0.1	0.29	0.390	0.610	0.0316	0.1090	0.280	0.179
5-*e*	0.2	0.2	0.38	0.353	0.647	0.0378	0.0994	0.282	0.153
5-*f*	0.2	0.3	0.47	0.345	0.655	0.0402	0.0856	0.248	0.131

[a]Given as a function of bh.
[b]Given as a function of h.
[c]Given as a function of bh^3.
[d]Given as a function of h^2.

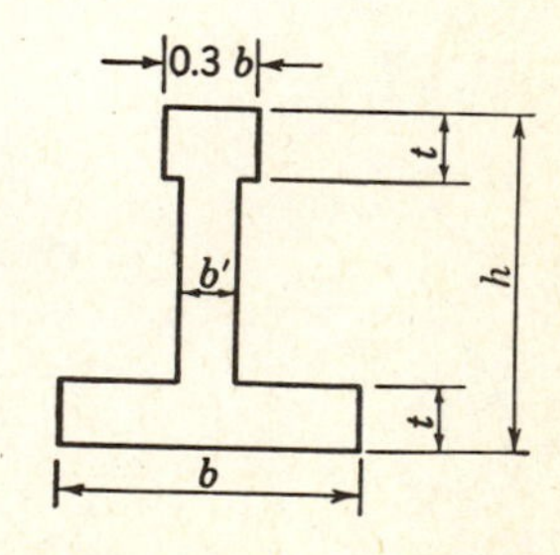

Cross section for Table C-5 constants.

Table C-6 Constants for Symmetrical I- and Box-Sections

Section	b'/b	t/h	A[a]	c_b[b]	c_t[b]	I[c]	r^2[d]	k_t[b]	k_b[b]
6-*a*	0.1	0.1	$0.28bh$	$0.500h$	$0.500h$	$0.0449bh^3$	$0.160h^2$	$0.320h$	$0.320h$
6-*b*	0.1	0.2	0.46	0.500	0.500	0.0671	0.146	0.292	0.292
6-*c*	0.1	0.3	0.64	0.500	0.500	0.0785	0.123	0.246	0.246
6-*d*	0.2	0.1	0.36	0.500	0.500	0.0492	0.137	0.274	0.274
6-*e*	0.2	0.2	0.52	0.500	0.500	0.0689	0.132	0.264	0.264
6-*f*	0.2	0.3	0.68	0.500	0.500	0.0791	0.117	0.234	0.234
6-*g*	0.3	0.1	0.44	0.500	0.500	0.0535	0.121	0.243	0.243
6-*h*	0.3	0.2	0.58	0.500	0.500	0.0707	0.122	0.244	0.244
6-*i*	0.3	0.3	0.72	0.500	0.500	0.0796	0.111	0.222	0.222
6-*j*	0.4	0.1	0.52	0.500	0.500	0.0577	0.111	0.222	0.222
6-*k*	0.4	0.2	0.64	0.500	0.500	0.0725	0.113	0.226	0.226
6-*l*	0.4	0.3	0.76	0.500	0.500	0.0801	0.105	0.211	0.211

[a] Given as a function of bh.
[b] Given as a function of h.
[c] Given as a function of bh^3.
[d] Given as a function of h^2.

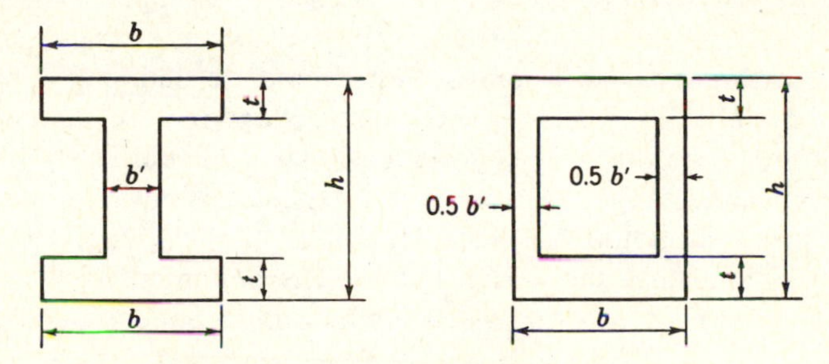

Cross section for Table C-6 constants.

APPENDIX D

PRESTRESSED CONCRETE LOSS OF PRESTRESS CALCULATIONS*

GENERAL METHOD—PCI COMMITTEE PRESTRESS LOSSES

2.4 Time-Dependent Losses (General)

Prestress losses due to steel relaxation and creep and shrinkage of concrete are interdependent and are time-dependent. To account for changes of these effects with time, a step-by-step procedure can be used with the time interval increasing with age of the concrete. Shrinkage from the time when curing is stopped until the time when the concrete is prestressed should be deducted from the total calculated shrinkage for posttensioned construction. It is recommended that a minimum of four time intervals be used as shown in Table 2.

When significant changes in loading are expected, time intervals other than those recommended should be used. Also, it is neither necessary, nor always desirable, to assume that the design live load is continually present. The four time intervals above are recommended for minimum noncomputerized calculations.

2.5 Loss Due to Creep of Concrete (*CR*)

2.5.1 Loss over each step

Loss over each time interval is given by

$$CR=(UCR)(SCF)(MCF)\times(PCR)(f_c) \tag{6}$$

where f_c is the net concrete compressive stress at the center of gravity of the prestressing force at time t_1, taking into account the loss of prestress force occurring over the preceding time interval.

*Sections from PCI Committee Report "Recommendations for Estimating Prestress Losses," *J. Prestressed Conc. Inst.*, July–August, 1975. Time-dependent losses by the General Method, and the Simplified Method equations.

Used with permission of the Prestressed Concrete Institute.

Calculations for Example 4-5(*b*) using equations from the PCI Committee Report (see summary of results in Example 4-5, Chapter 4, along with problem statement.)

The concrete stress f_c at the time t_1 shall also include change in applied load during the preceding time interval. Do not include the factor *MCF* for accelerated cured concrete.

Table 2 Time intervals

Step	Beginning time, t_1	End time, t
1	Pretensioned: anchorage of prestressing steel Posttensioned: end of curing of concrete	Age at prestressing of concrete
2	End of Step 1	Age = 30 days, or time when a member is subjected to load in addition to its own weight
3	End of Step 2	Age = 1 year
4	End of Step 3	End of service life

2.5.2 Ultimate creep loss

2.5.2.1 *Normal weight concrete (UCR)*

Moist cure not exceeding 7 days:

$$UCR = 95 - 20E_c/10^6 \geqslant 11 \qquad (7)$$

Accelerated cure:

$$UCR = 63 - 20E_c/10^6 \geqslant 11 \qquad (8)$$

2.5.2.2 *Lightweight concrete (UCR)*

Moist Cure not exceeding 7 days:

$$UCR = 76 - 20E_c/10^6 \geqslant 11 \qquad (9)$$

Accelerated cure:

$$UCR = 63 - 20E_c/10^6 \geqslant 11 \qquad (10)$$

Table 3 Creep factors for various volume-to-surface ratios

Volume to surface ratio, in.	Creep factor *SCF*
1	1.05
2	0.96
3	0.87
4	0.77
5	0.68
>5	0.68

Table 4 Creep factors for various ages of prestress and periods of cure

Age of prestress transfer, days	Period of cure, days	Creep factor, *MCF*
3	3	1.14
5	5	1.07
7	7	1.00
10	7	0.96
20	7	0.84
30	7	0.72
40	7	0.60

Table 5 Variation of creep with time after prestress transfer

Time after prestress transfer, days	Portion of ultimate creep, *AUC*
1	0.08
2	0.15
5	0.18
7	0.23
10	0.24
20	0.30
30	0.35
60	0.45
90	0.51
180	0.61
365	0.74
End of service life	1.00

Table 6 Shrinkage factors for various volume to surface ratios

Volume to surface ratio, in.	Shrinkage factor *SSF*
1	1.04
2	0.96
3	0.86
4	0.77
5	0.69
6	0.60

2.5.3 Effect of size and shape of member (SCF)

To account for the effect of size and shape of the prestressed members, the value of *SCF* in equation 6 is given in Table 3.

2.5.4 Effect of age at prestress and length of cure (MCF)

To account for effects due to the age at prestress of moist cured concrete and the length of the moist cure, the value of *MCF* in equation 6 is given in Table 4. The factors in this table do *not* apply to accelerated cured concretes nor are they applicable as shrinkage factors.

2.5.5 Variation of creep with time (AUC)

The variation of creep with time shall be estimated by the values given in Table 5. Linear interpolation shall be used between the values listed.

2.5.6 Amount of creep over each step (PCR)

The portion of ultimate creep over the time interval t_1 to t, *PCR* in equation 6, is given by the following equation:

$$PCR = (AUC)_t - (AUC)_{t_1} \tag{11}$$

2.6 Loss Due to Shrinkage of Concrete (*SH*)

2.6.1 Loss over each step

Loss over each time interval is given by

$$SH = (USH)(SSF)(PSH) \tag{12}$$

2.6.2 Ultimate loss due to shrinkage of concrete

The following equations apply to both moist cured and accelerated cured concretes.

2.6.2.1 *Normal weight concrete (USH)*

$$USH = 27{,}000 - 3000E_c/10^6 \tag{13}$$

but not less than 12,000 psi.

2.6.2.2 *Lightweight concrete (USH)*

$$USH = 41{,}000 - 10{,}000E_c/10^6 \tag{14}$$

but not less than 12,000 psi.

2.6.3 Effect of size and shape of member (SSF)

To account for effects due to the size and shape of the prestressed member, the value of SSF in equation 12 is given in Table 6.

2.6.4 Variation of shrinkage with time (AUS)

The variation of shrinkage with time shall be estimated by the values given in Table 7. Linear interpolation shall be used between the values listed.

2.6.5 Amount of shrinkage over each step (PSH)

The portion of ultimate shrinkage over the time interval t_1 to t, PSH in equation 12, is given by the following equation:

$$PSH = (AUS)_t - (AUS)_{t_1} \tag{15}$$

2.7 Loss Due to Steel Relaxation (*RET*)

Loss of prestress due to steel relaxation over the time interval t_1 to t may be estimated using the following equations. (For mathematical correctness, the value for t_1 at the time of anchorage of the prestressing steel shall be taken as $\frac{1}{24}$ of a day so that $\log t_1$ at this time equals zero.)

Table 7 Shrinkage coefficients for various curing times

Time after end of curing, days	Portion of ultimate shrinkage, AUS
1	0.08
3	0.15
5	0.20
7	0.22
10	0.27
20	0.36
30	0.42
60	0.55
90	0.62
180	0.68
365	0.86
End of service life	1.00

2.7.1 Stress-relieved steel

$$RET = f_{st}\{[\log 24t - \log 24_{t_1}]/10\} \times [f_{st}/f_{py} - 0.55] \quad (16)$$

where

$$f_{st}/f_{py} - 0.55 \geqslant 0.05$$
$$f_{py} = 0.85 f_{pu}$$

2.7.2 Low-relaxation steel

The following equation applies to prestressing steel given its low relaxation properties by simultaneous heating and stretching operations.

$$RET = f_{st}\{[\log 24t - \log 24t_1]/45\} \times [f_{st}/f_{py} - 0.55] \quad (17)$$

where

$$f_{st}/f_{py} - 0.55 \geqslant 0.05$$
$$f_{py} = 0.90 f_{pu}$$

2.7.3 Other prestressing steel

Relaxation of other types of prestressing steel shall be based upon manufacturer's recommendations supported by test data.

CHAPTER 3—SIMPLIFIED METHOD FOR COMPUTING PRESTRESS LOSSES

3.1 Scope

Computations of stress losses in accordance with the General Method can be laborious for a designer who does not have the procedure set up on a computer program. The Simplified Method is based on a large number of design examples in which the parameters were varied to show the effect of different levels of concrete stress, dead load stress, and other factors. These examples followed the General Method and the procedures given in the Design Examples.

3.2 Principles of the Simplified Method

3.2.1 Concrete stress at the critical location

Compute f_{cr} and f_{cds} at the critical location on the span. The critical location is the point along the span where the concrete stress under full live load is either in maximum tension or in minimum compression. If f_{cds} exceeds f_{cr} the simplified method is not applicable.

f_{cr} and f_{cds} are the stresses in the concrete at the level of the center of gravity of the tendons at the critical location. f_{cr} is the net stress due to the prestressing force plus the weight of the prestressed member and any other permanent loads on the member at the time the prestressing force is applied. The prestressing force used in computing f_{cr} is the force existing immediately after the prestress has been applied to the concrete. f_{cds} is the stress due to all permanent (dead) loads not used in computing f_{cr}.

Table 8 Simplified method equations for computing total prestress loss (*TL*)

Equation number	Concrete weight		Type of tendon			Tensioning		Equations
	NW	LW	SR	LR	BAR	PRE	POST	
N-SR-PRE-70	X		X			X		$TL = 33.0 + 13.8 f_{cr} - 4.5 f_{cds}$
L-SR-PRE-70		X	X			X		$TL = 31.2 + 16.8 f_{cr} - 3.8 f_{cds}$
N-LR-PRE-75	X			X		X		$TL = 19.8 + 16.3 f_{cr} - 5.4 f_{cds}$
L-LR-PRE-75		X		X		X		$TL = 17.5 + 20.4 f_{cr} - 4.8 f_{cds}$
N-SR-POST-68.5	X		X				X	$TL = 29.3 + 5.1 f_{cr} - 3.0 f_{cds}$
L-SR-POST-68.5		X	X				X	$TL = 27.1 + 10.1 f_{cr} - 4.9 f_{cds}$
N-LR-POST-68.5	X			X			X	$TL = 12.5 + 7.0 f_{cr} - 4.1 f_{cds}$
L-LR-POST-68.5		X		X			X	$TL = 11.9 + 11.1 f_{cr} - 6.2 f_{cds}$
N-BAR-POST-70	X				X		X	$TL = 12.8 + 6.9 f_{cr} - 4.0 f_{cds}$
L-BAR-POST-70		X			X		X	$TL = 12.5 + 10.9 f_{cr} - 6.0 f_{cds}$

Note: Values of TL, f_{cr}, and f_{cds} are expressed in ksi.

3.2.2 Simplified loss equations

Select the applicable equation from Table 8 or 9 substitute the values for f_{cr} and f_{cds} and compute TL or f_{se}, whichever is desired.

3.2.3 Basic parameters

The equations are based on members having the following properties:

1. Volume-to-surface ratio = 2.0.
2. Tendon tension as indicated in each equation.
3. Concrete strength at time prestressing force is applied:
   3500 psi for pretensioned members
   5000 psi for posttensioned members
4. 28-day concrete compressive strength = 5000 psi.
5. Age at time of prestressing:
   18 hours for pretensioned members
   30 days for posttensioned members
6. Additional dead load applied 30 days after prestressing.

Compare the properties of the beam being checked with Items 1 and 2. If there is an appreciable difference, make adjustments as indicated under Section 3.4.

Table 9 Simplified method equations for computing effective prestress (f_{se})

Equation number	Concrete weight NW	Concrete weight LW	Type of tendon SR	Type of tendon LR	Type of tendon BAR	Tensioning PRE	Tensioning POST	Equations
N-SR-PRE-70	X		X			X		$f_{se}=f_t-(33.0+13.8f_{cr}-11\ f_{cds})$
L-SR-PRE-70		X	X			X		$f_{se}=f_t-(31.2+16.8f_{cr}-13.5f_{cds})$
N-LR-PRE-75	X			X		X		$f_{se}=f_t-(19.8+16.3f_{cr}-11.9f_{cds})$
L-LR-PRE-75		X		X		X		$f_{se}=f_t-(17.5+20.4f_{cr}-14.5f_{cds})$
N-SR-POST-68.5	X		X				X	$f_{se}=f_{si}-(29.3+\ 5.1f_{cr}-\ 9.5f_{cds})$
L-SR-POST-68.5		X	X				X	$f_{se}=f_{si}-(27.1+10.1f_{cr}-14.6f_{cds})$
N-LR-POST-68.5	X			X			X	$f_{se}=f_{si}-(12.5+\ 7.0f_{cr}-10.6f_{cds})$
L-LR-POST-68.5		X		X			X	$f_{se}=f_{si}-(11.9+11.1f_{cr}-15.9f_{cds})$
N-BAR-POST-70	X				X		X	$f_{se}=f_{si}-(12.8+\ 6.9f_{cr}-10.5f_{cds})$
L-BAR-POST-70		X			X		X	$f_{se}=f_{si}-(12.5+10.9f_{cr}-15.7f_{cds})$

Note: Values of f_t, f_{si}, f_{cr}, and f_{cds} are expressed in ksi.

It was found that an increase in concrete strength at the time of prestressing or at 28 days made only a nominal difference in final loss and could be disregarded. For strength at prestressing less than 3500 psi or for 28-day strengths less than 4500 psi, an analysis should be made following Design Example 1.

Wide variations in Items 5 and 6 made only nominal changes in net loss so that further detailed analysis is needed only in extreme cases.

3.2.4 Computing f_{cr}

$$f_{cr}=A_s f_{si}/A_c+A_s f_{si}e^2/I_c-M'e/I_c \tag{18}$$

3.2.5 Tendon stress for pretensioned members

Except for members that are very heavily or very lightly* prestressed, f_{si} can be taken as follows:

For stress-relieved steel

$$f_{si}=0.90f_t \tag{19}$$

For low-relaxation steel

$$f_{si}=0.925f_t \tag{20}$$

3.2.6 Tendon stress for posttensioned members

Except for members that are very heavily or very lightly* prestressed, f_{si} can be taken as

$$f_{si}=0.95(T_o-FR) \tag{21}$$

3.3 Equations for Simplified Method

3.3.1 Total prestress loss

The equations in Table 8 give total prestress loss TL in ksi. This value corresponds to TL shown in the summaries of Design Examples 1 and 3.

3.3.2 Effective stress

The equations in Table 9 give effective stress in prestressing steel under dead load after losses. This value corresponds to f_{se} shown in the summary of Design Example 1.

As shown in the summary of Design Example 1, the stress existing in the tendons under dead load after all losses have taken place is the initial tension reduced by the amount of the total losses and increased by the stress created in the tendon by the addition of dead load after the member was prestressed. The increase in tendon stress due to the additional dead load is equal to $f_{cds}(E_s/E_c)$.

$$f_{se}=f_t-TL+f_{cds}(E_s/E_c) \tag{22}$$

3.3.3 Explanation of equation number

The equation number in Tables 8 and 9 defines the conditions for which each equation applies:

1. The first term identifies the type of concrete.
   N = normal weight = approximately 145 lb per cu ft
   L = lightweight = approximately 115 lb per cu ft
2. The second term identifies the steel in the tendon:
   SR = stress-relieved
   LR = low
   = relaxation
   BAR = high strength bar
3. The third term identifies the type of tensioning:
   PRE = pretensioned and is based on accelerated curing
   POST = posttensioned and is based on moist curing
4. The fourth term indicates the initial tension in percent of f_{pu}:
   For *pretensioned* tendons it is the tension at which the tendons are anchored in the casting bed before concrete is placed.

*When f_{cr} computed by Eq. (18) using the approximations for f_{si} is greater than 1600 psi or less than 800 psi the value of f_{si} should be checked as illustrated in Design Example 2.

For *posttensioned* tendons it is the initial tension in the tendon at the critical location in the concrete member after losses due to friction and anchor set have been deducted.

Notation For PCI Method

A_c = gross cross-sectional area of concrete member, sq in.
A_s = cross-sectional area of prestressing tendons, sq in.
ANC = loss of prestress due to anchorage of prestressing steel, psi
AUC = portion of ultimate creep at time after prestress transfer
AUS = portion of ultimate shrinkage at time after end of curing
CR = loss of prestress due to creep of concrete over time interval t_1 to t, psi
DEF = loss of prestress due to deflecting device in pretensioned construction, psi
e = tendon eccentricity measured from center of gravity of concrete section to center of gravity of tendons, in.
E_c = modulus of elasticity of concrete at 28 days taken as $33w^{3/2}\sqrt{f_c'}$, psi
E_{ci} = modulus of elasticity of concrete at time of initial prestress, psi
E_s = modulus of elasticity of steel, psi
ES = loss of prestress due to elastic shortening, psi
f_{cds} = concrete compressive stress at center of gravity of prestressing force due to all permanent (dead) loads not used in computing f_{cr}, psi
f_c = concrete compressive stress at center of gravity of prestressing steel, psi
f_c' = compressive strength of concrete at 28 days, psi
f_{ci}' = initial concrete compressive strength at transfer, psi
f_{cr} = concrete stress at center of gravity of prestressing force immediately after transfer, psi
f_{pu} = guaranteed ultimate tensile strength of prestressing steel, psi
f_{py} = stress at 1% elongation of prestressing steel, psi
f_{se} = effective stress in prestressing steel under dead load after losses
f_{si} = stress in tendon at critical location immediately after prestressing force has been applied to concrete
f_{st} = stress in prestressing steel at time t_1, psi
f_t = stress at which tendons are anchored in pretensioning bed, psi
FR = friction loss at section under consideration, psi
I_c = moment of inertia of gross cross section of concrete member, in.^{4}
K = friction wobble coefficient per foot of prestressing steel
l_{tx} = length of prestressing steel from jacking end to point x, ft
MCF = factor that accounts for the effect of age at prestress and length of moist cure on creep of concrete
M_{ds} = moment due to dead weight added after member is prestressed
M' = moment due to loads, including weight of member, at time prestress is applied to concrete

P = final prestress force in member after losses
P_o = initial prestress force in member
PCR = amount of creep over time interval t_1 to t
PSH = amount of shrinkage over time interval t_1 to t
RE = total loss of prestress due to relaxation of prestressing steel in pretensioned construction, psi
REP = total loss of prestress due to relaxation of prestressing steel in posttensioned construction, psi
RET = loss of prestress due to steel relaxation over time interval t_1 to t, psi
SCF = factor that accounts for the effect of size and shape of a member on creep of concrete
SH = loss of prestress due to shrinkage of concrete over time interval t_1 to t, psi
SSF = factor that accounts for the effect of size and shape of a member on concrete shrinkage
t = time at end of time interval, days
t_1 = time at beginning of time interval, days
T_o = steel stress at jacking end of posttensioning tendon, psi
T_x = steel stress at any point x, psi
TL = total prestress loss, psi
UCR = ultimate loss of prestress due to creep of concrete, psi per psi of compressive stress in the concrete
USH = ultimate loss of prestress due to shrinkage of concrete, psi
w = weight of concrete, lb per cu ft
α = total angular change of posttensioning tendon profile from jacking end to point x, radians
μ = friction curvature coefficient

Calculations For Example 4-5*

See Fig. 4-11 for details of the materials and the loading which occurs for this beam. Shown below are the calculations for the PCI Committee General Method (reference 1, Chapter 4) solution of losses as summarized in the table in part (*b*) of example 4-5.

To estimate the prestress at various stages assume the following basic data:

$$f_{ci}'=4500 \text{ psi } (31 \text{ N/mm}^2),\ E_{ci}=3.824 \text{ ksi } (26.4 \text{ kN/mm}^2),\ n'=7.2$$

$$f_c'=6000 \text{ psi } (41.4 \text{ N/mm}^2),\ E_c=4.415 \text{ ksi } (30.4 \text{ kN/mm}^2),\ n'=6.2$$

$$\frac{V}{S}=3,\ SSF=0.86,\ SCF=0.87$$

$$UCR=95-20\times 4.415=6.7 \ngeq 11. \text{ Use } UCR=11.$$

$$USH=27-3\left(\frac{4.415}{10^3}\right)=13.76 \text{ ksi}>12 \text{ ksi O.K.}$$

$$(UCR)(SCF)=9.57 \text{ ksi } (66.0 \text{ N/mm}^2)$$

$$(USH)(SSF)=11.83 \text{ ksi } (81.6 \text{ N/mm}^2)$$

Stage 1. From time we initially tension tendon to transfer at 48 hours
Relaxation

$$t_1=\tfrac{1}{24},\ t=2,\ f_{st}=0.75(270)=202.5 \text{ ksi } (1396 \text{ N/mm}^2)$$

$$f_{st}/f_{py}=(0.75)(270)/(0.85)(270)=0.882$$

$$RET=(202.5)\left[\frac{(\log 48-\log 1)}{10}\right](0.882-0.55)=11.30 \text{ ksi } (77.9 \text{ N/mm}^2)$$

Elastic shortening at transfer

$$M_G=\frac{(0.47)(65)^2(12)}{8}=2979 \text{ in.-k } (336.6 \text{ kN-m})$$

$$f_c(\text{due to } M_G)=2979\times\frac{13.2}{82{,}170}=0.479 \text{ ksi } (3.30 \text{ N/mm}^2)$$

*From section 4-11: Illustrating Computation of Loss of Prestress.

Assume $ES = 13.4$ ksi (92.4 N/mm^2),

$$f_{si} = 202.5 - 11.30 - 13.4 = 177.8 \text{ ksi } (1226 \text{ N/mm}^2)$$

$$P_0 = 177.8 \times 20 \times 0.153 = 544 \text{ k } (2.420 \text{ kN})$$

$$f_{cr} \text{ (due to } P_0) = \frac{544}{452} + \frac{544(13.2)^2}{82{,}170} = 2.357 \text{ ksi } (16.25 \text{ N/mm}^2)$$

$$f_{cr} = 2.357 - 0.479 = 1.878 \text{ ksi } (12.95 \text{ N/mm}^2)$$

$$ES = f_{cr} \times n' = 1.878 \times 7.2 = 13.5 \text{ ksi} \cong 13.4 \text{ ksi} \quad \text{O.K.}$$

No loss from shrinkage and creep in short time prior to transfer

$$SH = 0, \ CR = 0$$

Total losses

$$TL = 11.30 + 13.50 = 24.8 \text{ ksi } (171 \text{ N/mm}^2)$$

$f_{st} = 202.5 - 24.8 = 177.7$ ksi (1225 N/mm^2) (vs. 178.9 ksi = 12,234 N/mm^2 from PBEAM analysis, Table 4-8)

Stage 2: Time from transfer to 30 days (only $w_G = 0.470$ k/ft)

Relaxation

$$t_1 = 2, \ t = 30, f_{st} = 177.7 \text{ ksi } (1225 \text{ N/mm}^2), \ \frac{f_{st}}{f_{py}} = \frac{177.7}{(0.85)(270)} = 0.774$$

$$RET = (177.7)[(\log 720 - \log 48)/10](0.774 - 0.55) = 4.68 \text{ ksi } (32.3 \text{ N/mm}^2)$$

Creep

$$CR = (UCR)(SCF)(PCR)f_c = (9.57)(0.35)(1.878) = 6.29 \text{ ksi } (43.4 \text{ N/mm}^2)$$

(where $PCR = 0.35$ at 30 days)

Shrinkage

$$SH = (USH)(SSF)(PSH) = (11.83)(0.42) = 4.97 \text{ ksi } (34.3 \text{ N/mm}^2)$$

(where $PSH = 0.42$ at 30 days)

Total losses

$$TL = 4.68 + 6.29 + 4.97 = 15.94 \text{ ksi } (109.9 \text{ N/mm}^2)$$

$f_{st} = 177.7 - 15.94 = 161.76$ ksi (1115 N/mm^2)(vs. 161 ksi = 1110 N/mm^2 from PBEAM analysis, Table 4-8)

Stage 3: Time to end of 3 years (with 1 k/ft added at 30 days)

Relaxation

$$t_1 = 30, \ t = 30 \times 365 = 1095$$

$$M_{DL} = 1/8(1.0)(65)^2(12) = 6338 \text{ in.-kip } (716 \text{ kN-m})$$

$$f_c \text{ (due to } DL) = 6338 \times \frac{13.2}{82{,}170} = 1.018 \text{ ksi } (7.0 \text{ N/mm}^2)$$

$$f_{st} = 161.76 + 1.018(6.2) = 168.07 \text{ ksi } (1159 \text{ N/mm}^2),$$

$$f_{st}/f_{py} = \frac{168.07}{0.85 \times 270} = 0.732$$

$$RET = \left[(168.07)\frac{(\log 26{,}280 - \log 720)}{10}\right](0.732 - 0.55)$$

$$= 4.78 \text{ ksi } (33.0 \text{ N/mm}^2)$$

Creep

$$f_c = 2.357(161.76/177.7) - 0.479 - 1.018 = 0.649 \text{ ksi } (4.47 \text{ N/mm}^2)$$

$$CR = (9.57)(1 - 0.35)(0.649) = 4.04 \text{ ksi } (27.8 \text{ N/mm}^2)$$

Shrinkage

$$SH = (11.83)(1 - 0.42) = 6.86 \text{ ksi } (47.3 \text{ N/mm}^2)$$

Total losses

$$TL = 4.78 + 4.04 + 6.86 = 15.68 \text{ ksi } (108.1 \text{ N/mm}^2)$$

$$f_{st} = 161.76 - 15.68 + 1.018(6.2) = 152.4 \text{ ksi } (1051 \text{ N/mm}^2)(\text{vs. } 153.4 \text{ ksi} = 1058 \text{ N/mm}^2 \text{ from PBEAM analysis, Table 4-8})$$

Note that this $f_{st} = 152.4$ ksi (1051 N/mm^2) corresponds closely to the result of PBEAM analysis at 5 years which is 152.5 ksi (1051 N/mm^2). We do not actually have the accuracy to quarrel with the precision of these estimates for design purposes. These results seem to be very good at all stages.

APPENDIX E

PROBLEM STATEMENT

Flexural Analysis-Service Load

Problems in this group assume that concepts in Chapters 1, 2, and 5 dealing with materials and flexural analysis have been covered. Loss or prestress is given as part of problem statement (Problems 1–5) where required.

Problem 1(*a*)

Given Cross section A shown in Fig. E-1. Normal weight concrete 150 lb/ft^3 (23.6 kN/m^2), thus $w_g = 200$ lb/ft (2.92 kN/m). Allowable stresses: compression 3000 psi (20.7 N/mm^2), tension, zero.

Find: What F_0 (initial prestress force) and e would correspond to the full use of the allowable concrete stresses given? Show your check of top and bottom fiber stresses at support and at midspan sections. Indicate which section governs limiting F_0 and e.

Problem 1(*b*)

Given: Same information as Prob. 1(*a*), except allow 300 psi (2.07 N/mm^2) tension in concrete (instead of zero) (Fig. E-1).

Find: Same as Prob. 1(*a*). Compare results if you have solved F_0 and e with and without allowing tension.

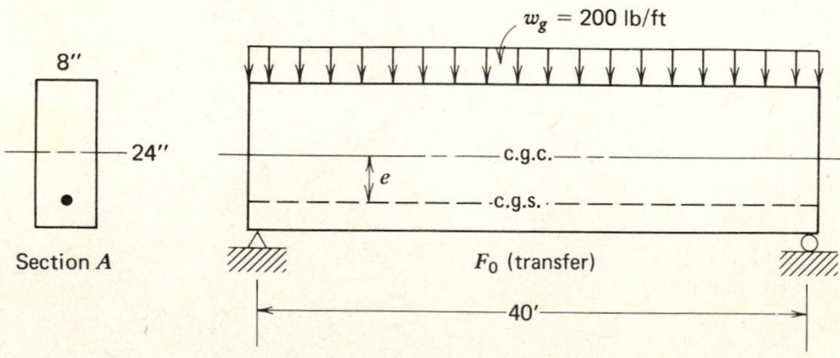

Fig. E-1. Problem 1, 3, and 4(*a*) (Problems 8 and 9-cross section only).

Problem 1(*c*)

Given: Beam shown in Fig. E-1 has $F_0 = 241$ k (1072 kN) (prestress at transfer) and $e = 5.56$ in. (141 mm). We are willing to allow 3000 psi (20.7 N/mm^2) compression and 490 psi (3.38 N/mm^2) tension in the concrete. After losses occur, assume $F = 0.80F_0$ for this beam.

Find: Are the stresses in excess of the given allowable values at transfer? What superimposed load, w_s, in addition to the beam weight, w_g, would be allowed based on these stresses?

Problem 2(*a*)

Given: Cross section B (Fig. E-2) with same area and w_g as section A of Problem 1. Allowable stresses: compression, 3000 psi (20.7 N/mm^2); tension, zero.

Find: What F_0 (initial prestress force) and e would fully utilize the allowable stresses given? Show center of top and bottom fiber stresses and indicate which section, support or midspan, governs the limiting F_0 and e.

Problem 2(*b*)

Given: Same information (Fig. E-2) as Problem 2(a), except allow 300 psi (2.07 N/mm^2) tension in concrete.

Find: Same as Problem 1(a). Compare results if you solved F_0 and e with and without allowing tension.

Problem 2(*c*)

Given: Beam shown in Fig. E-2. $F_0 = 241$ k (1072 kN) and $e = 8.65$ in. (220 mm). Allowable stresses: 3000 psi (20.7 N/mm^2) compression and

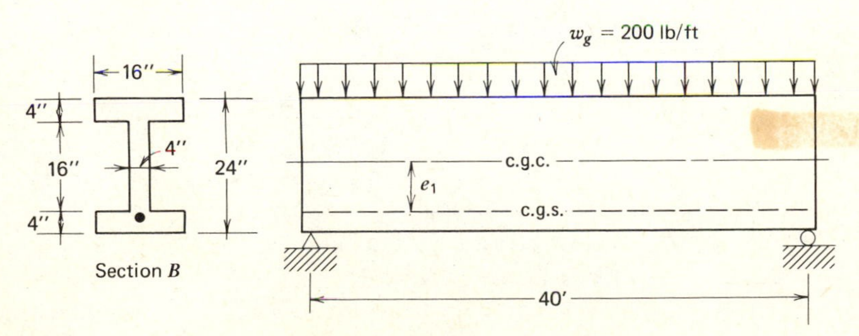

Fig. E-2. Problems 2, 3, and 4(*b*).

490 psi (3.38 N/mm^2) tension, and F (after losses) same as Prob. 1(c).

Find: Check stresses at transfer and find w_s same as Prob. 1(c).

Problem 3

Given: Sections A and B shown in Fig. E-1 and E-2 for same 40-ft (12.2 m) simple span. Efficiency of section can be expressed using two factors: concrete efficiency factor $= w_s/A$ with units $\dfrac{\text{lb/ft}}{\text{in.}^2}$ $\left(\dfrac{\text{N/m}}{\text{mm}^2}\right)$ and prestressing steel efficiency factor $= w_s/F_0$ with units $\dfrac{\text{lb/ft}}{\text{k}}$ $\left(\dfrac{\text{N/m}}{\text{kN}}\right)$.

Find: Compare the efficiency of the rectangular and I-shape section in terms of these two factor (a) with allowable compression 300 psi (20.7 N/mm^2) and tension zero, (b) with allowable compression 3000 psi (20.7 N/mm^2) and tension 490 psi (3.38 N/mm^2). Tabulate results and comment on the merits of the I-shape versus rectangular cross section as well as the benefit of allowable tension in the concrete.

Problem 4(a)

Given: Rectangular beam of Fig. E-1. $f'_{ci} = 5000$ psi (34.5 N/mm^2), $f'_c =$ 5500 psi (37.9 N/mm^2), $F = 0.80F_0$, allowable stresses following ACI Code.

Find: What F_0 and e would correspond to full use of the allowable stresses with straight strands? What w_s (superimposed load) in addition to w_g of 200 lb/ft (2.92 kN/m) could be allowed without overstress?

Problem 4(b)

Same statement as Problem 4(a) except refer to Fig. E-2 and use the I-shape section given.

Problem 5(a)

Given: Cross section A (Fig. E-1) with *draped tendon* shown in Fig. E-3. ACI Code allowable stresses. $F = 0.80F_0$ due to losses.

Find: What F_0 at transfer and e (e_1 at ends and $e_{℄}$) would fully utilize the allowable concrete stress? Is the $e_{℄}$ limited by the $g = 3$ in. (76.2 mm) min. for cover (Fig. E-3)? With your F and parabolic tendon layout, what superimposed load w_s could the beam carry without overstressing the concrete?

allowable stress

0.86. Ans.

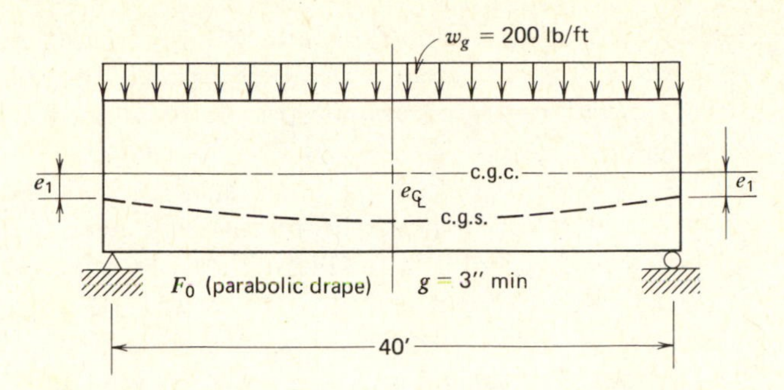

Fig. E-3. Problem 5.

Problem 5(*b*)

Same statement as Problem 5(a) except refer to Fig. E-2 for I-shape section and Fig. E-3 for *draped tendon* layout.

Loss of Prestress

Analysis for flexure and the material in Chapters 2 and 4 should be covered as background for the following problems.

Problem 6

Given: Beam of Fig. E-4.

$f'_{ci} = 4500$ psi (31.0 N/mm^2)

$f'_c = 6000$ psi (41.4 N/mm^2)

(Normal weight concrete.)

$f_{pu} = 270$ ksi (1862 N/mm^2)

(see Appendix B, Fig. B-1, for stress-strain curve for 7-wire strand)

Find: (a) If the strands (A_{ps} = 18-1/2 in. (12.7 mm) diameter strands) are initially stressed in a pretensioning bed to $0.70 f_{pu}$, find the

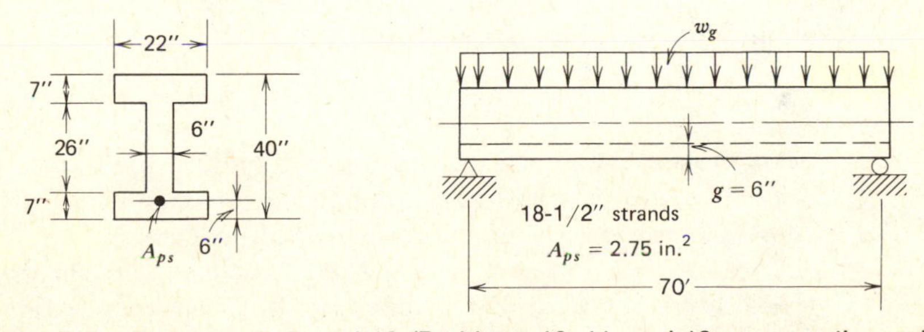

Fig. E-4. Problems 6, 7, and 13 (Problems 10, 11, and 12-cross section only).

elastic loss at support (where $M=0$) and at midspan immediately after transfer. Use gross section properties in making calculations and assume transfer occurs 48 h after initial stressing.

(*b*) Use the ACI Committee Recommendation to estimate total losses at midspan if the beam carries a permanent superimposed load $w_s = 1.0$ k/ft (14.6 kN/m) in addition to its own weight w_g as shown in Fig. E-4. Assume the average relative humidity to which the beam is exposed is 70%.

(*c*) Compare the calculated losses from (*b*) with the "lump sum" recommendation from AASHTO (Table 4-1) of about 45,000 psi (310 N/mm^2).

Problem 7

Given: Beam of Fig. E-4.
$f'_{ci} = 3500$ psi (24.1 N/mm^2)
$f'_c = 5000$ psi (34.5 N/mm^2)
(Normal weight concrete.)
$f_{pu} = 270$ ksi (1862 N/mm^2)
$\frac{1}{2}$ in. (12.7 mm) 7-wire strand

Find: (*a*) Estimate the loss of prestress at midspan for the beam if it carries a permanent load of 0.850 k/ft (12.4 kN/m) in addition to its own weight, w_g, for several years. Use the ACI Committee Recommendation of Chapter 4 and assume the average relative humidity is 75%.

(*b*) Follow the PCI simplified method using equations from Appendix D to estimate the loss of prestress.

(*c*) Compare calculated loss with lump sum recommendation of 45,000 psi (310 N/mm^2) from AASHTO (Table 4-1).

Moment-Curvature Relationship and Strength

Chapter 5 material on flexural analysis, especially sections 5-6 and 5-7 are essential background for the following problems.

Problem 8

Given: Beam cross section rectangular with $b=8$ in. (203 mm) and $d=19.0$ in. (483 mm). Strands are bonded to concrete.
$f'_c = 5000$ psi (34.5 N/mm^2) (normal weight concrete)
$A_{ps} = 0.765$ in^2 (494 mm^2) (5-1/2 in. strands), $f_{pu} = 270$ ksi (1862 N/mm^2) (Fig. B-1, Appendix B gives stress/strain curve)

$f_0 = 189$ ksi (1303 N/mm^2) (immediately after transfer)
$f_{se} = 155$ ksi (1069 N/mm^2) (after losses)
Straight strands with $e = 7.0$ in. (178 mm)

Find: (*a*) If tendons are bonded, find the stress and strains in the concrete over the depth of the beam due to the effective steel stress, f_{se} ($F_e = f_{se}A_{ps}$) Assume applied moment, $M = 0$ and $f_{se} = 155$ ksi. (1069 N/mm^2)

(*b*) Find M and ϕ which correspond to zero strain in concrete at level of steel.

(*c*) Find M and ϕ when top fiber concrete strain is 0.0015.

(*d*) Find M and ϕ when top fiber concrete strain is 0.003. Consider this to be the nominal moment strength for the section.

(*e*) Plot the M versus ϕ relationship for this cross section.

Problem 9

Given: same cross section as Problem 8.

Find: (*a*) Use the ACI Code equation for a *bonded beam* to estimate steel stress, f_{ps}, at ultimate moment. What is the ultimate moment capacity for the cross section using this estimate?

(*b*) If the member has *unbonded tendons*, find the minimum A_s which is required by ACI Code as bonded rebars. If this steel has $f_y = 60$ ksi (414 N/mm^2) and the strand has f_{ps} from the ACI Code Equation, find the ultimate moment capacity for the section. Assume the c.g.s. for the rebar is at $g = 3$ in. (76.2 mm), $e = 9$ in. (229 mm).

Problem 10 (Note: Problem 13 is shear strength analysis for this same beam.)

Given: *I*-shape beam of Fig. E-4
$f_c' = 5000$ psi (34.5 N/mm^2) (normal weight concrete)
$f_{pu} = 270$ ksi (1860 N/mm^2) (stress-strain curve Fig. B-1 Appendix B)
Bonded beam with $A_{ps} = 2.75$ in.^{2} (1774 mm^2) and $f_{se} = 155$ ksi (1069 N/mm^2).

Find: Determine the moment-curvature response of this cross section. Follow (*a*) through (*e*) as stated in Problem 8, but use the I-shape section of Fig. E-4 for this problem.

Problem 11

Given: I-shape beam of Prob. 10 (Fig. E-4) All materials for bonded beam same as Problem 10.

Find: Use the ACI Code equation for a bonded beam to estimate the steel stress at ultimate, f_{ps}. What is the moment strength which we might use in design if $\phi = 0.9$ for flexure and M_u is [0.9 × (calculated moment capacity of section)]?

Problem 12

Given: The following data from the moment-curvature analysis of an I-shape beam similar to that of Fig. E-4.

c (in.)	f_{ps} (ksi)	M (in.-k)		$\phi \times 10^{-5}$/in.	Remarks
	154	0	linear	−1.495	No applied moment
	166	12980	linear	+1.66	Zero conc. strain @ e
	~166+	13610	linear	~+1.82	Cracking
15.0	195	16100	nonlinear	+6.67	0.0010
11	225	19000	nonlinear	+13.5	0.0015
9.1	238	20,200	nonlinear	+22.0	0.0020
8.0	245	20,700	nonlinear	+31.2	0.0025
7.5	247	20,750	nonlinear	+40.0	0.0030 (ult.)

Find: If two loads applied at points 25 ft (7.62 m) from the end supports (10 ft (3.05 m) each side of midspan) produce the moment corresponding to 0.003 strain in concrete, find the magnitude of the load. Estimate the deflection under this load by using the curvature distribution along the span consistent with the moment-curvature response data given.

Shear Strength

The following problems require understanding of analysis for flexure as well as the shear strength analysis of Chapter 7.

Problem 13 (Note: Problem 10 is flexural strength analysis of this same beam.)

Given: Beam of Fig. E-4.
$f'_c = 5000$ psi (34.5 N/mm^2) (normal weight concrete)
$f_{se} = 155$ ksi (1069 N/mm^2) (steel stress after losses—bonded strands)

Find: (*a*) At section 1-1 located $h/2$ from support, find the shear strength contribution of the concrete (V_{cw}).
(*b*) At section 2-2 located 20 ft (6.10 m) from support, check to see which cracking mechanism, V_{cw} or V_{ci}, controls V_c. Evaluate both values and find V_c.
(*c*) At section 3-3 located 30 ft (9.14 m) from support, find V_c. Does V_{cw} or V_{ci} control? Why?

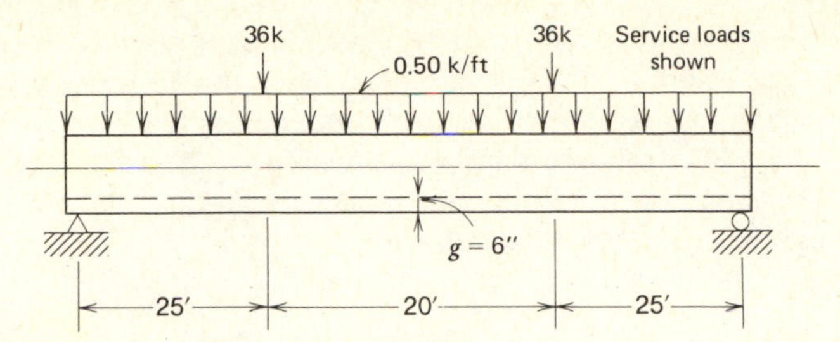

Fig. E-5. Problem 14. See cross section, Fig. E-4.

Problem 14

Given: Beam of Problem 13, Fig. E-4.
Loading—Fig. E-5 (service loads shown)
Materials—Problem 13
Apply ACI load factor $U = 1.4D + 1.7L$ to obtain ultimate loads for design.

Find: (*a*) Show the factored load shear together with the shear strength which the concrete can carry (Problem 13), and indicate shear for which stirrups must be designed. Tabulate values from analysis for V_{ci} and V_{cw} in a form similar to Fig. 7-15 and plot results in form similar to Fig. 7-16. Use stations 5 ft (1.52 m) apart for half the span plus an additional station $h/2$ from support.

(*b*) Find spacing required for No. 3 U-stirrups at a position just outside the 36-k (160-kN) load point. Assume $f_y = 50$ ksi (3448 N/mm^2). Check both shear strength requirements and minimum stirrup requirements of the ACI Code.

Composite Beams

Chapter 5 and 6 material on composite beams should be covered as background for the following problems.

Problem 15

Given: The precast, pretensioned tee section of Fig. E-6(*a*). Layout of draped strands shown in Fig. E-6(*b*). Normal weight concrete with $f'_{ci} = 4000$ psi (27.6 N/mm^2) and $f'_c = 5000$ psi (34.5 N/mm^2); $\frac{1}{2}$ in. (12.7 mm) diameter 7-wire strands with $f_{pu} = 270$ ksi (1862 N/mm^2). Assume F_0 based on $0.7 f_{pu}$ and $F = 0.78 F_0$.

Find: (*a*) If 16 strands are in two rows of 8 strands draped as shown [Fig. E-6(*b*)], find the eccentricities e_1 and e_2 which would make maximum use of the ACI allowable concrete stresses at transfer. The 8 strands are bundled vertically in the center 20 ft (6.10 m) But fan out to a 2-in. (50.8 mm) vertical distance between strands at the end support section.

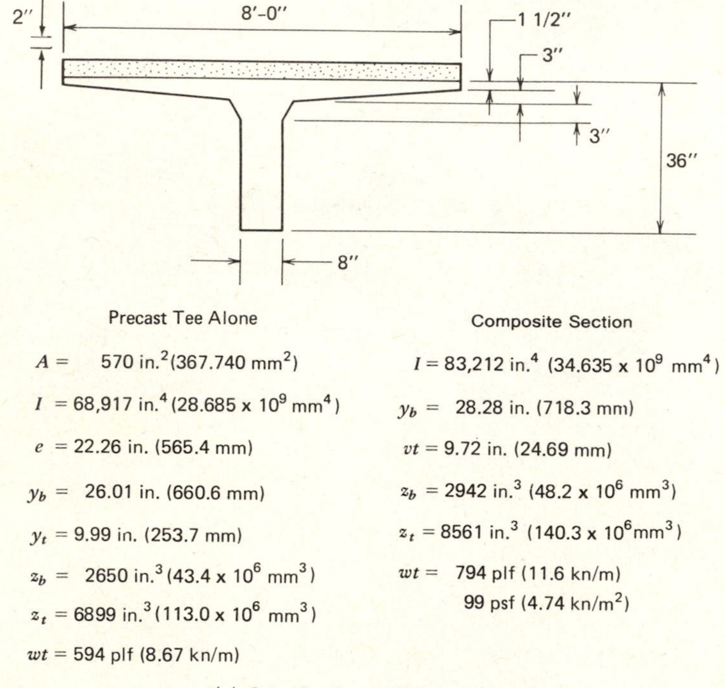

(*a*) Cross Section and Section Properties

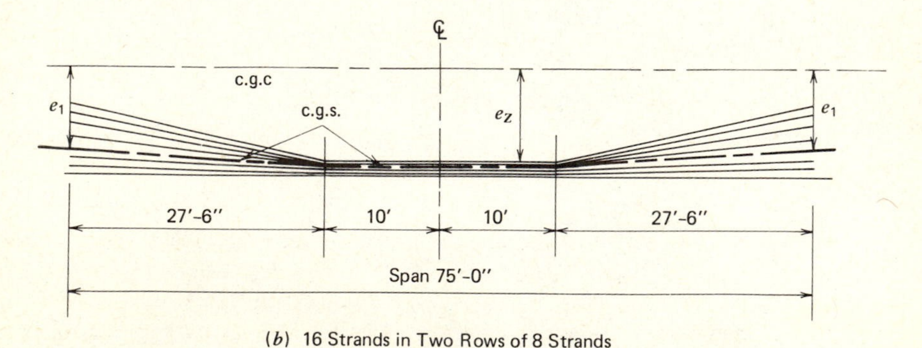

(*b*) 16 Strands in Two Rows of 8 Strands

Fig. E-6. Problem 15.

(b) What is the maximum superimposed service load (lb/ft^2) (N/m^2) which you could allow on this section without the 2 in. (50.8 mm) concrete topping? Limit stresses to ACI "after losses" values with $6\sqrt{f_c'}$ tension maximum.
(c) Repeat (b) for the section with the 2 in. (50.8 mm) composite topping considered assuming that it is added to the unshored beam and has $f_c' = 4000$ psi.

Problem 16

Given: A composite structural floor consists of precast, pretensioned Type II beams [Fig. 6-21(d)] which are *shored* during the construction of the 6-in.-thick cast-in-place composite slab which makes the total depth of the structure 42 in. The simple span for the composite structure is 60 ft. and the precast beams are spaced at a uniform spacing, S. Normal weight concrete: Beams, $f_{ci}' = 5000$ psi (34.5 N/mm^2), $f_c' = 7000$ psi (48.3 N/mm^2); composite slab, $f_c' =$ 4500 psi (31.0 N/mm^2). Service loads for the structure are (in addition to the self weight of the beams and slab): Additional dead load of 20 lb/ft^2 (958 N/m^2) and live load of 50 lb/ft^2 (2395 N/m^2).

Find:
(a) Based on the midspan section for the precast beam, compute the prestress F_0 (transfer) and e which would fully utilize the limiting stresses which ACI allows. Determine the number of strands and the strand pattern which would give your F_0 and e assuming that the strands have $0.7 f_{pu}$ just after transfer. Do we need to drape some strands?
(b) Assume prestress losses are 20% ($F = 0.8F_0$). Check the stresses over the depth of the composite beam under full service load with $S = 10$ ft (3.05 m) for the beam spacing. Compare the stresses with ACI Code allowable values. Could the spacing, S, be increased or should it be decreased based on allowable stresses?
(c) Check the horizontal shear stress between the precast beam and the composite slab at the end of the beam with $S = 10$ ft (3.05 m). With factored loads acting which cause maximum horizontal shear stress, does this stress exceed 350 psi (2.41 N/mm^2)? What concrete surface for top of precast beam and ties between beam and slab would you specify here?
(d) Check the composite beam with your F_0 and e from (a) and the maximum S (beam spacing) from (b) for flexural strength.

Continuous Beams

Chapter 10 should be covered as background for the following problems. The material in Chapter 11 is also helpful for analysis and design of these continuous beams.

Problem 17

Given: The idealized parabolic tendon layout of Fig. E-7 has a prestress force of 400 k (1779 kN) in the tendon throughout the entire length.

Find:
(*a*) What upward (balanced) loading along the beam *A* to *C* is produced by the tendon shown in Fig. E-7?
(*b*) Find the moment due to prestress at *B* and at midspan due to posttensioning this tendon. Is there secondary moment at *B* due to the posttensioning? Solve the external support reactions at *A*, *B*, and *C* due to this posttensioning.
(*c*) Locate the *C*-line at *B* with only the posttensioning of the tendon in Fig. E-7 and show the *C*-line location *A* to *C* on a sketch along with the c.g.s. layout.
(*d*) Develop an actual layout for the tendon and repeat (*a*), (*b*), and (*c*).

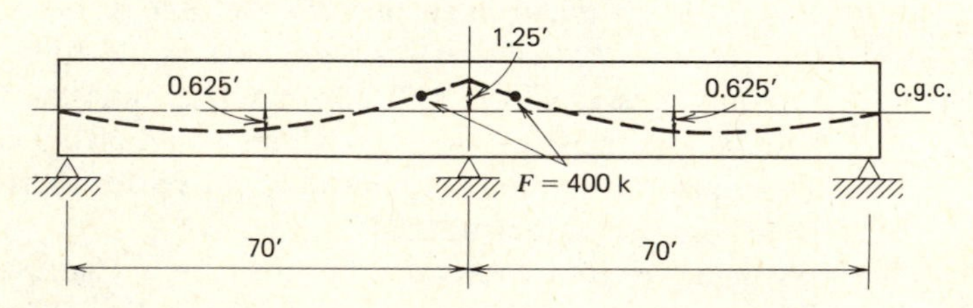

Fig. E-7. Problem 17: c.g.s. layout.

Problem 18

Given: A three-span continuous prestressed concrete beam (Fig. E-8) has a c.g.s. profile as shown, with an effective $F_e = 600$ k. $f'_c = 5$ ksi, $f_{pu} = 270$ ksi; $f_{se} = 162$ ksi. Bonded tendons. I_c of section $= 92{,}000$ in.^{4} $A_c = 792$ in.^{2} ($F_e = 2669$ kN, $f_{pu} = 1862$ N/mm^2, $f_{se} = 1117$ N/mm^2, $I_c = 38.3 \times 10^9$ mm^4, and $A_c = 511{,}000$ mm^2)

Find:
(*a*) Locate and show the *C*-line under the action of the prestress only. Compute the secondary moment over *B*, and find the reaction at *B*.

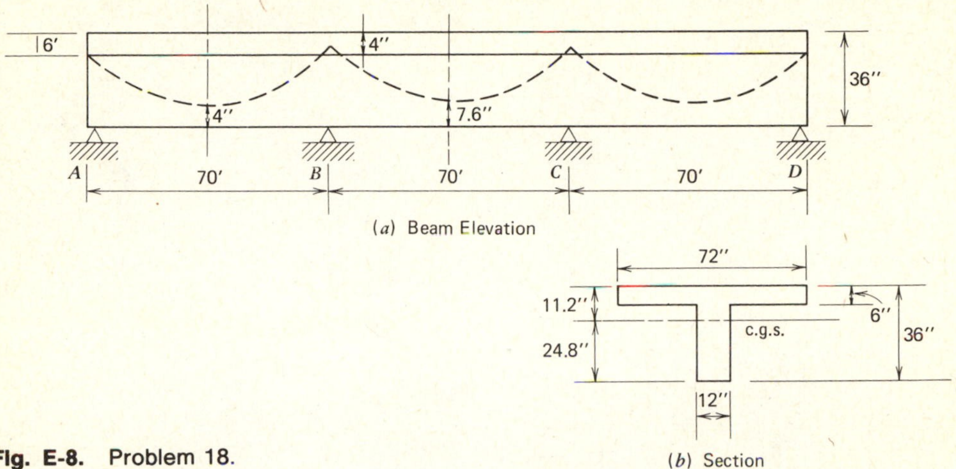

Fig. E-8. Problem 18.

(b) Under the beam's own weight, at 823 plf (12.0 kN/m), and the effect of prestress, compute the resulting extreme fiber stresses over support B, also the reaction at B. Find extreme fiber stresses at midspan AB and BC.

(c) If a uniform live load of 2.2 k/ft (32.1 kN/m) is applied to the beam, compute the live load moment at B and the resulting extreme fiber stresses due to prestress, beam weight, and live load. Discuss whether these stresses are permissible.

(d) Compute the ultimate moment capacity M_u of the beam section at B, using formula $f_{ps} = f_{pu}(1 - 0.5 f_{pu}/f_c')$ and other reasonable assumptions to select the A_{ps} for the tendon. Is the strength at B satisfactory for $1.4D + 1.7L$ loading?

(e) Develop an actual tendon layout and repeat (a), (b), and (c).

Slabs

Chapters 10 and 11 as well as Chapter 12 should be covered as background for the following problems.

Problem 19

Given: A one-way slab of 6 in. (152 mm) uniform thickness spans three successive 20-ft (6.10 m) spans. Posttensioned tendons will be draped to 1 in. (25.4 mm) from top of bottom fiber of the slab, using maximum eccentricity available at interior support and midspan sections. Tendons are unbonded and bonded reinforcement will be added (ACI Code). $f_c' = 4000$ psi (27.6 N/mm^2) for

3″

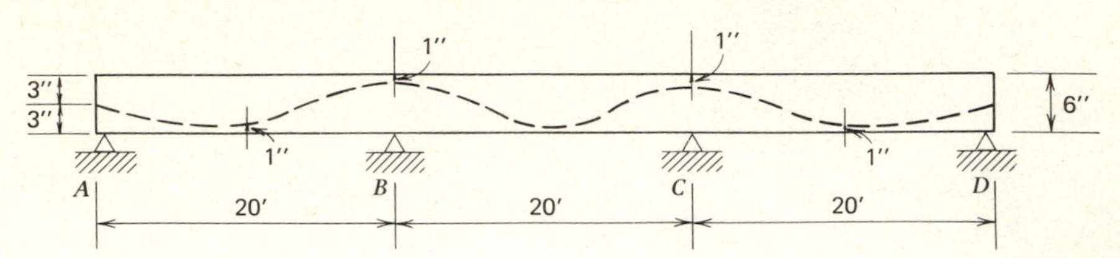

Fig. E-9. Problem 19.

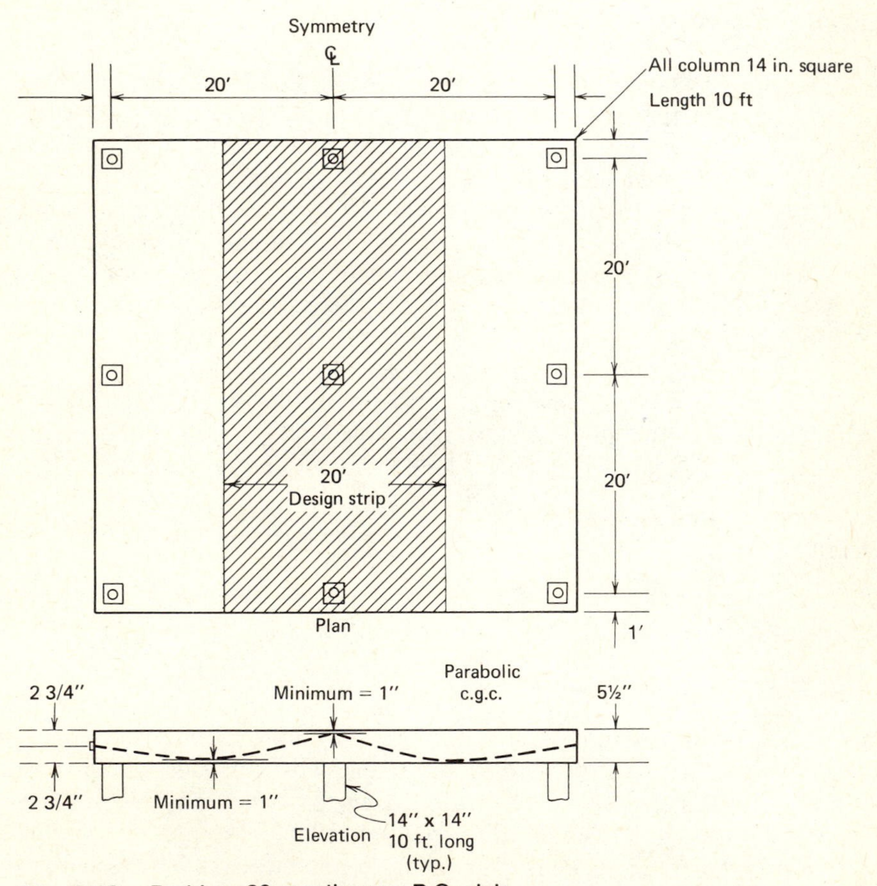

Fig. E-10. Problem 20; continuous P.C. slab.

normal weight concrete and $f_{pu} = 270$ ksi (1,862 kN/mm^2) for posttensioning strands. The service loads are slab load and live load of 50 lb/ft^2 (2395 N/m^2).

Find: (a) If the tendon is first assumed as idealized parabola in each span, find F in k/ft (kN/m) which will balance 85% of the slab weight on the span BC. What load is balanced in the outer spans? Assume F after loss is $0.6f_{pu}$.

(b) With your tendon from (a), what stresses occur at B with the loading which produces maximum moment at B? Use idealized layout for this calculation.

(c) Find the minimum rebar (bonded) which we must add along with this unbonded tendon and assume $f_y = 60$ ksi (414 N/mm^2). With this reinforcement plus the tendon (at same e) at B, is the strength satisfactory or is more than the minimum rebar needed?

Problem 20

Given: It is required to design a posttensioned concrete continuous slab, Fig. E-10 and to investigate its behavior and strength. The $5\frac{1}{2}$-in. (139.7 mm) slab has two spans of 20 ft (6.10 m) in each direction and is to be identically prestressed in both directions. It is of normal weight concrete @ 150 pcf (23.6 kN/m^3) with $f'_c = 4$ ksi (34.5 N/mm^2), $f_{pu} = 270$ ksi (1862 N/mm^2).

Find: (a) Design the prestressing force and the c.g.s. profile, using the beam method, to balance 72% of the weight of the slab. Distribute the tendons among the middle and column strips in some proper proportions.

(b) For the design in (a), compute approximately the actual stresses due to prestress and dead load.

(c) For a uniform live load of 50 psf (2395 N/m^2) over the entire area, compute the resulting stresses at critical points by the beam method.

(d) For the condition given in (c), what do you think are the actual stresses at critical points (reference 1)*.

(e) Compute the ultimate total load w_u (in psf (N/m^2) over entire area) using the beam method and assuming plastic action with moment redistribution allowed by ACI Code. What live load can be permitted, based on w_u, using proper load and phi factors?

*REF. 1 "Behavior of a Continuous Slab Prestressed in Two Directions," A.C. Scordelis, T.Y. Lin, and R. Itaya, J. Am. Conc. Inst., Dec. 1959, pp 441-459.

INDEX

AASHTO-PCI sections, 223-225
Abbreviations, appendix A
Air Entrainment, 49
Allowable stresses, for concrete, and for steel, 27-31
American Steel and Wire, 54, appendix B
Anchorage bearing plates, 274
Anchorage take-up, loss due to, 103
Arrangement of steel, 231

Bars, high-tensile, 57, 77, 82, appendix B
  reinforcing, *see* Nonprestressed reinforcements
B.B.R.V. system, 81, appendix B
Beam layouts, cantilever, 315
  continuous, 349
  fully continuous, 349
  partially continuous, 353
  post-tensioned, 308
  pre-tensioned, 307
  simple, 306
Beam section, actual examples of, 221
  constants for, appendix C
  shapes of, 218
Bearing at anchorage, 274, appendix B
Bending moments, primary and secondary, 360
Biaxial prestressing, 43, 429
Billner system, 78
Bond, at intermediate points, 267
  prestress transfer, 268
Bonded tendons, 26
Bridge girder, 563
  standard beams:
    AASHTO, 223, 224
    Washington State, 222
Bridge specifications, 37
Bridge types, 535
Building codes, 37
Building types, 520

Cable profiles and location, 309, 378
Cables, concordant and nonconcordant, 370
Cable stayed bridges, 535
Calcium chloride, 48
Camber, 290, 326
Cantilever layout, 315
Cast-in-place construction, 26
CCL anchorages, 71
Cement, high-early-strength, 48
  self-stressing, 51
Chemical prestressing, 51, 79
Circular prestressing, 467
Circumferential prestressing, 469
Collar for lift slabs, 454
Columns under eccentric load, 496
Composite construction, 26
Composite sections, analysis of, 178
  design of, 207
Compression members, 491
Concordancy of cables, 370
Concrete, admixtures for, 48
  air entrainment for, 49
  cover, 239
  creep strains in, 43, 51, 99
  curing of, 48
  elastic shortening of, 91
  elastic strain in, 42, 91
  modulus of elasticity of, 42
  modulus of rupture of, 42, 50
  Poisson's ratio of, 43, 50
  shearing strength of, 42
  shrinkage strains in, 46, 51, 100
  slump of, 41
  special manufacturing techniques, 48
  strain characteristics, 42
  strength requirements, 40
  tensile strength, 41
  water-cement ratio, 41
Connections, *see* Joints
Continuous beams, 345, 545
  load-balancing method, 399
Corrosion resistance, 34, 238
Cracking load, 30
  for composite sections, 179
Cracking moment, 144, 161

Cracking strength, continuous beams, 382
tension members, 465
Cracks in reinforced beams, 19
Criteria for prestressed concrete, 27, 32

Dams, 11
Decentering, 28
Definitions, 12, 24, appendix A
Deflections in beams, 290, 326
Depth-span ratios, 322
Design examples, 545
Dill, R. E., 3
Doehring, C. E. W., 2
Dome prestressing, 478
Dorland anchorages, 71
Dywadag bars, 77, appendix B

Economics of prestressed concrete, 35, 517
Effective prestress, 115
Elastic design, for composite section, 207
for continuous beams, 357
for flexure, 192, 195, 201, 203
for tension members, 461
vs. ultimate design, 216
Elasticity, *see* Modulus of elasticity
Electrical prestressing, 78
Elongation of tendons, 121
End-anchored tendons, 25, appendix B
End-block, reinforcement, 282
transverse tension at, 278

Falsework, 26
Fatigue strength, 61
Fiberglass, 58
Fire resistance, 38
Flat slabs, 412, 435-459, 575
Flexure, analysis for, 126
design for, 188
moment-curvature, 156
Formwork, 450, 599
Freyssinet, anchorages, 75, 76, appendix B
Eugene, 4, 13
jacks, 75
method of arch compensation, 11
system, 568
Frictional loss prestress, 107, 111, 123

Galvanized strands, 56
Grouting, 59
Guyon, Y., 279-280

Hose, corrugated sheet metal, 59
Hoyer, E., 4, 69

Initial prestress, 28, 87, 115
International Federation for Prestressing, 10, 61

Jacking stress, 28, 87, 115
Jacks, 73
Jackson, P. H., 2
Joints for precast members, 525

Layouts, *see* Beam layouts
Lee-McCall system, 83
Leonhardt system, 78
Lift slabs, *see* Slabs
Light-weight concrete, 49
Linear transformation, 370
Lin Tee, 220
Load-balancing method, 339
Load factors, 31
Loading, stages of, 27
Loss of prestress, 3, 87

Modulus of elasticity, of concrete, 42, 50
of steel, 54, 55, 57
Modulus of rupture, 42, 50
Moment-curvature analysis, 156

Non-end-anchored tendons, 25
Nonprestressed reinforcements, 329-342,
appendix B
Notations, appendix A

Over-reinforced beams, 148, 326

Partial prestress, 26, 203, 325, appendix A
Patents, 67
Pavements, 541
Piles, 504
Pipes joints of, 468
prestressed-concrete, 468
Plastic sheathing, 60
Poles, 543
Post-tensioning, 25, 73, appendix B
anchorages, 79, 80, 82, appendix B
couplers, 350, 352
jacks, 73
systems, appendix B
Precast construction, 26
Preflex method, 79
Preliminary design, 188, 213
Preload Company, 6
Prescon system, 74-76, 81, appendix B

Prestressed concrete, definition, 12
Prestressed Concrete Institute, 7, 9, 61
Prestressed reinforcements, appendix A. *See also* Steel; Tendons
Prestressed steel, 11
Prestressing, external and internal, 24
full and partial, 26, appendix A
linear and circular, 25, 467, appendix A
Prestressing bed, 69
Prestressing force, measurement of, 77
Prestressing systems, 67
addresses of some, appendix B
comparison of, 84
Pre-tensioning, 25
end anchorages, 68
strands, appendix B
systems, 68
Principal stresses, 246
Protection for prestressed reinforcement, 239

Reinforced concrete, combination of prestressed and, 343
cost comparison with, 517
prestressed *vs.*, 32
Retensioning, 28
Rigid frames, 405
Roof beam, 556

Safety, factor of, 31, 33
Secondary moments, 359
Segmental bridges, 535
Shapes of concrete section, 218, appendix C
Shear, 241
combined moment and, 263
inclined flexural cracking, V_{ci}, 257
principal tensile stress, 246
ultimate strength, 251
web reinforcement, 262
web shear cracking, V_{ew}, 254
Shorer system, 71
Sign conventions, 126
Slabs, continuous flat, 439
flat, 454
lift, 11, 449, 454
load-balancing method, 412, 452, 547, 577
one-way, 428, 545
tendon distribution, 441-449
test slabs, 443, 445, 447
two-way, 412, 435, 575
Span depth ratios, 322
Stages of loading, 27
Standard sections:
bridge beams, 222, 223, 224
double tee, 221
Lin tee, 220
piles, 507
Steam curing of concrete, 48
Steel, arrangement of, 231, 431, 443, 445, 447
chemical composition, 54
conduits for, 59
modulus of elasticity, 54
percentage, 148, 155
for prestressing, 52
Steiner, C. R., 3
Stirrups, 262, 265
Strands, high-tensile, 55
Stresses, in concrete, 126, 134
in steel, 27, 138
Stress-relieving process, 54
Structural types, 517

Tanks, prestressed-concrete, 6, 469, 474
Tendons, 13, appendix B. *See also* Steel
banded arrangement, 445, 447
distance between and protections for, 239
elongation of, 121
layout, 349, 408, 548
size, cover, and spacing of, 239, appendix B
specifications, appendix B
stressed at low level, 342
Tension at end block, 278
Tension members, 461, 465
Thin shells, 421, 478, 528
Ties, railroad, 540
Torsional strength, 284
Transfer, behavior and strength at, 183
length of, 268
prestress, 28
Transverse prestress, 428
Trusses, 524

Udall composite section, 180
Ultimate load, 30
composite section, 179
continuous beam, 382
Ultimate moment, 147, 173
Ultimate strength, in shear, 251
of beams, 213
of continuous beams, 382
of nonprestressed reinforcements, 335

of tension members, 465
Unbonded beams, 173
Unbonded tendons, 26
Under-reinforced beams, 148

Vertical prestresssing in tanks, 474
Vibration, internal and external, 48
Vibration in buildings, 323
VSL system, appendix B

Walnut Lane Bridge, 6
Web reinforcement, 262
Wires, 54, appendix B
stress-relieved, 54
Working load, 29

Yield point of steel, 54
Young's modulus, *see* Modulus of elasticity